Brauer Groups, Hopf Algebras and Galois Theory

K-Monographs in Mathematics

VOLUME 4

This book series is devoted to developments in the mathematical sciences which have links to *K*-theory. Like the journal *K-theory*, it is open to all mathematical disciplines.

K-*Monographs in Mathematics* provides material for advanced undergraduate and graduate programmes, seminars and workshops, as well as for research activities and libraries.

The series' wide scope includes such topics as quantum theory, Kac-Moody theory, operator algebras, noncommutative algebraic and differential geometry, cyclic and related (co)homology theories, algebraic homotopy theory and homotopical algebra, controlled topology, Novikov theory, transformation groups, surgery theory, Hermitian and quadratic forms, arithmetic algebraic geometry, and higher number theory.

Researchers whose work fits this framework are encouraged to submit book proposals to the Series Editor or the Publisher.

Series Editor:
A. Bak, *Dept. of Mathematics, University of Bielefeld, Postfach 8640, 33501 Bielefeld, Germany*

Editorial Board:
A. Connes, *Collège de France, Paris, France*
A. Ranicki, *University of Edinburgh, Edinburgh, Scotland, UK*

Brauer Groups, Hopf Algebras and Galois Theory

by

Stefaan Caenepeel

Free University of Brussels,
VUB,
Brussels, Belgium

KLUWER ACADEMIC PUBLISHERS
DORDRECHT / BOSTON / LONDON

A C.I.P. Catalogue record for this book is available from the Library of Congress.

ISBN 0-7923-4829-X

Published by Kluwer Academic Publishers,
P.O. Box 17, 3300 AA Dordrecht, The Netherlands.

Sold and distributed in the U.S.A. and Canada
by Kluwer Academic Publishers,
101 Philip Drive, Norwell, MA 02061, U.S.A.

In all other countries, sold and distributed
by Kluwer Academic Publishers,
P.O. Box 322, 3300 AH Dordrecht, The Netherlands.

Printed on acid-free paper

Dedicated to the memory of
Marleen Buysse
1961-1992

Contents

Preface

The development of the theory of the Brauer group of a field began around 1930; the underlying idea is to study properties of central simple algebras by considering them as representatives of elements of the Brauer group. It became clear that the Brauer group of a field is an important invariant, often reflecting arithmetic properties of the groundfield k. The Brauer group of a field continues to be an important research topic, we mention Amitsur's non-crossed product Theorem [5] and the Merkure'ev-Suslin Theorem [134] as more recent key results. As this volume goes to press, two other monographs dealing with division algebras and Brauer groups of fields are about to appear: [108], focussing on the generic construction of division algebras, and [148], discussing the representation theoretic aspects of the subject. For a more elementary exposition, we refer to classical texts such as [101, Chapter IV], [6, Chapter VIII] or the more recent [110].
Starting in the late fifties, a series of generalizations and variations of the Brauer group were introduced: the groundfield k was successively replaced by a commutative ring (Auslander-Goldman [12]), a scheme (Grothendieck [97]), a symmetric monoidal category (Pareigis [156]) and a braided monoidal category (Van Oystaeyen-Zhang [193]). This recent generalization includes almost all other known examples of Brauer groups, including the Brauer-Wall group and its generalizations, such as the Brauer-Long group.
The outline of this monograph is essentially the following. In Part I, we will discuss Auslander and Goldman's Brauer group of a commutative ring, with special focus on the (étale) cohomology theory and on Taylor's Brauer group of Azumaya algebras without a unit. Part II contains an introduction to the theory of Hopf algebras and Hopf-Galois theory. The important invariant that we are mainly interested in here is the group of Galois objects. The results of Parts I and II are interesting in its own right, but they are also developed as tools for explicit computations on the Brauer-Long group and its variations in Part III. Each of the three parts has a different flavour, so let us explain somewhat more of the philosophy behind each part.

The elements of the Brauer group are classes of Azumaya algebras. There are basically three different ways to introduce Azumaya algebras, and, in the theory of Auslander and Goldman, these three approaches turn out to be equivalent. This is not the case for the generalizations that we mentioned above. Auslander and Goldman introduce central separable algebras, modifying slightly Azumaya's notion of maximally central algebra [13]. Historically, this is the first approach, but it turned out afterwards that it is not the most natural one. A more conceptual definition is

the following: an Azumaya algebra A over a commutative ring R is a faithfully projective algebra, such that the enveloping algebra is isomorphic to the isomorphism ring over itself. This definition has the advantage that it can be translated into a categorical statement: we have an equivalence between the categories of R-modules and A-bimodules. Basic Theorems such as the Skolem-Noether Theorem follow immediately from this definition.
The third approach is due to Grothendieck [97]: Azumaya algebras can be viewed as twisted forms of matrix rings. This leads to the cohomological description of the Brauer group, generalizing the classical Theorem that every central simple algebra is equivalent in the Brauer group to a Galois crossed product.

In Part I, we will discuss Auslander and Goldman's Brauer group $\mathrm{Br}(R)$. Our purpose is twofold: the first aim is to introduce techniques, such as étale cohomology, descent theory and Morita theory, that will be useful in Parts II and III. We will also give a detailed discussion of Taylor's version of the Brauer group. As far as the author knows, this is the only version of the Brauer group that is not a special case of Van Oystaeyen and Zhang's Brauer group of a braided monoidal category. What is the importance of Taylor's Brauer group? We have mentioned above that the Brauer group of a commutative ring can be described in cohomological terms; the involved cohomology - étale cohomology - is more complicated than group cohomology, and the Brauer group embeds as a subgroup in the second étale cohomology group. A famous theorem due to Gabber [89] actually states that the Brauer group equals the torsion part of the second cohomology group. Taylor's Brauer group is bigger than the classical Brauer group, and it covers the full second étale cohomology group. Its elements are represented by algebras that do not necessarily have a unit.
At first sight our development of the theory of Taylor-Azumaya algebras is different from the one presented in Taylor's papers [175] and [160]. The reason is that Taylor follows the first approach to introducing Azumaya algebras, considering them as (generalized) central separable algebras. This is not so convenient, since the notion of separable algebra without a unit is technically rather complicated. Moreover, many of the relevant properties of Taylor-Azumaya algebras do not depend directly on the separability conditions. For the sake of completeness, we have included a separate chapter about separability conditions, but almost all the rest of the book is independent of the results of this Chapter. We will introduce Taylor-Azumaya algebras following the categorical approach, requiring that they fit into a strict Morita context, establishing a category equivalence between the categories of R-modules and (unital) A-bimodules. In Chapter 3, we will discuss the splitting Theorem, stating that a Taylor-Azumaya algebra is a twisted form of a so-called elementary algebra. This Theorem enables us to construct the embedding in the second étale cohomology group.
In Chapter 5, we introduce étale cohomology and its variations. In fact, we discuss the theory of sheaves over Grothendieck topologies (see [7] and [95]) in the particular situation where the topology is the étale or flat topology over the spectrum of a commutative ring R. The author is well aware of the fact that there is a small lack of beauty in the theory here: we have to translate some basically geometric notions into an algebraic language, and this might cause the loss of some geometric

intuition. On the other hand, once we have the machinery in the affine case, it turns out to be not so difficult to generalize many definitions and properties to arbitrary (noetherian) schemes, see Section 6.11.
We have managed to present most of the theory with a minimal use of the theory of étale algebras; actually, for the "functorial" part of the theory, one can work with the flat topology as well. In fact, we need the flat topologoy in the second and third part of this book). The more fundamental Theorems of Gabber and Taylor, and also the result that the Brauer group of a regular ring is equal to the full second cohomology group require deeper results of étale algebras, as they are based on Artin's Refinement Theorem [8]. This fundamental Theorem implies that the étale cohomology groups can be described as inductive limits of Čech cohomology groups (in our situation: Amitsur cohomology groups), a result that is used in the above mentioned Theorems. Let us also point out that it is not known whether the three results mentioned above hold for arbitrary schemes. We have contented ourselves with a simple formulation of Artin's Theorem, with a reference to the literature for its proof. The cohomological theory with its applications is then discussed extensively in Chapter 6. The author's favourite approach to constructing the embedding of the Brauer group in the second étale cohomology group is the Villamayor-Zelinsky approach (see [199]), and this is the one that is explained here. In Parts II and III of this book, some other applications of Villamayor and Zelinsky's method will be discussed. Other approaches to the embedding of the Brauer group are possible, we will give a brief survey of them.

In Part II of this book, we focus attention to another invariant of a commutative ring: given a Hopf algebra H, one can introduce the notion of H-Galois object. The definition is due to Chase and Sweedler in the case where H is faithfully projective, and it was generalized to the infinite case by Schneider. If H is the dual of a finite group ring, then one recovers classical Galois theory. We are mainly interested in the case where H is cocommutative: then isomorphism classes of H-Galois objects form a group $\mathrm{Gal}(R, H)$; the operation is a kind of dual version of the tensor product, called the cotensor product.
Our definition of Galois objects is also a categorical one: an H-comodule algebra A is an H-Galois object if the category of relative (A, H)-modules is equivalent to the category of R-modules. We have that twisted forms of H as an H-comodule algebra are H-Galois objects, but the converse does not hold this time. This leads to the introduction of the split part $\mathrm{Gal}^s(R, H)$ of the group of Galois objects. $\mathrm{Gal}^s(R, H)$ has a cohomological description, similar to the cohomological description of the Brauer group and the Picard group: it is the first flat cohomology group with values in the group of algebra maps from H to R, the grouplike elements of the dual Hopf algebras in the finite case.
In this case, the full group of Galois objects can be described by an exact sequence. This exact sequence, due to Early and Kreimer, and, independently, Yokogawa, involves a new type of cohomology introduced by Sweedler. In the case where H is not necessarily faithfully projective, this exact sequence still exists, but it then describes only a subgroup of the group of Galois objects, consisting of Galois objects with a so-called geometric normal basis. We point out that there exists a dual the-

ory, which is formally less complicated: if H is a commutative Hopf algebra, then one can introduce the group of H-Galois coobjects. Now the operation is the tensor product over H, the flat cohomology groups that we have to consider take values in the grouplike elements of H, and Sweedler cohomology is replaced by Harrison cohomology. In the case where H is finite, the two theories are each others dual, but in the infinite case they are different. We develop both theories parallel one to another in Chapters 8, 9 and 10. We have provided some explicit examples in Chapter 11. With an eye to Part III of our monograph, we have restricted attention to Hopf algebras that are faithfully projective, commutative and cocommutative. Apart from the group ring and its dual, we focus attention to the so-called Tate-Oort orders and monogenic Larson orders.

Another invariant of a commutative ring is the Brauer-Wall group. This invariant involves the cyclic group of order two, and, during the late sixties and early seventies, several generalizations to arbitrary finite abelian groups have been proposed. Long [125] introduced a Brauer group $\mathrm{BD}(R, H)$ of H-Azumaya algebras, where H is a faithfully projective, commutative and cocommutative Hopf algebra, and it turned out that all the previously suggested generalizations of the Brauer-Wall group happen to be special cases, or at least they are subgroups of Long's Brauer group. However, explicit computations remained scarce until the mid eighties, when DeMeyer and Ford [71] gave a complete description in the case where H is the group ring over the cyclic group of order 2. Their approach was very computational, and, in order to have some information in the general situation, one hopes for a general theory, comparable to the cohomological theory for the classical Brauer group. Such a theory was set up by the author and Beattie, first in the case where the Hopf algebra is a group ring [37], and later in the arbitrary case ([28] and [29]). Let us give an outline of the differences with the classical theory of the Brauer group.
An immediate observation is that the Brauer-Long group is not commutative: the operation is induced by the smash product, which is a non-commutative twisted version of the tensor product. Another more important observation is that the generalizations of the three possible characterizations of Azumaya algebras are not equivalent. The categorical definition turns out to be the most natural one, and it is also the one that appears in Long's paper [125]: let A be an H-dimodule algebra, that is, an algebra on which the Hopf algebra H acts and coacts. Then A is an H-Azumaya algebra if it is faithfully projective as an R-module, and if the two enveloping algebras $A\#\overline{A}$ and $\overline{A}\#A$ are respectively isomorphic to the endomorphism ring of A and its opposite. $\overline{A}$ is a kind of twisted opposite of A, we refer to Chapter 12 for the details.
Not every H-Azumaya algebra is a twisted form of an endomorphism ring of a faithfully projective H-dimodule. This phenomenon already appears when one considers the Brauer-Wall group; indeed, the Brauer-Wall group of an algebraically closed field of characteristic different from 2 is not trivial, it is equal to the cyclic group of order 2. As it was the case with the group of Galois objects, this leads to the introduction of the split part of the Brauer-Long group $\mathrm{BD}^s(R, H)$, consisting of classes of H-Azumaya algebra that can be split by a faithfully flat extension S of R. In a similar way, a discussion of separability conditions leads to the introduction

of H-separable and strongly H-separable algebras, and to another subgroup of the Brauer-Long group, called the strong part of the Brauer-Long group.
Our strategy is now the following: first we look at the split part of the Brauer-Long group, and we give a cohomological description. A modification of the arguments presented in part I shows that, as a set, $\mathrm{BD}^s(R, H)$ is the product of the Brauer group and the split parts of the groups of H-Galois objects and H^*-Galois objects. The multiplication rules can be described explicitly, and follow from the cohomology.
The second step is then to give a description of the cokernel of the natural inclusion of the split part into the full Brauer-Long group. As an application of the (generalized version of) the Rosenberg-Zelinsky exact sequence, one can show that this cokernel is contained in a kind of orthogonal subgroup of the Hopf automorphism group of the Hopf algebra $H \otimes H^*$. In the case where H is a group ring, with the order of the group G invertible in R, and R containing "enough" roots of unity, the cokernel is well-known. Therefore, the Brauer-Long group of a group ring is known, up to part of the multiplication rules. These multiplication rules are known in the case where G is a cyclic group of primary order. We can also carry out some explicit computations in the case where H is a Tate-Oort order. The details of all these computations are explained in Chapter 13.
In Chapters 12 and 13, our Hopf algebras are faithfully projective, commutative and cocommutative. In two recent papers (see [46] and [47]), it has been explained how the notion of H-Azumaya algebra can be extended to the situation where the Hopf algebra H has a bijective antipode, without any further condition. The main idea there is to replace H-dimodules by Yetter-Drinfel'd modules. These are also modules with an H-action and an H-coaction, but with a more complicated compatibility relation. Some of the results that are known for the Brauer-Long group generalize to this new Brauer group, but some more research needs to be done before we have a complete description. We give a brief account of the theory in Chapter 14.
Apart from Artin's Refinement Theorem, this book is written in an essentially self-contained way. We assume that the reader is familiar with some elementary notions from commutative algebra; the theory of Taylor's Brauer group will be more readable to those readers that have a certain familiarity with the classical theory of the Brauer group of a commutative ring, as explained for example in [75], [114] or [155].

Many have contributed directly or indirectly to this monograph. Here is an attempt to list people that supported me in various forms, through helpful ideas, stimulating discussions (sometimes late at night), reviews and referee reports (sometimes critical): Anthony Bak, Lindsay Childs, Frank DeMeyer, Françoise Grandjean, Raymond Hoobler, George Janelidze, Max Knus, Lieven Lebruyn, Gigel Militaru, Manuel Ojanguren, Şerban Raianu, Mitsuhiro Takeuchi, Francis Tilborghs, Michel Van den Bergh, Lucien Van Hamme, Alain Verschoren, Yinhuo Zhang and Shenglin Zhu. Special thanks go to Margaret Beattie (for telling me about the Brauer-Long group), Fred Van Oystaeyen (for telling me about the Brauer group), Roger Brynaert (for telling me about groups and rings), and Camille Caenepeel (for telling me about numbers, a long, long time ago).
My most enthusiastic supporters are Matthias and Frederik Caenepeel. Above all,

I want to thank my wife Lieve for her continuous love, patience and support, especially at the times when Matthias and/or Frederik were a little bit too enthusiastic.

Stefaan Caenepeel
July 30, 1997

Part I

The Brauer group of a commutative ring

Chapter 1

Morita theory for algebras without a unit

1.1 Morita contexts

Throughout this chapter, R will be a commutative ring with 1. $\otimes$ will be a shorter notation for $\otimes_R$. Let A be an R-algebra, not necessarily having a unit element. Given a left A-module M, and a right A-module N, the tensor product $N \otimes_A M$ is by definition the coequalizer

$$N \otimes A \otimes M \underset{I_N \otimes \psi_N}{\overset{\psi_N \otimes I_M}{\rightrightarrows}} N \otimes M \longrightarrow N \otimes_A M \longrightarrow 0$$

$\psi_N : N \otimes A \to N$ and $\psi_M : A \otimes M \to M$ represent the right action of A on N and the left action of A on M. A (left) A-module will be called *unital* if the canonical map $A \otimes_A M \to M$ is an isomorphism. Similar notions apply to right A-modules and A-bimodules. If A itself is unital as a left or right A-module, then we will call A a unital R-algebra. This implies that the multiplication map $m_A: A \otimes A \to A$ is surjective. If A has a unit, then every A-module is unital. For an R-algebra A, A-mod^u and mod^u-A will denote the categories of unital left and right A-modules.

Example 1.1.1 A ring A is called a *ring with local units* if every finite subset of A is contained in a subring of A of the form eAe, where e is an idempotent of A. Observe that e is a unit element of eAe. We claim that an R-algebra A with local units is unital in the above sense. To this end, it suffices to show that the sequence

$$A \otimes A \otimes A \underset{I_A \otimes m_A}{\overset{m_A \otimes I_A}{\rightrightarrows}} A \otimes A \xrightarrow{m_A} A \longrightarrow 0 \tag{1.1}$$

is exact. Take $a \in A$. There exists an idempotent $e \in A$ such that $a = exe$ for some $x \in A$. But then $a = m_A(e \otimes xe) \in \mathrm{Im}\,(m_A)$, and this shows that m_A is surjective. Now let $a_i, b_i \in A$ be such that $\sum_i a_i b_i = 0$. There exists an idempotent e in A such that $a_i, b_i \in eAe$. Observe that

$$(m_A \otimes I_A - I_A \otimes m_A)(\sum_i e \otimes a_i \otimes b_i) = \sum_i ea_i \otimes b_i - e \otimes a_i b_i = \sum_i a_i \otimes b_i$$

and this shows that the sequence is exact at $A \otimes A$.
In [2], a Morita theory for rings with local unit is developed. Our aim is to develop a Morita theory for unital algebras.

Definition 1.1.2 *Let A and B be unital R-algebras. A Morita context between A and B is a sextuple (A, B, P, Q, f, g) such that*

1. *P is a unital (A, B)-bimodule, and Q is a unital (B, A)-bimodule;*
2. *f: $P \otimes_B Q \to A$ is an (A, A)-bimodule homomorphism, and g: $Q \otimes_A P \to B$ is a (B, B)-bimodule homomorphism.*
3. *For all $p, p' \in P$ and $q, q' \in Q$, we have*

$$f(p \otimes q)p' = pg(q \otimes p')$$
$$g(q \otimes p)q' = qf(p \otimes q')$$

If f and g are isomorphisms, then we call the Morita context strict , and we will then say that A and B are Morita equivalent, and we will denote this by $A \sim B$.

Theorem 1.1.3 *Let A, B, P, Q be as in Definition 1.1.2, and suppose that*

$$h : A \longrightarrow P \otimes_B Q \quad \text{and} \quad g : Q \otimes_A P \longrightarrow B$$

are respectively an (A, A)-bimodule homomorphism and a (B, B)-bimodule homomorphism such that the diagrams

$$\begin{array}{ccc} P \otimes_B Q \otimes_A P & \xrightarrow{I_P \otimes g} & P \otimes_B B \\ \uparrow{\scriptstyle h \otimes I_P} & & \uparrow{\scriptstyle \cong} \\ A \otimes_A P & \xrightarrow{\cong} & P \end{array} \quad \text{and} \quad \begin{array}{ccc} Q \otimes_A P \otimes_B Q & \xrightarrow{g \otimes I_Q} & B \otimes_B Q \\ \uparrow{\scriptstyle I_Q \otimes h} & & \uparrow{\scriptstyle \cong} \\ Q \otimes_A A & \xrightarrow{\cong} & Q \end{array}$$

commute. Then the functor

$$P \otimes_B \bullet\colon B\text{-mod}^u \to A\text{-mod}^u$$

is a right adjoint of

$$Q \otimes_A \bullet\colon A\text{-mod}^u \to B\text{-mod}^u$$

A similar property holds for the categories of right unital A-modules and B-modules.

Proof Take $a \in A$, and write $h(a) = \sum_i p_i \otimes q_i \in P \otimes_B Q$. Then it follows from the commutativity of the two above diagrams that

$$\sum_i p_i g(q_i \otimes p) = ap \tag{1.2}$$
$$\sum_i g(q \otimes p_i) q_i = qa \tag{1.3}$$

for all $p \in P, q \in Q$.
Take $M \in A\text{-mod}^u$ and $N \in B\text{-mod}^u$, and define two maps

$$\mathrm{Hom}_A(M, P \otimes_B N) \underset{\Phi}{\overset{\Psi}{\rightleftarrows}} \mathrm{Hom}_B(Q \otimes_A M, N)$$

as follows:

$$\Psi(v):\ Q \otimes_A M \xrightarrow{I_Q \otimes v} Q \otimes_A P \otimes_B N \xrightarrow{g \otimes I_N} B \otimes_B N \cong N$$

for all $v \in \mathrm{Hom}_A(M, P \otimes_B N)$, and

$$\Phi(u):\ M \cong A \otimes_A M \xrightarrow{f^{-1} \otimes I_M} P \otimes_B Q \otimes_A M \xrightarrow{I_P \otimes u} P \otimes_B N$$

for any $u \in \mathrm{Hom}_B(Q \otimes_A M, N)$. These formulas can be rewritten as follows: take $m \in M$ and $q \in Q$, and write $v(m) = \sum_j x_j \otimes y_j$. Then

$$\Psi(v)(q \otimes m) = \sum_j g(q \otimes x_j) y_j \tag{1.4}$$

$$\Phi(u)(am) = \sum_i p_i \otimes u(q_i \otimes m) \tag{1.5}$$

Using equations (1.2-1.5), we can easily verify that Φ and Ψ are each others inverses:

$$\begin{aligned} \Psi(\Phi(u))(q \otimes am) &= \sum_j g(q \otimes p_i) u(q_i \otimes m) \\ &= \sum_j u(g(q \otimes p_i) q_i \otimes m) = u(qa \otimes m) \end{aligned}$$

and

$$\begin{aligned} \Phi(\Psi(v))(am) &= \sum_i p_i \otimes \Psi(v)(q_i \otimes m) \\ &= \sum_{i,j} p_i \otimes g(q_i \otimes x_j) y_j \\ &= \sum_{i,j} p_i g(q_i \otimes x_j) \otimes y_j \\ &= \sum_j a x_j \otimes y_j = av(m) = v(am) \end{aligned}$$

□

Theorem 1.1.4 (The Morita Theorem) *Let (A, B, P, Q, f, g) be a strict Morita context. Then the functors*

$$P \otimes_B \bullet:\ B\text{-mod}^u \to A\text{-mod}^u$$

and

$$Q \otimes_A \bullet:\ A\text{-mod}^u \to B\text{-mod}^u$$

are inverse equivalences. Similar properties hold for the categories of right unital A-modules and B-modules.

Proof Applying Proposition 1.1.3 with $h = f^{-1}$, we obtain that the two functors in question are adjoint functors. It therefore suffices to show that for all $M \in A\text{-mod}^u$ and $N \in B\text{-mod}^u$ the maps

$$\Phi(I_{Q\otimes_A M}):\ M \longrightarrow P\otimes_B Q\otimes_A M$$

and

$$\Psi(I_{P\otimes_B N}):\ Q\otimes_A P\otimes_B N \longrightarrow N$$

are isomorphisms. These maps are given by the following formulas (with notations as in the proof of 1.1.3.

$$\begin{aligned}\Phi(I_{Q\otimes_A M})(am) &= \sum_i p_i\otimes q_i\otimes m = f^{-1}(a)\otimes m\\ \Psi(I_{P\otimes_B N})(q\otimes p\otimes n) &= g(q\otimes p)n\end{aligned}$$

and they are isomorphisms since f and g are isomorphisms. □

Before stating some examples, let us first state the following important property; the proof is taken from [15, II.3.4].

Proposition 1.1.5 *Suppose that (A, B, P, Q, f, g) is a Morita context. Assume that A has a unit and that f is surjective. Then*

1. *f is bijective;*
2. *g induces bimodule isomorphisms*
$$P \cong \mathrm{Hom}_B(Q, B) \ \text{ and } \ Q \cong \mathrm{Hom}_B(P, B);$$
3. *the bimodule structures induce R-algebra isomorphisms*
$$A \cong \mathrm{End}_B(P) \cong \mathrm{End}_B(Q)^{\mathrm{op}};$$
4. *P and Q are A-generators;*
5. *P and Q are finitely generated and have a dual basis as B-modules. Consequently, P and Q are projective as B-modules.*
6. *We have the following isomorphisms of functors:*
$$\begin{cases} P\otimes_B \bullet \cong \mathrm{Hom}_B(Q, \bullet)\\ \bullet\otimes_B Q \cong \mathrm{Hom}_B(P, \bullet)\end{cases}$$

Proof 1) Since f is surjective, there exists $\sum_{i\in I} p_i\otimes q_i \in P\otimes_B Q$ such that $f(\sum_{i\in I} p_i\otimes q_i) = 1_A$. Using condition 2) of the above definition, we obtain for every $\sum_j p'_j\otimes q'_j \in$ Ker f:

$$\begin{aligned}\sum_j p'_j\otimes q'_j &= \sum_{j,i} p'_j\otimes q'_j f(p_i\otimes q_i)\\ &= \sum_{j,i} p'_j\otimes g(q'_j\otimes p_i)q_i\\ &= \sum_{j,i} f(p'_j\otimes q'_j)p_i\otimes q_i\\ &= 0\end{aligned}$$

2) Define h: $P \to \mathrm{Hom}_B(Q, B)$ by

$$h(p)(q) = g(q \otimes p)$$

If $h(p) = 0$, then

$$p = f(\sum_i p_i \otimes q_i)p = \sum_i p_i g(q_i \otimes p) = 0$$

hence h is injective. Next, take $\alpha \in \mathrm{Hom}_B(Q, B)$. Then for all $q \in Q$

$$\begin{aligned}
\alpha(q) &= \alpha(\sum_i q f(p_i \otimes q_i)) \\
&= \alpha(\sum_i g(q \otimes p_i) q_i) \\
&= \sum_i g(q \otimes p_i)\alpha(q_i) \\
&= \sum_i g(q \otimes p_i \alpha(q_i)) \\
&= h(\sum_i p_i \alpha(q_i))(q)
\end{aligned}$$

so h is surjective. The assertion for Q may be proved in a similar way.

3) Define k: $A \to \mathrm{End}_B(P)$ by

$$k(a)(p) = ap$$

for all $a \in A$ and $p \in P$. If $k(a) = 0$, then

$$a = af(\sum_i p_i \otimes q_i) = \sum_i f(ap_i \otimes q_i) = 0$$

so k is injective. Take $\alpha \in \mathrm{End}_B(P)$. Then for all $p \in P$:

$$\begin{aligned}
\alpha(p) &= \alpha(f(\sum_i p_i \otimes q_i)p) \\
&= \alpha(\sum_i p_i g(q_i \otimes p)) \\
&= \sum_i \alpha(p_i g(q_i \otimes p)) \\
&= \sum_i f(\alpha(p_i) \otimes q_i)p \\
&= k\big(f(\sum_i \alpha(p_i) \otimes q_i)\big)(p)
\end{aligned}$$

so k is surjective.

4) Define $h_i: P \to A$ by $h_i(p) = f(p \otimes q_i)$, for every $p \in P$ and $i \in I$. Then

$$\sum_{i \in I} h_i(p_i) = \sum_{i \in I} f(p_i \otimes q_i) = 1_A$$

and this means that P is an A-generator. In a similar way, we can show that Q is an A-generator.

5) For all $q \in Q$, we have

$$q = \sum_i q f(p_i \otimes q_i) = \sum_i g(q \otimes p_i) q_i,$$

hence $\{q_i, g(\bullet \otimes p_i) | i \in I\}$ is a dual basis for Q. Similarly, $\{p_i, f(q_i \otimes \bullet) | i \in I\}$ is a dual basis for P.

6) We know from 3) that $P \cong \mathrm{Hom}_B(Q, \bullet)$. Take $N \in B\text{-mod}^u$, and define

$$\alpha\colon \mathrm{Hom}_B(Q, B) \otimes_B N \longrightarrow \mathrm{Hom}_B(Q, N)$$

by

$$\alpha(\pi \otimes n)(q) = \pi(q) n$$

for all $\pi \in \mathrm{Hom}_B(Q, B)$, $n \in N$, $q \in Q$. From 4), we know that Q has a dual basis $\{q_i, f_i | i = 1, \ldots, n\}$ as a left B-module. This allows us to define a map

$$\beta\colon \mathrm{Hom}_B(Q, N) \longrightarrow P \otimes_B N$$

by

$$\beta(\psi) = \sum_i f_i \otimes \psi(q_i)$$

for all $\psi \in \mathrm{Hom}_B(Q, N)$. It now follows easily that

$$\begin{aligned}(\alpha \circ \beta)(\psi)(q) &= \alpha(\sum_i f_i \otimes \psi(q_i))(q) \\ &= \sum_i f_i(q)\psi(q_i) = \psi(q)\end{aligned}$$

and

$$\begin{aligned}(\beta \circ \alpha)(\pi \otimes n) &= \sum_i f_i \otimes \alpha(\pi \otimes n)(q_i) \\ &= \sum_i f_i \otimes \pi(q_i) n = \pi \otimes n\end{aligned}$$

Therefore α and β are each others inverses. The proof of the other assertion is similar. □

Let A and B be R-algebras with a unit. Recall that an A-module that is finitely generated, projective and an A-generator is called an *A-progenerator*. If A is commutative, then an A-progenerator is nothing else then a faithfully projective A-module (see for example [75]). If (A, B, P, Q, f, g) is a strict Morita context, then it follows from Proposition 1.1.5 that P and Q are A and B-progenerators. We also have the following "converse" property.

Proposition 1.1.6 *Suppose that B is an R-algebra with a unit, and that P is a right B-progenerator. Then*

$$(A = \mathrm{End}_B(P), B, P, Q = \mathrm{Hom}_B(P, B), f, g)$$

with

$$f:\ P\otimes_B Q\to A \quad\text{and}\quad g:\ Q\otimes_A P\to B$$

defined by

$$f(p\otimes q)(p')=pq(p') \quad\text{and}\quad g(q\otimes p)=q(p)$$

for all $p,p'\in P$ and $q\in Q$, is a strict Morita context.

Proof The bimodule structures of P and Q are given by the following formulas

$$apq=a(p)q \quad\text{and}\quad (bqa)(p)=q(a(p))b$$

for all $a\in A$, $b\in B, p\in P$ and $q\in Q$. It is easy to see that f and g are bimodule maps, and that

$$f(p\otimes q)p'=pq(p')=pg(q\otimes p')$$

and

$$\begin{aligned}(g(q\otimes p)q')(p') &= (q(p)q')(p')\\ &= q(p)q'(p')\\ &= q(pq'(p))\\ &= q((f(p\otimes q')(p'))\\ &= (qf(p\otimes q'))(p')\end{aligned}$$

and it follows that we have a Morita context. The fact that P is a B-generator implies that f is surjective, and from the fact that P has a finite dual basis as a right B-module, it follows that g is surjective. From Proposition 1.1.5, it follows that f and g are also injective. Thus our Morita context is strict. □

Proposition 1.1.7 *Let A,B,C,D be unital R-algebras. Then*
1) $A\sim A$;
2) if $A\sim B$, then $B\sim A$;
3) if $A\sim B$, and $B\sim C$, then $A\sim C$;
4) if $A\sim B$, and $C\sim D$, then $A\otimes C\sim B\otimes D$.

Proof The first two statements follow immediately. For the third one, suppose that we have two strict Morita contexts

$$(A,B,{}_AP_B,{}_BQ_A,f,g)$$

and

$$(B,C,{}_BM_C,{}_CN_B,h,k)$$

then

$$(A,C,P\otimes_B M,N\otimes_B Q,f\circ(I\otimes h\otimes I),k\circ(I\otimes g\otimes I))$$

is again a strict Morita context. Similarly, if

$$(A,B,P,Q,f,g)$$

and

$$(C,D,M,N,h,k)$$

are strict Morita contexts, then

$$(A\otimes C,B\otimes D,P\otimes M,Q\otimes N,f\otimes h,g\otimes k)$$

is a strict Morita context. We leave the details to the reader. □

1.2 Dual pairs and elementary algebras

A *dual pair* of R-modules $\underline{M} = (M, M', \mu)$ consists of two R-modules M and M' and a surjective map μ: $M' \otimes M \to R$. A morphism between two dual pairs $\underline{M}$ and $\underline{N}$ is a couple of maps $\underline{f} = (f, f')$ with f: $M \to N$ and f': $M' \to N'$ such that $\mu = \nu \circ (f' \otimes f)$. The category of dual pairs and dual pair isomorphisms will be denoted by $\underline{\underline{\mathrm{DP}}}(R)$. A typical example of such a dual pair is the following: let M be a *faithfully projective* R-module, that is, a finitely generated, faithful, projective R-module. Then $(M, M^*, \langle\bullet,\bullet\rangle)$, with the canonical pairing $\langle m^*, m\rangle = m^*(m)$, for all $m \in M$, $m^* \in M^*$, is a dual pair. If we denote $\underline{\underline{\mathrm{FP}}}(R)$ for the category of faithfully projective R-modules and R-module isomorphisms, then we obtain a natural functor

$$F: \underline{\underline{\mathrm{FP}}}(R) \longrightarrow \underline{\underline{\mathrm{DP}}}(R)$$

given by

$$F(M) = (M, M^*, \langle\bullet,\bullet\rangle)$$

and

$$F(f) = (f, (f^*)^{-1})$$

To any dual pair $\underline{M} = (M, M', \mu)$, we associate an associative R-algebra $A = E_R(\underline{M})$ as follows: as an R-module, $A = M \otimes M'$, and the multiplication is given by the rule

$$(m_1 \otimes m_1')(m_2 \otimes m_2') = \mu(m_1' \otimes m_2)(m_1 \otimes m_2').$$

We leave it to the reader to show that A is associative. We call A the *elementary algebra* associated to $\underline{M}$. E_R is a functor from $\underline{\underline{\mathrm{DP}}}(R)$ to associative R-algebras. Indeed, if $\underline{f}$: $\underline{M} \to \underline{N}$ is a morphism in $\underline{\underline{\mathrm{DP}}}(R)$, then $f \otimes f'$: $E_R(\underline{M}) \to E_R(\underline{N})$ is an R-algebra isomorphism. E_R preserves the tensor product:

$$E_R(\underline{M} \otimes \underline{N}) \cong E_R(\underline{M}) \otimes E_R(\underline{N})$$

since for $\underline{M}, \underline{N} \in \underline{\underline{\mathrm{DP}}}(R)$ the switch map

$$I \otimes \tau \otimes I: M' \otimes N' \otimes M \otimes N \to M' \otimes M \otimes N' \otimes N$$

is an isomorphism of R-algebras. Observe also that E_R preserves the composition: if $\underline{f} = (f, f') : \underline{M} \to \underline{N}$ and $\underline{g} = (g, g') : \underline{N} \to \underline{P}$ in $\underline{\underline{\mathrm{DP}}}(R)$, then

$$E_R(\underline{g} \circ \underline{f}) = (g' \circ f') \otimes (g \circ f) = (g' \otimes g) \circ (f' \otimes f) = E_R(\underline{g}) \circ E_R(\underline{f}).$$

Observe that for every $P \in \underline{\underline{\mathrm{FP}}}(R)$, we have:

$$E_R(F(P)) \cong \mathrm{End}_R(P)$$

as R-algebras.

Proposition 1.2.1 *An elementary R-algebra is unital.*

Proof Let $A = E_R(M, M', \mu)$. The fact that μ is surjective implies that the multiplication map m_A: $A \otimes A \to A$ is surjective. To finish the proof, we have to show that the sequence

$$A \otimes A \otimes A \xrightarrow{\lambda} A \otimes A \xrightarrow{m_A} A \to 0$$

with

$$\lambda(a \otimes b \otimes c) = ab \otimes c - a \otimes bc$$

is exact at $A \otimes A$. It is clear that $m_A \circ \lambda = 0$. Conversely, suppose that

$$x = \sum_i (m_i \otimes m_i') \otimes (n_i \otimes n_i') \in \operatorname{Ker} m_A,$$

that is

$$\sum_i \mu(m_i' \otimes n_i) m_i \otimes n_i' = 0.$$

Take $k = \sum_j q_j \otimes p_j \in M' \otimes M$ such that $\mu(k) = 1$. Then

$$\lambda(\sum_{i,j} m_i \otimes q_j \otimes p_j \otimes m_i' \otimes n_i \otimes n_i') = \mu(k)x - \sum_{i,j} \mu(m_i' \otimes n_i) m_i \otimes q_j \otimes p_j \otimes n_i' = x$$

hence $x \in \operatorname{Im}\lambda$. □

In general, $E_R(\underline{M})$ is not a unitary algebra. More precisely, we have (cf. [175, Prop.4.5]):

Proposition 1.2.2 *If $A = E_R(M, M', \mu)$ has a unit element, then M and M' are faithfully projective R-modules and $(M, M', \mu) \cong F(M)$ in $\underline{\underline{\mathrm{DP}}}(R)$.*

Proof Suppose that $1_A = \sum_{i=1}^n m_i \otimes n_i$ is a unit element of A. Choose $k = \sum_j q_j \otimes p_j \in M' \otimes M$ such that $\mu(k) = 1$. For all $m \in M$, $n \in M'$, we have

$$m \otimes n = 1_A(m \otimes n) = \sum_i \mu(n_i \otimes m) m_i \otimes n$$

hence, for all $m, p \in M$, $n \in M'$, we have

$$m \otimes n \otimes p = \sum_i \mu(n_i \otimes m) m_i \otimes n \otimes p$$

In $M \otimes M' \otimes M$, we therefore have the following identity:

$$m \otimes k = \sum_i \mu(n_i \otimes m) m_i \otimes k$$

Applying $I_M \otimes \mu$ to both sides, we obtain

$$m = \sum_i \mu(n_i \otimes m) m_i$$

Define f: $M' \to M^*$ by $\langle f(m'), m \rangle = \mu(m' \otimes m)$. For all $m \in M$ we then have that

$$m = \sum \langle f(n_i), m \rangle m_i$$

and $\{m_i, f(n_i) | i = 1, \ldots, n\}$ is a dual basis for M. Thus M is finitely generated and projective. Therefore, for all $m^* \in M^*$ we have :

$$m^* = \sum_i \langle m^*, m_i \rangle f(n_i) = f(\sum_i \langle m^*, m_i \rangle n_i),$$

and this implies that f is surjective. Similarly, we may show that for all $n \in M'$:

$$n = \sum_i \mu(n \otimes m_i) n_i = \sum_i \langle g(m_i), n \rangle n_i,$$

where $g\colon M \to M'^*$ is given by $\langle g(m), n \rangle = \mu(n \otimes m)$. Thus $\{n_i, g(m_i) | i = 1, \ldots, n\}$ is a dual basis for M', and g is surjective. It is clear that $f^* = g$, so f and g are injective, since g and f are surjective. Finally, assume that M is not faithful. Then there exists a nonzero idempotent e of R such that $eM = 0$. But then μ is not surjective, because $\operatorname{Im} \mu \subset (1 - e)R$. □

Theorem 1.2.3 *Take* $\underline{M} = (M, M', \mu) \in \underline{\underline{DP}}(R)$*, and let* $A = E_R(\underline{M})$*. Then we have a strict Morita context* (R, A, M', M, ϕ, I_A)*. The map* $\phi\colon M' \otimes_A M \to R$ *is given by* $\phi(m' \otimes m) = \mu(m' \otimes m)$*, for all* $m' \in M'$*,* $m \in M$*.*

Proof We define a left A-module structure on M and a right A-module on M' as follows:

$$\begin{aligned} (m \otimes m').n &= \mu(m' \otimes n)m \\ n'.(m \otimes m') &= \mu(n' \otimes m)m' \end{aligned}$$

for all $m, n \in M$, $m', n, \in N'$. An argument similar to the one used in the proof of Proposition 1.2.1 shows M is unital as a left A-module, and M' is unital as a right A-module. Obviously, I_A is an isomorphism. Let us show that ϕ is well-defined. We have the following commutative diagram with exact rows and columns

$$\begin{array}{ccccccc}
M' \otimes M \otimes M' \otimes M & \rightrightarrows & M' \otimes M & \xrightarrow{\pi} & M' \otimes_A M & \to & 0 \\
\downarrow{\scriptstyle \mu \otimes \mu} & & \downarrow{\scriptstyle \mu} & & \downarrow{\scriptstyle \phi} & & \\
R & \rightrightarrows & R & \to & R & \to & 0 \\
\downarrow & & \downarrow & & \downarrow & & \\
0 & & 0 & & 0 & &
\end{array}$$

In the top row we consider the maps

$$I_{M'} \otimes I_M \otimes \mu \text{ and } \mu \otimes I_{M'} \otimes I_M\colon M' \otimes M \otimes M' \otimes M \to M' \otimes M,$$

and $\pi\colon M' \otimes M \to M' \otimes_A M$ is the natural quotient map. In the bottom row all the maps considered are the identity on R. Now suppose that $\pi(\sum_i n_i \otimes m_i) = 0$. Then there exist $n'_j, n''_j \in M'$ and $m'_j, m''_j \in M$ such that

$$\sum_j (n'_j \otimes m'_j)\mu(n''_j \otimes m''_j) - \mu(n'_j \otimes m'_j)(n''_j \otimes m''_j) = \sum_i n_i \otimes m_i$$

Applying μ to both sides, we see that $\mu(\sum_i n_i \otimes m_i) = 0$, hence ϕ is well-defined. ϕ is surjective since μ is surjective. To show that ϕ is injective, consider $\sum_i n_i \otimes m_i \in \operatorname{Ker}\phi$. Then $\mu(\sum_i n_i \otimes m_i) = 0$, and

$$(I_{M'} \otimes I_M \otimes \mu - \mu \otimes I_{M'} \otimes I_M)(\sum_{i,j} n_i \otimes m_i \otimes q_j \otimes p_j) = \sum_i n_i \otimes m_i,$$

implying

$$\sum_i n_i \otimes m_i = 0$$

in $M' \otimes_A M$. This shows that ϕ is injective. We leave it to the reader to verify condition 3) in the definition of Morita contexts. □

1.3 Invertible modules

Let I be an R-module. Recall (cf. e.g. [75]) that the following conditions are equivalent:
1) I is projective of rank one;
2) $I \otimes I^* \cong R$;
3) I is invertible, that is, there exists an R-module J such that $I \otimes J \cong R$.
In the sequel, $\underline{\underline{\mathrm{Pic}}}(R)$ will be the full subcategory of $\underline{\underline{\mathrm{FP}}}(R)$ with invertible modules as objects. Similarly, $\underline{\underline{\mathrm{DPic}}}(R)$ will be the full subcategory of $\underline{\underline{\mathrm{DP}}}(R)$ with objects of the type (I, I', ϕ), with $I, I' \in \underline{\underline{\mathrm{Pic}}}(R)$. It follows easily from condition 2) above that any object of $\underline{\underline{\mathrm{DPic}}}(R)$ is isomorphic to an object of the form $(I, I^*, \langle\bullet,\bullet\rangle)$. This implies that

$$\mathrm{K}_0\underline{\underline{\mathrm{DPic}}}(R) \cong \mathrm{K}_0\underline{\underline{\mathrm{Pic}}}(R) \cong \mathrm{Pic}(R)$$

Recall that $\mathrm{Pic}(R)$, the *Picard group* of R is the group consisting of isomorphism classes of invertible R-modules. The group operation is induced by the tensor product. Also note that for $\underline{I} = (I, I', \phi) \in \underline{\underline{\mathrm{DPic}}}(R)$, the map

$$\phi \circ \tau\colon\ E_R(\underline{I}) = I \otimes I' \to R$$

is an isomorphism.

Consider two dual pairs $\underline{M}, \underline{N}$ and an isomorphism $\underline{f} = (f, f')\colon \underline{M} \to \underline{N}$ in $\underline{\underline{\mathrm{DP}}}(R)$. We have seen that $E(\underline{f}) = f \otimes f'\colon E_R(\underline{M}) \to E_R(\underline{N})$ is an isomorphism of R-algebras. In the next proposition we will describe to what extent an isomorphism between R-elementary algebra is of type $f \otimes f'$.

Proposition 1.3.1 *Let $\Phi\colon E_R(\underline{M}) \to E_R(\underline{N})$ be an isomorphism of elementary R-algebras. Then there exists $\underline{I} = (I, I', \phi) \in \underline{\underline{\mathrm{DPic}}}(R)$ and an isomorphism $\underline{f} = (f, f')\colon \underline{M} \to \underline{N} \otimes \underline{I}$ such that*

$$\Phi = E(\underline{f})\colon\ E_R(M) \to E_R(N \otimes I) \cong E_R(N) \otimes E_R(I) \otimes E_R(N),$$

$$\Phi = (I_N \otimes \phi \otimes I_{N'}) \circ (f \otimes f') \tag{1.6}$$

The quintuple (I, I', ϕ, f, f') *is unique in the following sense: suppose that* (J, J', ψ, g, g') *also satisfies condition (1.6). Then we have isomorphisms*

$$\alpha\colon I \to J \text{ and } \alpha'\colon I' \to J'$$

such that

$$g = (I_N \otimes \alpha) \circ f \quad ; \quad g' = (I_{N'} \otimes \alpha') \quad ; \quad \phi = (\alpha \otimes \alpha') \circ \psi.$$

Proof Write $\underline{M} = (M, M', \mu)$, $\underline{N} = (N, N', \nu)$, $A = E_R(\underline{M})$, $B = E_R(\underline{N})$. From the proof of Theorem 1.2.3, we know that M is a left unital A-module, and M' is a right unital A-module. Using Φ^{-1}, we can therefore put a left B-module structure on M and a right B-module structure on M' as follows:

$$\begin{aligned} (n \otimes n').m &= \Phi^{-1}(n \otimes n').m \\ m'.(n \otimes n') &= m'.\Phi^{-1}(n \otimes n') \end{aligned}$$

for $n \in N$, $m \in M$, $n' \in N'$, $m' \in M'$. With these definitions, $A = M' \otimes M$ is a unital (B, B')-bimodule, and Φ is a (B, B')-bimodule isomorphism. Using the Morita Theorem 1.1.4 and Theorem 1.2.3, we obtain that

$$\begin{aligned} M &\cong N \otimes (N' \otimes_B M) && \text{in } B\text{-mod} \\ M' &\cong (M' \otimes_B N) \otimes N' && \text{in mod-}B \end{aligned}$$

Write $I = N' \otimes_B M$, $I' = M' \otimes_B N$. Recall from Theorem 1.2.3 that $N' \otimes_B N \cong R$. Therefore we have an isomorphism

$$\phi = I_{N'} \otimes \Phi \otimes I_N\colon I \otimes I' = N' \otimes_B M \otimes M' \otimes_B N \to N' \otimes_B N \otimes N' \otimes_B N \cong R \otimes R = R$$

This implies that $I, I' \in \underline{\underline{\mathrm{Pic}}}(R)$. We also have isomorphisms

$$\begin{aligned} f&\colon M \to N \otimes I \\ f'&\colon M' \to I' \otimes N' \end{aligned}$$

given by

$$\begin{aligned} f^{-1}(n \otimes n' \otimes m) &= \Phi^{-1}(n \otimes n').m \\ f'^{-1}(m' \otimes n \otimes n') &= m'.\Phi^{-1}(n \otimes n') \end{aligned}$$

for $n \in N$, $m \in M$, $n' \in N'$, $m' \in M'$. We now claim that the map

$$(I_N \otimes \phi \otimes I_{N'}) \circ (f \otimes f')\colon M \otimes M' \to N \otimes I \otimes I' \otimes M' \to N \otimes N'$$

is equal to Φ. Indeed, consider $n \in N$, $n' \in N$ and $\sum_i q_i \otimes p_i \in N \otimes N'$ in the inverse image of 1 under ν. Then

$$\begin{aligned} &(f^{-1} \otimes f'^{-1})(I_N \otimes \phi \otimes I_{N'})^{-1}(n \otimes n') \\ &= (f^{-1} \otimes f'^{-1})(\sum_{i,j} n \otimes q_i \otimes \Phi^{-1}(p_i \otimes q_j) \otimes p_j \otimes n') \\ &= \sum_{i,j} \Phi^{-1}(n \otimes q_i)\Phi^{-1}(p_i \otimes q_j)\Phi^{-1}(p_j \otimes n') \\ &= \Phi^{-1}(\sum_{i,j}(n \otimes q_i)(p_i \otimes q_j)(p_j \otimes n')) \\ &= \Phi^{-1}(n \otimes n'). \end{aligned}$$

The uniqueness may be shown as follows: the map

$$g \circ f^{-1}\colon N \otimes I \to N \otimes J$$

is a B-module isomorphism, hence

$$I \cong N' \otimes_B N \otimes I \cong N' \otimes_B N \otimes J \cong J$$

in R-mod, and this yields an isomorphism $\alpha\colon I \to J$. The functorial properties imply that $I_N \circ \alpha = g \circ f^{-1}$. Similar arguments show that we have an isomorphism $\alpha'\colon I' \to J'$. Finally, consider the commutative diagram

$$\begin{array}{ccc} M \otimes M' & \stackrel{f \otimes f'}{\longrightarrow} & N \otimes I \otimes I' \otimes N' \\ \Big\downarrow{\scriptstyle g\otimes g'} & & \Big\downarrow{\scriptstyle I_N \otimes \phi \otimes I_{N'}} \\ N \otimes J \otimes J' \otimes N' & \stackrel{I_N \otimes \psi \otimes I_{N'}}{\longrightarrow} & N \otimes N' \end{array}$$

Applying the functor $N' \otimes_B \bullet \otimes_B N$ to this diagram, we obtain, after identifying $N' \otimes_B N \cong R$ that $\phi = (\alpha \otimes \alpha') \circ \psi$. We leave it to the reader to show that the map $\underline{f} = (f, \tau \circ f')$ defines an isomorphism $M \to N \otimes I$ in $\underline{\underline{\mathrm{DP}}}(R)$. □

1.4 Left modules versus bimodules

If two R-algebras A and B have a unit, then the categories A-mod-B and $A \otimes A^{\mathrm{op}}$-mod are obviously isomorphic. A similar result holds for the categories of unital (A, B)-bimodules and left unital $A \otimes B^{\mathrm{op}}$-modules, at least in the situation where A and B are unital R-algebras. This is somewhat more technical, and we will present some more detail in this Section.

If M is a unital (A, B)-bimodule, then we define $F(M) = M$ as an R-module, with an $A \otimes B^{\mathrm{op}}$-action given by the formula

$$(a \otimes b^{\mathrm{op}}) \cdot m = amb$$

for all $a \in A$, $b \in B$ and $m \in M$.

Lemma 1.4.1 *Let A and B be unital R-algebras, and M and N unital (A, B)-bimodules. Then*

$$\mathrm{Hom}_{(A,B)}(M, N) = \mathrm{Hom}_{A \otimes B^{\mathrm{op}}}(F(M), F(N))$$

Proof It is obvious that an (A, B)-linear map $f : M \to N$ is $A \otimes B^{\mathrm{op}}$-linear. Conversely, suppose that f is $A \otimes B^{\mathrm{op}}$-linear. Writing $m \in M$ under the form $m = \sum_i a_i m_i b_i$, we obtain

$$\begin{aligned} f(am) &= f(\sum_i a a_i m_i b_i) \\ &= \sum_i a a_i f(m_i) b_i \end{aligned}$$

$$\begin{aligned} &= \sum_i (aa_i \otimes b_i).f(m_i) \\ &= a(\sum_i (a_i \otimes b_i).f(m_i)) \\ &= a.f(\sum_i (a_i \otimes b_i)m_i) \\ &= a.f(m) \end{aligned}$$

so f is A-linear. In a similar way, we can show that f is B-linear. □

Lemma 1.4.2 *Let A and B be unital R-algebras, and M a unital (A,B)-bimodule. Then $F(M)$ is a unital $A \otimes B^{\text{op}}$-module. In particular, $A \otimes B^{\text{op}}$ is a unital R-algebra.*

Proof First we prove that, for any $A \otimes B^{\text{op}}$-module M, we have a natural isomorphism

$$(A \otimes B^{\text{op}}) \otimes_{A \otimes B^{\text{op}}} M \cong A \otimes_A M \otimes_B B \tag{1.7}$$

We define $\alpha : (A \otimes B^{\text{op}}) \otimes_{A \otimes B^{\text{op}}} M \longrightarrow A \otimes_A M \otimes_B B$ as follows:

$$\alpha((a \otimes b^{\text{op}}) \otimes m) = a \otimes m \otimes b$$

A straightforward computation shows that α is well-defined. α is bijective, its inverse is given by the formula

$$\alpha^{-1}(a \otimes m \otimes b) = (a \otimes b^{\text{op}}) \otimes m$$

If M is unital as a left A-module and as a right B-module, then $A \otimes_A M \otimes_B B \cong M$, and the Lemma follows. □

Proposition 1.4.3 *Let A and B be unital R-algebras. Then*

$$F : A\text{-mod}^{\text{u}}\text{-}B \longrightarrow A \otimes B^{\text{op}}\text{-mod}^{\text{u}}$$

is an isomorphism of categories.

Proof It follows from Lemmas 1.4.1 and 1.4.2 that F is a functor. We have to construct the inverse functor. Let M be a unital $A \otimes B^{\text{op}}$-module. On M, we define an A-action as follows:

$$\begin{aligned} A \otimes M &\cong A \otimes \Big((A \otimes B^{\text{op}}) \otimes_{A \otimes B^{\text{op}}} M\Big) \\ &\cong \Big(A \otimes (A \otimes B^{\text{op}})\Big) \otimes_{A \otimes B^{\text{op}}} M \\ &\xrightarrow{m_A \otimes I_{B^{\text{op}}} \otimes I_M} (A \otimes B^{\text{op}}) \otimes_{A \otimes B^{\text{op}}} M \cong M \end{aligned}$$

This means the following: take $a \in A$ and $m \in M$. We know that there exist $a_i \in A$, $b_i \in B$ and $m_i \in M$ such that $m = \sum_i (a_i \otimes b_i^{\text{op}}) \cdot m_i$. Then

$$am = \sum_i (aa_i \otimes b_i^{\text{op}}) \cdot m_i$$

In a similar way, we can introduce a right B-action on M. For $b \in B$, we have

$$mb = \sum_i (a_i \otimes (b_i b)^{\mathrm{op}}) \cdot m_i$$

Let $G(M)$ be M as an R-module, with the above introduced A and B-actions. Then $G(M)$ is an (A, B)-bimodule. Indeed, for $a \in A$, $b \in B$ and $m = \sum_i (a_i \otimes b_i^{\mathrm{op}}) \cdot m_i \in M$, we have

$$\begin{aligned} (am)b &= \sum ((aa_i \otimes b_i^{\mathrm{op}}) \cdot m_i)b \\ &= \sum_i (aa_i \otimes (b_i b)^{\mathrm{op}}) \cdot m_i \\ &= a(\sum_i (a_i \otimes (b_i b)^{\mathrm{op}}) \cdot m_i) \\ &= amb \end{aligned}$$

It follows from (1.7) that $G(M)$ is unital as A-module and as a B-module. Lemma 1.4.2 implies that G is a functor. It is clear that F and G are each others inverses. □

Chapter 2

Azumaya algebras and Taylor-Azumaya algebras

2.1 Central algebras, the separator and the trace map

Central algebras

Let R be a commutative ring, and A an R-algebra, not necessarily having a unit element. From now on, we will denote

$$A^e = A \otimes A^{\text{op}}$$

Obviously A is a left A^e-module; we define the *center* of A as follows:

$$\mathrm{Z}(A) = \mathrm{End}_{A^e}(A)$$

If A has a unit, then it is clear that

$$\mathrm{Z}(A) \cong \{a \in A | ab = ba, \text{ for all } b \in A\}$$

(map any α in the center to $\alpha(1_A)$). In general, we don't have this equality, there is even no natural embedding of $\mathrm{Z}(A)$ into A.

Lemma 2.1.1 *Suppose that $A^2 = A$. Then $\mathrm{Z}(A)$ is commutative, and A is a $\mathrm{Z}(A)$-algebra.*

Proof Take $\alpha, \beta \in \mathrm{Z}(A)$ and $a, b \in A$. Then

$$\begin{aligned}(\alpha \circ \beta)(ab) &= \alpha(\beta(ab)) = \alpha(a\beta(b)) = \alpha(a)\beta(b) \\ &= \beta(\alpha(a)b) = \beta(\alpha(ab)) = (\beta \circ \alpha)(ab))\end{aligned}$$

and the commutativity follows from the fact that we may write any $a \in A$ under the form $a = \sum_i a_i b_i$. It is clear that A is a $\mathrm{Z}(A)$-algebra under the action

$$\alpha.a = \alpha(a)$$

for any $\alpha \in Z(A)$, $a \in A$. □

We have a natural map $R \to Z(A)$. If this map is an isomorphism, then we say that A is a *central* R-algebra.

The Goldman module and the separator

Let A be an R-algebra. Recall that A may be viewed as a left A^e-module, as follows:

$$(b \otimes c).a = bac$$

for all $a, b, c \in A$. A viewed as a left A^e-module will be denoted by A_ℓ. Similarly, A_r will denote A with the right A^e-module structure

$$a.(b \otimes c) = cab$$

for all $a, b, c \in A$. With these conventions, $A_\ell \otimes A_r$ is an A^e-bimodule. A^e itself is also an A^e-bimodule, so we can consider

$$\Omega_A = \Omega = \mathrm{Hom}_{(A^e,A^e)}(A_\ell \otimes A_r, A^e),$$

that is, ω: $A \otimes A \to A \otimes A$ belongs to Ω if and only if for all $a, b, c, d \in A$:

$$\omega(cad \otimes b) = (c \otimes d)\omega(a \otimes b) \tag{2.1}$$
$$\omega(a \otimes dbc) = \omega(a \otimes b)(c \otimes d) \tag{2.2}$$

which can also be stated as follows: Ω preserves the two left A-actions on $A \otimes A$, while it interchanges the two right actions. We can restate (2.1) as follows: suppose that $\omega(a \otimes b) = \sum_i a_i \otimes b_i$. Then for all $c, d, f, g \in A$:

$$\omega(cad \otimes fbg) = \sum_i ca_ig \otimes fb_id \tag{2.3}$$

Ω is called the *Goldman module*; it is clear from (2.3) that we may also write

$$\Omega = \mathrm{Hom}_{(A^e,A^e)}(A^e, A_\ell \otimes A_r).$$

We define an A^e-bimodule map

$$\theta\colon\ A_\ell \otimes A_r \otimes \Omega \to A^e$$

called the *separator* as follows:

$$\theta(a \otimes b \otimes \omega) = \omega(a \otimes b)$$

The trace map

Fix $\omega \in \Omega$, and $b \in A$. We define a new map

$$\mathrm{tr}_\omega(b)\colon\ A \to A$$

by

$$\mathrm{tr}_\omega(b)(a) = (m_A \circ \omega)(a \otimes b)$$

Here m_A is the multiplication map $A \otimes A \to A$. With these notations, we have

Lemma 2.1.2 *Suppose that $A^2 = A$. Then for all $a, b, c \in A$ and $\omega \in \Omega$, we have*
1) $\mathrm{tr}_\omega(b) \in \mathrm{Z}(A)$;
2) $\mathrm{tr}_\omega\colon\ A \to \mathrm{Z}(A)$ *is* $\mathrm{Z}(A)$*-linear;*
3) $\mathrm{tr}_\omega(ab) = tr_\omega(ba)$;
4)

$$\mathrm{tr}_\omega(ab)(c) = \omega(c \otimes a).b = a.\omega(b \otimes c) \tag{2.4}$$

Proof For all $a, b, c, d \in A$, we have

$$\begin{aligned}\mathrm{tr}_\omega(b)(cad) &= (m_A \circ \omega)(cad \otimes b)\\ &= m_A((c \otimes d)\omega(a \otimes b))\\ &= cm_A(\omega(a \otimes b))d\\ &= c\mathrm{tr}_\omega(b)d\end{aligned}$$

proving 1). We can view $A_\ell \otimes A_r$ and A^e as right A^e-modules, and therefore also as (A^{op}, A)-bimodules, by Proposition 1.4.3, and ω is an (A^{op}, A)-bimodule isomorphism. Therefore, for all $a, b, c \in A$ we have that $\omega(c \otimes a) = \sum_i c_i \otimes a_i$ implies that

$$\begin{aligned}\omega(c \otimes ab) &= \sum_i c_i b \otimes a_i\\ \omega(c \otimes ba) &= \sum_i c_i \otimes ba_i\end{aligned}$$

hence

$$\begin{aligned}\mathrm{tr}_\omega(ab)(c) &= \sum_i c_i ba_i = \omega(c \otimes a).b\\ \mathrm{tr}_\omega(ba)(c) &= \sum_i c_i ba_i = a.\omega(b \otimes c)\end{aligned}$$

and this proves 3) and 4). To show 2), we take $\alpha \in \mathrm{Z}(A)$ and $a, b, c \in A$. Then

$$\begin{aligned}\mathrm{tr}_\omega(\alpha(ab))(c) &= \mathrm{tr}_\omega(\alpha(a).b)(c)\\ &= \omega(c \otimes b).\alpha(a)\\ &= \alpha(\omega(c \otimes b)a)\\ &= \alpha(\mathrm{tr}_\omega(ab)(c))\end{aligned}$$

and this implies 2) since $A^2 = A$. □

Using Lemma 2.1.2, we obtain a map

$$\text{tr}: A \otimes \Omega \to Z(A)$$

defined by $\text{tr}(a \otimes \omega) = \text{tr}_\omega(a)$. We call this map the *trace map.*

Proposition 2.1.3 *Let A be a faithful unital R-algebra, and consider a submodule $\Omega' \subset \Omega$ which is invertible as an R-module. If for all $\omega \in \Omega'$, $\text{tr}_\omega(A) \subset R$, then we have a Morita context*

$$(R, A^e, A_r \otimes \Omega', A_\ell, t, \theta)$$

where t: $(A_r \otimes \Omega') \otimes_{A^e} A_\ell \to R$ is defined by

$$t(a \otimes \omega \otimes b) = \text{tr}_\omega(ab)$$

for all $a, b \in A$, $\omega \in \Omega$.

Proof From the fact that A is faithful, it follows that $R \subset Z(A)$. t is well-defined, since for all $\omega \in \Omega', a \in A_r, c \in A_\ell, b \otimes d \in A^e$, we have, by lemma 2.1.2:

$$\begin{aligned} t(a.(b \otimes d) \otimes \omega \otimes c) &= t(dab \otimes \omega \otimes c) \\ &= \text{tr}_\omega(dabc) \\ &= \text{tr}_\omega(abcd) \\ &= t(a \otimes \omega \otimes bcd) \\ &= t(a \otimes \omega \otimes (b \otimes d).c) \end{aligned}$$

We also have to verify condition 3) of definition 1.1.2, that is, for all $\omega', \omega'' \in \Omega, a, b, c \in A$:

$$\begin{aligned} t((a \otimes \omega') \otimes b)(c \otimes \omega'') &= a.\theta(b \otimes c \otimes \omega') \otimes \omega'' \\ \theta(b \otimes c \otimes \omega')a &= bt(c \otimes \omega' \otimes a) \end{aligned}$$

The second identity amounts to

$$\omega'(b \otimes c)a = \text{tr}_{\omega'}(ac)b$$

which is exactly 4) of lemma 2.1.2. Observe that it suffices to show the first identity after localization at an arbitrary prime ideal p of R, that is, it suffices to consider the case where R is local. But if R is local, then Ω' is free of rank one, hence we may assume that $\Omega' = R\omega$, and therefore $\omega' = r\omega$, $\omega'' = s\omega$ for some $r, s \in R$. Then the first identity amounts to

$$rs\ \text{tr}_\omega(ab)(c) = rs\ a.\omega(b \otimes c)$$

which is again 4) of lemma 2.1.2. □

2.2 Taylor-Azumaya algebras

Definition 2.2.1 *A unital faithful R-algebra A is called an Taylor-Azumaya algebra if there exists an invertible R-submodule Ω' of Ω such that the Morita context of Proposition 2.1.3 is strict. If a Taylor-Azumaya algebra has a unit, then we call it an Azumaya algebra.*

Before stating an example, let us reformulate the definition. Suppose that A is a faithful R-algebra, and that I is an invertible R-module. Suppose that

$$\psi\colon\ A_\ell \otimes A_r \otimes I \to A^e$$

is an A^e-bimodule isomorphism. Then we have a monomorphism $\iota\colon\ I \to \Omega$ given by

$$\iota(x)(a\otimes b) = \psi(a\otimes b\otimes x) = \theta(a\otimes b\otimes \iota(x))$$

for all $a, b \in A$ and $x \in I$. Define for $x \in I$ and $b \in A$:

$$T(b\otimes x)\colon\ A \to A$$

by $T(b\otimes x)(a) = m_A \circ \psi(a\otimes b\otimes x)$. Then it follows easily that $T(b\otimes x) = \mathrm{tr}_{\iota(x)}(b) \in \mathrm{Z}(A)$. If for all $b \in A$, $T(b\otimes x) \in R \subset \mathrm{Z}(A)$, then we have that

$$m_A \circ \psi = I_A \otimes T$$

Since m_A and ψ are surjective, T is surjective. From Proposition 1.1.5, it then follows easily that the Morita context $(R, A^e, A_\ell, A_r \otimes \iota(I), t, \theta)$ is strict. Hence A is a Taylor-Azumaya algebra. We therefore have shown:

Proposition 2.2.2 *Let A be a faithful unital R-algebra. Then A is a Taylor-Azumaya algebra if and only if there exists an invertible R-module I and an A^e-bimodule isomorphism*

$$\psi\colon\ A_\ell \otimes A_r \otimes I \to A^e$$

such that the map $T\colon\ A\otimes I \to \mathrm{Z}(A)$ given by $T(b\otimes x)(a) = m_A \circ \psi(a\otimes b\otimes x)$ takes values in R.

Example 2.2.3 *Any R-elementary algebra is a Taylor-Azumaya algebra.*

Proof Let $A = E_R(M, M', \mu)$, and consider $\omega\colon\ A_\ell \otimes A_r \to A^e$ defined by

$$\omega((m\otimes m')\otimes(n\otimes n')) = (m\otimes n')\otimes(n\otimes m')$$

It is straightforward to verify that ω is an A^e-bimodule isomorphism. Furthermore

$$\mathrm{tr}_\omega(n\otimes n')(m\otimes m') = m_A(\omega((m\otimes m')\otimes(n\otimes n'))) = \mu(n'\otimes n)(m\otimes m')$$

such that $\mathrm{tr}_\omega = \mu\circ\tau$ is surjective. The module $\Omega' = R\omega$ therefore satisfies the conditions of Definition 2.2.1. □

As a consequence to Proposition 1.1.5, we immediately have the following properties:

Proposition 2.2.4 *Suppose that A is a Taylor-Azumaya algebra. Then we have the following*

1. $A_r \otimes \Omega' \cong \mathrm{Hom}_{A^e}(A_\ell, A^e)$
2. $A_\ell \cong \mathrm{Hom}_{A^e}(A_r \otimes \Omega', A^e)$.
3. $R = \mathrm{End}_{A^e}(A_\ell)$
4. $A^e \cong E_R(A_\ell, A_r \otimes \Omega', t)$

This entails that A is R-central, R-faithful, and that A^e is an R-elementary algebra. Furthermore, A_ℓ is a finitely generated A^e-module, having a dual basis. Therefore A_ℓ is A^e-projective.

Proof All the assertions follow from Proposition 1.1.5. We only have to prove the fourth statement. The map

$$\theta\colon\ A_\ell \otimes A_r \otimes \Omega \to A^e$$

induces a multiplicative structure on $A_\ell \otimes A_r \otimes \Omega$ as follows:

$$\begin{aligned}(a\otimes b\otimes\omega)(c\otimes d\otimes\omega') &= \theta^{-1}(\theta(a\otimes b\otimes\omega)\theta(c\otimes d\otimes\omega'))\\ &= \theta^{-1}(\theta(\theta(a\otimes b\otimes\omega).c\otimes d\otimes\omega'))\\ &= \omega(a\otimes b).c\otimes d\otimes\omega'\\ &= \mathrm{tr}_\omega(bc)a\otimes d\otimes\omega'\\ &= t(b\otimes\omega\otimes c)a\otimes d\otimes\omega'\end{aligned}$$

□

Proposition 2.2.5 *Let A be a faithful unital R-algebra. A is a Taylor-Azumaya algebra if and only if there exists an invertible R-module I such that the functors*

$$F\colon\ R\text{-mod} \to A^e\text{-mod}^u\colon\ N \to A_\ell \otimes N$$

and

$$G\colon\ A^e\text{-mod}^u \to R\text{-mod}\colon\ M \to (A_r \otimes I) \otimes_{A^e} M$$

are inverse equivalences.

Proof One implication follows immediately from the definition and Proposition 1.1.5. Conversely, suppose that F and G are inverse equivalences. Then

$$R = \mathrm{End}_R(R) = \mathrm{End}_{A^e}(F(R)) = \mathrm{End}_{A^e}(A_\ell)$$

hence A is central. We have isomorphisms

$$\psi\colon\ A_\ell \otimes A_r \otimes I \to A^e$$

in A^e-mod and

$$\phi\colon (A_r \otimes I) \otimes_{A^e} A_\ell \to R$$

in R-mod. Consider the square

$$\begin{array}{ccc} A_\ell \otimes A_r \otimes I \otimes_{A^e} A_\ell & \xrightarrow{\psi \otimes I} & A^e \otimes_{A^e} A_\ell \\ \downarrow{\scriptstyle I\otimes\phi} & & \downarrow{\scriptstyle \cong} \\ A_\ell \otimes R & \xrightarrow{\cong} & A^e \end{array}$$

Since all the maps in this square are isomorphisms, $(I \otimes \phi) \circ (\psi \otimes I)^{-1}$ is an A^e-isomorphism of A_ℓ, and is therefore given by multiplication by a unit $x \in R$, since A is central. Replacing ϕ by $x^{-1}\phi$, we may assume that the above square is commutative. This implies that

$$(I \otimes \phi)(c \otimes a \otimes \nu \otimes b) = (\psi \otimes I)(c \otimes a \otimes \nu \otimes b) \tag{2.5}$$

For all $\nu \in I$, define $\omega_\nu\colon A_\ell \otimes A_r \to A^e$ by

$$\omega_\nu(a \otimes b) = \psi(a \otimes b \otimes \nu).$$

Then

$$\omega_\nu(a \otimes b) = \theta(a \otimes b \otimes \omega_\nu)$$

and

$$\Omega' = \{\omega_\nu | \nu \in I\} \cong I$$

as an R-module. From (2.5), it follows that

$$\begin{aligned} \phi(a \otimes \nu \otimes b)c &= \psi(c \otimes a \otimes \nu).b \\ &= \omega_\nu(c \otimes a).b \\ &= \mathrm{tr}_{\omega_\nu}(ab)c \end{aligned}$$

Since A is a faithful R-algebra, this last assertion implies that

$$\phi(a \otimes \nu \otimes b) = \mathrm{tr}_{\omega_\nu}(ab)$$

and therefore Ω' satisfies the conditions of definition 2.2.1. □

Let us next show that the R-module Ω' occuring in Definition 2.2.1 is necessarily the whole of Ω.

Proposition 2.2.6 *Suppose that A is a Taylor-Azumaya algebra. Then $\Omega' = \Omega$.*

Proof It suffices to show that for all $p \in \mathrm{Spec}(R)$, $\Omega'_{A_p} = \Omega_{A_p}$. It is therefore enough to show the property for R local. In this case, we may assume that $\Omega' = R\omega$, with $\omega\colon A_\ell \otimes A_r \to A^e$ an isomorphism. Then the map

$$\mathrm{Hom}_{A^e,A^e}(A_\ell \otimes A_r, A_\ell \otimes A_r) \to \mathrm{Hom}_{A^e,A^e}(A_\ell \otimes A_r, A^e)\colon\ f \mapsto \omega \circ f$$

is an isomorphism. Now we have category equivalences between R-mod and A^e-mod and between mod-R and mod-A^e. Hence

$$\Omega \cong \mathrm{Hom}_{A^e,A^e}(A_\ell \otimes A_r, A_\ell \otimes A_r) = \mathrm{Hom}_{R,A^e}(A_r, A_r) = \mathrm{Hom}_{R,R}(R, R) = R \cong \Omega'.$$

□

In Example 2.2.3, Ω' happens to be free as an R-module. We will show later on that for any Taylor-Azumaya algebra A that is finitely generated as an R-module, Ω_A and is free of rank one. At the moment, we can only show that $\Omega \otimes \Omega$ is free of rank one:

Proposition 2.2.7 *Let A be a Taylor-Azumaya algebra. Then $\Omega \otimes \Omega \cong R$ as R-modules.*

Proof We know that $\Omega = \mathrm{Hom}_{A^e}(A_\ell \otimes A_r, A^e) = \mathrm{Hom}_{A^e}(A^e, A_\ell \otimes A_r)$. We therefore have a map ρ: $\Omega \otimes \Omega \to \mathrm{End}_{A^e}(A^e)$, given by

$$\rho(\omega \otimes \nu) = \nu \circ \omega.$$

It is easy to see that the following diagram is commutative:

$$\begin{array}{ccc} A \otimes A \otimes \Omega \otimes \Omega & \stackrel{\theta \otimes I_\Omega}{\longrightarrow} & A \otimes A \otimes \Omega \\ \downarrow{\scriptstyle I_{A \otimes A} \otimes \rho} & & \downarrow{\scriptstyle \theta} \\ A \otimes A \otimes R & \stackrel{\cong}{\longrightarrow} & A \otimes A \end{array}$$

Therefore ρ is surjective, since θ is surjective. Hence ρ is an isomorphism, since Ω is projective of rank one. □

Let us now give another characterization of Taylor-Azumaya algebras ; let A be a unital, faithful R-algebra, and consider the functors

$$\begin{aligned} F_\ell = A_\ell \otimes \bullet \quad &: \quad R\text{-mod} \to A^e\text{-mod}^u \\ G_\ell = \mathrm{Hom}_{A^e}(A_\ell, \bullet) \quad &: \quad A^e\text{-mod}^u \to R\text{-mod} \end{aligned}$$

It is easy to see that G_ℓ is a right adjoint of F_ℓ: for $M \in R$-mod, $N \in A^e$-mod^u, define

$$\mathrm{Hom}_{A^e}(A_\ell \otimes M, N) \underset{\Psi}{\overset{\Phi}{\rightleftarrows}} \mathrm{Hom}_R(M, \mathrm{Hom}_{A^e}(A_\ell, N))$$

as follows

$$\begin{aligned} \Phi(f)(m)(a) &= f(a \otimes m) \\ \Psi(g)(a \otimes m) &= g(m)(a) \end{aligned}$$

for all $f \in \mathrm{Hom}_{A^e}(A_\ell \otimes M, N)$, $g \in \mathrm{Hom}_R(M, \mathrm{Hom}_{A^e}(A_\ell, N))$, $m \in M$, $a \in A$. Then Φ and Ψ are well-defined, and each others inverses. In the next Proposition, we will show that A is Taylor-Azumaya if and only if F_ℓ and G_ℓ are inverse equivalences.

Proposition 2.2.8 *Let A be a unital, faithful R-algebra. Then the following assertions are equivalent:*

1. *A is a Taylor-Azumaya algebra;*
2. *the adjoint functors*
$$\begin{cases} F_\ell = A_\ell \otimes \bullet\colon\ R\text{-mod} \longrightarrow A^e\text{-mod}^u \\ G_\ell = \mathrm{Hom}_{A^e}(A_\ell, \bullet)\colon\ A^e\text{-mod}^u \longrightarrow R\text{-mod} \end{cases}$$
are inverse equivalences;
3. *the adjoint functors*
$$\begin{cases} F_r = \bullet \otimes A_r\colon\ R\text{-mod} \longrightarrow \text{mod}^u\text{-}A^e \\ G_r = \mathrm{Hom}_{A^e}(A_r, \bullet)\colon\ \text{mod}^u\text{-}A^e \longrightarrow R\text{-mod} \end{cases}$$
are inverse equivalences.

Proof 1. $\Rightarrow$ 2. follows from Proposition 2.2.5 and assertion 5. from Proposition 1.1.5
2. $\Leftrightarrow$ 3. is obvious, since $A^e\text{-mod}^u \cong \text{mod}^u\text{-}(A^e)^{\text{op}} \cong \text{mod}^u\text{-}A^e$.
2. $\Rightarrow$ 1. Write $P = G_\ell(A^e) = \mathrm{Hom}_{A^e}(A_\ell, A^e)$. The right A^e-module structure on A^e induces a right A^e-module structure on P. We will show that $G_\ell(\bullet) \cong P \otimes_{A^e} \bullet$. Take $M \in A^e\text{-mod}^u$. Then there exists a surjection $R^J \to G_\ell(M)$ in R-mod, for some index set J. Applying the functor F_ℓ, we obtain a surjection $A_\ell^J \to F_\ell(G_\ell(M)) \cong M$ of unital A^e-modules. If we take the composition with the multiplication map $m_A^J\colon (A^e)^J \to A_\ell^J$, then we obtain a surjection $(A^e)^J \to M$. Call K the kernel of this surjection. Invoking the fact that R-mod and $A^e\text{-mod}^u$ are equivalent, and using the fact that R-mod has kernels and cokernels, it follows that $A^e\text{-mod}^u$ has kernels and cokernels. More precisely, K is a unital A^e-module, hence we have a surjection $(A^e)^I \to K$ for some index set I. We therefore have an exact sequence of A^e-modules:
$$(A^e)^I \to (A^e)^J \to M \to 0 \tag{2.6}$$
Now we have a natural transformation
$$\Phi\colon\ P \otimes_{A^e} \bullet \longrightarrow \mathrm{Hom}_{A^e}(A_\ell, \bullet)$$
with $\Phi(M)\colon\ P \otimes_{A^e} M \to \mathrm{Hom}_{A^e}(A_\ell, M)$ given by
$$\Phi(M)(f \otimes m)(a) = f(a).m$$
for $f \in P = \mathrm{Hom}_{A^e}(A_\ell, A^e)$, $m \in M$, $a \in A_\ell$. Clearly $\Phi(A^e)$ is the identity on P, and, for any index set I, $\Phi((A^e)^I) = \Phi(A^e)^I$ is the identity on P^I. Applying this to (2.6), we obtain the following diagram with exact rows:
$$\begin{array}{ccccccc} P^I & \longrightarrow & P^J & \longrightarrow & P \otimes_{A^e} M & \longrightarrow & 0 \\ \downarrow\cong & & \downarrow\cong & & \downarrow\Phi(M) & & \\ P^I & \longrightarrow & P^J & \longrightarrow & \mathrm{Hom}_{A^e}(A_\ell, M) & \longrightarrow & 0 \end{array}$$

and it follows from the lemma of five that $\Phi(M)$ is an isomorphism.
We have seen already that condition 3. holds if 2. holds. Apply the functor G_r to P:

$$G_r(P) = \mathrm{Hom}_{A^e}(A_r, \mathrm{Hom}_{A^e}(A_\ell, A^e)) = \mathrm{Hom}_{A^e,A^e}(A_\ell \otimes A_r, A^e) = \Omega$$

and therefore

$$P \cong F_r(G_r(P)) = A_r \otimes \Omega.$$

If we can show that Ω is invertible as an R-module, then the conditions of Proposition 2.2.5 are fulfilled, and then it follows that A is an Taylor-Azumaya algebra, finishing the proof. To show that Ω is invertible, observe that the isomorphism

$$\psi\colon\ A_r \otimes \Omega \to P = \mathrm{Hom}_{A^e}(A_\ell, A^e)$$

is given by $\psi(a \otimes \omega)(b) = \omega(b \otimes a)$. The isomorphism

$$\varphi\colon\ A_\ell \otimes P \to A^e$$

is given by $\varphi(a \otimes f) = f(a)$, hence we obtain an isomorphism

$$A_\ell \otimes A_r \otimes \Omega \to A^e$$

as follows :

$$(\varphi \circ (I_{A_\ell} \otimes \psi))(b \otimes a \otimes \omega) = \omega(b \otimes a) = \theta(b \otimes a \otimes \omega)$$

Therefore the separator θ is an isomorphism. As in the proof of the Proposition 2.2.7, it follows that the map

$$I_{A \otimes A} \otimes \rho\colon\ A_\ell \otimes A_r \otimes \Omega \otimes \Omega \to A_\ell \otimes A_r \otimes R$$

is an isomorphism. Apply successively G_ℓ and G_r to this isomorphism. Then we obtain that $\Omega \otimes \Omega \cong R$, and Ω is invertible. □

Let us next focus on the case where A has a unit. The next Proposition will tell us that in this case we reobtain the classical definition of Azumaya algebra, cf. e.g. [114, III.5.1].

Proposition 2.2.9 *Let A be an R-algebra with a unit. Then the following assertions are equivalent:*

1. *A is an Azumaya algebra;*

2. *A is faithfully projective as an R-module, and the map*

$$F\colon A^e \to \mathrm{End}_R(A)$$

given by

$$F(a \otimes b)(x) = axb$$

is an isomorphism;

3. A is an R-central A^e-progenerator.

Proof 1 ⇒ 2. Since A has a unit, it follows from Proposition 1.1.5 that A is finitely generated and projective as an R-module. From 2.2.4, it follows that A is faithful, so A is a faithfully projective R-module. Consider the map k: $A_r \otimes \Omega' \to A_\ell^*$ given by

$$k(b\otimes\omega)(a) = \mathrm{tr}_\omega(ab) = f(a\otimes\omega\otimes b)$$

It follows from 1.1.5 that k is a bijection, since A has a unit element. We therefore have the following commutative diagram of A^e-bimodule homomorphisms:

$$\begin{array}{ccc} A_\ell\otimes A_r\otimes\Omega & \xrightarrow{\theta} & A^e \\ \downarrow{\scriptstyle I\otimes k} & & \downarrow{\scriptstyle F} \\ A_\ell\otimes A_\ell^* & \xrightarrow{\psi} & \mathrm{End}_R(A) \end{array}$$

Where ψ is the canonical isomorphism. The diagram is commutative since

$$\begin{aligned} F(\theta(a\otimes b\otimes\omega))(x) &= \omega(a\otimes b).x \\ &= \mathrm{tr}_\omega(bx)a \\ &= \mathrm{tr}_\omega(xb)a \\ &= k(b\otimes\omega)(x)a \\ &= \psi(a\otimes k(b\otimes\omega))x \end{aligned}$$

It follows that F is an isomorphism.
2 ⇒ 1. Consider the dual pair $(A, A^*, \langle\bullet,\bullet\rangle)$, cf. 1.2.2. By 1.2.3, we have a strict Morita context

$$(R, A\otimes A^*, A^*, A, \phi, I_{A\otimes A^*})$$

with $\phi(a^*, a) = < a^*, a >$. Now we have isomorphisms

$$A\otimes A^* \xrightarrow{\psi} \mathrm{End}_R(A) \xrightarrow{F^{-1}} A^e$$

hence a strict Morita context

$$(R, A^e, A^*, A_\ell, \phi, F^{-1}\circ\psi).$$

From 1.1.5, we know that A_ℓ is A^e-projective, and that $A^* = \mathrm{Hom}_{A^e}(A_\ell, A^e)$. Hence

$$A^*\otimes_{A^e} M = \mathrm{Hom}_{A^e}(A_\ell, A^e)\otimes_{A^e}\mathrm{Hom}_{A^e}(A^e, M) = \mathrm{Hom}_{A^e}(A_\ell, M) = M^A$$

and it follows from the Proposition 2.2.8 that A is an Azumaya algebra.
2 ⇒ 3. From Proposition 1.1.6, it follows that $(A^e \cong \mathrm{End}_R(A), R, A, A^e, A^*, f, g)$ is a strict Morita context. From Proposition 1.1.5, it then follows that A is an A^e-progenerator and that $R \cong \mathrm{End}_{A^e}(A) = Z(A)$.
3 ⇒ 2. From Proposition 1.1.6, it now follows that $(R \cong \mathrm{End}_{A^e}(A), A^e, A, \mathrm{Hom}_{A^e}(A, A^e), f, g$ is a strict Morita context. From Proposition 1.1.5, we deduce that A is an R-progenerator (that is, A is faithfully projective), and that $A^e \cong \mathrm{End}_R(A)$. □

Proposition 2.2.10 *If A and B are Taylor-Azumaya algebras, then $A\otimes B$ and A^{op} are also Taylor-Azumaya algebras.*

Proof This follows immediately from Proposition 1.1.7 and the definition of Taylor-Azumaya algebra. □

Proposition 2.2.11 *Suppose k is a field. Then A is a k-Azumaya algebra if and only if A is a central simple k-algebra.*

Proof Suppose that A is an Azumaya algebra. From the preceding Proposition, we know that A is a faithfully projective k-module, hence a finite dimensional k-vector space. A is k-central, by Proposition 2.2.4. From the fact that the categories A^e-mod and k-mod are equivalent, and the fact that k has only one ideal, it follows that A is simple. Hence A is central simple. Conversely, suppose that A is central simple. Then it is well-known, cf. e.g. [110, I.6.4] that the canonical map $F\colon A^e \to \mathrm{End}_R(A)$ is an isomorphism. Hence A is Azumaya, by the preceding Proposition. □

Azumaya algebras have the property that they may always be defined over a noetherian subring:

Proposition 2.2.12 *Let A be an Azumaya algebra. Then there exists a noetherian subring R_0 of R, and a subalgebra A_0 of A such that A_0 is R_0-Azumaya, and $A_0 \otimes_{R_0} R \cong A$.*

Proof It is well-known (cf. e.g. [114, Prop. I.2.8]) that a morphism $f : P \to Q$ between finitely generated projective modules can be defined over a noetherian subring R_0: there exist finitely generated projective R_0-modules P_0 and Q_0 such that $P_0 \otimes_{R_0} R \cong P$, $Q_0 \otimes_{R_0} R \cong Q$ and $f = f_0 \otimes I_R$. If A and B are R-algebras, and $f : A \to B$ is multiplicative, then we can take R_0 such that A_0 and B_0 are R_0 algebras and f_0 is multiplicative. If f is an isomorphism, then f_0 is also an isomorphism.
It suffices to apply this property to the isomorphism $A \otimes A^{\mathrm{op}} \to \mathrm{End}_R(A)$. □

Originally, Azumaya algebras were introduced as central separable algebras (cf. [12]), and the same idea was used to introduce Taylor-Azumaya algebras ([175]). In Chapter 4, we will discuss separable algebras without a unit, and we will show that the Taylor-Azumaya algebras that we introduced are exactly the central separable algebras. This theory is rather technical, and that is why we collected it in a separate Chapter, independent of the rest of this book. If A has a unit, then everything becomes less complicated, and we will give details below. We will show below that an Azumaya algebra is exactly a central separable algebra.
Recall that an R-algebra (with unit) is called *separable* if and only if one of the following equivalent conditions holds:
1) A is projective as an A^e-module;
2) the sequence

$$0 \longrightarrow \mathrm{Ker}\,(m_A) \longrightarrow A^e \xrightarrow{m_A} A \longrightarrow 0$$

splits as a sequence of R-modules;
3) A^e contains an element e such that $m_A(e) = 1$ and $\mathrm{Ker}\,(m_A)e = 0$.

e is an idempotent, called the *separability idempotent* of A. If φ_A is the right inverse of m_A, then $e = \varphi_A(1)$. For a detailed discussion of separable algebras, we refer to [75].
A is called *central separable* if A is central and separable. The following properties of central separable algebras are easy to show. If S is a commutative R-algebra, and A is a central separable R-algebra, then $S \otimes A$ is a central separable S-algebra. If A and B are central separable R-algebras, then A^{op}, $A \otimes B$ and A^e are also central separable R-algebras.
If A is a central separable R-algebra, then R embeds as a direct summand in A: a projection $t: A \to R$ in R-mod is given by the formula

$$t(a) = \sum_i a_i a b_i$$

where we wrote $e = \sum_i a_i \otimes b_i$ for the separability idempotent.
Another elementary property of separable algebras is the following.

Lemma 2.2.13 *Let A and B be R-algebras with a unit. Suppose that A is faithfully projective as an R-module and $A \otimes B$ is separable as an R-algebra. Then B is separable R-algebras.*

Proof Let $\eta: R \to A$ be the unit map. η is injective, so the dual map $\eta^*: A^* \to R$ is surjective, and splits since A and A^* are projective. This implies that $\eta: R \to A$ has a section.
Now $A = A_0 \oplus R$ as R-modules, and $A \otimes B = (A_0 \otimes B) \oplus B$ as B^e-modules. Now $A \otimes B$ is a direct factor of $(A \otimes B)^e \cong A^e \otimes B^e$ as an $A^e \otimes B^e$-module and as a B^e-module. Thus B is a direct factor of $A^e \otimes B^e$ as a B^e-module. $A^e \otimes B^e$ is projective as a B^e-module, so B is projective as a B^e-module. □

If A is a central separable algebra over a field k, then A is a central simple k-algebra. In fact, we have a structure Theorem for separable algebras over a field k, see [114, Theorem 3.1].

Lemma 2.2.14 *Let A a central separable R-algebra. If A is simple, then R is a field.*

Proof Let $I \neq 0$ be an ideal of R. Then IA is a two-sided ideal of A, and $IA = A$ since A is simple. Applying the projection t defined above, we find that

$$R = t(A) = t(IA) = It(A) = I$$

and it follows that R has no nontrivial ideals, and R is a field. □

Lemma 2.2.15 *Let A a central separable R-algebra. If M is a maximal two-sided ideal of A, then $M \cap R = I$ is a maximal ideal of R, and $M = IA$.*

Proof It is straightforward to show that A/M is central and separable as an R/I-algebra. Since M is maximal, A/M is simple, and R/I is a field by Lemma 2.2.14. Hence I is a maximal ideal of R.
$A/IA = A \otimes R/I$ is central separable as an R/I-algebra. R/I is a field, so A/IA is simple. But then IA is a maximal two-sided ideal of A. $IA \subset M$, so we necessarily have that $IA = M$. □

We are now able to give the following characterization of Azumaya algebras.

Proposition 2.2.16 *Let A be an R-algebra with unit. A is an Azumaya algebra if and only if A is a central separable algebra.*

Proof If A is Azumaya, then A is an R-central A^e-progenerator (see Proposition 2.2.9). Then A is A^e-projective and separable.
Conversely, suppose that A is central and separable. A is A^e-projective. If we can show that A is an A^e-generator, then A is an R-central A^e-progenerator, and our result follows from Proposition 2.2.9. Let

$$T = \{\sum_i f_i(x_i) \mid x_i \in A,\ f_i \in \mathrm{Hom}_{A^e}(A, A^e)\}$$

Take $f \in \mathrm{Hom}_{A^e}(A, A^e)$ and $x \in A$. Then

$$f(1) = f(m_A(e)) = (f \circ m_A)(e^2) = e(f \circ m_A)(e) = ef(1) \in eA^e$$

and

$$f(x) = (x \otimes 1)f(1) \in A^e eA^e$$

so $T \subset A^e eA^e$.
If we take $f(1) = 1$ and $x = 1$, then we see that $e \in T$. T is a two-sided ideal of A^e, so $T = A^e eA^e$.
Assume that $T \neq A^e$. Then T is contained in a maximal ideal M of A^e. A^e is a central separable R-algebra, and it follows from Lemma 2.2.15 that

$$A^e eA^e = T \subset M = (M \cap R)A^e$$

Applying m_A to both sides, we find

$$A \subset (M \cap R)A \quad \text{and} \quad A = (M \cap R)A$$

Now we apply t to both sides, and find that

$$R = t(A) = (M \cap R)t(A) = M \cap R$$

and

$$M = (M \cap R)A = RA = A$$

which is a contradiction. Thus $T = A^e$, and A is an A^e-progenerator. □

As an application, we have the following result.

Proposition 2.2.17 *Suppose that A and B are Azumaya algebras with a unit, and that $A \otimes B = C$ is an Azumaya algebra. Then A and B are also Azumaya algebras.*

Proof C is faithfully projective as an R-module, and the same property holds for A and B. From Lemma 2.2.13, it follows tat A and B are separable. Now $R = Z(C) = Z(A \otimes B) = Z(A) \otimes Z(B)$, so $Z(A)$ and $Z(B)$ are projective of rank one as R-modules. $Z(A) \subset Z(C) = R$, so we necessarily have that $Z(A) = R$. In a similar way, $Z(B) = R$, and A and B are central separable algebras, and therefore Azumaya algebras, by Proposition 2.2.16. □

2.3 The Rosenberg-Zelinsky exact sequence

Let A be an R-algebra with a unit. It is clear that a left A-endomorphism ν of A is given by right multiplication by $u = \nu(1)$. Similarly, a right A-module homomorphism ν' is given by left multiplication by $u' = \nu'(1)$. If A has no unit, then this property does not hold anymore, but we still have that $\nu \circ \nu' = \nu' \circ \nu$.

Definition 2.3.1 *An R-algebra isomorphism α of a unital R-algebra A is called inner if there exists a left A-endomorphism ν and a right A-endomorphism ν of A such that*

$$\alpha = \nu \circ \nu' = \nu' \circ \nu$$

and

$$\nu(a)\nu'(b) = ab$$

for all $a, b \in A$. $\mathrm{Inn}_R(A)$ will denote the subgroup of the group of R-automorphisms $\mathrm{Aut}_R(A)$ consisting of inner automorphisms.

If A has a unit, then this notion is the classical one. Indeed, with notations as above, the first condition of the Definition means that $\alpha(a) = uau'$. If we apply the second one to $a = b = 1$, then we obtain that $uu' = \nu(1)\nu'(1) = 1$.

Theorem 2.3.2 (Rosenberg-Zelinsky exact sequence) *Let A be an R-Azumaya algebra. Then we have an exact sequence:*

$$0 \longrightarrow \mathrm{Inn}_R(A) \longrightarrow \mathrm{Aut}_R(A) \xrightarrow{\Phi} \mathrm{Pic}(R)$$

Proof Recall from Proposition 1.4.3 that the categories A^e-mod^u and A-mod-A are equivalent. Invoking Proposition 1.1.5, we therefore have two inverse equivalent functors:

$$\begin{aligned} F = A \otimes \bullet &: \quad R\text{-mod} \longrightarrow A\text{-mod-}A \\ G = \mathrm{Hom}_{A,A}(A, \bullet) &: \quad A\text{-mod-}A \longrightarrow R\text{-mod} \end{aligned}$$

For all $M, M' \in R$-mod, we have:

$$F(M \otimes M') = A \otimes M \otimes M' \cong (A \otimes M) \otimes_A (A \otimes M') = F(M) \otimes_A F(M')$$

Since F and G are inverse equivalences, it follows also that for all $N, N' \in A$-mod-A:

$$G(N) \otimes G(N') \cong G(N \otimes_A N')$$

This also implies that F and G induce an isomorphism between $\mathrm{Pic}(R)$ and the group of invertible A-bimodules. Now let $\alpha, \beta, \gamma \in \mathrm{Aut}_R(A)$, and let $A_{\alpha,\beta}$ be equal to A as an R-module, and with A-bimodule structure given by

$$a \rightharpoonup b = \alpha(a)b \text{ and } b \leftharpoonup a = b\beta(a).$$

It is easy to prove that

$$A_{\alpha,\beta} \cong A_{\gamma\alpha,\gamma\beta} \text{ and } A_{1,\alpha} \otimes_A A_{1,\beta} \cong A_{1,\alpha\beta} \tag{2.7}$$

Therefore, $A_{\alpha,\beta}$ is an invertible A-bimodule, and the map

$$\Phi\colon \mathrm{Aut}_R(A) \longrightarrow \mathrm{Pic}(R)$$

given by $\Phi(\alpha) = [G(A_{1,\alpha})] = [\mathrm{Hom}_{A,A}(A, A_{1,\alpha})]$ is a homomorphism.

Suppose that α is inner. Then $\alpha = \nu \circ \nu'$ with ν and ν' as in the Definition. For all $a, b, c \in A$, we have:

$$\nu(abc) = \nu(a\nu(b)\nu'(c)) = a\nu(b)\nu(\nu'(c)) = a\nu(b)\alpha(c)$$

hence $\nu \in \mathrm{Hom}_{A^e}(A, A_{1,\alpha})$. Since ν is an isomorphism, the map

$$\nu \circ \bullet\colon R \cong \mathrm{Hom}_{A^e}(A, A) \to \mathrm{Hom}_{A^e}(A, A_{1,\alpha})$$

is an isomorphism, showing that $\Phi(\alpha) = 1$ in $\mathrm{Pic}(R)$. Conversely, suppose that $\mathrm{Hom}_{A^e}(A, A_{1,\alpha}) \cong R$. Then there exists an isomorphism $\nu\colon A \to A_{1,\alpha}$ of A-bimodules. Then ν is left A-linear, and for all $a, b \in A$ we have that $\nu^{-1}(ab) = \nu^{-1}(a)\alpha^{-1}(b)$. If we write $\nu' = \alpha \circ \nu^{-1}$, then it follows that $\nu'(ab) = \nu'(a)b$, so ν' is right A-linear. It is clear that $\nu \circ \nu' = \alpha$, and for all $a, b \in A$:

$$\begin{aligned}\alpha(ab) &= \alpha(a)\alpha(b) = \nu'(\nu(a))\nu(\nu'(b))\\ &= \nu'(\nu(a)\nu(\nu'(b))) = \nu'(\nu(\nu(a)\nu'(b)))\\ &= \alpha(\nu(a)\nu'(b))\end{aligned}$$

and therefore $ab = \nu(a)\nu'(b)$, proving that α is inner. □

Chapter 3

The Brauer group

3.1 Equivalent Taylor-Azumaya algebra

Let R be a commutative ring. If two Taylor-Azumaya algebras A and B are Morita equivalent, then we write $A \sim B$. We know from Proposition 1.1.7 that Morita equivalence is an equivalence relation. In the next Proposition, we will list some equivalent properties for two Taylor-Azumaya algebras to be Morita equivalent.

Proposition 3.1.1 *Let A and B be Taylor-Azumaya algebras. Then the following assertions are equivalent:*
1. $A \sim B$, that is, A is Morita equivalent to B;
2. $A \otimes B^{op}$ is R-elementary;
3. there exist $\underline{M}, \underline{N} \in \underline{\underline{\mathrm{DP}}}(R)$ such that $A \otimes E_R(\underline{M}) \cong B \otimes E_R(\underline{N})$.
If A and B are Azumaya algebras, then the above assertions are equivalent to
4. $A \otimes B^{op} \cong \mathrm{End}_R(Q)$ for some faithfully projective R-module Q;
3. there exist faithfully projective R-modules P and Q such that $A \otimes \mathrm{End}_R(P) \cong B \otimes \mathrm{End}_R(Q)$.

Proof 1. $\Rightarrow$ 2. If $A \sim B$, then $A \otimes B^{op} \sim B \otimes B^{op}$. By Proposition 2.2.4, $B \otimes B^{op}$ is R-elementary, and therefore Morita equivalent to R. Thus $A \otimes B^{op}$ is Morita equivalent to R, by Proposition 1.1.7 and therefore R-elementary.
2. $\Rightarrow$ 3. $A \otimes B^{op} \cong E_R(\underline{N})$, then $A \otimes B \otimes B^{op} \cong B \otimes E_R(\underline{N})$, and this implies 5. since $B \otimes B^{op}$ is R-elementary.
3. $\Rightarrow$ 1. Using Proposition 1.1.7, we obtain

$$A \cong A \otimes R \sim A \otimes E_R(\underline{M}) \cong B \otimes E_R(\underline{N}) \sim B \otimes R \cong B$$

Now assume that A and B have a unit.
2. $\Rightarrow$ 4. $A \otimes B^{op}$ is R-elementary and has a unit. By Proposition 1.2.2, it is isomorphic to the endomorphism ring of a faithfully projective module.
4. $\Rightarrow$ 5. $A \otimes \mathrm{End}_R(B) \cong A \otimes B \otimes B^{op} \cong B \otimes \mathrm{End}_R(Q)$.
5. $\Rightarrow$ 3. is trivial. □

Before we are able to state some more conditions for two Taylor-Azumaya algebras to be equivalent, we have to generalize the concept of elementary algebra. Suppose that B is a unital R-algebra, M' a unital left B-module, M a unital right B-module and $\mu: M' \otimes M \to B$ a surjective B-bimodule map. Then the R-algebra $A = M \otimes_B M'$ with multiplication given by the formula

$$(m \otimes m')(n \otimes n') = m \otimes \mu(m' \otimes n)n'$$

is called a *B-elementary algebra*. Obviously, if A is Morita equivalent to B, then A is B-elementary. Our aim is to show that the converse property holds if B is a Taylor-Azumaya algebra. First, let us show that a B-elementary algebra is a Taylor-Azumaya algebra.

Proposition 3.1.2 *Suppose that B is a Taylor-Azumaya algebra. Then any B-elementary algebra A is also a Taylor-Azumaya algebra.*

Proof We define a left A-module structure on M and a right A-module structure on M' as follows: for $m, n \in M$ and $m', n' \in M'$ let

$$(m \otimes m').n = m\mu(m' \otimes n) \tag{3.1}$$
$$n'.(m \otimes m') = \mu(n' \otimes m)m' \tag{3.2}$$

Consider the separator

$$\theta: B \otimes B \otimes \Omega \to B \otimes B$$

θ is an isomorphism, preserving the two left B-actions on B, and interchanging the two right B-actions on B. Apply $M \otimes_B \bullet$ to the second factor of each side. We then obtain an isomorphism

$$B \otimes (M \otimes_B B) \otimes \Omega = B \otimes M \otimes \Omega \to B \otimes (M \otimes_B B) = B \otimes M$$

Applying $\bullet \otimes_B M'$ to the second factor of the left hand side and to the first factor on the right hand side, we get an isomorphism

$$\tilde{\theta}: B \otimes (M \otimes_B M') \otimes \Omega = B \otimes A \otimes \Omega \to M' \otimes M \tag{3.3}$$

The left and right actions of A and B on each side are preserved in the following way:

left B-action on B	is turned into	left B-action on M'
left A-action on A	is turned into	left A-action on M
right B-action on B	is turned into	right B-action on M
right A-action on A	is turned into	right A-action on M'

Now consider the maps

$$B \otimes A \otimes \Omega \xrightarrow{\tilde{\theta}} M' \otimes M \xrightarrow{\mu} B$$

Apply $\bullet \otimes_B M'$ to B, M and B. Then we obtain

$$M' \otimes A \otimes \Omega \to M' \otimes (M \otimes_B M') \to M'$$

Now apply $M \otimes_B \bullet$ to the left of each side. Then we obtain

$$(M \otimes_B M') \otimes A \otimes \Omega \longrightarrow (M \otimes_B M') \otimes (M \otimes_B M') \stackrel{I\otimes\mu\otimes I}{\longrightarrow} (M \otimes_B M')$$

or

$$A_\ell \otimes A_r \otimes \Omega \stackrel{\psi}{\longrightarrow} A \otimes A \stackrel{m_A}{\longrightarrow} A \tag{3.4}$$

The map ψ is an isomorphism. For each $a \otimes \omega \in A \otimes \Omega$, consider the map $T(a \otimes \omega)$: $B \to B$, defined by

$$T(a \otimes \omega)(b) = (\mu \circ \tilde{\theta})(b \otimes a \otimes \omega)$$

Since $T(a \otimes \omega)$ is a B-bimodule homomorphism, $T(a \otimes \omega) \in R$ since B is R-central. From (3.3) and (3.4), it follows that for all $a' \in A$:

$$T(a \otimes \omega)a' = (m_A \circ \psi)(a' \otimes a \otimes \omega)$$

Hence T coincides with the map considered in Proposition 2.2.2, and T takes values in R. From Proposition 2.2.2, it follows that A is a Taylor-Azumaya algebra. □

In [175], Proposition 3.1.3 is stated for arbitrary Taylor-Azumaya algebras. The proof presented in [175] is only valid under the assumption that A is flat as an R-module. Nevertheless, the result is true even when A is not flat. We present a proof due to F. Grandjean (see [92]).

First we recall a result from category theory. Let $f, g : M \to N$ be two morphisms in R-mod, where R is an arbitrary ring. $q : N \to P$ is called the *absolute coequalizer* of f and g if $q \circ f = q \circ g$, q has a section r, g has a section s, and

$$f \circ s = r \circ q$$

An absolute coequalizer is always a coequalizer: take $n \in N$, and assume that $q(n) = 0$. Then

$$g(s(n)) - f(s(n)) = n - r(q(n)) = n$$

Proposition 3.1.3 *Suppose that A and B are Taylor-Azumaya algebras. The following assertions are equivalent:*
1) A and B are Morita equivalent;
2) A is a B-elementary algebra;
3) B is an A-elementary algebra.

Proof We have already remarked that 1) $\Longrightarrow$ 2) and 1) $\Longrightarrow$ 3) are trivial. We will prove 2) $\Longrightarrow$ 1). The proof of 3) $\Longrightarrow$ 1) then follows from the fact that Morita equivalence is symmetric.

Assume that $A = E_B(M, M', \mu) = M \otimes_B M'$ is B-elementary. We know from Proposition 2.2.4 that B is B^e-projective. In particular, the multiplication map $m_B : B^e \to B$ is split, we write φ_B for the splitting map. We use the following Sweedler type notation:

$$\varphi_B(b) = \sum b^1 \otimes b^2 \in B^e$$

for $b \in B$. Observe that $\sum b^1 b^2 = b$.
The map $\mu : M' \otimes M \to B$ is surjective, and therefore split as a B^e-module map. We write r for the splitting map, and

$$r(b) = \sum b^{M'} \otimes b^M$$

for any $b \in B$. Then $\sum \mu(b^{M'} \otimes b^M) = b$.
Consider the right B-action on M:

$$\psi^r_M : M \otimes B \longrightarrow M$$

ψ^r_M is split. The splitting map $t : M \to M \otimes B$ is given by the formula (see Lemma 4.1.3)

$$t(mb) = \sum mb^1 \otimes b^2$$

for all $m \in M$ and $b \in B$. In a similar way, the left B-action on M' is split. (3.1) and (3.1) define a left A-action ψ^l_M on M and a right A-action $\psi^r_{M'}$ on M'. We claim that the maps $\psi^r_{M'}$ and ψ^l_M are split. The map

$$s : M \xrightarrow{t} M \otimes B \xrightarrow{I_M \otimes r} M \otimes M' \otimes M \xrightarrow{\pi \otimes I_M} M \otimes_B M' \otimes M = A \otimes M$$

is a left inverse of ψ^l_M. For all $m \in M$ and $b \in B$, we have

$$\begin{aligned} \psi^l_M(s(mb)) &= \psi^l_M(\sum (mb^1 \otimes (b^2)^{M'}) \otimes (b^2)^M) \\ &= \sum (mb^1 \otimes (b^2)^{M'}) \cdot (b^2)^M \\ &= \sum mb^1 \mu((b^2)^{M'} \otimes (b^2)^M) \\ &= \sum mb^1 b^2 = mb \end{aligned}$$

It follows that M is unital as a left A-module. In a similar way, M' is unital as a right A-module.
We will now show that $\mu : M' \otimes M \to B$ is the absolute coequalizer of

$$\left\{ \begin{array}{l} \psi^r_{M'} \otimes I_M \\ I_{M'} \otimes \psi^l_M \end{array} \right. : M' \otimes A \otimes M \rightrightarrows M' \otimes M$$

Indeed, $\mu \circ (\psi^r_{M'} \otimes I_M) = \mu \circ (I_{M'} \otimes \psi^l_M)$, r is a section of μ and $I_{M'} \otimes s$ is a section of $I_{M'} \otimes \psi^l_M$. Finally, we have that

$$(\psi^r_{M'} \otimes I_M) \circ (I_{M'} \otimes s) = r \circ \mu$$

since

$$\begin{aligned} & (\psi^r_{M'} \otimes I_M)((I_{M'} \otimes s)(m' \otimes mb)) \\ = \ & (\psi^r_{M'} \otimes I_M)(m' \otimes (mb^1 \otimes (b^2)^{M'}) \otimes (b^2)^M) \\ = \ & \sum \mu(m' \otimes mb^1)(b^2)^{M'} \otimes (b^2)^M \\ = \ & \sum \mu(m' \otimes mb^1) r(b^2) = \sum r(\mu(m' \otimes mb^1) b^2) \\ = \ & \sum r(\mu(m' \otimes mb^1 b^2)) = (r \circ \mu)(m' \otimes mb) \end{aligned}$$

for all $m \in M$, $m' \in M'$ and $b \in B$.
Being the coequalizer of $\psi^r_{M'} \otimes I_M$ and $I_{M'} \otimes \psi^l_M$, B is isomorphic to $M' \otimes_A M$, and the isomorphism is the map $\overline{\lambda}$ induced by λ. It now follows easily that $(A, B, M, M', I_A, \overline{\lambda})$ is a strict Morita context. □

3.2 The (big) Brauer group

Now consider the set of Morita equivalence classes of finitely generated Taylor-Azumaya algebras. We restrict attention to algebras that are finitely generated as an R-module, because otherwise we run into logical trouble. We define a product on this set as follows: for two Taylor-Azumaya algebras A and B, we let

$$[A][B] = [A \otimes B]$$

It is clear from Proposition 3.1.1 that $[A \otimes B]$ is independent of the chosen representatives A and B. The tensor product is associative, and $[R]$ is a neutral element for the product. Moreover, every Morita equivalence class of Taylor-Azumaya algebras $[A]$ has an inverse: this inverse is represented by A^{op}, and this follows from Proposition 2.2.4. The commutative group we obtain is denoted $\mathrm{Br}'(R)$ and is called the *big Brauer group* of R. The subgroup $\mathrm{Br}(R)$ consisting of classes represented by Azumaya algebra is called the *Brauer group* of R. We can reformulate our definitions in K-theoretic language à la Bass: let $\underline{\underline{\mathrm{Az}}}'(R)$ be the category of R-Taylor-Azumaya algebras that are finitely generated as an R-module, and R-algebra isomorphisms. $\underline{\underline{\mathrm{Az}}}(R)$ will be the full subcategory of R-Azumaya algebras. Similarly, let $\underline{\underline{\mathrm{DP}}}'(R)$ be the full subcategory of $\underline{\underline{\mathrm{DP}}}(R)$ consisting of dual pairs (M, M', μ) such that $\mathrm{E}_R(M, M', \mu)$ is finitely generated as an R-module. We then have a commutative diagram of functors:

$$\begin{array}{ccc}
\underline{\underline{\mathrm{FP}}}(R) & \xrightarrow{\mathrm{End}_R} & \underline{\underline{\mathrm{Az}}}(R) \\
\downarrow & & \downarrow \\
\underline{\underline{\mathrm{DP}}}'(R) & \xrightarrow{\mathrm{E}_R} & \underline{\underline{\mathrm{Az}}}'(R)
\end{array}$$

$\mathrm{Br}(R)$ and $\mathrm{Br}'(R)$ are now given by the formulas

$$\mathrm{Br}(R) = \mathrm{Coker}\left(\mathrm{K}_0\mathrm{End}_R\colon \mathrm{K}_0\underline{\underline{\mathrm{FP}}}(R) \to \mathrm{K}_0\underline{\underline{\mathrm{Az}}}(R)\right) \tag{3.5}$$

$$\mathrm{Br}'(R) = \mathrm{Coker}\left(\mathrm{K}_0\mathrm{E}_R\colon \mathrm{K}_0\underline{\underline{\mathrm{DP}}}'(R) \to \mathrm{K}_0\underline{\underline{\mathrm{Az}}}'(R)\right) \tag{3.6}$$

The notations $\mathrm{Br}(R)$ and $\mathrm{Br}'(R)$ should be viewed as analogues of the notations $\mathrm{K}_0(R)$ and $\mathrm{K}_0'(R)$ for the Grothendieck groups of respectively the categories of finitely generated projective and finitely generated R-modules.

Proposition 3.2.1 *Let A be a Taylor-Azumaya algebra. Suppose that $e \in A$ is an idempotent element such that $e \notin mA$ for all $m \in \max(A)$. Then eAe is an Azumaya algebra Morita equivalent to A, and consequently $[A] \in \mathrm{Br}(R)$.*

Proof AeA is a unital left A^e-submodule of A. Therefore $AeA = IA$ for some ideal of R, since the categories R-mod and A^e-mod^u are equivalent. Since $e^3 = e \in AeA = IA$, it follows that $I = R$. We therefore have that $AeA = A$ From the fact that e is an idempotent, it follows that Ae and eA are unital as A-modules. Applying Proposition 4.3.4 to Ae and eA, we obtain that $[eAe] = [A]$ in $\mathrm{Br}'(R)$. e is a unit in eAe, so eAe is an Azumaya algebra. □

Proposition 3.2.2 *For a field k,* $\mathrm{Br}(k) = \mathrm{Br}'(k)$

Proof Let A be a finitely generated k-Taylor-Azumaya algebra. Then A is finite dimensional as a k-vector space. Suppose that $\dim A = n$. From classical linear algebra, it follows that n is the maximum possible nilpotency degree of a linear transformation of A. Therefore, if every element of A is nilpotent, then we can conclude that $A^n = 0$, contradictory to the fact that $A^2 = A$. We may therefore assume that there exists a non-nilpotent $a \in A$. Consider the linear map $L\colon A \to A$ given by left multiplication by a. Applying the Cayley-Hamilton theorem to L, we obtain

$$L^n + b_1 L^{n-1} + \cdots + b_{n-k} L^k = 0$$

with $b_{n-k} \neq 0$ for some $k \in \{0, 1, \ldots, n-1\}$, since L is not nilpotent. This means that we may write A as a direct sum of two vector spaces $A = V_1 \oplus V_2$, with $L^k V_1 = 0$; $L^k V_2 = V_2$; $\dim V_1 = k$; $\dim V_2 = n - k$. On V_2 we have, since the restriction of L^k to V_2 is invertible:

$$L^{n-k} + b_1 L^{n-k+1} + \cdots + b_{n-k} = 0$$

or

$$-b_{n-k}^{-1}(L^{n-k} + b_1 L^{n-k+1} + \cdots + b_{n-k-1} L) = I_{V_2}$$

Now consider the linear transformation

$$L' = (-b_{n-k}^{-1}(L^{n-k} + b_1 L^{n-k+1} + \cdots + b_{n-k-1} L))^k$$

Then

$$\begin{cases} L'(x) = x & \text{if } x \in V_2 \\ L'(x) = 0 & \text{if } x \in V_1 \end{cases}$$

Hence L' is a nonzero idempotent. Being a polynomial in L, L' is given by multiplication by some nonzero $b \in A$. Put $e = b^2$, then $e^2 = b^4 = b^2 b^2 = b b^2 = b^2 b = b^2 = e$, so e is a nonzero idempotent. The result now follows from Proposition 3.2.1. □

Following [175], we call an idempotent element $e \in A$ a *rank one idempotent* if $eAe = Re \cong R$. Observe that a rank one idempotent automatically has the property that $e \notin mA$ for all $m \in \max(R)$. Indeed, otherwise we would have that

$$Re = eAe = e^2 Ae \subset emAe = me,$$

which is a contradiction. Therefore, if a Taylor-Azumaya algebra has a rank one idempotent, then it represents an element of the Brauer group. In fact we have

Proposition 3.2.3 *If a Taylor-Azumaya algebra contains a rank one idempotent e, then A is an elementary algebra, and $[A] = 1$ in $\mathrm{Br}(R)$.*

Proof By Proposition 3.2.1, $A \sim eAe \cong R$. □

3.3 The splitting theorem for Taylor-Azumaya algebras

Let R be a commutative ring, and $f\colon R \to S$ a commutative R-algebra. If A is an R-Taylor-Azumaya algebra, then it follows easily from the definition that $A \otimes S$ is an S-Taylor-Azumaya algebra. If A is R-elementary, then $A \otimes S$ is S-elementary, hence we obtain a map $\mathrm{Br}'(f)\colon \mathrm{Br}'(R) \to \mathrm{Br}'(S)$. It is clear that this map behaves functorially, so we may view $\mathrm{Br}'(\bullet)$ as a covariant functor from commutative rings to abelian groups. It is clear that $\mathrm{Br}(\bullet)$ is a subfunctor of $\mathrm{Br}'(\bullet)$. If $A \otimes S$ is an S-elementary algebra, that is, if $[A \otimes S] = 1$ in $\mathrm{Br}'(S)$, then we say that the Taylor-Azumaya algebra A is *split* by S. The subgroup of $\mathrm{Br}'(R)$ of all Azumaya algebras split by S will be denoted by $\mathrm{Br}'(S/R)$:

$$\mathrm{Br}'(S/R) = \mathrm{Ker}\,(\mathrm{Br}'(R) \to \mathrm{Br}'(S)),$$

and similarly

$$\mathrm{Br}(S/R) = \mathrm{Ker}\,(\mathrm{Br}(R) \to \mathrm{Br}(S)).$$

The aim of this section is to show that for every R-Taylor-Azumaya algebra A, there exists a faithfully flat R algebra S such that A is split by S.
Recall that a *Henselian ring* (R,m) is a (commutative) local ring R with maximal ideal m such that every finite R-algebra A (with unit) is a direct product of local rings. Several characterizations are possible, we refer to [162] for details. One criterion is the following (cf. [162, I,Prop. 3-4]) : a local ring R is Henselian if and only if for every finite R-algebra A with a unit, there is a bijective correspondence between the idempotents of A and A/mA.
A Henselian ring is called *strictly Henselian* if R/m is a separably closed field.

Proposition 3.3.1 *If* (R,m) *is a Henselian ring, then* $\mathrm{Br}(R) = \mathrm{Br}'(R)$.

Proof Consider a finitely generated Taylor-Azumaya algebra A, representing $[A] \in \mathrm{Br}'(R)$. We will show that there exists an idempotent $e \notin mA$, and this will imply that $[A] \in \mathrm{Br}(R)$. $A/mA = \overline{A}$ is a finite dimensional R/m-Taylor-Azumaya algebra, and from the proof of Proposition 3.2.2, we know that this algebra contains a nonzero idempotent. Therefore $\overline{A}$ contains a nonzero idempotent, say $\overline{a}$ lifting to $a \in A$. We will lift $\overline{a}$ to an idempotent of A. Since R is local, Ω is free of rank one, and we may write $\Omega = R\omega$. Consider the separator

$$\theta\colon A \otimes A \otimes \Omega \to A \otimes A$$

and choose $u_i, v_i \in A$ such that

$$\theta(\sum_{i=1}^{n} u_i \otimes v_i \otimes \omega) = \omega(\sum_{i=1}^{n} u_i \otimes v_i) = a \otimes a$$

Then for all $b \in A$:

$$\sum_i u_i \mathrm{tr}_\omega(v_i b) = m_A(\omega(\sum_i u_i \otimes v_i b)) = aba$$

Write $f(b) = aba$, then $f = u \circ v$, with

$$\begin{cases} u\colon R^n \to A & \text{given by } u(r_1, \ldots, r_n) = \sum_i r_i u_i \\ v\colon A \to R^n & \text{given by } v(b) = (\mathrm{tr}_\omega(v_1 b), \ldots, \mathrm{tr}_\omega(v_n b)) \end{cases}$$

It follows easily that for $m \geq 1$:

$$f^m = u \circ w^{m-1} \circ v$$

with $w = v \circ u$ an endomorphism of R^n. By the Cayley-Hamilton theorem, there exists a monic polynomial $q \in R[X]$ such that $q(w) = 0$, one may take the characteristic polynomial of the matrix of the endomorphism w. Write

$$q(w) = w^n + \alpha_1 w^{n-1} + \cdots + \alpha_n = 0$$

Apply u to the left and v to the right of this equation. We then obtain

$$f^{n+1} + \alpha_1 f^n + \cdots + \alpha_n f = 0 \tag{3.7}$$

Now $f^i(a) = a^{2i+1}$. So if we apply (3.7) to a, we obtain

$$f^{n+1}(a) + \alpha_1 f^n(a) + \cdots + \alpha_n f(a) = a^{2n+3} + \alpha_1 a^{2n+1} + \cdots + \alpha_n a^3 = 0 \tag{3.8}$$

Consider the R-algebra

$$B = XR[X]/(X^{2n+3} + \alpha_1 X^{2n+1} + \cdots + \alpha_n X^3)$$

From (3.8), it follows that we have an algebra homomorphism

$$B = RX \oplus RX^2 \oplus \cdots \oplus RX^{2n+2} \to A$$

given by mapping X to a. Let C be the image of B into A. Being a quotient of a free R-module, C is projective, and therefore free as an R-module. C contains a, and is commutative, since B is commutative. The algebra $C_1 = R \oplus C$, with multiplication rule

$$(r, c)(r', c') = (rr', rc' + r'c + cc')$$

is finitely generated, free, commutative, and has a unit. Therefore, there is a bijective correspondence between the idempotents of C_1 and those of $C_1/mC_1 = k \oplus C/mC$. Take an idempotent $e_1 = (r, e) \in C_1$ lifting the idempotent $(0, \overline{a}) \in C_1/mC_1$. Then $r \in m$ and $r^2 = r$, hence $r = 0$. Therefore $e \in C \subset A$ is an idempotent such that $\overline{e} = \overline{a}$. □

Proposition 3.3.2 *Let (R, m) be a Henselian ring. Then the canonical map* $\mathrm{Br}(R) \to \mathrm{Br}(R/m)$ *is injective. Consequently* $\mathrm{Br}(R) = \mathrm{Br}'(R) = 1$ *if R is strictly Henselian.*

Proof Write $k = R/m$, and let A be an R-Azumaya algebra such that $[A/mA] = 1$ in $\mathrm{Br}(k)$. Then $A/mA \cong M_n(k)$ for some integer n. Let f be the matrix idempotent with a one in the top left corner and zeros elsewhere. Lift f to an element $a \in A$. Then $B = R[a]$ is a finite commutative R-algebra, hence the idempotent f may be lifted to an idempotent $e \in B \subset A$. Ae is a free R-module, since R is local, and we have an R-algebra homomorphism η: $A \to \mathrm{End}_R(Ae) = B$, given by

$$\eta(a)(be) = abe.$$

Suppose that $0 \neq a \in \mathrm{Ker}\ \eta$. Then, by Nakayama's Lemma, we have that $ma \neq Ra \subset \mathrm{Ker}\ \eta$, hence there exists $ra \in \mathrm{Ker}\ \eta$ with $ra \neq 0 \bmod mA$. This contradicts the fact that $\overline{\eta}$: $\overline{A} \to \mathrm{End}_k(\overline{A}f) \cong \overline{A}$ is an isomorphism. Therefore, η is injective. Similarly, if η is not surjective, then $B/\mathrm{Im}\ \eta \neq 0$, hence $m(B/\mathrm{Im}\ \eta) \neq B/\mathrm{Im}\ \eta$, by Nakayama's Lemma. Take $x \in B$ such that $\overline{x} \in B/\mathrm{Im}\ \eta \setminus m(B/\mathrm{Im}\ \eta)$. Then $x \notin \mathrm{Im}\ \eta$ and $x \notin mB$. This is again a contradiction. We therefore have an isomorphism η: $A \to \mathrm{End}_R(Ae)$, and this proves our result. □

Proposition 3.3.3 *Let (R, m) be a Henselian ring. Then every finitely generated elementary algebra A contains a rank one idempotent.*

Proof From the proof of Proposition 3.3.1, we know that A contains an idempotent $e \notin mA$. From Proposition 3.2.1, it follows that eAe is an Azumaya algebra equivalent to A, and $AeA = A$. eAe is therefore an elementary algebra with unit. It follows that eAe is the endomorphism ring of a faithfully projective R-module, which is free, since R is local. So eAe is a matrix ring. Now eAe contains an idempotent f such that $feAef = Rf$: take f the matrix with one in the top left corner and zero elsewhere. We now claim that $fAf = Rf$. Take $f \in A$, and write $f = ebe$ for some $b \in A$ (recall that $f \in eAe$). Then

$$faf = f^2af^2 = febeaebef \in feAef = Rf$$

□

Proposition 3.3.4 *Suppose that a Taylor-Azumaya algebra A has a rank one idempotent e. Then there exists a unique $\varpi \in \Omega$ such that $\varpi(e \otimes e) = e \otimes e$, or, equivalently, $\mathrm{tr}_{\varpi}(e) = 1$. ϖ is independent of the choice of e, and $\Omega \cong R\varpi$.*

Proof Let e be rank one idempotent, that is $eAe = Re \cong R$. Consider the separator θ: $A_\ell \otimes A_r \otimes \Omega \to A^e$, and recall that θ is an isomorphism. For all $\omega \in \Omega$, we have

$$\begin{aligned} \theta(e \otimes e \otimes \omega) &= \omega(e \otimes e) \\ &= \omega(e^3 \otimes e^3) \\ &= (e \otimes e)\omega(e \otimes e)(e \otimes e) \in eAe \otimes e = Re \otimes e \cong R \end{aligned}$$

hence there exists a unique number $\nu(\omega)$ such that

$$\omega(e \otimes e) = \nu(\omega)(e \otimes e)$$

This defines a map ν: $\Omega \to R$. Observe that ν may also be defined by the formula

$$\nu(\omega) = \mathrm{tr}_\omega(e)$$

Indeed,

$$\begin{aligned}\mathrm{tr}_\omega(e)e\otimes e &= m_A(\omega(e\otimes e))\otimes e\\ &= m_A(\nu(\omega)e\otimes e)\otimes e\\ &= \nu(\omega)e\otimes e\end{aligned}$$

We claim that ν is surjective. Since θ is an isomorphism, there exists

$$\sum_i a_i\otimes b_i\otimes\omega_i\in A\otimes A\otimes\Omega$$

such that

$$\theta(\sum_i a_i\otimes b_i\otimes\omega_i)=\sum_i\omega_i(a_i\otimes b_i)=e\otimes e$$

Then

$$\sum_i\omega_i(ea_ie\otimes eb_ie)=e^3\otimes e^3=e\otimes e$$

Since $eAe=Re$, we have $ea_ie=r_ie$, $eb_ie=s_ie$ for some $r_i,\ s_i\in R$. Hence

$$e\otimes e=\sum_i\omega_i(r_ie\otimes s_ie)=(\sum_i r_is_i\omega_i)(e\otimes e).$$

Now write $\varpi=\sum_i r_is_i\omega_i$. Then $\nu(\varpi)=1$, and ν is surjective. Since Ω and R are both projective of rank one, it follows that ν is an isomorphism. This shows that ϖ is the unique element in Ω satisfying $\nu(\varpi)=\mathrm{tr}_\varpi(e)=1$, or $\varpi(e\otimes e)=e\otimes e$.
We still have to show that ν is independent of the choice of e. It suffices to show that the localized map $\nu_p:\ \Omega\otimes R_p\cong\Omega_{A_p}\to R_p$ is independent of e, for every prime ideal p of R. Therefore, we can assume that R is local, with maximal ideal m. Let f be another rank one idempotent. Then for all $\omega\in\Omega$:

$$\begin{aligned}\theta(f\otimes e\otimes\omega) &= \omega(f\otimes e)\\ &= \omega(f^3\otimes e^3)\\ &= (f\otimes f)\omega(f\otimes e)(e\otimes e)\in fAe\otimes eAf\end{aligned}$$

and

$$\begin{aligned}m_A(\theta(f\otimes e\otimes\omega)) &= m_A(\omega(f\otimes e))\\ &= \mathrm{tr}_\omega(e)f\\ &= \nu(\omega)f\end{aligned}$$

Hence $m_A(\theta(f\otimes e\otimes\omega))$ runs through Rf if ω runs through Ω. Since

$$m_A(\theta(f\otimes e\otimes\omega))\in fAeeAf=fAeAf\subset fAf=Rf,$$

it follows that $Rf=fAeAf$. There exist $a_1,b\in A$ such that $fa_1eebf=xf$, with $x\in R\setminus m$. Indeed, otherwise $fAeAf\subset mf\neq Rf$. Let $a=x^{-1}a_1$, $u=fae\in fAe$, $v=ebf\in eAf$. Then $uv=f$ and

$$(vu)^2=vuvu=vfu=ebfffae=ebffae=vu\in eAffAe=Re,$$

hence $vu = xe$, for some idempotent $x \in R$. We have

$$xf = xuv = uxebf = uxeebf = uvuv = f^2 = f,$$

so $f = xf \in xA$. If $x \neq 1$, then xR is a proper ideal of R, and this contradicts the fact that f is a rank one idempotent. Hence $x = 1$, and $vu = e$. Finally

$$\mathrm{tr}_\omega(f) = \mathrm{tr}_\omega(uv) = \mathrm{tr}_\omega(vu) = \mathrm{tr}_\omega(e) = \nu(\omega)$$

and ν and ϖ are independent of the choice of the rank one idempotent e. □

Let S be a commutative R-algebra. Recall that S is called *locally of finite type* if for every $q \in \mathrm{Spec}(S)$, with $p = f^{-1}(q)$, there exists $g \in S \setminus q$ and $f \in R \setminus p$ such that S_g is finitely generated as an R_f-algebra, that is, $S_g \cong R_f[x_1, \ldots, x_n]/I$ for some ideal I of $R_f[x_1, \ldots, x_n]$.
S is called S is called an *étale R-algebra* if S is flat as an R-module, locally of finite type and separable as an R-algebra (this last assertion means that S is projective as an $S \otimes S$-module). Compare this definition to [162, Theorem 2, p.55] and [135, I.3.5]. Many other characterizations are possible, let us mention the following (cf. [162, Theorem 1, p.51]) : S is an étale R-algebra, if it is locally a *standard étale algebra*, that is, for any $q \in \mathrm{Spec}(S)$, with notations as above,

$$S_g \cong (R_f[x]/(P(x)))_h$$

for some polynomial $P(x) \in R_f[x]$ and $h \in R_f[x]/(P(x))$ such that P', the derivative of P is invertible in $(R_f[x]/(P(x)))_h$. For more properties of étale algebras, we refer to [162] and [135]. S is called an *étale covering* of R if S is étale and faithfully flat as an R-algebra.
Etale algebras may be used to construct the *Henselization* and the *strict Henselization* of a local ring. We refer to [162] and [135] for these constructions. In the proof of the following theorem, we will need the following description of the strict Henselization: let R be a commutative ring, and p a prime ideal of R. The strict Henselization R_p^{sh} of the localized ring R_p may be obtained as the direct limit

$$R_p^{sh} = \varinjlim S_\alpha$$

where the limit is taken over the filtering system of commutative diagrams of the form

$$\begin{array}{ccc} R & \xrightarrow{f_\alpha} & S_\alpha \\ & \searrow & \downarrow \\ & & k(p)^{sep} \end{array} \tag{3.9}$$

where f_α is étale, and where $k(p)^{sep}$ is a fixed separable closure of R_p/pR_p. R_p^{sh} satisfies a universal property, but the only fact that we will need in the proof of the next theorem is that R_p^{sh} is strictly Henselian.

Theorem 3.3.5 *Let A be an R-algebra that is finitely generated as an R-module. Then the following assertions are equivalent:*

1) A is a Taylor-Azumaya algebra;
2) $A \otimes S$ contains a rank one idempotent, for some étale covering S of R;
3) $A \otimes S$ is S-elementary, for some étale covering S of R;
4) $A \otimes S$ is S-elementary, for some faithfully flat R-algebra S which is locally of finite type;
5) $A \otimes S$ is S-elementary, for some faithfully flat R-algebra S;
6) $A \otimes S$ is an S-Taylor-Azumaya algebra, for some faithfully flat R-algebra S;
7) $A \otimes S$ contains a rank one idempotent, for some faithfully flat R-algebra S;
8) A is a faithful R-module, central and unital as an R-algebra, and there exists an A^e-bimodule isomorphism ω: $A_\ell \otimes A_r \to A^e$.

Proof 1 ⇒ 2. Let A be a Taylor-Azumaya algebra, and take $p \in \text{Spec}R$. Consider a commutative diagram of the filtering system (3.9), and write A_α for $A \otimes S_\alpha$ and A_p^{sh} for $A \otimes R_p^{sh}$. From Propositions 3.3.2 and 3.3.3, we know that $A \otimes R_p^{sh}$ contains a rank one idempotent e. Since

$$A \otimes R_p^{sh} = \varinjlim A_\alpha,$$

this e occurs already in one of the S_α. The natural map $S_\alpha \to S_\alpha e$ is a projection ; therefore, we can view $S_\alpha e$ as a direct factor of S_α. The map $R \to S_\alpha e$ is still étale. Replacing S_α by $S_\alpha e$, we therefore have

$$\begin{array}{ccccc} S_\alpha & \cong & S_\alpha e & \subset & eA_\alpha e \\ \downarrow & & \downarrow & & \downarrow \\ R_p^{sh} & \cong & R_p^{sh} e & = & eA_p^{sh} e \end{array}$$

$S_\alpha e \subset eA_\alpha e$, since $se = se^3 = esee \in eA_\alpha e$, for all $s \in S_\alpha$. Since A_α is finitely generated, there exists $\lambda \geq \alpha$ in the filtering system, such that $S_\lambda = eA_\lambda e$. This means that e is a rank one idempotent for A_λ.
We apply this construction to every $p \in \text{Spec}(R)$, and we call the obtained étale R-algebra $S(p)$. We thus obtain a set $\{S(p)|p \in \text{Spec}(R)\}$ of étale R-algebras such that the union of the images of the Spec($S(p)$) cover Spec(R), and such that every $A \otimes S(p)$ contains a rank one idempotent. By the quasicompactness of Spec(R), we can find a finite subset $\{p_1, \ldots, p_n\}$ of Spec(R) such that the images of the Spec(S_{p_i}) cover Spec(R). Then $S = S(p_1) \times \cdots \times S(p_n)$ is an étale covering of R, and $A \otimes S$ contains a rank one idempotent.
2 ⇒ 3 ⇒ 4 ⇒ 5 ⇒ 6. is trivial.
6 ⇒ 7. By 1 ⇒ 2, there exists an étale covering T of S such that $(A \otimes S) \otimes_S T = A \otimes T$ has a rank one idempotent. It is clear that T is a faithfully flat R-algebra.
7 ⇒ 8. Write $B = A \otimes S$. Then B is S-unital, S-faithful and S-central. The first assertion means that

$$B \otimes_S B \otimes_S B \rightrightarrows B \otimes_S B \to B \to 0$$

is exact, or

$$(A \otimes A \otimes A) \otimes S \rightrightarrows (A \otimes A) \otimes S \to A \otimes S \to 0$$

is exact. Since S is faithfully flat, this implies that

$$A \otimes A \otimes A \rightrightarrows A \otimes A \to A \to 0$$

is exact, so A is unital. A is faithful : take $r \in R$, and suppose that $rA = 0$. Then $(r \otimes 1)(A \otimes S) = 0$, hence $r \otimes 1 = 0$ in S, since $B = A \otimes S$ is S-faithful.
Let us next show that A is R-central. Suppose that $\alpha \in Z(A) = \mathrm{End}_{A^e}(A)$. Then $\alpha \otimes I$ belongs to the center of $A \otimes S$, hence $\alpha \otimes I$ is given by multiplication by some $s \in S$. To show that $s \in R$, we need to prove that $s \otimes 1 = 1 \otimes s$ in $S \otimes S$, cf. the Theorem of Faithfully Flat Descent of Elements B.0.6. Now multiplication by $s \otimes 1$ is exactly the endomorphism $\alpha \otimes I \otimes I$ of $A \otimes S \otimes S$, and the same is true for $1 \otimes s$. This implies that $1 \otimes s = s \otimes 1$, since $A \otimes S \otimes S$ is a faithful $S \otimes S$-module. Hence $s \in R$, and α is given by multiplication by s. Thus A is R-central.
Next, consider the canonical isomorphism

$$\varpi\colon\ (B_\ell \otimes_S B_r) = (A_\ell \otimes A_r) \otimes S \to B^e = A^e \otimes S$$

discussed in Proposition 3.3.4. We claim that ϖ descends to an isomorphism

$$\omega\colon\ A_\ell \otimes A_r \to A^e.$$

Applying the Theorem of Faithfully Flat Descent for homomorphisms, cf. Proposition B.0.7 this means that we have to show that $\varpi_2 = \varpi_3$. These are two isomorphisms

$$(A_\ell \otimes A_r) \otimes S \otimes S \to A^e \otimes S \otimes S$$

Now suppose that ϖ is constructed using a rank one idempotent $e \in A \otimes S$. Then it is clear that ϖ_2 and ϖ_3 are constructed using e_2 and e_3, which are rank one idempotents of $A \otimes S \otimes S$. As we have seen in Proposition 3.3.4, the construction is independent of the choice of the rank one idempotent, hence $\varpi_2 = \varpi_3$. Thus ϖ descends to an a homomorphism ω which is clearly an A^e-isomorphism, since S is faithfully flat as an R-module.
$8 \Rightarrow 1$. follows immediately from Proposition 2.2.2. □

Corollary 3.3.6 *Let A be a finitely generated Taylor-Azumaya algebra. Then the invertible R-module Ω is free of rank one.*

Proof In the proof of $6 \Rightarrow 7$. in Theorem 3.3.5, we referred to Proposition 2.2.2. In this Proposition, we have to take $I = R$. But we have seen in Proposition 2.2.6 that $I \cong \Omega$, and this finishes the proof. □

Observe that we did not only show that Ω is free of rank one, but that there exists a canonical generator for Ω, namely the unique isomorphism ω that is obtained from a rank one idempotent after tensoring up by a faithfully flat extension of R.
As an immediate corollary to Theorem 3.3.5, we obtain the following characterizations of Azumaya algebras.

Corollary 3.3.7 *Let A be an R-algebra. Then the following assertions are equivalent:*
1) A is an Azumaya algebra;
2) there exists an étale covering S of R and a faithfully projective S-module M such

that $A \otimes S \cong \mathrm{End}_S(M)$;
3) there exists a faithfully flat commutative R-algebra S and a faithfully projective S-module M such that $A \otimes S \cong \mathrm{End}_S(M)$;
4) $A \otimes S$ *is an S-Azumaya algebra for some faithfully flat commutative R-algebra S.*

3.4 The determinant map for an Azumaya algebra

If A is an Azumaya algebra, then Theorem 3.3.5 and Proposition 1.2.2 imply that we may find a faithfully flat R-algebra S such that $A \otimes S \cong \mathrm{End}_S(P)$, for some faithfully projective S-module P. If we assume that A is of constant rank, then we can replace S by another faithfully flat R-algebra such that P is free as an S-module. We then have an isomorphism β: $A \otimes S \to M_n(S)$. In the next proposition, we show that this isomorphism allows us to define a determinant map from A to R. The proof of 3.4.1 is taken from [114, IV.2].

Proposition 3.4.1 *With notations as above, the determinant*

$$\det_S\colon M_n(S) \longrightarrow S$$

descends to a map, called the determinant or the reduced norm

$$\det_R\colon A \longrightarrow S$$

$\det_R$ *is independent of the choice of* S *and* β.

Proof First, we prove the following assertion: if α: $M_n(R) \to M_n(R)$ is an isomorphism, then for any $f \in M_n(R)$, we have that $\det_R(f) = \det_R(\alpha(f))$. If α is inner, this is obvious. Otherwise we take a Zariski covering S of R such that $\alpha \otimes I_S$ is inner. This is always possible, cf. the Rosenberg-Zelinsky exact sequence 2.3.2. The statement then follows easily from the fact that S/R is faithfully flat.
Now consider two faithfully flat R-algebras S and T such that we have isomorphisms

$$\sigma\colon A \otimes S \to M_n(S) \text{ and } \rho\colon A \otimes T \to M_n(T)$$

Define the $S \otimes T$-isomorphism α such that the following diagram commutes in $S \otimes T$-mod:

$$\begin{array}{ccc} A \otimes S \otimes T & \xrightarrow{\sigma \otimes I_T} & M_n(S \otimes T) \\ \downarrow{\scriptstyle \tau_{12}} & & \downarrow{\scriptstyle \alpha} \\ S \otimes A \otimes T & \xrightarrow{I_S \otimes \sigma} & M_n(S \otimes T) \end{array}$$

It therefore follows that for $a \in A$

$$\begin{aligned} \det_S(\sigma(a \otimes I_S)) \otimes 1_T &= \det_{S \otimes T}(\sigma(a \otimes I_S) \otimes I_T) \\ &= \det_{S \otimes T}(\alpha(\sigma(a \otimes I_S) \otimes I_T)) \\ &= \det_{S \otimes T}(I_S \otimes \rho(a \otimes I_T)) \\ &= I_S \otimes \det_T(\rho(a \otimes I_T)) \end{aligned}$$

If we take $S = T$ and $\sigma = \rho$, then we obtain that

$$\det_S\sigma(a \otimes 1_S) \otimes 1_S = 1_S \otimes \det_S\sigma(a \otimes 1_S)$$

hence $\det_S\sigma(a \otimes 1_S) \in R$, by Corollary B.0.6.
If S and T are different R-algebras, then it also follows that

$$1_S \otimes \det_S(\sigma(a \otimes 1_S))1_T = \det_S(\sigma(a \otimes 1_S)) \otimes 1_T = 1_S \otimes \det_T(\rho(a \otimes 1_T))1_T$$

and therefore $\det_S(\sigma(a \otimes 1_S)) = \det_T(\rho(a \otimes 1_T))1_T$, since S/R is faithfully flat. It is therefore legitimate to write

$$\det_R(a) = \det_S(\sigma(a \otimes 1_S))$$

□

Observe that if the rank of A is not constant, we can still define the determinant map: consider orthogonal idempotents $e_1, \ldots, e_r$ of R such that the rank of Ae_i is constant over Re_i, and then write

$$\det_R = \oplus_i \det_{Re_i}$$

If A is an Azumaya algebra, then $A[T]$ is an $R[T]$-Azumaya algebra, and we can consider for each $a \in A$ the determinant

$$P(T) = \det_{R[T]}(T - a) \in R[T]$$

We call P the characteristic polynomial of a, and we have the following version of the Cayley-Hamilton theorem:

Theorem 3.4.2 (Cayley-Hamilton theorem) *P is a monic polynomial of degree $n = \sqrt{\mathrm{rk}_R(A)}$, the squareroot of the rank of A, and $P(a) = 0$.*

Proof Take S as in Proposition 3.4.1. Then

$$P(T) = \det_{R[T]}(T - a) = \det_{S[T]}(T - \sigma(a \otimes 1_S))$$

is a polynomial of degree n, and $P(a) = 0$ by the classical Cayley-Hamilton theorem, which holds over arbitrary commutative rings, cf. e.g. [155, 4.1] or [62, 11.2,Thm.3].
□

3.5 The splitting theorem for semilocal rings

It is well-known, cf. for example [101, chapter 4] or [110, 10.3] that, in case $R = k$ is a field, then every k-Azumaya algebra A can be split by a Galois extension ℓ of k. Over a commutative ring, this property does not hold in general. However, if R is semilocal and connected, then DeMeyer ([66]) proved that every Azumaya can be split by a Galois extension. Before we are able to prove DeMeyer's result, we need some preliminary lemmas. In this Section, all algebras will have a unit.

Lemma 3.5.1 *If A is a faithfully projective R-algebra, then R is a direct summand of A.*

Proof Let $\eta : R \to A$ be the unit map. It suffices to show that $\bullet\circ\eta$: $\mathrm{Hom}_R(A, R) \to \mathrm{Hom}_R(R, R) = R$ is surjective. If R is local, then A is free, and the assertion is trivial. The general assertion then follows from a local-global argument. □

Lemma 3.5.2 *Let A, B be R-algebras, and suppose that B is faithfully projective as an R-module. If $A \otimes B$ is separable over R, then A is separable over R.*

Proof From Lemma 3.5.1, it follows that $B = R \oplus B_0$ as R-modules. Therefore $A \otimes B = A \oplus (A \otimes B_0)$ as A-bimodules, or as A^e-modules. Thus A is an A^e-direct factor of $A \otimes B$. $A \otimes B$ is separable, and therefore a direct factor of $A^e \otimes B^e$, as an $A^e \otimes B^e$-module and a fortiori as an A^e-module. Therefore A is an A^e-direct summand of $A^e \otimes B^e$.
Now B^e is projective over R, so $A^e \otimes B^e$ is projective over A^e. Hence A is A^e-projective. □

Proposition 3.5.3 *Let A be a k-Azumaya algebra, where k is a field. Then there exists a maximal commutative subalgebra S of A such that S is separable and $S = k[\alpha]$ for some $\alpha \in A$.*

Proof First take $A = D$, a skewfield with center k. It is well-known that D contains a maximal commutative subalgebra ℓ which is separable over k, and which is a field. We refer to [110, 10.2] for details. If we write $\ell = k[\alpha]$, then the result follows for $A = D$.
For the general case, observe that $A = M_r(D)$ for some skewfield D, by Wedderburn's Theorem. First suppose that k is infinite. The set S of diagonal matrices with entries in $\ell = k[\alpha]$ is then a maximal commutative subalgebra of $A = M_r(D)$. Moreover, $S = k[\beta]$, where β is a diagonal matrix with r different entries.
If k is finite, then D is finite, so $k = D$. Let ℓ be the unique field extension of degree r over k. ℓ is separable, and ℓ embeds in $M_r(k) = \mathrm{End}_k(\ell)$. It is clear that ℓ is a maximal commutative subalgebra of A. □

Lemma 3.5.4 *Let A be an Azumaya algebra, and S a commutative subalgebra of A. If S is separable over R, then A is projective as an S-module.*

Proof Since S is R-separable, the sequence

$$0 \longrightarrow \mathrm{Ker}\ m_S \longrightarrow S \otimes S \longrightarrow S \longrightarrow 0$$

is split exact. Therefore the sequence

$$0 \longrightarrow A \otimes_S \mathrm{Ker}\ m_S \longrightarrow A \otimes_S S \otimes S = A \otimes S \longrightarrow A \otimes_S S = A \longrightarrow 0$$

and this implies that A is $A \otimes S$-projective. Now $A \otimes S$ is S-projective, so A is S-projective. □

Proposition 3.5.5 *Let A be an R-Azumaya algebra, and S a maximal commutative subalgebra. Then $A \otimes S \cong \mathrm{End}_S(A)$, where A is viewed as an S-module by right multiplication. If A is projective as an S-module (this is the case if S is a separable R-algebra, by lemma 3.5.4), then S is projective as an R-module, and in this case*

$$\mathrm{rk}_R(A) = \mathrm{rk}_S(A)^2 \quad \text{and} \quad \mathrm{rk}_R(S) = \mathrm{rk}_S(A)$$

Proof Consider the R-algebra $B = (A \otimes A^{\mathrm{op}})^{R\otimes S}$. It is clear that

$$A \cong A \otimes R \subset A \otimes S \subset (A \otimes A^{\mathrm{op}})^{R\otimes S} = B$$

so A is a subalgebra of B, and this makes B into an A-bimodule. Now

$$B^A \subset (A \otimes A^{\mathrm{op}})^{A\otimes R} = R \otimes A^{\mathrm{op}}$$

and

$$R \otimes S = A^A \otimes (A^{\mathrm{op}})^S \subset (A \otimes A^{\mathrm{op}})A \otimes S = B^A$$

so

$$S \cong R \otimes S \subset B^A \subset R \otimes A^{\mathrm{op}} \cong A^{\mathrm{op}}.$$

Thus B^A is a commutative subalgebra of A^{op} containing S, and $B^A \cong S$. From Proposition 2.2.8, it follows that $B \cong A \otimes B^A \cong A \otimes S$. Observe indeed that if A has a unit, then $M^A \cong \mathrm{Hom}_{A^e}(A_\ell, M)$ for an A-bimodule M.
Now let $F:\ A \otimes A^{\mathrm{op}} \to \mathrm{End}_R(A)$ be the canonical isomorphism given by $F(a \otimes b)(x) = axb$, for $a, b, x \in A$. We claim that $F(B) = \mathrm{End}_S(A)$. Let $\sum_i a_i \otimes b_i \in A^e$, and let $f = F(\sum_i a_i \otimes b_i)$. Then $\sum_i a_i \otimes b_i \in B$ if and only if for all $s \in S$:

$$(1 \otimes s)(\sum_i a_i \otimes b_i) = (\sum_i a_i \otimes b_i)(1 \otimes s)$$

or

$$\sum_i a_i \otimes sb_i = \sum_i a_i \otimes b_i s$$

or for all $x \in A$:

$$\begin{aligned} f(xs) &= \sum_i a_i x s b_i = F(\sum_i a_i \otimes s b_i)(x) \\ &= F(\sum_i a_i \otimes b_i s)(x) = \sum_i a_i x b_i s = f(x)s \end{aligned}$$

We can now conclude that

$$A \otimes S \cong B \cong \mathrm{End}_S(A)$$

proving the first assertion.
Now suppose that A is projective as an S-module. Then

$$\mathrm{rk}_R(A) = \mathrm{rk}_S(A \otimes S) = \mathrm{rk}_S(\mathrm{End}_S(A)) = \mathrm{rk}_S(A)^2$$

It follows from this that A is finitely generated as an S-module. Clearly A is S-faithful, since $S \subset A$, so A is faithfully projective as an S-module. Lemma 3.5.1 now tells us that S is a direct factor of A. Since A is R-projective, it follows that S is projective over R and

$$\mathrm{rk}_R(A) = \mathrm{rk}_S(A \otimes S) = \mathrm{rk}_S(A \otimes_S (S \otimes S)) = \mathrm{rk}_S(A)\mathrm{rk}_S(S \otimes S) = \mathrm{rk}_S(A)\mathrm{rk}_R(S)$$

□

Theorem 3.5.6 (DeMeyer [66]) *Let R be a semilocal ring. Then every Azumaya algebra A can be split by a faithfully projective separable extension S of R. If R is connected, then S can be taken Galois in the sense of [75].*

Proof It suffices to show the assertion for Azumaya algebras of constant rank. So assume that $\mathrm{rk}_R(A) = m^2$. Let $J(R)$ be the Jacobson radical of R. Then $R/J(R)$ is a direct sum of fields, say

$$\overline{R} = R/J(R) = \oplus_{i=1}^n k_i$$

and

$$\overline{A} = A/J(R)A = \oplus_{i=1}^n B_i$$

where each B_i is a central simple k_i-algebra of rank m^2. By 3.5.3, we may find a separable maximal commutative subalgebra $k_i[\beta_i]$ of B_i for each index i. Write $\beta =\sim_{i=1}^n \beta_i$, and lift β to some $\alpha \in A$. Now let S be the R-module generated by $\{1, \alpha, \alpha^2, \ldots, \alpha^{m-1}\}$. Now α satisfies its own characteristic polynomial, by the Cayley-Hamilton theorem 3.4.2. Therefore $\alpha^m \in S$, and S is an R-algebra.
Clearly $S/J(R)S = \overline{R}[\beta] = \oplus_{i=1}^n k_i[\beta_i]$ is separable as an $\overline{R}$-algebra. Since $J(R) \subset m$ for every $m \in \mathrm{Max}(R)$, this implies that S/mS is separable over R/mR, and S is separable over R, by [114, III.2.6]. From Lemma 3.5.4, it follows that A is S-projective. Then it follows from Proposition 3.5.5 that it suffices to show that S is a maximal commutative subalgebra of A, or that $A^S = S$.
Consider the S-algebra A^S. Then

$$A^S = S + (A^S \cap J(R)A) \tag{3.10}$$

since $S/J(R)S$ is a maximal commutative subalgebra of $A/J(R)A$. Write $B = \mathrm{End}_S(A)$, where A is an S-module by right multiplication. If we denote F for the canonical isomorphism $A \otimes A^{\mathrm{op}} \to \mathrm{End}_R(A)$, then we have

$$B = \mathrm{End}_S(A) \subset \mathrm{End}_R(A) = F(A \otimes A^{\mathrm{op}})$$

and

$$F(A \otimes S) \subset \mathrm{End}_S(A) = B$$

making B into an $A \otimes S$-bimodule. Since $A \otimes S$ is S-Azumaya, we have that (cf. Proposition 2.2.8) $B \cong A \otimes S \otimes_S B^{A \otimes S}$. Now

$$F^{-1}(B^{A\otimes S}) \subset (A \otimes A^{\mathrm{op}})^{A\otimes S} \subset R \otimes A^S$$

and

$$F(R \otimes A^S) \subset B^{A\otimes S}$$

hence $B^{A\otimes S} \cong A^S$, and it follows that $B = A \otimes S \otimes_S A^S$.
Now B is a separable S-algebra, and $A \otimes S$ is faithfully projective as an S-module, hence A^S is a separable S-algebra, by Lemma 3.5.2. Furthermore

$$\mathrm{Z}(A \otimes S) \otimes_S \mathrm{Z}(A^S) = S \otimes_S \mathrm{Z}(A^S) \subset \mathrm{Z}(B) = S$$

such that $\mathrm{Z}(A^S) = S$. A^S is a central separable S-algebra, and therefore an S-Azumaya algebra, by Proposition 2.2.16. We have a one to one correspondence between the ideals of S and the two-sided ideals of A^S (cf. Prop 2.2.8). Consider the two-sided ideal $A^S \cap J(R)A$ of A^S. We have that

$$A^S \cap J(R)A = (A^S \cap J(R)A \cap S)A^S = (S \cap J(R)A)A^S = J(R)A^S$$

From (3.10), it now follows that

$$A^S = S + J(R)A^S$$

and $A^S = S$ by Nakayama's Lemma.
Now we have a faithfully projective separable extension S of R that splits A. Suppose that R is connected. Then the image e of 1_R in S is an idempotent, and eS is still a faithfully projective separable extension of R splitting A. By the embedding Theorem [75, III.2.9], eS embeds in a Galois extension of R, and this proves our theorem. □

Chapter 4

Central separable algebras

It is well known, cf. e.g. [12], [114], [75], that an Azumaya algebra A is a central separable algebra, that is, A is central, and projective as an A^e-module. One may ask whether a similar approach is possible for algebras not having a unit; the answer is affirmative, and, in fact, this is how the big Brauer group was originally introduced by Taylor in [175]. In this Chapter, we will discuss the notion of separable algebra, and we will show that a central separable algebra in the sense of [175] is exactly the same as a Taylor-Azumaya algebra, as introduced in Chapter 2.
In Section 4.3, we will discuss some properties of Taylor-Azumaya algebras that happen to be flat as R-modules. The most striking property is that, over a noetherian ring, every Taylor-Azumaya algebra is Morita equivalent to a subalgebras that is finitely generated as an R-module.
The main aim of this technical Chapter is to explain how the theory that we developed in the previous Chapters is related to Taylor's papers [175] and [160]. The rest of this book is independent of the results of this Chapter, so the reader may skip it if he wants to.

4.1 Separable algebras

As in the previous Chapters, R will be a commutative ring, A an associative R-algebra, not necessarily having a unit. Note that if we view A^e as an A-bimodule, then the multiplication map $m_A: A^e \to A$ is an A-bimodule homomorphism.

Definition 4.1.1 *A splitting map φ for A is an A-bimodule homomorphism φ : $A \to A^e$ which is a left inverse for the multiplication map $m_A: A^e \to A$: $m_A \circ \varphi_A = I_A$.*

Proposition 4.1.2 *If A has a splitting map, then A is a unital R-algebra.*

Proof We have to show that $A \otimes_A A \cong A$. Recall that $A \otimes_A A = \operatorname{Coker} \lambda$, where $\lambda: A^{\otimes 3} \to A^{\otimes 2}$ is defined by

$$\lambda(a \otimes b \otimes c) = a \otimes bc - ab \otimes c$$

The multiplication map is surjective since $m_A(\varphi(a)) = a$. Hence it suffices to show that

$$A^{\otimes 3} \xrightarrow{\lambda} A^{\otimes 2} \xrightarrow{m_A} A \longrightarrow 0$$

is exact at $A^{\otimes 2}$. It is clear that $m_A \circ \lambda = 0$. For all $a, b \in A$, we have

$$\begin{aligned} &\varphi_A \circ m_A + \lambda \circ (I_A \otimes \varphi_A))(a \otimes b) \\ &= \varphi_A(ab) + a \otimes m_A(\varphi_A(b)) - (m_A \otimes I_A)(a \otimes \varphi_A(b)) \\ &= \varphi_A(ab) + a \otimes b - a.\varphi_A(b) \\ &= a \otimes b \end{aligned}$$

where we used the fact that φ_A is an A-bimodule homomorphism. Therefore, if $x \in \operatorname{Ker}(m_A)$, then $x = \lambda((I_A \otimes \varphi_A)(x) \in \operatorname{Im}\lambda$ □

If A is a projective object of A^e-mod, then A has a splitting map, but the converse is not necessarily true if A has no unit. If A has a unit, then a splitting map φ_A is completely determined by $\varphi_A(1)$. If $\varphi_A(1) = e$, then we have for all $a \in A$:

$$\varphi_A(a) = (a \otimes 1)e = (1 \otimes a)e$$

Moreover, e is an idempotent. Indeed, from the above equation, it follows that $\operatorname{Ker}(m_A)e = 0$. Now $e^2 - e = (e-1)e \in \operatorname{Ker}(m_A)e$, since $m_A(e) = 1$. Hence $e^2 = e$. We call e a *separability idempotent* for A.

Lemma 4.1.3 *If A has a splitting map, and if M is a unital left A-module, then the A-action ψ_M: $A \otimes M \to M$ has a right inverse φ_M.*

Proof We define φ_M as the composition of the following left A-module homomorphisms:

$$M \cong A \otimes_A M \overset{\varphi_A \otimes I_M}{\longrightarrow} A \otimes A \otimes_A M \cong M$$

It is easy to see that φ_M is a right inverse for ψ_M : take $m \in M$, and choose $a_i \in A$, $m_i \in M$ such that $m = \sum_i a_i m_i$. Let $\varphi_A(a) = \sum_j a_{ij} \otimes b_{ij}$. Then

$$\psi_M(\varphi_M(m)) = \psi_M(\sum_{i,j} a_{ij} \otimes b_{ij} m_i) = \sum_{i,j} a_{ij} b_{ij} m_i = \sum_i a_i m_i = m$$

□

Proposition 4.1.4 *If A has a splitting map, then the restriction of scalars functor A-mod$^u \longrightarrow R$-mod is separable in the sense of [145]. Recall that this means that, for every M, $N \in A$-mod^u, there exists a map $\phi_{M,N}$: $\operatorname{Hom}_R(M, N) \to \operatorname{Hom}_A(M, N)$ such that*

1. *$\phi_{M,N}(\alpha) = \alpha$ if α is A-linear;*

2. *for all $M, N, M', N' \in A\text{-mod}^u$ and for every commutative diagram*

$$\begin{array}{ccc} M & \xrightarrow{f} & N \\ \downarrow{\scriptstyle\alpha} & & \downarrow{\scriptstyle\beta} \\ M' & \xrightarrow{g} & N' \end{array}$$

in R-mod, with α, β R-linear, the diagram

$$\begin{array}{ccc} M & \xrightarrow{\phi_{M,N}(f)} & N \\ \downarrow{\scriptstyle\alpha} & & \downarrow{\scriptstyle\beta} \\ M' & \xrightarrow{\phi_{M,N}(g)} & N' \end{array}$$

commutes.

Proof (cf. [145, Prop.1.3]) Let $f: M \to N$ be R-linear. We define

$$\phi_{M,N}(f) = \psi_N \circ (I_A \otimes f) \circ \varphi_M$$

where φ_M is defined as in Proposition 4.1.3. Suppose first that f is A-linear. If $\varphi_M(m) = \sum_i a_i \otimes m_i$, then

$$\phi_{M,N}(f)(m) = \sum_i a_i f(m_i) = \sum_i f(a_i m_i) = f(m)$$

and this proves 1. Assume next that we have a commutative diagram in R-mod as in 2. Then the diagram

$$\begin{array}{ccccccc} M & \xrightarrow{\varphi_M} & A \otimes M & \xrightarrow{I \otimes f} & A \otimes N & \xrightarrow{\psi_N} & N \\ \downarrow{\scriptstyle\alpha} & & \downarrow{\scriptstyle I\otimes\alpha} & & \downarrow{\scriptstyle I\otimes\beta} & & \downarrow{\scriptstyle\beta} \\ M' & \xrightarrow{\varphi_{M'}} & A \otimes M & \xrightarrow{I \otimes g} & A \otimes N' & \xrightarrow{\psi'_N} & N' \end{array}$$

commutes since α and β are A-linear. This proves the second assertion. □

If A has a unit, then the converse of Proposition 4.1.4 also holds, we refer to [145, Prop.1.3] for details. Also observe that, if M, N are A-bimodules, and $f: M \to N$ is right A-linear, then $\phi_{M,N}$ is (A, A)-linear. To see this, take $f = g$ in 4.1.4, and let α and β be given by right multiplication by $a \in A$.

Lemma 4.1.5 *Suppose that A has a splitting map, and consider a short exact sequence*

$$0 \longrightarrow L \xrightarrow{\alpha} M \xrightarrow{\beta} N \longrightarrow 0$$

in $A\text{-mod}^u$. If the sequence is R-split, then it is also A-split.

Proof We know that the A-linear map $\beta: M \to N$ has an R-linear right inverse, say $g: N \to M$. Apply 4.1.4 to the square

$$\begin{array}{ccc} M & \xrightarrow{I_M} & M \\ \downarrow{\scriptstyle\beta} & & \downarrow{\scriptstyle I} \\ N & \xrightarrow{g} & M \end{array}$$

It follows that $\phi_{M,N}(\beta)$ is a right inverse of β. □

Lemma 4.1.6 *Suppose that A has a splitting map and that we have an R-split short exact sequence*

$$0 \longrightarrow L \xrightarrow{\alpha} M \xrightarrow{\beta} N \longrightarrow 0 \tag{4.1}$$

of left A-modules, such that M is unital and $AL = L$. Then L and N are also unital, and the sequence is also A-split.

Proof Recall that $A \otimes_A L = \text{Coker } \lambda$, where $\lambda : A \otimes A \otimes L \to A \otimes L$ is given by $\lambda(a \otimes b \otimes m) = a \otimes bm - ab \otimes m$. We claim that the short exact sequence

$$0 \longrightarrow \text{Im } \lambda \longrightarrow A \otimes L \longrightarrow A \otimes_A L \longrightarrow 0$$

is A-split. The inclusion $\text{Im}\lambda \to A \otimes L$ is split by the map $-\lambda \circ (\varphi_A \otimes I_L)$. Indeed, for $a, b \in A$, $m \in L$, we have

$$\begin{aligned}
&(-\lambda \circ (\varphi_A \otimes I_L) \circ \lambda)(a \otimes b \otimes m) \\
&= (-\lambda \circ (\varphi_A \otimes I_L))(a \otimes bm - ab \otimes m) \\
&= -\lambda(\varphi(a) \otimes bm - \varphi(a)b \otimes m) \\
&= a \otimes bm - \sum_i a_i \otimes a_i'bm + \sum_i a_i \otimes a_i'bm - a \otimes bm \\
&= a \otimes bm - a \otimes bm \\
&= \lambda(a \otimes b \otimes m)
\end{aligned}$$

Now consider the commutative diagram

$$\begin{array}{ccccccccc}
0 & \longrightarrow & A \otimes A \otimes L & \longrightarrow & A \otimes A \otimes M & \longrightarrow & A \otimes A \otimes N & \longrightarrow & 0 \\
 & & \downarrow & & \downarrow & & \downarrow & & \\
0 & \longrightarrow & A \otimes L & \longrightarrow & A \otimes M & \longrightarrow & A \otimes N & \longrightarrow & 0 \\
 & & \downarrow & & \downarrow & & \downarrow & & \\
0 & \longrightarrow & A \otimes_A L & \longrightarrow & A \otimes_A M & \longrightarrow & A \otimes_A N & \longrightarrow & 0 \\
 & & \downarrow & & \downarrow & & \downarrow & & \\
 & & 0 & & 0 & & 0 & &
\end{array} \tag{4.2}$$

In (4.2), the three columns are split exact, by the argument given above. The two top horizontal rows are exact, because (4.1) is R-split. A diagram chasing argument now immediately tells us that the third row

$$0 \longrightarrow A \otimes_A L \longrightarrow A \otimes_A M \longrightarrow A \otimes_A N \longrightarrow 0$$

is exact. Now look at the following diagram with exact rows

$$\begin{array}{ccccccccc}
0 & \longrightarrow & A \otimes_A L & \longrightarrow & A \otimes_A M & \longrightarrow & A \otimes_A N & \longrightarrow & 0 \\
 & & \downarrow & & \downarrow \cong & & \downarrow & & \\
0 & \longrightarrow & L & \longrightarrow & M & \longrightarrow & N & \longrightarrow & 0
\end{array}$$

to conclude that $A \otimes_A L \to L$ is injective. By assumption, this map is also surjective, so L is unital. The lemma of 5 then tells us that N is also unital. Now (4.1) is an R-split exact sequence of unital A-modules. By the preceding lemma, it is therefore also A-split. □

Lemma 4.1.7 *Suppose that A and B have splitting maps. Then A^{op} and $A \otimes B$ also have splitting maps. In particular, A^e has a splitting map.*

Proof Obvious. □

Lemma 4.1.8 *Suppose that an R-algebra A has a splitting map φ_A. For any $a \in A$, we define a map $\mu_A(a) : A \to A$ by $\mu_A(a)(b) = \varphi_A(b).a$, for all $b \in A$. We then have the following properties:*
1. For all $a \in A$, $\mu_A(a) \in Z(A)$;
2. $\mu_A : A \to Z(A)$ is a $Z(A)$-module homomorphism.

Proof For all $a, b, c, d \in A$, we have

$$\begin{aligned} \mu_A(a)((b\otimes d).c) &= \mu_A(a)(bcd) \\ &= \varphi(bcd).a \\ &= (b\otimes d).\varphi(c).a \\ &= (b\otimes d).\mu_A(a)(c) \end{aligned}$$

and this proves 1. To show 2., take $\alpha \in Z(A) = \mathrm{End}_{A^e}(A)$. Then for all $a, b \in A$:

$$\begin{aligned} \mu(\alpha(a))(b) &= \varphi_A(b).\alpha(a) \\ &= \alpha(\varphi_A(b).a) \\ &= \alpha(\mu_A(a)(b)) \\ &= (\alpha\circ\mu_A(a))(b) \end{aligned}$$

hence

$$\mu_A(\alpha(a)) = \alpha\circ\mu_A(a).$$

□

Now suppose for a moment that A has a unit and a splitting map. Then we can consider the separability idempotent $e = \sum_i e_i \otimes e_i'$. The map μ_A is then given by

$$\mu_A(a)(b) = \varphi_A(b).a = (1\otimes b)e.a = \sum_i e_i a e_i' b$$

and it follows that $\mu_A(a) = \sum_i e_i a e_i'$, and, in particular, $\mu_A(1) = 1$. Therefore, μ_A is surjective, and this justifies the following definition :

Definition 4.1.9 *An R-algebra A is called a separable R-algebra if A has a splitting map φ_A such that the corresponding map $\mu_A : A \to Z(A)$ is surjective.*

Proposition 4.1.10 *Suppose that the R-algebra A is separable. Then we have:*

1. *A is a projective object of A^e-mod;*
2. *for any proper ideal I of $Z(A)$, $IA \neq A$;*

3. $Z(A)$ *embeds as a direct summand into* A.

Proof Suppose that we have the following diagram of A^e-module homomorphisms, with the top row exact:

$$\begin{array}{ccccc} & & A & & \\ & & \downarrow{\scriptstyle\beta} & & \\ M & \xrightarrow{\alpha} & N & \longrightarrow & 0 \end{array}$$

Choose $e_0 \in A$ such that $\mu_A(e_0) = 1$, the identity in $Z(A) = \mathrm{End}_{A^e}(A)$, and $m_o \in M$ such that $\alpha(m_0) = \beta(e_0)$. Consider $\gamma:\ A \to M$ defined by

$$\gamma(a) = \varphi_A(a).m_0$$

Then γ is A^e-linear, the proof is similar to the proof of the first assertion of lemma 4.1.8. Furthermore, for all $a \in A$:

$$\begin{aligned} (\alpha \circ \gamma)(a) &= \alpha(\varphi_A(a).m_0) \\ &= \varphi_A(a)\alpha(m_0) \\ &= \varphi_A(a)\beta(e_0) \\ &= \beta(\varphi_A(a)e_0) \\ &= \beta(\mu_A(e_0)a) \\ &= \beta(a) \end{aligned}$$

hence γ lifts β, and this proves the first assertion.
Take $x \in Z(A) \setminus I$. We claim that $xe_0 \in A \setminus IA$. Otherwise $xe_0 = \sum_i x_i a_i$, with $x_i \in I,\ a_i \in A$, then

$$x = \mu_A(xe_0) = \mu_A(\sum_i x_i a_i) = \sum_i x_i \mu_A(a_i) \in I$$

which is a contradiction. This proves the second statement.
The map $\pi:\ A \to Z(A)e_0$ given by

$$\pi(a) = \mu_A(a)e_0$$

is a projection. Now $Z(A)e_0 \cong Z(A)$, since $xe_0 = 0$ implies that $x = \mu_A(xe_0) = 0$. Hence $Z(A)$ is a direct summand of A. □

4.2 Central separable algebras

Definition 4.2.1 *An* R*-algebra* A *is called a central separable algebra if it is separable and* R*-central.*

If the algebra A has a unit, then it is well-known that a central separable algebra is an Azumaya algebra, and conversely. We will show in this section that this property also holds for central separable algebras without a unit, and this will yield new characterizations for Taylor-Azumaya algebras.

Theorem 4.2.2 *An R-algebra A is central separable if and only if the following four properties hold:*

1. *A is central;*

2. $A^2 = A$;

3. *A is a projective object of A^e-mod;*

4. *For any proper ideal I of R, $IA \neq A$.*

Proof The fact that a central separable algebra satisfies conditions 1 to 4 follows immediately from Propositions 4.1.10 and 4.1.2. Before we can prove the converse, we need some lemmas.

Lemma 4.2.3 *Let k be a field, and A a nonzero central k algebra with a splitting map. Then $\varphi_A(A)(A^e) = A^e$ and A^e is central.*

Proof We will first show that

$$\varphi_A(A)A^e \neq 0 \tag{4.3}$$

To this end, consider $N = \{n \in A | anb = 0, \text{ for all } a, b \in A\}$. Suppose first that $N = 0$. Then clearly

$$\begin{aligned} N_1 &= \{n \in A | an = 0, \text{ for all } a \in A\} = 0 \\ N_2 &= \{n \in A | na = 0, \text{ for all } a \in A\} = 0 \end{aligned}$$

Let $\{e_i | i \in I\}$ be a basis for A as a k-vector space, and let $\pi_j : A \to k$ be the projection onto the j-th component, i.e.

$$\pi_j(\sum_i \alpha_i e_i) = \alpha_j$$

If $n \in A$ is such that $\pi_i(an) = 0$ for all $i \in I$ and $a \in A$, then $n = 0$, because $N_1 = 0$. Similarly, if $\pi_i(nb) = 0$ for all $i \in I$ and $b \in A$, then $n = 0$. Now take

$$x = \sum_{i,j} \alpha_{ij} e_i \otimes e_j \in A^e$$

and suppose that $x(a \otimes b) = 0$ for all $a, b \in A$. Then for all $a, b \in A$, we have

$$\sum_{i,j} \alpha_{ij} e_i a \otimes b e_j = 0$$

hence for all $k, \ell \in I$ and $a, b \in A$:

$$\sum_{i,j} \alpha_{ij} \pi_k(e_i a) \pi_\ell(b e_j) \pi_\ell(\sum_{i,j} \pi_k(\alpha_{ij} e_i a) b e_j) = 0$$

It follows that for all $k \in I$ and $a \in A$:

$$\sum_j \pi_k(\sum_i \alpha_{ij} e_i a) e_j = 0$$

and for all $j, k \in I$ and $a \in A$:

$$\pi_k(\sum_i \alpha_{ij} e_i a) = 0$$

For all $j \in I$, we now have

$$\sum_i \alpha_{ij} e_i = 0$$

and therefore $\alpha_{ij} = 0$ for all $i, j \in I$, and $x = 0$.
We have shown that $x(a \otimes b) = 0$ implies that $x = 0$. Suppose that $\varphi_A(A)A^e = 0$. Then it follows that $\varphi_A(a) = 0$ for all $a \in A$, hence $a = m_A(\varphi_A(a)) = 0$, hence $A = 0$, in contradiction with condition 4. on A. So (4.3) holds if $N = 0$.
If $N \neq 0$, then we apply the above arguments to A/N. Thus if $x \in A^e$ is such that $x(b \otimes c) = 0$ for all $b, c \in A$, then $x \in A \otimes N + N \otimes A$. Suppose that $\varphi_A(A)A^e = 0$. Then for all $a \in A$:

$$\varphi_A(a) = \sum_i a_i \otimes n_i + \sum_j m_j \otimes b_j \in A \otimes N + N \otimes A$$

Therefore for all $a, b, c \in A$:

$$\varphi_A(bac) = \sum_i ba_i \otimes n_i c + \sum_j bm_j \otimes b_j c$$

hence

$$bac = m_A(\varphi_A(bac)) = \sum_i ba_i n_i c + \sum_j bm_j b_j c = 0$$

since $m_j, n_i \in N$. This implies that $A = A^3 = 0$, again a contradiction. This finishes the proof of (4.3)
We will next show that A^e is R-central. Take $\alpha \in Z(A^e) = \mathrm{Hom}_{(A^e)^e}(A^e, A^e)$. For all $a \in A$, $f \in A^*$, consider the composition

$$\psi : A \xrightarrow{a\otimes\bullet} A \otimes A \xrightarrow{\alpha} A \otimes A \xrightarrow{f\otimes I} A$$

that is, $\psi(b) = (f \otimes I)(\alpha(a \otimes b)$. Then for all $b, c, d \in A$, we have (write $a = \sum_i a'_i a_i a''_i$ and $b = \sum_j b'_j b_j b''_j$)

$$\begin{aligned}
\psi(cbd) &= (f \otimes I)\alpha(a \otimes cbd) \\
&= \sum_{i,j} (f \otimes I)\alpha(a'_i a_i a''_i \otimes cb'_j b_j b''_j d) \\
&= \sum_{i,j} (f \otimes I)\alpha((a'_i \otimes b''_j d)(a_i \otimes b_j)(a''_i \otimes cb'_j)) \\
&= \sum_{i,j} (f \otimes I)((a'_i \otimes b''_j d)\alpha(a_i \otimes b_j)(a''_i \otimes cb'_j)) \\
&= c(\sum_{i,j} (f \otimes I)((a'_i \otimes b''_j)\alpha(a_i \otimes b_j)(a''_i \otimes b'_j))d \\
&= c\psi(b)d
\end{aligned}$$

hence $\psi \in Z(A) = k$. This means that ψ is given by multiplication by $r(a, f) \in k$. We therefore have for all $g \in A^*$ that

$$(f \otimes g)\alpha(a \otimes b) = r(a, f)g(b)$$

Now take $f = \pi_i$, $g = \pi_j$, with $i, j \in I$. Then it follows that

$$\alpha(a \otimes b)_{ij} = r_i(a)b_j \text{ for some } r_i(a) \in k$$

Here we wrote $\alpha(a\otimes b)_{ij}$ for the (i,j) component of $\alpha(a\otimes b)$ in the basis $\{e_i\otimes e_j | i, j \in I\}$ of A^e. In a similar way, we may show that

$$\alpha(a \otimes b)_{ij} = s_i(b)a_{ij} \text{ for some } s_i(b) \in k$$

Take $a = e_k$, $b = e_\ell$. Then we obtain

$$\alpha(e_k \otimes e_\ell)_{ij} = r_i(e_k)\delta_{jk} = s_j(e_\ell)\delta_{ik}$$

The only nonzero component of $\alpha(e_k \otimes e_\ell)$ is therefore the (k,ℓ) component, and

$$\alpha(e_k \otimes e_\ell) = r_k(e_k)e_k \otimes e_\ell = s_\ell(e_\ell)e_k \otimes e_\ell$$

It follows that all the $r_k(e_k)$ and $s_\ell(e_\ell)$ are equal. Denote $r_k(e_k) = s_\ell(e_\ell) = r$, then $\alpha(a \otimes b) = ra \otimes b$ for all $a, b \in A$. This proves that A^e is central.
Let us finally show that

$$\varphi_A(A)(A^e) = A^e \tag{4.4}$$

Since $A^e\varphi_A(A) = \varphi_A(A^e.A) = \varphi_A(A^3) = \varphi_A(A)$, it follows that $\varphi_A(A)(A^e)$ is a unital two-sided A^e-ideal. From Proposition 4.1.7, it follows that A^e and $(A^e)^e$ have splitting maps. Since $\varphi_A(A)A^e$ is a k-direct summand of A^e (k is a field), it follows from Proposition 4.1.5 that is also an $(A^e)^e$-direct summand of A^e. This means that there is a projection

$$\pi : A^e \to \varphi_A(A)A^e$$

which is an A^e-bimodule homomorphism. Since A^e is central, π is given by multiplication by an element $r \in k$. Since π is a projection, r is an idempotent. $r = 0$ is impossible, because of (4.3). Hence $r = 1$, and this implies (4.4). □

Lemma 4.2.4 *Suppose that A satisfies the four conditions of Theorem 4.2.2, and that I is a two-sided ideal of A. Then $Z(A/I)$ is a quotient of R.*

Proof Let $\alpha \in Z(A/I) = \mathrm{End}_{A^e}(A/I)$, and let $\overline{\alpha} = \alpha \circ \pi$, where $\pi : A \to A/I$ is the canonical surjection. Then the A^e-projectivity of A implies that $\overline{\alpha} : A \to A/I$ may be lifted to a map $\beta : A \to A$, which is A^e-linear, which means that $\beta \in Z(A) = \mathrm{End}_R(A^e)$.

$$\begin{array}{ccc} & & A \\ & \beta \swarrow & \downarrow \overline{\alpha} \\ A & \xrightarrow{\pi} & A/I \end{array}$$

It follows that the map $Z(A) = R \to Z(A/I)$ is surjective. □

We now finish the proof of Theorem 4.2.2. If A is A^e-projective, then A has a splitting map φ_A. The only thing we have to show, is that μ_A is surjective. Suppose that $\mu_A(A)$ is a proper ideal of R. Then $\mu_A(A)$ is contained in a maximal ideal m of R, and for all $a, b \in A$: $\mu(b)(a) = \varphi_A(a)b \in mA$, hence

$$\varphi_A(A).A \subset mA \tag{4.5}$$

Condition 4. of Theorem 4.2.2 implies that $\overline{A} = A/mA \neq 0$. From Lemma 4.2.4, it follows that $Z(\overline{A}) = R/m = k$. Define $\varphi_{\overline{A}}$: $\overline{A} \to \overline{A}^e$ by

$$\varphi_{\overline{A}}(\overline{a}) = \sum_i \overline{a}_i \otimes \overline{a}'_i$$

if

$$\varphi_A(a) = \sum_i a_i \otimes a'_i$$

$\varphi_{\overline{A}}$ is well-defined, since for $x \in M$, $a \in A$, we have that $\sum_i x\overline{a}_i \otimes \overline{a}'_i = 0$, and $\varphi_{\overline{A}}$ is clearly a splitting map for $\overline{A}$. Hence we may apply Lemma 4.2.3 to conclude that

$$\varphi_{\overline{A}}(\overline{A}).\overline{A}^e = \overline{A}^e$$

Applying the multiplication map, and using the fact that $\varphi_{\overline{A}}$ is a left $\overline{A}^e$-module homomorphism, we obtain

$$\varphi_{\overline{A}}(\overline{A}).\overline{A} = \overline{A}$$

From (4.5) it follows that $\varphi_{\overline{A}}(\overline{A}).\overline{A} = 0$. This is a contradiction! □

It is well known that a central separable algebra (with a unit) over a field k is a central simple algebra. In the following Proposition, this property is generalized to arbitrary central separable algebras.

Proposition 4.2.5 *Let A be a central separable algebra over a field k. Then A has no nontrivial unital two-sided ideals.*

Proof From Lemma 4.2.3, it follows that A^e is central. A^e has a splitting map, by Lemma 4.1.7, and it is clear that $\mu_{A^e} = \mu_A \otimes \mu_{A^{op}}$ is surjective. Hence A^e is a central separable algebra. Let I be a unital two-sided ideal of A. Since k is a field, the sequence

$$0 \longrightarrow I \longrightarrow A \longrightarrow A/I \longrightarrow 0$$

is k-split, and therefore A^e-split, by Lemma 4.1.5. We therefore have a projection π : $A \to I$, which is A^e-linear. This means that $\pi \in Z(A) = k$. Since π is an idempotent, $\pi = 0$ or $\pi = 1$. Hence $I = 0$ or $I = A$. □

Proposition 4.2.6 *Let A be a central separable R-algebra, and J a maximal two-sided ideal of A. Then $\mu_A(J)$ is a maximal ideal of R and $\overline{A} = A/\mu_A(J)A$ is a central separable k-algebra, with $k = R/\mu_A(J)$. Furthermore, $\mu_A(J)A = AJA$.*

Proof We will first show that $\mu_A(J)$ is a proper subset of R. Suppose that $a \in J$ is such that $\mu_A(a) = 1$. Then for all $b \in B$:

$$b = \mu_A(a)b = \varphi_A(b)a \in J$$

implying $J = A$. Take a maximal ideal m of R containing $\mu_A(J)$. Then $J+mA$ is a two-sided ideal of A containing J. $J+mA$ is a proper ideal since $\mu_A(J+mA) = m$, and therefore $J + mA = J$. But then $mA \subset J$, so $m = \mu_A(mA) \subset \mu_A(J)$, and $m = \mu_A(J)$. This finishes the proof of the first part.
Write $\overline{A} = A/\mu_A(J)A$. In the proof of Theorem 4.2.2, we have shown that $\overline{A}$ has a splitting map $\varphi_{\overline{A}}$. From Lemma 4.2.4, it follows that $\overline{A}$ is k-central. If $e_0 \in A$ is such that $\mu_A(e_0) = 1$, then clearly $\mu_{\overline{A}}(\overline{l}_0) = 1$, and this shows that $\overline{A}$ is a central separable k-algebra. From Proposition 4.2.5, it follows that $\overline{A}$ has no nontrivial unital two-sided ideals, and therefore the image of AJA in $\overline{A}$ is zero. Therefore

$$AJA \subset \mu_A(J)A \subset J,$$

and, since $A^2 = A$,

$$AJA = A^2JA^2 \subset A\mu_A(J)A^2 = A\mu_A(J)A \subset AJA$$

and it follows that $AJA = \mu_A(J)$. □

Corollary 4.2.7 *If A is a central separable R-algebra, then any two-sided ideal J of A is contained in a maximal two-sided ideal.*

Proof $\mu_A(J)$ is an ideal of R contained in a maximal ideal m. Let

$$I = \{a \in A|\varphi_A(b)a \in mA, \text{ for all } b \in A\}.$$

Then $mA \subset I$, and $mA = AIA$. I is a proper ideal, since $I = A$ would imply that $A = \varphi_A(A)A = \varphi_A(A)I \subset mA$, in contradiction with condition 4 of Theorem 4.2.2. If K is a two-sided ideal of A with $\mu_A(K) \subset m$, then $K \subset I$. Indeed, for all $a \in K$, we have that $\mu_A(a) = x \in m$, hence for all $b \in A$:

$$\varphi_A(b)a = \mu_A(a)b = xb \in mA,$$

and $a \in I$. This implies that I is maximal, because for any proper ideal K of A, $\mu_A(K)$ is a proper ideal of R, hence $I \subset K$ implies $m = \mu_A(I) \subset \mu_A(K)$, so $m = \mu_A(K)$, and $K \subset I$. We also have that $J \subset I$ since $\mu_A(J) \subset m$. □

Proposition 4.2.8 *Let A and B be central separable R-algebras. Then $A^{\text{op}}, A \otimes B$ and A^e are also central separable.*

Proof It is clear that A^{op} is a central separable R-algebra. we know from Lemma 4.1.7 that $A \otimes B$ has a splitting map. It is easy to see that $\mu_{A\otimes B}$ is given by

$$\mu_{A\otimes B}(a \otimes b) = \mu_A(a)\mu_B(b)$$

This implies that $\mu_{A\otimes B}$ maps onto R, and that $A \otimes B$ is R-central. □

We can now generalize Lemma 4.2.3 to central separable algebras over commutative rings:

Theorem 4.2.9 *For any central separable algebra A over a commutative ring R, we have that $\varphi_A(A)A^e = A^e$.*

Proof Suppose that $\varphi_A(A)A^e$ is a proper unital two-sided ideal of A^e. Using Corollary 4.2.7, we obtain a maximal two-sided ideal I of A^e such that $\varphi_A(A)A^e \subset I$. Invoking Propositions 4.2.6 and 4.2.8, we obtain that $A^eJA^e = \mu_{A^e}(J)A^e$, and $\mu_{A^e}(J)$ is a maximal ideal of R. Since $\varphi_A(A)A^e$ is unital,

$$\varphi_A(A)A^e \subset A^eJA^e = \mu_{A^e}(J)A^e.$$

Applying the multiplication map m_A, we obtain that

$$\varphi_A(A)A = \mu_A(\varphi_A(A)A^e) \subset \mu_{A^e}(J)A$$

Now $\varphi_A(A)A = A$, by the surjectivity of μ_A, hence we have a contradiction with condition 4 of theorem 4.2.2. □

Theorem 4.2.10 *For an R-algebra A, the following assertions are equivalent:*
1. A is a central separable R-algebra;
2. A is an R-Taylor-Azumaya algebra;
3. A is faithful as an R-module, and $\text{Im tr} = R$.

Proof 1. $\Longrightarrow$ 2. By Proposition 2.2.2, it suffices to show that the separator $\theta : A_\ell \otimes A_r \otimes \Omega \to A^e$ is bijective and that the trace map $\text{tr}: A \otimes \Omega \to R$ is surjective. We will first show that θ is surjective. For any $\alpha \in A^e$, we define $\omega_\alpha \in \Omega$ by

$$\omega_\alpha(a \otimes b) = \varphi_A(a)\alpha(\tau \circ \varphi_A)(b)$$

Let us verify that $\omega_\alpha \in \Omega$:

$$\begin{aligned}
\omega_\alpha(cad \otimes fbg) &= \varphi_A(cad)\ \alpha\ \tau(\varphi_A(fbg)) \\
&= (c \otimes d)\ \varphi_A(a)\ \alpha\ \tau((f \otimes g)\varphi_A(b)) \\
&= (c \otimes d)\ \varphi_A(a)\ \alpha\ \tau(\varphi_A(b))\ (g \otimes f) \\
&= (c \otimes d)\omega_\alpha(a \otimes b)(g \otimes f)
\end{aligned}$$

Now let a, b run through A, and α through A, then $\theta(a \otimes b \otimes \omega_\alpha)$ runs through $\varphi_A(A)A^e\tau(\varphi_A(A))$. Applying Theorem 4.2.9 to A and A^{op}, we obtain

$$\varphi_A(A)A^e\tau(\varphi_A(A)) = A^e\tau(\varphi_A(A)) = \tau(\varphi_A(A)((A^{\text{op}})^e) = \tau((A^{\text{op}})^e) = A^e$$

and therefore θ is surjective.
Recall that we have a map $\rho: \Omega \otimes \Omega \to Z(A) = R$ defined by $\rho(\omega \otimes \omega') = \omega \circ \omega'$, and consider the following diagram

$$\begin{array}{ccc}
A \otimes A \otimes \Omega \otimes \Omega & \xrightarrow{\theta \otimes I_\Omega} & A \otimes A \otimes \Omega \\
\downarrow \cong & & \downarrow \theta \\
A \otimes A \otimes \Omega \otimes \Omega & \xrightarrow{I_{A \otimes A} \otimes \rho} & A \otimes A
\end{array} \tag{4.6}$$

(4.6) is commutative, since for all $a, b \in A$ and $\omega, \omega' \in \Omega$:

$$(\theta \otimes I_\Omega)(\omega \otimes \omega' \otimes a \otimes b) = (\omega \circ \omega')(a \otimes b) = \rho(\omega \otimes \omega')a \otimes b,$$

where we used the fact that $\omega \circ \omega' \in Z(A) = R$.
Now suppose that ρ is not surjective. Then $\mathrm{Im}\rho \subset m \in \max(R)$, and

$$\mathrm{Im}(\rho \otimes I_{A\otimes A}) = \mathrm{Im}\theta(\theta \otimes I_\Omega) = m(A \otimes A) \neq A \otimes A$$

contradicting the surjectivity of θ. Hence ρ is surjective.
Choose $\sum_i \omega_i \otimes \omega_i' \in \Omega \otimes \Omega$ such that

$$\rho(\sum_i \omega_i \otimes \omega_i') = \sum_i \omega_i \circ \omega_i' = 1$$

Define $\sigma: R \to \Omega \otimes \Omega$ by

$$\sigma(r) = r \sum_i \omega_i \otimes \omega_i'.$$

Then $(\rho \circ \sigma)(r) = 1$, and

$$\begin{aligned}(\sigma \circ \rho)(\omega \otimes \omega') &= \sigma(\omega \circ \omega') \\ &= (\omega \otimes \omega')(\sum_i \omega_i \otimes \omega_i') \\ &= \sum_i \omega \circ \omega' \circ \omega_i \otimes \omega_i' \\ &= \sum_i \omega \otimes \omega' \circ \omega_i \circ \omega_i' \\ &= \omega \otimes \omega'\end{aligned}$$

Therefore ρ is an isomorphism. It now follows that θ is injective, since otherwise $\theta \circ (\theta \otimes I_\Omega$ and therefore $I_{A\otimes A} \otimes \rho$ is not surjective.
Finally, suppose that $\mathrm{Im}\,\mathrm{tr} \subset m \in \max(R)$. Recall that for all $\omega \in \Omega$, we have

$$m_A \circ \omega = I_A \otimes \mathrm{tr}_\omega$$

hence we have for all $a, b \in A$, and $\omega \in \Omega$:

$$(m_A \circ \theta)(a \otimes b \otimes \omega) = (m_A \circ \omega)(a \otimes b) = (I_A \otimes \mathrm{tr}_\omega)(a \otimes b) = (I_A \otimes \mathrm{tr})(a \otimes b \otimes \omega)$$

and it follows that $\mathrm{tr} \otimes I_A$ is surjective, contradicting the fact that $\mathrm{Im}(\mathrm{tr} \otimes I_A) = mA \neq A$.
2. $\Longrightarrow$ 3. This follows immediately from Definition 2.2.1.
3. $\Longrightarrow$ 1. Suppose that A is faithful and that the trace map maps $A \otimes \Omega$ onto R. We will show that A satisfies the four conditions listed in Theorem 4.2.2.
1. $A^2 = A$. Take $\sum b_i \otimes \omega_i \in A \otimes \Omega$ such that

$$\mathrm{tr}(\sum_i b_i \otimes \omega_i) = \sum_i \mathrm{tr}_{\omega_i}(b_i) = 1$$

We then have for all $a \in A$ that

$$a = \sum_i \mathrm{tr}_{\omega_i}(b_i)(a) = m_A(\sum_i \omega_i(a \otimes b)) \in A^2$$

and it follows that $A^2 = A$.
2. A is R-central. Recall (cf. Lemma 2.1.2) that $\mathrm{tr}_\omega : A \to Z(A)$ is $Z(A)$-linear. Choose $\alpha \in Z(A)$. Then

$$\begin{aligned}\mathrm{tr}(\sum_i \alpha(b_i)\otimes\omega_i) &= \sum_i \mathrm{tr}_{\omega_i}(\alpha(b_i))\\ &= \alpha\circ(\sum_i \mathrm{tr}_{\omega_i}(b_i)) = \alpha\end{aligned}$$

hence $\alpha \in R$, since tr takes values in R, and A is central.
3. Suppose that $IA = I$ for some ideal I of R. Then

$$R = \mathrm{tr}(A\otimes\Omega) = \mathrm{tr}(IA\otimes\Omega) = I\mathrm{tr}(A\otimes\Omega) = I.$$

4. A is A^e-projective. Choose $b_i, c_i \in A$ and $\omega_i \in \Omega$ such that $\sum_i \mathrm{tr}_{\omega_i}(b_ic_i) = 1$ (we already know that $A^2 = A$). Define $\phi_i : A \to A^e$ by

$$\phi_i(a) = \omega_i(a\otimes c_i).$$

Then ϕ_i is an A^e-module homomorphism, and

$$\begin{aligned}\sum_i \phi_i(a)b_i &= \sum_i (m_A\circ\omega_i)(a\otimes b_ic_i)\\ &= \sum_i \mathrm{tr}_{\omega_i}(b_ic_i)(a) = a\end{aligned}$$

Now consider the following diagram with bottom row exact in A^e-mod:

$$\begin{array}{ccccc} & & A & & \\ & & \downarrow{\scriptstyle\nu} & & \\ M & \xrightarrow{\mu} & N & \longrightarrow & 0\end{array}$$

Choose $m_i \in M$ such that $\mu(m_i) = \nu(b_i)$, and define $\gamma : A \to M$ by

$$\gamma(a) = \sum_i \phi_i(a).m_i$$

Then γ is A^e-linear, and

$$\begin{aligned}(\mu\circ\gamma)(a) &= \sum_i \phi_i(a)\mu(m_i)\\ &= \sum_i \phi_i(a)\nu(b_i)\\ &= \sum_i \nu(\phi_i(a)b_i)\\ &= \nu(a)\end{aligned}$$

it follows that $\mu\circ\gamma = \nu$, and A is A^e-projective. □

Here are some more characterizations of Taylor-Azumaya algebras:

Theorem 4.2.11 *For an R-algebra A, the following assertions are equivalent:*
1. A is a Taylor-Azumaya algebra;
2. A^e is an R-elementary algebra, and R is an R-module direct summand of A;
3. there exists an R-algebra B such that $A \otimes B$ is R-elementary, and R is an R-module direct summand of B.

Proof 1. $\Longrightarrow$ 2. It follows from Proposition 2.2.4 that A^e is R-elementary. We know that A is central separable, so from Proposition 4.1.10, it follows that $Z(A) = R$ is an R-module direct summand of A.
2. $\Longrightarrow$ 3. is trivial.
3. $\Longrightarrow$ 1. We will show that A satisfies the four conditions listed in Theorem 4.2.2. This will imply that A is central separable, hence that A is Taylor-Azumaya. In the sequel, we let $\pi: B \to R$ be a projection, and we take $b \in B$ such that $\pi(b) = 1$.
1. $A^2 = A$. Take $a \in A$, and $a_i, a'_i \in A$, $b_i, b'_i \in B$ such that $a \otimes b = \sum_i a_i a'_i \otimes b_i b'_i$. Then

$$a = a\pi(b) = \sum_i a_i a'_i \pi(b_i b'_i) \in A^2.$$

2. For every maximal ideal m of R, we have that $mA \neq A$. Indeed, if $mA = A$, then $mA \otimes B = A \otimes B$, in contradiction with the fact that $A \otimes B$ is Taylor-Azumaya.
3. A is A^e-projective. Suppose that in A^e-mod, we have the following diagram with exact bottom row :

$$\begin{array}{ccccc} & & A & & \\ & & \downarrow{\scriptstyle \beta} & & \\ M & \xrightarrow{\alpha} & N & \longrightarrow & 0 \end{array}$$

Then we have the following diagram in $A^e \otimes B^e$-mod:

$$\begin{array}{ccccc} & & A \otimes B & & \\ & & \downarrow{\scriptstyle \beta \otimes I_B} & & \\ M \otimes B & \xrightarrow{\alpha \otimes I_B} & N \otimes B & \longrightarrow & 0 \end{array}$$

The bottom row is exact in B^e-mod, since the functor $\bullet \otimes B$ is an equivalence of categories ; therefore, the bottom row is exact in $A^e \otimes B^e$-mod. Since $A \otimes B$ is $A^e \otimes B^e$-projective, there exists and $A^e \otimes B^e$-module homomorphism $\gamma: A \otimes B \to M \otimes B$ making the diagram commutative. Define $\delta: A \to M$ by

$$\delta(a) = ((I_A \otimes \pi) \circ \gamma)(a \otimes b)$$

for all $a \in A$. Then

$$\begin{aligned} (\alpha \circ \delta)(a) &= ((\alpha \otimes \pi) \circ \gamma)(a \otimes b) \\ &= (I_M \otimes \pi)(\alpha \otimes I_B)\gamma(a \otimes b) \\ &= (I_M \otimes \pi)(\beta \otimes I)(a \otimes b) = \beta(a) \end{aligned}$$

4. A is central. Suppose that $\alpha: A \to A$ lies in $Z(A)$. Then $\alpha \otimes I_B \in Z(A \otimes B) = R$. Write $\alpha \otimes I_B = r$, then for all $a \in A$ and $c \in B$:

$$\alpha(a) \otimes c = ra \otimes c$$

Now take $c = b$ and apply π to the second factor. Then we obtain $\alpha(a) = ra$, for all $a \in A$. □

4.3 Flat Taylor-Azumaya algebras

The big Brauer group $\mathrm{Br}'(R)$ classifies equivalence classes of finitely generated Taylor-Azumaya algebras. In this Section, we will discuss properties of flat Taylor-Azumaya algebras that are not necessarily finitely generated. One of the main results is that, in the noetherian case, every flat Taylor-Azumaya algebra is Morita equivalent to a finitely generated Taylor-Azumaya algebra. In [175], the results that follow are stated without the flatness assumption. It was pointed out to the author by F. Grandjean that we really need the assumption that the Taylor-Azumaya algebra in question is flat as an R-module. We begin with the following Lemma.

Lemma 4.3.1 *Let A be a Taylor-Azumaya algebra. If A is flat as an R-module, then A is flat as an A-module.*

Proof For any A-module M, we have an exact sequence

$$A \otimes A \otimes M \xrightarrow{\lambda} A \otimes M \longrightarrow A \otimes_A M \longrightarrow 0 \tag{4.7}$$

with $\lambda(a \otimes b \otimes m) = ab \otimes m - a \otimes bm$. A has a splitting map $\varphi_A : A \to A \otimes A$. φ_A is A-bilinear, and $m_A \circ \varphi_A = I_A$. We now claim that the map

$$\pi = \lambda \circ (\varphi I_M) : \; A \otimes M \longrightarrow \mathrm{Im}\,(\lambda)$$

is a projection. For $a \in A$, we introduce the Sweedler-type notation

$$\varphi_A(a) = \sum a^1 \otimes a^2$$

We will use this type of notation extensively in the second and third part of this book. For all $m \in M$, write

$$\varphi_A(a)m = \sum a^1 \otimes a^2 m$$

Then

$$\begin{aligned} \pi(a \otimes m) &= \lambda(\varphi_A(a) \otimes m) \\ &= a \otimes m - \varphi(a) \otimes m \\ &= a \otimes m - \sum a^1 \otimes a^2 m \end{aligned}$$

and

$$\begin{aligned} \pi^2(a \otimes m) &= a \otimes m - \sum a^1 \otimes a^2 m \\ &- \sum a^1 \otimes a^2 m + \sum \varphi_A(a^1) a^2 m \end{aligned}$$

Now

$$\sum \varphi_A(a^1) a^2 m = \varphi_A(\sum a^1 a^2) m = \varphi_A(a) m = \sum a^1 \otimes a^2 m$$

since φ_A is A-bilinear, so

$$\pi^2(a \otimes m) = a \otimes m - \sum a^1 \otimes a^2 m = \pi(a \otimes m)$$

and $\pi \circ \pi = \pi$. This implies that the exact sequence (4.7) is split.
Now let $N \to M$ be a monomorphism between A-modules. consider the diagram

$$\begin{array}{ccccccc} 1 & & 1 & & & & \\ \downarrow & & \downarrow & & & & \\ A\otimes A\otimes N & \xrightarrow{\lambda} & A\otimes N & \longrightarrow & A\otimes_A N & \longrightarrow & 1 \\ \downarrow & & \downarrow & & \downarrow & & \\ A\otimes A\otimes M & \xrightarrow{\lambda} & A\otimes M & \longrightarrow & A\otimes_A M & \longrightarrow & 1 \end{array}$$

The two rows are split, and the two columns are exact since A and $A \otimes A$ are flat as R-modules. It follows that $A \otimes_A N$ injects into $A \otimes_A M$. □

Corollary 4.3.2 *Let A be an R-flat Taylor-Azumaya algebra, and suppose that M is a unital left A-module. If N is a submodule of M such that $AN = N$, then L is unital as an A-module.*

Proof By assumption, the canónical map $A \otimes_A N \to N$ is surjective. From Lemma 4.3.1, we know that A is A-flat, and we have that $A \otimes_A N$ injects into $A \otimes_A M$. Thus we have

$$\begin{array}{ccc} A\otimes_A N & \hookrightarrow & A\otimes_A M \\ \downarrow & & \downarrow\cong \\ N & \hookrightarrow & M \end{array}$$

and this implies that $A \otimes_A N$ injects into N. □

Corollary 4.3.2 does not hold if A is not flat as an R-module! It is even not true that a principal ideal of a Taylor-Azumaya algebra is unital. This will be made clear in the next example.

Example 4.3.3 (Grandjean) Let $R = \mathbb{Z}$, and define

$$\pi : (\mathbb{Z}/2\mathbb{Z} \times \mathbb{Z}) \otimes_{\mathbb{Z}} \mathbb{Z} \longrightarrow \mathbb{Z}$$

by

$$\pi((\overline{z}_1, z_2) \otimes z_3) = z_2 z_3$$

for all $z_1, z_2, z_3 \in \mathbb{Z}$. After we identify $(\mathbb{Z}/2\mathbb{Z} \times \mathbb{Z}) \otimes_{\mathbb{Z}} \mathbb{Z}$ and $\mathbb{Z}/2\mathbb{Z} \times \mathbb{Z}$ by the canonical isomorphism, we obtain

$$\pi(\overline{z}_1, z_2) = z_2$$

It is clear that π is surjective, and $A = E_{\mathbb{Z}}(\mathbb{Z}, \mathbb{Z}/2\mathbb{Z} \times \mathbb{Z}, \pi)$ is a $\mathbb{Z}$-elementary algebra. A is unital, it is even a $\mathbb{Z}$-Taylor-Azumaya algebra. As a $\mathbb{Z}$-module,

$$A = (\mathbb{Z}/2\mathbb{Z} \times \mathbb{Z}) \otimes_{\mathbb{Z}} \mathbb{Z} \cong \mathbb{Z}/2\mathbb{Z} \times \mathbb{Z}$$

and the multiplication is given by the formula

$$\begin{aligned} (\overline{z}_1, t_1)(\overline{z}_2, t_2) &= \pi(\overline{z}_1, t_1)(\overline{z}_2, t_2) \\ &= (t_1 \overline{z}_2, t_1 t_2) \end{aligned}$$

Take $a = (\overline{0}, 2)$. Then $aA = 0 \times 2\mathbb{Z}$, and the canonical map

$$\alpha:\ aA \otimes_A A \longrightarrow A:\ (\overline{0}, 2t_1) \otimes (\overline{z}_2, t_2) \mapsto (\overline{0}, 2t_1t_2)$$

is surjective, but not injective. Indeed,

$$\alpha((\overline{0}, 2) \otimes (\overline{1}, 0)) = 0$$

while

$$(\overline{0}, 2) \otimes (\overline{1}, 0) \neq 0 \text{ in } aA \otimes_A A$$

To see this, consider the well-defined homomorphism

$$aA \otimes_A A \longrightarrow \mathbb{Z}/2\mathbb{Z} \times \mathbb{Z}:\ (\overline{0}, 2s) \otimes (\overline{z}, t) \mapsto s\overline{z}$$

This homomorphism sends $(\overline{0}, 2) \otimes (\overline{1}, 0)$ to $\overline{1} \in \mathbb{Z}/2\mathbb{Z}$.

Proposition 4.3.4 *Let A be a flat Taylor-Azumaya algebra. If M' is a unital left ideal of A, and M is a unital right ideal of A such that $M'M = A$, then MM' is a Taylor-Azumaya algebra which is Morita equivalent to A.*

Proof The map μ: $M' \otimes M \to A$ induced by the multiplication map is surjective. By Propositions 3.1.2 and 3.1.3, $M \otimes_A M'$ is a Taylor-Azumaya algebra Morita equivalent to A. Hence it suffices to show that the natural map π: $M \otimes_A M' \to MM'$ is a bijection. π is clearly surjective, so we only need to show that π is injective. We claim that the map

$$\psi_{M'}:\ A \otimes M' \to M':\ a \otimes m' \mapsto am'$$

has a right inverse

$$\varphi_{M'}:\ M' \to A \otimes M'$$

This can be seen as follows: we know from Proposition 2.2.4 that A is A^e-projective, hence the multiplication map m_A: $A \otimes A \to A$ has a right inverse φ_A: $A \to A \otimes A$. Let $\varphi_{M'}$ be the composition

$$M' \cong A \otimes_A M' \xrightarrow{\varphi_A} A \otimes A \otimes_A M' \cong A \otimes M'.$$

Here we use the fact that M' is unital. We leave it to the reader to show that $\varphi_{M'}$ is indeed a right inverse of $\psi_{M'}$. Now assume that

$$\pi(\sum_i m_i \otimes m'_i) = \sum_i m_i m'_i = 0,$$

for some $m_i \in M_i$, $m'_i \in M'_i$. Write

$$\varphi_{M'}(m'_i) = \sum_j a_{ij} \otimes m'_{ij}$$

for each index i. Then

$$\sum_{i,j} m_i a_{ij} \otimes m'_{ij} = \varphi_{M'}(\sum_i m_i m'_i) = 0$$

so

$$\sum_i m_i \otimes m_i' = \sum_{i,j} m_i \otimes a_{ij} m_{ij}' - m_i a_{ij} \otimes m_{ij}' \in \mathrm{Ker}\,(M \otimes M' \to M \otimes_A M')$$

and

$$\sum_i m_i \otimes m_i' = 0 \text{ in } M \otimes_A M'$$

□

We are now able to prove the following property.

Proposition 4.3.5 *Any flat Taylor-Azumaya algebra A is Morita equivalent to a subalgebra of A, which is also Taylor-Azumaya, and contained in a finitely generated R-submodule of A. If R is noetherian, it follows that any flat R-Taylor-Azumaya algebra is Morita equivalent to a finitely generated Taylor-Azumaya algebra.*

Proof Take $a, b \in A$ and write

$$\theta^{-1}(a \otimes b) = \sum_i x_i \otimes y_i \otimes \omega_i \in A \otimes A \otimes \Omega$$

This means that for all $c \in A$:

$$\begin{aligned} \sum_i \mathrm{tr}(y_i c \otimes \omega_i) x_i &= (m_A \circ \theta)(\sum_i x_i \otimes y_i c \otimes \omega_i) \\ &= m_A(\theta(\sum_i x_i \otimes y_i \otimes \omega_i)(c \otimes 1)) \\ &= m_A((a \otimes b)(c \otimes 1)) \\ &= acb \end{aligned}$$

Consequently, if $M = aA$, $M' = Ab$, then $MM' \subset \sum_i Rx_i$ is contained in a finitely generated R-submodule of A. Therefore, if M is a finitely generated right unital ideal, and M' is a finitely generated left unital ideal, then MM' is contained in a finitely generated R-submodule of A. Now choose $a_i, b_i, c_i \in A$, $\omega_i \in \Omega$ such that

$$\mathrm{tr}(\sum_i a_i b_i c_i \otimes \omega_i) = 1,$$

Choose $d \in A$, and write for any index i:

$$\omega_i(d \otimes b_i) = \sum_j x_{ij} \otimes y_{ij}$$

Then

$$\begin{aligned} d &= \mathrm{tr}(\sum_i a_i b_i c_i \otimes \omega_i) d \\ &= m_A\Big(\sum_i \omega_i(d \otimes b_i.(c_i \otimes a_i))\Big) \\ &= m_A\Big(\sum_{i,j} (x_{ij} \otimes y_{ij})(c_i \otimes a_i)\Big) \\ &= \sum_{i,j} x_{ij} c_i a_i y_{ij} \in (\sum_i A c_i)(\sum_i a_i A) \end{aligned}$$

Let $M' = \sum_i Ac_i$, $M = \sum_i a_i A$. By Corollary 4.3.2, M' and M are unital as respectively left and right A-modules. By Proposition 4.3.4, MM' is Taylor-Azumaya algebra, and $A \sim MM'$. MM' is contained in a finitely generated R-submodule of A. □

A field k is always noetherian, and every module over a field is flat, so we have the following Corollary.

Corollary 4.3.6 *If k is a field, then every Taylor-Azumaya algebra over k is equivalent to a finite dimensional Taylor-Azumaya subalgebra, and, consequently, to an Azumaya algebra.*

Lemma 4.3.7 *Suppose that A is a Taylor-Azumaya algebra, and let M and M' be as in Proposition 4.3.4. If e is a rank one idempotent in $B = MM'$, then it is a rank one idempotent in A.*

Proof Observe that $e = e^2 \in eB = eMM'$, and, similarly, $e \in MM'e$. Furthermore, $M^2 \subset MA = M$ and $M'^2 \subset M'$. Now

$$eAe \subset eMM'M'MMM'e \subset eMM'MM'e = eB^2e = eBe = R$$

and

$$R = eBe \subset eAe$$

proving the Lemma. □

Proposition 4.3.8 *Let R be a Henselian ring with maximal ideal m. Every Taylor-Azumaya algebra contains an equivalent Azumaya algebra.*

Proof As in the proof of Proposition 3.3.1, it suffices to show that there exists an idempotent $e \in A \setminus mA$. By Corollary 4.3.6, $A/mA = \overline{A}$ contains a finite dimensional equivalent R/m-Taylor-Azumaya algebra, containing a nonzero idempotent. Proceed as in the proof of Proposition 3.3.1. □

We are now able to show that Theorem 3.3.5 also holds for flat Taylor-Azumaya algebras.

Theorem 4.3.9 *Let A be an R-algebra that is flat as an R-module. Then the following assertions are equivalent:*

1) A is a Taylor-Azumaya algebra;

2) $A \otimes S$ contains a rank one idempotent, for some étale covering S of R;

3) $A \otimes S$ is S-elementary, for some étale covering S of R;

4) $A \otimes S$ is S-elementary, for some faithfully flat R-algebra S which is locally of finite type;

5) $A \otimes S$ is S-elementary, for some faithfully flat R-algebra S;

6) $A \otimes S$ is an S-Taylor-Azumaya algebra, for some faithfully flat R-algebra S;

7) $A \otimes S$ contains a rank one idempotent, for some faithfully flat R-algebra S;

8) A is a faithful R-module, central and unital as an R-algebra, and there exists an A^e-bimodule isomorphism ω: $A_\ell \otimes A_r \to A^e$.

Proof We restrict to giving the proof of 1) $\Rightarrow$ 2). The proof of all the other implications is identical to the proof of the corresponding implications in Theorem 3.3.5. The proof of 1) $\Rightarrow$ 2) is slightly more complicated then the the proof of 1) $\Rightarrow$ 2) in Theorem 3.3.5.
Let A be a Taylor-Azumaya algebra, and take $p \in \mathrm{Spec}R$. From Proposition 4.3.5, we know that A is equivalent to a Taylor-Azumaya algebra B, such that B is embedded in a finitely generated R-submodule M of A:

$$B \subset M \subset A$$

Consider a commutative diagram of the filtering system (3.9), and write A_α for $A \otimes S_\alpha$, B_α for $B \otimes S_\alpha$ and A_p^{sh} for $A \otimes R_p^{sh}$. We will show that B_α contains a rank one idempotent. From Lemma 4.3.7, it will then follow that this idempotent is also a rank one idempotent for A_α. From Propositions 3.3.2 and 3.3.3, we know that $B \otimes R_p^{sh}$ contains a rank one idempotent e, which is also a rank one idempotent for A_p^{sh}. Since

$$B \otimes R_p^{sh} = \varinjlim B_\alpha,$$

this e occurs already in one of the S_α. The natural map $S_\alpha \to S_\alpha e$ is a projection ; therefore, we can view $S_\alpha e$ as a direct factor of S_α. The map $R \to S_\alpha e$ is still étale. Replacing S_α by $S_\alpha e$, we therefore have

$$\begin{array}{ccccccccc} S_\alpha & \cong & S_\alpha e & \subset & eB_\alpha e & \subset & eM_\alpha e & \subset & eA_\alpha e \\ \downarrow & & \downarrow & & \downarrow & & \downarrow & & \downarrow \\ R_p^{sh} & \cong & R_p^{sh} e & = & eB_p^{sh} e & = & e(M \otimes R_p^{sh})e & = & eA_p^{sh} e \end{array}$$

$S_\alpha e \subset eB_\alpha e$, since $se = se^3 = esee \in eB_\alpha e$, for all $s \in S_\alpha$. Since M_α is finitely generated, there exists $\lambda \geq \alpha$ in the filtering system, such that $S_\lambda = eB_\lambda e$. This means that e is a rank one idempotent for B_λ, and therefore for A_λ.
We apply this construction to every $p \in \mathrm{Spec}(R)$, and we call the obtained étale R-algebra $S(p)$. We thus obtain a set $\{S(p) | p \in \mathrm{Spec}(R)\}$ of étale R-algebras such that the union of the images of the $\mathrm{Spec}(S(p))$ cover $\mathrm{Spec}(R)$, and such that every $A \otimes S(p)$ contains a rank one idempotent. By the quasicompactness of $\mathrm{Spec}(R)$, we can find a finite subset $\{p_1, \ldots, p_n\}$ of $\mathrm{Spec}(R)$ such that the images of the $\mathrm{Spec}(S_{p_i})$ cover $\mathrm{Spec}(R)$. Then $S = S(p_1) \times \cdots \times S(p_n)$ is an étale covering of R, and $A \otimes S$ contains a rank one idempotent. □

Remark 4.3.10 It is not known whether an arbitrary Taylor-Azumaya algebra is equivalent to a finitely generated Taylor-Azumaya algebra. In [92], $\mathrm{BR}(R)$ is introduced as the set of equivalence classes of Taylor-Azumaya algebras represented by a flat Taylor-Azumaya algebra. It follows from Proposition 4.3.5 that $\mathrm{BR}(R)$ is a set, and also that $\mathrm{BR}(R) \subset \mathrm{Br}'(R)$, if R is noetherian. It will follow from Theorem 6.3.8 that $\mathrm{BR}(R) \subset \mathrm{Br}'(R)$ even if R is not noetherian. If we would know that a finitely generated Taylor-Azumaya algebra is equivalent to a flat Taylor-Azumaya algebra, then it would follow that $\mathrm{BR}(R) = \mathrm{Br}'(R)$.

Chapter 5

Amitsur cohomology and étale cohomology

5.1 Grothendieck topologies

Definition 5.1.1 *(cf. [7]) A Grothendieck topology $\mathcal{T}$ consists of a category* $\underline{\underline{\text{cat}}}(\mathcal{T})$ *and a class* $\underline{\underline{\text{cov}}}(\mathcal{T})$ *consisting of families* $\{\phi_i : U_i \to U | i \in I\}$*. where the* ϕ_i *are morphisms in* $\underline{\underline{\text{cat}}}(\mathcal{T})$*.* $\{\phi_i : U_i \to U | i \in I\}$ *is called a* $\mathcal{T}$*-covering of* U*.* $\underline{\underline{\text{cat}}}(\mathcal{T})$ *and* $\underline{\underline{\text{cov}}}(\mathcal{T})$ *have to satisfy the following properties :*

1. *If* f *is an isomorphism in* $\underline{\underline{\text{cat}}}(\mathcal{T})$*, then* $\{f\} \in \underline{\underline{\text{cov}}}(\mathcal{T})$ *;*
2. *If* $\{\phi_i : U_i \to U | i \in I\} \in \underline{\underline{\text{cov}}}(\mathcal{T})$*, and, for every* $i \in I$*,* $\{\psi_{ij} : U_{ij} \to U_i | j \in I_i\} \in \underline{\underline{\text{cov}}}(\mathcal{T})$*, then* $\{\psi_{ij} \circ \phi_i : U_{ij} \to U | i \in I,\ j \in I_i\} \in \underline{\underline{\text{cov}}}(\mathcal{T})$ *; we call* $\{\psi_{ij}\}$ *a refinement of* $\{\phi_i\}$
3. *If* $\{\phi_i : U_i \to U | i \in I\} \in \underline{\underline{\text{cov}}}(\mathcal{T})$*, and* $V \to U$ *is a morphism in* $\underline{\underline{\text{cat}}}(\mathcal{T})$*, then the fibred products* $U_i \times_U V$ *exist in* $\underline{\underline{\text{cat}}}(\mathcal{T})$ *and* $\{U_i \times_U V \to V | i \in I\} \in \underline{\underline{\text{cov}}}(\mathcal{T})$*.*

The basic example of a Grothendieck topology is the following : consider a topological space X, and let $\underline{\underline{\text{cat}}}(\mathcal{T})$ be the category with open sets as objects and inclusions as morphisms. $\underline{\underline{\text{cov}}}(\mathcal{T})$ consists of open coverings of open subsets of X. Our main interest goes to the following examples of Grothendieck categories :

Examples 5.1.2 Let R be a commutative ring, The *Zariski topology* R_{Zar} is the Grothendieck topology defined by the topological space Spec(R). Observe that it follows from the quasicompactness of Spec(R) that every covering in R_{Zar} can be refined to a covering of the type $\{\text{Spec}(R_{f_i}) \to \text{Spec}(R) | i = 1, \ldots, n\}$, with $f_i \in R$. Such a covering is completely determined if we give a faithfully flat commutative R-algebra of the type $R' = \oplus_{i=1}^n R_{f_i}$. In fact this means that we can restrict attention to coverings given by a single map, and this observation leads to the following two examples:
Let $\underline{\underline{\text{cat}}}(R_{fl})$ be the dual of the category of (commutative) flat R-algebras, locally

of finite type, and flat morphisms of R-algebras. $\underline{\underline{\text{cov}}}(R_{fl})$ is the class of morphisms $f: S' \leftarrow S$ in $\underline{\underline{\text{cat}}}(R_{fl})$ such that S' is a faithfully flat S-algebra. In the sequel, we will write $S \to S'$ for an R-algebra homomorphism, or, equivalently, for a morphism from S' to S in $\underline{\underline{\text{cat}}}(R_{fl})$. We will call R_{fl} the *flat topology* on R.
In a similar way, the *étale topology* $R_{ét}$ on R is defined by $\underline{\underline{\text{cat}}}(R_{ét})$, (the dual of) the category of étale R-algebras and étale R-algebra morphisms and $\underline{\underline{\text{cov}}}(R_{ét})$ consisting of faithfully flat étale morphisms between R-algebras (also called étale coverings).

In the sequel, we will state some of the properties of a Grothendieck topology in the special situation of the above example. R_E will denote R_{Zar}, R_{fl} or $R_{ét}$. ; by an E-morphism, we will mean a morphism in $\underline{\underline{\text{cat}}}(R_E)$, and a covering of R_E will be called an E-covering.
Let $\mathcal{T}$ and $\mathcal{T}'$ be two Grothendieck topologies. A morphism $\mathcal{T} \to \mathcal{T}'$ consists of a functor $\underline{\underline{\text{cat}}}(\mathcal{T}') \to \underline{\underline{\text{cat}}}(\mathcal{T})$, that sends coverings to coverings. In the case where $\mathcal{T}$ and $\mathcal{T}'$ come from two topological spaces X and X', a morphisms $\mathcal{T} \to \mathcal{T}'$ correspond to continuous maps $f\ X \to X'$. The functor involved is then given by f^{-1}.
If $f:\ R \to R'$ is an algebra morphism, then a morphism $R'_E \to R_E$ is given by the functor

$$\tilde{f}:\ \underline{\underline{\text{cat}}}(R_E) \to \underline{\underline{\text{cat}}}(R'_E)$$

defined by $\tilde{f}(S) = R' \otimes S$, for every $S \in \underline{\underline{\text{cat}}}(R_E)$. Another example of a morphism of Grothendieck topologies is the following : the identity maps

$$\underline{\underline{\text{cat}}}(R_{Zar}) \to \underline{\underline{\text{cat}}}(R_{ét}) \to \underline{\underline{\text{cat}}}(R_{fl})$$

define morphisms

$$R_{fl} \to R_{ét} \to R_{Zar}$$

Definition 5.1.3 *A contravariant functor from $\underline{\underline{\text{cat}}}(R_E)$ to $\underline{\underline{Ab}}$ will be called a presheaf of abelian groups on R_E. A presheaf P on R_E is called a sheaf if for every E-covering S' of S, the following sequence is exact :*

$$1 \longrightarrow P(S) \longrightarrow P(S') \rightrightarrows P(S' \otimes_S S')$$

Here the two maps $P(S') \to P(S' \otimes_S S')$ are $P(\varepsilon_1)$ and $P(\varepsilon_2)$, where $\varepsilon_1(s) = 1 \otimes s$ and $\varepsilon_2(s) = s \otimes 1$. $\mathcal{P}(R_E)$ will denote the category of presheaves of abelian groups on R_E, and $\mathcal{S}(R_E)$ will be the full subcategory, with sheaves as objects.

From the fact that $\underline{\underline{Ab}}$ is an $Ab5$ and an $Ab4^*$ category with a generator, it follows easily that $\mathcal{P}(R_E)$ also satisfies this properties. As a consequence, $\mathcal{P}(R_E)$ has enough injectives, cf. Theorem A.1.6. We will show later that $\mathcal{S}(R_E)$ is also an $Ab5$, $Ab4^*$-category with a generator and enough injectives. First, we have to show that the natural functor $i:\ \mathcal{P}(R_E) \to \mathcal{S}(R_E)$ has a left adjoint. To this end, we first introduce Amitsur cohomology.

5.2 Amitsur cohomology

Let R be a commutative ring, and S a commutative R-algebra. Let P be a covariant functor from a full subcategory of the category of commutative R-algebras that contains all tensor powers $S^{\otimes n}$ of S. For $i = 1, \ldots, n+1$, we have maps

$$\varepsilon_i : S^{\otimes n} \to S^{\otimes n+1}$$

defined by

$$\varepsilon_i(s_1 \otimes \cdots \otimes s_n) = s_1 \otimes \cdots \otimes 1 \otimes \cdots \otimes s_n$$

We define a complex

$$0 \longrightarrow P(S) \xrightarrow{\Delta_0} P(S^{\otimes 2}) \xrightarrow{\Delta_1} P(S^{\otimes 3}) \xrightarrow{\Delta_2} \cdots \tag{5.1}$$

in $\underline{\underline{\mathrm{Ab}}}$ by

$$\Delta_n = \sum_{i=1}^{n+2} (-1)^{i-1} P(\varepsilon_i) : P(S^{\otimes(n+1)}) \to P(S^{\otimes(n+2)})$$

(5.1) is called the *Amitsur complex*, and it is denoted by $\mathcal{C}(S/R, P)$. As usual, we will write

$$Z^n(S/R, P) = \operatorname{Ker} \Delta_n \quad ; \quad B^n(S/R, P) = \operatorname{Im} \Delta_{n-1}$$
$$H^n(S/R, P) = Z^n(S/R, P)/B^n(S/R, P)$$

$H^n(S/R, P)$ will be called the n-th Amitsur cohomology group of P.

Remarks 5.2.1 1) If S is a Galois extension of R with group G in the sense of [75], and if P commutes with direct sums, then we have a natural isomorphism

$$H^n(S/R, P) \cong H^n(G, P(S)),$$

where $H^n(G, P(S))$ is the n-th group cohomology group. For a direct proof, we refer to [114, Prop. V.1.6]. We will prove a more general statement in Chapter 9 (see Propositions 9.1.1, 9.2.3 and 9.2.4).
2) In [7], *Čech cohomology* is defined over a Grothendieck topology. In fact, Amitsur cohomology is nothing else then Čech cohomology applied to the Grothendieck topology R_E.
3) Using Amitsur cohomology, the definition of a sheaf may be restated as follows : a presheaf P is a sheaf if and only if for every covering $S \to S'$ in R_E, $H^0(S'/S, P) = P(S)$.

Consider the presheaves $\mathbb{G}_a$ and $\mathbb{G}_m$ on R_E defined by

$$\begin{cases} \mathbb{G}_a(S) = S \text{ viewed as an abelian group} \\ \mathbb{G}_m(S) = \{x \in S | x \text{ invertible}\} \end{cases}$$

We call $\mathbb{G}_a(S)$ and $\mathbb{G}_m(S)$ the general additive and multiplicative group of S.

Proposition 5.2.2 *$\mathbb{G}_a$ and $\mathbb{G}_m$ are sheaves on R_E.*

Proof This follows immediately from the remark made above and faithfully flat descent for elements, cf. B.0.5. □

In the following Lemma, we will show that we can define the inductive limit of Amitsur cohomology groups over all objects of R_E. The proof of the following is taken from [114, V.1.7].

Lemma 5.2.3 *Let S, T be two faithfully flat R-algebras, and suppose that $f, g : S \to T$ are two homomorphisms of R-algebras. Then for every presheaf P on R_E, the induced maps $C(S/R, P) \to C(T/R, P)$ are homotopic, and, consequently, the induced maps $H^n(S/R, P) \to H^n(T/R, P)$ coincide.*

Proof We will construct maps $\Theta_n : P(S^{\otimes(n+1)}) \to P(T^{\otimes(n+1)})$, for $n \geq 1$, such that

$$\Delta_{n-1} \circ \Theta_n + \Theta_{n+1} \circ \Delta_n = P(g^{\otimes(n+1)}) - P(f^{\otimes(n+1)}). \tag{5.2}$$

Define, for $j = 1, 2, ...n$, $\theta_j^{(n)} : S^{\otimes(n+1)} \to T^{\otimes(n)}$ by

$$\theta_j^{(n)}(s_1 \otimes \cdots \otimes s_{n+1}) = f(s_1) \otimes \cdots \otimes f(s_j)g(s_{j+1}) \otimes \cdots \otimes g(s_{n+1}).$$

Let us look at $\theta_j^{(n+1)} \circ \varepsilon_i$, for $i = 1, ..., n+2$, $j = 1, ..., n+1$. For $1 \leq j \leq i-2$, we have

$$(\theta_j^{(n+1)} \circ \varepsilon_i)(s_1 \otimes \cdots \otimes s_{n+1}) = (\varepsilon_{i-1} \circ \theta_j^{(n)}(s_1 \otimes \cdots \otimes s_{n+1})$$

For $j = i$, we have

$$\begin{aligned}(\theta_j^{(n+1)} &\circ \varepsilon_i))(s_1 \otimes \cdots \otimes s_{n+1}) \\ &= f(s_1) \otimes \cdots \otimes f(s_{i-1}) \otimes f(1)g(s_i) \otimes \cdots \otimes g(s_{n+1}) \\ &= (f^{\otimes(i-1)} \otimes g^{\otimes(n+2-i)})(s_1 \otimes \cdots \otimes s_{n+1})\end{aligned}$$

For $j = i-1$, we have

$$\begin{aligned}(\theta_j^{(n+1)} &\circ \varepsilon_i))(s_1 \otimes \cdots \otimes s_{n+1}) \\ &= f(s_1) \otimes \cdots \otimes f(s_{i-1}) \otimes f(s_i)g(1) \otimes \cdots \otimes g(s_{n+1}) \\ &= (f^{\otimes(i)} \otimes g^{\otimes(n+1-i)})(s_1 \otimes \cdots \otimes s_{n+1})\end{aligned}$$

For $i+1 \leq j \leq n+1$, we have

$$(\theta_j^{(n+1)} \circ \varepsilon_i))(s_1 \otimes \cdots \otimes s_{n+1}) = (\varepsilon_i \circ \theta_{j-1}^{(n)}(s_1 \otimes \cdots \otimes s_{n+1})$$

Now define $\Theta_n : P(S^{\otimes(n+1)}) \to P(T^{\otimes(n)}$ by

$$\Theta_n = \sum_{i=1}^{n} (-1)^i P(\theta_i^{(n)}).$$

We then have

$$
\begin{aligned}
\Theta_{n+1} \circ \Delta_n &= \sum_{j=1}^{n+1} \sum_{i=1}^{n+2} (-1)^{i+j} P(\theta_j^{(n+1)} \circ \varepsilon_i) \\
&= \sum_{i=3}^{n+2} \sum_{j=1}^{i-2} (-1)^{i+j} P(\varepsilon_{i-1} \circ \theta_j^{(n)}) \\
&\quad - \sum_{i=2}^{n+2} P(f^{\otimes(i-1)} \otimes g^{\otimes(n+2-i)}) \\
&\quad + \sum_{i=1}^{n+1} P(f^{\otimes(i-1)} \otimes g^{\otimes(n+2-i)}) \\
&\quad + \sum_{i=1}^{n} \sum_{j=i+1}^{n+1} (-1)^{i+j} P(\varepsilon_i \circ \theta_{j-1}^{(n)}) \\
&= \sum_{i=2}^{n+1} \sum_{j=1}^{i-1} (-1)^{i+j} P(\varepsilon_i \circ \theta_j^{(n)}) \\
&\quad - P(f^{\otimes(n+1)}) + P(g^{\otimes(n+1)}) \\
&\quad - \sum_{i=1}^{n} \sum_{j=i}^{n} (-1)^{i+j} P(\varepsilon_i \circ \theta_j^{(n)}) \\
&= - \sum_{i=1}^{n+1} \sum_{j=1}^{n} (-1)^{i+j} P(\varepsilon_i \circ \theta_j^{(n)}) - P(f^{\otimes(n+1)}) + P(g^{\otimes(n+1)}) \\
&= -\Delta_{n-1} \circ \Theta_n - P(f^{\otimes(n+1)}) + P(g^{\otimes(n+1)})
\end{aligned}
$$

We leave it up to the reader to verify that the apparent problem in the last but one step with the indices i = 1 and i = n+1 is in fact no problem at all. □

Let Ω_E be the set of isomorphism classes of E-coverings of R. Define the following ordering on Ω_E : $[S] \leq [T]$ if and only if there exists an E-morphism from S to T. Then Ω_E is filtered : for every S, T in R_E, we clearly have that $[S] \leq [S \otimes T]$ and $[T] \leq [S \otimes T]$. In view of the preceding lemma, we have, for $[S] \leq [T]$, a well-defined homomophism $\Phi_{T,S}$: $H^n(S/R, P) \to H^n(T/R, P)$. Also $\Phi_{W,T} \circ \Phi_{T,S} = \Phi_{W,S}$ and $\Phi_{S,S}$ is the identity, so we can define

$$\check{H}^n(R_E, P) = \varinjlim H^n(S/R, P),$$

where the limit is taken over Ω_E.

5.3 The category of sheaves

We now prove the following theorem, which is a special case of [7, Th.II.1.1].

Theorem 5.3.1 *The natural embedding functor i : $\mathcal{S}(R_E) \to \mathcal{P}(R_E)$ has a left adjoint a : $\mathcal{P}(R_E) \to \mathcal{S}(R_E)$*

Proof A necessary (but not sufficient) condition for a presheaf P to be a sheaf is the following, in the sequel referred to as condition (*) : if $S \to S'$ is an E-covering, then the induced map $P(S) \to P(S')$ is injective. Consider a presheaf P. We will construct a presheaf P_1 satisfying condition (*). For any $U \in \underline{\underline{\text{cat}}}(R_E)$, let

$$P_0(U) = \{s \in P(U) | (P(f))(s) = 0 \text{ for some } E\text{-covering } f : U \to U'\}$$

and

$$P_1(U) = P(U)/P_0(U).$$

Clearly P_0 and P_1 are presheaves and P_1 satifies condition (*). Now define aP by

$$(aP)(S) = \check{H}^0(S_E, P_1)$$

for any $S \in \underline{\text{cat}}(R_E)$. Let us first show that aP is a sheaf. It is not difficult to see that aP is a presheaf, and that aP satifies condition (*). In order to have that aP is a sheaf, we therefore have to show the following : let $f : S \to S'$ be an E-covering, and suppose that $x' \in (aP)(S')$ satisfies

$$(aP)(\varepsilon_1)(x') = (aP)(\varepsilon_2)(x') \tag{5.3}$$

then we have to show that $x' \in \text{Im}\,(aP)(f)$. x' may be represented by an ordered pair (u', T'), where T' is an E-covering of S', and $u' \in H^0(T'/S', P_1) \subset P_1(T')$. Consider the following diagram

$$\begin{array}{ccccc}
S & \xrightarrow{f} & S' & \xrightarrow{g} & T' \\
\downarrow{\scriptstyle f} & & \downarrow{\scriptstyle \varepsilon_1} & & \downarrow{\scriptstyle \rho_1} \\
S' & \xrightarrow{\varepsilon_2} & S' \otimes_S S' & \xrightarrow{g_1} & S' \otimes_S T' \\
\downarrow{\scriptstyle g} & & \downarrow{\scriptstyle g_2} & & \downarrow{\scriptstyle g_2} \\
T' & \xrightarrow{\rho_2} & T' \otimes_S S' & \xrightarrow{g_1} & T' \otimes_S T'
\end{array}$$

with $\rho_1(t') = 1 \otimes t'$, $\rho_2(t') = t' \otimes 1$. It is clear that $g_2 \circ \rho_1 = \varepsilon_1$, $g_1 \circ \rho_2 = \varepsilon_2$. If we can show that $P_1(\varepsilon_1)(u') = P_1(\varepsilon_2)(u')$ in $P_1(T' \otimes_S T')$, then $u' \in H^o(T'/S, P_1)$, and then (u', T') represents an element of $(aP)(S) = \check{H}^0(S_E, P_1)$, which implies the result. Now $(aP)(\varepsilon_1)(x')$ is represented by

$$(P_1(\rho_1)(u'), S' \otimes_S T')$$

and $(aP)(\varepsilon_2)(x')$ is represented by

$$(P_1(\rho_2)(u'), T' \otimes_S S')$$

Here

$$\begin{aligned}
P_1(\rho_1)(u') &\in H^0(S' \otimes_S T'/S' \otimes_S S', P_1) \\
P_1(\rho_2)(u') &\in H^0(T' \otimes_S S'/S' \otimes_S S', P_1).
\end{aligned}$$

By (5.3), there must exist a common E-covering U of $S' \otimes_S T'$ and $T' \otimes S'$ such that the images of $P_1(\rho_1)(u')$ and $P_1(\rho_2)(u')$ are equal in $P_1(U)$. By the universal property of the tensor product, we must have a morphism

$$(S' \otimes_S T') \otimes_{S' \otimes_S S'} (T' \otimes_S S') = T' \otimes_S T' \to U,$$

which has to be an E-covering. Since P_1 satisfies condition (*), $P_1(T' \to T') \hookrightarrow P_1(U)$, hence the images of $P_1(\varepsilon_1)(u')$ and $P_1(\varepsilon_2)(u')$ in $P_1(T' \otimes_S T')$ coincide. Finally, we need to show that a is left adjoint to i, that is, for $P \in \mathcal{P}(R_E)$, $F \in \mathcal{S}(R_E)$, we have

$$\mathrm{Hom}_{\mathcal{P}(R_E)}(P, iF) \cong \mathrm{Hom}_{\mathcal{S}(R_E)}(aP, F). \tag{5.4}$$

Take a morphism $\Phi : P \to F$. Since F satisfies condition (*), we have a factorization

$$P \longrightarrow P_1 \xrightarrow{\Phi_1} F$$

We define $\Psi : aP \to F$ as follows : take $S \in \underline{\underline{\mathrm{cat}}}(R_E)$, and let $x' \in aP(S)$ be represented by (u, S'), with $f : S \to S'$ an E-covering. Then look at the following diagram with exact bottom row :

$$\begin{array}{ccccccc} & & & & P_1(S') & \rightrightarrows & P_1(S' \otimes_S S') \\ & & & & \downarrow{\scriptstyle \Phi_1(S')} & & \downarrow{\scriptstyle \Phi_1(S' \otimes_S S')} \\ 1 & \longrightarrow & F(S) & \longrightarrow & F_1(S') & \rightrightarrows & F_1(S' \otimes_S S') \end{array}$$

For $u \in H^0(S'/S, P_1)$ we find a unique $\Psi(aP)(S)(x) \in F(S)$.
Conversely, suppose that $\Psi : aP \to F$ is a morphism of sheaves. Consider $u \in P_1(S)$, and $x = [(u, S)] \in \check{H}^0(S_E, P_1)$. Define $\Phi_1(F)(S)(u) = \psi(aP)(S)(x)$. Then Φ_1 is a morphism $P_1 \to F$. We define Φ to be the composition of Φ_1 and the canonical map $P \to P_1$. We leave it to the reader to show that the correspondence between Φ and Ψ is a bijection. □

Remark 5.3.2 *The sheaf aP is called the sheaf associated to P. Observe that we have the following universal property : we have a canonical map of presheaves $\pi : P \to aP$ such that for every sheaf F, and any morphism $\Phi : P \to F$, we have a unique morphism $\Psi : aP \to F$ such that $\Phi = \Psi \circ \pi$.*

For later use, let us make the following observation :

Lemma 5.3.3 *If P is a torsion presheaf (that is, a presheaf of torsion groups), then aP is a torsion sheaf.*

Proof This follows easily from the proof of the preceding theorem : for every $S \in \underline{\underline{\mathrm{cat}}}(R_E)$, $P_1(S)$ is torsion, since it is a quotient of a torsion group. For every covering $S \to T$, $H^0(T/S, P_1)$ is a torsion group, being a subgroup of $P_1(T)$. Therefore $aP(S) = \check{H}^0(S_E, P_1)$ is torsion. □

Lemma 5.3.4 *The functor i preserves monics ; a preserves epics and monics.*

Proof That i preserves monics and a preserves epics follows from the fact that a is a left adjoint of i. Suppose that $\varphi : P \to P'$ is monic in $\mathcal{P}(R_E)$. First let us show that $\varphi_1 : P_1 \to P_1'$ is also monic in $\mathcal{P}(R_E)$. So take $U \in \underline{\underline{\mathrm{cat}}}(R_E)$, and suppose that

$s \in P(U)$, $s' = \varphi(U)(s) \in P'_0(U)$. There exists an E-covering $f : U \to V$ such that $P'(f)(s') = 0$. Looking at the following diagram, with all the maps injections

$$\begin{array}{ccc} P(U) & \longrightarrow & P'(U) \\ \downarrow & & \downarrow \\ P(V) & \longrightarrow & P'(V) \end{array}$$

we conclude that $P(f)(s) = 0$, hence $s \in P_0(U)$. Consequently φ_1 is monic. Now the following diagram is commutative :

$$\begin{array}{ccc} H^0(V/U, P_1) & \longrightarrow & H^0(V/U, P'_1) \\ \subset & & \subset \\ P_1(V) & \hookrightarrow & P'_1(V) \end{array}$$

Therefore $H^0(V/U, P_1) \hookrightarrow H^0(V/U, P'_1)$. Taking the inductive limit over the E-coverings V of U, we conclude that $(aP)(U) \hookrightarrow (aP')(U)$, hence $aP \to aP'$ is monic in $\mathcal{P}(R_E)$, and a fortiori in $\mathcal{S}(R_E)$, since i preserves monics. □

Theorem 5.3.5 *$\mathcal{S}(R_E)$ is an Ab5 and Ab3*-category. a is an exact functor, and i is a left exact functor.*

Proof Observe first that the constant sheaf $\{0\}$ is a zero object for $\mathcal{S}(R_E)$. Consider $\varphi : F \to G$ in $\mathcal{S}(R_E)$. It may be shown easily that the presheaf kernel $\mathrm{Ker}_{\mathcal{P}(R_E)}(\varphi)$ is a sheaf, and that this kernel is also the sheaf kernel $\mathrm{Ker}_{\mathcal{S}(R_E)}(\varphi)$ of φ. Concerning the cokernel, we claim that $\mathrm{Coker}_{\mathcal{S}(R_E)}(\varphi) = a(\mathrm{Coker}_{\mathcal{S}(R_E)}(\varphi))$. Indeed, suppose that we have (let $\gamma = \mathrm{Coker}_{\mathcal{S}(R_E)}(\varphi)$):

$$\begin{array}{ccccc} F & \xrightarrow{\varphi} & G & \xrightarrow{a(\gamma)} & aQ \\ & & \downarrow{\scriptstyle \beta} & & \\ & & X & & \end{array} \tag{5.5}$$

in $\mathcal{S}(R_E)$, with $\beta \circ \varphi = 0$. Then we have in $\mathcal{P}(R_E)$:

$$\begin{array}{ccccc} F & \xrightarrow{\varphi} & G & \xrightarrow{\gamma} & Q \\ & & \downarrow{\scriptstyle \beta} & & \\ & & X & & \end{array} \tag{5.6}$$

Hence there exists a map $Q \to X$ making (5.6) commutative. The associated map $aQ \to aX = X$ then makes (5.5) commutative. It now follows from Lemma 5.3.4 that a is exact and that i is left exact.
Similar arguments show that every monic is a kernel : if $\varphi : F \to G$ is monic in $\mathcal{S}(R_E)$, then it is monic in $\mathcal{P}(R_E)$, and therefore a presheaf kernel, and also a sheaf kernel. Next suppose that $\varphi : F \to G$ is epic in $\mathcal{S}(R_E)$. If P is the presheaf cokernel of φ, then we have an exact sequence $F \to G \to P \to 0$ in $\mathcal{P}(R_E)$. Since $\mathcal{P}(R_E)$ is abelian, we can extend this to an exact sequence $P' \to F \to G \to P \to 0$ in $\mathcal{P}(R_E)$. Taking associated sheaves, we obtain the following exact sequence in $\mathcal{S}(R_E)$:

$$aP' \to F \to G \to 0$$

showing that $F \to G$ is a cokernel.
Next, let us show that $\mathcal{S}(R_E)$ has arbitrary direct products and coproducts. Let $\{F_i | i \in I\}$ be a family of sheaves, and define $F, G \in \mathcal{P}(R_E)$ by $F(S) = \prod_i F_i(S)$ and $G(S) = \coprod_i F_i(S)$. We claim that F and G are sheaves. Indeed, since the F_i's are sheaves, we have exact sequences in $\underline{\underline{Ab}}$

$$1 \longrightarrow F_i(S) \longrightarrow F_i(S') \rightrightarrows F_i(S' \otimes_S S')$$

for every $i \in I$. Since direct products are left exact, it follows that the sequence

$$1 \longrightarrow \prod_i F_i(S) \longrightarrow \prod_i F_i(S') \rightrightarrows \prod_i F_i(S' \otimes_S S')$$

is exact, hence F is a sheaf. Since $\underline{\underline{Ab}}$ is *Ab*4, direct sums in $\underline{\underline{Ab}}$ are also left exact, hence the sequence

$$1 \longrightarrow \coprod_i F_i(S) \longrightarrow \coprod_i F_i(S') \rightrightarrows \coprod_i F_i(S' \otimes_S S')$$

is exact, hence G is a sheaf. It then follows easily that F and G are respectively the direct product and direct sum of $\{F_i | i \in I\}$ in $\mathcal{S}(R_E)$, and we have shown that $\mathcal{S}(R_E)$ is *Ab*3 and *Ab*3*. We therefore also know that $\mathcal{S}(R_E)$ has direct inductive and inverse projective limits. To find the direct inductive (resp. inverse projective) limit in $\mathcal{S}(R_E)$, construct the direct inductive (resp. inverse projective) limit in $\mathcal{P}(R_E)$: this is a sheaf, and this may be seen as follows : repeat the arguments above for the direct sum and product, and invoke the fact that the category $\underline{\underline{Ab}}$ is *Ab*5, and that inverse projective limits are left exact. It now follows easily that $\mathcal{S}(R_E)$ is *Ab*5 : let $\{F_i | i \in I\}$, $\{G_i | i \in I\}$ be two directed systems in $\mathcal{S}(R_E)$, and $f_i : F_i \to G_i$ monic in $\mathcal{S}(R_E)$. Then the f_i are monic in $\mathcal{P}(R_E)$, since i is left exact. Hence

$$f = \varinjlim f_i$$

is monic in $\mathcal{P}(R_E)$, since $\mathcal{P}(R_E)$ is *Ab*5. It follows that f is monic in $\mathcal{S}(R_E)$, since a is exact. □

Corollary 5.3.6 *Let $\varphi : F \to G$ be a morphism of sheaves. φ is monic if and only if $\varphi(S) : F(S) \to G(S)$ is a monomorphism for every $S \in \underline{\text{cat}}(R_E)$. φ is epic if and only if for every $S \in \underline{\text{cat}}(R_E)$ and $x \in G(S)$, there exists an E-covering $f : S \to S'$ such that $\varphi(f)(x) \in \text{Im}(\varphi(S'))$.*

Proof The first statement follows from the fact that both i and a are left exact. For the second statement, consider the presheaf cokernel H defined by $H(S) = \text{Coker}\,\varphi(S)$. The sheaf cokernel is then aH. If φ is epic, then $aH(S) = \check{H}^0(S_E, H) = 0$ for every S, and this is easily seen to be equivalent to the condition in the corollary.
□

In order to conclude that $\mathcal{S}(R_E)$ has enough injective objects, we still miss one ingredient (cf. Theorem A.1.6 : we need a generator for $\mathcal{S}(R_E)$, or, in view of the remark made after definition A.1.5, a family of generators. To this end, we first define a contravariant functor $\mathbb{Z}_\bullet \underline{\underline{\text{cat}}}(R_E) \to \mathcal{S}(R_E)$ as follows : for $T, U \in \underline{\underline{\text{cat}}}(R_E)$, we denote by $\text{Mor}_E(T,U)$ the set of morphisms from T to U in $\underline{\underline{\text{cat}}}(R_E)$. For example, $\text{Mor}_{fl}(T,U)$ is the set of flat R-algebra morphisms from T to U. We then define $\mathbb{Z}_T$ by

$$\mathbb{Z}_T(U) = \bigoplus_{\text{Mor}_E(T,U)} \mathbb{Z},$$

the direct sum of $|\text{Mor}_E(T,U)|$ copies of $\mathbb{Z}$, and, for $a:\ U \to V$ in $\underline{\underline{\text{cat}}}(R_E)$, $\mathbb{Z}_T(a)$ is given by $\mathbb{Z}_T(a)(f) = a \circ f$, for all $f \in \text{Mor}_E(T,U)$; observe that it suffices to define $\mathbb{Z}_T(a)$ on the $\mathbb{Z}$-basis $\text{Mor}_E(T,U)$ of $\mathbb{Z}_T(U)$. Furthermore, for $\delta :\ T \to T'$ in $\underline{\underline{\text{cat}}}(R_E)$, $F(\delta) = \mathbb{Z}_\delta$ is given by

$$\mathbb{Z}_\delta(U)(f') = f' \circ \delta$$

for all $\delta \in \text{Mor}_E(T,U)$. We claim that $\mathbb{Z}_T$ is a sheaf, that is, for every E-covering $S \to S'$, the sequence

$$1 \to \bigoplus_{\text{Mor}_E(T,S)} \mathbb{Z} \to \bigoplus_{\text{Mor}_E(T,S')} \mathbb{Z} \to \bigoplus_{\text{Mor}_E(T,S' \otimes_S S')} \mathbb{Z}$$

is exact. This follows from the theorem of faithfully flat descent for elements, cf. B.0.6 : suppose that $f :\ T \to S'$ is such that $\varepsilon_1 \circ f = \varepsilon_2 \circ f :\ T \to S' \otimes_S S'$; then f takes values in S, hence $f \in \text{Mor}_E(T,S')$.

Lemma 5.3.7 *For all* $P \in \mathcal{P}(R_E)$, $\text{Hom}_{\mathcal{P}(R_E)}(\mathbb{Z}_T, P) \cong P(T)$.

Proof Define maps

$$\Lambda\colon \text{Hom}(\mathbb{Z}_T, P) \to P(T) \ ;\ \Omega :\ P(T) \to \text{Hom}(\mathbb{Z}_T, P)$$

by

$$\Lambda(\psi) = \psi(T)(I_T)$$

for all $\psi \in \text{Hom}(\mathbb{Z}_T, P)$; here I_T is the identity on T. For $p \in P(T)$, $\Omega(p)$ is defined by $\Omega(p)(U) :\ \mathbb{Z}_T(U) \to P(U)$, given by

$$\Omega(p)(U)(f) = P(f)(p)$$

for all $f \in \text{Mor}_E(T,U)$. Let us check that Ω and Λ are each others inverses : For all $p \in P(T)$, we have :

$$(\Lambda \circ \Omega)(p) = \Lambda(\Omega(p)) = \Omega(p)(T)(I_T) = P(I_T)(p) = p$$

For all $\psi \in \text{Hom}(\mathbb{Z}_T, P)$, we will show that $(\Omega \circ \Lambda)(\psi) = \psi$. For all $U \in \underline{\underline{\text{cat}}}(R_E)$, and for all $f \in \text{Mor}_E(T,U)$, we have that

$$(\Omega \circ \Lambda)(\psi)(U)(f) = \Omega(\psi(T)(I_T))(U)(f) = P(f)(\psi(T)(I_T)) = (P(f) \circ \psi(T))(I_T),$$

so it suffices to show that

$$(P(f) \circ \psi(T))(I_T) = \psi(U)(f).$$

This follows from the commutativity of the following diagram :

$$\begin{array}{ccc} \mathbb{Z}_T(T) & \xrightarrow{\psi(T)} & P(T) \\ \downarrow{\scriptstyle \mathbb{Z}_T(f)} & & \downarrow{\scriptstyle P(f)} \\ \mathbb{Z}_T(U) & \xrightarrow{\psi(U)} & P(U) \end{array}$$

Indeed, $(\psi(U) \circ \mathbb{Z}_T(f))(I_T) = \psi(U)(f)$. □

Remark that in Lemma 5.3.7, Ω is functorial in P, that is, for any morpism $f : P \to Q$ of presheaves, we have a commutative diagram

$$\begin{array}{ccc} \mathrm{Hom}(\mathbb{Z}_T, P) & \xrightarrow{\Lambda} & P(T) \\ \downarrow{\scriptstyle f\circ\bullet} & & \downarrow{\scriptstyle f(T)} \\ \mathrm{Hom}(\mathbb{Z}_T, Q) & \xrightarrow{\Lambda} & Q(T) \end{array}$$

Indeed, for every $\psi \in \mathrm{Hom}(\mathbb{Z}_T, P)$,

$$f(T)(\Lambda(\psi)) = f(T)(\psi(T)(I_T)) = (f \circ \psi)(T)(I_T) = \lambda(f \circ \psi)$$

We are now able to show

Proposition 5.3.8 *Let X be the set of isomorphism classes in $\underline{\underline{\mathrm{cat}}}(R_E)$. Then $\{\mathbb{Z}_T : [T] \in X\}$ is a family of generators for $\mathcal{S}(R_E)$.*

Proof Take $F, G \in \mathcal{S}(R_E)$, and $a \neq b : F \to G$. Then there exists $T \in \underline{\underline{\mathrm{cat}}}(R_E)$ such that $a(T) \neq b(T) : F(T) \to G(T)$. Applying Ω, and taking into account the functorial property of Ω mentioned above, we obtain that the maps

$$a \circ \bullet \text{ and } b \circ \bullet : \mathrm{Hom}(\mathbb{Z}_T, F) \to \mathrm{Hom}(\mathbb{Z}_T, G)$$

are not equal. Hence there exists $\psi \in \mathrm{Hom}(\mathbb{Z}_T, F)$ such that $a \circ \psi \neq b \circ \psi$, finishing the proof. □

Corollary 5.3.9 *$\mathcal{S}(R_E)$ has enough injectives.*

Proof This follows from 5.3.5, 5.3.8 and A.1.6. □

5.4 Direct and inverse image sheaves and presheaves

Let $f : R \to S$ be an algebra homomorphism. As we have seen, f defines a morphism $S_E \to R_E$ given by the functor $\tilde{f} : \underline{\underline{\mathrm{cat}}}(R_E) \to \underline{\underline{\mathrm{cat}}}(S_E)$, $\tilde{f}(U) = S \otimes U$. We now associate the following functor to f:

$$f_p : \mathcal{P}(S_E) \to \mathcal{P}(R_E)$$

defined by

$$f_p(P') = P'(S \otimes \bullet)$$

for all $P' \in \mathcal{P}(S_E)$. $f_p(P')$ is called the *direct image presheaf* of P'. Let us show that f_p has a left adjoint f^p. $f^p(P)$ will be called the *inverse image presheaf* of P.

Theorem 5.4.1 *f_p has a left adjoint f^p*

Proof Take $V \in \underline{\underline{\mathrm{cat}}}(S_E)$, and consider the category $\mathcal{C}_V$ with objects commutative diagrams of the form

$$\begin{array}{ccc} R & \xrightarrow{f} & S \\ \downarrow & & \downarrow \\ U & \xrightarrow{g} & V \end{array}$$

such that $R \to U$ is in $\underline{\underline{\mathrm{cat}}}(R_E)$. A morphism of this object to another one

$$\begin{array}{ccc} R & \xrightarrow{f} & S \\ \downarrow & & \downarrow \\ U_1 & \xrightarrow{g_1} & V \end{array}$$

is given by a map $h: U \to U_1$ such that $g = g_1 \circ h$. Consider $P \in \mathcal{P}(R_E)$ and define $f^p(P) \in \mathcal{P}(S_E)$ by

$$f^p(P)(V) = \varinjlim P(U)$$

where the limit is taken over the category $\mathcal{C}_V$.
If $V \to W$ is a morphism in $\underline{\underline{\mathrm{cat}}}(S_E)$, then we have a natural functor $\mathcal{C}_V \to \mathcal{C}_W$ given by

$$\begin{array}{ccc} R & \xrightarrow{f} & S \\ \downarrow & & \downarrow \\ U & \xrightarrow{g} & V \end{array} \quad \mapsto \quad \begin{array}{ccccc} R & & \xrightarrow{f} & & S \\ \downarrow & & & & \downarrow \\ U & \xrightarrow{g} & V & \longrightarrow & W \end{array}$$

Therefore, we have a natural map $f^p(V) \to f^p(W)$, making f^p into a presheaf. Let us show that f^p is a left adjoint of f_p.
Suppose that $P \in \mathcal{P}(R_E)$, $P' \in \mathcal{P}(S_E)$. We have to show that

$$\mathrm{Hom}_{\mathcal{P}(R_E)}(P, f_p P') \cong \mathrm{Hom}_{\mathcal{P}(S_E)}(f^p P, P')$$

Take $u: P \to f_p P'$. We will define $\psi(u): f^p P \to P'$. For all $U \in \underline{\underline{\mathrm{cat}}}(R_E)$, we have a map $u(U): P(U) \to f_p P'(U) = P'(S \otimes U)$. Take $V \in \underline{\underline{\mathrm{cat}}}(S_E)$, and let $x \in f^p P(V)$ be represented by $x \in P(U)$, for some diagram

$$\begin{array}{ccc} R & \xrightarrow{f} & S \\ \downarrow & & \downarrow \\ U & \xrightarrow{g} & V \end{array}$$

Consider the map

$$\alpha: S \otimes U \xrightarrow{I \otimes g} S \otimes V \longrightarrow V$$

and let $\psi(u)(V)(x)$ be the image of x under the composition

$$P(U) \xrightarrow{u(U)} P'(U \otimes S) \xrightarrow{P'(\alpha)} P'(V)$$

Conversely, for a map $v: f^p P \to P'$, we define $\varphi(v): P \to f_p P'$ as follows : take $U \in \underline{\underline{\mathrm{cat}}}(R_E)$. Then $x \in P(U)$ represents an element of $(f^p P)(S \otimes U)$: we consider the diagram

$$\begin{array}{ccc} R & \xrightarrow{f} & S \\ \downarrow & & \downarrow \\ U & \longrightarrow & S \otimes U \end{array}$$

We define $\varphi(v)(U)(x) = v(S \otimes U)(x)$.
Let us check that φ and ψ are each others inverses. With notations as above, we have that

$$\varphi(\psi(u))(x) = \psi(u)(S \otimes U)(x)$$

Take $V = S \otimes U$ in the definition of ψ. Then α is the identity, so $P'(\alpha)$ is the identity on $P'(S \otimes U)$, and $\psi(u)(S \otimes U) = u$. Therefore $\varphi \circ \psi = I$.
Conversely, take $v : f^p P \to P'$. The map $\psi(\varphi(v))$ is now defined by

$$\begin{array}{ccccc} P(U) & \xrightarrow{\varphi(v)(U)} & P'(S \otimes U) & \xrightarrow{P'(\alpha)} & P'(U) \\ x & \mapsto & v(S \otimes U)(x) & \mapsto & v(V)(x) \end{array}$$

and it follows that $(\psi \circ \varphi)(v) = v$. □

Proposition 5.4.2 *Let $f : R \to S$ be an R-algebra. Then f_p is exact and f^p is right exact.*

Proof The first assertion is trivial. The second one follows from the fact that direct limits are exact on $\underline{\underline{Ab}}$. □

Proposition 5.4.3 *Let $f : R \to S$ be an R-algebra. If F is a sheaf, then $f_p F$ is also a sheaf. The restriction of f_p to the category of sheaves therefore gives a left exact functor*

$$f_* = a \circ f_p \circ i : \mathcal{S}(S_E) \to \mathcal{S}(R_E)$$

The functor

$$f^* = a \circ f^p \circ i : \mathcal{S}(R_E) \to \mathcal{S}(S_E)$$

is right exact and is a left adjoint of f_*

Proof Suppose that $F \in \mathcal{S}(S_E)$ is a sheaf. If $U \to V$ is a covering in $\underline{\underline{\text{cov}}}(R_E)$, then $S \otimes U \to S \otimes V$ is a covering in $\underline{\underline{\text{cov}}}(S_E)$, and we have an exact sequence

$$0 \to F(S \otimes U) \to F(S \otimes V) \rightrightarrows F((S \otimes V) \otimes_S (S \otimes V)) = F(S \otimes V \otimes V)$$

or

$$0 \to f_p F(U) \to f_p F(V) \rightrightarrows f_p F(V \otimes V)$$

and $f_p F$ is a sheaf. Furthermore

$$\begin{aligned} \text{Hom}_{\mathcal{S}(R_E)}(F, f_* F') &= \text{Hom}_{\mathcal{P}(R_E)}(i(F), f_p F') \\ &= \text{Hom}_{\mathcal{P}(S_E)}(f^p(i(F)), iF') = \text{Hom}_{\mathcal{S}(S_E)}(a f^p(i(F)), F') \end{aligned}$$

and it follows that f^* is a left adjoint of f_*. □

$f_*(F)$ is called the *direct image sheaf* of F; $f^*(F')$ is called the *inverse image sheaf* of F'. In the following proposition, we look at the special situation where $R \to S$ is in $\underline{\underline{\text{cat}}}(R_E)$.

Proposition 5.4.4 *Suppose that $S \in \underline{\underline{\mathrm{cat}}}(R_E)$. Then the functor f^p is given by the formula*

$$f^p(V) = P(V)$$

for all $V \in \underline{\underline{\mathrm{cat}}}(S_E)$. f^* is then an exact functor, and we write $f^* = f_{|S}$. Moreover f^p and f^* have an exact left adjoint.

Proof For all $V \in \underline{\underline{\mathrm{cat}}}(S_E)$, the diagram

$$\begin{array}{ccc} R & \longrightarrow & S \\ \downarrow & & \downarrow \\ V & \xrightarrow{I_V} & V \end{array}$$

is a final object of $\mathcal{C}_V$, and this immediately implies the first statement. It is then also clear that f^p is a sheaf if F is a sheaf, and that f^p is exact. We already now that f^* is right exact. $f^* = a \circ f^p \circ i$ is left exact, since it is a composition of left exact functors.

A left adjoint $f_! : \mathcal{P}(S_E) \to \mathcal{P}(R_E)$ of f^p may be constructed as follows : for $U \in \underline{\underline{\mathrm{cat}}}(R_E)$, consider the set of all morphisms $\phi : S \to U$ in R_E, and let U_ϕ be U considered as an object of $\underline{\underline{\mathrm{cat}}}(S_E)$ via ϕ. We now define

$$(f_!P')(U) = \bigoplus_{\phi:\, S \to U} P'(U_\phi)$$

for every $P' \in \mathcal{P}(S_E)$. An isomorphism

$$\mathrm{Hom}_{\mathcal{P}(R_E)}(f_!P', P) \cong \mathrm{Hom}_{\mathcal{P}(S_E)}(P', f^pP)$$

is now constructed as follows: if $u : P' \to f^pP$, then we have maps

$$u(U_\phi) : P'(U_\phi) \longrightarrow f^pP(U_\phi) = P(U)$$

for any $U \in \underline{\underline{\mathrm{cat}}}(R_E)$ and $\phi : S \to U$, and therefore a map

$$\varphi(u)(U) : (f_!P')(U) \to P(U)$$

Conversely, suppose given

$$v(U) : (f_!P')(U) = \bigoplus P'(U_\phi) \to P(U)$$

for every $U \in \underline{\underline{\mathrm{cat}}}(R_E)$. For $g : S \to V \in \underline{\underline{\mathrm{cat}}}(S_E)$, let $\psi(v)(V)$ be the composition

$$P'(V_g) \to \bigoplus P'(V_\phi) \to P(V)$$

It is straightforward to verify that φ and ψ are each others inverses, and that $f_!$ is exact. A left adjoint of f^* is given by the composition

$$\mathcal{S}(S_E) \xrightarrow{i} \mathcal{P}(S_E) \xrightarrow{f_!} \mathcal{P}(P_E) \xrightarrow{a} \mathcal{S}(R_E)$$

which is right exact (being a left adjoint) and left exact (being the composition of left exact functors). □

5.5 Stalks in the étale topology

In this Section, we will work on the étale site. The reader who wishes to avoid arguments involving étale algebras should skip this Section and take Corollary 5.5.8 for granted.

Suppose that k is a separably closed field. Then the only étale k-algebras are the direct sums of a finite number of copies of k. If F is a sheaf on $k_{ét}$, then $F(k^n) = F(k)^n$, so F is completely determined by $F(k)$, and the functor $\mathcal{S}(k_{ét}) \to \underline{\underline{Ab}}$ mapping F to $F(k)$ is an equivalence of categories.

Now let R be a commutative ring, and take $p \in \text{Spec}(R)$. Let $k(p)^{sep} = k$ be a separable closure of the residue class field $k(p) = R_p/pR_p$, and consider the canonical map

$$u(p): \ R \longrightarrow k(p)^{sep} = k$$

and the inverse image $u(p)^p(P)$ for every presheaf P.

Lemma 5.5.1 *$u(p)^p(P)$ is a sheaf on $k(p)^{sep}_{ét}$.*

Proof From the construction in the proof of Theorem 5.4.1, it follows that

$$u(p)^p(P)(k^n) = \varinjlim P(S),$$

where the limit runs over all commutative diagrams of the form

$$\begin{array}{ccc} R & \xrightarrow{u(p)} & k \\ \big\downarrow \scriptstyle{ét} & & \big\downarrow \\ S & \xrightarrow{g} & k^n \end{array}$$

It suffices to take the limit over the cofinal subcategory consisting of diagrams of the form

$$\begin{array}{ccc} R & \xrightarrow{u(p)} & k \\ \big\downarrow \scriptstyle{ét} & & \big\downarrow \\ S^n & \xrightarrow{g} & k^n \end{array}$$

to conclude that

$$u(p)^p P(k^n) = (u(p)^p P(k))^n$$

finishing the proof. □

We now define the *stalk* of a presheaf P at $p \in \text{Spec}(R)$ in the étale topology as

$$P_{\overline{p}} = u(p)^p P(k(p)^{sep}) = \varinjlim P(S)$$

where the limit runs over diagrams of the form

$$\begin{array}{ccc} R & \xrightarrow{u(p)} & k(p)^{sep} \\ \big\downarrow \scriptstyle{ét} & \nearrow & \\ S & & \end{array}$$

From Lemma 5.5.1, it follows that $u(p)^p(P)$ is completely determined by $P_{\overline{p}}$. Let us now show that the stalks of a presheaf and its associated sheaf coincide. We first need a Lemma.

Lemma 5.5.2 *Let $\pi : R \to S$ be a ring homomorphism. Then for every presheaf P on R_E, we have that $\pi^* aP = a\pi^p P$.*

Proof For every sheaf $F \in \mathcal{S}(R_E)$, we have that

$$\begin{aligned} \mathrm{Hom}_{\mathcal{S}(S_E)}(\pi^* aP, F) &= \mathrm{Hom}_{\mathcal{S}(R_E)}(aP, \pi_* F) \\ &= \mathrm{Hom}_{\mathcal{P}(R_E)}(P, i\pi_* F) \\ &= \mathrm{Hom}_{\mathcal{P}(R_E)}(P, \pi_p iF) \\ &= \mathrm{Hom}_{\mathcal{P}(S_E)}(\pi^p P, iF) \\ &= \mathrm{Hom}_{\mathcal{S}(S_E)}(a\pi^p P, F) \end{aligned}$$

and this implies that $\pi^* aP = a\pi^p P$. □

Corollary 5.5.3 *Let $P \in \mathcal{P}(R_{ét})$, and $p \in \mathrm{Spec}(R)$. Then $(aP)_{\overline{p}} \cong P_{\overline{p}}$. Conversely, if for all $p \in \mathrm{Spec}(R)$, $P_{\overline{p}} = 0$, then $aP = 0$.*

Proof From Lemmas 5.5.1 and 5.5.2, it follows that

$$\begin{aligned} (aP)_{\overline{p}} &= u(p)^p aP(k(p)^{sep}) = u(p)^* aP(k(p)^{sep}) \\ &= au(p)^p P(k(p)^{sep}) = u(p)^p P(k(p)^{sep}) = P_{\overline{p}} \end{aligned}$$

proving the first statement.
Let $R \to S$ be étale, and consider $s \in P(S)$. Take $q \in \mathrm{Spec}(S)$ and $p = f^{-1}(q)$. We then find

$$\begin{array}{ccc} R & \longrightarrow & k(p) \\ \downarrow & & \uparrow \\ S & \xrightarrow{ét} & T(q) \end{array}$$

such that the image $s_{|T(q)}$ of s in $T(q)$ is 0. A finite number of the $T(q)$ will cover $\mathrm{Spec}(S)$, so, taking their direct sum, we find an étale covering $S \to T$ such that $s_{|T} = 0$. This means that the presheaf P_0 constructed in the proof of 5.3.1 equals P, and this implies that $aP = 0$. □

Let $f : R \to S$ be étale, and consider $q \in \mathrm{Spec}(S)$, $p = f^{-1}(q) \in \mathrm{Spec}(R)$. Then we have

$$\begin{array}{ccc} R & \longrightarrow & k(p)^{sep} \\ \downarrow f & \nearrow & \\ S & & \end{array}$$

For $s \in F(S)$, we let $s_{\overline{p}}$ be the image of F in $F_{\overline{p}}$.

Lemma 5.5.4 *If $s \neq 0$ in $F(S)$, then there exists $q \in \mathrm{Spec}(S)$ such that $s_{\overline{p}} \neq 0$ in $F_{\overline{p}}$.*

Proof If $s_{\overline{p}} = 0$, then we have $S \to T(q)$ in $R_{\acute{e}t}$ with

$$\begin{array}{ccc} R & \longrightarrow & k(p)^{sep} \\ \downarrow & & \uparrow \\ S & \xrightarrow{\acute{e}t} & T(q) \end{array}$$

such that the image of s in $F(T(q))$ is zero. If $s_{\overline{p}} = 0$ for all q, then we find a $T(q)$ for every q, and the images of the $\mathrm{Spec}(T(q))$ in $\mathrm{Spec}(S)$ cover the whole of $\mathrm{Spec}(S)$. Thus a finite number of $T(q_1), \ldots, T(q_n)$ are sufficient to cover $\mathrm{Spec}(S)$, and $S \to T = T(q_1) \times \cdots \times T(q_n)$ is an étale covering of S. The image of s in $F(T)$ is zero, so $s = 0$. □

Theorem 5.5.5 *Let F, F', $F'' \in \mathcal{S}(R_{\acute{e}t})$. Then the following are equivalent:*

1. $0 \to F' \to F \to F'' \to 0$ *is exact in* $\mathcal{S}(R_{\acute{e}t})$*;*
2. $0 \to F'_{\overline{p}} \to F_{\overline{p}} \to F''_{\overline{p}} \to 0$ *is exact in* $\underline{\underline{Ab}}$*, for every* $p \in \mathrm{Spec}(R)$.

Consequently two sheaves are equal if their stalks coincide.

Proof 1) ⇒ 2). Let $P = \mathrm{Coker}_{\mathcal{P}(R_{\acute{e}t})}(F \to F'')$. Then $F \to F''$ is surjective if and only if $aP = 0$. We have the following exact sequences:

$$F \to F'' \to P \to 0$$

in $\mathcal{P}(R_{\acute{e}t})$;

$$u(p)^p F \to u(p)^p F'' \to u(p)^p P \to 0$$

in $\mathcal{P}(k(p)^{sep}_{\acute{e}t})$ (since $u(p)^p$ is exact, cf. 5.4.2);

$$F_{\overline{p}} \to F''_{\overline{p}} \to P_{\overline{p}} \to 0$$

in $\underline{\underline{Ab}}$ (taking global sections). By 5.5.3, $P_{\overline{p}} = (aP)_{\overline{p}} = 0$ if $F \to F''$ is surjective. Hence $F_{\overline{p}} \to F''_{\overline{p}}$ is surjective.
The left exactness of i implies that $0 \to F' \to F \to F''$ is exact in $\mathcal{P}(R_{\acute{e}t})$. Therefore for every $R \to S$ étale the sequence $0 \to F'(S) \to F(S) \to F''(S)$ is exact. Now $0 \to F'_{\overline{p}} \to F_{\overline{p}} \to F''_{\overline{p}}$ is exact, since inductive limits preserve injective maps.
2) ⇒ 1). Take $f: R \to S$ in $R_{\acute{e}t}$, and consider $s \in F'(S)$. Suppose that the image of s in $F(S)$ is zero. For any $q \in \mathrm{Spec}(S)$, with $p = f^{-1}(q)$, the image $s_{\overline{p}}$ in $F_{\overline{p}}$ is zero, hence $s_{\overline{p}} = 0$, since $F'_{\overline{p}} \to F_{\overline{p}}$ is injective. From Lemma 5.5.4, it then follows that $s = 0$, and therefore $F' \to F$ is injective.
Next, take $s \in F(S)$, and suppose that the image of s in $F''(S)$ is zero. Then for all $q \in \mathrm{Spec}(S)$, with $p = f^{-1}(q)$, we have that $s_{\overline{p}} \in F'_{\overline{p}} \subset F_{\overline{p}}$. Then we find $T(q)$ such that

$$\begin{array}{ccc} R & \longrightarrow & k(p)^{sep} \\ \downarrow & & \uparrow \\ S & \xrightarrow{\acute{e}t} & T(q) \end{array}$$

such that the image $s_{|T(q)}$ of s in $F(T(q))$ lies in $F'(T(q))$. Restrict to a finite number of $T(q)$, such that the union of their spectra cover $\mathrm{Spec}(S)$, and let T be the finite direct product of these $T(q)$. Then $s_{|T} \in F'(T) \subset F(T)$. Now since F and F' are sheaves, and $S \to T$ is an étale covering, we have the following commutative diagram with exact rows and columns:

$$\begin{array}{ccccccc} & & 1 & & 1 & & 1 \\ & & \downarrow & & \downarrow & & \downarrow \\ 1 & \longrightarrow & F'(S) & \longrightarrow & F'(T) & \rightrightarrows & F'(T \otimes_S T) \\ & & \downarrow & & \downarrow & & \downarrow \\ 1 & \longrightarrow & F(S) & \longrightarrow & F(T) & \rightrightarrows & F(T \otimes_S T) \end{array}$$

A diagram chasing argument now shows that $s \in F'(S)$. Thus $1 \to F' \to F \to F''$ is exact.

Finally, let us show that $F \to F''$ is surjective. As in the proof of 1) $\Rightarrow$ 2), let

$$P = \mathrm{Coker}_{\mathcal{P}(R_{\text{ét}})}(F \to F'').$$

We have seen that

$$F_{\overline{p}} \to F''_{\overline{p}} \to P_{\overline{p}} \to 0$$

is exact, hence $P_{\overline{p}} = 0$, hence $aP = 0$, by 5.5.3. Therefore $F \to F''$ is surjective. □

Lemma 5.5.6 *Let $\pi : R \to S$ be a ring homomorphism, $p \in \mathrm{Spec}(R)$, $f : R \to R_p^{sh}$ the canonical map, $\tilde{S} = S \otimes R_p^{sh}$ and $g = I \otimes f : S \to \tilde{S}$, $\tilde{\pi} : R_p^{sh} \to \tilde{S}$, such that we have*

$$\begin{array}{ccc} R & \xrightarrow{f} & R_p^{sh} \\ \downarrow{\scriptstyle \pi} & & \downarrow{\scriptstyle \tilde{\pi}} \\ S & \xrightarrow{g} & \tilde{S} \end{array}$$

Then for any sheaf $F \in \mathcal{S}(S_{\text{ét}})$, we have that

$$(\pi_* F)_{\overline{p}} = g^* F(\tilde{S})$$

Proof On one hand,

$$g^* F(\tilde{S}) = \varinjlim F(U')$$

over diagrams of the form

$$\begin{array}{ccc} S & \xrightarrow{g} & \tilde{S} \\ \downarrow{\scriptstyle \text{ét}} & \nearrow & \\ U & & \end{array} \tag{5.7}$$

On the other hand,

$$(\pi_* F)_{\overline{p}} = \varinjlim (\pi_* F)(U) = \varinjlim F(U \otimes S)$$

over diagrams of the form

$$\begin{array}{ccc} R & \longrightarrow & \overline{k}(p)^{sep} \\ \downarrow{\scriptstyle \text{ét}} & \nearrow & \\ U & & \end{array} \tag{5.8}$$

Now

$$R_p^{sh} = \varinjlim U,$$

over the diagrams (5.8). Since U is finitely generated as an R-algebra, the maps $U \to k(p)^{sep}$ factor through R_p^{sh}, so we may as well take the limit over

$$\begin{array}{ccc} R & \longrightarrow & \tilde{R}_p^{sh} \\ \downarrow_{\acute{e}t} & \nearrow & \\ U & & \end{array} \tag{5.9}$$

It therefore suffices to show that, after tensoring up over S, the diagrams (5.9) are cofinal in (5.8). Take a diagram (5.7). Now

$$\tilde{S} = S \otimes \varinjlim V = \varinjlim S \otimes V,$$

over (5.8). Since U' is finitely generated as an R-algebra, $U' \to \tilde{S}$ factors through $S \otimes V$ for some V. Hence

$$\begin{array}{ccc} S & \xrightarrow{g} & \tilde{S} \\ \downarrow_{\acute{e}t} & & \uparrow \\ U' & \xrightarrow{\acute{e}t} & S \otimes V \end{array}$$

which is exactly what we need. □

Lemma 5.5.7 *Let $\pi : R \to S = R/I$ be a surjective ring homomorphism, $I =$ Ker π. Then for every $F \in \mathcal{S}(S_{\acute{e}t})$, we have*

$$(\pi_* F)_{\overline{p}} = \begin{cases} 0 & \text{if } I \not\subset p; \\ F_{\overline{p}} & \text{if } I \subset p \text{ and } p = \pi^{-1}(q). \end{cases}$$

Proof It is well-known that Spec$(R/I) = V/I$. If $p \notin V(I)$, then $R_p^{sh} \otimes R/I = 0$, so $(\pi_* F)_{\overline{p}} = g^* F(R_p^{sh} \otimes R/I) = 0$.
If $I \subset p$, then $R/I \otimes R_p^{sh} = (R/I)_q^{sh}$. Then the map g of Lemma 5.5.6 is the canonical map $R/I \to (R/I)_q^{sh}$, and

$$(\pi_* F)_{\overline{p}} = g^* F((R/I)_q^{sh}) = F_{\overline{q}}.$$

□

Corollary 5.5.8 *Let $\pi : R \to R/I$ be surjective. Then the direct image functor $\pi_* : \mathcal{S}((R/I)_{\acute{e}t}) \to \mathcal{S}(R_{\acute{e}t})$ is exact.*

Proof Suppose that $0 \to F' \to F \to F'' \to 0$ is exact in $\mathcal{S}((R/I)_{\acute{e}t})$. By Theorem 5.5.5, this is equivalent to $0 \to F'_{\overline{q}} \to F_{\overline{q}} \to F''_{\overline{q}} \to 0$ being exact for every $q \in$ Spec(R/I). By Lemma 5.5.7. Take $p \in$ Spec(R). If $I \subset p$, take $q \in$ Spec(R/I) such that $p = \pi^{-1}(q)$. From Lemma 5.5.7, it follows that

$$0 \longrightarrow (\pi_* F')_{\overline{p}} \longrightarrow (\pi_* F)_{\overline{p}} \longrightarrow (\pi_* F'')_{\overline{p}} \longrightarrow 0 \tag{5.10}$$

is exact. If $I \not\subset p$, then (5.10) consists of zeroes, and is therefore exact. In any case, (5.10) is exact, and Theorem 5.5.5 implies that

$$0 \to \pi_* F' \to \pi_* F \to \pi_* F'' \to 0$$

is exact. □

Remark 5.5.9 *Corollary 5.5.8 may be generalized as follows :* $\pi_* : (S_{ét}) \to (R_{ét})$ *is exact if* $\pi : R \to S$ *is finitely generated as an R-module. For details, we refer to [135, II.3.6].*

5.6 Etale cohomology

Definition 5.6.1 *Let* $\Gamma : \mathcal{S}(R_E) \to \underline{\underline{Ab}}$ *be the "global section" functor, that is*

$$\Gamma(F) = F(R)$$

for all $F \in \mathcal{S}(R_E)$. *We define the n-th E-cohomology group of F to be the n-th right derived functor of* Γ *at F :*

$$H^n(R_E, F) = R^n\Gamma(F).$$

If $E = (ét)$ or (fl), then we call these groups resp. the *étale* or *flat cohomology groups* on R. Let $f : R \to S$ be a morphism in $\underline{\underline{\text{cat}}}(R_E)$. Consider the functor $\Gamma_S : \mathcal{S}(R_E) \to \underline{\underline{Ab}}, \Gamma_S(F) = F(S)$, and its derived functors $R^n\Gamma_S$. Alternatively, we can also consider the inverse image sheaf $f^*F = F_{|S}$ on S_E, and therefore the cohomology groups $H^n(S_E, F_{|S})$. The next Proposition tells us that these two types of cohomology coincide, and that they also determine the derived functors of the functor $i : \mathcal{P}(R_E) \to \mathcal{S}(R_E)$.

Lemma 5.6.2 *With notations as above, we have isomorphisms*

$$H^n(S_E, F_{|S}) \cong R^n\Gamma_S(F) \cong R^n i(F)(S)$$

Proof We have seen in Proposition 5.4.4 that f^* is an exact functor, and that it has an exact left adjoint. Therefore f^* preserves injectives, and we can apply Theorem A.2.2 to

$$\mathcal{S}(R_E) \xrightarrow{f^*} \mathcal{S}(S_E) \xrightarrow{\Gamma} \underline{\underline{Ab}}$$

The first statement follows. For $F \in \mathcal{S}(R_E)$, consider an injective resolution

$$F \to X^0 \to X^1 \to \cdots$$

Then $R^n i(F)(S) = \text{Ker}\,(X^n(S) \to X^{n+1}(S))/\text{Im}(X^{n-1}(S) \to X^n(S)) = R^n\Gamma_S(F)$ □

Our next aim will be to relate étale and flat cohomology groups to Amitsur cohomology groups. Consider the following functors (in the sequel, E will be $(ét)$ or (fl)) :

$$\mathcal{S}(R_E) \xrightarrow{i} \mathcal{P}(R_E) \xrightarrow{g} \underline{\underline{Ab}},$$

where i is as above, and g is defined by $g(F) = H^0(S/R, F)$, the 0-th Amitsur cohomology group. If F is a sheaf, then $g(F) = F(R)$, and therefore $g \circ i = \Gamma$.

Proposition 5.6.3 *With notations as above,* $R^n g = H^n(S/R, \bullet)$.

Before proving proposition 5.6.3, we need to investigate further the sheaves $\mathbb{Z}_T$ introduced above.

Lemma 5.6.4 *Let* $P \in \mathcal{P}(R_E)$. *Then* P *is isomorphic to* $\mathrm{Hom}_{\mathcal{P}(R_E)}(\mathbb{Z}_\bullet, P)$.

Proof Consider an E-morphism $\delta : T \to T'$. This induces maps

$$P(\delta) : P(T) \to P(T')$$

and

$$\Delta = \mathrm{Hom}(\mathbb{Z}_\delta, P) : \mathrm{Hom}(\mathbb{Z}_T, P) \to \mathrm{Hom}(\mathbb{Z}_{T'}, P)$$

We have to show that they induce a commutative diagram

$$\begin{array}{ccc} \mathrm{Hom}(\mathbb{Z}_T, P) & \xrightarrow{\Delta} & \mathrm{Hom}(\mathbb{Z}_{T'}, P) \\ \downarrow{\scriptstyle\Lambda} & & \downarrow{\scriptstyle\Lambda} \\ P(T) & \xrightarrow{P(\delta)} & P(T') \end{array}$$

Take $\psi \in \mathrm{Hom}(\mathbb{Z}_T, P)$. Then $(P(\delta) \circ \Lambda)(\psi) = P(\delta)(\psi(T)(I_T))$, and

$$\begin{aligned} (\Lambda \circ \Delta)(\psi) &= \Lambda(\psi \circ \Delta) \\ &= (\psi \circ \Delta)(T')(I_{T'}) \\ &= \psi(\Delta(T'))(I_{T'}) \\ &= (\psi(T') \circ \Delta(T'))(I_{T'}) \\ &= \psi(T')(\delta) \end{aligned}$$

The fact that $\psi(T')(\delta) = P(\delta)(\psi(T)(I_T))$ follows from the commutativity of the following diagram:

$$\begin{array}{ccc} \mathbb{Z}_T(T) & \xrightarrow{\psi(T)} & P(T) \\ \downarrow{\scriptstyle \mathbb{Z}_T(\delta)} & & \downarrow{\scriptstyle P(\delta)} \\ \mathbb{Z}_T(T') & \xrightarrow{\psi(T')} & P(T') \end{array}$$

□

Lemma 5.6.5 *If* P *is an injective object of* $\mathcal{P}(R_E)$, *then for all* $n > 0$, $H^n(S/R, P) = 0$.

Proof We have to show that the sequence

$$P(S) \xrightarrow{\Delta_0} P(S^{\otimes 2}) \xrightarrow{\Delta_1} P(S^{\otimes 3}) \xrightarrow{\Delta_2} \cdots$$

is exact. From the preceding two lemmas, it follows that this is equivalent to the exactness of

$$\mathrm{Hom}(\mathbb{Z}_S, P) \longrightarrow \mathrm{Hom}(\mathbb{Z}_{S^{\otimes 2}}, P) \longrightarrow \mathrm{Hom}(\mathbb{Z}_{S^{\otimes 3}}, P) \longrightarrow \cdots$$

From the fact that P is injective, it follows that it suffices to show that the sequence

$$\cdots \longrightarrow \mathbb{Z}_{S^{\otimes 3}} \longrightarrow \mathbb{Z}_{S^{\otimes 2}} \longrightarrow \mathbb{Z}_S$$

is exact in $\mathcal{P}(R_E)$, or, for all $T \in \underline{\underline{\text{cat}}}(R_E)$:

$$\cdots \longrightarrow \mathbb{Z}_{S^{\otimes 3}}(T) \longrightarrow \mathbb{Z}_{S^{\otimes 2}}(T) \longrightarrow \mathbb{Z}_S(T)$$

is exact in $\underline{\underline{Ab}}$. It can be shown easily that

$$\text{Mor}_E(S^{\otimes n}, T) \cong \text{Mor}_E(ST)^n.$$

Indeed, for an E-morphism $f: S^{\otimes n} \to T$, define $f_i : S \to T$ by $f_i(s) = f(1 \otimes \cdots \otimes s \otimes \cdots \otimes 1)$. Since f is an algebra homomorphism, it follows that

$$f(s_1 \otimes \cdots \otimes s_n) = f_1(s_1) \cdots f_n(s_n).$$

Let $M = \text{Mor}_E(ST)$. Then the above complex can be written as

$$\cdots \xrightarrow{d_2} \oplus_{M^3} \mathbb{Z} \xrightarrow{d_1} \oplus_{M^2} \mathbb{Z} \xrightarrow{d_0} \oplus_M \mathbb{Z}$$

What are the maps d_n? It suffices to define

$$d_{n-2}: \oplus_{M^n} \mathbb{Z} \to \oplus_{M^{n-1}} \mathbb{Z}$$

on the $\mathbb{Z}$-basis M_n. The map $\varepsilon_i : S^{\otimes(n-1)} \to S^{\otimes n}$ induces the map

$$\delta_i : \text{Mor}_E(S^{\otimes n}, T) \to \text{Mor}_E(S^{\otimes(n-1)}, T)$$

defined by

$$\delta_i(f) = f \circ \varepsilon_i.$$

Therefore, for $f = (f_1, \ldots, f_n) \in M^{n-2}$,

$$\delta_i(f) = (f_1, \cdots, f_{i-1}, f_{i+1}, \cdots, f_n),$$

and d_{n-2} is defined by

$$d_{n-2}(f_1, \ldots, f_n) = (f_2, \ldots, f_n) - (f_1, f_3, \cdots, f_n) + \cdots \pm (f_1, \ldots, f_{n-1}).$$

We define a contracting homotopy

$$s_{n-2}: \oplus_{M^n} \mathbb{Z} \to \oplus_{M^{n+1}} \mathbb{Z}$$

by

$$s_{n-2}(f_1, \ldots, f_n) = (1, f_1, \ldots, f_n).$$

It is straightforward to check that $d_{n-1} \circ s_{n-2} + s_{n-3} \circ d_{n-2}$ is the identity on $\oplus_{M^n} \mathbb{Z}$.
□

Proof of Proposition 5.6.3 We have to show that $H^n(S/R, \bullet)$ satisfies the three conditions listed in A.2. The first condition is obvious, and the second one follows from Lemma 5.6.5. Consider an exact sequence in $\mathcal{P}(R_E)$:

$$0 \longrightarrow P_1 \xrightarrow{\alpha} P_2 \xrightarrow{\beta} P_3 \longrightarrow 0.$$

Then we have to show that we have a long exact sequence

$$\begin{aligned} 0 &\longrightarrow H^0(S/R, P_1) \longrightarrow H^0(S/R, P_2) \longrightarrow H^0(S/R, P_3) \\ &\longrightarrow H^1(S/R, P_1) \longrightarrow H^1(S/R, P_2) \longrightarrow H^1(S/R, P_3) \\ &\longrightarrow H^2(S/R, P_1) \longrightarrow \cdots \end{aligned}$$

The proof is straightforward ; let us restrict to giving the definition of the map

$$\gamma : H^i(S/R, P_3) \to H^{i+1}(S/R, P_1).$$

Take $p_3 \in Z^i(S/R, P_3) \subset P_3(S^{\otimes(i+1)})$. Then there exists a $p_2 \in P_2(S^{\otimes(i+1)})$ such that

$$\beta(S^{\otimes(i+1)})(p_2) = p_3.$$

Now

$$\beta(S^{\otimes(i+2)})(\Delta_i(p_2)) = \Delta_i(p_3) = 0,$$

so there exists $p_1 \in P_1(S^{\otimes(i+2)})$ such that

$$\alpha(S^{\otimes(i+2)})(p_1) = \delta_i(p_2).$$

Now

$$\alpha(S^{\otimes(i+3)})(\Delta_{i+1}(p_1)) = \Delta_{i+1}(\Delta_i(p_2)) = 0,$$

hence p_1 is a cocycle, since $\alpha(S^{\otimes(i+3)})$ is injective. We define $\gamma([p_3]) = [p_1]$. Standard arguments show that γ is well-defined and that the above complex is exact. □

Lemma 5.6.6 *If F is an injective object of $\mathcal{S}(R_E)$, then $i(F)$ is injective in $\mathcal{P}(R_E)$. Consequently i takes injective objects of $\mathcal{S}(R_E)$ to g-acyclics.*

Proof Suppose that

$$0 \longrightarrow P_1 \xrightarrow{\alpha} P_2 \xrightarrow{\beta} P_3 \xrightarrow{0}$$

is exact in $\mathcal{P}(R_E)$. Then, by exactness of the functor a,

$$0 \longrightarrow aP_1 \xrightarrow{a\alpha} aP_2 \xrightarrow{a\beta} aP_3 \xrightarrow{0}$$

is exact in $\mathcal{S}(R_E)$, so, by injectivity of F,

$$0 \longrightarrow \mathrm{Hom}_{\mathcal{S}(R_E)}(aP_3, F) \longrightarrow \mathrm{Hom}_{\mathcal{S}(R_E)}(aP_2, F) \longrightarrow \mathrm{Hom}_{\mathcal{S}(R_E)}(aP_1, F) \longrightarrow 0$$

is exact. Now since a is the left adjoint of i,

$$\mathrm{Hom}_{\mathcal{S}(R_E)}(aP_i, F) \cong \mathrm{Hom}_{\mathcal{P}(R_E)}(P_i, iF),$$

since a is the left adjoint of i, so we obtain an exact sequence in $\underline{\underline{Ab}}$:

$$0 \longrightarrow \mathrm{Hom}_{\mathcal{P}(R_E)}(P_3, iF) \longrightarrow \mathrm{Hom}_{\mathcal{P}(R_E)}(P_2, iF) \longrightarrow \mathrm{Hom}_{\mathcal{P}(R_E)}(P_1, iF) \longrightarrow 0$$

□

We are now able to apply Theorem A.2.2 to

$$\mathcal{S}(R_E) \xrightarrow{i} \mathcal{P}(R_E) \xrightarrow{H^0(S/R,\bullet)} \underline{\underline{Ab}}$$

For $F \in \mathcal{S}(R_E)$, we obtain the following exact sequences :

$$\begin{aligned} 0 &\longrightarrow H^1(S/R, C^q) \longrightarrow H^{q+1}(R_E, F) \longrightarrow H^0(S/R, H^{q+1}(\bullet, F)) \\ &\longrightarrow H^2(S/R, C^q) \longrightarrow H^1(S/R, C^{q+1}) \longrightarrow H^1(S/R, H^{q+1}(\bullet, F)) \\ &\longrightarrow \cdots \\ &\longrightarrow H^{p+1}(S/R, C^q) \longrightarrow H^p(S/R, C^{q+1}) \longrightarrow H^p(S/R, H^{q+1}(\bullet, F)) \\ &\longrightarrow \cdots \end{aligned} \tag{5.11}$$

In particular, we have maps $H^{p+1}(S/R, C^q) \longrightarrow H^p(S/R, C^{q+1})$ and an injection $H^1(S/R, C^q) \hookrightarrow H^{q+1}(R_E, F)$. If we remark that $C_0 = F$, then we obtain a chain of maps

$$H^p(S/R, F) \to H^{p-1}(S/R, C^1) \to \cdots \to H^1(S/R, C^{p-1}) \to H^p(R_E, F).$$

Therefore, in view of the remarks following Lemma 5.2.3, we have a map

$$\check{H}^p(R_E, F) \to H^p(R_E, F).$$

If $E = (\acute{e}t)$, we have the following result :

Theorem 5.6.7

$$\check{H}^p(R_{\acute{e}t}, F) \cong H^p(R_{\acute{e}t}, F).$$

Proof The proof is an easy exercise, if one applies Theorem 5.6.8, due to M. Artin. □

Theorem 5.6.8 (Artin's Refinement Theorem) *Let R be a noetherian ring. Suppose that $S \in \underline{\text{cat}}(R_{\acute{e}t})$, and that $S^{\otimes n} \to T$ is an étale covering. Then there exists an étale covering S' of S such that we have a factorization*

$$S^{\otimes n} \to T \to S'^{\otimes n}.$$

The proof of this fundamental Theorem requires a detailed study of étale algebras, we refer to [8], or to the forthcoming [38]. Many proofs of Theorems about the Brauer group are easy to prove if one applies Artin's Theorem, let us mention Knus and Ojanguren's proof of the fact that $\text{Br}(R)$ is torsion (cf. [114]), or the construction of the map $\text{Br}(R) \hookrightarrow H^2(R_{\acute{e}t}, \mathbb{G}_m)$ (cf. [114], [155]). In the forthcoming chapters, we will work on the flat site, and we will have to look out for ways to escape Artin's Theorem. For $p = 1$, Theorem 5.6.7 always holds :

Proposition 5.6.9 *For any sheaf F on R_E, we have*

$$\check{H}^1(R_E, F) \cong H^1(R_E, F)$$

Proof We have to show that $\check{H}^1(R_E, F)$ satisfies the conditions A.2. It follows from Lemmas 5.6.4 and 5.6.5 that $\check{H}^1(R_E, F) = 0$ if F is an injective sheaf. Let

$$0 \longrightarrow F_1 \xrightarrow{\alpha} F_2 \xrightarrow{\beta} F_3 \longrightarrow 0$$

be an exact sequence in $\mathcal{S}(R_E)$. Then we have a sequence

$$0 \longrightarrow F_1(R) \xrightarrow{\alpha(R)} F_2(R) \xrightarrow{\beta(R)} F_3(R) \xrightarrow{\gamma} \check{H}^1(R_E, F_1) \longrightarrow \check{H}^1(R_E, F_2) \longrightarrow \check{H}^1(R_E, F_3). \tag{5.12}$$

The map γ may be defined as follows : take $z \in F_3(R)$. Then there exists an E-covering $f: R \to S$ such that $F_3(f)(z) \in \operatorname{Im} \beta(S)$, say $F_3(f)(z) = \beta(S)(y)$. Now $\beta(S)(\Delta_0(y)) = 0$, so $\Delta_0(y) \in \operatorname{Im} \alpha(S^{\otimes 2})$, say $\Delta_0(y) = \alpha(S^{\otimes 2})(x)$. Define $\gamma(z) = [x]$. Standard arguments show that x is a cocycle, that γ is well-defined and that the sequence is exact. □

Corollary 5.6.10 *Let F be a sheaf on R_{fl}. Then we have natural inclusions*

$$H^n(R_{\text{Zar}}, F) \subset H^n(R_{\text{ét}}, F) \subset H^n(R_{\text{fl}}, F)$$

Proof First consider the case $n = 1$. If $S \to S'$ is a covering in R_E, then the map $H^1(S/R, F) \to H^1(S'/R, F)$ is injective (this follows from the fact that both of them inject into $H^1(R_E, F)$. Taking the unions over respectively all Zariski coverings, étale coverings and faithfully flat extensions of R, we therefore obtain that

$$\check{H}^n(R_{\text{Zar}}, F) \subset \check{H}^n(R_{\text{ét}}, F) \subset \check{H}^n(R_{\text{fl}}, F)$$

and the result follows from Proposition 5.6.9. The general case now follows from the fact that

$$H^n(R_E, F) \cong H^1(R_E, C_{n-1})$$

□

Remark 5.6.11 *We leave it to the reader to define a map*

$$\gamma: \check{H}^n(R_{\text{ét}}, F_3) \to \check{H}^{n+1}(R_{\text{ét}}, F_1),$$

using Artin's Refinement Theorem.

Proposition 5.6.12 *Let $F \in \mathcal{S}(R_E)$, and suppose that $x \in H^n(R_E, F)$. Then there exists an E-covering $R \to S$ such that the image of x in $H^n(S_E, F)$ is trivial.*

Proof Since $H^n(R_E, F) \cong H^1(R_E, C_{n-1})$, it follows that it suffices to prove the Proposition for $n = 1$. Invoking Proposition 5.6.9, it follows that it suffices to show that, for any E-covering S, and $[u] \in H^1(S/R, F)$, there exists an E-covering T of R such that the image of $[u]$ in $H^1(T \otimes S/T, F)$ is trivial. Take $T = S$. Then $u \in F(S \otimes S)$ lies in the kernel of the map Δ_1 in the complex

$$F(S) \xrightarrow{\Delta_0} F(S^{\otimes 2}) \xrightarrow{\Delta_1} F(S^{\otimes 3}) \xrightarrow{\Delta_2} \cdots$$

Here $\Delta_0 = F(\varepsilon_1) - F(\varepsilon_2)$, $\Delta_1 = F(\varepsilon_1) - F(\varepsilon_2) + F(\varepsilon_3)$. Tensoring up by S at the left, we obtain

$$F(S^{\otimes 2}) \xrightarrow{\Delta_0} F(S^{\otimes 3}) \xrightarrow{\Delta_1} F(S^{\otimes 4}) \xrightarrow{\Delta_2} \cdots$$

where now $\Delta_0 = F(\varepsilon_2) - F(\varepsilon_3)$, $\Delta_1 = F(\varepsilon_2) - F(\varepsilon_3) + F(\varepsilon_4)$. The image of u in $F(S^{\otimes 3})$ is

$$F(\varepsilon_1)(u) = F(\varepsilon_2)(u) - F(\varepsilon_3)(u)$$

hence the image of u is a coboundary. □

Proposition 5.6.13 *The map $\check{H}^2(R_E, F) \to H^2(R_E, F)$ is injective.*

Proof Consider the exact sequence (5.11) in the cases $q = 0$ and $q = 1$:

$$\begin{array}{ccccccc} 0 & \longrightarrow & H^1(S/R, F) & \longrightarrow & H^1(R_E, F) & \longrightarrow & H^0(S/R, H^1(\bullet_E, F)) \\ & \xrightarrow{\alpha} & H^2(S/R, F) & \xrightarrow{\beta} & H^1(S/R, C_1) & \longrightarrow & \cdots \\ 0 & \longrightarrow & H^1(S/R, C_1) & \xrightarrow{\gamma} & H^2(R_E, F) & \longrightarrow & \cdots \end{array}$$

Suppose that $x \in H^2(S/R, F)$ is in the kernel of $\gamma \circ \beta$. Then $x \in \operatorname{Ker} \beta$, hence $x \in \operatorname{Im} \alpha$, say $x = \alpha(y)$. But y is represented by an element of $H^1(S_E, F)$. Replacing S by an E-covering, y, and therefore x, becomes trivial, by Proposition 5.6.12. Hence $[x] = 1$ in $\check{H}^2(R_E, F)$. □

As an application, let us show the following:

Proposition 5.6.14 *Suppose that F is a torsion sheaf, that is, a sheaf of torsion groups, on R_E. Then for all $q \geq 0$ the cohomology group $H^q(R_E, F)$ is torsion.*

Proof For $q = 1$, the result follows immediately from Proposition 5.6.9, and the fact that the Amitsur cohomology groups of a torsion (pre)sheaf are torsion. If we work over the étale topology, then the result for $q \geq 1$ follows from Theorem 5.6.7. If one wants to avoid Artin's refinement theorem, or if one wants the result for the flat topology, then one has to use another description of the cohomology groups. It was shown by Verdier in [9] that derived functor cohomology may be described as an inductive limits of generalized Čech cohomology groups. This groups are easily seen to be torsion if taken for a torsion (pre)sheaf. The construction is explained in our particular situation in [45, Sec. V.2]. □

5.7 Flabby sheaves

Consider an R-algebra $f : R \to S$. We have seen (Proposition 5.4.3) that the functor $f_* : \mathcal{S}(S_E) \to \mathcal{S}(R_E)$ is left exact, so we can consider its right derived functors. To study these derived functors, we have to introduce the notion of *flabby sheaf.*

Definition 5.7.1 *A sheaf F is called flabby if it is i-acyclic, where i is the natural functor $\mathcal{S}(R_E) \to \mathcal{P}(R_E)$, or, equivalently, if F is Γ_S-acyclic for every $S \in \underline{\underline{\text{cat}}}(R_E)$.*

Observe that a flabby sheaf is Γ-acyclic ($\Gamma = \Gamma_R$), and that the class of flabby sheaves satisfies the three conditions of Lemma A.2.1. Indeed, the first condition follows from the fact that an injective sheave is necessarily flabby. Condition 2 follows from the fact that derived functors commute with finite direct sums (since the finite direct sum injective resolutions is the injective resolution of the finite direct sum). Finally, suppose that

$$0 \longrightarrow F' \longrightarrow F \longrightarrow F'' \longrightarrow 0$$

is exact in $\mathcal{S}(R_E)$, with F and F' flabby. Then we have the following exact sequence in $\mathcal{P}(R_E)$:

$$\cdots R^n iF' = 0 \longrightarrow R^n iF = 0 \longrightarrow R^n iF'' \longrightarrow R^{n+1} iF' = 0 \longrightarrow \cdots$$

implying $R^n iF'' = 0$ for all $n \geq 1$. Therefore F'' is flabby, and this establishes condition 3.
We now give a characterization of flabby sheaves using Amitsur cohomology.

Theorem 5.7.2 *For any $F \in \mathcal{S}(R_E)$, the following conditions are equivalent:*

1. *F is flabby;*
2. *$H^n(T/S, C_q) = 0$ for every $S \to T$ in $\underline{\underline{\text{cov}}}(R_E)$, $n \geq 1$ and $q \geq 0$;*
3. *$H^n(T/S, F) = 0$ for every $S \to T$ in $\underline{\underline{\text{cov}}}(R_E)$ and $n \geq 1$;*
4. *$\check{H}^n(S_E, F_{|S}) = 0$ for every $S \in \underline{\underline{\text{cat}}}(R_E)$ and $n \geq 1$.*

Here $C^q = \ker(X^q \to X^{q+1})$ in an injective resolution of F, as usual.

Proof From Lemma 5.6.6, it follows that we may apply Theorem A.2.2 to

$$\mathcal{S}(R_E) \xrightarrow{i} \mathcal{P}(R_E) \xrightarrow{H^0(T/S,\bullet)} \underline{\underline{Ab}}$$

We have that $H^0(T/S, \bullet) \circ i = \Gamma_S$, and $(R^n\Gamma)H^0(T/S, \bullet) = H^n(T/S, \bullet)$ (the proof is similar to the proof of 5.6.3). We therefore have exact sequences, for $q \geq 0$:

$$\begin{array}{ccccccc} 0 & \longrightarrow & H^1(T/S, C_q) & \longrightarrow & H^{q+1}(S_E, F_{|S}) & \longrightarrow & H^0(S/R, H^{q+1}(\bullet, F_{|\bullet})) \\ & \longrightarrow & H^2(T/S, C_q) & \longrightarrow & H^1(T/S, C_{q+1}) & \longrightarrow & H^1(T/S, H^{q+1}(\bullet, F_{|\bullet})) \\ & \longrightarrow & \cdots & & & & \end{array}$$

If F is flabby, then $H^{q+1}(\bullet, F_{|\bullet}) = 0$, hence

$$\begin{array}{rcl} H^1(T/S, C_q) & = & 0 \\ H^{n+1}(T/S, C_q) & \cong & H^n(T/S, C_{q+1}) \end{array}$$

and this implies that $H^n(T/S, C_q) = 0$ for all $n \geq 1$, $q \geq 0$. This proves $1 \Rightarrow 2$. $2 \Rightarrow 3$ is trivial (take $q = 0$) and $3 \Rightarrow 4$ follows after taking direct limits over all coverings of S. To prove $4 \Rightarrow 1$, we consider the functor

$$\check{H}^0(S_E, \bullet) : \mathcal{P}(R_E) \to \underline{\underline{Ab}}$$

We claim that $R^n\check{H}^0(S_E, \bullet) = \check{H}^n(S_E, \bullet)$. Indeed, if a presheaf P is injective, then for $n > 0$

$$\check{H}^n(S_E, P) = \varinjlim H^n(T/S, P) = 0$$

by Proposition 5.6.4. Furthermore, if

$$0 \longrightarrow P' \longrightarrow P \longrightarrow P'' \longrightarrow 0$$

is an exact sequence of presheaves, then we have a long exact sequence of Amitsur cohomology groups

$$\cdots \longrightarrow H^n(T/S, P') \longrightarrow H^n(T/S, P) \longrightarrow H^n(T/S, P'') \longrightarrow H^{n+1}(T/S, P') \longrightarrow \cdots$$

and this sequence stays exact after we take direct limits over all coverings T of S. Next, consider the functor

$$\check{H}^0(\bullet_E, \bullet) : \mathcal{P}(R_E) \to \mathcal{P}(R_E)$$

We then have that $R^n\check{H}^0(\bullet_E, \bullet) = \check{H}^n(\bullet_E, \bullet)$ (the proof is similar to the proof of 5.6.2). We can therefore apply Theorem A.2.2 to

$$\mathcal{S}(R_E) \xrightarrow{i} \mathcal{P}(R_E) \xrightarrow{H^0(\bullet_E, \bullet)} \mathcal{P}(R_E)$$

resulting in long exact sequences

$$\begin{array}{cccccc} 0 & \longrightarrow & \check{H}^1(\bullet, C_q) & \longrightarrow & R^{q+1}i(F) & \longrightarrow & \check{H}^0(\bullet, R^{q+1}i(F)) \\ & \longrightarrow & \check{H}^2(\bullet, C_q) & \longrightarrow & \check{H}^1(\bullet, C_{q+1}) & \longrightarrow & \check{H}^1(\bullet, R^{q+1}i(F)) \\ & \longrightarrow & \cdots & & & & \end{array}$$

Now suppose that for all $n > 0$ we have that $\check{H}^n(\bullet, F) = 0$. We then have (using Proposition 5.6.9)

$$R^1i(F) = H^1(\bullet, F) = \check{H}^1(\bullet, F) = 0$$

Now we proceed by induction on n. Suppose that for a given $n \geq 1$ the following two conditions are fulfilled for all $m \leq n$ and $p \geq 1$:

$$\begin{aligned} H^m(\bullet, F) &= R^m i(F) = 0 \\ \check{H}^{p+1}(\bullet, C_{m-2}) &\cong H^p(\bullet, C_{m-1}) \end{aligned}$$

These conditions are fulfilled if $n = 1$. If they are fulfilled for n, then we consider the above exact sequence in the case where $n = q - 1$. Then for all $p \geq 1$, we obtain

$$\check{H}^{p+1}(\bullet, C_{n-1}) \cong H^p(\bullet, C_n)$$

and the second condition is fulfilled for $n+1$. Also

$$\begin{aligned} R^{n+1}i(F) &= H^{n+1}(\bullet, F) \\ &= H^1(\bullet, C_n) \\ &= \check{H}^1(\bullet, C_n) \\ &= \check{H}^{n+1}(\bullet, F) \end{aligned}$$

and the first condition is also fulfilled for $n+1$. □

Corollary 5.7.3 *Let $f: R \to S$ be a commutative R-algebra, and consider $f_*: \mathcal{S}(S_E) \to \mathcal{S}(R_E)$. If F' is flabby, then $f_*(F')$ is also flabby. Consequently flabby sheaves are Γ-acyclic, and we can apply Theorem A.2.2 to*

$$\mathcal{S}(S_E) \xrightarrow{f_*} \mathcal{S}(R_E) \xrightarrow{\Gamma} \underline{\underline{Ab}}$$

Proof Take $U \to V \in \underline{\underline{\text{cov}}}(R_E)$. Inspecting the definition of Amitsur cohomology, we easily see that the complexes defining $H^n(V/U, f_*F)$ and $H^n(S \otimes V/S \otimes U, F')$ are isomorphic, hence $H^n(V/U, f_*F) = H^n(S \otimes V/S \otimes U, F') = 0$ if F is flabby, so f_*F is flabby. The other statements now follow immediately. □

Corollary 5.7.4 *Suppose that $f: R \to S$ is in $\underline{\underline{\text{cat}}}(R_E)$, and consider $f^*: \mathcal{S}(R_E) \to \mathcal{S}(S_E)$. If F is flabby, then $f^*(F)$ is also flabby. Consequently flabby sheaves are f^*-acyclic, and we can apply Theorem A.2.2 to*

$$\mathcal{S}(R_E) \xrightarrow{f^*} \mathcal{S}(S_E) \xrightarrow{\Gamma} \underline{\underline{Ab}}$$

Proof If $U \to V \in \underline{\underline{\text{cov}}}(S_E)$, then $U \to V \in \underline{\underline{\text{cov}}}(R_E)$, hence $H^n(V/U, F) = H^n(V/U, f^*F) = 0$, and $f^*F$is flabby. The other statements now follow immediately. □

The direct image sheaf functor $f_*: \mathcal{S}(S_E) \to \mathcal{S}(R_E)$ is left exact, so we can consider its derived functors. They may be described as follows:

Proposition 5.7.5 *Let $f: R \to S$ be a ring homomorphism. Then*

$$R^q f_* = a f_p(R^q i),$$

that is, for $F \in \mathcal{S}(S_E)$, $R^q f_* F$ is the sheaf associated to the presheaf P on R_E given by

$$P(U) = H^q(U \otimes S, F_{|U \otimes S})$$

Proof We know that $f_* = a \circ f_p \circ i$, cf. Proposition 5.4.3. Therefore we can apply Theorem A.2.2 to

$$f_*: \mathcal{S}(S_E) \xrightarrow{i} \mathcal{P}(S_E) \xrightarrow{a \circ f_p} \mathcal{S}(R_E)$$

Now a and f_p are exact, and the result follows. □

Chapter 6

Cohomological interpretation of the Brauer group

6.1 Cohomology with values in the category of invertible modules

We start this Section with a cohomological interpretation of the Picard group :

Proposition 6.1.1 *Let $R \to S$ be an E-covering. Then* $\mathrm{Pic}(S/R) \cong H^1(S/R, \mathbb{G}_m)$. *Consequently* $\mathrm{Pic}(R) \cong H^1(R_E, \mathbb{G}_m)$ *for* $E = (\acute{e}t)$, (fl), or (Zar).

Proof The first statement is an easy application of descent theory : suppose $x \in \mathbb{G}_m(S^{\otimes 2})$ represents $[x] \in H^1(S/R, \mathbb{G}_m)$. Then $x_1 x_3 = x_2$, so multiplication by x defines a descent datum $m(x) : S^{\otimes 2} \to S^{\otimes 2}$. The descended module I is of rank one, since $S \otimes I \cong S$. It is straightforward to show that the map $H^1(S/R, \mathbb{G}_m) \to \mathrm{Pic}(S/R)$ defined by sending $[x]$ to $[I]$ is a well-defined isomorphism. From the fact that the Picard group of a local ring is trivial, it follows that every element of the Picard group may be split by an E-covering. Therefore

$$\mathrm{Pic}(R) = \bigcup \mathrm{Pic}(S/R) \cong \check{H}^1(R_E, \mathbb{G}_m) \cong H^1(R_E, \mathbb{G}_m)$$

where the union runs over all E-coverings S of R. □

In this Chapter, we will discuss the relationship between $\mathrm{Br}(R)$, $\mathrm{Br}'(R)$ and $H^2(R_{\acute{e}t}, \mathbb{G}_m)$. To this end, we need an explicit description of the second cohomology group; we will present a construction due to Villamayor and Zelinsky, cf. [199], [209]. Other possible approaches will be discussed briefly later on.

Recall that $\mathbb{G}_m$ is a sheaf. Let C_i be defined as in Theorem A.2.2, that is :

$$C_i = \mathrm{Ker}\,(X_i \to X_{i+1})$$

where

$$1 \to \mathbb{G}_m \to X_0 \to X_1 \to \cdots$$

is an injective resolution of $\mathbb{G}_m$ in $\mathcal{S}(R_E)$. Clearly $\mathbb{G}_m = C_0$. We do not know any description for C_i, with $i > 0$, but we can describe $H^n(S/R, C_1)$, for any E-covering S, and for any $n \geq 0$. For $n \geq 1$, we define a functor

$$\delta_{n-1} : \underline{\underline{\mathrm{Pic}}}(S^{\otimes n}) \to \underline{\underline{\mathrm{Pic}}}(S^{\otimes(n+1)})$$

by

$$\delta_{n-1}(I) = I_1 \otimes_{S^{\otimes(n+1)}} I_2^* \otimes_{S^{\otimes(n+1)}} I_3 \otimes_{S^{\otimes(n+1)}} \cdots \otimes_{S^{\otimes(n+1)}} I_{n+1}^{(*)}$$

for $I \in \underline{\underline{\mathrm{Pic}}}(S^{\otimes n})$), and

$$\delta_{n-1}(f) = f_1 \otimes (f_2^*)^{-1} \otimes \cdots \otimes (f_{n+1}^{(*)})^{\pm 1}$$

for $f : I \to J$ in $\underline{\underline{\mathrm{Pic}}}(S^{\otimes n})$. Observe that it is necessary to restrict to isomorphisms for the morphisms at this place. From the fact that the map

$$\Delta_n \circ \Delta_{n-1} : S^{\otimes n} \to S^{\otimes(n+2)}$$

is the zero map (Amitsur's complex is a complex !), it follows that there is a natural isomorphism

$$\lambda_I : \delta_n(\delta_{n-1}(I)) \to S^{\otimes(n+2)}$$

This may be made explicit as follows : for $I \in \underline{\underline{\mathrm{Pic}}}(S^{\otimes n})$), write $\varphi : I \otimes_{S^{\otimes n}} I^* \to S^{\otimes n}$ for the natural isomorphism. We then have that

$$\begin{aligned} \delta_n \delta_{n-1}(I) &= \bigotimes_{j=1}^{n+2} \bigotimes_{i=1}^{n+1} (I_{ij})^{(-1)^{i+j}} \\ &= \bigotimes_{j=1}^{n+2} \bigotimes_{i=1}^{j-1} I_{ij} \otimes (I_{ij})^* \cong S^{\otimes(n+2)} \end{aligned}$$

where the connecting isomorphism is $\lambda_I = \bigotimes_{j=1}^{n+2} \bigotimes_{i=1}^{j-1} \varphi_{ij}$. If we identify $I \otimes I^*$ and $S^{\otimes n}$ using φ, then λ_I is the identity. We will extensively use this identification in the sequel.

It is clear that δ_n is a cofinal product preserving functor. The exact sequence (C.8) takes the form

$$\mathbb{G}_m(S^{\otimes n}) \xrightarrow{\Delta_{n-1}} \mathbb{G}_m(S^{\otimes n+1}) \xrightarrow{\kappa_{n-1}} \mathrm{K}_0 \underline{\Psi\delta}_{n-1} \xrightarrow{\theta_{n-1}} \mathrm{Pic}(S^{\otimes n}) \xrightarrow{\Delta'_{n-1}} \mathrm{Pic}(S^{\otimes n+1}) \tag{6.1}$$

Δ_{n-1} and Δ'_{n-1} are the maps in the Amitsur complexes of $\mathbb{G}_m$ and Pic. The category $\underline{\Psi\delta}_{n-1}$ may be described as follows:

Objects: (I, α), with $I \in \underline{\underline{\mathrm{Pic}}}(S^{\otimes n})$ and $\alpha : \delta_{n-1}(I) \to S^{\otimes n+1}$ an isomorphism;

Morphisms: $f : (I, \alpha) \to (J, \beta)$ is an $S^{\otimes n}$-module isomorphism $f : I \to J$ such that $\beta \circ \delta_{n-1}(f) = \alpha$.

Product: $(I, \alpha)(J, \beta) = (I \otimes_{S^{\otimes n}} J, \alpha \otimes \beta)$.

Every element in $\mathrm{K}_0 \underline{\Psi\delta}_{n-1}$ is represented by an object of the category $\underline{\Psi\delta}_{n-1}$. The maps κ_{n-1} and θ_{n-1} are given by the following formulas. For $x \in \mathbb{G}_m(S^{\otimes n+1})$, we let

$$\kappa_{n-1}(x) = (S^{\otimes n}, x) \tag{6.2}$$

(here x means "multiplication by x"). For $(I,\alpha) \in \underline{\Psi\delta}_{n-1}$, we let

$$\theta_{n-1}[(I,\alpha)] = [I] \tag{6.3}$$

We will now define two additional maps:

$$D'_{n-1} : \operatorname{Pic}(S^{\otimes n}) \longrightarrow \mathrm{K}_0\underline{\Psi\delta}_n$$

is defined by the formula

$$D'_{n-1}([I]) = [(\delta_{n-1}(I), \lambda_I)] \tag{6.4}$$

and

$$D_{n-1} : \mathrm{K}_0\underline{\Psi\delta}_{n-1} \longrightarrow \mathbb{G}_m(S^{\otimes n+2})$$

is defined by

$$D_{n-1}[(I,\alpha)] = \delta_n(\alpha) \circ \lambda_I^{-1} \tag{6.5}$$

By this we mean the following: $\delta_n(\alpha)\circ\lambda_I^{-1}$ is an isomorphism

$$S^{\otimes n+2} \xrightarrow{\lambda_I^{-1}} \delta_n(\delta_{n-1}(I)) \xrightarrow{\delta_n(\alpha)} \delta_n(S^{\otimes n+1}) = S^{\otimes n+2}$$

and is given by multiplication by a unit in $S^{\otimes n+2}$. This unit is the image of $[(I,\alpha)]$. We can now combine the maps $\Delta_n, \Delta'_n, D_n, D'_n, \kappa_n$ and θ_n in the following diagram.

$$\begin{array}{ccccc}
1 & & & & \\
\downarrow & & & & \\
\mathbb{G}_m(S) & & & & 1 \\
\downarrow{\scriptstyle \Delta_0} & & & & \downarrow \\
\mathbb{G}_m(S^{\otimes 2}) & \xrightarrow{\kappa_0} & \mathrm{K}_0\underline{\Psi\delta}_0 & \xrightarrow{\theta_0} & \operatorname{Pic}(S) \\
\downarrow{\scriptstyle \Delta_1} & \swarrow{\scriptstyle D_0} & & \swarrow{\scriptstyle D'_0} & \downarrow{\scriptstyle \Delta'_0} \\
\mathbb{G}_m(S^{\otimes 3}) & \xrightarrow{\kappa_1} & \mathrm{K}_0\underline{\Psi\delta}_1 & \xrightarrow{\theta_1} & \operatorname{Pic}(S^{\otimes 2}) \\
\downarrow{\scriptstyle \Delta_2} & \swarrow{\scriptstyle D_1} & & \swarrow{\scriptstyle D'_1} & \downarrow{\scriptstyle \Delta'_1} \\
\mathbb{G}_m(S^{\otimes 4}) & \xrightarrow{\kappa_2} & \mathrm{K}_0\underline{\Psi\delta}_2 & \xrightarrow{\theta_2} & \operatorname{Pic}(S^{\otimes 3}) \\
\downarrow{\scriptstyle \Delta_3} & \swarrow{\scriptstyle D_2} & & \swarrow{\scriptstyle D'_2} & \downarrow{\scriptstyle \Delta'_2} \\
\vdots & & \vdots & & \vdots
\end{array} \tag{6.6}$$

The first column is the Amitsur complex of $\mathbb{G}_m$, and the last one is the Amitsur complex of Pic. The cranks in (6.6) are the exact sequences (6.1). Furthermore

$$D_n \circ D'_{n-1} = 1 \tag{6.7}$$

Indeed,

$$\begin{aligned}
D_n\left(D'_{n-1}([I])\right) &= D_n[(\delta_{n-1}(I), \lambda_I)] \\
&= \delta_n(\lambda_I) \circ \lambda_{\delta_{n-1}(I)^{-1}} = 1
\end{aligned}$$

for all $I \in \underline{\underline{\mathrm{Pic}}}(S^{\otimes n})$. The parallelograms in (6.6) anticommute:

$$(\kappa_n \circ D_{n-1})[(I,\alpha)] = (D'_{n-1} \circ \theta_{n-1})[(I,\alpha)]^{-1} \tag{6.8}$$

for all $(I,\alpha) \in \underline{\Psi\delta}_{n-1}$. Indeed, on one hand we have that

$$\begin{aligned} D'_{n-1}\Big(\theta_{n-1}[(I,\alpha)]\Big) &= D'_{n-1}([I]) \\ &= [(\delta_{n-1}(I), \lambda_I)] \end{aligned}$$

while, on the other hand

$$\begin{aligned} \kappa_n\Big(D_{n-1}[(I,\alpha)]\Big) &= \kappa_n(\delta_n(\alpha) \circ \lambda_I^{-1}) \\ &= \Big[(S^{\otimes n}, \delta_n(\alpha) \circ \lambda_I^{-1})\Big] \\ &= \Big[(S^{\otimes n}, \lambda_I \circ \delta_n(\alpha)^{-1})\Big]^{-1} \end{aligned}$$

Now $\alpha:\ \delta_{n-1}(I) \to S^{\otimes n}$ defines an isomorphism

$$(\delta_{n-1}(I), \lambda_I) \longrightarrow \Big[(S^{\otimes n}, \lambda_I \circ \delta_n(\alpha)^{-1})\Big]$$

since the diagram

$$\begin{array}{ccc} \delta_n(\delta_{n-1}(I)) & \xrightarrow{\delta_n(\alpha)} & S^{\otimes n+1} \\ \downarrow{\scriptstyle \lambda_I} & \swarrow{\scriptstyle \lambda_I \circ \delta_n(\alpha)^{-1}} & \\ S^{\otimes n+1} & & \end{array}$$

commutes.

The triangles in (6.6) commute:

$$\theta_n \circ D'_{n-1} = \Delta'_{n-1} \tag{6.9}$$

$$D_n \circ \kappa_n = \Delta_{n+1} \tag{6.10}$$

Indeed,

$$\begin{aligned} (\theta_n \circ D'_{n-1})([I]) &= \theta_n[(\delta_{n-1}(I), \lambda_I)] \\ &= [\delta_{n-1}(I)] = \Delta'_{n-1}([I]) \end{aligned}$$

for all $I \in \underline{\underline{\mathrm{Pic}}}(S^{\otimes n})$, and,

$$(D_n \circ \kappa_n)(x) = D_n[(S^{\otimes n+1}, x)] = \Delta_{n+1}(x)$$

for all $x \in \mathbb{G}_m(S^{\otimes n+2})$.

We now define *Amitsur cohomology with values in the category of invertible modules* $\underline{\underline{\mathrm{Pic}}}$ as follows:

$$H^n(S/R, \underline{\underline{\mathrm{Pic}}}) = \mathrm{Ker}\,(D_n)/\mathrm{Im}\,(D'_{n-1}) \tag{6.11}$$

Remarks 6.1.2 1) Another way to introduce $H^n(S/R, \underline{\underline{\mathrm{Pic}}})$ is the following. For $n \geq 0$, we consider the category Ω_n, with objects (I,α) where $I \in \underline{\underline{\mathrm{Pic}}}(S^{\otimes n})$ and $\alpha:\ \delta_{n-1}(I) \to S^{\otimes(n+1)}$ an isomorphism such that

$$\delta_n(\alpha) = \lambda_I$$

or

$$\delta_n(\alpha) = I_{S^{\otimes(n+2)}}$$

if we make the above mentioned identification. A morphism from (I, α) to (J, β) in Ω_n will be an isomorphism $f: I \to J$ in $\underline{\underline{\mathrm{Pic}}}(S^{\otimes n})$ such that $\alpha = \beta \circ \delta_{n-1}(f)$. We have a functor

$$\delta_{n-1}: \underline{\underline{\mathrm{Pic}}}(S^{\otimes n}) \to \Omega_{n+1}$$

given by $\delta_{n-1}(I) = (\delta_{n-1}(I), \lambda_I)$. It is now easy to see that

$$H^n(S/R, \underline{\underline{\mathrm{Pic}}}) = \mathrm{Coker}\,(\mathrm{K}_0\delta_{n-1}\ \mathrm{K}_0\underline{\underline{\mathrm{Pic}}}(S^{\otimes n}) \to \mathrm{K}_0\Omega_{n+1}),$$

2) Villamayor and Zelinsky (cf. [199]) use the notation

$$E_n = H^{n-1}(S/R, \underline{\underline{\mathrm{Pic}}})$$

3) The above construction may be extended easily to some other types of cohomology. We leave it to the reader to define group cohomology with values in the category $\underline{\underline{\mathrm{Pic}}}$. In Part II, we will consider Harrison and Sweedler cohomology with values in a category, and in part III, we will consider cohomology with values in the categories of comodules or dimodules.

Theorem 6.1.3 (Villamayor-Zelinsky) *We have a long exact sequence of cohomology groups*

$$\begin{array}{lllllll} 1 & \longrightarrow & H^1(S/R, \mathbb{G}_m) & \xrightarrow{\alpha_1} & H^0(S/R, \underline{\underline{\mathrm{Pic}}}) & \xrightarrow{\beta_1} & H^0(S/R, \mathrm{Pic}) \\ & \xrightarrow{\gamma_1} & H^2(S/R, \mathbb{G}_m) & \xrightarrow{\alpha_2} & H^1(S/R, \underline{\underline{\mathrm{Pic}}}) & \xrightarrow{\beta_2} & H^1(S/R, \mathrm{Pic}) \\ & \xrightarrow{\gamma_2} & \cdots & & & & \\ & \xrightarrow{\gamma_{n-1}} & H^n(S/R, \mathbb{G}_m) & \xrightarrow{\alpha_n} & H^{n-1}(S/R, \underline{\underline{\mathrm{Pic}}}) & \xrightarrow{\beta_n} & H^{n-1}(S/R, \mathrm{Pic}) \\ & \xrightarrow{\gamma_n} & \cdots & & & & \end{array} \qquad (6.12)$$

Proof The maps in the sequence are defined as follows.

$$\alpha_n: H^n(S/R, \mathbb{G}_m) \longrightarrow H^{n-1}(S/R, \underline{\underline{\mathrm{Pic}}})$$

is induced by the map κ_{n-1};

$$\beta_n: H^{n-1}(S/R, \underline{\underline{\mathrm{Pic}}}) \longrightarrow H^{n-1}(S/R, \mathrm{Pic})$$

is induced by the map θ_{n-1};

$$\gamma_n: H^{n-1}(S/R, \mathrm{Pic}) \longrightarrow H^{n+1}(S/R, \mathbb{G}_m)$$

is defined as follows: if $I \in \underline{\mathrm{Pic}}(S^{\otimes n})$ represents an element in $H^{n-1}(S/R, \mathrm{Pic})$, then $[I] \in \mathrm{Ker}\,(\Delta'_{n-1}) = \mathrm{Im}\,(\theta_{n-1})$. Take $x \in \mathrm{K}_0\underline{\Psi\delta}_{n-1}$ in the inverse image of $[I]$ under θ_{n-1} and define $\gamma_n([I]) = D_{n-1}(x)$.
Diagram chasing arguments show that the maps α_n, β_n and γ_n are well-defined and that the sequence (6.12) is exact. □

Theorem 6.1.4 (Villamayor-Zelinsky) *Let S be an E-covering of R. Then*

$$H^0(S/R, \underline{\underline{\mathrm{Pic}}}) \cong H^1(R_E, \mathbb{G}_m) \cong \mathrm{Pic}(R)$$

and

$$H^n(S/R, \underline{\underline{\mathrm{Pic}}}) \cong H^n(S/R, C_1)$$

for all $n \geq 1$.

Proof We define a map $\xi:\ \mathrm{Pic}(R) \to H^0(S/R, \underline{\underline{\mathrm{Pic}}})$ as follows:

$$\xi([I]) = [(I \otimes S, \alpha)]$$

where

$$\alpha:\ \delta_0(I) = (I \otimes S \otimes S) \otimes_{S^{\otimes 2}} (S \otimes I^* \otimes S) \longrightarrow S^{\otimes 2}$$

is given by

$$\alpha((x \otimes 1 \otimes 1) \otimes (1 \otimes x^* \otimes 1)) = \langle x^*, x \rangle$$

for all $x \in I$ and $x^* \in I^*$. It is easy to verify that ξ is a well-defined homomorphism. For $n \geq 1$, we will define a map

$$\xi_n:\ H^n(S/R, C_1) \longrightarrow H^n(S/R, \underline{\underline{\mathrm{Pic}}})$$

Let

$$0 \longrightarrow \mathbb{G}_m \longrightarrow X^0 \longrightarrow X^1 \longrightarrow \cdots$$

be an injective resolution of the sheaf $\mathbb{G}_m$. We then have an exact sequence of sheaves

$$0 \longrightarrow \mathbb{G}_m \longrightarrow X^0 \longrightarrow C_1 \longrightarrow 0$$

Take $c \in C_1(S^{\otimes n+1})$. There exists a covering T of $S^{\otimes n+1}$ such that the image of c in $C_1(T)$ lies in the image of the map $X^0(T) \to C_1(T)$. Let $x \in X^0(T)$ be an inverse image of the image of c in $C_1(T)$, and consider the commutative diagram.

$$\begin{array}{ccccccc}
0 & \longrightarrow & \mathbb{G}_m(T \otimes_{S^{\otimes n+1}} T) & \xrightarrow{\theta} & X^0(T \otimes_{S^{\otimes n+1}} T) & \xrightarrow{\theta} & C_1((T \otimes_{S^{\otimes n+1}} T) \\
 & & \uparrow & & \uparrow & & \uparrow \\
0 & \longrightarrow & \mathbb{G}_m(T) & \xrightarrow{\theta} & X^0(T) & \longrightarrow & C_1(T) \\
 & & \uparrow & & \uparrow & & \uparrow \\
0 & \longrightarrow & \mathbb{G}_m(S^{\otimes n+1}) & \longrightarrow & X^0(S^{\otimes n+1}) & \xrightarrow{\theta} & C_1(S^{\otimes n+1})
\end{array} \tag{6.13}$$

All the horizontal maps in the diagram are denoted by θ, all the vertical ones by Δ. A diagram chasing argument shows that $\Delta(x) \in \mathrm{Ker}\,(\theta)$, hence $\Delta(x) = \theta(u)$ for some $u \in \mathbb{G}_m(T \otimes_{S^{\otimes n+1}} T)$. Observing (6.13), we see that u is a cocycle in $Z^1(X/S^{\otimes n+1}, \mathbb{G}_m)$, defining $I \in \underline{\underline{\mathrm{Pic}}}(S^{\otimes n+1})$.
Now suppose that $c \in Z^1(S/R, C_1)$ is a cocycle. Write

$$T_i = T \otimes_{\varepsilon_i, S^{\otimes n+1}} S^{\otimes n+2}$$

and

$$\delta_n(T) = T_1 \otimes_{S^{\otimes n+1}} T_2 \otimes_{S^{\otimes n+1}} \cdots \otimes_{S^{\otimes n+1}} T_{n+1}$$

Then $\delta_n(T)$ is a covering of $S^{\otimes n+2}$, and we have a commutative diagram

$$\begin{array}{ccccccc}
0 \to & \mathbb{G}_m(\delta_n(T)^{\otimes_{S^{\otimes n+2}} 2}) & \xrightarrow{\theta} & X^0(\delta_n(T)^{\otimes_{S^{\otimes n+2}} 2}) & \xrightarrow{\theta} & C_1((\delta_n(T) \otimes_{S^{\otimes n+2}} \delta_n(T)) \\
& \uparrow & & \uparrow & & \uparrow \\
0 \to & \mathbb{G}_m(\delta_n(T)) & \xrightarrow{\theta} & X^0(\delta_n(T)) & \longrightarrow & C_1(\delta_n(T)) \\
& \uparrow & & \uparrow & & \uparrow \\
0 \to & \mathbb{G}_m(S^{\otimes n+2}) & \longrightarrow & X^0(S^{\otimes n+2}) & \xrightarrow{\theta} & C_1(S^{\otimes n+2})
\end{array}$$

and we have a map δ_n from the diagram (6.13) to the above diagram. If c is a cocycle, then $\theta(\delta_n(x)) = 1$ in $C_1(\delta_n(T))$. Thus $\delta_n(x) = \theta(v)$, with $v \in \mathbb{G}_m(\delta_n(x))$. We find that $\Delta(v) = \delta_n(u)$, meaning that $\delta_n(u)$ is a coboundary in $Z^1(\delta_n(T)/S^{\otimes n+2}, \mathbb{G}_m)$. This implies that $\delta_n(I)$ is isomorphic to $S^{\otimes n+2}$, and we find an isomorphism

$$\alpha: \delta_n(I) \longrightarrow S^{\otimes n+2}$$

Similar arguments show that $\delta_{n+1}(\alpha) = \lambda_I$, and we define

$$\xi([c]) = [(I, \alpha)]$$

We leave it to the reader to show that

$$\xi_n : H^n(S/R, C_1) \longrightarrow H^n(S/R, \underline{\underline{\mathrm{Pic}}})$$

is a well-defined homomorphism. The proof is complete if we can show that ξ and ξ_n are injective and surjective. There are two different ways to do this, they have in common that they are complicated if one wants to check all the necessary details. The first approach is to prove directly that ξ and ξ_n are isomorphisms. This involves descent arguments.

Applying Proposition 6.1.1, Theorem A.2.2 may be written as (take $q = 0$):

$$\begin{array}{ccccccc}
1 & \longrightarrow & H^1(S/R, \mathbb{G}_m) & \longrightarrow & H^1(R_E, \mathbb{G}_m) & \longrightarrow & H^0(S/R, \mathrm{Pic}) \\
& \longrightarrow & H^2(S/R, \mathbb{G}_m) & \longrightarrow & H^1(S/R, C_1) & \longrightarrow & H^1(S/R, \mathrm{Pic}) \\
& \longrightarrow & \cdots & & & & \\
& \longrightarrow & H^n(S/R, \mathbb{G}_m) & \longrightarrow & H^{n-1}(S/R, C_1) & \longrightarrow & H^{n-1}(S/R, \mathrm{Pic}) \\
& \longrightarrow & \cdots & & & &
\end{array} \tag{6.14}$$

If we show that the maps ξ and ξ_n define a map from the exact sequence (6.14) to (6.12), then we are done, by the Lemma of 5. □

6.2 The Brauer group versus the second cohomology group

Theorem 6.2.1 *Let S be a faithfully flat R-algebra. Then we have a natural monomorphism*

$$\mathrm{Br}'(S/R) \hookrightarrow H^1(S/R, \underline{\underline{\mathrm{Pic}}}).$$

If S is faithfully projective as an R-module, then

$$\mathrm{Br}(S/R) \cong \mathrm{Br}'(S/R) \cong H^1(S/R, \underline{\underline{\mathrm{Pic}}}).$$

Proof We proceed in several steps. Let us first recall the following notations : If M is an R-module, then we write $M_1 = S \otimes M$, $M_2 = M \otimes S$, $M_{11} = M_{12} = S \otimes S \otimes M$, $M_{22} = M_{23} = M \otimes S \otimes S$, etc. Similar notations are used for morphisms : if $f : M \to N$ in R-mod, then $f_1 = I_S \otimes f : M_1 \to N_1$, $f_2 = f \otimes I_S : M_2 \to N_2$, etc.

Step 1 Take $[A] \in \mathrm{Br}'(S/R)$. We then have an S-algebra isomorphism $\rho : A \otimes S = A_2 \to E_S(Q)$, for some $\underline{Q} \in \underline{\underline{DP}}(S)$. Let

$$\Phi = \rho_3 \circ \tau_3 \circ \rho_1^{-1} : E_{S^{\otimes 2}}(S \otimes Q) \to E_{S^{\otimes 2}}(Q \otimes S).$$

We have a commutative diagram

$$\begin{array}{ccc} A_{13} = S \otimes A \otimes S & \xrightarrow{\rho_1} & E_{S^{\otimes 2}}(S \otimes Q) \\ \downarrow{\scriptstyle \tau \otimes I} & & \downarrow{\scriptstyle \Phi} \\ A_{23} = A \otimes S \otimes S & \xrightarrow{\rho_3} & E_{S^{\otimes 2}}(Q \otimes S) \end{array} \tag{6.15}$$

From Proposition 1.3.1, it follows that $\Phi = E(f)$ for some $f : S \otimes \underline{Q} \otimes_{S^{\otimes 2}} \underline{I} \to \underline{Q} \otimes S$, with $I \in \underline{\underline{\mathrm{Pic}}}(S^{\otimes 2})$ and $\underline{I} = (I, I^*, \varphi)$. We now have the following isomorphism in $\underline{\underline{\mathrm{Pic}}}(S^{\otimes 2})$:

$$f_3 \circ \varphi_1^{-1} : Q_1 \xrightarrow{\cong} Q_1 \otimes_{S^{\otimes 2}} I \otimes_{S^{\otimes 2}} I^* \to Q_2 \otimes_{S^{\otimes 2}} I^*$$

which we still denote by f (in fact we identify - again - $I \otimes_{S^{\otimes 2}} I^*$ and $S^{\otimes 2}$ using φ). Then we have

$$f : Q_1 \to Q_2 \otimes_{S^{\otimes 2}} I^*.$$

Using this identification, we may consider

$$f_2^{-1} \circ (f_3 \otimes I) \circ (f_1 \otimes I \otimes I) : Q_{11} \otimes_{S^{\otimes 3}} I_1 \otimes_{S^{\otimes 3}} I_3 \otimes_{S^{\otimes 3}} I_2^* \to Q_{11}$$

This composition will be denoted also by $f_2^{-1} \circ f_3 \circ f_1$. Observe that

$$E(f_2^{-1} \circ f_3 \circ f_1) = E(f_2^{-1}) \circ E(f_3) \circ E(f_1) = \Phi_2^{-1} \circ \Phi_3 \circ \Phi_1 = I,$$

hence, from the uniqueness property in Proposition 1.3.1, it follows that we have an isomorphism in $\underline{\underline{\mathrm{Pic}}}(S^{\otimes 3})$:

$$\alpha : \delta_1(I) = I_1 \otimes_{S^{\otimes 3}} I_3 \otimes_{S^{\otimes 3}} I_2^{-1} \to S^{\otimes 3}$$

such that the diagram

$$\begin{array}{ccc} Q_{11} \otimes \delta_1(I) & \xrightarrow{f_2^{-1} f_3 f_1} & Q_{11} \\ \downarrow{\scriptstyle I \otimes \alpha} & & \downarrow{\scriptstyle I} \\ Q_{11} \otimes S^{\otimes 3} & \xrightarrow{\cong} & Q_{11} \end{array}$$

commutes. Hence $f_2^{-1} f_3 f_1 = I \otimes \alpha$.

Step 2 (I, α) represents an element of $Z^1(S/R, \underline{\mathrm{Pic}})$. Indeed, $f_2^{-1} \circ f_3 \circ f_1 = I \otimes \alpha$. Observe that

$$\begin{aligned} & I \otimes \delta_2(\alpha) \\ = \ & (f_2^{-1} \circ f_3 \circ f_1)_4^{-1} \circ (f_2^{-1} \circ f_3 \circ f_1)_3 \circ (f_2^{-1} \circ f_3 \circ f_1)_2^{-1} \circ (f_2^{-1} \circ f_3 \circ f_1)_1 : \\ & Q_{111} \otimes_{S^{\otimes 4}} \delta_2(\delta_1(I)) \to Q_{111} \end{aligned}$$

is the identity after we identify $\delta_2(\delta_1(I))$ with $S^{\otimes 4}$ using λ_I, hence $I \otimes \delta_2(\alpha) = I$, and $\delta_2(\alpha) = I$. Write $\theta(A,\rho) = [(I,\alpha)] \in H^1(S/R, \underline{\underline{\mathrm{Pic}}})$. Applying the uniqueness property in Proposition 1.3.1, we find that $[(I,\alpha)]$ is independent of the choice of f and I.

Step 3 $\theta(A,\rho) = [(I,\alpha)]$ is independent of the choice of ρ. Let $\mu: A \otimes S \to E_S(M)$ be another choice. Then $\mu\rho^{-1}: E_S(Q) \to E_S(M)$ can be written as $\mu\rho^{-1} = E_S(g)$, with $g: Q \otimes_S J \to M$, $J \in \underline{\underline{\mathrm{Pic}}}(S)$. Consider the commutative diagram

$$\begin{array}{ccccc} A_{13} & \xrightarrow{\rho_1} & E_2(Q_1) & \xrightarrow{(\mu\rho^{-1})_1} & E_2(M_1) \\ \downarrow{\scriptstyle \tau_3} & & \downarrow{\scriptstyle \Phi} & & \downarrow{\scriptstyle \Psi} \\ A_{23} & \xrightarrow{\rho_3} & E_2(Q_2) & \xrightarrow{(\mu\rho^{-1})_2} & E_2(M_2) \end{array}$$

Then $\Psi = E(g_2 \circ f \circ g_1^{-1}) = E(h)$ with

$$h = g_2 \circ f \circ g_1^{-1}: M_1 \otimes_{S^{\otimes 2}} J_1^* \otimes_{S^{\otimes 2}} I \otimes_{S^{\otimes 2}} J_2 = M_1 \otimes_{S^{\otimes 2}} I \otimes_{S^{\otimes 2}} \delta_0(J^*) \to M_2$$

Now

$$\begin{aligned} & h_2^{-1} \circ h_3 \circ h_1 \\ = \ & g_{12} \circ f_2^{-1} \circ g_{22}^{-1} \circ g_{23} \circ f_{33} \circ g_{13}^{-1} \circ g_{21} \circ f_1 \circ g_{11}^{-1} \\ = \ & g_{12} \circ f_2^{-1} \circ f_3 \circ f_1 \circ g_{11}^{-1} \\ = \ & (I \otimes \alpha)(g_{12} \circ g_{11}^{-1}) \\ = \ & (I \otimes \alpha) \end{aligned}$$

hence

$$\theta(A,\mu) = [(I \otimes_{S^{\otimes 2}} \delta_0(J^{-1}), \alpha)] = [(I, a)]$$

Step 4 If $A = E_R(P)$ for some $P \in \underline{\underline{DP}}(R)$, then $\theta(A,\rho) = [(S^{\otimes 2}, I)]$. This is obvious : take the natural map $\rho: E_R(P) \otimes S \to E_S(P \otimes S)$. Then $\Phi = E(I)$.

Step 5 $\theta(A \otimes B, \rho \otimes \mu) = \theta(A,\rho)\theta(B,\mu)$. Let Φ, f, Q, I, α be defined as above, and introduce in a similar way Ψ, g, M, J, β for B. Then

$$\Phi \otimes \Psi = E(f) \otimes E(g) = E(f \otimes g),$$

with

$$f \otimes g: (Q \otimes M)_1 \otimes_{S^{\otimes 2}} I \otimes_{S^{\otimes 2}} J \to (Q \otimes M)_2,$$

and

$$\begin{aligned} & (f \otimes g)_2^{-1} \circ (f \otimes g)_3 \circ (f \otimes g)_1 \\ = \ & f_2^{-1} \circ f_3 \circ f_1 \otimes g_2^{-1} \circ g_3 \circ g_1 \\ = \ & I \otimes \alpha \otimes \beta : \\ & Q_{11} \otimes_{S^{\otimes 3}} M_{11} \otimes_{S^{\otimes 3}} \delta_1(I) \otimes_{S^{\otimes 3}} \delta_1(J) \to Q_{11} \otimes M_{11} \end{aligned}$$

and Step 5 follows.

From Step 4 and Step 5, it follows that $\theta(A,\rho)$ is independent from the choice of A in $[A]$, so we have defined a well-defined homomorphism

$$i: \mathrm{Br}'(S/R) \to H^1(S/R, \underline{\underline{\mathrm{Pic}}})$$

given by

$$i([A]) = \theta(A, \rho).$$

Step 6 i is injective. Suppose $\theta(A,\rho) = (\delta_0(J), \lambda_J)$, with $J \in \underline{\underline{\mathrm{Pic}}}(S)$. Then $\Phi = E(f)$, with

$$f:\ Q_1 \otimes_{S^{\otimes 2}} J_1 \otimes_{S^{\otimes 2}} J_2^{-1} \to Q_2,$$

or

$$f:\ Q_1 \otimes_{S^{\otimes 2}} J_1 \to Q_2 \otimes_{S^{\otimes 2}} J_2$$

It is clear that f is a descent datum, so there exists $P \in \underline{\underline{DP}}(R)$ such that

$$\begin{array}{ccc} P_{13} & \longrightarrow & (Q\otimes J)_1 \\ \downarrow{\scriptstyle \tau_3} & & \downarrow{\scriptstyle f} \\ P_{23} & \longrightarrow & (Q\otimes J)_2 \end{array}$$

commutes. Applying E_R to this diagram, and invoking the uniqueness property in the Theorem of faithfully flat descent, we obtain that $A \cong E_R(P)$ as R-algebras.

Step 7 Suppose that S is faithfully projective as an R-module. Take $[(I,\alpha)] \in H^1(S/R, \underline{\underline{\mathrm{Pic}}})$, and consider the $S^{\otimes 3}$-module isomorphism

$$\beta = \alpha \otimes I_{I_2} :\ I_1 \otimes_{S^{\otimes 3}} I_3 \longrightarrow I_2$$

Let $P = I$ as an R-module, with S-action given by

$$s \cdot x = (s \otimes 1)x$$

for all $s \in S$ and $x \in I$. We will say that "$P = I$, with S acting on the first factor". With similar conventions, we can say that
$P_1 = I_1$, with $S^{\otimes 2}$ acting on the first two factors;
$P_2 = I_3$, with $S^{\otimes 2}$ acting on the first and the third factor;
$P_1 \otimes_{S^{\otimes 2}} I = I_1 \otimes_{S^{\otimes 3}} I_3$ with $S^{\otimes 2}$ acting on the first two factors.
We have an isomorphism of $S^{\otimes 2}$-modules

$$f = \tau_1 \circ \beta :\ P_1 \otimes_{S^{\otimes 2}} I \xrightarrow{\beta} I_2 \xrightarrow{\tau_1} P_2$$

We point out that f is $S^{\otimes 2}$-linear, but that the two factors β and τ_1 are not $S^{\otimes 2}$-linear. Consider the map

$$f_2^{-1} \circ f_3 \circ f_1 :\ P_{11} \otimes_{S^{\otimes 3}} I_1 \otimes_{S^{\otimes 3}} I_3 \longrightarrow P_{11} \otimes_{S^{\otimes 3}} I_2$$

Observe that

$$\begin{aligned} f_1 &= \tau_{12} \circ \beta_1 \\ f_3 &= \tau_{14} \circ \beta_4 = \tau_{14} \circ \tau_{12} \circ \beta_3 \tau_{12} \\ f_2 &= \tau_{14} \circ \tau_{12} \circ \beta_2 \end{aligned}$$

and

$$\begin{aligned} f_2^{-1} \circ f_3 \circ f_1 &= \beta_2^{-1} \circ \tau_{12} \circ \tau_{14} \circ \tau_{14} \circ \tau_{12} \circ \beta_3 \circ \tau_{12} \circ \tau_{12} \circ \beta_1 \\ &= \beta_2^{-1} \circ \beta_3 \circ \beta_1 \end{aligned}$$

Now $\beta_2^{-1} \circ \beta_3 \circ \beta_1$, or, more precisely, $\beta_2^{-1} \circ \beta_3 \circ (\beta_1 \otimes I_{I_33}) = I_{I_11} \circ \beta_4$, because $\delta_1(\alpha) = \lambda_I$. This means that $f_2^{-1} \circ f_3 \circ f_1$ induces the identity of $\mathrm{End}_{S^{\otimes 3}}(P_{11})$. Therefore

$$\mathrm{End}(f): \ \mathrm{End}_{S^{\otimes 2}}(P_1) \longrightarrow \mathrm{End}_{S^{\otimes 2}}(P_2)$$

is a descent datum. The descended algebra A is an Azumaya algebra, and it is not difficult to show that $i([A]) = [(I, \alpha)]$. □

Corollary 6.2.2 *We have natural embeddings*

$$\mathrm{Br}(R) \hookrightarrow \mathrm{Br}'(R) \hookrightarrow H^2(R_{\text{ét}}, \mathbb{G}_m) \hookrightarrow H^2(R_{\text{fl}}, \mathbb{G}_m)$$

Proof From the previous Theorem, Theorem 6.1.4 and (5.11) (in the case $q = 1$) it follows that

$$\mathrm{Br}'(S/R) \hookrightarrow H^1(S/R, C_1) \cong H^1(S/R, \underline{\underline{\mathrm{Pic}}}) \hookrightarrow H^1(R_{\text{ét}}, C_1) \cong H^2(R_{\text{ét}}, \mathbb{G}_m)$$

for any faithfully flat R-algebra S. From Theorem 3.3.5, it follows that every Taylor-Azumaya algebra A can be split by an étale covering S of R. Therefore

$$\mathrm{Br}'(R) = \cup_S \mathrm{Br}'(S/R) \hookrightarrow \cup_S H^1(S/R, C_1) \cong H^2(R_{\text{ét}}, \mathbb{G}_m)$$

and this proves the second embedding. The third one has already been established in Corollary 5.6.10. □

Corollary 6.2.3 (Chase-Rosenberg exact sequence) *Suppose that S is a faithfully projective R-algebra. Then we have a long exact sequence*

$$1 \to H^1(S/R, \mathbb{G}_m) \to \mathrm{Pic}(R) \to H^0(S/R, \mathrm{Pic}) \to H^2(S/R, \mathbb{G}_m)$$
$$\to \mathrm{Br}(S/R) \to H^1(S/R, \mathrm{Pic}) \to H^3(S/R, \mathbb{G}_m) \tag{6.16}$$

Proof Take the long exact sequence (6.12) and apply 6.1.1 and 6.2.1. □

The above sequence was first obtained by Chase and Rosenberg in [50]. An elementary proof, with an explicit description of all the connecting maps, has been given by Knus in [113].

Suppose that S is a Galois extension of R with group G. We have seen (Lemma 5.2.1) that

$$H^n(S/R, \mathbb{G}_m) \cong H^n(G, \mathbb{G}_m(S)) \quad \text{and} \quad H^n(S/R, \mathrm{Pic}) \cong H^n(G, \mathrm{Pic}(S))$$

In this case, the exact sequence (6.16) takes the form

$$1 \to H^1(G, \mathbb{G}_m(S)) \to \mathrm{Pic}(R) \to \mathrm{Pic}(S)^G \to H^2(G, \mathbb{G}_m(S))$$
$$\to \mathrm{Br}(S/R) \to H^1(G, \mathrm{Pic}(S)) \to H^3(G, \mathbb{G}_m(S)) \tag{6.17}$$

and is called the Chase-Harrison-Rosenberg sequence. A detailed prove of the exactness of the sequence in this case may be found in [75]. In particular, the map $H^2(G, \mathbb{G}_m(S)) \to \mathrm{Br}(S/R)$ may be described as follows: the class in $\mathrm{Br}(S/R)$ corresponding to a cocycle $f: G \times G \to \mathbb{G}_m(S)$ is represented by the algebra

$$A = (S, G, f) = \oplus_{\sigma \in G} S u_\sigma \tag{6.18}$$

with

$$(s u_\sigma)(t u_\tau) = f(\sigma, \tau) s(\sigma \cdot t) u_{\sigma\tau}$$

As a special case of (6.17), we obtain *Hilbert's Theorem 90* and the *Crossed Product Theorem.*

Corollary 6.2.4 (Hilbert 90) *Suppose that S is a Galois extension of R with group G and that* $\mathrm{Pic}(S) = 1$. *Then*

$$\mathrm{Br}(S/R) \cong H^2(G, \mathbb{G}_m(S))$$

Corollary 6.2.5 (Crossed Product Theorem) *Suppose that S is a Galois extension of R with group G and that* $\mathrm{Pic}(R) = 1$. *Then* $H^2(G, \mathbb{G}_m(S)) = 1$

In the special situation where S is a cyclic Galois extension of R, we have an easy formula that allows us to compute $H^2(G, \mathbb{G}_m(S))$. First we need the following Lemma.

Lemma 6.2.6 *Let S be a Galois extension of R with group G, and consider two cocycles $f, g \in Z^2(G, \mathbb{G}_m(S))$. Then f and g are cohomologous if and only if there exists an isomorphism φ of S-modules and R-algebras*

$$\varphi:\ (S, G, f) = \oplus_{\sigma \in G} S u_\sigma \longrightarrow (S, G, g) = \oplus_{\sigma \in G} S v_\sigma$$

such that $v_\sigma = h(\sigma)\varphi(u_\sigma)$ with $h(\sigma) \in \mathbb{G}_m(S)$. In this case $g = (\Delta_1 h) f$.

Proof Suppose first that $g = (\Delta_1 h) f$, that is,

$$g(\sigma, \tau) = (\sigma \cdot h(\tau)) h(\sigma\tau)^{-1} h(\sigma) f(\sigma, \tau)$$

for all $\sigma, \tau \in G$. Define

$$\varphi:\ (S, G, f) = \oplus_{\sigma \in G} S u_\sigma \longrightarrow (S, G, g) = \oplus_{\sigma \in G} S v_\sigma$$

by $\varphi(u_\sigma) = h(\sigma)^{-1} v_\sigma$. It is clear that φ is an S-module isomorphism. φ is also multiplicative since

$$\begin{aligned}
\varphi(u_\sigma)\varphi(u_\tau) &= h(\sigma)^{-1} v_\sigma h(\tau)^{-1} v_\tau \\
&= h(\sigma)^{-1} \left(\sigma \cdot h(\tau)^{-1}\right) g(\sigma, \tau) v_{\sigma\tau} \\
&= h(\sigma\tau)^{-1} f(\sigma, \tau) v_{\sigma\tau} \\
&= f(\sigma, \tau) \varphi(u_{\sigma\tau}) \\
&= \varphi(u_\sigma u_\tau)
\end{aligned}$$

Conversely, suppose that the automorphism φ is given. Then

$$\begin{aligned} g(\sigma,\tau)v_{\sigma\tau} &= v_\sigma v_\tau \\ &= \varphi\left(h(\sigma)u_\sigma h(\tau)u_\tau\right) \\ &= \varphi\left(h(\sigma)(\sigma\cdot h(\tau))f(\sigma,\tau)u_{\sigma\tau}\right) \\ &= h(\sigma)(\sigma\cdot h(\tau))f(\sigma,\tau)\varphi(u_{\sigma\tau}) \\ &= h(\sigma)(\sigma\cdot h(\tau))f(\sigma,\tau)h(\sigma\tau)^{-1}v_{\sigma\tau} \\ &= (\Delta_1 h)(\sigma,\tau)f(\sigma,\tau)u_{\sigma\tau} \end{aligned}$$

and it follows that

$$g = (\Delta_1 h)f$$

□

If S is a Galois extension of R with group G, then the norm map $N_{S/R}:\ S\to R$ is given by the formula

$$N_{S/R}(x) = \prod_{\sigma\in G}\sigma\cdot x \tag{6.19}$$

$N_{S/R}(x)$ is also given by the determinant of the endomorphism of S given by left multiplication by x. We refer to Section 8.5 for details. Using $N_{S/R}$, we can describe $Z^1(G,\mathbb{G}_m(S))$ and $H^2(G,\mathbb{G}_m(S))$ explicitly.

Proposition 6.2.7 *Let S be a Galois extension of R with cyclic group $G=\{1,\sigma,\ldots,\sigma^{n-1}\}$. Then*

$$Z^1(G,\mathbb{G}_m(S)) \cong \mathrm{Ker}\,(N_{S/R}) \tag{6.20}$$

Assume that $\mathrm{Pic}(R)=1$. *For all $a\in\mathrm{Ker}\,(N_{S/R})$, there exists $b\in\mathbb{G}_m(S)$ such that $a=\sigma(b)b^{-1}$.*

Proof Take $f\in Z^1(G,\mathbb{G}_m(S))$, and write $f(\sigma)=a$. The cocycle relation implies that

$$f(\sigma^k) = a(\sigma\cdot a)\cdots(\sigma^{k-1}\cdot a)$$

for all $k\geq 1$. Thus f is completely determined by a, and, in particular,

$$f(1) = f(\sigma^n) = N_{S/R}(a)\in R$$

and

$$f(\sigma) = f(\sigma^{n+1}) = a(\sigma\cdot N_{S/R}(a)) = aN_{S/R}(a)$$

and $N_{S/R}(a)=1$. So $a\in\mathrm{Ker}\,(N_{S/R}$, and we obtain a one to one correspondence between the cocycles in $Z^1(G,\mathbb{G}_m(S))$ and $\mathrm{Ker}\,(N_{S/R}$.
The second part follows immediately from Corollary 6.2.4. It is actually the original formulation of Hilbert 90. □

Proposition 6.2.8 *Let S be a Galois extension of R with cyclic group $G=\{1,\sigma,\ldots,\sigma^{n-1}\}$. Then*

$$H^2(G,\mathbb{G}_m(S)) \cong \mathbb{G}_m(R)/N_{S/R}(\mathbb{G}_m(S)) \tag{6.21}$$

Proof For $a \in \mathbb{G}_m(R)$, we define a cocycle

$$f_a : G \times G : \mathbb{G}_m(S) : (\sigma^i, \sigma^j) \mapsto \begin{cases} 1 & \text{if } i+j < n \\ a & \text{if } i+j \geq n \end{cases}$$

with $0 \leq i+j \leq n-1$. It is readily verified that f_a is a cocycle. Define $\alpha : \mathbb{G}_m(R) \to H^2(G, \mathbb{G}_m(S))$ by $\alpha(a) = [f_a]$, for all $a \in \mathbb{G}_m(R)$.
We claim that α is surjective. Take an arbitrary (normalized) cocycle $f \in Z^2(G, \mathbb{G}_m(S))$. Then $[f] = [f_a]$, with

$$a = \prod_{m=1}^{n-1} f(\sigma^m, \sigma)$$

It is clear that $a \in R$, since a is invariant under the G-action. Now write $(S, G, f) = \oplus_{i=0}^{n-1} Sv_{\sigma^i}$. Then

$$\begin{aligned} v_\sigma^2 &= f(\sigma, \sigma) v_{\sigma^2} \\ v_\sigma^3 &= f(\sigma,\sigma) f(\sigma^2, \sigma) v_{\sigma^3} \\ &\vdots \\ v_\sigma^{n-1} &= \prod_{m=1}^{n-2} f(\sigma^m, \sigma) v_{\sigma^{n-1}} \\ v_\sigma^n &= \prod_{m=1}^{n-1} f(\sigma^m, \sigma) v_{\sigma^n} = a \end{aligned}$$

and define $\varphi : (S, G, f_a) \to (S, G, f)$ by $\varphi(u_{\sigma_i}) = v_\sigma^i$ for $i = 0, \ldots, n-1$. It is clear that φ is an isomorphism of S-modules and R-algebra, and f and f_a are cohomologous by Lemma 6.2.6.
The proof will be finished if we can show that Ker $(\alpha) = N_{S/R}(S)$. Take $a = N_{S/R}(b)$, with $b \in S$, and write

$$(S, G, 1) = \bigoplus_{i=0}^{n-1} Su_\sigma$$

Consider

$$\varphi : (S, G, f_a) \longrightarrow (S, G, 1) : u_{\sigma^i} \mapsto (bv_\sigma)^i$$

for $i = 0, 1, \ldots, n-1$. Then

$$\begin{aligned} \varphi(u_\sigma)^n &= (bv_\sigma)^n \\ &= b(\sigma \cdot b)(\sigma^2 \cdot b) \cdot (\sigma^{n-1} \cdot b) v_\sigma^n \\ &= N_{S/R}(b) = a = \varphi(u_\sigma^n) \end{aligned}$$

and it follows that φ is an S-module R-algebra isomorphism, and $a \in$ Ker (α), by Lemma 6.2.6.
Conversely, suppose that $a \in$ Ker (α). Then $f_a = \Delta_1 h$ for some $h : G \to \mathbb{G}_m(S)$, and we can apply Lemma 6.2.6. We obtain an isomorphism $\varphi : (S, G, f_a) \to (S, G, 1)$. Now write $b = h(\sigma)^{-1}$. Then

$$a = \varphi(u_\sigma^n) = \varphi(u_\sigma)^n = (bv_\sigma)^n = N_{S/R}(b)$$

and $a \in N_{S/R}(S)$. □

If k is a field, then it is very well known that every Azumaya algebra can be split by a Galois extension, and it then follows from 6.2.5 that

$$\mathrm{Br}(k) = \bigcup_{\ell/k \text{ Galois}} H^2(\mathrm{Gal}(\ell/k), \mathbb{G}_m(\ell))$$

We can now extend this property to connected semilocal rings:

Corollary 6.2.9 *Let R be a connected semilocal ring. Then*

$$\mathrm{Br}(R) = \bigcup_{S/R \text{ Galois}} H^2(\mathrm{Gal}(S/R), \mathbb{G}_m(S))$$

Proof The property follows immediately from the fact that every Azumaya algebra over a connected semilocal ring can be split by a Galois extension, cf. Theorem 3.5.5, from corollary 6.2.5 and the fact that the Picard group of a semilocal ring is trivial. □

Remarks 6.2.10 Other approaches to the definition of the injection $\mathrm{Br}(R) \to H^2(R_E, \mathbb{G}_m)$ are possible ; the most elementary approach may be found in [114, Ch. V] ; the disadvantage of this approach is that it makes use of Artin's Refinement Theorem, and this is an obstruction for generalizing the map i to other situations (for example to the Brauer group of a scheme, or to the relative Brauer group of Van Oystaeyen and Verschoren, see further). Another possibility is to apply hypercohomology, as introduced in [9]. This construction was applied by the author and F. Van Oystaeyen in order to describe the $\mathbb{Z}$-graded Brauer group (cf. [45]). We will discuss this construction briefly in a later Section. Finally, it is also possible to use Giraud's non-abelian cohomology (cf. [91]). A sketch of this approach is given by Milne in [135, Chapter IV].

6.3 Taylor's theorem

In Section 6.2, we have seen that Taylor's Brauer group $\mathrm{Br}'(R)$ embeds into the second étale cohomology group $H^2(R_{\acute{e}t}, \mathbb{G}_m)$. We will now prove that $\mathrm{Br}'(R)$ covers the full étale cohomology group. In Section 6.7, we will prove that the classical Brauer group $\mathrm{Br}(R)$ is equal to the torsion part of $H^2(R_{\acute{e}t}, \mathbb{G}_m)$.
We need some preliminary results about étale algebras. Let R be a noetherian ring, and let I and J be ideals of R. Recall that

$$V_R(I) = V(I) = \{p \in \mathrm{Spec}(R) | I \subset p\}$$

$$D_R(I) = D(I) = \mathrm{Spec}(R) \setminus V(I)$$

The $V(I)$ and $D(I)$ are respectively the closed and open subsets of $\mathrm{Spec}(R)$ for the Zariski topology. Recall also that

$$D(IJ) = D(I) \cap D(J)$$

and

$$V(I) \subset V(J) \iff D(J) \subset D(I) \iff \sqrt{J} \subset \sqrt{I}$$

For a principal ideal $I = (f)$ of R, we have that

$$D(f) = \{p \in \mathrm{Spec}(R) | f \notin p\}$$

and the $D(f)$ form a basis for the open subsets in the Zariski topology.
A ring homomorphism $\varphi: R \to S$ is called an *open immersion* if the induced map

$$\mathrm{Spec}(S) \longrightarrow \mathrm{Spec}(R): \; q \mapsto \varphi^{-1}(q)$$

embeds $\mathrm{Spec}(S)$ as an open subset of $\mathrm{Spec}(R)$ and if, for each $q \in \mathrm{Spec}(S)$, the canonical map $R_{\varphi^{-1}(q)} \to S_q$ is an isomorphism. A typical example of an open immersion is a localization $R \to R_f$ at an element f of R.
If $R \to S$ is an open immersion, and I is an ideal of R such that $D(I) = \mathrm{Spec}(S)$, the the kernel K of the map $R \to S$ is given by the formula

$$K = \mathrm{Ker}\,(R \to S) = \{x \in R | xI^n = 0 \text{ for some integer } n\}$$

If R is noetherian, then we can choose I in such a way that $I \subset S$. Indeed, K is finitely generated as an R-module, so we can find an integer n such that $KI^n = 0$. From the Artin-Rees Lemma ([27, III.3.1, Corollary 2]), it follows that

$$K \cap I^{n+m} \subset KI^n = 0$$

for some integer m, and I^{n+m} injects into S. Observe now that $D(I^{n+m}) = D(I) = \mathrm{Spec}(S)$, and I can be replaced by I^{n+m}.
Suppose now that S is a commutative R-algebra, not necessarily having a unit. If is faithfully projective as an R-module, then the map

$$\tau: \; S \longrightarrow \mathrm{End}_R(S) \cong S \otimes S^* \longrightarrow R$$

is called the *standard trace* on S. Here the map $S \to \mathrm{End}_R(S)$ is given by left multiplication by elements of S, while $S \otimes S^* \to R$ is the canonical pairing. In the sequel, our problem will be that we will work with étale coverings S of R, and they are not necessarily faithfully projective as R-modules. To overcome this difficulty, we proceed as follows.
A commutative R-algebra S is called *quasi-finite* if for every prime p of R the $k(p)$-algebra $S \otimes k(p)$ is finite dimensional as a $k(p)$-vector space. $k(p) = R_p/pR_p$ is the residue class field at the prime ideal p. If S has a unit, then S is quasi-finite if and only if the inverse image in $\mathrm{Spec}(S)$ of every $p \in \mathrm{Spec}(R)$ is finite. An étale R-algebra S is always quasi-finite.
If S is a quasi-finite R-algebra, then a map $\tau: \; S \to R$ is called a *trace map* if the induced map

$$\tau \otimes I_{k(p)}: \; S \otimes k(p) \longrightarrow k(p)$$

equals the standard trace map on $S \otimes k(p)$, for every prime ideal p of R.

Lemma 6.3.1 *Let S be a quasi-finite R-algebra and suppose that R and S are noetherian rings with unit. Then every S-ideal I is quasi-finite as an R-algebra.*

Proof I is finitely generated as an S-module, because S is noetherian. Therefore $I \otimes k(p)$ is finitely generated as an $S \otimes k(p)$-module, and also as a $k(p)$-module, since $S \otimes k(p)$ is finitely generated as a $k(p)$-module. □

Lemma 6.3.2 *Let R and S be as in Lemma 6.3.1. If $J \subset I$ are ideals of S, and if $\tau : I \to R$ is a trace map, then $\tau_{|J} : J \to R$ is a trace map.*

Proof From Lemma 6.3.1, we know that I and J are quasi-finite as R-algebras. We have an exact sequence of R-modules

$$0 \longrightarrow J \xrightarrow{i} I \longrightarrow I/J \longrightarrow 0$$

and therefore an exact sequence of $k(p)$-vector spaces

$$0 \longrightarrow K \longrightarrow J \otimes k(p) \xrightarrow{i} I \otimes k(p) \longrightarrow I/J \otimes k(p) \longrightarrow 0$$

Take $x \in J$. x acts on I and J by left multiplication, and the induced action on $I/J \otimes k(p)$ clearly disappears. The induced action on $J \otimes k(p)$ factorizes as

$$J \otimes k(p) \quad \overset{i \otimes I_{k(p)}}{\longrightarrow} \quad I \otimes k(p) \quad \overset{x \otimes I_{k(p)}}{\longrightarrow} \quad J \otimes k(p)$$

and this implies that the induced action on K also disappears. Write $f = i \otimes I_{k(p)}$, and choose bases for the $k(p)$-vector spaces as follows:

$\{v_1, \ldots, v_n\}$ for K
$\{w_1, \ldots, w_m, v_1, \ldots, v_n\}$ for $J \otimes k(p)$
$\{f(w_1), \ldots, f(w_m), x_1, \ldots, x_k\}$ for $I \otimes k(p)$
$\{\overline{x}_1, \ldots, \overline{x}_k\}$ for $I/J \otimes k(p)$

Since $x \otimes 1$ acts trivially on K, the matrix representing the left multiplication by $x \otimes 1 \in J \otimes k(p)$ is then of the following form:

$$\begin{pmatrix} M_{m \times m} & 0_{m \times n} \\ 0_{n \times m} & 0_{n \times n} \end{pmatrix}$$

$f(x \otimes 1)$ acts trivially on $I/J \otimes k(p)$, so the matrix representing the left multiplication by $f(x \otimes 1) \in I \otimes k(p)$ is

$$\begin{pmatrix} M_{m \times m} & 0_{m \times k} \\ 0_{k \times k} & 0_{k \times k} \end{pmatrix}$$

Both matrices have the same trace. Since the trace of the second one equals the image of $\tau(x)$ in $k(p)$, so does the trace of the first one. □

Definition 6.3.3 *Let $f : R \to S$ be a commutative R-algebra, with R and S noetherian. We say that the R-algebra S is of trace type if f factorizes as*

$$R \xrightarrow{g} T \xrightarrow{i} S$$

with T finitely generated as an R-module and $i : T \to S$ an open immersion, with the additional condition that we have an ideal I of T such that $D(I) = \mathrm{Spec}(S)$ and there exists a trace map $\tau : I \to R$.

Observe that I is finitely generated as an R-module (R is noetherian) and therefore quasi-finite as an R-algebra.

Our next aim is to show that an étale covering may always be refined to an étale covering of trace type. This follows basically from Zariski's Main Theorem (see the proof of Lemma 6.3.5).

Lemma 6.3.4 *Suppose that R and S are noetherian rings. If $R \to S$ is of trace type, and $S \to U$ is an open immersion, then $R \to U$ is of trace type.*

Proof Let I, T and τ be as in Definition 6.3.3. The composition $T \to S \to U$ is an open immersion. Let J be an ideal of T such that $D(J) = \text{Spec}(U)$. Then

$$D(IJ) = D(I) \cap D(J) = \text{Spec}(S) \cap \text{Spec}(U) = \text{Spec}(U)$$

and $\tau_{|IJ} : |IJ \to R$ is a trace, by Lemma 6.3.2. □

Lemma 6.3.5 *Suppose that R and S are noetherian rings, and let $\alpha \in R$. If $R_\alpha \to S$ is of trace type, then $R \to S$ is of trace type.*

Proof According to Definition 6.3.3, we have a factorization

$$R_\alpha \xrightarrow{g} T \xrightarrow{i} S$$

with g finite, and i an open immersion. We also have an ideal I of T such that $D(I) = \text{Spec}(S)$ and $\tau : I \to U$ is a trace map.
Now T is finitely generated and quasi-finite as an R-algebra. By Zariski's Main Theorem (cf. e.g. [162, p. 42, Corollaire 2]), $R \to T$ may be factored as

$$R \xrightarrow{h} A \xrightarrow{j} T$$

with A finitely generated as an R-module, and $A \to T$ an open immersion. Now look at the following diagram

$$\begin{array}{ccccc} & & R_\alpha & & & \\ & \nearrow & & \searrow^{g} & & \\ R & & & & T & \xrightarrow{i} & S \\ & \searrow_{h} & & {}_{j}\nearrow & & \\ & & A & & & \end{array}$$

Observe that $D_T(I) \subset D_A(j^{-1}(I))$. Take an ideal J' of A such that $D_A(J') = D_T(I)$, and let $J = J'j^{-1}(I)$. then

$$D_A(J) = D_A(J') \cap D_A(j^{-1}(I)) = D_T(I) = \text{Spec}(S)$$

and

$$j(J) \subset I$$

We have the following situation

$$\begin{array}{ccccccc} & & & & J & \xrightarrow{j} & I \\ & & & & & & \downarrow{\scriptstyle\tau} \\ 0 & \longrightarrow & K & \longrightarrow & R & \longrightarrow & R_\alpha \end{array}$$

and we want to define a map $\tau_1 : J \to R$. We have to get rid of the kernel and the cokernel of the open immersion $R \to R_\alpha$. The kernel K is given by the formula

$$K = \{x \in R | x\alpha^n = 0 \text{ for some integer } n\}$$

K is finitely generated as an R-module, since R is noetherian. Consequently, there exists an integer n_1 such that $K\alpha^{n_1} = 0$. By the Artin-Rees Lemma ([27, III.3.1, Corollary 2]), there exists an integer n_2 such that

$$(\alpha^{n_1})^{n_2} \cap K \subset \alpha^{n_1} K = 0$$

and we can conclude that

$$\alpha^{n_1+n_2}(R/K) = \alpha^{n_1+n_2} R/K \cong \alpha^{n_1+n_2} R$$

and we may regard $\alpha^{n_1+n_2}R$ as a submodule of U.
Now J is finitely generated as an A-module, since R (and A) are noetherian. Therefore $(\tau \circ j)(J) \subset U$ is finitely generated as an R-module, and as an R/K-module. Thus there exists an integer n_3 such that

$$(\tau \circ j)(\alpha^{n_3} J) = \alpha^{n_3}(\tau \circ j)(J) \subset R/K$$

Now put $m = n_1 + n_2 + n_3$. Then

$$(\tau \circ j)(\alpha^m J) \subset \alpha^{n_1+n_2} R/K \cong \alpha^{n_1+n_2} R \subset R$$

and we have a map

$$\tau_1 = \tau \circ j : \alpha^m J \longrightarrow R$$

Furthermore $D_A(\alpha^m J) = D_T(\alpha^m I) = D_T(I) = \mathrm{Spec}(S)$ since α is invertible in T.
Let us finally show that $\tau_1 : \alpha^m J \longrightarrow R$ is a trace. Take a prime ideal p of R.
If $p \in D(\alpha)$, that is, $\alpha \notin p$, then α is invertible in R_p and $k(p)$ and

$$\alpha^m J \otimes k(p) \cong J \otimes k(p) \cong I \otimes k(p) \cong \alpha^m I \otimes k(p)$$

Thus $\tau_1 \otimes I_{k(p)} \cong \tau \otimes I_{k(p)}$ equals the standard trace map.
If $p \in V(\alpha)$, that is, $\alpha \in p$, then $R_\alpha \otimes k(p) = 0$, and $\tau_1 \otimes I_{k(p)} = 0$. Moreover $\alpha^m J \otimes R_p/pR_p = 0$ once that $m > 0$. So if we make sure that $m > 0$, then the standard trace is also zero. □

Theorem 6.3.6 (Taylor's Refinement Theorem) *Let R be noetherian ring, and $R \to S$ an étale covering. Then there exists an étale covering $S \to S'$ such that $R \to S'$ is of trace type.*

Proof Any étale R-algebra S is locally of standard type (see [162, p. 51, Théorème 1]). This means that for all $q \in \mathrm{Spec}(S)$, there exist $f \in S \setminus q$ and $h \in R \setminus (R \cap q)$ such that

$$S_f \cong (R_h[X]/P(X))_g$$

with P a monic polynomial such that P' is invertible in $(R_h[X]/P(X))_g$. Now choose $q_1, \dots q_n \in \mathrm{Spec}(S)$ such that the corresponding $f_1, \dots, f_n \in S$ are such that $\mathrm{Spec}(S) = D(f_1) \cup \dots \cup D(f_n)$. Then $S \to S' = S_{f_1} \times \dots \times S_{f_n}$ is an étale covering, and from the two previous Lemmas, it follows that each $R \to S_{f_i}$ is of trace type, and therefore $R \to S'$ is of trace type. □

Lemma 6.3.7 *Let $R \to S$ be an étale covering of trace type, and let I, T and τ be as in Definition 6.3.3. Then $\tau : I \to R$ is surjective.*

Proof It suffices to show that the map $\tau_p : I \otimes R_p \to R_p$ is surjective, for all $p \in \mathrm{Spec}(R)$. If τ_p is not surjective, then $\mathrm{Im}(\tau_p) \subset pR_p$ and $\tau \otimes I_{k(p)} : I \otimes k(p) \to k(p)$ is the zero map. Let $q_1, \dots q_n$ be the prime ideals of S lying above p. There are only finitely many such ideals since S is étale and therefore quasi-finite over R, and there is at least one such ideal since S/R is faithfully flat. Then

$$S \otimes k(p) = \bigoplus_{i=1}^{n} S_q \otimes_{R_p} k(p)$$

and $S_q \otimes_{R_p} k(p)$ is a finite separable field extension of $k(p)$. For a finite separable field extension l/k, we know that the trace map $l \to k$ is different from 0. Therefore the standard trace $S_q \otimes_{R_p} k(p) \to k(p)$ is not zero, and $\tau \otimes I_{k(p)} : I \otimes k(p) \to k(p)$ is not zero since it has to coincide with the standard trace. □

Theorem 6.3.8 (Taylor) *Let R be a commutative ring. Then*

$$\mathrm{Br}'(R) \cong H^2(R_{\acute{e}t}, \mathbb{G}_m)$$

Proof Suppose first that R is a noetherian ring. Using Artin's Refinement Theorem 5.6.8, we may write (Theorem 5.6.7):

$$H^2(R_{\acute{e}t}, \mathbb{G}_m) = \varinjlim H^2(S/R, \mathbb{G}_m)$$

where the limit runs over all étale coverings S of R. In view of Taylor's Refinement Theorem 6.3.6, we may restrict attention to étale coverings of trace type. So take an étale covering $R \to S$ of trace type, and consider a cocycle

$$u \in Z^2(S/R, \mathbb{G}_m) \subset \mathbb{G}_m(S^{\otimes 3})$$

representing a class $[u] \in H^2(R_{\acute{e}t}, \mathbb{G}_m)$. According to Definition 6.3.3, we have a factorization

$$R \xrightarrow{g} T \xrightarrow{i} S$$

with T finitely generated as a g-module and i an open immersion. We have an ideal I of T such that $D(I) = \mathrm{Spec}(S)$ and a trace map $\tau : I \to R$. As it has been explained at the beginning of this Section, we can choose I in such a way that $I \subset S$. Now consider the maps

$$\left\{\begin{array}{l} p_1 \\ p_2 \end{array}\right. : T\otimes S \longrightarrow T\otimes S\otimes S : t\otimes s \mapsto \left\{\begin{array}{l} t\otimes 1\otimes s \\ t\otimes s\otimes 1 \end{array}\right.$$

Let $S\otimes \underline{S}\otimes S$ be $S^{\otimes 3}$ considered as a $T\otimes S$-module, with T and S acting respectively on the first and third factor. Then $I\otimes \underline{S}\otimes S$ is a $T\otimes S$-submodule of $S\otimes \underline{S}\otimes S$, and we can consider the $T\otimes S$-modules

$$\begin{aligned} M &= (I\otimes \underline{R}\otimes S)\cap u^{-1}(I\otimes \underline{S}\otimes S) \\ M' &= (I\otimes \underline{R}\otimes S)\cap u(I\otimes \underline{S}\otimes S) \end{aligned}$$

and let

$$\begin{aligned} M_i &= M \otimes_{T\otimes S,p_i} (T\otimes S\otimes S) \\ M_i' &= M' \otimes_{T\otimes S,p_i} (T\otimes S\otimes S) \end{aligned}$$

for $i = 1,2$. With this notation, we mean that $T\otimes S\otimes S$ is considered as a $T\otimes S$-module via p_i. M_1 and M_2 are $T\otimes S\otimes S$-submodules of $S\otimes S\otimes S$, and we claim that

$$u_2M_1 = M_2 \ \text{ and } \ u_2M_2' = M_1' \tag{6.22}$$

We write

$$u = \sum u^1\otimes u^2\otimes u^3 \in S^{\otimes 3} \ \text{ and } \ u^{-1} = \sum v^1\otimes v^2\otimes v^3 \in S^{\otimes 3}$$

Recall the notations

$$u_1 = \sum 1\otimes u^1\otimes u^2\otimes u^3,\ u_2 = \sum u^1\otimes 1\otimes u^2\otimes u^3 \in S^{\otimes 4},\ldots$$

We have that

$$\begin{aligned} M_1 &= (I\otimes \underline{R}\otimes S\otimes S)\cap u_3^{-1}(I\otimes \underline{S}\otimes S\otimes S) \\ M_2 &= (I\otimes \underline{R}\otimes S\otimes S)\cap u_4^{-1}(I\otimes \underline{S}\otimes S\otimes S) \end{aligned}$$

Now take

$$x = \sum_i v^1x_i\otimes v^2s_i\otimes s_i'\otimes v^3s_i'' \in M_1$$

with $x_i \in I$ and $s_i', s_i', s_i'' \in S$. From the fact that $x\in I\otimes \underline{R}\otimes S\otimes S$ and Corollary B.0.8, it follows that

$$x = \sum_i v^1x_i\otimes 1\otimes v^2s_is_i'\otimes v^3s_i''$$

and

$$\begin{aligned} u_2x &= \sum_i u^1v^1x_i\otimes 1\otimes u^2v^2s_is_i'\otimes u^3v^3s_i'' \\ &= \sum_i x_i\otimes 1\otimes s_is_i'\otimes s_i'' \in (I\otimes \underline{R}\otimes S\otimes S) \end{aligned}$$

Now u is a cocycle, hence $u_2u_3^{-1} = u_4^{-1}u_1$, and

$$\begin{aligned} u_2u_3^{-1}(I\otimes \underline{S}\otimes S\otimes S) &= u_4^{-1}u_1(I\otimes \underline{S}\otimes S\otimes S) \\ &= u_4^{-1}(I\otimes \underline{S}\otimes S\otimes S) \end{aligned}$$

and this proves that $u_2M_1 \subset M_2$. In a similar way, we can prove that $u_2^{-1}M_2 \subset M_1$ and $u_2M_2' = M_1'$. By extension of scalars, $\tau \otimes I_S : I \otimes S \to S$ is a trace. This trace restricts to a trace $MM' \to S$, since MM' is a $T \otimes S$-subideal of $I \otimes S$, which is surjective by Lemma 6.3.7. The composition

$$\mu : M' \otimes M \to MM' \to S$$

is also a surjection, and we obtain a dual pair $\underline{M} = (M, M', \mu)$ of S-modules, after we consider M and M' as S-modules via restriction of scalars. T is finitely generated as an R-module. Since R is noetherian, the submodule I of T is also finitely generated, and it follows that $I \otimes R \otimes S$ is finitely generated as an S-module. M and M' are submodules of $I \otimes R \otimes S$, and they are therefore also finitely generated as S-modules. Furthermore, the map

$$(u_2, u_2^{-1}) : \underline{M}_1 \longrightarrow \underline{M}_2$$

is an isomorphism of dual pairs over $S \otimes S$. To prove this, we need to check that

$$\mu_1 = \mu_2 \circ (u_2 \otimes u_2^{-1})$$

Take $x \in M_1$ and $y \in M_2$, and write, with notations as above,

$$\begin{aligned} x &= \sum_i v^1 x_i \otimes 1 \otimes v^2 s_i s_i' \otimes v^3 s_i'' \\ y &= \sum_j u^1 y_j \otimes 1 \otimes u^2 t_j t_j' \otimes u^3 t_j'' \end{aligned}$$

with $x_i, y_j \in I$ and $s_i', s_i', s_i'', t_j', t_j', t_j'' \in S$. Then

$$\begin{aligned} \mu_1(x \otimes y) &= (\tau \otimes I_{S\otimes S}) \Big(\sum_{i,j} v^1 x_i u^1 y_j \otimes 1 \otimes v^2 s_i s_i' u^2 t_j t_j' \otimes v^3 s_i'' u^3 t_j'' \Big) \\ &= (\tau \otimes I_{S\otimes S}) \Big(\sum_{i,j} x_i y_j \otimes 1 \otimes s_i s_i' t_j t_j' \otimes s_i'' t_j'' \Big) \\ &= \sum_{i,j} \tau(x_i y_j) \otimes 1 \otimes s_i s_i' t_j t_j' \otimes s_i'' t_j'' \\ &= \mu_2 \Big((\sum_i x_i \otimes 1 \otimes s_i s_i' \otimes s_i'') \otimes (\sum_j y_j \otimes 1 \otimes t_j t_j' \otimes t_j'') \Big) \\ &= \mu_2(u_2 x \otimes u_2^{-1} y) \\ &= (\mu_2 \circ (u_2 \otimes u_2^{-1}))(x \otimes y) \end{aligned}$$

(u_2, u_2^{-1}) induces an isomorphism

$$\Phi : E_{S\otimes S}(\underline{M}_1) \longrightarrow E_{S\otimes S}(\underline{M}_2)$$

and the isomorphism

$$\Phi_2^{-1} \circ \Phi_3 \circ \Phi_1 : E_{S\otimes S\otimes S}(\underline{M}_{11}) \longrightarrow E_{S\otimes S\otimes S}(\underline{M}_{11})$$

is induced by

$$\begin{aligned}\left(u_{24}^{-1}u_{25}u_{23}, u_{24}u_{25}^{-1}u_{23}^{-1}\right) &= \left(u_{32}^{-1}u_{42}u_{22}, u_{32}u_{42}^{-1}u_{22}^{-1}\right)\\ &= \left((u_2u_3^{-1}u_4)_2, (u_2u_3^{-1}u_4)_2^{-1}\right)\\ &= (u_{12}, u_{12}^{-1})\end{aligned}$$

From the fact that $\underline{M}_{11} \subset I \otimes R \otimes S^{\otimes 3}$, it follows that $\Phi_2^{-1} \circ \Phi_3 \circ \Phi_1$ is the identity on $E_{S\otimes S\otimes S}(\underline{M}_{11})$, and Φ is a descent datum. Thus $E_S(\underline{M})$ descends to a Taylor-Azumaya algebra A. A is finitely generated as an R-module, because M, M' and $E_S(\underline{M})$ are finitely generated as S-modules.
A careful inspection of the first part of the proof of Theorem 6.2.1 yields that the image of $[A]$ in $H^2(R_{ét}, \mathbb{G}_m)$ is represented by $(S^{\otimes 2}, u) \in Z^1(S/R, \underline{\underline{\mathrm{Pic}}})$, or by $[u] \in H^2(S/R, \mathbb{G}_m)$.
In the case where R is not necessarily noetherian, our result follows from Theorem 6.3.9. □

Theorem 6.3.9 *Let R be a commutative ring. For every $x \in H^n(R_{ét}, \mathbb{G}_m)$, there exists a noetherian subring R_0 of R such that x is defined over R_0, in other words,*

$$H^n(R_{ét}, \mathbb{G}_m) = \varinjlim H^n(R_{0_{ét}}, \mathbb{G}_m)$$

where the limit runs over all noetherian subrings R_0 of R.

Proof From Artin's Refinement Theorem 5.6.8, we know that

$$H^n(R_{ét}, \mathbb{G}_m) = \varinjlim H^n(S/R, \mathbb{G}_m)$$

where the limit runs over all étale coverings S of R. Let x be represented by $u \in Z^n(S/R, \mathbb{G}_m) \subset \mathbb{G}_m(S^{\otimes n+1})$. We invoke again the fact that any étale R-algebra S is locally of standard type. As in the proof of Theorem 6.3.6, we find a Zariski covering $S' = \prod_{i=1}^{m} S_{f_i}$ of S such that every S_{f_i} is a standard étale algebra:

$$S_{f_i} \cong (R_{h_i}[X]/P_i(X))_{g_i}$$

with P_i' invertible in S_{f_i}. It now suffices to observe that S_{f_i} can be defined over a noetherian subring R_0 of R:

$$S_{f_i} = S_i \otimes_{R_0} R \tag{6.23}$$

with S_i a standard étale algebra. It is not difficult to see that we can choose R_0 in such a way that (6.23) for $i = 1, \ldots, m$, and

$$S' = \left(\prod_{i=1}^{m} S_i\right) \otimes_{R_0} R$$

$S_0 \prod_{i=1}^{m} S_i$ is an étale covering of R_0, and the image of u in $Z^1(S'/R, \mathbb{G}_m)$ lies in the image of $Z^1(S_0/R_0, \mathbb{G}_m)$ in $Z^1(S'/R, \mathbb{G}_m)$. □

6.4 Verschoren's construction and Takeuchi's exact sequence

In Section 6.1, we have discussed the exact sequence (6.12), connecting the Amitsur cohomology groups with values in $\mathbb{G}_m$ and Pic, and we have seen that this sequence can be identified with the sequence (6.14), thus providing a description of the cohomology groups $H^n(S/R, C_1)$, for $n > 0$.
In this Section, we will see that there is a similar exact sequence, now connecting Amitsur cohomology groups with values in Pic and Br (or Br′). We will first construct two categories $\underline{\underline{A}}(R)$ and $\underline{\underline{A}}'(R)$ such that

$$\mathrm{Br}(R) = \mathrm{K}_0\underline{\underline{A}}(R) \;\; ; \;\; \mathrm{Br}'(R) = \mathrm{K}_0\underline{\underline{A}}'(R)$$

$$\mathrm{Pic}(R) = \mathrm{K}_1\underline{\underline{A}}(R) = \mathrm{K}_1\underline{\underline{A}}'(R)$$

The construction of these categories is due to Verschoren [197].
The objects of $\underline{\underline{A}}'(R)$ are finitely generated Taylor-Azumaya algebras To define the morphisms, we proceed as follows. For two Taylor-Azumaya algebras A and B, let $\Delta(A, B)$ consist of all triples $(\underline{P}, u, \underline{Q})$, with $\underline{P}, \underline{Q} \in \underline{\underline{\mathrm{DP}}}'(R)$ and

$$u: \; A \otimes E_R(\underline{P}) \longrightarrow B \otimes E_R(\underline{Q})$$

an isomorphism of R-algebras. Two elements $(\underline{P}, u, \underline{Q})$ and $(\underline{M}, v, \underline{N})$ in $\Delta(A, B)$ are called equivalent if there exist $\underline{U}, \underline{V} \in \underline{\underline{\mathrm{DP}}}'(R)$ and isomorphisms

$$\underline{f}: \; \underline{P} \otimes \underline{U} \longrightarrow \underline{M} \otimes \underline{V} \;\; \text{and} \;\; \underline{g}: \; \underline{Q} \otimes \underline{U} \longrightarrow \underline{N} \otimes \underline{V}$$

such that the following diagram commutes

$$\begin{array}{ccc} A \otimes E_R(\underline{P}) \otimes E_R(\underline{U}) & \xrightarrow{u \otimes I} & B \otimes E_R(\underline{Q}) \otimes E_R(\underline{U}) \\ \downarrow{\scriptstyle I_A \otimes E_R(\underline{f})} & & \downarrow{\scriptstyle I_B \otimes E_R(\underline{g})} \\ A \otimes E_R(\underline{M}) \otimes E_R(\underline{V}) & \xrightarrow{v \otimes I} & B \otimes E_R(\underline{N}) \otimes E_R(\underline{V}) \end{array} \tag{6.24}$$

Here we identified $E_R(\underline{P}) \otimes E_R(\underline{U})$ and $E_R(\underline{P} \otimes \underline{U})$ by the canonical isomorphism.
If two elements of $\Delta(A, B)$ are equivalent, then we denote this by

$$(\underline{P}, u, \underline{Q}) \sim (\underline{M}, v, \underline{N})$$

We now put

$$\mathrm{Hom}_{\underline{\underline{A}}'(R)}(A, B) = \Delta(A, B)/\sim$$

and we define the composition of two morphisms as follows. For $(\underline{P}, u, \underline{Q}) \in \Delta(A, B)$ and $(\underline{M}, v, \underline{N}) \in \Delta(B, C)$, we put

$$[(\underline{M}, v, \underline{N})] \circ [(\underline{P}, u, \underline{Q})] = \Big[\big(\underline{P} \otimes \underline{M}, (v \otimes I_{E_R(\underline{Q})}) \circ (u \otimes I_{E_R(\underline{M})}), \underline{N} \otimes \underline{Q}\big)\Big]$$

It is easy to check that this composition is independent of the chosen representatives in $\Delta(A, B)$ and $\Delta(B, C)$. The tensor product defines a product on the category $\underline{\underline{A}}'(R)$.
Replacing Taylor-Azumaya algebras by classical Azumaya algebras, dual pairs by faithfully projective modules and elementary algebras by endomorphism rings in the above construction, we obtain the category $\underline{\underline{A}}(R)$.

Proposition 6.4.1 (Verschoren [197]) *With notations as above, we have for any commutative ring R that*

$$\mathrm{Br}(R) = \mathrm{K}_0\underline{\underline{A}}(R) \quad ; \quad \mathrm{Br}'(R) = \mathrm{K}_0\underline{\underline{A}}'(R) \tag{6.25}$$

$$\mathrm{Pic}(R) = \mathrm{K}_1\underline{\underline{A}}(R) = \mathrm{K}_1\underline{\underline{A}}'(R) \tag{6.26}$$

Proof We will prove the statements about the category $\underline{\underline{A}}'(R)$, and leave the ones about $\underline{\underline{A}}(R)$ to the reader.

<u>Computation of $\mathrm{K}_0\underline{\underline{A}}'(R)$.</u>

From the construction of $\underline{\underline{A}}'(R)$, it is clear that two Taylor-Azumaya algebras are isomorphic in $\underline{\underline{A}}'(R)$ if and only if they are Brauer equivalent. Therefore the isomorphism classes in $\underline{\underline{A}}'(R)$ correspond to the elements of the big Brauer group. Elementary algebras are isomorphic to R in $\underline{\underline{A}}'(R)$, and this implies that, for every Taylor-Azumaya algebra A, the inverse of $[A]$ in $\mathrm{K}_0\underline{\underline{A}}'(R)$ is represented by $[A^{\mathrm{op}}]$, since we know that $A \otimes A^{\mathrm{op}}$ is an elementary algebra. Thus every class in $\mathrm{K}_0\underline{\underline{A}}'(R)$ is represented by a Taylor-Azumaya algebra, and $\mathrm{K}_0\underline{\underline{A}}'(R)$ is nothing else then the set of isomorphism classes in $\underline{\underline{A}}'(R)$, with operation induced by the tensor product, that means that it is not else then Taylor's Brauer group.

<u>Computation of $\mathrm{K}_1\underline{\underline{A}}'(R)$.</u>

According to (C.6), $\mathrm{K}_1\underline{\underline{A}}'(R) = \mathrm{Aut}_{\underline{\underline{A}}'(R)}(R)$. We will define an isomorphism

$$F: \ \mathrm{Aut}_{\underline{\underline{A}}'(R)}(R) \longrightarrow \mathrm{Pic}(R) = \mathrm{K}_0\underline{\underline{\mathrm{DPic}}}(R)$$

Let $(\underline{M}, v, \underline{N})$ represent an (iso)morphism in of R in $\underline{\underline{A}}'(R)$. Then $v: \ E_R(\underline{M}) \to E_R(\underline{N})$ is an isomorphism of R-algebras. Applying Proposition 1.3.1, we obtain

$$\underline{I} = (I, I', \phi) \in \underline{\underline{\mathrm{DPic}}}(R) \text{ and } \underline{\psi} = (\psi, \psi'): \ \underline{M} \to \underline{N} \otimes \underline{I}$$

such that v is induced by $\underline{\psi}$. We define

$$F([(\underline{M}, v, \underline{N})]) = [I]$$

Let us show that F is well-defined, that is, $[I]$ is independent of the chosen representative of the morphism in $\underline{\underline{A}}'(R)$. Suppose that $(\underline{M}, v, \underline{N}) \sim (\underline{P}, u, \underline{Q})$ in $\Delta(R, R)$. The diagram (6.24) now takes the form

$$\begin{array}{ccc} E_R(\underline{P}) \otimes E_R(\underline{U}) & \xrightarrow{u \otimes I} & E_R(\underline{Q}) \otimes E_R(\underline{U}) \\ \downarrow E_R(\underline{f}) & & \downarrow E_R(\underline{g}) \\ E_R(\underline{M}) \otimes E_R(\underline{V}) & \xrightarrow{v \otimes I} & E_R(\underline{N}) \otimes E_R(\underline{V}) \end{array}$$

Suppose that u is induced by $\underline{\varphi} = (\varphi, \varphi'): \ \underline{P} \to \underline{Q} \otimes \underline{J}$. Then $v \otimes I$ is induced by

$$\underline{g} \circ (\underline{psi} \otimes I_{\underline{U}}) \circ \underline{f}^{-1}: \ \underline{M} \otimes \underline{V} \longrightarrow \underline{N} \otimes \underline{V} \otimes \underline{I}$$

and also by

$$\varphi \otimes I_{\underline{V}}: \ \underline{M} \otimes \underline{V} \longrightarrow \underline{N} \otimes \underline{V} \otimes \underline{J}$$

From the uniqueness property in Proposition 1.3.1, it follows that $uII \cong \underline{J}$ and $I \cong J$.
Now we define

$$G:\ \mathrm{Pic}(R) \longrightarrow \mathrm{Aut}_{\underline{\underline{A}}'(R)}(R)$$

by

$$G([I]) = [(R, \varphi, \mathrm{End}_R(I) = E_R(I, I^*, \varphi)]$$

where $\varphi:\ R \to \mathrm{End}_R(I)$ is the canonical isomorphism. It is clear that G is well-defined and that F and G are each others inverses. □

Now we proceed as in Section 6.1. Let S be a faithfully flat R-algebra. For $n \geq 1$, we define a functor

$$\delta'_{n-1}:\ \underline{\underline{A}}'(S^{\otimes n}) \longrightarrow \underline{\underline{A}}'(S^{\otimes n+1})$$

by

$$\delta'_{n-1}([A]) = A_1 \otimes_{S^{\otimes n+1}} \otimes A_2^{\mathrm{op}} \otimes_{S^{\otimes n+1}} \cdots \otimes_{S^{\otimes n+1}} A_{n+1}^{(\mathrm{op})}$$

for any $S^{\otimes n}$-Taylor-Azumaya algebra A and

$$\delta_{n-1}(f) = f_1 \otimes f_2 \otimes \cdots \otimes f_{n+1}$$

for $f = [(\underline{M}, u, \underline{N})]:\ A \to B$ in $\underline{\underline{A}}'(S^{\otimes n})$. From the fact that the Amitsur complex is a complex, we derive that there exists a natural isomorphism

$$\lambda_A:\ \delta'_n(\delta'_{n-1}(A)) \longrightarrow S^{\otimes n+2}$$

in $\underline{\underline{A}}'(S^{\otimes n+2})$. It is of course rather complicated to describe λ_A explicitly. But δ_n is cofinal and product preserving, and the exact sequence (C.8) takes the form

$$\mathrm{Pic}(S^{\otimes n}) \xrightarrow{\Delta'_{n-1}} \mathrm{Pic}(S^{\otimes n+1}) \xrightarrow{\kappa'_{n-1}} \mathrm{K}_0\underline{\Psi\delta'}_{n-1} \xrightarrow{\theta'_{n-1}} \mathrm{Br}'(S^{\otimes n}) \xrightarrow{\Delta''_{n-1}} \mathrm{Br}'(S^{\otimes n+1}) \tag{6.27}$$

The exact sequences (6.1) and (6.27) fit together into an eight term long exact sequence. Δ'_{n-1} and Δ''_{n-1} are the maps in the Amitsur complexes of Pic and Br′. The objects in the category $\underline{\Psi\delta'}_{n-1}$ are twotuples of the form (A, α), with $A \in \underline{\underline{A}}'(S^{\otimes n})$ and $\alpha:\ \delta_{n-1}(A) \to S^{\otimes n+1}$ an isomorphism in $\underline{\underline{A}}'(S^{\otimes n+1})$. We also have that

$$\kappa'_{n-1}([I]) = \big[(S^{\otimes n}, G[I])\big]$$

for all $I \in \underline{\underline{\mathrm{Pic}}}(S^{\otimes n+1})$, where G is defined as in the proof of Proposition 6.4.1. Furthermore

$$\theta'_{n-1}[(A, \alpha)] = [A]$$

for all $[(A, \alpha)] \in \mathrm{K}_0\underline{\Psi\delta'}_{n-1}$. Exactly as in Section 6.1, we can define two more maps

$$E'_{n-1}:\ \mathrm{Br}'(S^{\otimes n}) \longrightarrow \mathrm{K}_0\underline{\Psi\delta'}_n:\ [A] \mapsto [(\delta'_{n-1}(A), \lambda_A)$$
$$E'_n:\ \mathrm{K}_0\underline{\Psi\delta'}_{n-1} \longrightarrow \mathrm{Pic}(S^{\otimes n+2}):\ [(A, \alpha)] \mapsto F(\delta'_n(\alpha) \circ \lambda_A^{-1})$$

where F is defined as in the proof of Proposition 6.4.1. All these maps may be combined in the following diagram, which is the analogue of (6.6). Actually (6.6) and (6.28) can be fit together to a big diagram.

$$
\begin{array}{ccccccc}
1 & & & & & & \\
\downarrow & & & & & & \\
\mathrm{Pic}(S) & & & & & & 1 \\
\downarrow{\scriptstyle \Delta'_0} & & & & & & \downarrow \\
\mathrm{Pic}(S^{\otimes 2}) & \xrightarrow{\kappa'_0} & \mathrm{K}_0\underline{\Psi\delta_0} & & \xrightarrow{\theta'_0} & & \mathrm{Br}'(S) \\
\downarrow{\scriptstyle \Delta'_1} & \swarrow{\scriptstyle E_0} & & & \swarrow{\scriptstyle E'_0} & & \downarrow{\scriptstyle \Delta'_0} \\
\mathrm{Pic}(S^{\otimes 3}) & \xrightarrow{\kappa'_1} & \mathrm{K}_0\underline{\Psi\delta_1} & & \xrightarrow{\theta'_1} & & \mathrm{Br}'(S^{\otimes 2}) \\
\downarrow{\scriptstyle \Delta'_2} & \swarrow{\scriptstyle E_1} & & & \swarrow{\scriptstyle E'_1} & & \downarrow{\scriptstyle \Delta'_1} \\
\mathrm{Pic}(S^{\otimes 4}) & \xrightarrow{\kappa'_2} & \mathrm{K}_0\underline{\Psi\delta_2} & & \xrightarrow{\theta'_2} & & \mathrm{Br}'(S^{\otimes 3}) \\
\downarrow{\scriptstyle \Delta'_3} & \swarrow{\scriptstyle E_2} & & & \swarrow{\scriptstyle E'_2} & & \downarrow{\scriptstyle \Delta'_2} \\
\vdots & & \vdots & & & & \vdots
\end{array}
\qquad (6.28)
$$

The first and last column are the Amitsur complexes of Pic and Br′. The parallelograms in (6.28) anticommute, while the triangles commute. Also $E_n \circ E'_{n-1} = 1$. The arguments are formal analogues of the arguments presented in Section 6.1. We define *Amitsur cohomology with values in the category* $\underline{\underline{A}}'$ by the formula

$$
H^n(S/R, \underline{\underline{A}}') = \mathrm{Ker}\,(E_n)/\mathrm{Im}\,(E'_{n-1}) \qquad (6.29)
$$

By a diagram chasing argument, we now obtain the following result.

Theorem 6.4.2 *With notations as above, we have a long exact sequence of cohomology groups*

$$
\begin{array}{llllll}
1 & \longrightarrow & H^1(S/R,\mathrm{Pic}) & \xrightarrow{\alpha_1} & H^0(S/R,\underline{\underline{A}}') & \xrightarrow{\beta_1} \quad H^0(S/R,\mathrm{Br}') \\
 & \xrightarrow{\gamma_1} & H^2(S/R,\mathrm{Pic}) & \xrightarrow{\alpha_2} & H^1(S/R,\underline{\underline{A}}') & \xrightarrow{\beta_2} \quad H^1(S/R,\mathrm{Br}') \\
 & \xrightarrow{\gamma_2} & \cdots & & & \\
 & \xrightarrow{\gamma_{n-1}} & H^n(S/R,\mathrm{Pic}) & \xrightarrow{\alpha_n} & H^{n-1}(S/R,\underline{\underline{A}}') & \xrightarrow{\beta_n} \quad H^{n-1}(S/R,\mathrm{Br}') \\
 & \xrightarrow{\gamma_n} & \cdots & & &
\end{array}
\qquad (6.30)
$$

Remarks 6.4.3 1) Replacing the category $\underline{\underline{A}}'$ by $\underline{\underline{A}}$ in the above construction, and replacing the big Brauer group by the classical Brauer group, we obtain the definition of Amitsur cohomology groups with values in the category $\underline{\underline{A}}$ and the long exact sequence

$$
\begin{array}{llllll}
1 & \longrightarrow & H^1(S/R,\mathrm{Pic}) & \xrightarrow{\alpha_1} & H^0(S/R,\underline{\underline{A}}) & \xrightarrow{\beta_1} \quad H^0(S/R,\mathrm{Br}) \\
 & \xrightarrow{\gamma_1} & H^2(S/R,\mathrm{Pic}) & \xrightarrow{\alpha_2} & H^1(S/R,\underline{\underline{A}}) & \xrightarrow{\beta_2} \quad H^1(S/R,\mathrm{Br}) \\
 & \xrightarrow{\gamma_2} & \cdots & & & \\
 & \xrightarrow{\gamma_{n-1}} & H^n(S/R,\mathrm{Pic}) & \xrightarrow{\alpha_n} & H^{n-1}(S/R,\underline{\underline{A}}) & \xrightarrow{\beta_n} \quad H^{n-1}(S/R,\mathrm{Br}) \\
 & \xrightarrow{\gamma_n} & \cdots & & &
\end{array}
\qquad (6.31)
$$

(6.31) is originally due to Takeuchi, see [178].
2) Applying Taylor's Theorem 6.3.8, the exact sequence (A.3) may be written as (take $q = 1$)

$$\begin{array}{lllll} 1 & \longrightarrow & H^1(S/R, C_1) & \longrightarrow & H^2(R_{\acute{e}t}, \mathbb{G}_m)) \cong \mathrm{Br}'(R) & \longrightarrow & H^0(S/R, \mathrm{Br}') \\ & \longrightarrow & H^2(S/R, C_1) & \longrightarrow & H^1(S/R, C_2) & \longrightarrow & H^1(S/R, \mathrm{Br}') \\ & \longrightarrow & H^2(S/R, C_1) & \longrightarrow & H^1(S/R, C_2) & \longrightarrow & H^1(S/R, \mathrm{Br}') \\ & \longrightarrow & \cdots & & & & \end{array} \tag{6.32}$$

Unfortunately, the sequences (6.32) and (6.30) may not be identified, so we do not obtain an analog of Villamayor and Zelinsky's Theorem 6.1.4. In particular, we do not have an explicit description of $H^n(S/R, C_2)$. In order to obtain such a description, we would need a category $\mathcal{C}$ such that $\mathrm{K}_0\mathcal{C} = \mathrm{Br}'(R)$ and $\mathrm{K}_1\mathcal{C} = C_1(R)$. We do not have such a category, we even do not know what $C_1(R)$ is.

6.5 The Brauer group is torsion

Theorem 6.5.1 $\mathrm{Br}(R)$ *is torsion.*

Proof First recall from [15, IX.3] that we have a determinant functor

$$\det_R = \det : \underline{\underline{\mathrm{P}}}(R) \to \underline{\underline{\mathrm{Pic}}}(R)$$

obtained by taking exterior powers, that is, if $\mathrm{rk}(M) = n$, then $\det(M) = \Lambda^n(M)$. This determinant satisfies the following properties: we have natural isomorphisms

$$\det(M \oplus N) = \det(M) \otimes \det(N), \tag{6.33}$$

and, if $f : M \to M'$, $g : N \to N'$ in $\underline{\underline{\mathrm{P}}}(R)$, then

$$\det(f \oplus g) = \det(f) \otimes \det(g). \tag{6.34}$$

If $f : R^n \to R^n$ is an isomorphism, then $\det(f)\ \det(R^n) = R \to \det(R^n) = R$ is nothing else then multiplication by the "classical" determinant of the matrix of f. The determinant also commutes with base extension : if S is an R-algebra, then $\det_S(M \otimes S) = \det_R(M) \otimes S$ and $\det_S(f \otimes I_S) = \det(f) \otimes I_S$. Now take an R-Azumaya algebra A of constant rank n. We will show that $[A]^n = 1$ in $\mathrm{Br}(R)$. The result will then follow easily, after decomposing R into a direct sum $\oplus_{i \in I} Re_i$, with e_i idempotent and A_{e_i} an R_{e_i}-Azumaya algebra of constant rank, for all i. We use the notations of Theorem 6.2.1. Take a faithfully flat R-algebra S such that

$$A \otimes S \cong \mathrm{End}_S(Q) \cong E_S(Q, Q^*, \langle \bullet, \bullet \rangle),$$

with $Q \in \underline{\underline{\mathrm{FP}}}(S)$. Replacing S by another faithfully flat extension, we may assume that Q is free, hence $Q = S^n$. As in step 1 of the proof of 6.2.1, we consider the map

$$\Phi : E_{S^{\otimes 2}}(Q_1) = M_n(S \otimes S) \to E_{S^{\otimes 2}}(Q_2) = M_n(S \otimes S)$$

induced by

$$f:\ Q_1\otimes I\cong I^n\to Q_2\cong (S\otimes S)^n$$

for some $I\in \underline{\underline{\mathrm{Pic}}}(S\otimes S)$. Consider

$$\det(f)\ \det_{S^{\otimes 2}}(I^n)=I^{\otimes n}\to det_{S^{\otimes 2}}(S\otimes S)^n=S\otimes S.$$

Recall that

$$f_2^{-1}f_3f_1=I\otimes\alpha:\ (S\otimes S\otimes S)^n\otimes_{S^{\otimes 3}}\delta_1(I)\to (S\otimes S\otimes S)^n,$$

or

$$f_2^{-1}f_3f_1=\alpha^n:\ (\delta_1(I))^n\to (S\otimes S\otimes S)^n,$$

and, in view of (6.34)

$$\det(f_2^{-1}f_3f_1)=\alpha^{\otimes n}:\ (\delta_1(I))^{\otimes n}\to S\otimes S\otimes S. \tag{6.35}$$

Observe that $(I^n)^{\otimes n}$ is a direct product of n^n copies of $I^{\otimes n}$:

$$(I^n)^{\otimes n}\cong (I^{\otimes n})^{n^n},$$

hence

$$f^{\otimes n}:\ (I^{\otimes n})^{n^n}\to (S\otimes S)^{n^n}.$$

$f^{\otimes n}$ induces $\Phi^{\otimes n}$. Consider the composition

$$h=f^{\otimes n}\circ(\det(f^{-1}))^{n^n}\ (S\otimes S)^{n^n}\to (I^{\otimes n})^{n^n}\to (S\otimes S)^{n^n}.$$

h still induces $\Phi^{\otimes n}$: from $E_{S^{\otimes 2}}(\det f)=I$, it follows that $E_{S^{\otimes 2}}((\det f)^{n^n})=I$. Then apply the fact that $E(g\circ f)=E(g)\circ E(f)$. Now $h_i=(\det f_i^{-1})^{n^n}\circ f_i^{\otimes n}$, hence

$$h_2^{-1}h_3h_1=(\det(f_2^{-1}f_3f_1)^{-1})\circ(\alpha^n)^{\otimes n}=I$$

using (6.35). We leave it to the reader to show that we may commute $\det f_i$ and f_j. Now $h:\ (S\otimes S)^{n^n}\to (S\otimes S)^{n^n}$ is a descent datum, hence by the faithfully flat descent theorem, there exists a faithfully projective R-module Q such that $Q\otimes S\cong S^{n^n}$ and the diagram

$$\begin{array}{ccc} S\otimes P\otimes S & \xrightarrow{I\otimes\eta} & S\otimes S^{n^n}\\ \downarrow{\scriptstyle \tau\otimes I} & & \downarrow{\scriptstyle h}\\ P\otimes S\otimes S & \xrightarrow{\eta\otimes I} & S^{n^n}\otimes S \end{array}$$

commutes. Take $E_{S^{\otimes 2}}$ of this diagram and compare it to the commutative diagram

$$\begin{array}{ccc} S\otimes A^{\otimes n}\otimes S & \longrightarrow & \mathrm{End}_{S^{\otimes 2}}(S\otimes S^{n^n})\\ \downarrow{\scriptstyle \tau\otimes I} & & \downarrow{\scriptstyle \Phi^{\otimes n}}\\ A^{\otimes n}\otimes S\otimes S & \longrightarrow & \mathrm{End}_{S^{\otimes 2}}(S^{n^n}\otimes S) \end{array}$$

The uniqueness property in Theorem B.0.13 then implies that $A^{\otimes n}\cong \mathrm{End}_R(P)$. □

In case I is trivial in $\mathrm{Pic}(S\otimes S)$, our proof coincides with the first part of the proof presented by Knus and Ojanguren in [114]. If I is not trivial, they present two ways

to escape : if we accept to use Artin's refinement theorem, then we can reduce the proof to the case where I is trivial, and we are done ; otherwise, we have to use a more complicated construction, given in part two of their proof. Another approach is presented by Milne in [135], but he uses Giraud's non-abelian cohomology. Yet another proof was given by Saltman in [166].
We now have $\mathrm{Br}(R) \subset \mathrm{Br}'(R)_{\mathrm{tors}}$. The reader will immediately ask the following question : do we have equality ? The cohomological version of this question is the following : do we have $\mathrm{Br}(R) \cong H^2(R_{\acute{e}t}, \mathbb{G}_m)_{\mathrm{tors}}$? An affirmative answer was been given by Gabber in [89], but other versions of the proof were given by Hoobler, cf. [103] and by by Knus and Ojanguren, cf. [116]. In what follows, we will present a proof that follows closely the Knus-Ojanguren proof. The first step will be to construct a Mayer-Vietoris sequence for the cohomology groups on R_E.

6.6 The Mayer-Vietoris exact sequence

Theorem 6.6.1 (Mayer-Vietoris exact sequence) *Suppose that $f, g \in R$ are such that $(f,g) = 1$. Then we have a long exact sequence of cohomology groups, for every $F \in \mathcal{S}(R_E)$:*

$$\begin{array}{lllll} 0\to & F(R) & \to & F(R_f)\oplus F(R_g) & \to & F(R_{fg}) \\ \to & H^1(R_E,F) & \to & H^1((R_f)_E,F_{|R_f})\oplus H^1((R_g)_E,F_{|R_g}) & \xrightarrow{d_1} & H^1((R_{fg})_E,F_{|R_{fg}}) \\ \to & \cdots & & & & \\ \to & H^i(R_E,F) & \to & H^i((R_f)_E,F_{|R_f})\oplus H^i((R_g)_E,F_{|R_g}) & \xrightarrow{d_i} & H^i((R_{fg})_E,F_{|R_{fg}}) \\ \to & \cdots & & & & \end{array}$$

The maps $d_i : H^i((R_f)_E, F_{|R_f}) \oplus H^i((R_g)_E, F_{|R_g}) \to H^i((R_{fg})_E, F_{|R_{fg}})$ are defined as follows : for $a \in H^i((R_f)_E, F_{|R_f})$, $b \in H^i((R_g)_E, F_{|R_g})$, let $d_i(a,b) = a - b$. Here we use the same notation for a (resp. b), and the image of a (resp. b) in $H^i((R_{fg})_E, F_{|R_{fg}})$. R_f is the localization of R at the multiplicative set $\{1, f, f^2, \ldots\}$.

Proof We proceed in two steps. Write $\alpha_f : R \to R_f$ for the localization map, and S_f as a shorter notation for $R_f \otimes S$. Consider the sheaf $F_f = (\alpha_f^* \circ \alpha_{f*})(F)$. Then for every $S \in \underline{\underline{cat}}(R_E)$, $F_f(S) = F(S_f)$. If $(f,g) = 1$, then we have an exact sequence of sheaves

$$0 \to F \to F_f \oplus F_g \to F_{fg} \to 0.$$

Indeed, let $S \in \underline{\underline{cat}}(R_E)$, then $S \to S' = S_f \times S_g$ is a Zariski covering, and therefore an E-covering, and, because F is a sheaf, an exact sequence

$$0 \to F(S) \to F(S') \rightrightarrows F(S' \otimes_S S')$$

Let us investigate this sequence in detail. We introduce the following notations :

$$\alpha_f : S \to S_f,\ \alpha_g : S \to S_g,\ \beta_g : S_f \to S_{fg},\ \beta_f : S_g \to S_{fg},$$

for the natural maps. Observe that

$$S' \otimes_S S' = (S_f \times S_g) \otimes_S (S_f \times S_g) = S_f \times S_{fg} \times S_{gf} \times S_g$$

and that

$$\begin{aligned} F(\varepsilon_1)(a,b) &= (a, F(\beta_f)(b), F(\beta_g(a)), b) \\ F(\varepsilon_2)(a,b) &= (a, F(\beta_g)(a), F(\beta_f(b)), b) \end{aligned}$$

Therefore, if $F(\beta_g)(a) - F(\beta_f)(b) = 0$, then $F(\varepsilon_1)(a,b) = F(\varepsilon_2)(a,b)$, and (a,b) lies in the image of $F(S) \to F(S')$. Therefore the sequence

$$0 \to F(S) \to F_f(S) \times F_g(S) \to F_{fg}(S)$$

is exact, and it only remains to be shown that $F_f \oplus F_g \to F_{fg}$ is surjective in $\mathcal{S}(R_E)$. Take $c \in F_{fg}(S)$. It suffices to show that the image of c in $F_{fg}(S')$ lies in the image of $(F_f \oplus F_g)(S')$ for some E-covering S' of S. Take $S' = S_f \times S_g$, as above. Then $F_{fg}(S') = F(S_{fg}) \times F(S_{fg})$, and the image of c is (c,c). The map

$$(F_f \oplus F_g)(S') = F_f(S) \times F(S_{fg}) \times F(S_{fg}) \times F_g(S) \to F_{fg}(S') = F(S_{fg}) \times F(S_{fg})$$

is given by $(a,b,c,d) \to (a-c, b-d)$, and we see that $(0,c,-c,0)$ is mapped to (c,c).
It now follows that we have a long exact sequence

$$\begin{array}{ccccccc} 0 & \longrightarrow & F(R) & \longrightarrow & F_f(R) \oplus F_g(R) & \longrightarrow & F_{fg}(R) \\ & \longrightarrow & H^1(R_E, F) & \longrightarrow & H^1(R_E, F_f) \oplus H^1(R_E, F_g) & \xrightarrow{d_1} & H^1(R_E, F_{fg}) \\ & \longrightarrow & \cdots & & & & \\ & \longrightarrow & H^i(R_E, F) & \longrightarrow & H^i(R_E, F_f) \oplus H^i(R_E, F_g) & \xrightarrow{d_1} & H^i(R_E, F_{fg}) \\ & \longrightarrow & \cdots & & & & \end{array}$$

and the Theorem will follow if we can show that $H^i((R_f)_E, F_{|R_f}) = H^i(R_E, F_f)$. Combining Proposition 5.6.2 and Corllary 5.7.3, we see that we can apply Theorem A.2.2 to the situation

$$\mathcal{S}(R_E) \xrightarrow{\alpha_f^* \circ \alpha_{f*}} \mathcal{S}(R_E) \xrightarrow{\Gamma} \underline{\underline{Ab}}$$

Observe that $\Gamma \circ \alpha_f^* \circ \alpha_{f*} = \Gamma_{R_f}$. Now α_{f*} is not right exact, but the composition $\alpha_f^* \circ \alpha_{f*}$ is exact. This can be easily seen as follows: if $\varphi : F \to F''$ is surjective in $\mathcal{S}(R_E)$, and $S \in \underline{cat}(R_E)$, then for every $x \in F''(S_f)$, there exists an E-covering $S_f \to T$, such that the image of x in $F''(T)$ is in $\varphi(T)$. Now T is an R_f-algebra, so $T \cong R_f \otimes T = T_f$ and the image of x is in the image of $\varphi_f(T) : F(T) = F(T_f) = F_f(T) \to F''(T) = F''(T_f) = F''_f(T)$, and therefore φ_f is surjective.
Applying Theorem A.2.2 and Lemma 5.6.2, we now see that

$$H^n(R_E, F_f) = R^n\Gamma_{R_f} = H^n(R_{f_e}, F_{|R_f})$$

finishing our proof. □

A direct proof of 6.6.1 for the étale site, based on Artin's refinement theorem 5.6.8 may be found in [116].

Corollary 6.6.2 *Let $f, g \in R$ be such that $(f,g) = 1$. Then we have an exact sequence*

$$\begin{array}{rlllll} 1 & \to \mathbb{G}_m(R) & \to & \mathbb{G}_m(R_f) \times \mathbb{G}_m(R_g) & \to & \mathbb{G}_m(R_{fg}) \\ & \to \mathrm{Pic}(R) & \to & \mathrm{Pic}(R_f) \times \mathrm{Pic}(R_g) & \to & \mathrm{Pic}(R_{fg}) \\ & \to \mathrm{Br}'(R) & \to & \mathrm{Br}'(R_f) \times \mathrm{Br}'(R_g) & \to & \mathrm{Br}'(R_{fg}) \end{array}$$

If we replace $\mathrm{Br}'(.)$ by $\mathrm{Br}(.)$, then we do not always obtain an exact sequence : the exactness at $\mathrm{Br}(.)$ is a problem. Partial results have been obtained by Knus and Ojanguren, we refer to [115]. A different procedure to overcome this problem is the following : consider $f, g \in R$ such that $(f,g) = 1$. A *relative Azumaya datum* consists of a 5-tuple $\Delta = (A, B, P, Q, \xi)$, with A an R_f-Azumaya algebra, B an R_g-Azumaya algebra, $P, Q \in \underline{\underline{FP}}(R_{fg}))$ and ξ an R_{fg}-algebra isomorphism

$$\xi : A_g \otimes_{R_{fg}} \mathrm{End}_{R_{fg}}(P) \to B_f \otimes_{R_{fg}} \mathrm{End}_{R_{fg}}(Q)$$

(this implies that $[A_g] = [B_f]$ in $\mathrm{Br}(R_{fg})$). An isomorphism $(\alpha, \beta, \sigma, \rho)\ \Delta \to \Delta'$ consists of isomorphisms $\alpha : A \to A'$, $\beta : B \to B'$, $\sigma : P \to P', \rho : Q \to Q'$ such that

$$(\beta \otimes \mathrm{End}(\rho)) \circ \xi = \xi' (\alpha \otimes \mathrm{End}(\sigma)).$$

A relative Azumaya datum of the form $\mathcal{E} = (\mathrm{End}_R(M), \mathrm{End}_R(N), P, Q, \mathrm{End}(\varphi))$, with

$$\varphi : M \otimes_{R_f} R_{fg} \otimes_{R_{fg}} P \to N \otimes_{R_g} R_{fg} \otimes_{R_{fg}} Q$$

will be called trivial. Two data Δ and Δ' are called equivalent if there exist trivial data $\mathcal{E}$, $\mathcal{E}'$ such that $\Delta \otimes \mathcal{E} \cong \Delta' \otimes \mathcal{E}'$ (the tensor product is defined in the obvious way). The set of equivalence classes forms a group under the operation defined by the tensor product. This group will be denoted by $\mathrm{Br}(R_f, R_g, R_{fg})$. A similar procedure for the Brauer' group leads to $\mathrm{Br}'(R_f, R_g, R_{fg})$. Observe that we have natural morphisms $\mathrm{Br}(R) \to \mathrm{Br}(R_f, R_g, R_{fg})$ and $\mathrm{Br}'(R) \to \mathrm{Br}'(R_f, R_g, R_{fg})$.

Proposition 6.6.3 *Let $f, g \in R$ be such that $(f,g) = 1$. Then we have the following commutative diagram with exact rows :*

$$\begin{array}{ccccccc}
1 & \longrightarrow & \mathbb{G}_m(R) & \longrightarrow & \mathbb{G}_m(R_f) \times \mathbb{G}_m(R_g) & \longrightarrow & \mathbb{G}_m(R_{fg}) \\
 & & \downarrow{=} & & \downarrow{=} & & \downarrow{=} \\
1 & \longrightarrow & \mathbb{G}_m(R) & \longrightarrow & \mathbb{G}_m(R_f) \times \mathbb{G}_m(R_g) & \longrightarrow & \mathbb{G}_m(R_{fg}) \\
 & \longrightarrow & \mathrm{Pic}(R) & \longrightarrow & \mathrm{Pic}(R_f) \times \mathrm{Pic}(R_g) & \longrightarrow & \mathrm{Pic}(R_{fg}) \\
 & & \downarrow{=} & & \downarrow{=} & & \downarrow{=} \\
 & \longrightarrow & \mathrm{Pic}(R) & \longrightarrow & \mathrm{Pic}(R_f) \times \mathrm{Pic}(R_g) & \longrightarrow & \mathrm{Pic}(R_{fg}) \\
 & \longrightarrow & \mathrm{Br}(R_f, R_g, R_{fg}) & \longrightarrow & \mathrm{Br}(R_f) \times \mathrm{Br}(R_g) & \longrightarrow & \mathrm{Br}(R_{fg}) \\
 & & \downarrow{j} & & \downarrow{\subset} & & \downarrow{\subset} \\
 & \longrightarrow & \mathrm{Br}'(R) & \longrightarrow & \mathrm{Br}'(R_f) \times \mathrm{Br}'(R_g) & \longrightarrow & \mathrm{Br}'(R_{fg})
\end{array}$$

The maps $\mathrm{Br}(R) \to Br'(R_f, R_g, R_{fg}) \to \mathrm{Br}'(R)$ *are injective, and*

$$\mathrm{Br}'(R) \cong Br'(R_f, R_g, R_{fg})$$

Proof In the top row, all the maps are the obvious ones, only the definition of $d: \text{Pic}(R_{fg}) \to \text{Br}(R_f, R_g, R_{fg})$ needs to be stated. For $I \in \text{Pic}(R_{fg})$, define $d([I]) = \Delta$ with $\Delta = (R_f, R_g, R_{fg}, I, \xi)$, where ξ is the canonical map

$$\xi: R_{fg} \to \text{End}_{R_{fg}}(I)$$

A routine computation then shows that the top row is exact. A similar computation applies to the bottom row, with $\text{Br}'(R)$ replaced by $\text{Br}'(R_f, R_g, R_{fg})$, and we have obvious maps from the top row to the bottom row. Comparing this to 6.6.2, it follows that $\text{Br}'(R) \cong \text{Br}'(R_f, R_g, R_{fg})$, using the lemma of 5. The lemma of 5 then also implies that $j: \text{Br}(R_f, R_g, R_{fg}) \to \text{Br}'(R)$ is injective. We therefore have a commutative diagram

$$\begin{array}{ccc} \text{Br}(R) & \longrightarrow & \text{Br}(R_f, R_g, R_{fg}) \\ \downarrow & & \downarrow \\ \text{Br}'(R) & \xrightarrow{\cong} & \text{Br}'(R_f, R_g, R_{fg}) \end{array}$$

hence $\text{Br}(R) \to \text{Br}(R_f, R_g, R_{fg})$ is injective. □

6.7 Gabber's theorem

Some results on the K-theory of projective modules

For any (semi)group G, let $G(R)$ be the (semi)group of continuous functions from $\text{Spec}(R)$ to G. The rank is then a functor

$$\text{rk}: \underline{\text{P}}(R) \to \mathbb{N}(R)$$

restricting to

$$\text{rk}: \underline{\underline{\text{FP}}}(R) \to \mathbb{N}_0(R)$$

These induce group homomorphisms

$$\text{rk}: \text{K}_0(R) \to \mathbb{Z}(R) \text{ and } \text{rk}: \text{K}_0\underline{\underline{\text{FP}}}(R) \to \mathbb{Q}^+(R)$$

These rank functions both have a section. For example, a section σ for the second map may be defined as follows : take $q \in \mathbb{Q}^+(R)$. If e is an idempotent such that q takes constant value $\frac{m}{n}$ on $\text{Spec}(eR)$, then $e\sigma(q) = [eR^m]/[eR^n]$ in $\text{K}_0\underline{\underline{\text{FP}}}(R)$. We define

$$\rho: \text{K}_0\underline{\underline{\text{FP}}}(R) \to \text{K}_0\underline{\underline{\text{FP}}}(R)$$

by $\rho(x) = x/\sigma(\text{rk}(x))$. It is then clear that $\text{rk}(\rho(x)) = 1$ for all $x \in \text{K}_0\underline{\underline{\text{FP}}}(R)$. Elements of rank 1 in $\text{K}_0\underline{\underline{\text{FP}}}(R)$ satisfy the following remarkable property :

Lemma 6.7.1 *if $x \in \text{K}_0\underline{\underline{\text{FP}}}(R)$ is such that $\text{rk}(x) = 1$, then x has a unique n-th root in $\text{K}_0\underline{\underline{\text{FP}}}(R)$.*

Proof Since the rank function is split, we can write

$$\mathrm{K}_0(R) = \mathrm{Rk}_0(R) \oplus \mathbb{Z}(R),$$

where $\mathrm{Rk}_0(R)$ is the kernel of the rank function. The rank extends to a function $\mathbb{Q}\otimes_{\mathbb{Z}}\mathrm{K}_0(R) \to \mathbb{Q}(R)$. Let $U^+(\mathbb{Q}\otimes_{\mathbb{Z}}\mathrm{Rk}_0(R))$ denote the set of elements in $\mathbb{Q}\otimes_{\mathbb{Z}}\mathrm{Rk}_0(R)$ with strictly positive rank. According to [15, IX.7.1], we have an isomorphism

$$\mathrm{K}_0\underline{\underline{\mathrm{FP}}}(R) \cong U^+(\mathbb{Q}\otimes_{\mathbb{Z}}\mathrm{Rk}_0(R)) \cong \mathbb{Q}^+(R) \oplus (\mathbb{Q}\otimes_{\mathbb{Z}}\mathrm{Rk}_0(R))$$

with the multiplicative structure on $\mathbb{Q}^+(R)$ and the additive structure on $\mathbb{Q}\otimes_{\mathbb{Z}} \mathrm{Rk}_0(R)$. The image of an element of rank one takes the form $(1, x)$, and has $(1, \frac{1}{x})$ as unique n-th root. □

In the proof of the next proposition, we will also use the fact that the free modules are cofinal in the category $\underline{\underline{\mathrm{FP}}}(R)$ (cf. [15, IX.4.6]). This implies that every $x \in \mathrm{K}_0\underline{\underline{\mathrm{FP}}}(R)$ may be represented by $[P]/[R^m]$ for some $P \in \underline{\underline{\mathrm{FP}}}(R)$ and $m \in \mathbb{N}_0$, or that $[R^m]x$ may be represented by $P \in \underline{\underline{\mathrm{FP}}}(R)$ for some $m \in \mathbb{N}_0$.

Proposition 6.7.2 *Let $f, g \in R$ be such that $(f, g) = 1$, and consider $[D] \in \mathrm{Br}'(R)_{\mathrm{tors}}$. If $[D_f] \in \mathrm{Br}(R_f)$, and $[D_g] \in \mathrm{Br}(R_g)$, then $[D] \in \mathrm{Br}(R)$.*

Proof Take $[D] \in Br'(R)_{\mathrm{tors}}$, and suppose that $[D_f] \in \mathrm{Br}(R_f)$, and $[D_g] \in \mathrm{Br}(R_g)$. Inspecting the commutative diagram in Proposition 6.6.3, we conclude that $[D] = j([\Delta])$ for some relative Azumaya datum $\Delta = (A, B, P, Q, \xi)$. Let us assume that the ranks of A, B, P, Q are all constant; we leave it to the reader to point out that this is no restriction. If $[D]^n = 1$ in $\mathrm{Br}'(R)$, then $[\Delta]^n = 1$ in $\mathrm{Br}'(R_f, R_g, R_{fg})$, hence

$$\Delta^{\otimes n} = (A^{\otimes n}, B^{\otimes n}, P^{\otimes n}, Q^{\otimes n}, x^{\otimes n}) \cong (\mathrm{End}_R(M), \mathrm{End}_R(N), P^{\otimes n}, Q^{\otimes n}, \mathrm{End}(\omega)).$$

Consider $x = [M_g \otimes_{R_{fg}} P^{\otimes n}] = [N_f \otimes_{R_{fg}} Q^{\otimes n}] \in \mathrm{K}_0\underline{\underline{\mathrm{FP}}}(R_{fg})$, and let $y = \rho(x)$. Then, according to Lemma 6.7.1, y^{-1} has a unique n-th root z in $\mathrm{K}_0\underline{\underline{\mathrm{FP}}}(R_{fg})$, and

$$(z\rho([P]))^n = y^{-1}\rho([P])^n = \rho([M_g])^{-1}$$

hence

$$z\rho([P]) = \rho([M_g])^{-1/n} \in \mathrm{Im}\,(\mathrm{K}_0\underline{\underline{\mathrm{FP}}}(R_f) \to \mathrm{K}_0\underline{\underline{\mathrm{FP}}}(R_{fg})).$$

It is clear that $\sigma(\mathrm{rk}([P]) \in \mathrm{Im}\,(\mathrm{K}_0\underline{\underline{\mathrm{FP}}}(R_f) \to \mathrm{K}_0\underline{\underline{\mathrm{FP}}}(R_{fg}))$, hence

$$z[P] \in \mathrm{Im}\,(\mathrm{K}_0\underline{\underline{\mathrm{FP}}}(R_f) \to \mathrm{K}_0\underline{\underline{\mathrm{FP}}}(R_{fg})),$$

and, similarly,

$$z[Q] \in \mathrm{Im}\,(\mathrm{K}_0\underline{\underline{\mathrm{FP}}}(R_f) \to \mathrm{K}_0\underline{\underline{\mathrm{FP}}}(R_{fg})).$$

Using the fact that the free modules are cofinal in the category of faithfully projective modules, we find an integer h such that $[R^h_{fg}]z = [H]$, $[R^h_{fg}]z[P] = [P'_g]$, $[R^h_{fg}]z[Q] = [Q'_f]$ with $H \in \underline{\underline{\mathrm{FP}}}(R_{fg})$, $P' \in \underline{\underline{\mathrm{FP}}}(R_f)$, $Q' \in \underline{\underline{\mathrm{FP}}}(Rg)$. We then have

$$[H \otimes P] = [P'_g] \text{ and } [H \otimes Q] = [Q'_f]$$

in $K_0\underline{\underline{FP}}(R_f)$, hence there exists $m \in \mathbb{N}$ such that

$$\begin{aligned} H^m \otimes P &\cong (H \otimes P)^m \cong (P'_g)^m \\ H^m \otimes Q &\cong (H \otimes Q)^m \cong (Q'_f)^m \end{aligned}$$

Replace h above by hm ; then we obtain

$$H \otimes P \cong P'_g \text{ and } H \otimes Q \cong Q'_f.$$

In $\mathrm{Br}'(R_f, R_g, R_{fg})$, we have

$$\begin{aligned} [\Delta] &= [(A, B, P, Q, \xi) \otimes (R_f, R_g, H, H, I)] \\ &= [(A, B, P'_g, Q'_f, \xi)] = [(A', B', R_{fg}, R_{fg}, \xi)] \end{aligned}$$

with $A' = A \otimes \mathrm{End}_{R_f}(P')$, $B' = B \otimes \mathrm{End}_{R_g}(Q')$. Since $\xi : A'_g \to B'_f$ is an isomorphism, it follows that there exists an Azumaya algebra C such that $C_f \cong A'$, $C_g \cong B'$ (this may be seen if we apply the faithfully flat descent theorem to $S = R_f \times R_g$ and the S-algebra $A \times B$). Hence

$$[\Delta] = [(C_f, C_g, R_{fg}, R_{fg}, I)] \in \mathrm{Im}\,(\mathrm{Br}'(R) \to \mathrm{Br}'(R_f, R_g, R_{fg}))$$

proving the assertion. □

Lemma 6.7.3 *Suppose that* $[A] \in \mathrm{Br}'(R)_{\mathrm{tors}}$. *If* $[A_m] \in \mathrm{Br}(R_m)$ *for all* $m \in \mathrm{Max}(R)$, *then* $[A] \in \mathrm{Br}(R)$.

Proof First fix $m \in \mathrm{Max}(R)$, and let S be a faithfully flat R-algebra such that

$$[A] \in \mathrm{Br}'(S/R) \cong H^1(S/R, \underline{\underline{\mathrm{Pic}}}).$$

Let $c = [(I, \alpha)]$ be the image of $[A]$ in $H^1(S/R, \underline{\underline{\mathrm{Pic}}})$, cf. Theorem 6.2.1. Let B be an R_m-Azumaya algebra such that $[B] = [A_m]$ in $\mathrm{Br}'(R_m)$. Then $[B]$ maps to c_m, the canonical image of c in $H^1((R_m \otimes S)/R_m, \underline{\underline{\mathrm{Pic}}})$. Inspecting the construction in the proof of Theorem 6.2.1 we see that we may find $f \in R \setminus m$ and an R_f-Azumaya algebra $B(f)$ such that $[B(f)]$ maps to c_f in $H^1(S_f/R_f, \underline{\underline{\mathrm{Pic}}})$. But then $[B(f)] = [A_f]$ in $\mathrm{Br}'(S/R)$ and $[A_f] \in \mathrm{Br}'(S/R)$.
Let $\Sigma = \{f \in R | [A_f] \in \mathrm{Br}(R)\}$. Clearly if $a \in R$ and $f \in \Sigma$, then $af \in \Sigma$. Also, if $f, g \in \Sigma$ then $f + g \in \Sigma$ (apply Proposition 6.7.2 to the ring R_{f+g}). Hence Σ is an ideal, and for every $m \in \mathrm{Max}(R)$, there exists $f \in R \setminus m$ lying in Σ. It follows that $S = R$, hence $1 \in R$, proving the result. □

Theorem 6.7.4 (Gabber's Theorem) *For any commutative ring* R, *we have*

$$\mathrm{Br}(R) \cong \mathrm{Br}'(R)_{\mathrm{tors}} \cong H^2(R_{fl}, \mathbb{G}_m)_{\mathrm{tors}} \cong H^2(R_{\acute{e}t}, \mathbb{G}_m)_{\mathrm{tors}}.$$

Proof By Theorem 6.5.1 and Proposition6.7.2, it suffices to prove that for R local, $\mathrm{Br}'(R)_{\mathrm{tors}} \subset \mathrm{Br}(R)$. Take $[A] \in \mathrm{Br}'(R)_{\mathrm{tors}}$, and let S be an étale covering of R splitting A. We know (cf. [162]) that S is locally of standard étale type, that is, if

p is a prime ideal of S lying above m, then $S_p \cong (R[X]/(P))_{m_1}$, with $P(X)$ a monic polynomial in $R[X]$, m_1 a maximal ideal in $C_1 = R[X]/(P)$ and P' invertible in $(C_1)_{m_1}$. Write $P(X) = X^N + a_1X^{N-1} + \cdots + a_N$, and let

$$C = R[t_1, ..., t_N]/I,$$

with I the ideal generated by $\{s_i(t_1, \ldots, t_N) - (-1)^i a_i | i = 1, \ldots, N\}$. Here s_i is the i-th symmetric function. C contains N roots of the polynomial P (the images of the t_i), and the symmetric group S_N acts as a group of R-automorphisms on C, with

$$\sigma(t_i) = t_{\sigma(i)}$$

for $\sigma \in S_N$. C is projective of rank $N!$ as an R-module and we have a map $C_1 \to C$, defined by sending X to t_1. S_N permutes the maximal ideals of C (they all lie above m, since R is local). Take a maximal ideal m' of C lying above m_1 in C_1. By the remarks made above, we have, for any other maximal ideal m'' of C that $C_{m'} \cong C_{m''}$, the isomorphism being induced by any permutation in S_N that maps m' to m''. We now have the following commutative diagram of R-algebras :

$$\begin{array}{ccccc} R & \longrightarrow & C_1 & \longrightarrow & C \\ \downarrow & & \downarrow & & \downarrow \\ S_p & \cong & (C_1)_{m_1} & \longrightarrow & C_{m'} \cong C_{m''} \end{array}$$

Consider the image of $[A]$ in $\mathrm{Br}'(C)$. Since A is split by S, the image of $[A]$ in $\mathrm{Br}'(S_p)$, and hence in $\mathrm{Br}'(C_{m''})$ is trivial for every $m'' \in \mathrm{Max}(C)$, and therefore $[A \otimes C_{m''}] \in \mathrm{Br}(C_{m''})$. By the preceding lemma, we therefore have that $[A \otimes C] \in \mathrm{Br}(C)$. Finally C is a semilocal ring, hence every C-Azumaya algebra may be split by a faithfully projective extension D of C, cf. Theorem 3.5.5. Consequently D splits A, and $[A \otimes D] \in \mathrm{Br}'(D/R) \cong \mathrm{Br}(D/R)$, since D/R is faithfully projective, and applying Theorem 6.2.1. □

6.8 The Brauer group modulo a nilpotent ideal

As an application, we will now show that the Brauer group remains unaffected if we divide out a nilpotent ideal. This result is originally due to DeMeyer (cf. [68]) ; his proof is fully algebraic, and is based on results of Ingraham (cf. [106]). The idea behind the following cohomological proof was communicated to the author by R. Hoobler.

Theorem 6.8.1 *If $\mathcal{N}$ is a nilpotent ideal of R, and $\overline{R} = R/\mathcal{N}$, then for all $q > 0$, we have that*

$$H^q(R_{ét}, \mathbb{G}_m) = H^q(\overline{R}_{ét}, \mathbb{G}_m).$$

Proof First suppose that $\mathcal{N}^2 = 0$, and write $\pi : R \to \overline{R}$ for the canonical surjection. We will also write, for $R \to S$ étale, $\overline{S} = S/\mathcal{N}S$. Now consider the following sheaves on $R_{ét}$: $F_{\mathcal{N}}$, given by

$$F_{\mathcal{N}}(S) = \mathcal{N}S,$$

and $\pi_*\mathbb{G}_{m,\overline{R}}$ given by

$$\pi_*\mathbb{G}_{m,\overline{R}}(S) = \mathbb{G}_m(\overline{S}).$$

We then have an exact sequence of sheaves, even of presheaves:

$$0 \longrightarrow F_{\mathcal{N}} \xrightarrow{\alpha} \mathbb{G}_m \longrightarrow \pi_*\mathbb{G}_{m,\overline{R}} \longrightarrow 1 \tag{6.36}$$

Here α is given by $\alpha(S)(ns) = 1 + ns \in \mathbb{G}_m(S)$, for $R \to S$ étale, $s \in S$, $n \in \mathcal{N}$. From Proposition B.0.4, it follows that for any étale covering $S \to T$ in $\underline{\underline{cov}}(R_{ét})$, the Amitsur cohomology groups

$$H^q(T/S, F_{\mathcal{N}}) = 0$$

disappear for $q > 0$. Hence $F_{\mathcal{N}}$ is flabby, by Theorem 5.7.2, and therefore Γ_R-acyclic, by definition 5.7.1. Thus

$$H^q(R_{ét}, F_{\mathcal{N}}) = 0$$

and it follows from the long exact sequence associated to (6.36) that

$$H^q(R_{ét}, \pi_*\mathbb{G}_{m,\overline{R}}) \cong H^q(R_{ét}, \mathbb{G}_m).$$

From Corollary 5.5.8, we know that π_* is exact, hence for all $q \geq 1$

$$\begin{aligned} H^q(\overline{R}_{ét}, \mathbb{G}_{m,\overline{R}}) &= R^q(\Gamma_R \circ \pi_*)(\mathbb{G}_{m,\overline{R}}) \\ &= R^q\Gamma_R(\pi_*\mathbb{G}_{m,\overline{R}}) \\ &= H^q(R_{ét}, \pi_*\mathbb{G}_{m,\overline{R}}) \end{aligned}$$

and this implies the result in the case where $\mathcal{N}^2 = 0$.
The theorem now follows easily for a nilpotent ideal $\mathcal{N}$: if $\mathcal{N}^n = 0$, then $(\mathcal{N}^p)^2 = 0$ if $2p \geq n$. But then $H^q(R_{ét}, \mathbb{G}_m) \cong H^q((R/\mathcal{N}^p)_{ét}, \mathbb{G}_m)$ and $\mathcal{N}/\mathcal{N}^p$ is nilpotent of order smaller then n in $R/\mathcal{N}^p$, so we may proceed using induction on n. □

Corollary 6.8.2 *If $\mathcal{N}$ is a nilpotent ideal of R, then*

$$\begin{aligned} \mathrm{Pic}(R) &\cong \mathrm{Pic}(R/\mathcal{N}) \\ \mathrm{Br}'(R) &\cong \mathrm{Br}'(R/\mathcal{N}) \\ \mathrm{Br}(R) &\cong \mathrm{Br}(R/\mathcal{N}). \end{aligned}$$

Proof The first two statements follow immediately from Theorems 6.8.2, 6.1.1 and 6.3.8. The last part now follows from Gabber's Theorem 6.7.4, after taking torsion parts. □

Corollary 6.8.3 *Let R be an Artinian ring. Then* $\mathrm{Br}(R) = \mathrm{Br}'(R) \cong H^2(R_{ét}, \mathbb{G}_m)$.

Proof In view of corollary 6.8.2, it suffices to show the result for a reduced Artinian ring. But a reduced Artinian ring is a finite direct sum of fields, by the decomposition theorem for Artinian rings, and the result follows from Proposition 3.2.2. □

6.9 The Brauer group of a regular ring

We have already seen that $\mathrm{Br}(R) = \mathrm{Br}'(R)$ if R is a field (13.2.2), a Henselian ring (13.3.1) or an Artinian ring (6.8.3). The aim of this Section is to prove that this property also holds when R is a regular ring. We will also show that $\mathrm{Br}(R)$ embeds into the Brauer group of the field of fractions of R, if R is a regular domain. We will use the following well-known properties from commutative algebras, we refer to the literature for the proof. In this Section, it is essential that we work over the étale topology.

Proposition 6.9.1 *A noetherian domain R is a unique factorization domain if and only if every prime ideal of height one is a principal ideal.*

Proof See [131, Theorem 47]. □

Proposition 6.9.2 (Auslander-Buchsbaum) *A regular local ring is a unique factorization domain.*

Proof See [131, Theorem 48]. □

Proposition 6.9.3 *An étale extension of a regular ring is regular.*

Proof See [135, I.3.17] or [162, p. 75]. □

Proposition 6.9.4 *Let k be a field. Then for every sheaf F on $k_{\acute{e}t}$, and for every $q \geq 1$, the cohomology group $H^q(k_{\acute{e}t}, F)$ is torsion. Furthermore*

$$H^1(k_{\acute{e}t}, F) = \cup_{\ell/k \text{ Galois}} H^1(\mathrm{Gal}(\ell/k), F(\ell)).$$

Proof Let C^q be the q-th kernel of an injective resolution of F, as usual. Since $H^q(k_{\acute{e}t}, F) = H^1(k_{\acute{e}t}, C^q)$, it suffices to show the statement for $q = 1$. From the exact sequence (5.11) in the case $q = 0$, we know that for any étale covering S of k,

$$H^1(S/k, F) = \mathrm{Ker}\,(H^1(k_{\acute{e}t}, F) \to H^1(S_{\acute{e}t}, F))$$

Now an étale covering S of k is of the form

$$S = \ell_1 \times \ell_2 \times \cdots \times \ell_n,$$

where each ℓ_i is a separable finite field extension of k. The normal closure $\overline{\ell}_i$ of ℓ_i is a Galois extension of k. Put $\overline{S} = \overline{\ell}_1 \times \overline{\ell}_2 \times \cdots \times \overline{\ell}_n$, then

$$H^1(S_{\acute{e}t}, F) = \oplus_i H^1(\overline{\ell}_{i,\acute{e}t}, F)$$

and consequently

$$\begin{aligned} H^1(S/k,F) &= \operatorname{Ker}(H^1(k_{\acute{e}t},F)\to H^1(S_{\acute{e}t},F)) \\ &\subset H^1(\overline{S}/k,F)=\operatorname{Ker}(H^1(k_{\acute{e}t},F)\to H^1(\overline{S}_{\acute{e}t},F)) \\ &= \cap_i \operatorname{Ker}(H^1(k_{\acute{e}t},F)\to H^1(\overline{\ell}_{i,\acute{e}t},F)) \\ &= \cap_i H^1(\overline{\ell}_i/k,F) \end{aligned}$$

Consequently every $x \in H^1(k_{\acute{e}t}, F)$ can be split by a Galois extension ℓ of k, it suffices to take one of the ℓ_i above. Therefore

$$\begin{aligned} H^1(k_{\acute{e}t},F) &= \cup_{\ell/k \text{ Galois}} \operatorname{Ker}(H^1(k_{\acute{e}t},F)\to H^1(\ell_{\acute{e}t},F) \\ &= \cup_{\ell/k \text{ Galois}} H^1(\ell/k,F) \\ &= \cup_{\ell/k \text{ Galois}} H^1(\operatorname{Gal}(\ell/k),F(\ell)) \end{aligned}$$

is torsion, since $\operatorname{Gal}(\ell/k)$ is a finite group. □

Now let $p \in \operatorname{Spec}(R)$, and denote $i(p):\ R \to R_p/pR_p = k(p)$. We then have the following result (cf. [97, II.1.1]).

Lemma 6.9.5 *For any sheaf F_p on $k(p)_{\acute{e}t}$ and $q \geq 1$, $R^q i(p)_* F_p$ is a sheaf of torsion groups, and $H^q(R_{\acute{e}t}, i(p)_*F_p)$ is a torsion group.*

Proof Recall from Proposition 5.7.5 that $R^q i(p)_* F_p$ is the sheaf associated to the presheaf

$$U \longrightarrow H^q(U\otimes k(p), F_{p|U\otimes k(p)}).$$

Now $U \otimes k(p)$ is a finite direct product of finite separable field extensions of $k(p)$:

$$U\otimes k(p) = \ell_1 \times \ell_2 \times \cdots \times \ell_n,$$

hence

$$H^q(U\otimes k(p), F_{p|U\otimes k(p)}) = \bigoplus_{i=1}^{q} H^q(\ell_{i,\acute{e}t}, F_{|\ell_i})$$

is torsion, by Proposition 6.9.4. Therefore $R^q i(p)_* F_p$ is torsion, by Lemma 5.3.3, and this proves the first statement.
From Corollary 5.7.3, it follows that we can apply Theorem A.2.2 to

$$\mathcal{S}(k(p)_{\acute{e}t}) \xrightarrow{i(p)_*} \mathcal{S}(R_{\acute{e}t}) \xrightarrow{\Gamma} \underline{\underline{Ab}}$$

resulting in a long exact sequence (take $q = 0$ in (A.4))

$$\begin{array}{ccccccc} 0 & \longrightarrow & H^1(R_{\acute{e}t}, i(p)_*F_p) & \longrightarrow & H^1(k(p)_{\acute{e}t}, F_p) & \longrightarrow & R^1 i(p)_*F_p(R) \\ & \longrightarrow & H^2(R_{\acute{e}t}, i(p)_*F_p) & \longrightarrow & H^1(R_{\acute{e}t}, i(p)_*C^1) & \longrightarrow & H^1(R, R^1 i(p)_*F_p) \\ & \longrightarrow & H^3(R_{\acute{e}t}, i(p)_*F_p) & \longrightarrow & H^2(R_{\acute{e}t}, i(p)_*C^1) & \longrightarrow & H^2(R, R^1 i(p)_*F_p) \\ & \longrightarrow & \cdots & & & & \end{array}$$

By Proposition 6.9.4, the second term in the sequence is torsion, hence $H^1(R_{\acute{e}t}, i(p)_*F_p)$ is torsion. Since this holds for every sheaf F_p, the fifth term in the sequence is

also torsion. We already have seen that the third term is torsion, and therefore $H^2(R_{\acute{e}t}, i(p)_*F_p)$ is torsion.
From Proposition 5.6.14, it follows that $H^q(R_{\acute{e}t}, (R^1i(p)_*)F_p)$ is torsion for any q. It is then easy to conclude that all terms in the above exact sequence are torsion. □

For any commutative ring S, let $Q(S)$ be the total ring of fractions of S, that is, the localization of S at the set of all non-zero divisors of S. Then Q is a sheaf of rings on $R_{\acute{e}t}$, and $\mathbb{G}_m \circ Q$ is a sheaf of abelian groups. With these notations, we have

Proposition 6.9.6 *(cf. [97, II.1.2]) If R is a regular ring, then for $q \geq 1$, the cohomology group $H^q(R_{\acute{e}t}, \mathbb{G}_m \circ Q)$ is torsion.*

Proof First observe that it follows from Propositions 6.9.2 and 6.9.3 that R is reduced. Indeed, all localizations of R are unique factorization domains, and are therefore reduced. Therefore every étale R-algebra S is reduced (use the fact that S is also regular, or invoke [162, Prop.1, p.74]). We claim that $Q(S)$ is an artinian ring. Let $p \in \text{Spec}(Q(S))$; by Auslander-Buchsbaum $Q(S)_p$ is a unique factorization domain, in which every non-zero-divisor is invertible. Thus $Q(S)_p$ is a field, so p is a minimal prime of S. Now $Q(S)$ is reduced and artinian, and it is therefore a direct sum of fields, cf. e.g. [207, Theorem 3, p.205]. This implies that

$$Q(S) = S_{q_1} \times \cdots \times S_{q_r}$$

where $q_1, \ldots, q_r$ are the minimal prime ideals of S. Therefore

$$(\mathbb{G}_m \circ Q)(S) = i(p_1)_*(\mathbb{G}_{m|R_{p_1}})(S) \oplus \cdots \oplus i(p_m)_*(\mathbb{G}_{m|R_{p_m}})(S)$$

where $p_1, \ldots, p_m$ are the minimal primes of R. Hence

$$H^q(R_{\acute{e}t}, \mathbb{G}_m \circ Q) = \oplus_{j=1}^m H^q(R_{\acute{e}t}, i(p_m)_*(\mathbb{G}_{m|R_{p_m}})$$

is torsion by Lemma 6.9.5. □

Now we introduce the sheaf of *Cartier divisors* Div_R as the sheaf cokernel of the inclusion $\mathbb{G}_m \to \mathbb{G}_m \circ Q$. We then have the following exact sequence of sheaves on $R_{\acute{e}t}$:

$$1 \longrightarrow \mathbb{G}_m \longrightarrow \mathbb{G}_m \circ Q \longrightarrow \text{Div}_R \tag{6.37}$$

Let R be a regular local ring. Then the localizations R_p at primes of height one are discrete valuation rings. Let v_p be the valuation on R_p. The sheaf of *Weil divisors* on $R_{\acute{e}t}$ is by definition the sheaf

$$\text{D}_R = \oplus_{p \in X_1(R)} i(p)_*\mathbb{Z}$$

where $i(p): R \to k(p)$ is the canonical map, $\mathbb{Z}$ is the constant sheaf on $k(p)_{\acute{e}t}$ and $X_1(R)$ is the set of prime ideals of rank one of R. Note that $\text{D}_R(S)$ is nothing else then the free abelian group on $X_1(S)$:

$$\text{D}_R(S) = \oplus_{q \in X_1(S)} \mathbb{Z}q$$

We will show below that, for R regular, the Weil divisors and the Cartier divisors are isomorphic:

Theorem 6.9.7 *If R is regular, then $\mathrm{D}_R = \mathrm{Div}_R$.*

Proof Let $R \to S$ be étale. As we have seen, $Q(S)$ is a finite direct product of fields. For a prime of height one q of S, the discrete valuation ring S_q is a subring of one of these fields, and this field is the localization of S at the minimal prime r that is contained in q. We call this field K_r, and we write

$$X_{1,r} = \{q \in X_1(S) | r \subset q\}$$

Then for every $f \in \mathbb{G}_m(K)$,the number of $q \in X_{1,r}$ such that $v_q(f) \neq 0$ is finite. Indeed, let $f = a/b$, with $a \in R$, $b \in R \setminus r$. If $v_q(f) < 0$, then $b \in q$, and this is possible only for a finite number of q. Similarly, if $v_q(f) > 0$, then $a \in q$, leading to the same conclusion. We therefore have a well-defined map

$$\mathbb{G}_m(K_r) \longrightarrow \oplus_{q \in X_{1,r}} \mathbb{Z}q$$

mapping f to $\sum v_q(f)q$, and this leads to a map

$$\alpha(S) : \mathbb{G}_m(Q(S)) = \mathbb{G}_m(\oplus_{r \in X_0(S)} K_r) \longrightarrow \oplus_{q \in X_1(S)} \mathbb{Z}q = \mathrm{D}_R(S)$$

It is obvious that Ker $\alpha(S) = \mathbb{G}_m(S)$, and that α defines a map between sheaves. We therefore have a short exact sequence of sheaves

$$1 \longrightarrow \mathbb{G}_m \longrightarrow \mathbb{G}_m \circ Q \longrightarrow \mathrm{D}_R$$

We claim that $\alpha(S)$ is surjective if S is a unique factorization domain. Indeed, in this case every $q \in X_1(S)$ is principal, let s_q be a generator for q. Then

$$\alpha(S)(\sum s(q)^{n_q}) = \sum n_q q$$

so $\alpha(S)$ is surjective. We can now prove that α is surjective as a map between sheaves: take $p \in \mathrm{Spec}(S)$, and fix $d \in \mathrm{D}_R(S)$. Then $\mathbb{G}_m(Q(S_p)) \longrightarrow \mathrm{D}_R(S_q)$ is surjective, since S_p is a unique factorization domain. We can therefore find $a \in S$ such that the image of d in $\mathrm{D}_R(S_a)$ is in the image of $\mathbb{G}_m(Q(S_a))$. A finite number of the $\mathrm{Spec}(S_a)$ will cover $\mathrm{Spec}(S)$, leading to a Zariski covering T of S, with the image of d in $\mathrm{D}_R(T)$ lying in $\mathrm{Im}(\mathbb{G}_m(Q(T)) \to \mathrm{D}_R(T)$. We therefore have an exact sequence

$$1 \longrightarrow \mathbb{G}_m \longrightarrow \mathbb{G}_m \circ Q \longrightarrow \mathrm{D}_R \longrightarrow 1$$

This shows that D_R is the sheaf cokernel of $\mathbb{G}_m \to \mathbb{G}_m \circ Q$, hence $\mathrm{D}_R \cong \mathrm{Div}_R$. □

Corollary 6.9.8 *(cf. [97, Prop. II.1.4]) If R is a regular ring, then for all $q \geq 2$, the cohomology group $H^q(R_{ét}, \mathbb{G}_m)$ is torsion.*

Proof (6.37) results in a long exact sequence

$$\begin{array}{ccccccc} 1 & \longrightarrow & \mathbb{G}_m(R) & \longrightarrow & \mathbb{G}_m(Q(R)) & \longrightarrow & \mathrm{Div}_R(R) \\ & \longrightarrow & H^1(R_{ét}, \mathbb{G}_m) & \longrightarrow & H^1(R_{ét}, \mathbb{G}_m \circ Q) & \longrightarrow & H^1(R_{ét}, \mathrm{Div}_R) \\ & \longrightarrow & H^2(R_{ét}, \mathbb{G}_m) & \longrightarrow & H^2(R_{ét}, \mathbb{G}_m \circ Q) & \longrightarrow & H^2(R_{ét}, \mathrm{Div}_R) \\ & \longrightarrow & \cdots & & & & \end{array} \qquad (6.38)$$

From Lemma 6.9.5, Proposition 6.9.6 and Theorem 6.9.7, it follows that $H^q(R_{ét}, \mathbb{G}_m \circ Q)$ and

$$H^q(R_{ét}, \mathrm{Div}_R) = H^q(R_{ét}, \mathrm{D}_R) = H^q(R_{ét}, \oplus_{p \in X_1} i(p)_* \mathbb{Z})$$

are torsion. The result then follows immediately. □

Corollary 6.9.9 *If R is a regular ring, then* $\mathrm{Br}(R) = \mathrm{Br}'(R) \cong H^2(R_{ét}, \mathbb{G}_m)$.

Proof Combine Theorems 6.3.8, 6.7.4 and 6.9.8. □

As another application of the fact that the Cartier and Weil divisors are isomorphic over a regular ring, let us prove the following:

Theorem 6.9.10 *If R is a regular domain with field of fractions K, then* $\mathrm{Br}(R) \to \mathrm{Br}(K)$ *is monomorphic.*

Proof If R is a domain, then $Q(R) = K$. We know from the above arguments that

$$\mathrm{Div}_R = \mathrm{D}_R = \oplus_{p \in X_1} i(p)_* \mathbb{Z}$$

We will first consider the cohomology groups of $i(p)_*\mathbb{Z}$. Apply Theorem A.2.2 to

$$\mathcal{S}(k(p)_{ét}) \xrightarrow{i(p)_*} \mathcal{S}(R_{ét}) \xrightarrow{\Gamma} \underline{\underline{Ab}}$$

to obtain a long exact sequence

$$0 \longrightarrow H^1(R_{ét}, i(p)_*\mathbb{Z}) \longrightarrow H^1(k(p)_{ét}, \mathbb{Z}) \longrightarrow (R^1 i(p)_*)\mathbb{Z} \longrightarrow \cdots \quad (6.39)$$

Now

$$\begin{aligned} H^1(k(p)_{ét}, \mathbb{Z}) &= \cup_{\ell/k(p) \text{ Galois}} H^1(\mathrm{Gal}(\ell/k(p)), \mathbb{Z}) \\ &= \cup_{\ell/k(p) \text{ Galois}} \mathrm{Hom}(\mathrm{Gal}(\ell/k(p)), \mathbb{Z}) = 0 \end{aligned}$$

Consequently $H^1(R_{ét}, i(p)_*\mathbb{Z}) = 0$, and

$$H^1(R_{ét}, \mathrm{Div}_R) = 0 \quad (6.40)$$

Finally, apply Theorem A.2.2 to

$$\mathcal{S}(K_{ét}) \xrightarrow{g_*} \mathcal{S}(R_{ét}) \xrightarrow{\Gamma} \underline{\underline{Ab}}$$

where $g : R \to K$ is the natural inclusion, in fact $g_* = \mathbb{G}_m \circ Q$. The long exact sequence (A.4) now takes the form

$$\begin{array}{cccccc} 0 & \longrightarrow & H^1(R_{ét}, g_*\mathbb{G}_m) & \longrightarrow & H^1(K_{ét}, \mathbb{G}_m) = \mathrm{Pic}(K) = 1 & \longrightarrow & (R^1 g_*)(\mathbb{G}_m) \\ & \longrightarrow & H^2(R_{ét}, g_*\mathbb{G}_m) & \longrightarrow & \mathrm{Ker}\,(H^2(K_{ét}, \mathbb{G}_m) \to (R^2 g_*)(\mathbb{G}_m)(R)) & \longrightarrow & \cdots \end{array} \quad (6.41)$$

Now by Proposition 5.7.5, $(R^1 g_*)\mathbb{G}_m$ is the sheaf associated to the presheaf

$$U \longrightarrow H^1(U \otimes K, \mathbb{G}_m) = \mathrm{Pic}(U \otimes K) =$$

so $(R^1g_*)\mathbb{G}_m = 1$, and

$$H^2(R_{\acute{e}t}, g_*\mathbb{G}_m) \hookrightarrow H^2(K_{\acute{e}t}, \mathbb{G}_m) = \mathrm{Br}(K) \qquad (6.42)$$

Now it follows from (6.38) and (6.40) that

$$H^2(R_{\acute{e}t}, \mathbb{G}_m) = \mathrm{Br}(R) \hookrightarrow H^2(R_{\acute{e}t}, g_*\mathbb{G}_m)$$

and therefore

$$\mathrm{Br}(R) \hookrightarrow \mathrm{Br}(K)$$

□

Remark 6.9.11 The original proof of Theorem 6.9.10 (cf. [12, Theorem 7.2]) is purely algebraic and uses the theory of maximal orders. A detailed version of it may also be found in [155, Theorem 6.19]. If R is not regular, then $\mathrm{Br}(R) \to \mathrm{Br}(K)$ may fail to be regular, even if R is a unique factorization domain. An elementary example is the following (cf. [12, 155]): take $R = \mathbb{R}[x,y]/(x^2+y^2)$, and let $A = \mathbb{H}[x,y]/(x^2+y^2)$. Then A is central and separable, and therefore Azumaya. The image of $[A]$ in $\mathrm{Br}(R)$ is nontrivial, so $[A] \neq 1$ in $\mathrm{Br}(R)$. In $K = Q(R)$, $x^2/y^2 = -1$, hence $K \otimes A \cong M_2(K)$, and $[K \otimes A] = 1$in $\mathrm{Br}(K)$. For a more detailed study of the map $\mathrm{Br}(R) \to \mathrm{Br}(K)$, we refer to [72].
If R is regular of dimension at most two, for example, if R is a Dedekind domain, then we have the following result:

Theorem 6.9.12 *If R is a regular ring of dimension at most two, then*

$$\mathrm{Br}(R) = \cap_{p\in X_1(R)}\mathrm{Br}(R_p) \subset \mathrm{Br}(K)$$

Proof We refer to [155, Theorem 6.33] □

6.10 Further results and examples

In this Section we present a survey of some explicit computations of the Brauer group. It would lead us to far to go into all the details, and this is why we refer to the literature for most of the proofs.

Local fields

We have seen that the Brauer group of an algebraically closed field (more generally, of a strictly Henselian ring) is trivial. In particular, $\mathrm{Br}(\mathbb{C}) = 1$. Using Proposition 6.2.8, we easily obtain that

$$\mathrm{Br}(\mathbb{R}) = \mathrm{Br}(\mathbb{C}/\mathbb{R}) = H^2(\mathrm{Gal}(\mathbb{C}/\mathbb{R}), \mathbb{G}_m(\mathbb{C})) = \mathbb{G}_m(\mathbb{R})/N_{\mathbb{C}/\mathbb{R}}(\mathbb{G}_m(\mathbb{C})) = \mathbb{G}_m(\mathbb{R})/\mathbb{R}_+ = \mathbb{Z}/2\mathbb{Z}$$

The nontrivial element in $\mathrm{Br}(\mathbb{R})$ is represented by the quaternions $\mathbb{H} = (\mathbb{C}, C_2, -1)$. One of Wedderburn's Theorems states that every finite division ring is commutative

(see for example [110, Satz 11.2]), and this implies that the Brauer group of a finite field is trivial: $\mathrm{Br}(\mathbb{F}_q) = 1$.
If K is a local field (for example, $K = \mathbb{Q}_p$), then

$$\mathrm{Br}(K) = \mathbb{Q}/\mathbb{Z} \tag{6.43}$$

We give a brief description of the isomorphism, and refer to [110, §16] for further details. Let R be the ring of integers of K, m the maximal ideal of R, $k = R/mR$ the residue class field, $q = \#(k)$, and π a generator of the maximal ideal m. It is well-known that the splitting field K_n of the polynomial $X^{q^n-1} - 1$ is a cyclic Galois extension of K, with Galois group of order n. Furthermore $\mathrm{Gal}(K_n/K) \cong \mathrm{Gal}(k_n/k)$, where k_n is the residue class field of K_n. The K-automorphism $\varphi_n : K_n \to K_n$ that corresponds to the k-automorphism $k_n \to k_n : x \mapsto x^q$ is called the Frobenius automorphism of K_n, and φ_n generates $G = \mathrm{Gal}(K_n/K)$. We now have that

$$\mathrm{Br}(K_n/K) \cong \mathbb{Z}/n\mathbb{Z} \cong \frac{1}{n}\mathbb{Z}/\mathbb{Z}$$

Here k/n corresponds to the cyclic algebra (F_n, G, f_{π^k}). Finally

$$\mathrm{Br}(K) = \bigcup_{n\geq 1} \mathrm{Br}(K_n/K) \cong \bigcup_{n\geq 1} \frac{1}{n}\mathbb{Z}/\mathbb{Z} = \mathbb{Q}/\mathbb{Z}$$

A more general result is due to Witt ([202]): if K is a complete discretely valued field with perfect residue class field k, then $\mathrm{Br}(K)$ fits into an exact sequence

$$0 \longrightarrow \mathrm{Br}(k) \longrightarrow \mathrm{Br}(K) \longrightarrow \mathrm{Hom}_{\mathrm{cont}}(\mathrm{Gal}(\overline{k}/k, \mathbb{Q}/\mathbb{Z} \longrightarrow 0 \tag{6.44}$$

where $\overline{k}$ is the algebraic closure of k. A proof of (6.44) may be found in [168, XII, §3, Theorem 2]

Number fields

Let K be a number field, that is, a finite field extension of $\mathbb{Q}$, and let R be the ring of integers in K. Discrete valuations on K correspond to non-zero prime ideals of R, and the completion K_v of K with respect to the non-archimedean distance on K induced by the valuation v is a local field. Thus $\mathrm{Br}(K_v) = \mathbb{Q}/\mathbb{Z}$.
An archimedean valuation on K is an embedding $\varphi : K \hookrightarrow \mathbb{C}$. The formula $d(x, y) = |\varphi(x - y)|$ defines a distance on K, and the completion K_φ of K with respect to this distance is either $\mathbb{R}$ or $\mathbb{C}$. The Brauer group of this completion is either $\frac{1}{2}\mathbb{Z}/\mathbb{Z}$ or $\{1\}$, and we regard them as subgroups of $\mathbb{Q}/\mathbb{Z}$.
Now let V be the set of discrete and archimedean valuations on K. For every $v \in V$, we have a map

$$\mathrm{Br}(K) \longrightarrow \mathrm{Br}(K_v)$$

and one can show that the image of a class in $\mathrm{Br}(K)$ in $\mathrm{Br}(K_v)$ is trivial for all but a finite number of v. We obtain a map

$$\mathrm{Br}(K) \longrightarrow \oplus_{v\in V} \mathrm{Br}(K_v)$$

Applying Class Field Theory, we may show that this maps fits into an exact sequence

$$0 \longrightarrow \mathrm{Br}(K) \longrightarrow \oplus_{v\in V} \mathrm{Br}(K_v) \xrightarrow{\text{sum}} \mathbb{Q}/\mathbb{Z} \longrightarrow 0 \tag{6.45}$$

For more details, we refer to [158, Theorem 18.5].

Number rings

The exact sequence (6.45) together with Theorem 6.9.12 can be used to compute the Brauer group of the ring of integers R of a number field K. The clue result is that we have an exact sequence

$$0 \longrightarrow \mathrm{Br}(R_p) \longrightarrow \mathrm{Br}(K) \longrightarrow \mathrm{Br}(K_p) \tag{6.46}$$

for every nonzero prime ideal of R. Here K_p is the completion with respect to the valuation corresponding to p. A proof of (6.46) may be found in [155, Prop. 6.34]. From (6.46), it follows that we have an exact sequence

$$0 \longrightarrow \cap_{p\in X_1(R)} \mathrm{Br}(R_p) \longrightarrow \mathrm{Br}(K) \longrightarrow \oplus_{p\in X_1(R)} \mathrm{Br}(K_p) \tag{6.47}$$

Using Theorem 6.9.12, this sequence takes the form

$$0 \longrightarrow \mathrm{Br}(R) \longrightarrow \mathrm{Br}(K) \longrightarrow \oplus_{p\in X_1(R)} \mathrm{Br}(K_p) \tag{6.48}$$

Now it is clear that

$$\oplus_{v\in V} \mathrm{Br}(K_p) = \oplus_{p\in X_1(R)} \mathrm{Br}(K_p) \oplus \mathrm{Br}(\mathbb{R})^r \oplus \mathrm{Br}(\mathbb{C})^s$$

With r the number of embeddings of R (or K) in $\mathbb{R}$. It follows that

$$\mathrm{Br}(R) = \begin{cases} (\mathbb{Z}/2\mathbb{Z})^{r-1} & \text{if } r \geq 1 \\ 1 & \text{if } r = 0 \end{cases} \tag{6.49}$$

In particular

$$\mathrm{Br}(\mathbb{Z}) = 1 \quad \text{and} \quad \mathrm{Br}(\mathbb{Z}[\sqrt{2}]) = \mathbb{Z}/2\mathbb{Z} \tag{6.50}$$

The exact sequence (6.48) is a special case of the following exact sequence due to Auslander and Brumer [11]. This exact sequence also generalizes Witt's result (6.44). Let R be a semilocal Dedekind domain, and suppose that $k(p) = R/p$ is perfect for all $p \in X_1(R) = \max(R)$. If K is the field of quotients of R, then we have an exact sequence

$$0 \longrightarrow \mathrm{Br}(R) \longrightarrow \mathrm{Br}(K) \longrightarrow \oplus_{p\in X_1(R)} \mathrm{Hom}_{\mathrm{cont}}(\mathrm{Gal}(\overline{k(p)}/k(p)), \mathbb{Q}/\mathbb{Z}) \longrightarrow 0 \tag{6.51}$$

Function fields

Let K be a field. We say that K satisfies condition (C_1) if every homogeneous polynomial in $K[x_1, \ldots, x_n]$ of degree less then n has a nontrivial zero.

Proposition 6.10.1 *If K satisfies condition (C_1), then* $\mathrm{Br}(K) = 1$.

Proof (following [155]). Let D be a central division ring over K, of dimension $n^2 > 1$. Let $a_1, \ldots, a_{n^2}$ be a basis of D over K, and consider the reduced norm (or determinant) $\det : D \to K$.

$$f(X_1, \ldots, X_{n^2}) = \det(\sum_{i=1}^{n^2} X_i a_i)$$

is a homogeneous polynomial of degree n^2 in the X_i's. Now $n > 1$, so $n < n^2$, and f has a nontrivial zero. Thus there exists a nonzero $d \in D$ such that $\det(d) = 0$. Now d has an inverse $d^{-1} \in D$, and $\det(d)\det(d^{-1}) = \det(dd^{-1}) = 1$, which is a contradiction. □

Proposition 6.10.2 *If k is an algebraically closed field then $K = k(X)$ satisfies (C_1).*

Proof (following [155]). Let k be an algebraically closed field, and suppose that f is a homogeneous polynomial in $k(X)[X_1, \ldots, X_n]$ of degree $d < n$. Multiplying f by a suitable polynomial in $k[X]$, we find a homogeneous polynomial with coefficients in $k[X]$. It suffices to show that this new polynomial, which we still call f, has a nontrivial zero.
To this end, we have to find n polynomials

$$x_i = \sum_{j=0} m y_{ij} X^j$$

such that $(x_1, \ldots, x_n)$ is a zero of f. We treat the y_{ij} as the new variables. If we substitute the x_i in the equation $f(x_1, \ldots, x_n) = 0$, then our equation reduces to an equation of the form

$$f_0(Y) + f_1(Y)X + \cdots + f_{dmr+r}X^{dm+r} = 0$$

Here r is the maximal degree of the coefficients of f, Y is the vector with entries $\{y_{ij} | 1 \leq i \leq n,\ 0 \leq j \leq m\}$, and the f_i's are homogeneous polynomials in the y_{ij}'s. Our equation now reduces to the simultaneous equations

$$\begin{aligned} f_0(y_{10}, \ldots, y_{nm}) &= 0 \\ f_1(y_{10}, \ldots, y_{nm}) &= 0 \\ &\vdots \\ f_{dm+r}(y_{10}, \ldots, y_{nm}) &= 0 \end{aligned}$$

This is a set of $dm+r+1$ simultaneous homogeneous equations in $n(m+1)$ variables in k. For $m > (r-n+1)/(n-d)$, there are more variables than forms, and then there is always a nontrivial solution, since the field k is algebraically closed (see [208, p. 209]). □

If a field K satisfies C_1, then every algebraic extension of K satisfies C_1; we refer to [155, Lemma 8.3] for the proof. It follows from this result and Propositions 6.10.1 and 6.10.2 that every algebraic extension of the function field of an algebraically closed field is trivial. This result is known as *Tsen's Theorem* .

Polynomial rings

Lemma 6.10.3 *Let R be a commutative ring, and write $\mathcal{B}'(R[x])$ for the cokernel of the map* $\mathrm{Br}(R) \to \mathrm{Br}(R[x])$. *Then we have a split exact sequence*

$$1 \longrightarrow \mathrm{Br}(R) \longrightarrow \mathrm{Br}(R[x]) \longrightarrow \mathcal{B}'(R[x]) \longrightarrow 1 \qquad (6.52)$$

Proof The natural map $i: R \to R[x]$ has a left inverse p, mapping x to 0. The composition

$$\mathrm{Br}(R) \xrightarrow{\mathrm{Br}(i)} \mathrm{Br}(R[x]) \xrightarrow{\mathrm{Br}(p)} \mathrm{Br}(R) \longrightarrow 1$$

is the identity, so $\mathrm{Br}(i)$ has a right inverse, and this shows that (6.52) is split. □

When is $\mathrm{Br}(i)$ an isomorphism? We will first investigate the case where $R = k$ is a field.

Lemma 6.10.4 *Let A be a $k[x]$-Azumaya algebra. Then $[A] \in \mathrm{Im}(\mathrm{Br}(i))$ if and only if A can be split by a $k[x]$-algebra $l[x]$ with l a Galois extension of k.*

Proof One implication is obvious. Suppose conversely that A is split by $l[x]$, with l a Galois extension of k. Then $l[x]$ is a Galois extension of $k[x]$, with the same Galois group G, and $\mathrm{Pic}(l[x]) = 1$, since $l[x]$ is a PID. Therefore

$$[A] \in \mathrm{Br}(l[x]/k[x]) \cong H^2(G, \mathbb{G}_m(l[x])) = H^2(G, \mathbb{G}_m(l)$$

Thus A is equivalent to a crossed product

$$B = \bigoplus_{\sigma \in G} l[x]u_\sigma = \left(\bigoplus_{\sigma \in G} lu_\sigma\right)[x] = C[x]$$

with C a central simple k-algebra. This proves the Lemma. □

Theorem 6.10.5 (Auslander-Goldman [12]) *The map* $\mathrm{Br}(i): \mathrm{Br}(k) \to \mathrm{Br}(k[x])$ *is an isomorphism if and only if k is perfect.*

Proof Assume first that k is perfect. It suffices to show that $\mathrm{Br}(i)$ is surjective (Lemma 6.10.3). Let $\overline{k}$ be the algebraic closure of k. From Propositions 6.10.1 and 6.10.2, we know that $\mathrm{Br}(\overline{k}(x)) = 1$. Now $\overline{k}[x]$ is regular, so $\mathrm{Br}(\overline{k}[x])$ injects into $\mathrm{Br}(\overline{k}(x))$ and is therefore trivial. Take a $k[x]$-Azumaya algebra A. A is split by $\overline{k}[x]$, and by therefore by $l[x]$ for some finite field extension of k. l is separable over k, since k is perfect, and the normalization l' of l is a Galois extension of k. $l'[x]$ splits A, and it follows from Lemma 6.10.4 that $A \in \mathrm{Br}(i)$.
Conversely, suppose that k is a non-perfect field. Let p be the characteristic of k and take $c \in k \setminus k^p$. Write $R = k[x]$ and $S = R[y]/(y^p - y - x)$. Then S is a cyclic Galois extension of R, with Galois group $G = C_p$ (see Theorem 11.3.4). Consider the cyclic algebra $A = (S, G, f_c)$. We claim that A represents a non-trivial element in $\mathrm{Br}(R)$. To this end, we consider $K = k(x)$ and $L = K \otimes_R S = K[y]/(y^p - y - x)$. It will be sufficient to show that $K \otimes_R A = (L, G, f_c)$ is nontrivial in $\mathrm{Br}(K)$, or, equivalently, that c is not a norm in $N_{L/K}(L)$ (see Proposition 6.2.8).
An element of L can be written under the form s/r, with $s \in S$ and $r \in R = k[x]$. s can be written in a unique way as

$$s = a_0 + a_1 y + \cdots + a_{p-1} y^{p-1}$$

where the a_i lie in $R = k[x]$. Now suppose that

$$c = N_{L/K}(s/r) = \frac{N_{L/K}(s)}{N_{L/K}(r)} = \frac{N_{L/K}(s)}{r^p}$$

It follows that the polynomial r^p divides $N_{L/K}(s) = s(\sigma \cdot s)(\sigma^2 \cdot s) \cdots (\sigma^{p-1} \cdot s)$ in $k[x]$. Consequently r divides s in S, and we can assume that $r = 1$.
As we have remarked earlier, the norm of s is also given by the determinant of the endomorphism of S (or L) given by left multiplication by s. $\{1, y, \ldots, y^{p-1}\}$ is a free basis of S over R, and the matrix M representing the left multiplication by s with respect to this basis is

$$M = \begin{pmatrix} a_0 & xa_{p-1} & xa_{p-2} & \cdots & xa_1 \\ a_1 & a_0 + a_{p-1} & xa_{p-1} + a_{p-2} & \cdots & xa_2 + a_1 \\ a_2 & a_1 & a_0 + a_{p-1} & \cdots & xa_3 + a_2 \\ a_3 & a_2 & a_1 & \cdots & xa_4 + a_3 \\ \vdots & \vdots & \vdots & & \vdots \\ a_{p-2} & a_{p-3} & a_{p-4} & \cdots & xa_{p-1} + a_{p-2} \\ a_{p-1} & a_{p-2} & a_{p-3} & \cdots & a_0 + a_{p-1} \end{pmatrix}$$

Let N be the maximum of the degrees of the polynomials $a_0, a_1, \ldots, a_{p-1}$, and let m be the maximal index for which $\deg(a_m) = N$. Write $a_m = \alpha_m x^N + \cdots$. The coefficient of highest degree in $\det(M)$ is then

$$\alpha_m^p x^{m+pN}$$

and

$$m + pN = \deg(\det(M)) = \deg(c) = 0$$

and it follows that $m = N = 0$. This means that a_0 is a constant, and that all the other a_i equal zero. But then

$$\det(M) = a_0^p = c$$

which is impossible since c is not a p-th power, by assumption. Thus c is not a norm, and $[A]$ represents a nontrivial element in $\mathrm{Br}(R)$.
Finally, we show that the image of $[A]$ in $\mathrm{Br}(k)$ is trivial. Write $T = k[y]/(y^p - y) = S \otimes_R k$. T is not a field, but, as an R-algebra, it is still a Galois extension of k. Now

$$A \otimes_R k = (T, G, f_c)$$

An easy computation shows that $N_{T/k}(c + (1-c)y^{p-1}) = c$, proving that $A \otimes_R k = 1$ in $\mathrm{Br}(k)$. □

Theorem 6.10.6 (Auslander-Goldman [12]) *Suppose that k is a field of characteristic p, and that l is a finite separable field extension of k. Then the map $f: \mathrm{Br}(k[x]) \to \mathrm{Br}(l[x])$ restricts to a monomorphism $f' : \mathcal{B}'(k[x]) \hookrightarrow \mathcal{B}'(l[x])$. Furthermore, every element in $\mathcal{B}'(k[x])$ has order a power of p.*

Proof From Lemma 6.10.4, it follows that $l[x]$ does not split any nontrivial element of $\mathcal{B}'(k[x])$, and therefore the restriction of f to $\mathcal{B}'(k[x])$ is injective. Now look at the commutative diagram

$$\begin{array}{ccc} \mathrm{Br}(k[x]) & \longrightarrow & \mathrm{Br}(k) \\ \downarrow & & \downarrow \\ \mathrm{Br}(l[x]) & \longrightarrow & \mathrm{Br}(l) \end{array}$$

Observing that $\mathcal{B}'(k[x]) = \mathrm{Ker}\,(\mathrm{Br}(k[x]) \to \mathrm{Br}(k))$, it follows that the image of $\mathcal{B}'(k[x])$ under f is contained in $\mathcal{B}'(l[x])$.
Now let $[A]$ be a nontrivial element in $\mathcal{B}'(k[x])$. From the proof of Lemma 6.10.4, it follows that $[A]$ can be split by a $k[x]$-algebra of the form $\tilde{l}[x]$, where $\tilde{l}$ is a finite field extension of k. $\tilde{l}$ is not necessarily separable over over k, since k is not necessarily perfect. Consider a maximal separable subfield l of $\tilde{l}$. The restriction f' of f to $\mathcal{B}'(k[x])$ is injective, by the first part of the proof, and $g : \mathrm{Br}(l[x]) \to \mathrm{Br}(l(x))$ is also injective, since $l[x]$ is regular. Therefore $g(f([A]))$ in $\mathrm{Br}(l(x))$ and $[A]$ in $\mathcal{B}'(k[x])$ have the same order. $g(f([A]))$ is split by $\tilde{l}(x)$, which is a purely inseparable extension of $l(x)$. The order of $g(f([A]))$ divides $[\tilde{l}(x) : l(x)]$ which is a primary number. □

Let us now consider polynomial rings over commutative rings. As an immediate consequence of Theorem 6.10.5, we have the following result.

Corollary 6.10.7 *Let R be a regular domain of characteristic zero. Then the canonical map* $\mathrm{Br}(R) \to \mathrm{Br}(R[x])$ *is an isomorphism.*

Proof Let K be the field of fractions of R and consider the commutative diagram

$$\begin{array}{ccc} \mathrm{Br}(R[x]) & \xrightarrow{\alpha} & \mathrm{Br}(R) \\ \downarrow{\scriptstyle\gamma} & & \downarrow \\ \mathrm{Br}(K[x]) & \xrightarrow{\beta} & \mathrm{Br}(K) \end{array}$$

K has characteristic zero and is perfect, and β is injective by Theorem 6.10.5. $R[x]$ is regular, because R is regular (see [143]), and therefore γ is also regular, by Theorem 6.9.10. It follows that α is also injective, proving the result. □

The next two results are generalizations of Theorem 6.10.5. We omit the proofs.

Proposition 6.10.8 (DeMeyer [67, Theorem 8]) *If R is a von Neumann regular ring, then* $\mathrm{Br}(R) \cong \mathrm{Br}(R[x])$ *if and only if R/m is perfect for every maximal ideal m of R. In particular, if R is a Boolean ring, then* $\mathrm{Br}(R) \cong \mathrm{Br}(R[x])$.

Proposition 6.10.9 (Hoobler [104, 1.6]) *If n is invertible in R, then the natural map* ${}_n\mathrm{Br}(R) \cong {}_n\mathrm{Br}(R[x])$ *is an isomorphism.*

Remark 6.10.10 For completeness sake, we mention the following result, giving the relation between the Picard group of a commutative ring R and the polynomial rings $R[x_1, \ldots, x_n]$. A ring R is called *seminormal* if it is reduced and whenever $b, c \in R$ satisfy $b^3 = c^2$, there is an $a \in R$ with $a^2 = b$ and $a^3 = c$. A result of Swan [170] states that, for any commutative ring R, the following conditions are equivalent.

- $\mathrm{Pic}(R) = \mathrm{Pic}(R[x_1, \ldots, x_n])$, for some $n \geq 1$;
- $\mathrm{Pic}(R) = \mathrm{Pic}(R[x_1, \ldots, x_n])$, for every $n \geq 1$;
- R_{red} is seminormal.

Laurent polynomial rings

Consider a Laurent-polynomial ring $R[x, x^{-1}]$. The Brauer group of $R[x, x^{-1}]$ has been investigated by several authors, see for example [69], [128], [104] and [83]. The most general result is due to Ford [84]. Suppose that R is a connected ring, and that n is invertible in R. Then the natural map

$$\mathrm{Pic}(R[x]) \longrightarrow \mathrm{Pic}(R[x, x^{-1}])$$

is injective ([84, (13)]). If R is a normal domain, then

$$\mathrm{Pic}(R) = \mathrm{Pic}(R[x]) = \mathrm{Pic}(R[x, x^{-1}])$$

(see [104, Prop. 2.1]). Now write

$$C = \mathrm{Pic}(R[x, x^{-1}])/\mathrm{Pic}(R[x])$$

Then we have a natural exact sequence

$$1 \longrightarrow nC \longrightarrow C \longrightarrow \mathrm{Gal}(R, RC_n^*)$$

where $\mathrm{Gal}(R, RC_n^*)$ is the group of Galois extensions of R with Galois group C_n (see Chapter 10). The main result in [84] is that we have a split exact sequence

$$1 \longrightarrow {}_n\mathrm{Br}(R) \longrightarrow {}_n\mathrm{Br}(R[x, x^{-1}]) \longrightarrow \mathrm{Gal}(R, (RC_n)^*)/(C/nC) \qquad (6.53)$$

Take a cyclic Galois extension S of R. The corresponding element in $\mathrm{Br}(R[x, x^{-1}])$ is represented by the Azumaya algebra

$$A = S\#R[\sqrt[n]{t}, 1/\sqrt[n]{t}] = (S, C_n, f_t)$$

We refer to Section 13.4 for the definition of the smash product $\#$; see also [85].

Power series rings

For any commutative ring R, the natural map

$$\mathrm{Br}(R) \longrightarrow \mathrm{Br}(R[[x]])$$

is an isomorphism. This result is due to DeMeyer [67, Cor. 4], and follows from the following fact (see [67, Theorem 2]): if $I \subset J(R)$ is an ideal of R, and R is I-adically complete, then the natural map

$$\mathrm{Br}(R) \longrightarrow \mathrm{Br}(R/I)$$

is an isomorphism. In the special case where R is local and I is the maximal idela of R, this result also follows from Proposition 3.3.2, together with the fact that a complete local ring is Henselian.

Geometric examples

Let $A = \mathbb{C}[x,y]/(y^2 - x^2(x+1))$ be the affine coordinate ring of the nodal cubic curve $y^2 - x^2(x+1) = 0$. For any integer $n \geq 1$, consider the polynomial $f_n = zy^{n-1} - x^n \in \mathbb{C}[x,y,z]$, and consider the following subring B_n of $\mathbb{C}[x,y,z]$ localized at f_n:

$$B_n = \{\frac{g}{f_n^r} \in \mathbb{C}[x,y,z]_{f_n} | g \text{ is homogeneous of degree } rn\}$$

In [83], Ford proves that

$$\mathrm{Br}(A \otimes_{\mathbb{C}} (B_{n_1} \times \cdots \times B_{n_r})) = C_{n_1} \times \cdots \times C_{n_r}$$

and this allows him to conclude that every finite abelian group is the Brauer group of a commutative ring.
In [76], the Brauer group of a real curve is investigated. If I is an ideal of coheight one in $\mathbb{R}[T_1, \ldots, T_n]$, then

$$\mathrm{Br}(\mathbb{R}[T_1, \ldots, T_n]/I) = (\mathbb{Z}/2\mathbb{Z})^s$$

where s is the number of real components of the curve $X = \mathrm{Spec}(R)$.
The Brauer group of a surface over an algebraically closed field is studied in [70]. Suppose that k is an algebraically closed field of characteristic zero and R is a normal two-dimensional positively graded k-algebra of finite type with $R_0 = k$. If the only singularities on Spec(A) are rational, then $\mathrm{Br}(R) = 1$ (see [70, Theorem 9]).
We end this Section with the following recent result, due to Hongnian Li ([123]). Set k be the real or complex field, and suppose that $k \subset R \subset k[x,y]$, with R normal and $k[x,y]$ integral over R. Then A is isomorphic to $\mathbb{C}[x,y]^H$ where H is a finite subgroup of $\mathrm{Aut}_{\mathbb{R}}(\mathbb{C}[x,y])$, and $\mathrm{Br}(R) = \mathrm{Br}(k)$.

6.11 The Brauer group of a scheme and further generalizations

Schemes and sheaves of modules

We briefly recall the definition of a scheme, and some of its elementary properties. For a more detailed discussion, we refer to Hartshorne's book [100]. In the sequel, we will follow Hartshorne's notations.

Let R be a commutative ring. The Zariski topology makes the set of prime ideals Spec(R) into a topological set. The closed sets for the Zariski topology are

$$V(I) = \{p \in \mathrm{Spec}(R) | I \subset p\}$$

where I runs through all ideals of R. The open sets are denoted $D(I) = \mathrm{Spec}(R) \setminus V(I)$. To any R-module M, a sheaf of abelian groups $\widetilde{M}$ on Spec(R) is defined as follows: for an open set determined by a principal ideal (f), one puts

$$\widetilde{M}(D(f)) = \Gamma(D(f),\ \widetilde{M}) = M \otimes R_f$$

For an arbitrary open set U of Spec(R), the description of $\widetilde{M}(U)$ is somewhat more complicated: $\widetilde{M}(U)$ is the set of functions

$$s:\ U \mapsto \coprod_{p \in U} M \otimes R_p$$

such that $s(p) \in M \otimes R_p$ for all $p \in U$ and such that the following condition holds: for all $p \in U$, there exist $f \in R \setminus p$ and $m \in M$ such that for all $q \in U$:

$$f \in R \setminus q(\text{ or } q \in D(f)\ ) \Longrightarrow\ s(q) = \frac{m}{f} \in M \otimes R_q$$

$\tilde{R}$ is a sheaf of commutative rings on Spec(R). Since the stalk $\tilde{R}_p$ at a prime ideal p is nothing else then the localized ring R_p, (Spec(R), $\tilde{R}$) is a so-called *locally ringed space*, and we call (Spec(R), $\tilde{R}$) an *affine scheme*. The sheaf $\widetilde{M}$ is then a sheaf of $\tilde{R}$-modules.

A locally ringed space $(X, \mathcal{O}_X)$ is called a *scheme* if every point $x \in X$ has an open neighborhood U such that the restriction $(U, \mathcal{O}_{X|U})$ is isomorphic to an affine scheme, that is, it is isomorpic to (Spec(R), $\tilde{R}$) for some commutative ring R. We then call U an open affine neighbourhood of x.

A scheme X is called *locally noetherian* if, for every open affine subset U of X, the restriction $(U, \mathcal{O}_{X|U})$ is isomorphic to the spectrum of a noetherian ring. If X is locally noetherian and quasi-compact, then we say that X is *noetherian.* In order to avoid some technical difficulties, we assume from now on that X is a noetherian scheme.

A sheaf of $\mathcal{O}_X$-modules $\mathcal{M}$ is called *quasi-coherent* if every point $x \in X$ has an open affine neighbourhood $U = \mathrm{Spec}R$ such that the restriction $\mathcal{M}_{|U}$ is isomorphic to $\widetilde{M}$ for some R-module M. In this case, $M = \Gamma(U, \mathcal{M})$. If M is a finitely generated R-module, for every affine open U, then we say that $\mathcal{M}$ is *coherent* It is not difficult to show that $\mathcal{M}$ is quasi-coherent if and only if $\mathcal{M}_{|U} = \widetilde{M}$, with $M = \Gamma(U, \mathcal{M})$, for every open affine subset $U = \text{Spec}(R)$ of X (cf. [100, II.5.4]). As an immediate application, we have a category equivalence between R-mod and the category of quasi-coherent sheaves of Spec(R)-modules, and a category equivalence between the the category of finitely generated R-modules and the category of coherent sheaves of Spec(R)-modules. The equivalence is defined by the functor sending an R-module M to $\widetilde{M}$. The inverse is given by sending a coherent sheaf $\mathcal{M}$ to its global section $\Gamma(\text{Spec}(R), \mathcal{M})$ (see [100, II.5.5]).

Suppose that $\mathcal{M}$ and $\mathcal{M}'$ are quasi-coherent sheaves of $\mathcal{O}_X$-modules. Recall from [100, p. 109] that $\mathcal{M}\otimes_{\mathcal{O}_X}\mathcal{M}'$ is the sheaf associated to the presheaf $U \to \Gamma(U, \mathcal{M})\otimes_{\Gamma(U,\mathcal{O}_X)} \Gamma(U, \mathcal{M}')$. From [100, Prop. II.5.2], it follows that

$$\Gamma(U, \mathcal{M} \otimes_{\mathcal{O}_X} \mathcal{M}') = \Gamma(U, \mathcal{M}) \otimes_R \Gamma(U, \mathcal{M}')$$

if $U = \text{Spec}(R)$ is an affine open subset of X.

The presheaf $\mathcal{H}om_{\mathcal{O}_X}(\mathcal{M}, \mathcal{M}')$ given by

$$\Gamma(U, \mathcal{H}om_{\mathcal{O}_X}(\mathcal{M}, \mathcal{M}')) = \text{Hom}_{\mathcal{O}_{X|U}}(\mathcal{M}_{|U}, \mathcal{M}'_{|U}))$$

is a sheaf ([100, p. 109]). If $U = \text{Spec}(R)$ is affine, then

$$\Gamma(U, \mathcal{H}om_{\mathcal{O}_X}(\mathcal{M}, \mathcal{M}')) = \text{Hom}_R(M, M') \tag{6.54}$$

where $M = \Gamma(U, \mathcal{M})$ and $M = \Gamma(U, \mathcal{M}')$, and this implies that $\mathcal{H}om_{\mathcal{O}_X}(\mathcal{M}, \mathcal{M}')$ is quasi-coherent.

Consider an epimorphism $\phi : \mathcal{M} \to \mathcal{M}'$ in the category of quasi-coherent sheaves over X. From [100, Prop. II.5.2], it then follows that $\phi(U) : \Gamma(U, \mathcal{M}) \to \Gamma(U, \mathcal{M}')$ is surjective for any affine open subset U of X.

A (quasi-coherent) sheaf of $\mathcal{O}_X$-algebras $\mathcal{A}$ is an algebra in the category of quasi-coherent $\mathcal{O}_X$-modules. This means that we have a morphism of sheaves

$$m_{\mathcal{A}} : \mathcal{A} \otimes_{\mathcal{O}_X} \mathcal{A} \longrightarrow \mathcal{A}$$

satisfying the associativity condition

$$m_{\mathcal{A}} \circ (I_{\mathcal{A}} \otimes m_{\mathcal{A}}) = m_{\mathcal{A}} \circ (m_{\mathcal{A}} \otimes I_{\mathcal{A}})$$

We then have that $\Gamma(U, \mathcal{A})$ is a $\Gamma(U, \mathcal{O}_X)$-algebra, for every open subset U of X. In the case where all the $\Gamma(U, \mathcal{A})$ have a unit, we say that $\mathcal{A}$ is an $\mathcal{O}_X$-algebra with a unit. This can also be stated as follows: there exists a morphism of sheaves $\eta_{\mathcal{A}} : \mathcal{O}_X \to \mathcal{A}$ such that

$$m_{\mathcal{A}} \circ (I_{\mathcal{A}} \otimes \eta_{\mathcal{A}}) = m_{\mathcal{A}} \circ (\eta_{\mathcal{A}} \otimes m_{\mathcal{A}}) = I_{\mathcal{A}}$$

Now let $\mathcal{A}$ be an $\mathcal{O}_X$-algebra, possibly without a unit. A (left) $\mathcal{A}$-module $\mathcal{M}$ is a quasi-coherent sheaf of $\mathcal{O}_X$-modules, together with a map

$$\psi_{\mathcal{M}} : \mathcal{A} \otimes_{\mathcal{O}_X} \mathcal{M} \longrightarrow \mathcal{M}$$

satisfying the condition

$$\psi_{\mathcal{M}} \circ (I_{\mathcal{A}} \otimes \psi_{\mathcal{M}}) = \psi_{\mathcal{M}} \circ (m_{\mathcal{A}} \otimes I_{\mathcal{M}})$$

We then have that $\Gamma(U,\mathcal{M})$ is a $\Gamma(U,\mathcal{A})$-module, for every open subset U of X. Right modules and bimodules are introduced in a similar way.
Now let $\mathcal{M}$ be a right $\mathcal{A}$-module, and $\mathcal{N}$ a left $\mathcal{A}$-module. By definition, the tensor product $\mathcal{M}\otimes_{\mathcal{A}}\mathcal{N}$ is the coequalizer of the maps

$$\begin{cases}\psi_{\mathcal{M}}\otimes I_{\mathcal{N}}\\ I_{\mathcal{M}}\otimes\psi_{\mathcal{N}}\end{cases}:\ \mathcal{M}\otimes_{\mathcal{O}_X}\mathcal{A}\otimes_{\mathcal{O}_X}\mathcal{N}\rightrightarrows\mathcal{M}\otimes_{\mathcal{O}_X}\mathcal{N}\longrightarrow\mathcal{M}\otimes_{\mathcal{A}}\mathcal{N}\longrightarrow 0$$

For $U=\mathrm{Spec}(R)\subset X$ an affine open subset, we obtain that

$$\Gamma(U,\mathcal{M})\otimes_{\Gamma(U,\mathcal{A})}\Gamma(U,\mathcal{N})=\Gamma(U,\mathcal{M}\otimes_{\mathcal{A}}\mathcal{N})$$

We call $\mathcal{N}$ a unital $\mathcal{A}$-module if the natural map $\mathcal{A}\otimes_{\mathcal{A}}\mathcal{N}\to\mathcal{N}$ is an isomorphism. To this end, it suffices that $\Gamma(U,\mathcal{N})$ is a unital $\Gamma(U,\mathcal{A})$-module for every open affine subset $U=\mathrm{Spec}(R)$ of X.
Now consider two left $\mathcal{A}$-modules $\mathcal{M}$ and $\mathcal{N}$. A morphism $\varphi:\ \mathcal{M}\to\mathcal{N}$ in the category of $\mathcal{O}_X$-modules is called $\mathcal{A}$-linear if the maps

$$\Gamma(U,\varphi):\ \Gamma(U,\mathcal{M})\longrightarrow\Gamma(U,\mathcal{N})$$

are $\Gamma(U,\mathcal{A})$-linear for all open subsets U of X. The set $\mathrm{Hom}_{\mathcal{A}}(\mathcal{M},\mathcal{N})$ of all $\mathcal{A}$-linear morphism $\mathcal{M}\to\mathcal{N}$ is an $\mathcal{O}_X$-module. The presheaf on X_E sending an open subset U of X to $\mathrm{Hom}_{\mathcal{A}|U}(\mathcal{M}_{|U},\mathcal{N}_{|U})$ is a sheaf, and we will denote this sheaf by $\mathcal{H}om_{\mathcal{A}}(\mathcal{M},\mathcal{N})$. If $U=\mathrm{Spec}(R)$ is an open affine subset of X, then

$$\Gamma(U,\mathcal{H}om_{\mathcal{A}}(\mathcal{M},\mathcal{N})=\mathrm{Hom}_{\Gamma(U,\mathcal{A})}(\Gamma(U,\mathcal{M}),\Gamma(U,\mathcal{N}))$$

The definition of (Taylor's) Brauer group can be generalized to schemes, in such a way that the Brauer group of an affine scheme Spec(R) coincides with the Brauer group of R, and many properties of the Brauer group of a commutative ring can be generalized to schemes. For example, we have a cohomological description of the Brauer group of a scheme. In this Section, we will give a brief survey, and we refer to the literature (cf. [97], [135]) for more details.

Etale cohomology

Let X and Y be schemes. Recall that a morphism of schemes $f:\ Y\to X$, is a morphism from Y to X as locally ringed spaces. A morphism $f:\ Y\to X$ is called flat if for every $y\in Y$, $\mathcal{O}_{Y,y}$ is flat as an $\mathcal{O}_{X,f(y)}$-module.
f is called *locally of finite type* if X can be covered by open affine subsets $U_i=\mathrm{Spec}(R_i)$ such that, for each i, $f^{-1}(U_i)$ can be covered by open affine subsets $V_{ij}=\mathrm{Spec}(S_{ij})$ such that every S_{ij} is finitely generated as an R_{ij}-algebra.
F is called *finite* if X can be covered by open affine subsets $U_i=\mathrm{Spec}(R_i)$ such that $f^{-1}(U_i)=\mathrm{Spec}(S_i)$ is affine and S_i is finitely generated as an R_i-module.
f is called *unramified* if for all $y\in Y$, $\mathcal{O}_{Y,y}/m_x\mathcal{O}_{Y,y}$ is a finite separable field extension of the field $\mathcal{O}_{X,x}/m_x\mathcal{O}_{X,x}=k(x)$. Here m_x is the maximal ideal of the local ring $\mathcal{O}_{X,x}$.
f is called *étale* if f is flat, locally of finite type and unramified. One can show that

these notions are consistent with the corresponding notions for commutative algebras. For example, a commutative R-algebra S is étale if and only if corresponding morphism of affine schemes $\mathrm{Spec}(S) \to \mathrm{Spec}(R)$ is étale.

Fix a scheme X. A Grothendieck topology $X_{\text{ét}}$ is defined as follows: $\underline{\underline{\mathrm{cat}}}(X_{\text{ét}})$ is the category of all X-schemes Y whose structure morphism $f: Y \to X$ is étale, and étale morphisms; $\underline{\underline{\mathrm{cov}}}(X_{\text{ét}})$ consists of families of étale morphisms of schemes $\{g_i: U_i \to Y | i \in I\}$ such that $\bigcup_{i\in I} g_i(U_i) = Y$. Replacing étale morphisms by flat morphisms that are locally of finite type, we obtain the Grothendieck topology X_{fl}. Now let $E = (\text{ét})$ or (fl), and let P be a presheaf on X_E, that is, a contravariant functor from $\underline{\underline{\mathrm{cat}}}(X_E)$ to $\underline{\underline{\mathrm{Ab}}}$. Take a covering $\mathcal{U} = \{g_i: U_i \to X | i \in I\} \in \underline{\underline{\mathrm{cov}}}(X_E)$. We now define the *Čech complex* $C(\mathcal{U}/X, P)$ of $\mathcal{U}$ as follows: for any $n+1$-tuple $\mathbf{i} = (i_1, i_2, \ldots, i_{n+1}) \in I^{n+1}$, we write

$$U_{\mathbf{i}} = \prod_{j=1}^{n+1} U_{i_j}$$

and the canonical projection

$$p_j: U_{\mathbf{i}} \longrightarrow U_{(i_1,\ldots,\hat{i}_j,\ldots,i_{n+1})}$$

Using the factorial properties of P, we obtain maps

$$\mathrm{res}_j = P(p_j): P(U_{(i_1,\ldots,\hat{i}_j,\ldots,i_{n+1})}) \longrightarrow P(U_{\mathbf{i}})$$

Write

$$C^n(\mathcal{U}/X, P) = \prod_{\mathbf{i}\in I^{n+1}} U_{\mathbf{i}}$$

and define the map

$$\Delta_n: C^n(\mathcal{U}, P) \longrightarrow C^{n+1}(\mathcal{U}, P)$$

as follows: for $s = (s_{\mathbf{i}} | \mathbf{i} \in I^{n+1})$, the component $(\Delta_n s)_{(i_1,\ldots,i_{n+2})}$ of $\Delta_n s$ is given by

$$(\Delta_n s)_{(i_1,\ldots,i_{n+2})} = \sum_{j=1}^{n+2} (-1)^{j+1} \mathrm{res}_j(s_{(i_1,\ldots,\hat{i}_j,\ldots,i_{n+1})})$$

A straightforward computation shows that

$$0 \longrightarrow C^0(\mathcal{U}, P) \xrightarrow{\Delta_0} C^1(\mathcal{U}, P) \xrightarrow{\Delta_1} \cdots$$

is a complex. The cohomology groups

$$H^n(\mathcal{U}/X, P) = \mathrm{Ker}\,(\Delta_n)/\mathrm{Im}\,(\Delta_{n-1})$$

are called the *Čech cohomology groups* of P with respect to the covering $\mathcal{U}$.

Recall that we assume that our base scheme X is noetherian, and therefore quasi-compact. This means that the E-covering $\mathcal{U}$ can always be restricted to a finite covering $\mathcal{V} = \{U_i \to X | i = 1, \ldots, n\}$. Now consider the covering $\mathcal{W} = \{U_1 \times \cdots \times U_n = U \to X\}$. It is not difficult to show that

$$H^n(\mathcal{V}/X, P) = H^n(\mathcal{W}/X, P)$$

Of course the formal definition of the Čech complex simplifies if we consider it with respect to a covering consisting of only one scheme. In the special case where $U = \mathrm{Spec}(S)$ and $X = \mathrm{Spec}(R)$ are both affine, a comparison between the Čech complex $\mathcal{C}(\{U\}/X, P)$ and the Amitsur complex $\mathcal{C}(S/R, P)$ (see Section 5.2) learns that both complexes coincide, and therefore

$$H^n(\{\mathrm{Spec}(S)\}/\mathrm{Spec}(R), P) = H^n(S/R, P) \tag{6.55}$$

and we may view Čech cohomology as a generalization of Amitsur cohomology.
A presheaf P on X_E is called a sheaf if for any $U \in \underline{\underline{\mathrm{cat}}}(X_E)$ and for every covering $\mathcal{U} = \{U_i \to U | i \in I\}$ in $\underline{\underline{\mathrm{cov}}}(X_E)$, the canonical morphism $P(U) \to H^0(\mathcal{U}/U, P)$ is an isomorphism. Arguments that are similar to the ones exhibited in Sections 5.1 and 5.2 show that $\mathcal{P}(X_E)$ and $\mathcal{S}(X_E)$, respectively the categories of presheaves and sheaves on X_E are $Ab5$, $Ab4^*$-categories with a generator and enough injectives. Furthermore, the inclusion functor $i: \mathcal{P}(X_E) \to \mathcal{S}(X_E)$ has a left adjoint. The n-th right derived functor of the global section functor

$$\Gamma(X, \bullet): \mathcal{S}(X_E) \longrightarrow \underline{\underline{\mathrm{Ab}}}: F \mapsto \Gamma(X, F)$$

is denoted by $H^n(X_E, \bullet)$. The exact sequence (5.11) can be generalized in a straightforward way to Čech cohomology over schemes. Let F be a sheaf. The full Čech cohomology group $\check{H}^n(X_E, F)$ is by definition the inductive limit of the $\check{H}^n(\mathcal{U}/X, F)$, where $\mathcal{U}$ runs over all E-coverings $\mathcal{U}$ of X. We have a natural map

$$\check{H}^n(X_E, F) \longrightarrow H^n(X_E, F)$$

This map is bijective if $n = 0, 1$ and injective for $n = 2$. For $n = 0$, this follows immediately from the definition of a sheaf. The proofs in the cases $n = 0$ and $n = 1$ is similar to the proofs of Propositions 5.6.9 and 5.6.13. Theorem 5.6.7 generalizes to quasicompact schemes X with the following additional property: every finite subset of X is contained in an affine subset of X. In this case

$$\check{H}^n(X_{\acute{e}t}, F) \cong H^n(X_{\acute{e}t}, F) \tag{6.56}$$

for all n. We refer to [8] for further details.

Azumaya algebras

Let $(X, \mathcal{O}_X)$ be a scheme. A quasi-coherent sheaf $\mathcal{A}$ of $\mathcal{O}_X$-algebras is called an $\mathcal{O}_X$-(Taylor)-Azumaya algebra if for every $x \in X$, there exists an affine open neighbourhood U of x such that $\Gamma(U, \mathcal{A})$ is a (Taylor)-Azumaya $\Gamma(U, \mathcal{O}_X)$-algebra.
A dual pair of $\mathcal{O}_X$-modules is by definition a triple $\underline{\mathcal{M}} = (\mathcal{M}, \mathcal{M}', \mu)$, where $\mathcal{M}$ and $\mathcal{M}'$ are quasi-coherent $\mathcal{O}_X$-modules and $\mu: \mathcal{M}' \otimes_{\mathcal{O}_X} \mathcal{M} \to \mathcal{O}_X$ is an epimorphism in the category of quasicoherent $\mathcal{O}_X$-modules. We then can form the sheaf of $\mathcal{O}_X$-algebras $\mathcal{A} = \mathcal{M} \otimes_{\mathcal{O}_X} \mathcal{M}'$. The product is defined by the map

$$I_{\mathcal{M}} \otimes \mu \otimes I_{\mathcal{M}'}: \mathcal{M} \otimes_{\mathcal{O}_X} \mathcal{M}' \otimes_{\mathcal{O}_X} \mathcal{M} \otimes_{\mathcal{O}_X} \mathcal{M}' \to \mathcal{M} \otimes_{\mathcal{O}_X} \mathcal{M}'$$

We now claim that $\mathcal{A}$ is an $\mathcal{O}_X$-Taylor-Azumaya algebra. Take an affine open subset $U = \mathrm{Spec}(R)$ of X. Using [100, Prop. II.5.4], we obtain that

$$\mathcal{M}_{|U} \cong \tilde{M} \text{ and } \mathcal{M}'_{|U} \cong \tilde{M'}$$

where $M = \Gamma(U, \mathcal{M})$ and $M' = \Gamma(U, \mathcal{M}')$ are R-modules. From the observations made above, it follows that $\Gamma(U, \mathcal{A}) = M \otimes M'$, and the map $\mu(U) : M' \otimes M \to R$ is surjective. Thus $\Gamma(U, \mathcal{A}) = E_R(M, M', \mu(U))$ is an R-Taylor-Azumaya algebra, and this proves that $\mathcal{A} = E_{\mathcal{O}_X}(\underline{\mathcal{M}})$ is an $\mathcal{O}_X$-Taylor-Azumaya algebra. $\mathcal{A}$ is called an elementary $\mathcal{O}_X$-algebra.

Now take a *locally free sheaf* $\mathcal{M}$ of $\mathcal{O}_X$-modules of finite positive rank. This means that for any affine open subset $U = \mathrm{Spec}(R)$ of X, $\Gamma(U, \mathcal{M})$ is faithfully projective as a $\Gamma(U, \mathcal{O}_X)$-module. We define the dual locally free sheaf $\check{\mathcal{M}}$ to be the sheaf $\mathcal{H}om_{\mathcal{O}_X}(\mathcal{M}, \mathcal{O}_X)$ (see [100, p. 123]). For any open affine subset $U = \mathrm{Spec}(R)$ of X, we have that

$$\Gamma(U, \check{\mathcal{M}}) = \Gamma(U, \mathcal{M})^*$$

and

$$\Gamma(U, \mathcal{M}) \otimes_R \Gamma(U, \check{\mathcal{M}}) \cong \mathrm{End}_R(\Gamma(U, \mathcal{M})) \cong \Gamma(U, \mathcal{E}nd_{\mathcal{O}_X}(\mathcal{M}))$$

and this shows that

$$\mathcal{M} \otimes_{\mathcal{O}_X} \check{\mathcal{M}} \cong \mathcal{E}nd_{\mathcal{O}_X}(\mathcal{M})$$

Since $\mathcal{E}nd_{\mathcal{O}_X}(\mathcal{M})$ is locally an endomorphism ring, it is an Azumaya $\mathcal{O}_X$-algebra.

As in the affine case, we have several characterizations of (Taylor-)Azumaya algebras. We list some of them in the next two Propositions.

Proposition 6.11.1 *Let X be a noetherian scheme, and let $\mathcal{A}$ be a quasi-coherent sheaf of $\mathcal{O}_X$-algebras. Then the following assertions are equivalent.*

1) $\mathcal{A}$ is an $\mathcal{O}_X$-Azumaya algebra;

2) for every open affine subset $U = \mathrm{Spec}(R)$ of X, $\Gamma(U, \mathcal{A})$ is an Azumaya R-algebra;

3) there exists an étale covering $\{g_i : V_i \to X | i \in I\}$ of X such that, for each $i \in I$, $\mathcal{A} \otimes_{\mathcal{O}_X} \mathcal{O}_{V_i}$ is isomorphic to $\mathcal{E}nd_{\mathcal{O}_{V_i}}(\mathcal{M}_i)$ for some sheaf of $\mathcal{O}_{V_i}$-modules that is locally free of finite positive rank;

4) there exists a flat covering $\{g_i : V_i \to X | i \in I\}$ of X such that, for each $i \in I$, $\mathcal{A} \otimes_{\mathcal{O}_X} \mathcal{O}_{V_i}$ is isomorphic to $\mathcal{E}nd_{\mathcal{O}_{V_i}}(\mathcal{M}_i)$ for some sheaf of $\mathcal{O}_{V_i}$-modules that is locally free of finite positive rank;

5) $\mathcal{A}$ is locally free of finite positive rank as an $\mathcal{O}_X$-module, and the canonical map

$$\mathcal{A} \otimes_{\mathcal{O}_X} \mathcal{A}^{\mathrm{op}} \longrightarrow \mathcal{E}nd_{\mathcal{O}_X}(\mathcal{A})$$

is an isomorphism.

Proof 1) $\Longrightarrow$ 2). Let $U = \mathrm{Spec}(R)$ be an open affine subset of X. For any $p \in \mathrm{Spec}(R) \subset X$, there exists an open affine neighbourhood V of p such that $\Gamma(V, \mathcal{A})$ is an Azumaya $\Gamma(V, \mathcal{O}_X)$-algebra. We can take V such that $V = D(f) \subset \mathrm{Spec}(R) = U$ for some $f \in R$. Now $\Gamma(V, \mathcal{A}) = \Gamma(U, \mathcal{A})_f$ is an R_f-Azumaya algebra. Choose $p_1, \ldots, p_n \in \mathrm{Spec}(R)$ such that the corresponding $D(f_1), \ldots, D(f_n)$ cover $\mathrm{Spec}(R)$. Then $S = R_{f_1} \times \cdots \times R_{f_n}$ is a faithfully flat R-algebra, and $\Gamma(V, \mathcal{A}) \otimes S$ is an

S-Azumaya algebra. From Corollary 3.3.7, it follows that $\Gamma(V, \mathcal{A})$ is an R-Azumaya algebra.

2) $\Longrightarrow$ 3). Let the elements of X be indexed by an index set I. For every $x_i \in X$, we take an open affine neighbourhood $U_i = \mathrm{Spec}(R_i)$ of x_i. Then $\Gamma(U_i, \mathcal{A})$ is an R_i-Azumaya algebra, and, by Corollary 3.3.7, there exists an étale covering S_i of R_i such that $\Gamma(U_i, \mathcal{A}) \otimes_{R_i} S_i \cong \mathrm{End}_{S_i}(M_i)$ for some faithfully projective S_i-module M_i. Now let $V_i = \mathrm{Spec}(S_i)$, and $\mathcal{M}_i = \widetilde{M}_i$. We have étale morphisms

$$V_i = \mathrm{Spec}(S_i) \to U_i = \mathrm{Spec}(R_i) \to X$$

and the V_i cover X. Finally

$$\mathcal{A} \otimes_{\mathcal{O}_X} \mathcal{O}_{U_i} \cong \mathcal{A} \otimes_{\mathcal{O}_X} \mathcal{O}_{X_i} \otimes_{\mathcal{O}_{X_i}} \mathcal{O}_{U_i} \cong \mathcal{A}_{|X_i} \otimes_{\mathcal{O}_{X_i}} \mathcal{O}_{U_i} \cong \mathcal{E}nd_{\mathcal{O}_{V_i}}(\mathcal{M}_i)$$

3) $\Longrightarrow$ 4) is trivial.

4) $\Longrightarrow$ 2) Let $Y = \mathrm{Spec}(R)$ be an open affine subset of X, and let $U_i = V_i \times_X Y$ (we refer to [100, p. 87] for the definition of the fibred product). Then $\{U_i \to Y | i \in I\}$ is a covering of Y in the flat topology. Now consider a set of open affine subsets $\{Z_{ij} = \mathrm{Spec}(S_{ij}) | j \in J_i\}$ of U_i that cover U_i. We have the following morphisms of schemes:

$$\begin{array}{ccccc} & & & V_i & \\ & & \nearrow & & \searrow \\ Z_{ij} & \longrightarrow & U_i & & X \\ & & \searrow & & \nearrow \\ & & & Y & \end{array}$$

and we have

$$\begin{aligned} (\mathcal{A} \otimes_{\mathcal{O}_X} \mathcal{O}_Y) \otimes_{\mathcal{O}_Y} \mathcal{O}_{Z_{ij}} &\cong \mathcal{A} \otimes_{\mathcal{O}_X} \mathcal{O}_{Z_{ij}} \\ &\cong (\mathcal{A} \otimes_{\mathcal{O}_X} \mathcal{O}_{V_i}) \otimes_{\mathcal{O}_{V_i}} \mathcal{O}_{Z_{ij}} \\ &\cong \mathcal{E}nd_{\mathcal{O}_{V_i}}(\mathcal{M}_i) \otimes_{\mathcal{O}_{V_i}} \mathcal{O}_{Z_{ij}} \\ &\cong \mathcal{E}nd_{\mathcal{O}_{Z_{ij}}}(\mathcal{M}_i \otimes_{\mathcal{O}_{V_i}} \mathcal{O}_{Z_{ij}}) \end{aligned}$$

Now take global sections of both sides. In view of the fact that Y and Z_{ij} are affine, we obtain

$$\Gamma(Y, \mathcal{A}) \otimes_R S_{ij} \cong \mathrm{End}_{S_{ij}} M_{ij}$$

with $M_{ij} = \Gamma(Z_{ij}, \mathcal{M}_i \otimes_{\mathcal{O}_{V_{ij}}})$ a faithfully projective S_{ij}-module. Now take a finite number of the S_{ij} such that the images of the corresponding $\mathrm{Spec}(S_{ij})$ in Y cover Y. This is possible since $Y = \mathrm{Spec}(R)$ is quasi-compact. Let S be the direct product of these S_{ij}'s, and M the direct product of the corresponding M_{ij}. Then S is a faithfully flat R-algebra, M is a faithfully projective S-module, and $\Gamma(Y, \mathcal{A}) \otimes_R S \cong \mathrm{End}_S(M)$. It follows from Corollary 3.3.7 that $\Gamma(Y, \mathcal{A})$ is an R-Azumaya algebra.

2) $\Longrightarrow$ 1) is trivial.

2) $\Longrightarrow$ 5) Let $\mathcal{A}$ be an $\mathcal{O}_X$-Azumaya algebra, and take $U = \mathrm{Spec}(R) \subset X$ an affine open subset. Then $\Gamma(U, \mathcal{A})$ is an R-Azumaya algebra, and it follows from Proposition 2.2.9 that $\Gamma(U, \mathcal{A})$ is faithfully projective as an R-module. Since this

holds for any open affine subset U of X, it follows that $\mathcal{A}$ is locally free of finite positive rank as an $\mathcal{O}_X$-module. Moreover

$$\Gamma(U,\mathcal{A})\otimes\Gamma(U,\mathcal{A})^{\mathrm{op}}\cong \mathrm{End}_R(\Gamma(U,\mathcal{A})) \tag{6.57}$$

Since (6.57) holds for any open affine subset of X, it follows that

$$\mathcal{A}\otimes_{\mathcal{O}_X}\mathcal{A}^{\mathrm{op}}\cong \mathcal{E}nd_{\mathcal{O}_X}(\mathcal{A}) \tag{6.58}$$

5) $\Longrightarrow$ 2) Assume that $\mathcal{A}$ is locally free of finite positive rank, and that (6.58) holds. Then for every open affine subset $U=\mathrm{Spec}(R)$ of X, we have that $\Gamma(U,\mathcal{A})$ satisfies (6.57) and is faithfully projective as an R-module. From Proposition 2.2.9, it follows that $\Gamma(U,\mathcal{A})$ is an R-Azumaya algebra. □

Proposition 6.11.2 *Let X be a noetherian scheme, and let $\mathcal{A}$ be a quasi-coherent sheaf of $\mathcal{O}_X$-algebras. Then the following assertions are equivalent.*
1) $\mathcal{A}$ is an $\mathcal{O}_X$-Taylor-Azumaya algebra;
2) for every open affine subset $U=\mathrm{Spec}(R)$ of X, $\Gamma(U,\mathcal{A})$ is an R-Taylor-Azumaya algebra.
3) there exists an étale covering $\{g_i: V_i\to X|i\in I\}$ of X such that, for each $i\in I$, $\mathcal{A}\otimes_{\mathcal{O}_X}\mathcal{O}_{V_i}$ is isomorphic to $E_{\mathcal{O}_{V_i}}(\underline{\mathcal{M}}_i)$ for some dual pair of quasi-coherent sheaves of $\mathcal{O}_{V_i}$-modules $\underline{\mathcal{M}}_i$.
4) there exists a flat covering $\{g_i: V_i\to X|i\in I\}$ of X such that, for each $i\in I$, $\mathcal{A}\otimes_{\mathcal{O}_X}\mathcal{O}_{V_i}$ is isomorphic to $E_{\mathcal{O}_{V_i}}(\underline{\mathcal{M}}_i)$ for some dual pair of quasi-coherent sheaves of $\mathcal{O}_{V_i}$-modules $\underline{\mathcal{M}}_i$.

Proof The proof is similar to the proof of Proposition 6.11.2, taking Theorem 3.3.5 into account. □

Let $\mathcal{A}$ be a coherent $\mathcal{O}_X$-Taylor-Azumaya algebra, and let $\mathcal{A}^e=\mathcal{A}\otimes_{\mathcal{O}_X}\mathcal{A}^{\mathrm{op}}$. Then $\mathcal{A}^e$ may be viewed as an $\mathcal{A}^e$-bimodule. Let $\mathcal{A}_l$ and $\mathcal{A}_r$ be $\mathcal{A}$ viewed respectively as a left and right $\mathcal{A}^e$-module. Now let

$$\Upsilon=\mathcal{H}om_{(\mathcal{A}^e,\mathcal{A}^e)}(\mathcal{A}^e,\mathcal{A}_l\otimes_{\mathcal{O}_X}\mathcal{A}_r)$$

Let $U=\mathrm{Spec}(R)\subset X$ be an open affine subset of X, and write $A=\Gamma(U,\mathcal{A})$. Then

$$\Omega=\Gamma(U,\Upsilon)=\mathrm{Hom}_{(A^e,A^e)}(A^e,A_l\otimes A_r) \tag{6.59}$$

We know from Corollary 3.3.6 that Ω is free of rank one as an R-module. Since this holds for every open affine subset $\mathrm{Spec}(R)$ of X, Υ is locally free of rank one as an $\mathcal{O}_X$-module, we have in fact that Υ is isomorphic to $\mathcal{O}_X$ as an $\mathcal{O}_X$-module. Now define a map

$$\tau:\ (\mathcal{A}_r\otimes_{\mathcal{O}_X}\Upsilon)\otimes_{\mathcal{A}^e}\mathcal{A}_l\longrightarrow\mathcal{O}_X$$

as follows: for each open affine subset $U=\mathrm{Spec}(R)$ of X, $\Gamma(U,\tau)$ is the map t defined in Proposition 2.1.3. τ is then completely determined, since the open affine subsets of X form a basis for the topology on X. From Propositions 2.2.4 and 2.2.6, it follows that t is surjective and that

$$A^e \cong E_R(A_l, A_r \otimes \Omega, t)$$

Since this holds for all open affine subsets $\mathrm{Spec}(R) = U$ of X, we have the following result:

Proposition 6.11.3 *Let X be a noetherianscheme, and $\mathcal{A}$ a coherent $\mathcal{O}_X$-Taylor-Azumaya algebra. Then the map τ defined above is an epimorphism of $\mathcal{O}_X$-modules, and*

$$\mathcal{A}^e \cong E_{\mathcal{O}_X}(\mathcal{A}_l, \mathcal{A}_r \otimes_{\mathcal{O}_X} \Upsilon, \tau) \tag{6.60}$$

We are now able to define the Brauer group of a noetherian scheme. Two $\mathcal{O}_X$-Azumaya algebras $\mathcal{A}$ and $\mathcal{B}$ over a noetherian scheme X are called equivalent if

$$\mathcal{A} \otimes_{\mathcal{O}_X} \mathcal{E}nd_{\mathcal{O}_X}(\mathcal{M}) \cong \mathcal{B} \otimes_{\mathcal{O}_X} \mathcal{E}nd_{\mathcal{O}_X}(\mathcal{N})$$

for some $\mathcal{O}_X$-modules $\mathcal{M}$ and $\mathcal{N}$ that are locally free of finite positive rank. In a similar way, two coherent $\mathcal{O}_X$-Azumaya algebras $\mathcal{A}$ and $\mathcal{B}$ are called equivalent if

$$\mathcal{A} \otimes_{\mathcal{O}_X} E_{\mathcal{O}_X}(\underline{\mathcal{M}}) \cong \mathcal{B} \otimes_{\mathcal{O}_X} E_{\mathcal{O}_X}(\underline{\mathcal{N}})$$

for some dual pairs of coherent $\mathcal{O}_X$-modules $\underline{\mathcal{M}}$ and $\underline{\mathcal{N}}$. The equivalence classes of $\mathcal{O}_X$-Azumaya algebras form a group, which is a group under the operation induced by the tensor product $\otimes_{\mathcal{O}_X}$. This is the Brauer group $\mathrm{Br}(X)$ of the scheme X. In a similar way, the set of equivalence classes of coherent $\mathcal{O}_X$-Taylor-Azumaya algebras forms a group, denoted by $\mathrm{Br}'(X)$. It is clear that

$$\mathrm{Br}(\mathrm{Spec}(R)) \cong \mathrm{Br}(R) \quad \text{and} \quad \mathrm{Br}'(\mathrm{Spec}(R)) \cong \mathrm{Br}'(R) \tag{6.61}$$

for any noetherian commutative ring R.
Proceeding as in Sections 6.1 and 6.2, we can prove the following result.

Theorem 6.11.4 *For any noetherian scheme X, we have natural monomorphisms*

$$\mathrm{Br}(X) \hookrightarrow \mathrm{Br}'(X) \hookrightarrow H^2(X_{\text{ét}}, \mathbb{G}_m)$$

Without proof, we mention the following properties, generalizing Theorems 6.3.8 and 6.7.4.

Theorem 6.11.5 (Gabber [89]) *Let X be a scheme, and suppose that $X = U \cup V$, where U, V and $U \cap V$ are open affine subsets of X. Then*

$$\mathrm{Br}(X) \cong H^2(X_{\text{ét}}, \mathbb{G}_m)_{\text{tors}}$$

Theorem 6.11.6 (Taylor [160]) *Let X be a noetherian scheme, and suppose that every finite subset of X is contained in an open affine subset of X. Then*

$$\mathrm{Br}'(X) \cong H^2(X_{\text{ét}}, \mathbb{G}_m)$$

For further results about the Brauer group of a scheme, we refer to [97] and [135]. Let us also mention that some explicit computations about the (cohomological) Brauer group of a *toric variety* have been carried out recently by DeMeyer, Ford, Miranda and Reignier ([73], [74] and [77]).
The definition of the Brauer group can be generalized further. B. Auslander ([10]) introduced the Brauer group of an arbitrary ringed space $(X, \mathcal{A})$, and Grothendieck ([97]) defined the Brauer group of a locally ringed topos.
A categorical approach is due to Pareigis ([156]). For a symmetric monoidal category $\mathcal{C}$, he defines two Brauer groups $\text{Br}_1(\mathcal{C})$ and $\text{Br}_2(\mathcal{C})$. If R is a commutative ring, then $\text{Br}_1(R\text{-mod}) = \text{Br}_2(R\text{-mod}) = \text{Br}(R)$, and this property is due to the fact that, over a commutative ring R, an algebra is Azumaya if and only if it is central and separable.
In a recent paper, Van Oystaeyen and Zhang generalize Pareigis' definitions to the situation where $\mathcal{C}$ is a braided monoidal category (see [193]). Special cases of their very general construction that are not covered by Pareigis' construction are the Brauer-Long group (see Part III) and the Brauer group of Yetter-Drinfel'd module algebras ([46], [47] and [194]).
In another recent paper, Vitale introduces Taylor's Brauer group a a symmetric monoidal category, we refer to [189].

The relative Brauer group

Let R be a commutative ring, and $\sigma : R\text{-mod} \to R\text{-mod}$ an idempotent kernel functor, that is, a subfunctor of the identity functor such that $\sigma(M/\sigma(M)) = 0$ for all $M \in R$-mod. Let

$$\begin{aligned}\mathcal{T}_\sigma &= \{M \in R\text{-mod} \mid \sigma(M) = M\} \quad \text{(the } \sigma\text{-torsion modules)}\\ \mathcal{F}_\sigma &= \{M \in R\text{-mod} \mid \sigma(M) = 0\} \quad \text{(the } \sigma\text{-torsion free modules)}\end{aligned}$$

$(\mathcal{T}_\sigma, \mathcal{F}_\sigma)$ is a so-called *hereditary torsion theory*, and, in fact, there is a one-to-one correspondence between idempotent kernel functors and hereditary torsion theories. An R-module E is called σ-closed if for every exact diagram

$$\begin{array}{ccccccccc} 0 & \longrightarrow & M' & \xrightarrow{i} & M & \longrightarrow & M'' & \longrightarrow & 0 \\ & & \downarrow f & & & & & & \\ & & E & & & & & & \end{array}$$

in R-mod with $E'' \in \mathcal{T}_\sigma$, there exists a unique $g : M \to E$ making the diagram commutative. (R, σ)-mod, the full subcategory of R-mod consisting of σ-closed modules, is a Grothendieck category, and the inclusion functor $i_\sigma : (R, \sigma)\text{-mod} \to R\text{-mod}$ has a left adjoint a_σ. The functor

$$Q_\sigma = i_\sigma \circ a_\sigma : R\text{-mod} \longrightarrow R\text{-mod}$$

is called the localization functor associated to σ. σ can be described explicitly: let $\mathcal{L}(\sigma)$ be the set of ideals of R such that $R/L \in \mathcal{T}_\sigma$. $\mathcal{L}(\sigma)$ is called the Gabriel topology corresponding to σ, and we have that

$$Q_\sigma(M) = \varinjlim \text{Hom}_R(I, M/\sigma(M))$$

where the limit runs over all ideals I in $\mathcal{L}(\sigma)$.
Assume now that R is σ-noetherian. This means that R satisfies the ascending chain condition on σ-closed ideals. Let $\mathcal{C}(\sigma)$ be the set of ideals of R that are maximal with respect to the property of not being contained in $\mathcal{L}(\sigma)$. Then $\mathcal{C}(\sigma)$ consists of prime ideals, and it can be shown that σ is completely determined by $\mathcal{C}(\sigma)$.
An R-module is called σ-finitely presented if there exists a finitely presented R-module M' and a map $u: M' \to M$ such that both $\mathrm{Ker}(u)$ and $\mathrm{Coker}(u)$ are in $\mathcal{T}_\sigma$.
An R-algebra is called a *σ-Azumaya algebra* if it is σ-finitely presented, σ-closed and if for every $p \in \mathcal{C}(\sigma)$, there exists $f \in R \setminus p$ such that A_f is an R_f-Azumaya algebra. Again, we have several characterizations, and we refer to [192] for a more detailed discussion. One can define an equivalence relation on the class of σ-Azumaya algebras (the "trivial" σ-Azumaya algebras are the endomorphism rings of the so-called σ-progenerators), and the set of equivalence classes forms the *relative Brauer group* $\mathrm{Br}(R,\sigma)$. The operation is not the tensor product this time, but the modified tensor product $\perp_\sigma$:

$$A \perp_\sigma B = Q_\sigma)A \otimes B)$$

In the special case where R is a Krull domain, and $\mathcal{C}(\sigma)$ is the set of height one prime ideal of R, the relative Brauer group turns out to be Yuan's reflexive Brauer group (cf. [206]).

The Brauer group of a quasi-affine scheme

Now let R be noetherian, and let I be an ideal of R. Then $D(I)$ is an open subset of $\mathrm{Spec}(R)$, and the scheme $(D(I), \widetilde{R}_{|D(I)})$ is called a *quasi-affine scheme* . In this Section, we will explain how the Brauer group of a quasi-affine scheme is related to Van Oystaeyen's and Verschoren's relative Brauer group.
Let σ_I be the idempotent kernel functor with $\mathcal{C}(\sigma_I)$ the set of prime ideals of R that are maximal in $D(I)$. The associated Gabriel topology then consists of the ideals L of R such that $\sqrt{L} = \sqrt{I}$. The associated localization functor provides another tool to describe the sheaf of $\widetilde{R}$-modules $\widetilde{M}$: it can be shown that

$$Q_{\sigma_I}(M) = \Gamma(D(I), \widetilde{M})$$

for any R-module M. Now let $\mathcal{A}$ be a $D(I)$-Azumaya algebra. Then $\Gamma(D(I), \mathcal{A})$ can be made into an R-module, and $Q_{\sigma_I}(\Gamma(D(I), \mathcal{A}))$ is a σ_I-Azumaya algebra (see [192, Cor. IV.2.6]). Conversely, if A is a σ_I-Azumaya algebra, then $\widetilde{A}_{|D(I)}$ is a $D(I)$-Azumaya algebra. Thus there is a one-to-one correspondence between σ_I-Azumaya algebras and $D(I)$-Azumaya algebras, and it can be shown that this correspondence induces an isomorphism of the Brauer groups (cf. [192, Prop. IV.2.8]):

Proposition 6.11.7 *Let I be an ideal in a noetherian commutative ring R. Then*

$$\mathrm{Br}(X(I)) \cong \mathrm{Br}(R, \kappa_I)$$

Remarks 6.11.8 1) Proposition 6.11.7 provides a "ring theoretical" description of the Brauer group of a quasi-affine scheme. A similar description can be given for

the Brauer group of a projective scheme, we refer to [192] for further details. An additional complication here is the fact that one has to consider a graded version of the relative Brauer group. For a detailed discussion of the graded Brauer group, we refer to [45]
2) It is possible that $D(I)$ is an affine open subset of R; in this case, we call $D(I) = \mathrm{Spec}(S) \to \mathrm{Spec}(R)$ an affine open immersion. This happens for example if $I = (f)$ is a principal ideal: we then have that $S = R_f$. In general, we have that $D(I)$ is an open affine subset of $\mathrm{Spec}(R)$ if the kernel functor σ_I is a so-called T-functor.
A kernel functor σ is called a T-functor if the associated localization functor Q_σ is exact and commutes with direct sums, or, equivalently, if the full subcategories (R,σ)-mod and $Q_\sigma(R)$-mod of R-mod are identical (see e.g. [192, Th. I.1.3]) for other characterizations). As a direct consequence of this property, one obtains that

$$\mathrm{Br}(R,\sigma) = \mathrm{Br}(Q_\sigma(R))$$

if σ is a T-functor. If we apply this to the situation where $\sigma = \sigma_I$, we obtain that

$$\mathrm{Br}(R,\sigma_I) = \mathrm{Br}(Q_{\sigma_I}(R))$$

if $D(I)$ is an affine subset of $\mathrm{Spec}(R)$. This is of course obvious!

Part II

Hopf algebras and Galois theory

Chapter 7

Hopf algebras

In this chapter, we will recall the definition of a Hopf algebra, and we will give some elementary properties and examples. For a more detailed discussion, we refer to the literature, e.g. [171], [1], [137]. A nice introduction to the subject is George Bergman's paper [25]. Note that in the monographs cited above, only Hopf algebras over fields are considered; many properties may be generalized directly to Hopf algebras over commutative rings. In the sequel, R will always be a commutative ring.

7.1 Algebras, coalgebras and Hopf algebras

Algebras

Remark that the definition of an R-algebra can be restated as follows: an R-module A together with R-module homomorphisms

$$\eta_A = \eta : R \to A \text{ and } m_A = m : A \otimes A \to A$$

is called an R-algebra (with unit) if

$$m \circ (m \otimes I) = m \circ (I \otimes m) \tag{7.1}$$

$$m \circ (\eta \otimes I) = m \circ (I \otimes \eta) = I \tag{7.2}$$

Here $I = I_A$ is the identity map on A. We will usually write

$$m(a \otimes b) = ab \text{ and } \eta(1) = 1_A = 1$$

(7.1) and (7.2) express that A is associative and have a unit, and can be rewritten in their usual form:

$$(ab)c = a(bc) \text{ and } 1_A a = a 1_A$$

for all $a, b, c \in A$. (7.1.2) and (7.1.3) tell us that the following diagrams are commutative:

$$\begin{array}{ccc} A \otimes A \otimes A & \xrightarrow{m \otimes I} & A \otimes A \\ \downarrow{\scriptstyle I \otimes m} & & \downarrow{\scriptstyle m} \\ A \otimes A & \xrightarrow{m} & A \end{array}$$

$$\begin{array}{ccccccc} R\otimes A & \xrightarrow{\eta\otimes I} & A\otimes A & \quad & A\otimes A & \xrightarrow{I\otimes\eta} & A\otimes A \\ & \cong\searrow & \downarrow m & & & \cong\searrow & \downarrow m \\ & & A & & & & A \end{array}$$

Remark that A is commutative if and only if

$$m = m \circ \tau \tag{7.3}$$

where $\tau = \tau_A : A \otimes A \to A \otimes A$ is the switch map. $m = m_A$ is called the multiplication map, and $\eta = \eta_A$ is called the unit map. The index A will be omitted if no confusion is possible.

Coalgebras

The definition of an R-algebra can be dualized in the following way: an R-module C together with R-module homomorphisms

$$\varepsilon = \varepsilon_C : C \to R \text{ and } \Delta_C = \Delta : C \to C \otimes C$$

is called an *R-coalgebra* (with counit)

$$(\Delta \otimes I) \circ \Delta = (I \otimes \Delta) \circ \Delta \tag{7.4}$$

$$(\varepsilon \otimes I) \circ \Delta = (I \otimes \varepsilon) \circ \Delta = I \tag{7.5}$$

ε is called the *counit map* or *augmentation map.* The *coassociativity* law 7.1.8 is equivalent to the commutativity of the following diagram:

$$\begin{array}{ccc} C & \xrightarrow{\Delta} & C\otimes C \\ \downarrow\Delta & & \downarrow I\otimes\Delta \\ C\otimes C & \xrightarrow{\Delta\otimes I} & C\otimes C\otimes C \end{array}$$

(7.5) is equivalent to the commutativity of the following diagrams:

$$\begin{array}{ccccccc} C & & & \quad & C & & \\ \downarrow\Delta & \searrow\cong & & & \downarrow\Delta & \searrow\cong & \\ C\otimes C & \xrightarrow{\varepsilon\otimes I} & R\otimes C & & C\otimes C & \xrightarrow{I\otimes\varepsilon} & R\otimes C \end{array}$$

A coalgebra C is called *cocommutative* if

$$\tau \circ \Delta = \Delta \tag{7.6}$$

Take $c \in C$. Then $\Delta(c) = \sum_i c_i \otimes c_i'$ for some $c_i, c_i' \in C$. Sweedler introduced the following notation:

$$\Delta(c) = \sum c_{(1)} \otimes c_{(2)}$$

From the coassociativity, it then follows that

$$\sum c_{(1)_{(1)}} \otimes c_{(1)_{(2)}} \otimes c_{(2)} = ((\Delta \otimes I) \circ \Delta)(c) = ((I \otimes \Delta) \circ \Delta)(c) = \sum c_{(1)} \otimes c_{2_{(1)}} \otimes c_{2_{(2)}}$$

and therefore we simply write

$$\Delta^2(c) = ((\Delta \otimes I) \circ \Delta)(c) = \sum c_{(1)} \otimes c_{(2)} \otimes c_{(3)}$$

Similarly

$$\Delta^3(c) = ((\Delta \otimes I \otimes I)(\Delta \otimes I) \circ \Delta)(c) = \sum c_{(1)} \otimes c_{(2)} \otimes c_{(3)} \otimes c_{(4)}$$

etc. The counit property can now be written as follows:

$$\sum \varepsilon(c_{(1)})c_{(2)} = \sum \varepsilon(c_{(2)})c_{(1)} = c$$

for all $c \in C$. The cocommutativity (7.6) is equivalent to

$$\sum c_{(1)} \otimes c_{(2)} = \sum c_{(2)} \otimes c_{(1)}$$

An R-linear map $f: C \to D$ between R-coalgebras is called an *R-coalgebra homomorphism* if

$$\Delta_D \circ f = (f \otimes f) \circ \Delta_C \text{ and } \varepsilon_D \circ f = \varepsilon_C$$

or, equivalently, for all $c \in C$:

$$\sum f(c)_{(1)} \otimes f(c)_{(2)} = \sum f(c_{(1)}) \otimes f(c_{(2)}) \quad (7.7)$$

$$\varepsilon(f(c)) = \varepsilon(c) \quad (7.8)$$

If C and D are coalgebras, then $C \otimes D$ is also a coalgebra:

$$\Delta_{C\otimes D} = (I \otimes \tau \otimes I) \circ (\Delta_C \otimes \Delta_D)$$

$$\varepsilon_{C\otimes D} = \varepsilon_C \otimes \varepsilon_D$$

or

$$\Delta(c \otimes d) = \sum c_{(1)} \otimes c_{(2)} \otimes d_{(1)} \otimes d_{(2)}$$

$$\varepsilon(c \otimes d) = \varepsilon(c)\varepsilon(d)$$

for all $c \in C,\ d \in D$.
If (C, Δ, ε) is a coalgebra, then the dual maps

$$C^* \otimes C^* \to (C \otimes C)^* \xrightarrow{\Delta^*} C^* \text{ and } \varepsilon^*: R = R^* \to C^*$$

define an algebra structure on C^*. The multiplication $*$ on C^* may then be written as follows:

$$(f * g)(c) = \sum f(c_{(1)})g(c_{(2)})$$

for all $c \in C,\ f, g \in C^*$. This multiplication is usually called the *convolution*. If C is cocommutative, then C^* is commutative.
In a dual way, if (A, m, η) is an R-algebra, then we can define a coalgebra structure on A^*, at least if A is faithfully projective as an R-module. Indeed, under this condition we have that $(A \otimes A)^* \cong A^* \otimes A^*$, and the dual maps

$$m^*: A^* \to (A \otimes A)^* \cong A^* \otimes A^* \text{ and } \eta^*: R \to A^*$$

satisfy all the necessary conditions. Observe that for $f \in A^*$, we have that

$$m^*(f) = \Delta_{A^*}(f) = \sum f_{(1)} \otimes f_{(2)}$$

if and only if for all $a, b \in A$ we have

$$f(ab) = \sum f_{(1)}(a) \otimes f_{(2)}(b)$$

Also

$$\eta^*(f) = \varepsilon_{A^*}(f) = f(1)$$

Examples 7.1.1 *1) Let X be a set, and RX the free R-module generated by X. The maps*

$$\Delta : RX \to RX \otimes RX \quad \text{and} \quad \varepsilon : RX \to R$$

defined by

$$\Delta(x) = x \otimes x \quad \text{and} \quad \varepsilon(x) = 1$$

for all $x \in X$ make RX into a coalgebra.
2) Let $X = \{c_0, c_1, c_2, \ldots, \}$. Another coalgebra structure may be defined on RX, by putting

$$\Delta(c_n) = \sum_{i=0}^{n} c_i \otimes c_{n-i} \quad \text{and} \quad \varepsilon(c_n) = \delta_{n0}$$

Bialgebras

Proposition 7.1.2 *Let $H \in R$-mod be at once an R-algebra and an R-coalgebra, with structure maps m, η, Δ and ε. Then the following assertions are equivalent:*

1. *m and η are R-coalgebra homomorphisms ;*
2. *Δ and ε are R-algebra homomorphisms ;*
3. *for all $g, h \in H$, we have*

$$\Delta(gh) = \sum g_{(1)}h_{(1)} \otimes g_{(2)}h_{(2)} \quad ; \quad \varepsilon(gh) = \varepsilon(g)\varepsilon(h)$$

and

$$\Delta(1) = 1 \otimes 1 \quad ; \quad \varepsilon(1) = 1$$

Proof Consider the following conditions

$$\Delta \circ m = (m \otimes m) \circ (I \otimes \tau \circ I) \circ (\Delta \otimes \Delta) \tag{7.9}$$
$$\Delta \circ \eta = \eta \otimes \eta \tag{7.10}$$
$$\varepsilon \circ m = \varepsilon \otimes \varepsilon \tag{7.11}$$
$$\varepsilon \circ \eta = I \tag{7.12}$$

Then it is clear that
m is an R-coalgebra homomorphism $\Longleftrightarrow$ (7.9) and (7.11) hold;

η is an R-coalgebra homomorphism $\Longleftrightarrow$ (7.10) and (7.12) hold;
Δ is an R-algebra homomorphism $\Longleftrightarrow$ (7.9) and (7.10) hold;
ε is an R-algebra homomorphism $\Longleftrightarrow$ (7.11) and (7.12) hold.
The equivalence of 1) and 2) now follows immediately. It is also clear that 3) is equivalent to (7.9-7.12). □

If H satisfies the equivalent conditions of Proposition 7.1.2, then we call H an R-*bialgebra.* If H and K are bialgebras, then a map $f: H \to K$ which is at once an R-algebra and an R-coalgebra homomorphism is called an R-bialgebra homomorphism.

The convolution product and Hopf algebras

Let C be an R-coalgebra, and A an R-algebra. We can now define a product $*$ on $\mathrm{Hom}_R(C, A)$ as follows: for $f, g: C \to A$, let $f * g$ be defined by

$$f * g = m_A \circ (f \otimes g) \circ \Delta_C \tag{7.13}$$

or

$$(f * g)(c) = \sum f(c_{(1)})g(c_{(2)}) \tag{7.14}$$

$*$ is called the *convolution product.* We have already met the convolution product above, in the special case where $A = R$. Observe that $*$ is associative, and that for any $f \in \mathrm{Hom}_R(C, A)$, we have that

$$f * (m_A \circ \varepsilon_C) = (m_A \circ \varepsilon_C) * f = f$$

so the convolution product makes $\mathrm{Hom}_R(C, A)$ into an R-algebra with unit.
Now suppose that H is a bialgebra, and take $H = A = C$ in the above construction. If the identity $I = I_H$ of H has a convolution inverse $S = S_H$, then we say that H is a Hopf algebra. $S = S_H$ is called the *antipode* of H. The antipode therefore satisfies the following property: $S * I = I * S = \eta \circ \varepsilon$ or

$$\sum h_{(1)}S(h_{(2)}) = \sum S(h_{(1)})h_{(2)} = \eta(\varepsilon(h)) \tag{7.15}$$

for all $h \in H$. A bialgebra homomorphism $f: H \to K$ is called a Hopf algebra homomorphism if $S_K \circ f = f \circ S_H$. In the sequel, we will denote

$$\begin{aligned} \mathrm{Hopf}_R(H, K) &= \{f : H \to K | f \text{ is a Hopf algebra homomorphism}\} \\ \mathrm{Aut}_{\mathrm{Hopf}}(H, H) &= \{f : H \to H | f \text{ is a Hopf algebra isomorphism}\} \end{aligned}$$

A Hopf algebra that is faithfully projective as an R-module will be called a finite Hopf algebra. If H is a finite Hopf algebra, then H^* is also a faithfully projective Hopf algebra. In the following proposition, we list some elementary properties of Hopf algebras, cf. e.g. [1, 2.1.4].

Proposition 7.1.3 *Let H be a Hopf algebra. For all $h, g \in H$, we have*

$$S(gh) = S(h)S(g) \tag{7.16}$$
$$S \circ \eta = \eta \tag{7.17}$$
$$\varepsilon \circ S = \varepsilon \tag{7.18}$$
$$\Delta(S(h)) = \sum S(h_{(2)}) \otimes S(h_{(1)}) \tag{7.19}$$

If H is commutative or cocommutative, then $S \circ S = I$.

Proof (7.16) Define $\nu, \rho \in \mathrm{Hom}_R(H \otimes H, H)$ by

$$\nu(g \otimes h) = S(h)S(g) \quad \text{and} \quad \rho(g \otimes h) = S(gh)$$

for all $g, h \in H$. A straightforward computation shows that

$$m * \rho = \rho * m = \eta \circ (\varepsilon \otimes \varepsilon) = m * \nu = \nu * m$$

hence ν and ρ are both convolution inverses of the multiplication map m, and therefore they are equal.
(7.17) $1 = \eta(\varepsilon(1)) = (I * S)(1) = I(1)S(1) = S(1)$
(7.18) $\varepsilon(h) = \varepsilon((\eta \circ \varepsilon)(h)) = \sum \varepsilon(S(h_{(1)})h_{(2)}) = \sum \varepsilon(S(h_{(1)}))\varepsilon(h_{(2)}) = \varepsilon(S(h))$
(7.19) Consider $\nu = \tau \circ (S \otimes S) \circ \Delta$, $\rho = \Delta \circ S \in \mathrm{Hom}_R(H, H \otimes H)$. It is straightforward to show that

$$\Delta * \nu = \nu * \Delta = (\eta \otimes \eta) \circ \varepsilon = \Delta * \rho = \rho * \Delta$$

so ρ and ν are both convolution inverse of Δ and therefore $\rho = \nu$, and (7.19) follows. Finally, suppose that H is commutative or cocommutative. Then for all $h \in H$:

$$\sum S(h_{(2)})h_{(1)} = (\eta \circ \varepsilon)(h)$$

and therefore

$$\begin{aligned}(S * (S \circ S))(h) &= \sum S(h_{(1)})S(S(h_{(2)})) \\ &= S(\sum S(h_{(2)})h_{(1)}) \\ &= S((\eta \circ \varepsilon)(h)) = \eta(\varepsilon(h))\end{aligned}$$

hence $S \circ S$ is a convolution inverse of S, so $S \circ S = I$. □

There are examples of (noncommutative, noncocommutative) Hopf algebras with an antipode that is not bijective. A Hopf algebra with bijective antipode is sometimes called a *quantum group*.
Let $H = (H, m, \eta, \Delta, \varepsilon, S)$ be a Hopf algebra, and suppose that the antipode S is bijective. It is straightforward to show that $H^{\mathrm{op}} = (H^{\mathrm{op}}, m \circ \tau, \eta, \Delta, \varepsilon, S^{-1})$, $H^{\mathrm{cop}} = (H^{\mathrm{cop}}, m, \eta, \tau \circ \Delta, \varepsilon, S^{-1})$ and $H^{\mathrm{opcop}} = (H^{\mathrm{opcop}}, m \circ \tau, \eta, \tau \circ \Delta, \varepsilon, S)$ are also Hopf algebras.

Example 7.1.4 Let S be a semigroup. The semigroup algebra RS with the coalgebrastructure given by $\Delta(s) = s \otimes s$ and $\varepsilon(s) = 1$ for all $s \in S$ is a (cocommutative) bialgebra. If G is a group, then RG is a cocommutative Hopf algebra. The antipode is given by $S(g) = g^{-1}$ for all $g \in G$, and RG is commutative if and only if G is commutative. If G and H are groups, then $\mathrm{Hopf}(RG, RH) = \mathrm{Hom}(G, H)$. This example motivates the following definition.

Definition 7.1.5 *An element $g \in H$ is called grouplike if $\Delta(g) = g \otimes g$ and $\varepsilon(h) = 1$. We denote*

$$\mathrm{G}(H) = \{g \in H | g \text{ is grouplike}\}$$

Proposition 7.1.6 $\mathrm{G}(H)$ *is a subgroup of* $\mathbb{G}_m(H)$.

Proof We leave it to the reader to show that $\mathrm{G}(H)$ is closed under multiplication, and that $gS(g) = S(g)g = 1$ if g is grouplike. □

Remark 7.1.7 *If H is finite Hopf algebra, then $h^* \in H^*$ is grouplike if and only if it h^* is an R-algebra homomorphism: $h^*(gh) = h^*(g)h^*(h)$ and $h^*(1) = 1$.*

Proposition 7.1.8 *Let H, K be Hopf algebras. Then* $\mathrm{G}(H) \times \mathrm{G}(K) \cong \mathrm{G}(H \otimes K)$.

Proof We clearly have an embedding $\mathrm{G}(H) \times \mathrm{G}(K) \to \mathrm{G}(H \otimes K)$, mapping (h, k) to $h \otimes k$. Let us show that this embedding is surjective. Take $\sum_i h_i \otimes k_i \in \mathrm{G}(H \otimes K)$. Then

$$\begin{aligned} h &= \sum_i \varepsilon(k_i) h_i \in \mathrm{G}(H) \\ k &= \sum_i \varepsilon(h_i) k_i \in \mathrm{G}(K) \end{aligned}$$

From the fact that $\sum_i h_i \otimes k_i$ is grouplike, it follows that

$$\sum_i h_{i_{(1)}} \otimes k_{i_{(1)}} \otimes h_{i_{(2)}} \otimes k_{i_{(2)}} = \sum_{i,j} h_i \otimes k_i \otimes h_j \otimes k_j$$

Applying $I_H \otimes \varepsilon_K \otimes \varepsilon_H \otimes I_K$ to both sides, we obtain

$$\sum_i h_i \otimes k_i = h \otimes k$$

□

We have seen above that for a group G, the group ring RG is a Hopf algebra. If G is finite, then $(RG)^*$ is also a Hopf algebra, we will write $(RG)^* = GR$. As an R-module, GR is the free R-module with basis $\{v_\sigma | \sigma \in G\}$, where v_σ is defined by

$$v_\sigma(u_\tau) = \langle v_\sigma, u_\tau \rangle = \delta_{\sigma,\tau}.$$

The Hopf structure on GR is given by the following formulas:

$$\begin{aligned} v_\sigma v_\tau &= \delta_{\sigma,\tau} v_\sigma \\ \eta(1) &= \sum_{\sigma \in G} v_\sigma \\ \Delta(v_\sigma) &= \sum_{\tau \in G} v_{\sigma\tau^{-1}} \otimes v_\tau \\ \varepsilon(v_\sigma) &= \delta_{\sigma,e} \\ S(v_\sigma) &= v_{\sigma^{-1}} \end{aligned}$$

for all $\sigma, \tau \in G$. Here e is the unit element of G.
If G is a finite abelian group, then GR is again a group ring, if the order of G is invertible in R, and if G contains enough roots of unity. Before we show this, let us recall the following elementary Lemma.

Lemma 7.1.9 *Let R be a commutative ring in which the positive integer n is not a zero-divisor. If R contains a primitive n-th root of unity η, then*

$$(1-\eta)(1-\eta^2)\cdots(1-\eta^{n-1}) = n \tag{7.20}$$

If the greatest common divisor of n and i equals 1, then $1+\eta+\eta^2+\cdots+\eta^{i-1}$ is invertible in R.

Proof In $R[x]$, we have that

$$x^n - 1 = (x-1)(x-\eta)\cdots(x-\eta^{n-1})$$

and

$$x^{n-1}+x^{n-2}+\cdots+1 = (x-\eta)\cdots(x-\eta^{n-1})$$

(7.20) follows after we take $x=1$. If $(n,i)=1$, then

$$\{1,\eta,\eta^2,\ldots,\eta^{n-1}\} = \{1,\eta^i,\eta^{2i},\ldots,\eta^{(n-1)i}\}$$

and

$$(x-1)(x-\eta)\cdots(x-\eta^{n-1}) = (x-1)(x-\eta^i)\cdots(x-\eta^{(n-1)i})$$

and

$$\frac{x-\eta^i}{x-\eta}\,\frac{x-\eta^{2i}}{x-\eta^2}\cdots\frac{x-\eta^{(n-1)i}}{x-\eta^{n-1}} = 1$$

Take $x=1$. The first factor of the left hand side is $1+\eta+\eta^2+\cdots+\eta^{i-1}$, and divides 1. □

Theorem 7.1.10 *Let R be a connected commutative ring, and G a finite abelian group such that $|G|$ is invertible in R, and such that R has a primitive $\exp(G)$-th root of unity. Then we have an isomorphism*

$$RG \cong R(G^*) \cong GR$$

Proof Recall that $G^* = \mathrm{Hom}(G, \mathbb{G}_m(R))$. If R has enough roots of unity, then $G \cong G^*$. Consider the map

$$f:\ R(G^*) \to (RG)^*$$

defined as follows: for $\sigma^* \in G^*$, we define $f(u_{\sigma^*})$ by

$$\langle f(u_{\sigma^*}), u_\tau\rangle = \langle \sigma^*, \tau\rangle.$$

We will show that f is an isomorphism of Hopf algebras. It is clear that f is a Hopf algebra homomorphism. Indeed, it preserves multiplication, and also comultiplication, since u_{σ^*} and $f(u_{\sigma^*})$ are both grouplike elements. $f(u_{\sigma^*})$ is grouplike because it is a multiplicative map. It suffices now to show that f is bijective, in the case where $G = C_q$, the cyclic group of order q, where q is a primary number. Indeed, since G is a finite abelian group, we have

$$G = C_{q_1} \times C_{q_2} \times \cdots \times C_{q_r} \quad ; \quad RG = RC_{q_1} \otimes RC_{q_2} \otimes \cdots \otimes RC_{q_r}$$
$$G^* = C^*_{q_1} \times C^*_{q_2} \times \cdots \times C^*_{q_r} \quad ; \quad (RG)^* = (RC_{q_1})^* \otimes (RC_{q_2})^* \otimes \cdots \otimes (RC_{q_r})^*$$

From now on, let us assume that $G = C_q =< \sigma >$, with q a primary number. Let η be a primitive q-th root of unity in R, and define $\sigma^* \in G^*$ by $\langle \sigma^*, \sigma \rangle = \eta$. In the sequel, we will write $u = f(u_{\sigma^*}) \in (RG)^*$. It is clear that $G^* =< \sigma^* >\cong G$.
To show that f is bijective, it will be sufficient to show that $\{u^i | i = 0, 1, \ldots, q-1\}$ is a basis of the free R-module $GR = (RG)^*$, or, equivalently, that every v_{σ^k} may be written in a unique way as a linear combination of the u^i. This comes down to showing that the equation

$$\sum_{i=0}^{q-1} \alpha_i u^i = v_{\sigma^k}$$

has a unique solution for every k. This is equivalent to showing that the linear system

$$\sum_{i=0}^{q-1} \alpha_i \langle u^i, u_{\sigma^j} \rangle = \langle v_{\sigma^k}, u_{\sigma^j} \rangle$$

or

$$\sum_{i=0}^{q-1} \alpha_i \eta^{ij} = \delta_{kj} \tag{7.21}$$

has a unique solution for every $k \in \{0, 1, \ldots, q-1\}$. The determinant D of (7.21) is a Vandermonde determinant and is equal to

$$D = \prod_{0<i<j<q} (\eta^i - \eta^j)$$

To show that D is invertible, it suffices to show that every factor of D is invertible. Dividing by the powers of η, it follows that it suffices to show that for all $i = 1, 2, \ldots, q-1$, $1 - \eta^i$ is invertible, or, equivalently, that

$$\prod_{0<i<q} (1 - \eta^i) = q$$

is invertible. q is invertible by assumption. □

Tate-Oort algebras and monogenic Larson orders

Let R be a commutative ring in which the prime number p is not a zero-divisor. R_p will be the ring R localized at the multiplicative set $\{p, p^2, p^3, \ldots\}$. Take a positive integer d, write $q = p^d$, and suppose that I and J are ideals of R such that

$$IJ^{p^{d-1}(p-1)} = pR \tag{7.22}$$

and write

$$K = \frac{1}{p} IJ^{p^{d-1}(p-1)-1} \subset R_p \tag{7.23}$$

Then K is an R-submodule of R_p, $JK = R$, and $R \subset K$. Now consider the Rees ring

$$\check{R}[Ky] = R \oplus Ky \oplus K^2y^2 \oplus K^3y^3 \oplus \cdots \subset R_p[y]$$

Lemma 7.1.11 *With notation as above, we have*

$$K^q((1+y)^q - 1) \subset \check{R}[Ky] \tag{7.24}$$

Proof We have to show that

$$K^q \binom{q}{i} y^i \subset \check{R}[Ky]$$

or

$$K^q \binom{q}{i} \subset K^i$$

for $i = 1, 2, \ldots q$. For any integer i, let $v_p(i)$ be the p-adic valuation of i. A result of Kummer ([120, p. 116]) states that the p-adic valuation of $\binom{m+n}{m}$ is equal to the number of carries that occur when the numbers m and n are added in their p-adic Hensel expansion. As a special case, it follows that

$$v_p \binom{q}{i} = d - v_p(i) \tag{7.25}$$

For two ideals M and N of R, we will write $M|N$ if there exists an ideal P of R such that $MP = N$.
Take $i \in \{1, \ldots, p^{d-1}\}$. Then $p^{m-1} \leq i < p^m$ for some $m \in \{1, 2 \ldots, d\}$, and $v_p(i) \leq m - 1$. Then $q - i \leq q - p^{m-1}$ and, using (7.25), we obtain that

$$\begin{aligned} J^{q-i} \ &| \ J^{p^d - p^{m-1}} = J^{p^d - p^{d-1}} J^{p^{d-1} - p^{d-2}} \cdots J^{p^m - p^{m-1}} \\ &| \ (J^{p^d - p^{d-1}})^{d-m+1} \ | \ (p^{d-m+1}) \\ &| \ (p^{d - v_p(i)}) \ | \ (\binom{q}{i}) \end{aligned}$$

We have that

$$\binom{q}{i} R = MJ^{q-i}$$

for some ideal M of R, and

$$K^q \binom{q}{i} = MJ^{q-i}K^q = MK^i \subset K^i$$

□

Consider the R-algebra

$$H_J = \check{R}[Ky] / \big(K^q((1+y)^q - 1)\big) \tag{7.26}$$

As an R-module, $H_J = R \oplus Ky \oplus \cdots \oplus K^{q-1}y^{q-1}$, and therefore H_J is faithfully projective as an R-module. It is clear that

$$R_p \otimes \check{R}[Ky] = R_p[y]$$

and

$$R_p \otimes H_J = R_p[y]/((1+y)^q - 1)$$

Now write $u = y + 1$. Then $R_p[y] = R_p[u]$ and

$$R_p \otimes H_J = R_p[u]/(u^q - 1) = R_pC_q$$

is the cyclic group ring of rank q. R_pC_q is a Hopf algebra over C_q. We already know that H_J is a commutative R-algebra, and it will follow that H_J is a Hopf algebra if we can show that the comultiplication, antipode and counit of R_pC_q respectively restrict to maps

$$H_J \to H_J \otimes H_J,\ H_J \to R,\ H_J \to H_J$$

Since these maps are multiplicative, it suffices to verify this for ky, with $k \in K$. For all $k \in K$, we have

$$\begin{aligned}\Delta(ky) &= k\Delta(u-1) = k(u \otimes u - 1 \otimes 1)\\ &= k(y \otimes y + y \otimes 1 + 1 \otimes y) \in H_J \otimes H_J\\ \varepsilon(ky) &= k\varepsilon(u-1) = 0 \in R\end{aligned}$$

Finally, in H_J, we have that $u^{-1} = u^{q-1}$, and therefore

$$\begin{aligned}S(ky) &= kS(u-1) = k(u^{q-1} - 1) = k(u-1)\sum_{i=0}^{q-2} u^i\\ &= ky\sum_{i=0}^{q-2} u^i \in H_J\end{aligned}$$

We summarize the above results as follows.

Theorem 7.1.12 *Let R be a commutative ring in which the prime number p is not a zero divisor. Let $q = p^d$, and J an ideal of R such that $J^{p^d - p^{d-1}}$ divides the ideal pR. If K is the inverse of I in the lattice of R-submodules of R_p, then*

$$H_J = \check{R}[Ky]/\big(K^q((1+y)^q - 1)\big)$$

is a faithfully projective, commutative, cocommutative Hopf algebra. The comultiplication and antipode are defined by the formulas

$$\Delta(u) = u \otimes u,\ \varepsilon(u) = 1,\ S(u) = u^{q-1}$$

with $u = 1 + y$.
If $J = R$, then $H_J = RC_q$.
$R_p \otimes H_J \cong R_pC_q$.
If R is connected, then $G(H_J) = \{1, u, u^2, \ldots, u^{q-1}\}$.

Proof The first statement has been proved above. The second and third statement follow immediately. The fourth statement follows from the fact $G(R_pC_q) = \{1, u, u^2, \ldots, u^{q-1}\}$. □

Remarks 7.1.13 1) Suppose that $I = Ra$ and $J = Rb$ are principal ideals. Then

$$ab^{p^d-p^{d-1}} = p \text{ and } K = \frac{R}{b}$$

so

$$H_b = R\Big[\frac{y}{b}\Big]/\Big(\frac{1}{b^q}((y+1)^q - 1)\Big) = R[x]/\Big(\frac{1}{b^q}((bx+1)^q - 1)\Big)$$

where we have written $y = bx$. H_b is a free Hopf algebra of rank q.
2) Suppose that R is a domain, and let K be its field of fractions. A Hopf algebra H is called a *Hopf algebra order* in KC_q if $K \otimes H \cong KC_q$. Some authors, cf. e.g. [164], call such a Hopf algebra *generically split*. Tate and Oort [174] classified Hopf algebra orders of rank p over commutative rings that are algebras over the ring $\Lambda_p = \mathbb{Z}[\zeta, p^{-1}(p-1)^{-1}] \cap \mathbb{Z}_p$ (here ζ is a primitive $p-1$-th root of unity, and $\mathbb{Z}_p$ is the ring of p-adic integers). Hopf algebras of rank p are sometimes called *Tate-Oort algebras*.
In [121] Hopf algebras of the form

$$H = R\Big[\frac{u-1}{b_1}, \frac{u^2-1}{b_n} \cdots \frac{u^n-1}{b_n}\Big]$$

where $u^{p^n} = 1$, and the b_i satisfy certain conditions. These Hopf algebras are Hopf algebra orders in KC_{p^n} and are called *Larson orders*. In the situation where $(u-1)/b_1$ generates H, we call H a *monogenic Larson order*. Monogenic Larson orders are special cases of the Hopf algebras discussed in Theorem 7.1.12. In [93], a new class of Hopf algebra orders in RC_{p^2} is introduced. Larson orders are special cases of these Hopf algebras, which are called *Greither orders*. These Greither orders have been investigated further by Underwood ([186] and [187]). In [187], a classification of all the Hopf algebras orders in KC_{p^3} is given, under the assumption that R contains a primitive p^3-th root of unity.
Raynaud [163] gives a class of Hopf algebra orders in $KC_p \times KC_p \times \cdots \times KC_p$.
Another aspect of the theory that we have to mention here is the relation between Hopf algebras and formal group laws. In fact, one can construct Hopf algebras as the kernel of an isogeny between n-dimensional polynomial formal group laws. In the 1-dimensional case (the degree of the polynomial is then always 2), this can be used to construct the rank p^n-Hopf algebras of Theorem 7.1.12, in the case where the groundring R is local (see [56] and [59] for more details). In the higher dimensional case, one obtains new examples of Hopf algebra orders in $KC_p \times KC_p \times \cdots \times KC_p$, see the paper by Childs and Sauerberg in [60]. In another paper (by Childs and Greither) in the same monograph [60] it is shown that one can also construct Hopf algebra orders in $KC_p \times KC_p \times \cdots \times KC_p$ via iterated extensions of Tate-Oort algebras.

The dual of a Tate-Oort order

Assume from now on that $q = p$ is a prime number and that R contains a primitive p-th root of unity η. From Lemma 7.1.9, we know that $p = (\eta - 1)^{p-1}\xi$, with ξ a unit in R. It follows that if b^{p-1} divides p, then it divides $(\eta - 1)^{p-1}$ also. In the sequel, we will assume that a factorization

$$ab = \eta - 1$$

is given. We can consider the Hopf algebras H_a and H_b, and our aim is to show that $H_b^* \cong H_a$. In particular, it will follow from the fact that $H_1 \cong RC_p$ that $H_{\eta-1} \cong RC_p^*$.

We have seen above that

$$H_b = R[x]/(f_b(x))$$

with $f_b(x) = b^{-p}(u^p - 1)$ and $u = 1 + bx$. In a similar way, let

$$H_a = R[z]/(f_a(z))$$

with $f_a(z) = a^{-p}(v^p - 1)$ and $v = 1 + az$. We know that

$$R_p \otimes H_a \cong R_p \otimes H_b \cong R_p C_p$$

and from Theorem 7.1.10, it follows that

$$R_p \otimes H_a \cong \mathrm{Hom}_{R_p}(R_p \otimes H_b, R_p) \cong R_p \otimes H_b^*$$

as Hopf algebras. The connecting isomorphism Φ may be described as follows (see the proof of Theorem 7.1.10). We keep in mind the fact that $\{1, v, v^2, \ldots, v^{p-1}\}$ and $\{1, u, u^2, \ldots, u^{p-1}\}$ are bases for respectively $R_p \otimes H_a$ and $R_p \otimes H_b$ (not for H_a and H_b!). Φ is defined by the formula

$$\langle \Phi(v^i), u^j \rangle = \eta^{ij} \tag{7.27}$$

Our strategy is to show that Φ restricts to an isomorphism

$$\Phi' : H_a \longrightarrow H_b^*$$

First, we show that, for $h \in H_a$, $\Phi(h)$ restricts to a map $H_b \to R$. For $h = 1$, this is trivial ($\Phi(1) = \varepsilon$), so it suffices to show this for $h = z, z^2, \ldots, z^{p-1}$. First take $h = z$. Then $\langle \Phi(z), 1 \rangle = \langle \Phi(v) - \varepsilon, 1 \rangle = 0 \in R$, and

$$\begin{aligned}
\langle \Phi(z), x^i \rangle &= \frac{1}{a}\langle \Phi(v) - \varepsilon, x^i \rangle \\
&= \frac{1}{a}(\langle \Phi(v), x^i \rangle - \langle \varepsilon, x^i \rangle) \\
&= \frac{1}{a}\langle \Phi(v), \frac{u-1}{b}\rangle^i \\
&= \frac{(\eta - 1)^i}{ab^i} = a^{i-1} \in R
\end{aligned}$$

We have therefore shown that $\Phi(z)$ restricts to a map $H_b \to R$. The same property follows immediately for $\Phi(z^i) = \Phi(z)^i$ (Φ is an algebra map). Thus Φ restricts to a homomorphism $\Phi' : H_a \longrightarrow H_b^*$. Now $\{1, z, \dots, z^{p-1}\}$ is a basis for H_a. In order to show that Φ' is an isomorphism, it suffices to show that $\{\varepsilon, \Phi(z) = \zeta, \dots, \Phi(z^{p-1}) = \zeta^{p-1}\}$ is a basis for H_b^*. Consider the following basis of H_b^*:

$$\{\varepsilon, \pi_x, \pi_{x^2}, \dots, \pi_{x^{p-1}}\}$$

where π_{x^i} is the natural projection on Rx^j:

$$\langle \pi_{x^i}, x_j \rangle = \delta_{ij}$$

We have to show that $\pi_x, \pi_{x^2}, \dots, \pi_{x^{p-1}}$ are in a unique way linear combinations of $\zeta, \zeta^2, \dots, \zeta^{p-1}$. We will proceed as follows. Write

$$Z = \begin{pmatrix} \zeta \\ \zeta^2 \\ \vdots \\ \zeta^{p-1} \end{pmatrix} \quad \text{and} \quad \Pi = \begin{pmatrix} \pi_x \\ \pi_{x^2} \\ \vdots \\ \pi_{x^{p-1}} \end{pmatrix}$$

and consider the maps ζ_i and π_{x^i} as maps $R_p \otimes H_b \to R_p$. From the fact that $\{\varepsilon, \pi_x, \pi_{x^2}, \dots, \pi_{x^{p-1}}\}$ is a basis for H_b^* and $(R_p \otimes H_b^*)$, it follows that the ζ_i are in a unique way linear combinations of the π_{x^j}, and that the coefficients are in R. Consequently we can write

$$Z = M\Pi \tag{7.28}$$

where M is a $(p-1) \times (p-1)$-matrix with entries in R.
It now suffices to show that the determinant of M is invertible in R. To this end, we will compute M explicitly. This works in two steps. Consider the maps

$$\psi_i : R_p \otimes H_b \longrightarrow R_p$$

given by

$$\langle \psi_i, u^j \rangle = \delta_{ij}$$

for $i, j = 1, \dots, p-1$, and write

$$\Psi = \begin{pmatrix} \psi_1 \\ \psi_2 \\ \vdots \\ \psi_{p-1} \end{pmatrix}$$

$\{\varepsilon, \psi_1, \psi_2, \dots, \psi_{p-1}\}$ is a basis for $\mathrm{Hom}_{R_p}(H_b, R_p)$ (but not for H_b^*!). For every $j, n \in \{1, \dots, p-1\}$, we have

$$x^n = \frac{1}{b^n} \sum_{k=0}^{n} \binom{n}{k} (-1)^{n-k} u^k$$

and therefore

$$\langle \psi_j, x^n \rangle = \begin{cases} \dfrac{1}{b^n}(-1)^{n-j}\dbinom{n}{k} & \text{if } j \le n \\ 0 & \text{if } j > n \end{cases}$$

and it follows that

$$\psi_j = \sum_{n=j}^{p-1} \frac{(-1)^{n-j}}{b^n} \binom{n}{j} \pi_{x^n}$$

In matrix notation

$$\Psi = BA\Pi$$

with

$$B = \begin{pmatrix} 1 & -1 & 1 & \cdots & -1 \\ 0 & \binom{2}{2} & -\binom{3}{2} & \cdots & \binom{p-1}{2} \\ 0 & 0 & \binom{3}{3} & \cdots & -\binom{p-1}{3} \\ \vdots & \vdots & \vdots & & \vdots \\ 0 & 0 & 0 & \cdots & \binom{p-1}{p-1} \end{pmatrix} \text{ and } A = \begin{pmatrix} b^{-1} & 0 & 0 & \cdots & 0 \\ 0 & b^{-2} & 0 & \cdots & 0 \\ 0 & 0 & b^{-3} & \cdots & 0 \\ \vdots & \vdots & \vdots & & \vdots \\ 0 & 0 & 0 & \cdots & b^{-(p-1)} \end{pmatrix}$$

Clearly $\det(B) = 1$ and $\det(A) = b^{-p(p-1)/2}$.
We will now express the ζ^i as linear combinations of the ψ_j. We have that

$$\zeta = \sum_{i=1}^{p-1} \langle \zeta, u^i \rangle \psi_i = \sum_{i=1}^{p-1} \frac{\eta^i - 1}{a} \psi_i = b \sum_{i=1}^{p-1} s_i \psi_i$$

where $s_i = \dfrac{\eta^i - 1}{\eta - 1} = 1 + \eta + \eta^2 + \cdots + \eta^{i-1}$. Using the fact that the ψ_i are idempotents, it follows that

$$\zeta^k = b^k \sum_{i=1}^{p-1} s_i^k \psi_i$$

In matrix notation, we have

$$Z = DC\Psi$$

with

$$D = \begin{pmatrix} b & 0 & \cdots & 0 \\ 0 & b^2 & \cdots & 0 \\ \vdots & \vdots & & \vdots \\ 0 & 0 & \cdots & b^{p-1} \end{pmatrix} \text{ and } C = \begin{pmatrix} s_1 & s_2 & \cdots & s_{p-1} \\ s_1^2 & s_2^2 & \cdots & s_{p-1}^2 \\ \vdots & \vdots & & \vdots \\ s_1^p & s_2^p & \cdots & s_{p-1}^p \end{pmatrix}$$

It follows that

$$Z = DCBA\Pi = M\Pi$$

Now $\det(D) = b^{p(p-1)/2} = \det(A)^{-1}$, and $\det(M) = \det(C)$. If we factor out $s_1 \cdots s_{p-1}$, then $\det(C)$ is a Vandermonde determinant:

$$\det(C) = s_1 \cdots s_{p-1} \prod_{0<i<j<p} (s_j - s_i)$$

From Lemma 7.1.9, we know that $s_1, s_2, \cdots s_{p-1}$ are invertible in R. Furthermore

$$\begin{aligned} s_j - s_i &= \frac{1}{\eta - 1}(\eta^j - 1 - (\eta^i - 1)) \\ &= \eta^i \frac{\eta^j - 1}{\eta - 1} = \eta^i s_j \end{aligned}$$

is invertible in R. Therefore $\det(M) = \det(C)$ is invertible in R, and we have shown the following result.

Theorem 7.1.14 *Let R be a commutative ring in which p is not a zero-divisor, and suppose that R contains a p-th root of unity η. If $ab = \eta - 1$, then the Hopf algebras $H_b = R[x]/f_b(x)$ and $H_a = R[z]/f_a(z)$ are each others dual. The duality is given by the formula*

$$\langle z, x \rangle = 1$$

Theorem 7.1.14 can be generalized to non-free Hopf algebras.

Theorem 7.1.15 *Let R be a commutative ring in which p is not a zero-divisor, and suppose that R contains a p-th root of unity η. If I and J are ideals such that $IJ = (\eta - 1)R$, then the Hopf algebras*

$$H_J = \check{R}[Ky]/\big(K^p((1+y)^p - 1)\big) \quad \text{and} \quad H_I = \check{R}[Lt]/\big(L^p((1+t)^p - 1)\big)$$

are each others dual. The duality is given by the formula

$$\langle y, t \rangle = \eta - 1$$

Proof We use the following notations

$$u = 1 + y \qquad K = \frac{1}{p} I^{p-1} J^{p-2}$$

$$v = 1 + t \qquad L = \frac{1}{p} I^{p-2} J^{p-1}$$

Observe also that

$$JK = R, \quad IL = R \quad \text{and} \quad (IJ)^{p-1} = pR$$

If $J = bR$ and $I = aR$ are principal ideals, then the result follows from the preceding Theorem. In the general situation, we still have that

$$R_p \otimes H_I \cong R_p \otimes H_J^*$$

and the connecting isomorphism Φ is still given by (7.27). We claim that Φ restricts to a map $\Phi' : H_I \to H_J^*$. To this end, it suffices to show that, for $i = 1, \ldots, p-1$ and $l_i \in L^i$, the map $\Phi(l_i t^i)$ restricts to a map $H_J \to R$. As in the proof of Theorem 7.1.14, we first consider the case $i = 1$. Then

$$\langle \Phi(t), 1 \rangle = \langle \Phi(v) - \varepsilon, 1 \rangle = 0 \in R$$

and, for all $k_i \in K^i$,

$$\begin{aligned} \langle \Phi(l_1 t), k_i y^i \rangle &= l_1 k_i \langle \Phi(v) - \varepsilon, y^i \rangle \\ &= l_1 k_i \langle \Phi(v), (u-1)^i \rangle \\ &= l_1 k_i (\eta - 1)^i \in LK^i(\eta - 1)^i = I^{i-1} L^i K^i (\eta - 1)^i = I^{i-1} \subset R \end{aligned}$$

The same property follows immediately for $\Phi(l_i y^i) = l_i \Phi(y)^i$, since Φ preserves the multiplication.

From the preceding Theorem, we know that Φ' becomes an isomorphism after we localize at a prime ideal of R. Hence Φ' is an isomorphism, and this finishes the proof. □

7.2 Modules and comodules

In this section, H will be a Hopf algebra over a commutative ring R. At some places, we will make the additional assumption that H is finite, commutative or cocommutative.

Modules

Let A be an R-algebra. Recall that a left A-module M is an R-module M together with a map

$$\psi = \psi_M : \ A \otimes M \to M$$

satisfying

$$\psi(ab \otimes m) = \psi(a \otimes \psi(b \otimes m)) \tag{7.29}$$

$$\psi(\eta(r) \otimes m) = rm \tag{7.30}$$

for all $a, b \in A$, $m \in M$ and $r \in R$. In the sequel, we will often write

$$\psi(a \otimes m) = a \rightharpoonup m$$

In a similar way, we define right A-modules. If M is a right A-module, with structure map ψ, then we will write

$$\psi(m \otimes a) = m \leftharpoonup a$$

An R-module M that is at once a left A-module and a right B-module is called an (A, B)-bimodule if

$$(a \rightharpoonup m) \leftharpoonup b = a \rightharpoonup (m \leftharpoonup b)$$

for all $a \in A$, $b \in B$, $m \in M$. If M is a right A-module, and N is a left A-module, then $M \otimes_A N$ is the coequalizer of $I_M \otimes \psi_N$ and $\psi_M \otimes I_N$:

$$M \otimes A \otimes N \rightrightarrows M \otimes N \longrightarrow M \otimes_A N$$

If M is a (B, A)-bimodule, then $M \otimes_A N$ is a left B-module.
Now let H be a Hopf algebra. If M and N are (left) H-modules, then $M \otimes N$ and $\mathrm{Hom}_R(M, N)$ are also H-modules. The H-module structure is given by the following formulas:

$$h \rightharpoonup (m \otimes n) = \sum (h_{(1)} \rightharpoonup m) \otimes (h_{(2)} \rightharpoonup n) \tag{7.31}$$

$$(h \rightharpoonup f)(m) = \sum h_{(1)} \rightharpoonup (f(S(h_{(2)}) \rightharpoonup m)) \tag{7.32}$$

for all $m \in M$, $n \in N$, $h \in H$, $f : \ M \to N$. (7.32) can be rewritten as follows. For each $h \in H$, we define a map

$$\psi_h : \ M \to M$$

by

$$\psi_h(m) = \psi(h \otimes m) = h \rightharpoonup m$$

Using this notation, for $f : \ M \to N$, we have

$$h \rightharpoonup f = \sum \psi_{h_{(1)}} \circ f \circ \psi_{S(h_{(2)})} \tag{7.33}$$

We will also write

$$M^H = \{m \in M | h \rightharpoonup m = \varepsilon(h)m \text{ for all } h \in H\}$$

Module algebras and module coalgebras

Now suppose that A is a (left) H-module and an R-algebra. We say that A is a left *H-module algebra* if for all $h \in H$ and $a, b \in A$, we have that $h \rightharpoonup 1 = \varepsilon(h)$ and $h \rightharpoonup (ab) = \sum (h_{(1)} \rightharpoonup a)(h_{(2)} \rightharpoonup b)$.

Now suppose that H is cocommutative. Then the tensor product $A \otimes B$ of two H-module algebras is again an H-module algebra. If M is an H-module, then the endomorphism ring $\mathrm{End}_R(M)$ is an H-module algebra, we leave verification of the details to the reader.

A left *H-module coalgebra* is an R-module C such that C is a left H-module and a R-coalgebra with the additional properties

$$\Delta_C(h \rightharpoonup c) = \sum (h_{(1)} \rightharpoonup c_{(1)}) \otimes (h_{(2)} \rightharpoonup c_{(2)}) \tag{7.34}$$

$$\varepsilon_C(h \rightharpoonup c) = \varepsilon_H(h)\varepsilon_C(c) \tag{7.35}$$

Let G be a group, and X a left G-set, that is, we have an action

$$G \times X \to X : \ (g, x) \mapsto g \cdot x$$

satisfying

$$h \cdot (g \cdot x) = (hg) \cdot x$$

for all $g, h \in G$ and $x \in X$. Then the coalgebra RX is a left RG-module coalgebra. Right H-module algebras and right H-module coalgebras are defined in a similar way. If C is a right H-module coalgebra, then C^* is a left H-module algebra. The left action of H on C^* is given by the formula

$$\langle h \rightharpoonup c^*, d \rangle = \langle c^*, d \leftharpoonup h \rangle$$

for all $c^* \in C^*$, $d \in C$ and $h \in H$.

Comodules

Let C be a coalgebra. A (right) *C-comodule* is an R-module M together with a map $\rho = \rho_C : \ M \to M \otimes C$ satisfying the following two conditions:

$$(\rho \otimes I_C) \circ \rho = (I_M \otimes \Delta_C) \circ \rho \tag{7.36}$$

$$(I_C \otimes \varepsilon_C) \circ \rho = I_M \tag{7.37}$$

We will also say that C *coacts* on M. For the map ρ, we introduce the Sweedler notation

$$\rho(m) = \sum m_{(0)} \otimes m_{(1)}$$

(7.36) justifies the notation

$$((\rho \otimes I_C) \circ \rho)(m) = \sum m_{(0)} \otimes m_{(1)} \otimes m_{(2)}$$

and (7.37) can be rewritten as

$$\sum \varepsilon_C(m_{(1)}) m_{(0)} = m$$

A map $f : M \to N$ between two comodules M and N is called a *C-comodule homomorphism* or a *C-colinear map* if

$$\rho_N \circ f = (f \otimes I_C) \circ \rho_M$$

or

$$\rho_N(f(m)) = \sum f(m_{(0)}) \otimes m_{(1)}$$

for all $m \in M$. The category of (right) C-comodules and C-colinear maps is denoted by comod-C. Recall that C^* is an R-algebra, and observe that we have a natural functor

$$F : \text{comod-}C \to C^*\text{-mod}$$

defined as follows: for $M \in C^*$-mod, we let $F(M)$ be equal to M as an R-module, with C^*-module structure given by

$$c^* \rightharpoonup m = \sum \langle c^*, m_{(1)} \rangle m_{(0)}$$

for all $m \in M$ and $c^* \in C^*$.

Proposition 7.2.1 *If C is faithfully projective as an R-module, then F is an isomorphism of categories.*

Proof We will construct an inverse G of F. Let $\{c_i, c_i^* | i = 1, \ldots, n\}$ be a dual basis of C. Recall that this means that for every $c \in C$ and $c^* \in C^*$ we have

$$c^* = \sum_i \langle c^*, c_i \rangle c_i^* \quad \text{and} \quad c = \sum_i \langle c_i^*, c \rangle c_i$$

Let $M \in C^*$-mod, and let $G(M) = M$ as an R-module, with C-coaction given by

$$\rho(m) = \sum_i (c_i^* \rightharpoonup m) \otimes c_i$$

We claim that $G(M)$ is a C-comodule. We have to show that ρ satisfies equations (7.36, 7.37). First observe that for all $m \in M$:

$$(I \otimes \varepsilon)\rho(m) = \sum_i \varepsilon(c_i)(c_i^* \rightharpoonup m) = \varepsilon \rightharpoonup m = m$$

proving (7.37). Equation (7.36)

$$(\rho \otimes I_C) \circ \rho = (I_M \otimes \Delta_C) \circ \rho$$

follows from the fact that for all $c^*, d^* \in C^*$ and $m \in M$

$$\begin{aligned}
&\big((I_M \otimes c^* \otimes d^*) \circ (\rho \otimes I_C) \circ \rho\big)(m) \\
&= (I_M \otimes c^* \otimes d^*)\Big(\sum_{i,j} \big((c_j^* * c_i^*) \rightharpoonup m \otimes c_j \otimes c_i\big)\Big) \\
&= \sum_{i,j} \langle c^*, c_j \rangle \langle d^*, c_i \rangle (c_j^* * c_i^*) \rightharpoonup m \\
&= (c^* * d^*) \rightharpoonup m
\end{aligned}$$

$$\begin{aligned}
&= \sum_i (c^* * d^*)(c_i)c_i^* \rightharpoonup m\\
&= \sum_i (c^* \otimes d^*)\Delta(c_i)c_i^* \rightharpoonup m\\
&= (I_M \otimes c^* \otimes d^*)\big(\sum_i c_i^* \rightharpoonup m \otimes \Delta(c_i)\big)\\
&= (I_M \otimes c^* \otimes d^*) \circ (I_M \otimes \Delta_C) \circ \rho(m)
\end{aligned}$$

We leave it to the reader to show that F and G are each others inverses. □

Now let H be a Hopf algebra. The tensor product $M \otimes N$ of two H-comodules is again an H-comodule. The structure map $\rho_{M\otimes N}$ is given by

$$\rho_{M\otimes N}(m \otimes n) = \sum m_{(0)} \otimes n_{(0)} \otimes m_{(1)}n_{(1)} \tag{7.38}$$

for all $m \in M$ and $n \in N$. This structure map is compatible with the functor $F:$ comod-$C \to C$-mod defined above.
If H is a faithfully projective Hopf algebra, then we can use Proposition 7.2.1 to define an H-comodule structure on $\mathrm{Hom}_R(M, N)$. Indeed, M and N are H^*-modules, and therefore H^* acts on $\mathrm{Hom}_R(M, N)$ as follows:

$$\begin{aligned}
(h^* \rightharpoonup f)(m) &= \sum h^*_{(1)} \rightharpoonup f(S^*(h^*_{(2)}) \rightharpoonup m)\\
&= \sum \langle S^*(h^*_{(2)}), m_{(1)}\rangle h^*_{(1)} \rightharpoonup f(m_{(0)})\\
&= \sum \langle h^*_{(2)}, S(m_{(1)})\rangle \langle h^*_{(1)}, f(m_{(0)})_{(1)}\rangle f(m_{(0)})_{(0)}\\
&= \sum \langle h^*, f(m_{(0)})_{(1)} S(m_{(1)})\rangle f(m_{(0)})_{(0)}
\end{aligned}$$

for all $m \in M,\ h^* \in H^*,\ f \in \mathrm{Hom}_R(M, N)$.
The map

$$\rho:\ \mathrm{Hom}_R(M, N) \to \mathrm{Hom}_R(M, N) \otimes H \cong \mathrm{Hom}_R(M, N \otimes H)$$

is therefore given by

$$\rho(f)(m) = \sum f(m_{(0)})_{(0)} \otimes f(m_{(0)})_{(1)} S(m_{(1)}) \tag{7.39}$$

for all $m \in M$ and $f \in \mathrm{Hom}_R(M, N)$. This may also be expressed as follows:

$$\rho(f) = \sum f_{(0)} \otimes f_{(1)} \in \mathrm{Hom}_R(M, N) \otimes H$$

if and only if for all $m \in M$ we have

$$\sum f_{(0)}(m) \otimes f_{(1)} = \sum f(m_{(0)})_{(0)} \otimes f(m_{(0)})_{(1)} S(m_{(1)}) \tag{7.40}$$

For later use, let us examine the special case where $N = R$. For $m^* \in M^*$ we now have that

$$\rho(m^*) = \sum m^*_{(0)} \otimes m^*_{(1)}$$

if and only if for all $m \in M$

$$\sum \langle m^*_{(0)}, m\rangle m^*_{(1)} = \sum \langle m^*, m_{(0)}\rangle S(m_{(1)}) \tag{7.41}$$

Similar definitions and properties apply to left H-comodules. For a left C-comodule M, we will use the Sweedler notation

$$\rho_M(m) = \sum m_{(-1)} \otimes m_{(0)} \in C \otimes M$$

Let C and D be coalgebras. A (C, D)-bicomodule is an R-module which is at once a left C-comodule and a right D-comodule such that

$$(\rho_C \otimes I_D) \circ \rho_D = (I_C \otimes \rho_D) \circ \rho_C \tag{7.42}$$

(7.42) justifies the following version of Sweedler's notation:

$$((\rho_C \otimes I_D) \circ \rho_D)(m) = ((I_C \otimes \rho_D) \circ \rho_C)(m) = \sum m_{(-1)} \otimes m_{(0)} \otimes m_{(1)}$$

Observe that the map Δ_C makes C itself in a C-bicomodule.

The cotensor product

Suppose that M is a right C-comodule and that N is a left C-comodule. We define the *cotensor product* $M\Box_C N$ of M and N as the equalizer of the maps $\rho_M \otimes I_N$ and $I_M \otimes \rho_N$

$$0 \longrightarrow M\Box_C N \longrightarrow M \otimes N \rightrightarrows M \otimes C \otimes N \tag{7.43}$$

$M\Box_C N$ is an R-module, but in some situations, it also has a comodule structure. This is discussed in the following Proposition:

Proposition 7.2.2 *Suppose that M is a right C-comodule, N is a (C, D)-bicomodule, and that the coalgebra D is flat as an R-module. Then the map $I_M \otimes \rho_N$ puts a right D-comodule structure on $M\Box_C N$. Similarly, if M is a (D, C)-bicomodule, N a left C-comodule and D flat as an R-module, then $M\Box_C N$ is a left D-comodule. If M is a (D', C)-bicomodule and if N is a (C, D)-bicomodule, then $M\Box_C N$ is a (D', D)-bicomodule, if D and D' are both R-flat.*

Proof We only prove the first statement. Consider $\sum_i m_i \otimes n_i \in M\Box_C N$, that is

$$\sum_i m_{i_{(0)}} \otimes m_{i_{(1)}} \otimes n_i = \sum_i m_i \otimes n_{i_{(-1)}} \otimes n_{i_{(0)}} \tag{7.44}$$

The only nontrivial thing we have to show is that

$$x = \sum_i m_i \otimes n_{i_{(0)}} \otimes n_{i_{(1)}} \in M\Box_C N \otimes D$$

From the fact that D is R-flat and the exactness of (7.42), it follows that we have an exact sequence

$$0 \longrightarrow M\Box_C N \otimes D \longrightarrow M \otimes N \otimes D \rightrightarrows M \otimes C \otimes N \otimes D$$

It therefore suffices to show that

$$(I_M \otimes \rho_N \otimes I_D)(x) = (\rho_M \otimes I_N \otimes I_D)(x)$$

or

$$\sum_i m_i \otimes n_{i_{(-1)}} \otimes n_{i_{(0)}} \otimes n_{i_{(1)}} = \sum_i m_{i_{(0)}} \otimes m_{i_{(1)}} \otimes n_{i_{(0)}} \otimes n_{i_{(1)}},$$

and this follows easily if we apply $I_M \otimes I_C \otimes \rho_N^r$ to (7.44) (The upper index r stands for the right D-comodule structure). □

If C is cocommutative, then it makes no sense to distinguish between left and right C-comodules, and in this case the cotensor product of two comodules is again a comodule.
If A is an R-algebra with a unit, then it is well-known that every A-module M is unital, that is $A \otimes_A M \cong M$. Below, we present the dual property.

Proposition 7.2.3 *If M is a right C-comodule, with C a coalgebra (with counit), then every C-comodule M is counital, that is*

$$M \square_C C \cong M$$

Proof It is easy to see that $\rho_M(m) \in M\square_C C$, for every $m \in M$. Now the maps $\rho_M : M \to M\square_C C$ and $I_M \otimes \varepsilon_C : M\square_C C \to M$ are each others inverses. Indeed,

$$(I_M \otimes \varepsilon_C)(\rho_M(m)) = m$$

for all $m \in M$. If $\sum_i m_i \otimes c_i \in M\square_C C$, then

$$\sum_i m_{i_{(0)}} \otimes m_{i_{(1)}} \otimes c_i = \sum_i m_i \otimes c_{i_{(1)}} \otimes c_{i_{(2)}}$$

and

$$\begin{aligned} \rho_M(I_M \otimes \varepsilon_C)(\sum_i m_i \otimes c_i) &= \sum_i m_{i_{(0)}} \otimes m_{i_{(1)}} \otimes \varepsilon(c_i) \\ &= \sum_i m_i \otimes c_{i_{(1)}} \otimes \varepsilon(c_{i_{(2)}}) \\ &= \sum_i m_i \otimes c_i \end{aligned}$$

□

Comodule algebras and comodule coalgebras

Let H be a Hopf algebra, and suppose that A is at once a (right) H-comodule and an R-algebra. We call A a right *H-comodule algebra* if m_A and η_A are H-colinear, that is

$$\begin{aligned} \rho(ab) &= \sum a_{(0)}b_{(0)} \otimes a_{(1)}b_{(1)} \\ \rho(1) &= 1 \otimes 1 \end{aligned}$$

for all $a, b \in A$. A left H-comodule algebra is defined in a similar way.
Now suppose that H is cocommutative. If A and B are H-module algebras, then

$A \otimes B$ is again an H-comodule algebra. If H is finite, and M is an H-comodule, then $\mathrm{End}_R(M)$ is an H-comodule algebra.
If H is finite, then the functor F from Proposition 7.2.1 defines an isomorphism between the categories of right H-comodule algebras and left H^*-module algebras.

The basic example of a comodule algebra is the following: let $H = RG$, where G is a group. An RG-comodule algebra A is nothing else then a G-graded R-algebra, and the relation between the G-grading and the RG-coaction on A is the following:

$$\deg(a) = \sigma \iff \rho(a) = a \otimes \sigma$$

To make the picture complete, let us state the definition of a right *H-comodule coalgebra.* This is a coalgebra C that is also a right H-comodule such that

$$\sum c_{[0]_{(1)}} \otimes c_{[0]_{(2)}} \otimes c_{[1]} = \sum c_{(1)_{[0]}} \otimes c_{(2)_{[0]}} \otimes_{(1)_{[1]}} c_{(2)_{[1]}}$$

and

$$\sum \varepsilon_C(c_{[0]})c_{[1]} = \varepsilon_C(c)1_H$$

for all $c \in C$. Here the indices between square brackets refer to the comodule structure, while the ones between brackets refer to the comultiplication on C. We leave it to the reader to state the definition of a left H-comodule algebra.

The smash product

Let H be a Hopf algebra, A a left H-module algebra, and B a left H-comodule algebra. The *smash product* $A\#B$ is the R-algebra defined as follows. $A\#B = A \otimes B$ as an R-module, and the multiplication rule is given by

$$(a\#b)(c\#d) = \sum a(b_{(-1)} \rightharpoonup c)\#b_{(0)}d \tag{7.45}$$

A straightforward computation show that $A\#B$ is an associative algebra. If A and B have a unit, then $1_A\#1_B$ is a unit for $A\#B$. The original formulation of this definition of the smash product goes back to [90].

Coinvariants

Let H be a Hopf algebra, and M a (right) H-comodule. We call

$$M^{\mathrm{co}H} = \{m \in M | \rho(m) = m \otimes 1\}$$

the submodule of *coinvariants* of M.

Chapter 8

Galois objects

8.1 Relative Hopf modules and Galois objects

Let R be a commutative ring and H a Hopf algebra with bijective antipode that is faithfully flat as an R-module. Suppose also that A is a right H-comodule algebra. We call M a (right) (A, H)-*Hopf module* if M is a right H-comodule and a right A-module, such that

$$\rho_M(ma) = \sum m_{(0)}a_{(0)} \otimes m_{(1)}a_{(1)} \tag{8.1}$$

for all $a \in A$ and $m \in M$. In the sequel, $\mathcal{M}(H)_A^H$ will denote the category of right (A, H)-Hopf modules and A-linear H-colinear maps. In a similar way, we call M a left (A, H)-Hopf module if M is a left H-comodule and a left A-module such that

$$\rho(am) = \sum m_{(-1)}S^{-1}(a_{(1)}) \otimes a_{(0)}m_{(0)} \tag{8.2}$$

for all $a \in A$ and $m \in M$. ${}_A^H\mathcal{M}(H)$ will denote the category of left (A, H)-Hopf modules and A-linear H-colinear maps. A two-sided (A, H)-Hopf module M is an R-module that is at once an A-bimodule and an H-bicomodule such that (8.1, 8.2) hold and furthermore

$$\begin{aligned}
\rho_M^r(am) &= \sum am_{(0)} \otimes m_{(1)} \\
\rho_M^l(ma) &= \sum m_{(-1)} \otimes m_{(0)}a
\end{aligned}$$

for all $a \in A$ and $m \in M$.

Examples 8.1.1 1) $A \otimes H$ is a two-sided (A, H)-Hopf module. The structure maps are the following:

$$(a \otimes h)b = \sum ab_{(0)} \otimes hb_{(1)} \tag{8.3}$$

$$\rho_{A\otimes H}^r(a \otimes h) = \sum a \otimes h_{(1)} \otimes h_{(2)} \tag{8.4}$$

$$b(a \otimes h) = ba \otimes h \tag{8.5}$$

$$\rho_{A\otimes H}^l(a \otimes h) = \sum h_{(1)}S^{-1}(a_{(1)}) \otimes a_{(0)} \otimes h_{(2)} \tag{8.6}$$

for all $a, b \in A$ and $h \in H$.
2) $H \otimes A$ is a two-sided (A, H)-Hopf module. The structure maps are the following:

$$
\begin{aligned}
(h \otimes a)b &= h \otimes ab && (8.7)\\
\rho^r_{H\otimes A}(h \otimes a) &= \sum h_{(1)} \otimes a_{(0)} \otimes h_{(2)}a_{(1)} && (8.8)\\
b(h \otimes a) &= \sum (hS^{-1}(b_{(1)})) \otimes b_{(0)}a && (8.9)\\
\rho^l_{H\otimes A}(h \otimes a) &= \sum h_{(1)} \otimes h_{(2)} \otimes a && (8.10)
\end{aligned}
$$

for all $a, b \in A$ and $h \in H$. We leave it to the reader to prove the six compatibility relations in each of our two examples. Furthermore, observe that $\nu : H\otimes A \longrightarrow A\otimes H$ defined by

$$\nu(h \otimes a) = \sum a_{(0)} \otimes ha_{(1)}$$

is an isomorphism of two-sided Hopf modules. Its inverse is given by

$$\nu^{-1}(a \otimes h) = \sum hS^{-1}(b_{(1)}) \otimes b_{(0)}$$

In Proposition 8.1.2, we will show that relative Hopf modules are related to the smash product. We will need the following right-left version of the smash product: let A be a right H-comodule algebra and B a left H-module algebra, and define $A^{\rm op}\#B$ by the multiplication rule

$$(a\#b)(c\#d) = \sum c_{(0)}a\#(c_{(1)}\rightharpoonup b)d \tag{8.11}$$

We leave it to the reader to show that $A^{\rm op}\#B$ is an associative algebra.
We already know that H^* is an algebra; observe that H^* can be made into a left H-module algebra as follows.

$$\langle h\rightharpoonup h^*, k\rangle = \langle h^*, kh\rangle$$

for all $h, k \in H$ and $h^* \in H^*$.

Proposition 8.1.2 *With notations as above, we have a functor*

$$\mathcal{M}(H)^H_A \longrightarrow A^{\rm op}\#H^*\text{-mod}$$

This functor is an isomorphism of categories if H is a finite Hopf algebra.
If $S^2 = I$, then we also have a functor

$${}^H_A\mathcal{M}(H) \longrightarrow \text{mod-}A^{\rm op}\#H^*$$

this functor is an isomorphism of categories if H is finite.
If H is finite and $S^2 = I$, then two-sided (A, H)-Hopf modules correspond to $A^{\rm op}\#H^*$-bimodules.

Proof Let M be a right (A, H)-Hopf module. Consider the following left $A^{\rm op}\#H^*$-action on M:

$$(a\#h^*) \cdot m = (h^* \cdot m)a = \sum \langle h^*, m_{(1)}\rangle m_{(0)}a$$

It is easy to see that M is an $A^{\rm op}\#H^*$-module: for all $a,b\in A$ and $h^*,k^*\in H^*$, we have that

$$\begin{aligned}
&((a\#h^*)(b\#k^*))\cdot m\\
&= (\sum b_{(0)}a\#(b_{(1)}\rightharpoonup h^*)*k^*)\cdot m\\
&= \sum\langle (b_{(1)}\rightharpoonup h^*)*k^*, m_{(1)}\rangle m_{(0)}b_{(0)}a\\
&= \sum\langle h^*, m_{(1)}b_{(1)}\rangle\langle k^*, m_{(2)}\rangle m_{(0)}b_{(0)}a\\
&= (a\#h^*)\cdot(\sum\langle k^*, m_{(1)}\rangle m_{(0)}b_{(0)})\\
&= (a\#h^*)\cdot((b\#k^*)\cdot m)
\end{aligned}$$

Suppose that H is finite, and suppose that M is a left $A^{\rm op}\#H^*$-module. The formulas

$$\begin{aligned}
ma &= (a\#\varepsilon)\cdot m\\
h^*\cdot m &= (1\#h^*)\cdot m
\end{aligned}$$

make M into a left H^*-module and a right A-module. M is therefore also a right H-comodule, see Proposition 7.2.1. We have to check that (8.1) is fulfilled. For all $a\in A$, $h^*\in H^*$ and $m\in M$, we have

$$\begin{aligned}
h^*\cdot(ma) &= (1\#h^*)\cdot((a\#\varepsilon)\cdot m)\\
&= \sum(a_{(0)}\#(a_{(1)}\rightharpoonup h^*))\cdot m\\
&= \sum(a_{(0)}\#\varepsilon)(1\#(a_{(1)}\rightharpoonup h^*)m\\
&= ((a_{(1)}\rightharpoonup h^*)\cdot m)a_{(0)}\\
&= \sum\langle h^*, m_{(1)}a_{(1)}\rangle m_{(0)}a_{(0)}
\end{aligned}$$

and (8.1) follows.

The proof of the second part is similar: let M be a left (A,H)-Hopf module and define a right $A^{\rm op}\#H^*$-action on M as follows:

$$m\cdot(a\#h^*) = (am)\leftharpoonup h^* = \sum\langle h^*, m_{(-1)}S(a_{(1)})\rangle a_{(0)}m_{(0)}$$

We leave it to the reader to check that M is a right $A^{\rm op}\#H^*$-module. If H is finite, and M is a right $A^{\rm op}\#H^*$-module, then we define a left (A,H)-module structure on H as follows:

$$am = m\cdot(a\#\varepsilon)\ \text{ and }\ m\leftharpoonup h^* = m\cdot(1\#h^*)$$

For all $a\in A$, $h^*\in H^*$ and $m\in M$, we have

$$\begin{aligned}
a(m\leftharpoonup h^*) &= (m\cdot(1\#h^*))(a\#\varepsilon)\\
&= m\cdot(a_{(0)}\#(a_{(1)}\rightharpoonup h^*))\\
&= \sum\langle a_{(1)}\rightharpoonup h^*, (a_{(0)}m)_{(-1)}\rangle(a_{(0)}m)_{(0)}
\end{aligned}$$

Now $a(m\leftharpoonup h^*) = \sum\langle h^*, m_{(-1)}\rangle am_{(0)}$, and it follows that

$$\sum(a_{(0)}m)_{(-1)}a_{(1)}\otimes(a_{(0)}m)_{(0)} = \sum m_{(-1)}\otimes am_{(0)}$$

It now follows that

$$\begin{aligned}\rho(am) &= \sum (a_{(0)}m)_{(-1)}a_{(1)}S(a_{(2)} \otimes (a_{(0)}m)_{(0)} \\ &= \sum m_{(-1)}S(a_{(1)} \otimes a_{(0)}m_{(0)}\end{aligned}$$

and (8.2) holds since $S = S^{-1}$. The third part of the Theorem now follows immediately. □

Proposition 8.1.3 *Let R, H and A be as above, and consider the functors*

$$\begin{aligned}F:\ R\text{-mod} \to \mathcal{M}(H)_A^H &:\ N \mapsto A \otimes N \\ G:\ \mathcal{M}(H)_A^H \to R\text{-mod} &:\ M \mapsto M^{\mathrm{co}H}\end{aligned}$$

Then G is a rightadoint to F.

Proof For $N \in R$-mod and $M \in \mathcal{M}(H)_A^H$, we define maps

$$\mathrm{Hom}_R(N, M^{\mathrm{co}H}) \underset{\beta}{\overset{\alpha}{\rightleftarrows}} \mathrm{Hom}_A^H(A \otimes N, M)$$

by

$$\alpha(f)(a \otimes n) = f(n)a \quad \text{and} \quad \beta(g)(n) = g(1 \otimes n)$$

It is straightforward to check that α and β are each others inverses. □

Remarks 8.1.4 1) Propositions 8.1.2 and 8.1.3 are special cases of more general statements. If H is a Hopf algebra, A an H-comodule algebra and C an H-module coalgebra, then we can define the notion of (H, A, C)-Hopf module. This coincides with our notion of Hopf module in the case where $C = H$. We refer to [78] and to [50] for more details. In [50, Theorem 1.3] pairs of adjoint functors between categories of Hopf modules are discussed.

2) If H is a finite Hopf algebra, then we can apply Proposition 8.1.2, and we obtain the following pair of adjoint functors:

$$\begin{aligned}F:\ R\text{-mod} \to (A^{\mathrm{op}}\#H^*)\text{-mod} &:\ N \mapsto A \otimes N \\ G:\ (A^{\mathrm{op}}\#H^*)\text{-mod} \to R\text{-mod} &:\ M \mapsto M^{H^*}\end{aligned}$$

3) For $N \in R$-mod and $M \in \mathcal{M}(H)_A^H$ we now have natural maps

$$\psi_N:\ N \to G(F(N)) = A^{\mathrm{co}H} \otimes N:\ n \mapsto 1 \otimes n$$

and

$$\phi_M:\ F(G(M)) = A \otimes M^{\mathrm{co}H} \to M:\ a \otimes m \mapsto ma$$

Definition 8.1.5 *With notations as above, an H-comodule algebra A is called an H-Galois object if the two functors F and G of Proposition 8.1.3 are inverse equivalences, or, equivalently, if ψ_N and ϕ_M are isomorphisms for all $N \in R$-mod and $M \in \mathcal{M}(H)_A^H$.*

We will now establish some necessary and sufficient conditions for an H-comodule algebra to be an H-Galois object. First of all, observe that it is clear that ψ_N is an isomorphism for all $N \in R$-mod if and only if

$$A^{\mathrm{co}H} = R \tag{8.12}$$

For the map ϕ_M, the situation is more complicated. Recall that $A \otimes H$ with A-action and H-coaction defined by

$$\begin{aligned} (a \otimes h) \leftharpoonup b &= \sum ab_{(0)} \otimes hb_{(1)} \\ \rho_{A \otimes H}(a \otimes h) &= \sum a \otimes h_{(1)} \otimes h_{(2)} \end{aligned}$$

is an (A, H)-Hopf module. Now observe that $H^{\mathrm{co}H} = R$ (use the maps $H^{\mathrm{co}H} \to R:$ $h \to \varepsilon(h)$ and $R \to H^{\mathrm{co}H} : x \to \eta(x)$), and therefore

$$F(G(A \otimes H)) = F(A) = A \otimes A$$

The map $\gamma = \phi_{A \otimes H} : F(G(A \otimes H)) = A \otimes A \to A \otimes H$ is given by the formula

$$\gamma(a \otimes b) = \sum ab_{(0)} \otimes b_{(1)} \tag{8.13}$$

A necessary condition for A to be an H-Galois object is therefore that $S^{\mathrm{co}H} = R$ and that γ is an isomorphism. In the next Theorem, we will show that this condition is almost sufficient.

Theorem 8.1.6 *Let H be a Hopf algebra which is faithfully flat as an R-module, and suppose that the antipode of H is bijective. For a right H-comodule algebra A, the following conditions are equivalent:*

1. *A is an H-Galois object;*
2. • *$A^{\mathrm{co}H} = R$;*
   • *$\gamma = \phi_{A \otimes H} : A \otimes A \to A \otimes H : a \otimes b \mapsto \sum ab_{(0)} \otimes b_{(1)}$ is an isomorphism;*
   • *A is flat as an R-module.*

Proof Observe that the Hopf module structure on $F(A) = A \otimes A$ is given by

$$(a \otimes b) \leftharpoonup c = a \otimes bc \quad \text{and} \quad \rho(a \otimes b) = \sum a \otimes b_{(0)} \otimes b_{(1)}$$

With these notations, γ is a map in $\mathcal{M}(H)_A^H$.
We have seen above that if A is an H-Galois object, then $S^{\mathrm{co}H} = R$ and γ is an isomorphism. In this case, F is an equivalence of categories, and therefore an exact functor. This means that A is a flat R-module.
Observe that we can make A into a left (A, H)-Hopf module, by defining

$$a \cdot b = ab \quad \text{and} \quad \rho^\ell(a) = \sum S^{-1}(a_{(1)}) \otimes a_{(0)}$$

indeed, for any $a, n \in A$, we have

$$\rho^\ell(an) = \sum S^{-1}(a_{(1)}n_{(1)}) \otimes a_{(0)}n_{(0)} = \sum S^{-1}(n_{(1)})S^{-1}(a_{(1)}) \otimes a_{(0)}n_{(0)}$$

and (8.2) is satisfied.

Now take $M \in {}^H_A \mathcal{M}(H)$ and $N \in \mathcal{M}(H)^H_A$. We may now consider the tensor product $N \otimes_A M$ and the cotensor product $N \square_H M$, and we have natural maps

$$N \square_H M \to N \otimes M \to N \otimes_A M$$

The image of $N \square_H M$ in $N \otimes_A M$ is called the *bitensor product* (cf. [50]) of N and M, and it is denoted by

$$N \otimes^H_A M = \mathrm{Im}\,(N \square_H M \to N \otimes_A M)$$

We now claim that

$$N \otimes^H_A A \cong N \square_H R = N^{\mathrm{coH}} \qquad (8.14)$$

Here R is viewed as an H-algebra via ε. First we will show that the map

$$f :\ N \otimes_A A \overset{\cong}{\longrightarrow} N$$

restricts to

$$f :\ N \otimes^H_A A \longrightarrow N^{\mathrm{coH}}$$

Take $\sum_i n_i \otimes a_i \in N \square_H A$. Then

$$\sum_i n_{i_{(0)}} \otimes n_{i_{(1)}} \otimes a_i = \sum_i n_i \otimes S^{-1}(a_{i_{(1)}}) \otimes a_{i_{(0)}}$$

Apply ρ_A to the last factor:

$$\sum_i n_{i_{(0)}} \otimes n_{i_{(1)}} \otimes a_{i_{(0)}} \otimes a_{i_{(1)}} = \sum_i n_i \otimes S^{-1}(a_{i_{(2)}}) \otimes a_{i_{(0)}} \otimes a_{i_{(1)}}$$

Now let the third factor act on the first one, and the fourth one on the second one

$$\sum_i n_{i_{(0)}} a_{i_{(0)}} \otimes n_{i_{(1)}} a_{i_{(1)}} = \sum_i n_i a_{i_{(0)}} \otimes S^{-1}(a_{i_{(2)}}) a_{i_{(1)}} = \sum_i n_i a_i \otimes 1$$

and this shows that

$$\sum_i n_i a_i = f(\sum_i n_i \otimes a_i) \in N^{\mathrm{coH}}$$

Now define

$$g :\ N^{\mathrm{coH}} \longrightarrow N \otimes^H_A A$$

by

$$g(n) = n \otimes 1$$

then

$$f \circ g = I_{N^{\mathrm{coH}}}$$

and for all $\sum_i n_i \otimes a_i \in N \square_H A$

$$(g \circ f)(\sum_i n_i \otimes a_i) = \sum_i n_i a_i \otimes 1 = \sum_i n_i \otimes a_i$$

in $N \otimes^H_A A$, and this proves (8.14).

We now claim that

$$(N \otimes^H_A A) \otimes A \cong N \otimes^H_A (A \otimes A) \qquad (8.15)$$

We start with the exact sequence

$$0 \longrightarrow N\Box_H A \longrightarrow N \otimes A \rightrightarrows N \otimes H \otimes A$$

Now A is flat, and this implies that the sequence

$$0 \longrightarrow (N\Box_H A) \otimes A \longrightarrow N \otimes A \otimes A \rightrightarrows N \otimes H \otimes A \otimes A$$

Comparing this sequence to the sequence defining $N\Box_H(A \otimes A)$, we obtain that

$$(N\Box_H A) \otimes A \cong N\Box_H(A \otimes A)$$

Now

$$\begin{aligned} (N \otimes_A^H A) \otimes A &= \mathrm{Im}\,(N\Box_H A \to N \otimes_A A) \otimes A \\ &= \mathrm{Im}\,((N\Box_H A) \otimes A \to N \otimes_A A \otimes A) \\ &= \mathrm{Im}\,(N\Box_H(A \otimes A) \to N \otimes_A A \otimes A) \\ &= N \otimes_A^H (A \otimes A) \end{aligned}$$

proving (8.15). Now observe that

$$\begin{aligned} F(G(N)) &= (N \otimes_A^H A) \otimes A \\ &\cong N \otimes_A^H (A \otimes A) \\ &\cong N \otimes_A^H (A \otimes H) \end{aligned}$$

We will now show that

$$N \otimes_A^H (A \otimes H) \cong N \tag{8.16}$$

First recall that $A \otimes H$ is a two-sided (A, H)-Hopf module. The Hopf module structure is given by (8.3-8.6), in particular

$$a(b \otimes h) = ab \otimes h \quad \text{and} \quad \rho^\ell(a \otimes h) = h_{(1)}S^{-1}(a_{(1)}) \otimes a_{(0)} \otimes h_{(2)}\ .$$

As above, we can show that the map

$$f:\ N \otimes_A (A \otimes H) \to N \otimes H$$

restricts to

$$f:\ N \otimes_A^H (A \otimes H) \to N\Box_H H = N$$

The inverse g of f is given by the formula

$$g(n) = \sum n_{(0)} \otimes (1 \otimes n_{(1)})$$

and this proves (8.16). Combining the above statements, we obtain an isomorphism $F(G(N)) \cong N$. It is not difficult to show that this isomorphism is exactly the map ϕ_N. □

We will now restate Theorem 8.1.6 in a slightly different way. First, we need a Lemma.

Lemma 8.1.7 *Suppose that A is a faithfully flat R-algebra, and $M \in R$-mod. Then the sequence*

$$1 \longrightarrow M \xrightarrow{\varepsilon} A \otimes M \xrightarrow{\delta} A \otimes A \otimes M$$

with

$$\varepsilon(m) = 1 \otimes m \text{ and } \delta(a \otimes m) = a \otimes 1 \otimes m - 1 \otimes a \otimes m$$

is exact.

Proof If A is commutative, then our result is nothing else then faithfully flat descent for elements, cf. Proposition B.0.4. The proof easily extends to the noncommutative case. Since A is faithfully flat, it suffices to show that

$$1 \longrightarrow A \otimes M \xrightarrow{I\otimes\varepsilon} A \otimes A \otimes M \xrightarrow{I\otimes\delta} A \otimes A \otimes A \otimes M$$

is exact. First observe that $I \otimes \varepsilon$ is injective: if $(I \otimes \varepsilon)(\sum_i a_i \otimes m_i) = \sum_i a_i \otimes 1 \otimes m_i$, then $\sum_i a_i \otimes m_i = 0$ (multiply the two first factors).
Take $\sum_i a_i \otimes b_i \otimes m_i \in \text{Ker}\,(I \otimes \delta)$. Then

$$\sum_i a_i \otimes 1 \otimes b_i \otimes m_i = \sum_i a_i \otimes b_i \otimes 1 \otimes m_i$$

Multiplying the first two factors, we obtain

$$\sum_i a_i \otimes b_i \otimes m_i = \sum_i a_i b_i \otimes 1 \otimes m_i = (I \otimes \varepsilon)(\sum_i a_i b_i \otimes m_i)$$

□

Theorem 8.1.8 *Let H be a Hopf algebra which is faithfully flat as an R-module, and suppose that the antipode of H is bijective. For an H-comodule algebra A, the following conditions are equivalent:*

1. *A is an H-Galois object;*
2. *$\gamma = \phi_{A\otimes H} : A \otimes A \to A \otimes H : a \otimes b \mapsto \sum ab_{(0)} \otimes b_{(1)}$ is an isomorphism and A is faithfully flat as an R-module.*

Proof The implication $1 \Rightarrow 2$ is trivial. $2 \Rightarrow 1$ will follow if we can show that $A^{\text{co}H} = R$. Take $a \in A^{\text{co}H}$. Then $\gamma(a \otimes 1) = a \otimes 1 = \gamma(1 \otimes a)$, and $a \otimes 1 = 1 \otimes a$. From Lemma 8.1.7, it now follows that $a \in R$. □

If A is commutative, then we have an alternative approach to Theorem 8.1.8; this will also clarify the connection to Faithfully Flat Descent.

Proposition 8.1.9 *Let S be a commutative H-comodule algebra. If the map γ is an isomorphism, then the categories $\mathcal{M}(H)_S^H$ of Hopf modules and $\mathcal{D}(S)$ of descent data (see Proposition B.0.12) are equivalent.*

Proof Let M be a (right) S-module. For $m \in M$ and $x = \sum_i s_i \otimes t_i \in S \otimes S$ or $S \otimes H$, we write $mx = \sum_i ms_i \otimes t_i$.
Suppose that (M, u) is a descent datum. We define an H-coaction on M as follows. If $u(1 \otimes m) = \sum_i m_i \otimes s_i \in M \otimes S$, then

$$\rho(m) = \sum_i m_i s_{i_{(0)}} \otimes s_{i_{(0)}} = \sum_i m_i \gamma(1 \otimes s_i)$$

We leave it to the reader to verify that ρ is an H-coaction, and that M is an (S, H)-Hopf module.
Conversely, if M is an (S, H)-Hopf module, then we define an $S \otimes S$-linear map $u: M_1 \to M_2$ by

$$u(1 \otimes m) = \sum m_{(0)} \gamma^{-1}(1 \otimes m_{(1)})$$

It can be shown that u is a descent datum. We leave it to the reader to the prove that the above constructions are functorial, and that they establish an equivalence of categories. □

We finish this Section with the following remarkable property. In the case where H is finite, it is due to Beattie, cf. [17, Lemma 1.1].

Proposition 8.1.10 *Let A and B be H-Galois objects. Then any H-comodule algebra homomorphism $f: A \to B$ is an isomorphism.*

Proof We make B into an object of $\mathcal{M}(H)_A^H$ as follows.

$$\begin{aligned} \rho(b) &= \sum b_{(0)} \otimes b_{(1)} \\ b \cdot a &= bf(a) \end{aligned}$$

for all $a \in A$ and $b \in B$. A straightforward computation shows that (8.1) holds. In 8.1.4, we have seen that there is an isomorphism

$$\phi_B: \ F(G(B)) = A \otimes B^{\mathrm{co}H} = A \otimes R = A \longrightarrow B$$

given by

$$\phi_B(a) = \phi_B(a \otimes 1_B) = 1_B \cdot a = 1_B f(a) = f(a)$$

and $f = \phi_B$ is an isomorphism. □

8.2 Galois objects and graded ring theory

Let G be an arbitrary group and R a commutative ring. Recall that a *G-graded R-algebra* A is an R-algebra that can be written as a direct sum of R-modules

$$A = \oplus_{\sigma \in G} A_\sigma$$

with $A_\sigma A_\tau \subset A_{\sigma\tau}$ for all $\sigma, \tau \in G$. As we already pointed out in the previous Section, a G-graded R-algebra is nothing else then an RG-comodule algebra.
An A-module is called G-graded if it can be written as a direct sum of A_e-modules

$$M = \oplus_{\sigma \in G} M_\sigma$$

with $A_\sigma M_\tau \subset M_{\sigma\tau}$ for all $\sigma, \tau \in G$. A right G-graded A-module is nothing else then a right (A, RG)-Hopf module. The relation between the RG-coaction and the G-grading is given by the following:

$$\rho_M(m) = \sum m_\sigma \otimes u_\sigma$$

if and only if

$$m = \sum m_\sigma \text{ with } m_\sigma \in M_\sigma$$

For full detail on graded ring theory, we refer the reader to [146]. Recall in particular that a G-graded algebra A is *strongly graded* if $A_\sigma A_\tau = A_{\sigma\tau}$ for all $\sigma, \tau \in G$. We now have the following:

Proposition 8.2.1 *A G-graded R-algebra A is an RG-Galois object if and only if $A_e = R$ and A is strongly graded.*

Proof This follows immediately from [146, Theorem A.1.3.4]. For completeness sake, let us give a proof here. Suppose first that A is an RG-Galois object. It follows immediately that $A_e = A^{\mathrm{co}RG} = R$. Furthermore, the map

$$\gamma: A \otimes A \to A \otimes RG: a \otimes b \mapsto \sum_{\sigma \in G} ab_\sigma \otimes u_\sigma$$

is an isomorphism. Take $\sigma \in G$. Then γ restricts to an isomorphism

$$\gamma_\sigma: A \otimes A_\sigma \to A \otimes Ru_\sigma \cong A$$

Now take $\tau \in G$. γ_σ restricts to a monomorphism

$$\gamma_{\tau,\sigma}: A_\tau \otimes A_\sigma \to A_{\tau\sigma}$$

and we are done if we can show that $\gamma_{\tau,\sigma}$ is surjective, that is, we have to show that the inverse image of $\gamma^{-1}(A_{\tau\sigma} \otimes Ru_\sigma) \subset A_\tau \otimes A_\sigma$.
Take $a \in A_{\tau\sigma}$, and let $\sum_i a_i \otimes b_i \in A \otimes A_\sigma$ be its inverse image under γ_σ. Then

$$\sum_i a_i b_i = a$$

Take $\tau' \neq \tau$ in G. Then

$$\sum_i (a_i)_{\tau'} b_i = (\sum_i a_i b_i)_{\tau'\sigma} = a_{\tau'\sigma} = 0$$

and therefore

$$\sum_i (a_i)_\tau b_i = a$$

and $\sum_i (a_i)_\tau \otimes b_i$ is an inverse of a in $A_\tau \otimes A_\sigma$.
Conversely, if A is strongly graded and $A_e = R$, then it follows easily that the maps $\gamma_{\tau,\sigma}$, γ_σ and γ described above are isomorphisms. It follows also that $A_\sigma \otimes A_{\sigma^{-1}} \cong A_e = R$, and therefore A_σ is an invertible R-module. But then A_σ and $A = \oplus_\sigma A_\sigma$ are flat as R-modules. The result now follows from Proposition 8.1.6. □

8.3 Galois objects and Morita theory

In this Section, we will consider the case where H is faithfully projective as an R-module. Suppose that A is an H-Galois object. As we have seen in Section 8.1, the categories $\mathcal{M}(H)_A^H$ and $A^{\mathrm{op}}\#H^*$-mod are isomorphic, and the pair of adjoint functors:

$$\begin{aligned} F:\ R\text{-mod} \to (A^{\mathrm{op}}\#H^*)\text{-mod} &:\ N \mapsto N\otimes A \\ G:\ (A^{\mathrm{op}}\#H^*)\text{-mod} \to R\text{-mod} &:\ M \mapsto M^{H^*} \end{aligned}$$

is an equivalence of categories. From the Eilenberg-Watts Theorem (cf. [15, II.3.1]), it follows that this category equivalence is given by a strict Morita context

$$(R, A^{\mathrm{op}}\#H^*, P, Q, f, g)$$

Let us construct P, Q, f and g. From the Eilenberg-Watts Theorem, it follows that

$$P = G(A^{\mathrm{op}}\#H^*) = (A^{\mathrm{op}}\#H^*)^{H^*} \text{ and } Q = F(R) = A$$

Observe that P is an $(R, A^{\mathrm{op}}\#H^*)$-bimodule. Indeed, $x \in P = (A^{\mathrm{op}}\#H^*)^{H^*}$ if and only if $(h^*\#1)x = \langle h^*, 1\rangle x$ for all $h^* \in H^*$, and this implies clearly that $xy \in P$ for all $y \in A^{\mathrm{op}}\#H^*$.
$Q = A$ is an $(A^{\mathrm{op}}\#H^*, R)$-bimodule, the left action of $A^{\mathrm{op}}\#H^*$ is given by the formula

$$(a\#h^*)B = \sum \langle h^*, B_{(1)}\rangle B_{(0)}A$$

The map g may be described easily.

$$g:\ Q\otimes P = A\otimes (A^{\mathrm{op}}\#H^*)^{H^*} = F(G(A^{\mathrm{op}}\#H^*)) \longrightarrow A^{\mathrm{op}}\#H^*$$

is nothing else then the map $\phi_{A^{\mathrm{op}}\#H^*}$, and is therefore given by

$$g(s\otimes x) = xs = x(s\otimes \varepsilon)$$

The inverse of the map

$$f:\ (A^{\mathrm{op}}\#H^*)^{H^*} \otimes_{A^{\mathrm{op}}\#H^*} A \longrightarrow R$$

is given by $f^{-1}(1) = 1\otimes 1$. To find a formula for f itself, we have to invoke Proposition 1.1.5. Part 3 of this Proposition tells us that the map

$$\Phi:\ A^{\mathrm{op}}\#H^* \longrightarrow \mathrm{End}_R(A)$$

given by

$$\Phi(a\#h^*)(b) = (a\#h^*)\cdot b = \sum \langle h^*, b_{(1)}\rangle b_{(0)}a$$

is an isomorphism. From part 2 of the same Proposition, it follows that there is an isomorphism

$$\psi:\ (A^{\mathrm{op}}\#H^*)^{H^*} \longrightarrow A^* = \mathrm{Hom}_R(A, R)$$

A formula for ψ can be found as follows: the strict Morita contexts $(R, A^{\rm op}\#H^*, (A^{\rm op}\#H^*)^{H^*}, A, f, g)$ and $(R, {\rm End}_R(A), A^*, A, \tilde f, \tilde g)$ are isomorphic, and the diagram

$$\begin{array}{ccc} A\otimes(A^{\rm op}\#H^*)^{H^*} & \stackrel{g}{\longrightarrow} & A^{\rm op}\#H^* \\ \downarrow{\scriptstyle I\otimes\psi} & & \downarrow{\scriptstyle \Phi} \\ A\otimes A^* & \stackrel{\tilde g}{\longrightarrow} & {\rm End}_R(A) \end{array}$$

commutes. We therefore have, for all $a \in A$ and $x \in (A^{\rm op}\#H^*)^{H^*}$

$$\begin{aligned} \psi(x)(a) &= \tilde g((I\otimes\psi)(1\otimes x))(a) \\ &= \Phi(g(1\otimes x))(a) \\ &= \Phi(x)(a) \end{aligned}$$

We can now also construct the map f from the commutativity of the diagram

$$\begin{array}{ccc} (A^{\rm op}\#H^*)^{H^*}\otimes A & \stackrel{f}{\longrightarrow} & R \\ \downarrow{\scriptstyle \psi\otimes I} & & \downarrow{\scriptstyle =} \\ A^*\otimes_{{\rm End}_R(A)} A & \stackrel{\tilde f}{\longrightarrow} & R \end{array}$$

We obtain that

$$\begin{aligned} f(x\otimes a) &= \tilde f((\psi\otimes I)(x\otimes a)) \\ &= \psi(x)(a) = \Phi(x)(a) \end{aligned}$$

We have shown one implication of the following Theorem.

Theorem 8.3.1 *Let H be a finite Hopf algebra, and suppose that A is an H-comodule algebra. Then the following are equivalent*

1. *A is an H-Galois object;*
2. - *$A^{H^*} = R$;*
   - *The sextuple $(R, A^{\rm op}\#H^*, (A^{\rm op}\#H^*)^{H^*}, A, f, g)$ with f and g defined by the formulas*

$$f:\ (A^{\rm op}\#H^*)^{H^*}\otimes_{A^{\rm op}\#H^*} A \to R\ :\ x\otimes a\mapsto \Phi(x)(a) \quad (8.17)$$
$$g:\ A\otimes(A^{\rm op}\#H^*)^{H^*} \to A^{\rm op}\#H^*\ :\ a\otimes x\mapsto x(a\otimes\varepsilon) \quad (8.18)$$

   is a strict Morita context;
3. - *A is faithfully projective as an R-module;*
   - *the map $\Phi:\ A^{\rm op}\#H^* \to {\rm End}_R(A)$ given by*

$$\Phi(a\#h^*)(b) = \sum\langle h^*, b_{(1)}\rangle b_{(0)}a$$

   is an isomorphism.

Proof 1) ⇒ 2) was shown above.
2) ⇒ 3). First let us show that the map g in the statement is well-defined. To this end, one shows that the map Φ restricts to a map

$$\psi:\ (A^{\rm op}\#H^*)^{H^*}\to A^*$$

Take $x=\sum_i a_i\#h_i^*\in(A^{\rm op}\#H^*)^{H^*}$. Then for all $h^*\in H^*$

$$\langle h^*,1\rangle x=(1\#h^*)(\sum_i a_i\#h_i^*)=\sum_i(h^*_{(2)}\rightharpoonup a_i)\#h^*_{(1)}h_i^*$$

Apply Φ to both sides, and evaluate at $a\in A$. Then

$$\begin{aligned}\langle h^*,1\rangle\Phi(x)(a)&=\sum_i((h^*_{(1)}h_i^*)\rightharpoonup a)(h^*_{(2)}\rightharpoonup a_i)\\ &= h^*\rightharpoonup((h_i^*\rightharpoonup a)a_i)=h^*\rightharpoonup(\Phi(x)(a))\end{aligned}$$

hence $\Phi(x)(a)\in A^{H^*}=R$, and $\Phi(x)\in A^*$.
From Proposition 1.1.5, it follows that Φ is an isomorphism (part 3) and that A is projective as an R-module (part 4).
3) ⇒ 1). A is faithfully projective, and therefore faithfully flat as an R-module. In view of Proposition 8.1.8, we only need to show that

$$\gamma:\ A\otimes A\to A\otimes H:\ a\otimes b\mapsto\sum ab_{(0)}\otimes b_{(1)}$$

is an isomorphism. Since A and H are faithfully projective, it suffices to show that

$$\Gamma=\mathrm{Hom}(\gamma,I):\ \mathrm{Hom}_A(A\otimes A,A)\longrightarrow\mathrm{Hom}_A(A\otimes H,A)$$

is bijective. From the hom-tensor relations, it follows that we have isomorphisms (as R-modules)

$$\begin{aligned}\pi_1:&\ A^{\rm op}\#H^*\longrightarrow\mathrm{Hom}_A(A\otimes H,A)\\ \pi_2:&\ \mathrm{End}_R(A)\longrightarrow\mathrm{Hom}_A(A\otimes A,A)\end{aligned}$$

given by

$$\begin{aligned}\pi_1(a\#h^*)(b\#h)&=\langle h^*,h\rangle ba\\ \pi_2(f)(b\otimes c)&=bf(c)\end{aligned}$$

Now the diagram

$$\begin{array}{ccc} A^{\rm op}\#H^* & \stackrel{\Phi}{\longrightarrow} & \mathrm{End}_R(A)\\ \downarrow{\scriptstyle\pi_1} & & \downarrow{\scriptstyle\pi_2}\\ \mathrm{Hom}_A(A\otimes A,A) & \stackrel{\Gamma}{\longrightarrow} & \mathrm{Hom}_A(A\otimes H,A)\end{array}$$

commutes. Indeed, for all $a,b,c\in A$ and $h^*\in H^*$ we have

$$\begin{aligned}(\pi_2\circ\Phi)(a\#h^*)(b\otimes c)&= b\Phi(a\#h^*)(c)\\ &=\sum\langle h^*,c_{(1)}\rangle bc_{(0)}a\\ &=\pi_1(a\#h^*)(\sum bc_{(0)}\otimes c_{(1)})\\ &=(\Gamma\circ\pi_1)(a\#h^*)(b\otimes c)\end{aligned}$$

The fact that π_1,π_2 and Φ are bijective now implies that Γ and γ are bijective. □

The reduced norm

Let A be an H-Galois object, and suppose that H is faithfully projective as an R-module. Then we can consider the map

$$N_{A/R}: A \longrightarrow A^{\rm op}\#H^* \xrightarrow{\Phi} {\rm End}_R(A) \xrightarrow{\det} R \tag{8.19}$$

where the first map maps a to $a\#\varepsilon$. The image of a in ${\rm End}_R(A)$ is given by left multiplication by a. The determinant map (or reduced norm map) was defined in Section 3.4.

8.4 Galois extensions

So far, we have been working with H-comodule algebras. We can also develop a theory for H-module algebras. In this Section, we will show that this gives us nothing new: if there exists an H-Galois extension A, then A and H are both faithfully projective, and A is an H^*-Galois object.

Let A be a left H-module algebra. An R-module M is called a right-left (A, H)-module if M is a right A-module and a left H-module such that

$$h \rightharpoonup (m \cdot a) = \sum (h_{(1)} \rightharpoonup m) \cdot (h_{(2)} \rightharpoonup a) \tag{8.20}$$

Left-left (A, H)-modules have been studied in [44]. ${}_H\mathcal{M}(H)_A$ will be the category o right-let (A, H)-modules and A-linear H-linear maps. As we will see below, (A, H)-modules are related to modules over a smash product. The version of the smash product that we need this time is the following: let A be a left H-module algebra, and B a right H-comodule algebra. $A^{\rm op}\#B$ is defined by the multiplication rule

$$(a\#b)(c\#d) = \sum (b_{(1)} \rightharpoonup c)a\#b_{(0)}d \tag{8.21}$$

Proposition 8.4.1 *Let A be a left H-module algebra. Then the categories ${}_H\mathcal{M}(H)_A$ and $A^{\rm op}\#H$-*mod *are isomorphic.*

Proof If M is an (A, H)-module, then we define a left $A^{\rm op}\#H$-module structure on M by the rule

$$(a\#h) \cdot m = (h \rightharpoonup m) \cdot a$$

Conversely, if M is an $A^{\rm op}\#H$-module, then

$$\begin{aligned} m \cdot a &= (a\#1) \cdot m \\ h \rightharpoonup m &= (1\#h) \cdot m \end{aligned}$$

make M into an (A, H)-module. We leave details to the reader. □

If H is finite, then (A, H)-modules are in one to one correspondence with (A, H^*)-Hopf modules.

Proposition 8.4.2 *Suppose that H is finite. With notations as above, A is a right H^*-comodule algebra, and B is a left H^*-module algebra. In this case, the smash products defined by the formulas (8.11) and (8.21) are isomorphic. The categories ${}_H\mathcal{M}(H)_A$ and $\mathcal{M}(H^*)_A^{H^*}$ are isomorphic.*

Proof According to (8.21),

$$\begin{aligned}(a\#b)(c\#d) &= \sum(b_{(1)}\rightharpoonup c)a\#b_{(0)}d\\ &= \sum\langle c_{(1)}, b_{(1)}\rangle c_{(0)}a\#b_{(0)}d\\ &= \sum c_{(0)}a\#(c_{(1)}\rightharpoonup b)d\end{aligned}$$

which is exactly (8.11). The second statement now follows immediately from Propositions 8.4.1 and 8.1.2. □

Proposition 8.4.3 *Let A be a left H-module algebra, and consider the functors*

$$\begin{aligned}F &: R\text{-mod}\longrightarrow {}_H\mathcal{M}(H)_A : N\mapsto A\otimes N\\ G &: {}_H\mathcal{M}(H)_A\longrightarrow R\text{-mod} : M\mapsto M^H\end{aligned}$$

where $M^H = \{m \in M | h\rightharpoonup m = \varepsilon(h)m$ for all $h \in H\}$.
G is a right adjoint of F. If H is finite, then F and G are isomorphic to the functors F and G of Proposition 8.1.3 (with H replaced by H^*). If F and G are inverse equivalences, then we call A an H-Galois extension of R. If A is an H-Galois extension of R, then both A and H are finitely generated and projective, and then A is an H^*-Galois object.

Proof For $N \in R$-mod and $M \in {}_H\mathcal{M}(H)_A$, we define maps

$$\mathrm{Hom}_R(N, M^H)\underset{\beta}{\overset{\alpha}{\rightleftarrows}}\mathrm{Hom}_{A,H}(A\otimes N, M)$$

by

$$\alpha(f)(a\otimes n) = f(n)a \quad\text{and}\quad \beta(g)(n) = g(1\otimes n)$$

It is straightforward to check that α and β are each others inverses.
Suppose that F and G are inverse equivalences. Since ${}_H\mathcal{M}(H)_A \cong A^{\mathrm{op}}\#H$-mod, it follows from the Eilenberg-Watts Theorem that R and $A^{\mathrm{op}}\#H$ are connected via a strict Morita context $(R, A^{\mathrm{op}}\#H, P, Q, f, g)$. As in Section 8.3, it follows that $P = G(A^{\mathrm{op}}\#H) = (A^{\mathrm{op}}\#H)^H$ and $Q = F(R) = A$. From Proposition 1.1.5, it follows that $Q = A$ is faithfully projective and that $A^{\mathrm{op}}\#H \cong \mathrm{End}_R(A)$. The connecting isomorphism

$$\Phi:\ A^{\mathrm{op}}\#H\longrightarrow\mathrm{End}_R(A)$$

is given by

$$\Phi(a\#h)(b) = (h\rightharpoonup b)(a)$$

for all $a, b \in A$ and $h \in H$. Therefore $A\otimes H$ is faithfully projective as an R-module, and H is faithfully projective as an R-module. □

8.5 Galois objects and classical Galois theory

Let G be a finite group and R a commutative ring, and let $H = RG$. Recall that $H^* = GR = \oplus_\sigma Rv_\sigma$, with multiplication rule

$$v_\sigma v_\tau = \delta_{\sigma,\tau} v_\sigma$$

We now claim that a G-Galois extension in the classical sense, cf. e.g. [75] (over a commutative ring) or [101] (over a field), is nothing else then an RG-Galois extension, or, invoking Proposition 8.4.3, a GR-Galois object. Indeed,

$$A^{\mathrm{op}} \# RG = \oplus_{\sigma \in G} Au_\sigma$$

with multiplication rule

$$(au_\sigma)(bu_\tau) = \sigma(b)au_{\sigma\tau}$$

The map

$$\Phi: \ A^{\mathrm{op}} \# RG = \oplus_{\sigma \in G} Au_\sigma \longrightarrow \mathrm{End}_R(A)$$

is given by

$$\Phi(au_\sigma)(b) = \sum_{\tau \in G} \langle u_\sigma, v_\tau \rangle \tau(b) a = \sigma(b) a$$

Theorem 8.3.1 now allows us to conclude that A is a GR-Galois object if and only if A is faithfully projective as an R-module and the map Φ described above is an isomorphism.

In [75, Prop. 1.2, p.80], equivalent conditions for a commutative R-algebra S on which G acts as a group of automorphisms are given. One of them (namely condition 3) is exactly the condition above. From condition 1) of the same theorem, it follows that this extends the classical notion of Galois field extension.

The map $\gamma: \ A \otimes A \to A \otimes GR$ can be described explicitly. The GR-coaction on A induced by the G-action on A is given by the formula

$$\rho(a) = \sum_{\sigma \in G} (\sigma \rightharpoonup a) \otimes v_\sigma \tag{8.22}$$

Indeed, for all $\tau \in G$, we have that

$$\sum_{\sigma \in G} \langle v_\sigma, u_\tau \rangle (\sigma \rightharpoonup a) = \tau \rightharpoonup a$$

The map γ is therefore given by the formula

$$\gamma(a \otimes b) = \sum_{\sigma \in G} a(\sigma \rightharpoonup b) \otimes v_\sigma \tag{8.23}$$

Let A be a commutative G-Galois extension. We will give a formula for the norm map $N_{A/R}$. For $a \in A$, we have to find the determinant of the element in $\mathrm{End}_R(A)$ given by left multiplication by a. We first a base extension by A. $\{v_\sigma | \sigma \in G\}$ is a basis for $A \otimes GR$ as an A-module. Furthermore

$$(1 \otimes a) \cdot v_\sigma = \sigma(a) \otimes v_\sigma$$

and the matrix representing the left multiplication by $1 \otimes a$ with respect to the basis $\{v_\sigma | \sigma \in G\}$ is the diagonal matrix with entries $\{\sigma \rightharpoonup a | \sigma \in G\}$. The determinant of this matrix is

$$N_{A/R}(a) = \prod_{\sigma \in G} \sigma(a) \tag{8.24}$$

(8.24) is consistent with (6.19).

8.6 Integrals

Let H be a faithfully projective Hopf algebra. We know that H is an H-Galois object, and that we have a category equivalence between the categories R-mod and $\mathcal{M}(H)^H_H$. For any (H, H)-Hopf module, we have that the map

$$\phi_M : H \otimes M^{\text{co}H} \longrightarrow M : h \otimes m \mapsto (m \leftharpoonup h)$$

is an isomorphism.
Replace H by H^*, which is also a Hopf algebra. We claim that H may be viewed as an object of $\mathcal{M}(H^*)^{H^*}_{H^*}$. The structure maps are the following.

$$h \leftharpoonup h^* = \sum \langle h^*, S(h_{(2)}) \rangle h_{(1)} \tag{8.25}$$

for all $h \in H$ and $h^* \in H^*$, and

$$\rho(h) = \sum h_{(0)} \otimes h_{(1)} \in H \otimes H^*$$

if and only if

$$kh = \sum \langle h_{(1)}, k \rangle h_{(0)} \tag{8.26}$$

for all $k \in H$.
We already know from Section 7.2 that (8.25) makes H into a right H^*-module, and that (8.26) defines a right H^*-comodule structure on H. To verify that $H \in \mathcal{M}(H^*)^{H^*}_{H^*}$, we have to show that

$$\rho(h \leftharpoonup h^*) = \sum (h_{(0)} \leftharpoonup h^*_{(1)}) \otimes (h_{(1)} * h^*_{(2)})$$

or, equivalently, for all $k \in H$:

$$k(h \leftharpoonup h^*) = \sum \langle h_{(1)} * h^*_{(2)}, k \rangle (h_{(0)} \leftharpoonup h^*_{(1)})$$

We compute easily that the right hand side amounts to

$$\begin{aligned}
& \sum \langle h_{(1)}, k_{(1)} \rangle \langle h^*_{(2)}, k_{(2)} \rangle (h_{(0)} \leftharpoonup h^*_{(1)}) \\
= & \sum \langle h^*_{(2)}, k_{(2)} \rangle ((k_{(1)} h) \leftharpoonup h^*_{(1)}) \quad \text{(by (8.26))} \\
= & \sum \langle h^*_{(2)}, k_{(3)} \rangle \langle h^*_{(1)}, S(k_{(2)} h_{(2)}) \rangle k_{(1)} h_{(1)} \quad \text{(by (8.25))} \\
= & \sum \langle h^*_{(3)}, k_{(3)} \rangle \langle h^*_{(1)}, S(h_{(2)}) \rangle \langle h^*_{(2)}, S(k_{(2)}) \rangle k_{(1)} h_{(1)} \\
= & \sum \langle h^*_{(2)}, S(k_{(2)}) k_{(3)} \rangle \langle h^*_{(1)}, S(h_{(2)}) \rangle k_{(1)} h_{(1)} \\
= & \sum \langle h^*, S(h_{(2)}) \rangle k h_{(1)} \\
= & \; k(h \leftharpoonup h^*)
\end{aligned}$$

as needed. We now define

$$\int_H^l = H^{\mathrm{co}H^*} = \{x \in H | \rho(x) = x \otimes \varepsilon\}$$
$$= \{x \in H | hx = \varepsilon(h)x, \text{ for all } h \in H\} \tag{8.27}$$

The elements of $\int_H^l$ are called the *left integrals* of H. They satisfy the following property.

Theorem 8.6.1 *Let H be a faithfully projective Hopf algebra (with bijective antipode). Then we have an R-module isomorphism*

$$V : \int_H^l \otimes H^* \mapsto H : x \otimes h^* \mapsto \sum \langle h^*, S(x_{(2)})\rangle x_{(1)}$$

Furthermore, $\int_H^l$ is an invertible R-module.

Proof The first part follows immediately if we apply the category equivalence between R-mod and $\mathcal{M}(H^*)_{H^*}^{H^*}$ described above.
From the fact that H and H^* are faithfully projective R-modules, it follows that $\int_H^l$ is also faithfully projective (see [15]). A count of ranks shows that the rank of $\int_H^l$ is one, and $\int_H^l$ is invertible. □

Remark 8.6.2 Let us restate Theorem 8.6.1, with the rôles of H and H^* interchanged. We obtain that

$$\int_{H^*}^l = \{h^* \in H^* | h^* * x^* = \sum \langle h^*, 1\rangle x^*, \text{ for all } h^* \in H^*\}$$

We have an isomorphism

$$V : \int_{H^*}^l \otimes H \longrightarrow H^*$$

defined by

$$\langle V(x^* \otimes h), k\rangle = \sum \langle h, S^*(x^*_{(2)})\rangle \langle x^*_{(1)}, k\rangle$$
$$= \langle x^*, kS(h)\rangle \tag{8.28}$$

Now fix $x^* \in \int_{H^*}^l$, and define $v : H \to H^*$ by

$$v(h) = V(x^* \otimes h)$$

We claim that

$$\sum \langle v(h), k_{(2)}\rangle k_{(1)} = \sum \langle v(h_{(1)}), k\rangle h_{(2)} \tag{8.29}$$

for all $h, k \in H$. Indeed, for all $h^* \in H^*$, we have that

$$\begin{aligned}
\sum \langle v(h), k_{(2)}\rangle \langle h^*, k_{(1)}\rangle &= \sum \langle x^*, k_{(2)}S(h)\rangle \langle h^*, k_{(1)}\rangle \\
&= \sum \langle h^*, k_{(1)}S(h_{(2)})h_{(3)}\rangle \langle x^*, k_{(2)}S(h_{(1)})\rangle \\
&= \sum \langle h^*_{(1)}, k_{(1)}S(h_{(2)})\rangle \langle h^*_{(2)}, h_{(3)}\rangle \langle x^*, k_{(2)}S(h_{(1)})\rangle \\
&= \sum \langle h^*_{(1)} * x^*, kS(h_{(1)})\rangle \langle h^*_{(2)}, h_{(2)}\rangle \\
&= \sum \langle h^*_{(1)}, 1\rangle \langle x^*, kS(h_{(1)})\rangle \langle h^*_{(2)}, h_{(2)}\rangle \\
&= \langle v(h_{(1)}), k\rangle \langle h^*, h_{(2)}\rangle
\end{aligned}$$

Example 8.6.3 Let G be a finite group, and take the canonical dual basis $\{u_\sigma, v_\sigma | \sigma \in G\}$ of RG. We then have that $\langle u_\sigma, v_\tau\rangle = \delta_{\sigma,\tau}$, and it is easy to prove that $\int_{RG}^l = Rx$, with

$$x = \sum_{\sigma\in G} u_\sigma$$

and $V: \int_{RG}^l \otimes (RG)^* \longrightarrow RG$ is given by the formula

$$V(x \otimes v_\sigma) = u_{\sigma^{-1}}$$

In a similar way, we find easily that

$$\int_{(RG)^*}^l = Rv_e$$

where e is the neutral element of G. The isomorphism V is now the following

$$V: \int_{(RG)^*}^l \otimes RG \longrightarrow (RG)^* : \; v_e \otimes u_\sigma \mapsto v_\sigma$$

8.7 Galois coobjects

The results in this Section may be viewed as formal duals of the results of the previous Section. In fact, the results of this Section applied to a faithfully projective Hopf algebra H are equivalent to the ones of the previous Section applied to H^*.
Let R be a commutative ring, and H a Hopf algebra that is faithfully flat as an R-module. Let C be a left H-module coalgebra. Then a left (H,C)-Hopf module M is an R-module that is a left H-module and a left C-comodule such that

$$\rho_M(h \cdot m) = \sum h_{(1)} \rightharpoonup m_{(-1)} \otimes h_{(2)} m_{(0)} \tag{8.30}$$

for all $m \in M$ and $h \in H$. In the sequel, ${}_H^C\mathcal{M}(H)$ will denote the category of left (H,C)-Hopf modules and H-linear C-colinear maps.
If H and C are both faithfully projective as R-modules, then H^* is a Hopf algebra, and C^* is an H^*-comodule algebra. We then have an isomorphism of categories

$${}_H^C\mathcal{M}(H) \cong \mathcal{M}(H^*)_{H^*}^{C^*}$$

and this will make the results of this Section equivalent to the ones of the previous Section.

Proposition 8.7.1 *With notations as above, consider the functors*

$$F: {}_H^C\mathcal{M}(H) \longrightarrow R\text{-mod} \; : \; M \mapsto R \otimes_H M = \overline{M}$$
$$G: R\text{-mod} \longrightarrow {}_H^C\mathcal{M}(H) \; : \; N \mapsto C \otimes N$$

Then G is a right adjoint to F.

Proof R is an H-module via the map ε. In fact, $\overline{M} = M/\mathrm{Ker}\,\varepsilon M$, and in $\overline{M}$ we have the following identity:

$$\overline{hm} = \varepsilon(h)\overline{m}$$

for all $h \in H$ and $m \in M$. For any $M \in {}^C_H\mathcal{M}(H)$ and $N \in R$-mod we consider the maps

$$\begin{aligned}\alpha &: \mathrm{Hom}^C_H(M, C \otimes N) \longrightarrow \mathrm{Hom}_R(\overline{M}, N)\\ \beta &: \mathrm{Hom}_R(\overline{M}, N) \longrightarrow \mathrm{Hom}^C_H(M, C \otimes N)\end{aligned}$$

given by

$$\begin{aligned}\alpha(f)(\overline{m}) &= (\varepsilon_C \otimes I_N)(f(m))\\ \beta(g)(m) &= \sum m_{(-1)} \otimes g(\overline{m_{(0)}})\end{aligned}$$

for all $f \in \mathrm{Hom}_{{}^C_H\mathcal{M}(H)}(M, C \otimes N)$, $g \in \mathrm{Hom}_R(\overline{M}, N)$ and $m \in M$. The map α is well-defined: take $m \in M$ and $h \in H$, and suppose that $f(m) = \sum_i c_i \otimes n_i$. Then

$$f(hm) = \sum_i h \rightharpoonup c_i \otimes n_i$$

and

$$\alpha(f)(\overline{hm}) = \sum_i \varepsilon_H(h)\varepsilon_C(c_i)m_i = \alpha(f)(\varepsilon_H(h)\overline{m})$$

Let us now show that α and β are each others inverses. Take $f \in \mathrm{Hom}_{{}^C_H\mathcal{M}(H)}(M, C \otimes N)$ and $m \in M$ and suppose again that $f(m) = \sum_i c_i \otimes n_i$. The fact that f is H-colinear means that

$$\sum m_{(-1)} \otimes f(m_{(0)}) = \sum_i c_{i_{(1)}} \otimes c_{i_{(2)}} \otimes n_i$$

and therefore

$$\sum m_{(-1)} \otimes (\varepsilon_C \otimes I_N)(f(m_{(0)})) = \sum c_i \otimes n_i = f(m)$$

It now follows that

$$\begin{aligned}\beta(\alpha(f))(m) &= \sum m_{(-1)} \otimes \alpha(f)(\overline{m_{(0)}})\\ &= \sum m_{(-1)} \otimes (\varepsilon_C \otimes I_N)(f(m_{(0)}))\\ &= f(m)\end{aligned}$$

and this shows that β is a left inverse of α. We also have for all $g \in \mathrm{Hom}_R(\overline{M}, N)$ and $m \in M$ that

$$\begin{aligned}\alpha(\beta(g))(\overline{m}) &= (\varepsilon_C \otimes I_N)(\beta(g)(m))\\ &= (\varepsilon_C \otimes I_N)(\sum m_{(-1)} \otimes g(\overline{m_{(0)}}))\\ &= g(\overline{m})\end{aligned}$$

and it follows that β is also a right inverse of α. □

Remarks 8.7.2 1) As is the case with Proposition 8.1.3, Proposition 8.7.1 is a special case of [50, Theorem 1.3].
2) From the adjointness of the functors F and G in Proposition 8.7.1, it follows that for all $M \in {}_H^C\mathcal{M}(H)$ and $N \in R$-mod we have natural maps

$$\begin{aligned} \psi_M &: M \longrightarrow G(F(M)) = C \otimes \overline{M} \\ \phi_N &: F(G(N)) = \overline{C} \otimes N \longrightarrow N \end{aligned}$$

given by

$$\begin{aligned} \psi_M(m) &= \sum m_{(-1)} \otimes \overline{m_{(0)}} \\ \phi_N(\sum_i \overline{c}_i \otimes n_i) &= \sum_i \varepsilon(c_i) n_i \end{aligned}$$

Definition 8.7.3 *With notations as above, an H-module coalgebra C is called an H-Galois coobject if the functors F and G from Proposition 8.7.1 are inverse equivalences, or, equivalently, if ψ_M and ϕ_N are isomorphisms for all $M \in {}_H^C\mathcal{M}(H)$ and $N \in R$-mod.*

We will now establish some necessary and sufficient conditions for an H-module coalgebra to be an H-Galois coobject. It is clear that ϕ_N is an isomorphism for all $N \in R$-mod if and only if the canonical map

$$\phi_C : \overline{C} \longrightarrow R : \overline{c} \mapsto \varepsilon(c)$$

is an isomorphism.
Observe that $H \otimes C$ can be given the structure of left (H, C)-Hopf module as follows:

$$\begin{aligned} k(h \otimes c) &= kh \otimes c \\ \rho_{H\otimes C}(h \otimes c) &= \sum h_{(1)} \rightharpoonup c_{(1)} \otimes h_{(2)} \otimes c_{(2)} \end{aligned}$$

for all $h, k \in H$ and $c \in C$. It is readily verified that condition 8.30 is satisfied:

$$\begin{aligned} \rho_{H\otimes C}(kh \otimes c) &= \sum k_{(1)} h_{(1)} \rightharpoonup c_{(1)} \otimes k_{(2)} h_{(2)} \otimes c_{(2)} \\ &= \sum k_{(1)} (h \otimes c)_{(-1)} \otimes k_{(2)} (h \otimes c)_{(0)} \end{aligned}$$

A necessary condition for M to be an H-Galois coobject is therefore that $\delta = \psi_{H\otimes C}$ is an isomorphism. Let us describe δ. First we remark that $F(H \otimes C) = \overline{H} \otimes C = C$, since $\overline{H} \cong R$. Indeed, the maps

$$I \otimes \varepsilon_H : \overline{H} = R \otimes_H H \longrightarrow R \quad \text{and} \quad \eta \otimes 1 : R \longrightarrow \overline{H} = R \otimes_H H$$

are well-defined and each others inverses.
Now $G(F(H \otimes C)) = C \otimes C$, where H acts and C coacts on the first factor:

$$\begin{aligned} h(c \otimes d) &= h \rightharpoonup c \otimes d \\ \rho_{C\otimes C}(c \otimes d) &= \sum c_{(1)} \otimes c_{(2)} \otimes d \end{aligned}$$

$\delta = \psi_{H\otimes C}$ is given by the formula

$$\delta(h \otimes c) = \sum (h_{(1)} \rightharpoonup c_{(1)}) \otimes \varepsilon(h_{(2)}) c_{(2)} = \sum (h \rightharpoonup c_{(1)}) \otimes c_{(2)}$$

The following Theorem is a dual analog of Theorem 8.1.6.

Theorem 8.7.4 *Let H be a Hopf algebra that is faithfully flat as an R-module, and suppose that the antipode of H is bijective. For a left H-comodule algebra C, the following conditions are equivalent:*

1. *C is an H-Galois coobject;*
2. * $\overline{C} = R$;
   * $\delta = \psi_{H\otimes C} : H\otimes C \longrightarrow C\otimes C : h\otimes c \mapsto \sum (h\rightharpoonup c_{(1)})\otimes c_{(2)}$ *is an isomorphism;*
   * *C is flat as an R-module.*

Proof It is clear that 1) implies 2). We have already shown above that $\overline{C} \cong R$ and that δ is an isomorphism. From the fact that G is an equivalence of categories, it follows that G is exact, and therefore C is flat as an R-module.
To prove that 2) implies 1), we proceed as in the proof of Theorem 8.1.6. A right (H, C)-Hopf module is a right H-module N that is also a right C-comodule such that

$$\rho_N(n.h) = \sum n_{(0)}h_{(2)} \otimes S^{-1}(h_{(1)})\rightharpoonup n_{(1)}$$

for all $n \in N$ and $h \in H$. The category of right (H, C)-Hopf modules is denoted by $\mathcal{M}(H)^C_H$. We can make C into a right (H, C)-Hopf module by putting

$$\rho(C) = \sum c_{(1)} \otimes c_{(2)} \quad \text{and} \quad c\leftharpoonup h = S^{-1}(h)\rightharpoonup c$$

Indeed, for all $c \in C$ and $h \in H$, we have that

$$\rho(c\leftharpoonup h) = \rho(S^{-1}(h)\rightharpoonup c) = \sum (S^{-1}(h_{(2)})\rightharpoonup c_{(1)}) \otimes (S^{-1}(h_{(1)})\rightharpoonup c_{(2)})$$

As in the proof of Theorem 8.1.6, we can consider, for $M \in {}^C_H\mathcal{M}(H)$ and $N \in \mathcal{M}(H)^C_H$ the bitensor product

$$N \otimes^C_H M = \mathrm{Im}\,(N\Box_C M \to N \otimes_H M)$$

We now claim that

$$C \otimes^C_H M \cong R \otimes_H M = \overline{M} \tag{8.31}$$

To show this, consider the isomorphism

$$f : M \longrightarrow C\Box_C M : m \mapsto \sum m_{(-1)} \otimes m_{(0)}$$

The inverse of f is given by the formula

$$f^{-1}(\sum_i c_i \otimes m_i) = \sum \varepsilon_C(c_i)m_i$$

Consider $m \in M$ and $h \in H$. Then

$$\begin{aligned} f(h.m) &= \sum h_{(1)}\rightharpoonup m_{(-1)} \otimes h_{(2)}m_{(0)} \\ &= m_{(-1)}\leftharpoonup S(h_{(1)}) \otimes h_{(2)}m_{(0)} \end{aligned}$$

and

$$f(\varepsilon(h)m) = \sum \varepsilon(h)m_{(-1)} \otimes m_{(0)} = \sum m_{(-1)} \otimes S(h_{(1)})h_{(2)}m_{(0)}$$

and it follows that the images of $f(h.m)$ and $f(\varepsilon(h)m)$ in $C \otimes_H^C M$ are equal. We therefore have a well-defined map

$$\overline{f}: \overline{M} \longrightarrow C \otimes_H^C M$$

Now define a map

$$\overline{g}: C \otimes_H M \longrightarrow \overline{M}$$

by

$$\overline{g}(c \otimes m) = \varepsilon_C(c)\overline{m}$$

It is clear that $\overline{g}$ is well-defined, since

$$\begin{aligned} \overline{g}(d \leftharpoonup h \otimes m) &= \overline{g}(S^{-1}(h)d \otimes m) \\ &= \varepsilon_C(S^{-1}(h)d)\overline{m} \\ &= \varepsilon_H(h)\varepsilon_C(d)\overline{m} \\ &= \varepsilon_C(d)\overline{h.m} \\ &= \overline{g}(d \otimes h, m) \end{aligned}$$

$\overline{g}$ restricts to a map $C \otimes_H^C M \to \overline{M}$, which is immediately verified to be the inverse of $\overline{f}$, and this proves (8.31). Now observe that

$$C \otimes (C \otimes_H^C M) \cong (C \otimes C) \otimes_H^C M$$

The proof is identical to the proof of (8.15), now using the fact that C is flat. We now have that

$$\begin{aligned} G(F(M)) = C \otimes \overline{M} &\cong C \otimes (C \otimes_H^C M) \\ &\cong (C \otimes C) \otimes_H^C M \cong (H \otimes C) \otimes_H^C M \end{aligned}$$

Let us examine the right Hopf module structure on $H \otimes C$ induced by the one on $C \otimes C$ using δ. The right Hopf module structure on $C \otimes C$ is the one on the second factor:

$$\begin{aligned} \rho^r(c \otimes d) &= \sum c \otimes d_{(1)} \otimes d_{(2)} \\ (c \otimes d) \leftharpoonup h &= c \otimes (S^{-1}(h) \rightharpoonup d) \end{aligned}$$

The right Hopf module structure on $H \otimes C$ is now the following:

$$\begin{aligned} \rho^r(h \otimes c) &= \sum h \otimes c_{(1)} \otimes c_{(2)} \\ (h \otimes c) \leftharpoonup k &= hk_{(2)} \otimes (S^{-1}(k_{(1)}) \rightharpoonup c) \end{aligned}$$

It is readily verified that this makes $H \otimes C$ into a right (H, C)-Hopf module and that δ is right H-linear and C-colinear. Let us check the right H-linearity:

$$\begin{aligned} \delta(h \otimes c) \leftharpoonup k &= (\sum h \otimes c_{(1)} \otimes c_{(2)}) \leftharpoonup k \\ &= \sum h \rightharpoonup c_{(1)} \otimes S^{-1}(k) \rightharpoonup c_{(2)} \end{aligned}$$

and

$$\begin{aligned}\delta((h\otimes c)\leftharpoonup k) &= \sum hk_{(2)}\rightharpoonup(S^{-1}(k_{(1)})_{(1)}\rightharpoonup c_{(1)})\otimes(S^{-1}(k_{(1)})_{(2)}\rightharpoonup c_{(2)})\\ &= \sum hk_{(3)}S-1(k_{(2)})\rightharpoonup c_{(1)}\otimes S^{-1}(k_{(1)})\rightharpoonup c_{(2)}\\ &= \sum h\rightharpoonup c_{(1)}\otimes S^{-1}(k)\rightharpoonup c_{(2)}\end{aligned}$$

We will now show that

$$G(F(M))\cong(H\otimes C)\otimes_H^C M\cong M$$

The map

$$f:\ H\otimes m\longrightarrow(H\otimes C)\square_C M:\ h\otimes m\mapsto h\otimes m_{(-1)}\otimes m_{(0)}$$

induces a map

$$\overline{f}:\ H\otimes_H M=M\stackrel{(}{\longrightarrow}H\otimes C)\otimes_H^C M$$

We leave it to the reader to verify that $\overline{f}$ is well-defined. The inverse of $\overline{f}$ is given by the formula

$$\overline{f}^{-1}(\sum h_i\otimes c_i\otimes m_i)=\sum h_i m_i\varepsilon_C(c_i)$$

It is straightforward to show that the isomorphism $\overline{f}$ is exactly ψ_M. □

We will now give a dual version of Theorem 8.1.8. First, we will need a dual version of lemma 8.1.7.

Lemma 8.7.5 *Let C be an R-coalgebra that is faithfully flat as an R-module. Then for any $M\in R$-mod, the sequence*

$$C\otimes C\otimes M\stackrel{\delta}{\longrightarrow}C\otimes M\stackrel{\varepsilon_C\otimes I_M}{\longrightarrow}M\longrightarrow 0 \tag{8.32}$$

with δ defined by the formula

$$\delta(c\otimes d\otimes m)=\varepsilon(c)d\otimes m-\varepsilon(d)c\otimes m$$

is exact. In particular, it follows that C is cofaithful, this means that ε_C is surjective.

Proof The fact that C is a faithfully flat R-module implies that it suffices to show that the sequence

$$C\otimes C\otimes C\otimes M\stackrel{I_C\otimes\delta}{\longrightarrow}C\otimes C\otimes M\stackrel{I_C\otimes\varepsilon_C\otimes I_M}{\longrightarrow}C\otimes M\longrightarrow 0$$

is exact. It is clear that the sequence is a complex. $I_C\otimes\varepsilon_C\otimes I_M$ is surjective:

$$\sum_i c_i\otimes m_i=(I_C\otimes\varepsilon_C\otimes I_M)(\sum_i c_{i_{(1)}}\otimes c_{i_{(2)}}\otimes m_i)$$

Suppose that

$$\sum_i c_i\otimes d_i\otimes m_i\in\operatorname{Ker}(I_C\otimes\varepsilon_C\otimes I_M)$$

Then

$$\sum_i c_i \otimes \varepsilon(d_i) m_i = 0$$

and

$$\sum_i c_{i_{(1)}} \otimes c_{i_{(2)}} \otimes \varepsilon(d_i) m_i = 0$$

Now we have that

$$(I_C \otimes \delta)(\sum_i c_{i_{(1)}} \otimes c_{i_{(2)}} \otimes d_i \otimes m_i)$$

$$= \sum_i c_i \otimes d_i \otimes m_i - \sum_i c_{i_{(1)}} \otimes c_{i_{(2)}} \otimes \varepsilon(d_i) m_i$$

$$= \sum_i c_i \otimes d_i \otimes m_i$$

$\square$

Theorem 8.7.6 *Let H be a Hopf algebra which is faithfully flat as an R-module, and suppose that the antipode of H is bijective. For a left H-module coalgebra C, the following conditions are equivalent:*

1. *C is an H-Galois coobject;*

2. *$\delta : H \otimes C \to C \otimes C : h \otimes c \mapsto \sum (h{\rightharpoonup}c_{(1)}) \otimes c_{(2)}$ is an isomorphism and C is faithfully flat as an R-module.*

Proof 1) $\Rightarrow$ 2) follows immediately from Theorem 8.7.4. To prove the converse, it suffices to show that the map

$$\phi_C : \overline{C} {\longrightarrow} R : \overline{c} \mapsto \varepsilon_C(c)$$

is an isomorphism. We have seen in Lemma 8.7.5 that ε_C is surjective, and this proves that ϕ_C is surjective.

Let us show that Φ_C is injective. If $\varepsilon_C(c) = 0$, then it follows from Lemma 8.7.5 that

$$c = \sum_i (\varepsilon_C(c_i) d_i - \varepsilon_C(d_i) c_i)$$

for some $c_i, d_i \in C$. Now take

$$\sum_j h_j \otimes e_j = \delta^{-1}(\sum_i c_i \otimes d_i)$$

It now follows that

$$\sum_i c_i \otimes d_i = \delta(\sum_j h_j \otimes e_j) = \sum_j h_j {\rightharpoonup} e_{j_{(1)}} \otimes e_{j_{(2)}}$$

and

$$c = \sum_i (\varepsilon_C(c_i) d_i - \varepsilon_C(d_i) c_i) = \sum_j (\varepsilon_H(h_i) e_i - h_i {\rightharpoonup} e_i$$

and this means that $\overline{c} = 0$ in $\overline{C}$.

$\square$

Corollary 8.7.7 *Let H be a Hopf algebra which is faithfully flat as an R-module. Then H viewed as a left H-module coalgebra is an H-Galois coobject.*

Proof We only have to show that the map

$$\delta : H \otimes H \longrightarrow H \otimes H : \ h \otimes k \mapsto \sum hk_{(1)} \otimes k_{(2)}$$

has an inverse. This inverse is given by the formula

$$\delta^{-1}(h \otimes k) = \sum hS(k_{(1)}) \otimes k_{(2)}$$

□

Chapter 9

Cohomology over Hopf algebras

9.1 Sweedler cohomology

Let H be a cocommutative Hopf algebra, and S a commutative faithfully flat H-module algebra. Let $P \in \mathcal{P}(R_{fl})$ be a covariant functor from flat R-algebras to abelian groups. For $n \in \mathbb{N}$, we consider $S_n = \mathrm{Hom}_R(H^{\otimes n}, S)$. We put $S_0 = \mathrm{Hom}_R(R, S) \cong S$. Then S_n is a commutative R-algebra under the convolution product: for $f, g \in S_n$ and $x \in H^{\otimes n}$, $f * g$ is defined by

$$(f * g)(x) = \sum f(x_{(1)})g(x_{(2)})$$

For $i = 0, 1, \ldots, n+1$, consider the maps

$$\varepsilon_i : \ S_n = \mathrm{Hom}_R(H^{\otimes n}, S) \to S_{n+1} = \mathrm{Hom}_R(H^{\otimes n+1}, S)$$

given by

$$\begin{aligned}
\varepsilon_0(f)(h_1 \otimes h_2 \otimes \ldots \otimes h_{n+1}) &= h_1 \rightharpoonup f(h_2 \otimes \cdots \otimes h_{n+1}) \\
\varepsilon_i(f)(h_1 \otimes h_2 \otimes \cdots \otimes h_{n+1}) &= f(h_1 \otimes \cdots h_i h_{i+1} \cdots \otimes h_{n+1}) \\
&\qquad \text{for } i = 1, \ldots, n \\
\varepsilon_{n+1}(f)(h_1 \otimes h_2 \otimes \cdots \otimes h_{n+1}) &= \varepsilon(h_{n+1}) f(h_1 \otimes \cdots \otimes h_n)
\end{aligned}$$

Now define

$$\Delta_n : \ P(S_n) = P(\mathrm{Hom}_R(H^{\otimes n}, S)) \to P(S_{n+1}) = P(\mathrm{Hom}_R(H^{\otimes n+1}, S))$$

by

$$\Delta_n = \sum_{i=0}^{n+1} (-1)^i P(\varepsilon_i)$$

A straightforward computation then shows that we have a complex

$$0 \longrightarrow P(\mathrm{Hom}_R(R, S)) = P(S) \xrightarrow{\Delta_0} P(\mathrm{Hom}_R(H, S)) = P(S_1) \xrightarrow{\Delta_1} P(S_2) \xrightarrow{\Delta_2} \cdots$$

We will write

$$\begin{aligned} Z^n(H,S,P) &= \operatorname{Ker}\Delta_n \\ B^n(H,S,P) &= \operatorname{Im}\Delta_{n-1} \\ H^n(H,S,P) &= Z^n(H,S,P)/B^n(H,S,P) \end{aligned}$$

We call $H^n(H,S,P)$ the n-th *Sweedler cohomology group* with values in P.

Sweedler cohomology with values in the category of invertible modules

In Chapter 1, we have defined Amitsur cohomology with values in the category $\underline{\underline{\mathrm{Pic}}}$. The same construction applies to Sweedler cohomology. Observe that the maps $\varepsilon_i : S_n \to S_{n+1}$ defined above are R-algebra maps. Then S_{n+1} is in $n+2$ different ways an S_n-algebra. For any S_n-module M, let $\varepsilon_i(M)$ be the S_{n+1}-module $M \otimes_{\varepsilon_i} S_{n+1}$. For $i = 0, 1, \ldots, n+1$, we therefore have a functor

$$\varepsilon_i : S_n\text{-mod} \longrightarrow S_{n+1}\text{-mod}$$

and this allows us to define a functor

$$\delta_n : \underline{\underline{\mathrm{Pic}}}(S_n) \longrightarrow \underline{\underline{\mathrm{Pic}}}(S_{n+1})$$

by

$$\delta_n(I) = \varepsilon_0(I) \otimes_{S_{n+1}} \varepsilon_1(I)^* \otimes_{S_{n+1}} \cdots \otimes_{S_{n+1}} \varepsilon_{n+1}(I)^{(*)}$$

and

$$\delta_n(f) = \varepsilon_0(f) \otimes (\varepsilon_1(f)^*)^{-1} \otimes \cdots (\varepsilon_{n+1}(f)^{(*)})^{\pm 1}$$

From the fact that the natural map $S_n \to S_{n+2}$ is an isomorphism, it follows that we have a natural map

$$\lambda_I : (\delta_{n+1} \circ \delta_n)(I) \longrightarrow S_{n+2}$$

We will identify λ_I to the identity on S_{n+2} - the argument is identical to the one used in the case of Amitsur cohomology.
Now for $n \geq 0$, we consider the category Ω_n. An object of Ω_n is a pair (I, α), with $I \in \underline{\underline{\mathrm{Pic}}}(S_n)$ and $\alpha : \delta_n(I) \to S_{n+1}$ such that

$$\delta_n(\alpha) = \lambda_I = I_{S_{n+2}}$$

A morphism from (I, α) to (J, β) in Ω_n will be an isomorphims $f : I \to J$ such that $\alpha = \beta \circ \delta_n(f)$. We now have a functor

$$\partial_n : \underline{\underline{\mathrm{Pic}}}(S_n) \longrightarrow \Omega_{n+1}$$

given by

$$\partial_n(I) = (\delta_n I, \lambda_I)$$

We now define

$$H^n(H,S,\underline{\underline{\mathrm{Pic}}}) = \operatorname{Coker}(K_0\partial_{n-1} : \mathrm{Pic}(S_{n-1}) \to K_0\Omega_n)$$

We call $H^n(H,S,\underline{\underline{\mathrm{Pic}}})$ the n-th Sweedler cohomology group with values in the category $\underline{\underline{\mathrm{Pic}}}$.

Sweedler cohomology versus group cohomology

Take the special case where $H = RG$, G being an arbitrary group. As we have seen before, an H-module algebra is then just a G-module algebra. G^n is a free basis for $H^{\otimes n}$, and we have a natural isomorphism

$$\mathrm{Hom}_R(H^{\otimes n}, S) \cong K^n(G,S)$$

Here $K^n(G,S)$ is the set of all maps from G^n to S. Sweedlers complex now takes the following form:

$$\varepsilon_i:\ K^n(G,S) \longrightarrow K^{n+1}(G,S)$$

with

$$\begin{aligned}
\varepsilon_0(f)(\sigma_1,\ldots,\sigma_{n+1}) &= \sigma_1(f(\sigma_2,\ldots,\sigma_{n+1}))\\
\varepsilon_i(f)(\sigma_1,\ldots,\sigma_{n+1}) &= f(\sigma_1,\ldots,\sigma_i\sigma_{i+1},\ldots,\sigma_{n+1})\\
&\qquad\qquad \text{for } i = 1,\ldots,n\\
\varepsilon_{n+1}(f)(\sigma_1,\ldots,\sigma_{n+1}) &= f(\sigma_1,\ldots,\sigma_n)
\end{aligned}$$

It is clear that Sweedlers complex is nothing else then the classical group cohomology complex with values in $P(S)$. Therefore group cohomology and Sweedler cohomology coincide.

Proposition 9.1.1 *Let G be a group and S a G-module algebra. If P commutes with direct sums, then*

$$H^n(H,S,P) \cong H^n(G,P(S))$$

Normalized cocycles

Let $f \in \mathbb{G}_m((H\otimes H)^*) \in Z^2(H,R,\mathbb{G}_m)$. The cocycle relation can be rewritten as follows.

$$f(h\otimes k)\varepsilon(l) = \sum f(h_{(1)}\otimes k_{(1)}l_{(1)})f^{-1}(h_{(2)}k_{(2)}\otimes l_{(2)})f(k_{(3)}\otimes l_{(3)})$$

for all $h,k,l \in H$. Take $k = 1$. Then we obtain

$$f(h\otimes 1)\varepsilon(l) = \varepsilon(h)f(1\otimes l)$$

for $h,k \in H$. Now take subsequently $h = 1$ and $l = 1$ in this formula. It follows that

$$f(h\otimes 1) = f(1\otimes h) = \varepsilon(h)f(1\otimes 1) \tag{9.1}$$

for all $h \in H$. The cocyle f is called *normalized* if $f(1\otimes 1) = 1$. In this case

$$f(h\otimes 1) = f(1\otimes h) = \varepsilon(h)$$

for all $h \in H$.
Let

$$\begin{aligned}
\mathbb{G}_m^n((H\otimes H)^*) &= \{f \in \mathbb{G}_m((H\otimes H)^*) | f(1\otimes h) = f(h\otimes 1) = \varepsilon(h) \text{ for all } h \in H\}\\
\mathbb{G}_m^n(H^*) &= \{f \in \mathbb{G}_m(H^*) | f(1) = 1\}
\end{aligned}$$

Lemma 9.1.2 *Every element in $H^2(H, R, \mathbb{G}_m)$ can be represented by a normalized cocycle. Sweedler's complex*

$$\mathbb{G}_m(H^*) \xrightarrow{\Delta_1} \mathbb{G}_m((H \otimes H)^*) \xrightarrow{\Delta_2} \mathbb{G}_m((H^{\otimes 3})^*)$$

restricts to a complex

$$\mathbb{G}_m^n(H^*) \xrightarrow{\Delta_1^n} \mathbb{G}_m^n((H \otimes H)^*) \xrightarrow{\Delta_2^n} \mathbb{G}_m((H^{\otimes 3})^*)$$

and

$$H^2(H, R, \mathbb{G}_m) \cong \operatorname{Ker}(\Delta_2^n)/\operatorname{Im}(\Delta_1^n)$$

Proof Let $f \in (H \otimes H)^*$ be a cocycle. $f(1 \otimes 1)$ is an invertible element of R, since $1 \otimes 1$ is grouplike in H. The map

$$g: \ H \longrightarrow R: \ h \mapsto \varepsilon(h) f(1 \otimes 1)^{-1}$$

is convolution invertible, its inverse is given by the formula

$$g^{-1}(h) = \varepsilon(h) f(1 \otimes 1)$$

Now

$$(\Delta_1 g)(h \otimes k) = \sum \varepsilon(h_{(1)}) \varepsilon(k_{(1)}) \varepsilon(h_{(2)} k_{(2)}) f(1 \otimes 1)^{-1} = \varepsilon(hk) f(1 \otimes 1)^{-1}$$

and it follows easily that

$$(\Delta_1 g * f)(1 \otimes 1) = 1$$

and $\Delta_1 g * f$ is normalized. Observe finally that a coboundary $\Delta_1 g$ is normalized if and only if $g(1)g(1)g^{-1}(1) = g(1) = 1$. The second part now follows immediately. □

9.2 Harrison cohomology

Consider now a commutative Hopf algebra H and a commutative faithfully flat H-comodule algebra S. For $i = 0, 1, \ldots, n+1$, consider the maps

$$\varepsilon_i: \ S \otimes H^{\otimes n} \longrightarrow S \otimes H^{\otimes n+1}$$

defined as follows:

$$\begin{aligned}
\varepsilon_0(s \otimes h_1 \otimes \cdots \otimes h_n) &= \rho(s) \otimes h_1 \otimes \cdots \otimes h_n \\
\varepsilon_i(s \otimes h_1 \otimes \cdots \otimes h_n) &= s \otimes h_1 \otimes \cdots \otimes \Delta(h_i) \otimes \cdots \otimes h_n \\
&\qquad \text{for } i = 1, \ldots, n \\
\varepsilon_{n+1}(s \otimes h_1 \otimes \cdots \otimes h_n) &= s \otimes h_1 \otimes \cdots \otimes h_n \otimes 1
\end{aligned}$$

Let P be a covariant functor from flat commutative R-algebras to abelian groups, and define

$$\Delta_n: \ P(S \otimes H^{\otimes n}) \longrightarrow P(S \otimes H^{\otimes n+1})$$

by

$$\Delta_n = \sum_{i=0}^{n+1}(-1)^i P(\varepsilon_i)$$

We obtain a complex

$$0 \longrightarrow P(S) \xrightarrow{\Delta_0} P(S \otimes H) \xrightarrow{\Delta_1} P(S \otimes H^{\otimes 2}) \xrightarrow{\Delta_2} \cdots$$

and its cohomology groups are denoted by $H^n_{\mathrm{Harr}}(H,S,P)$. We call $H^n_{\mathrm{Harr}}(H,S,P)$ the n-th *Harrison cohomology* group with values in P.

Harrison cohomology with values in the category of invertible modules

In Section 6.1, we have introduced Amitsur cohomology with values in the category of invertible modules. We will now apply the same construction to Harrison cohomology.

The maps ε_i are algebra maps, and therefore $S \otimes H^{\otimes n+1}$ is an $S \otimes H^{\otimes n}$-algebra via $\varepsilon_0, \varepsilon_1, \ldots, \varepsilon_{n+1}$. For $M \in S \otimes H^{\otimes n}$-mod, we write $\varepsilon_i(M) = M \otimes_{\varepsilon_i} (S \otimes H^{\otimes n+1})$. This defines $n+2$ functors

$$\varepsilon_i : \underline{\underline{\mathrm{Pic}}}(S \otimes H^{\otimes n}) \longrightarrow \underline{\underline{\mathrm{Pic}}}(S \otimes H^{\otimes n+1})$$

Define

$$\delta_n : \underline{\underline{\mathrm{Pic}}}(S \otimes H^{\otimes n}) \longrightarrow \underline{\underline{\mathrm{Pic}}}(S \otimes H^{\otimes n+1})$$

by

$$\delta_n(I) = \varepsilon_0(I) \otimes_{S\otimes H^{\otimes n}} \varepsilon_1(I)^* \otimes_{S\otimes H^{\otimes n}} \cdots \otimes_{S\otimes H^{\otimes n}} \varepsilon_{n+1}(I)^{(*)}$$

and

$$\delta_n(f) = \varepsilon_0(f) \otimes (\varepsilon_1(f)^*)^{-1} \otimes \cdots (\varepsilon_{n+1}(f)^{(*)})^{\pm 1}$$

We have a natural isomorphism

$$\lambda_I : (\delta_{n+1} \circ \delta_n)(I) \longrightarrow S \otimes H^{\otimes n+2}$$

of $S \otimes H^{\otimes n+2}$-modules. δ_n is cofinal and product preserving, and the exact sequence (C.8) takes the form

$$\mathbb{G}_m(S \otimes H^{\otimes n}) \xrightarrow{\Delta_n} \mathbb{G}_m(S \otimes H^{\otimes n+1}) \xrightarrow{\kappa_n} \mathrm{K}_0\underline{\Psi\delta_n} \xrightarrow{\theta_n} \mathrm{Pic}(S \otimes H^{\otimes n}) \xrightarrow{\Delta'_n} \mathrm{Pic}(S \otimes H^{\otimes n+1}) \tag{9.2}$$

Δ_n and Δ'_n are the maps in the Harrison complexes of $\mathbb{G}_m$ and Pic. $\underline{\Psi\delta_n}$ is the following category.

Objects: (I,α), with $I \in \underline{\underline{\mathrm{Pic}}}(S \otimes H^{\otimes n})$ and $\alpha : \delta_n(I) \to S \otimes H^{\otimes n+1}$ an $S \otimes H^{\otimes n+1}$-linear isomorphism.

Morphisms: a morphism $f : (I,\alpha) \to (J,\beta)$ is an $S \otimes H^{\otimes n}$-linear isomorphism $I \to J$ such that $\beta \circ \delta_n(f) = \alpha$.

Product: $(I,\alpha)(J,\beta) = (I \otimes_{S\otimes H^{\otimes n}} J, \alpha \otimes \beta)$.

We leave it to the reader to describe the maps κ_n and θ_n. We have maps

$$D'_n : \mathrm{Pic}(S \otimes H^{\otimes n}) \longrightarrow \mathrm{K}_0\underline{\Psi\delta_n}$$
$$D_n : \mathrm{K}_0\underline{\Psi\delta_n} \longrightarrow \mathbb{G}_m(S \otimes H^{\otimes n})$$

defined by the formulas

$$\begin{aligned} D'_n[I] &= [(\delta_n(I), \lambda_I)] \\ D_n[(I,\alpha)] &= \delta_{n+1}(\alpha) \circ \lambda_I^{-1} \end{aligned}$$

The following diagram is the analog of (6.5)

$$\begin{array}{ccccc}
1 & & & & \\
\downarrow & & & & \\
\mathbb{G}_m(S) & & & & 1 \\
\downarrow \Delta_0 & & & & \downarrow \\
\mathbb{G}_m(S\otimes H) & \xrightarrow{\kappa_0} & \mathrm{K}_0\underline{\Psi\delta_0} & \xrightarrow{\theta_0} & \mathrm{Pic}(S) \\
\downarrow \Delta_1 & \swarrow D_0 & & \swarrow D'_0 & \downarrow \Delta'_0 \\
\mathbb{G}_m(S\otimes H^{\otimes 2}) & \xrightarrow{\kappa_1} & \mathrm{K}_0\underline{\Psi\delta_{(1)}} & \xrightarrow{\theta_1} & \mathrm{Pic}(S\otimes H) \\
\downarrow \Delta_2 & \swarrow D_1 & & \swarrow D'_1 & \downarrow \Delta'_1 \\
\mathbb{G}_m(S\otimes H^{\otimes 3}) & \xrightarrow{\kappa_2} & \mathrm{K}_0\underline{\Psi\delta_2} & \xrightarrow{\theta_2} & \mathrm{Pic}(S\otimes H^{\otimes 2}) \\
\downarrow \Delta_3 & \swarrow D_2 & & \swarrow D'_2 & \downarrow \Delta'_2 \\
\vdots & & \vdots & & \vdots
\end{array} \tag{9.3}$$

The first and third column are the Harrison complexes of $\mathbb{G}_m$ and Pic, and the cranks are the exact sequences (9.2). Arguments similar to the ones in Section 6.1 show that $D_n \circ D'_{n-1} = 1$ and that the parallelograms in (9.3) anticommute. We define

$$H^n_{\mathrm{Harr}}(H, S, \underline{\underline{\mathrm{Pic}}}) = \mathrm{Ker}\,(D_n)/\mathrm{Im}\,(D'_{n-1}) \tag{9.4}$$

Theorem 9.2.1 *Let H be a commutative Hopf algebra, and S a commutative faithfully flat H-comodule algebra. We have a long exact sequence of cohomology groups*

$$\begin{array}{ccccccc}
1 & \longrightarrow & H^1_{\mathrm{Harr}}(H,S,\mathbb{G}_m) & \xrightarrow{\alpha_1} & H^0_{\mathrm{Harr}}(H,S,\underline{\underline{\mathrm{Pic}}}) & \xrightarrow{\beta_1} & H^0_{\mathrm{Harr}}(H,S,\mathrm{Pic}) \\
 & \xrightarrow{\gamma_1} & H^2_{\mathrm{Harr}}(H,S,\mathbb{G}_m) & \xrightarrow{\alpha_2} & H^1_{\mathrm{Harr}}(H,S,\underline{\underline{\mathrm{Pic}}}) & \xrightarrow{\beta_2} & H^1_{\mathrm{Harr}}(H,S,\mathrm{Pic}) \\
 & \xrightarrow{\gamma_2} & \cdots & & & & \\
 & \xrightarrow{\gamma_{n-1}} & H^n_{\mathrm{Harr}}(H,S,\mathbb{G}_m) & \xrightarrow{\alpha_n} & H^{n-1}_{\mathrm{Harr}}(H,S,\underline{\underline{\mathrm{Pic}}}) & \xrightarrow{\beta_n} & H^{n-1}_{\mathrm{Harr}}(H,S,\mathrm{Pic}) \\
 & \xrightarrow{\gamma_n} & \cdots & & & &
\end{array} \tag{9.5}$$

Proof Similar to the proof of Theorem 6.1.3 □

Remark 9.2.2 If $S = R$, with trivial coaction, then

$$H^0_{\mathrm{Harr}}(H, R, \mathrm{Pic}) = 0$$

and

$$H^0_{\mathrm{Harr}}(H, R, \underline{\underline{\mathrm{Pic}}}) = H^1_{\mathrm{Harr}}(H, R, \mathbb{G}_m) = G(H)$$

Harrison cohomology versus Sweedler cohomology

Proposition 9.2.3 *Let H be a faithfully projective cocommutative Hopf algebra, and suppose that S is a commutative faithfully projective H-module algebra. Then the Sweedler complex for H and the Harrison complex for H^* are isomorphic, and therefore*

$$H^n(H,S,P) \cong H^n_{\mathrm{Harr}}(H^*,S,P)$$

Proof Observe that if S is an H-module algebra, then it is also an H^*-comodule algebra. Since S and H are faithfully projective, we have a natural isomorphism

$$S \otimes (H^*)^{\otimes n} \longrightarrow \mathrm{Hom}(H^{\otimes n}, S)$$

and it is straightforward to check that this establishes an isomorphism between the Sweedler complex and the Harrison complex. □

Harrison cohomology versus Amitsur cohomology

Proposition 9.2.4 *Let H be a commutative faithfully flat Hopf algebra, and suppose that S is a commutative H-Galois object. Then the Harrison complex and the Amitsur complex are isomorphic, and, in particular,*

$$H^n(S/R,P) \cong H^n_{\mathrm{Harr}}(H,S,P) \quad \text{and} \quad H^n(S/R,\underline{\underline{\mathrm{Pic}}}) \cong H^n_{\mathrm{Harr}}(H,S,\underline{\underline{\mathrm{Pic}}})$$

Proof Since S is an H-Galois object, we have an isomorphism

$$\gamma:\ S \otimes S \longrightarrow S \otimes H:\ s_1 \otimes s_2 \mapsto \sum s_1 s_{2_{(0)}} \otimes s_{2_{(1)}}$$

For any $n \geq 1$, we now define

$$\gamma_n : S^{\otimes n+1} \longrightarrow S \otimes H^{\otimes n}$$

recursively as follows:

$$\gamma_1 = \gamma \quad \text{and} \quad \gamma_n = (\gamma \otimes I_{H^{\otimes n-1}}) \circ (I_S \otimes \gamma_{n-1})$$

It is clear that γ_n is a bijection. Recall that the Amitsur complex is defined using maps

$$\epsilon_i:\ S^{(n+1)} \to S^{(n+2)}:\ s_1 \otimes \cdots \otimes s_{n+1} \mapsto s_1 \otimes \cdots \otimes 1 \otimes \cdots \otimes s_{n+1}$$

the 1 is inserted in the i-th position. Now we claim that the diagram

$$\begin{array}{ccc} S^{\otimes n+1} & \xrightarrow{\epsilon_i} & S^{\otimes n+2} \\ \downarrow{\scriptstyle \gamma_n} & & \downarrow{\scriptstyle \gamma_{n+1}} \\ S \otimes H^{\otimes n} & \xrightarrow{\varepsilon_{i-1}} & S \otimes H^{\otimes n+1} \end{array}$$

commutes for $i = 1, \dots, n+2$. This may be checked easily using the formula

$$\begin{aligned}\gamma_n(s_1 \otimes \cdots \otimes s_n) \\ = \sum s_1 s_{2_{(0)}} s_{3_{(0)}} \cdots s_{n+1_{(0)}} \otimes s_{2_{(1)}} s_{3_{(1)}} \cdots s_{n+1_{(1)}} \otimes s_{3_{(2)}} \cdots s_{n+1_{(2)}} \otimes \cdots \otimes s_{n+1_{(n)}}\end{aligned}$$

This now implies that the Amitsur complex and the Harrison complex are isomorphic, and the result follows. □

Now let H be a cocommutative faithfully projective Hopf algebra, and suppose that S is a commutative H^*-Galois object. S is a right H^*-comodule algebra, and therefore a left H-module algebra. Combining Theorem 6.2.1, Corollary 6.2.3, Proposition 9.2.3 and Proposition 9.2.4, we now obtain:

Proposition 9.2.5 *Let H and S be as above. We then have an isomorphism*

$$\mathrm{Br}(S/R) \cong H^1(H, S, \underline{\underline{\mathrm{Pic}}})$$

The Chase-Rosenberg exact sequence takes the form

$$\begin{array}{cccccc} 1 & \longrightarrow & H^1(H,S,\mathbb{G}_m) & \longrightarrow & \mathrm{Pic}(R) & \longrightarrow & H^0(H,S,\mathrm{Pic}) \\ & \longrightarrow & H^2(H,S,\mathbb{G}_m) & \xrightarrow{\gamma} & \mathrm{Br}(S/R) & \longrightarrow & H^1(H,S,\mathrm{Pic}) \\ & \longrightarrow & H^3(H,S,\mathbb{G}_m) & & & & \end{array}$$

If $\mathrm{Pic}(S) = \mathrm{Pic}(S \otimes S) = 0$, then the map γ is an isomorphism, and

$$\mathrm{Br}(S/R) \cong H^2(H, S, \mathbb{G}_m)$$

This property is called the crossed product theorem

Let us describe the map γ explicitly. Take a Sweedler cocycle $\sigma \in Z^2(H, R, \mathbb{G}_m)$. This means that $\sigma : H \otimes H \to \mathbb{G}_m(S)$ is a convolution invertible map satisfying the cocycle relation

$$\sum (h_{(1)} \rightharpoonup \sigma(k_{(1)} \otimes \ell_{(1)}))\sigma(h_{(2)} \otimes k_{(2)}\ell_{(2)}) = \sum \sigma(h_{(1)}k_{(1)} \otimes \ell)\sigma(h_{(2)} \otimes k_{(2)}) \quad (9.6)$$

An Azumaya algebra $A = S\#_\sigma H$ may now be constructed as follows: let A be equal to $S \otimes H$ as an R-module, and define a multiplication on $S \otimes H$ as follows:

$$(s\#_\sigma h)(t\#_\sigma k) = \sum \sigma(h_{(2)} \otimes k_{(1)}) s(h_{(1)} \rightharpoonup t)\#_\sigma h_{(3)}k_{(2)} \quad (9.7)$$

A is called the crossed product of S, H and σ. Using (9.6), we can show easily that the multiplication rule (9.7) is associative. We claim that

$$1_A = \sigma^{-1}(1,1)\#_\sigma 1$$

is a unit element of A. To show this, take $k = 1$ in the cocycle relation (9.6). We obtain that

$$\sum (h_{(1)} \rightharpoonup \sigma(1 \otimes \ell_{(1)}))\sigma(h_{(2)} \otimes \ell_{(2)}) = \sum \sigma(h_{(1)} \otimes \ell)\sigma(h_{(2)} \otimes 1)$$

Taking the convolution product of both sides with σ^{-1}, the convolution inverse of σ, we obtain:

$$\sigma(h \otimes 1)\varepsilon(\ell) = h \rightharpoonup \sigma(1 \otimes \ell)$$

Taking $h = 1$ and $\ell = 1$ in the above equations, we obtain

$$\sigma(1 \otimes \ell) = \varepsilon(\ell)\sigma(1 \otimes 1) \text{ and } \sigma(h \otimes 1) = h \rightharpoonup \sigma(1 \otimes 1) \tag{9.8}$$

We now compute easily that

$$\begin{aligned}
(s\#_\sigma h)1_A &= \sum s(h_{(1)} \rightharpoonup \sigma^{-1}(1 \otimes 1))\sigma(h_{(2)} \otimes 1)\#_\sigma h_{(3)} \\
&= \sum s(h_{(1)} \rightharpoonup \sigma^{-1}(1 \otimes 1))(h_{(2)} \rightharpoonup \sigma(1 \otimes 1))\#_\sigma h_{(3)} \\
&= s\#_\sigma h \\
1_A(s\#_\sigma h) &= \sigma^{-1}(1 \otimes 1)s\sigma(1 \otimes h_{(1)})\#_\sigma h_{(2)} \\
&= \sigma^{-1}(1 \otimes 1)s\varepsilon(h_{(1)})\sigma(1 \otimes 1)\#_\sigma h_{(2)} \\
&= s\#_\sigma h
\end{aligned}$$

The cocycle σ is called *normal* if $\sigma(1 \otimes 1) = 1$. It follows from 9.8 that a normal cocycle has the property

$$\sigma(h \otimes 1) = \sigma(1 \otimes h) = \varepsilon(h)$$

for all $h \in H$. The unit of A is then given by $1_A = 1_S \otimes 1_H$. It may be shown easily that every cocycle is cohomologous to a normalized cocycle. Suppose that $\sigma \in Z^2(H, S, \mathbb{G}_m)$ is a cocycle. Consider the map $\tau : H \to S$ given by

$$\tau(h) = \varepsilon(h)\sigma^{-1}(1 \otimes 1)$$

It is clear that τ is convolution invertible. Its inverse is given by

$$\tau^{-1}(h) = \varepsilon(h)\sigma(1 \otimes 1)$$

Now

$$\begin{aligned}
\Delta_1(\tau)(h \otimes k) &= (h \rightharpoonup \sigma^{-1}(1 \otimes 1))\sigma(1 \otimes 1)\sigma^{-1}(1 \otimes 1)\varepsilon(k) \\
&= \varepsilon(k)(h \rightharpoonup \sigma^{-1}(1 \otimes 1)) \\
\Delta_1(\tau)(1 \otimes 1) &= \sigma^{-1}(1 \otimes 1)
\end{aligned}$$

Therefore

$$(f * \Delta_1(g))(1 \otimes 1) = \sigma(1 \otimes 1)\sigma^{-1}(1 \otimes 1) = 1$$

and $f * \Delta_1(g)$ is a normalized cocycle cohomologous to f.
We now claim that if $\sigma = \Delta_1(\tau)$ is a coboundary, then

$$S\#_\sigma H \cong S\#H \cong \mathrm{End}_R(S) \tag{9.9}$$

The second isomorphism is given in Theorem 8.3.1. To prove the first one, we first observe that

$$\sigma(h \otimes k) = \sum (h_{(1)} \rightharpoonup \tau(k_{(1)}))\tau^{-1}(h_{(2)}k_{(2)})\tau(h_{(3)})$$

Now define

$$f:\ S\#H \longrightarrow S\#_\sigma H$$

by

$$f(s\#h)\sum s\tau^{-1}(h_{(1)})\#_\sigma h_{(2)}$$

It is clear that f is R-linear and bijective. The inverse of f is given by

$$f^{-1}(s\#_\sigma h) = \sum s\tau(h_{(1)})\#h_{(2)}$$

f is a multiplicative map:

$$\begin{aligned}
f(s\#h)f(t\#k) &= \sum\left(s\tau^{-1}(h_{(1)})\#_\sigma h_{(2)}\right)\left(t\tau^{-1}(k_{(1)})\#_\sigma k_{(2)}\right)\\
&= \sum s\tau^{-1}(h_{(1)})(h_{(2)}\rightharpoonup(t\tau^{-1}(k_{(1)})))(h_{(3)}\rightharpoonup\tau^{-1}(k_{(2)}))\tau^{-1}(h_4k_{(3)})\tau(h_5)\#h_6k_4\\
&= \sum s(h_{(1)}\rightharpoonup t)\tau^{-1}(h_{(2)}k_{(1)})\#h_{(3)}k_{(2)}\\
&= f\left(\sum s(h_{(1)}\rightharpoonup t)\#h_{(2)}k\right)\\
&= f((s\#h)(t\#k))
\end{aligned}$$

We can now show that $A = S\#_\sigma H$ is an Azumaya algebra split by S. We know that the image of an Amitsur cocycle in $Z^2(S/R, \mathbb{G}_m)$ in $Z^2(S\otimes S/S, \mathbb{G}_m)$ is a coboundary. In view of Propositions 9.2.3 and 9.2.4, the same is true for a Sweedler cocycle: if $\sigma \in Z^2(H, S\mathbb{G}_m)$ is a cocycle, then the map

$$\sigma\otimes I:\ (H\otimes S)\otimes_S(H\otimes S) \cong (H\otimes H)\otimes S \longrightarrow S\otimes S:\ (h\otimes s)\otimes(k\otimes t)\mapsto \sigma(h\otimes k)\otimes st$$

is a coboundary in $Z^2(S\otimes H, S\otimes S, \mathbb{G}_m)$. Now it is clear that

$$A\otimes S \cong (S\otimes S)\#_{\sigma\otimes I}(H\otimes S) \cong (S\otimes S)\#(H\otimes S) \cong \mathrm{End}_R(S\otimes S)$$

and this means that A is an Azumaya algebra split by S.
Given two cocycles $\sigma,\tau \in Z^2(H, S, \mathbb{G}_m)$, we can show that the Azumaya algebras $(S\#_\sigma H)\otimes(S\#_\tau H)$ and $S\#_{\sigma*\tau}H$ are Brauer equivalent. We therefore have a well-defined group homomorphism

$$\gamma:\ H^2(H, S, \mathbb{G}_m)\longrightarrow \mathrm{Br}(S/R):\ [f]\mapsto S\#_\sigma H$$

A long computation shows that the diagram

$$\begin{array}{ccc}
H^2(S/R, \mathbb{G}_m) & \xrightarrow{\gamma'} & \mathrm{Br}(S/R)\\
\downarrow\cong & & \downarrow I\\
H^2(H, S, \mathbb{G}_m) & \xrightarrow{\gamma} & \mathrm{Br}(S/R)
\end{array}$$

is almost commutative, in fact, if $u \in Z^2(S/R, \mathbb{G}_m)$ is an Amitsur cocycle, and σ is the corresponding Sweedler cocycle, then $\gamma'([u])$ and $\gamma([\sigma])$ are each others inverses in the Brauer group. Here γ' is one of the maps in the Chase-Rosenberg sequence 6.2.3.
If $H = RG$, with G a finite group, then an H-Galois object is nothing else then a classical G-Galois extension in the sense of [75]. As we have seen, Sweedler

cohomology then corresponds to ordinary group cohomology. Recall that a map $\sigma : G \times G \to \mathbb{G}_m(S)$ is a 2-cocycle if

$$g.\sigma(h,k)\sigma(g,hk) = \sigma(gh,k)\sigma(g,h)$$

The crossed product $A = S\#_\sigma RG$ may now be described as follows:

$$A = \oplus_{g\in G} Su_g$$

with multiplication rule

$$(su_g)(tu_h) = \sigma(g,h)s(g.t)u_{gh}$$

We leave it to the reader to show that this formula follows from (9.7). The Chase-Rosenberg sequence in the special case where $H = RG$, and S a G-Galois extension was first given in [49]. An explicit construction of all the maps and the proof of the exactness of the sequence in this case may be found in [75]. The fact that the Chase-Rosenberg sequence can also be written in the Hopf algebra case was first observed in [140].

Chapter 10

The group of Galois (co)objects

10.1 Galois coobjects and Harrison cohomology

Throughout this Section, H will be a commutative Hopf algebra that is faithfully flat as an R-module. We will show that the set of isomorphism classes of H-Galois coobjects forms a group under the operation induced by the tensor product over H, and we will describe this group using Harrison cohomology.
If C and D are two H-module coalgebras, then $C \otimes_H D$ is again an H-module coalgebra. The action and comultiplication are given by the formulas

$$h \rightharpoonup (c \otimes d) = h \rightharpoonup c \otimes d = c \otimes h \rightharpoonup d$$

and

$$\Delta(c \otimes d) = \sum (c_{(1)} \otimes d_{(1)}) \otimes (c_{(2)} \otimes d_{(2)})$$

We leave it to the reader to verify that Δ is well-defined. Obviously, H is itself an H-module coalgebra, and we have an H-module coalgebra isomorphism

$$H \otimes_H C \longrightarrow C : \ h \otimes c \mapsto h \rightharpoonup c$$

for every H-module coalgebra C.

Proposition 10.1.1 *Suppose that C and D are two H-Galois coobjects. Then $C \otimes_H D$ is also an H-Galois coobject.*

Proof As we have seen above, $C \otimes_H D$ is an H-module coalgebra. Let us first show that $C \otimes_H D$ is faithfully flat as an R-module. Using Theorems 8.7.4 and 8.7.6, we obtain that

$$C \otimes_H D \otimes D \cong C \otimes_H H \otimes D \cong C \otimes D$$

is faithfully flat as an R-module. Using the fact that D is faithfully flat as an R-module, we obtain that $C \otimes_H D$ is faithfully flat as an R-module. The map

$$\delta : \ H \otimes C \otimes_H D \longrightarrow C \otimes_H D \otimes C \otimes_H D : \ h \otimes c \otimes d \mapsto \sum h \rightharpoonup c_{(1)} \otimes d_{(1)} \otimes c_{(2)} \otimes d_{(2)}$$

is an isomorphism. Indeed, observe that we have natural isomorphisms

$$H \otimes C \otimes_H D \cong H \otimes_H H \otimes C \otimes_H D \cong (H \otimes C) \otimes_{H\otimes H} (H \otimes D)$$

We may therefore view δ as the map

$$\delta: \ (H \otimes C) \otimes_{H\otimes H} (H \otimes D) \longrightarrow C \otimes_H D \otimes C \otimes_H D$$

given by

$$\delta = \tau_{23} \circ (\delta_C \otimes \delta_D)$$

and this map is an isomorphism. From Theorem 8.7.6, it now follows that $C \otimes_H D$ is an H-Galois coobject. □

Theorem 10.1.2 *Suppose that H is a Hopf algebra that is faithfully flat as an R-module, and let C be an H-Galois coobject. Then there exists an H-Galois coobject D such that $C \otimes_H D \cong H$ as H-module coalgebras. D is given by the following data:*

$D = C^{\mathrm{cop}}$ as a coalgebra;
$h \overset{D}{\rightharpoonup} d = S(h) \overset{C}{\rightharpoonup} d$ for all $h \in H$ and $d \in D$.

Proof Let $K = H^{\mathrm{cop}}$ as a coalgebra, with H-action given by $h \overset{k}{\rightharpoonup} d = S(h)k$ for all $h, k \in H$. Then the antipode $S: \ H \to K$ is an isomorphism of H-module coalgebras. It therefore suffices to show that $C \otimes_H D \cong K$ as H-module coalgebras. Since $\overline{C} = R$, R is the coequalizer of the maps

$$\begin{cases} \varepsilon_H \otimes I_C \\ \psi_C \end{cases} : \ H \otimes C \rightrightarrows C \xrightarrow{\varepsilon_C} R \longrightarrow 0$$

Now H is flat as an R-module, and this implies that H is the coequalizer of the maps

$$\begin{cases} I_H \otimes \varepsilon_H \otimes I_C \\ I_H \otimes \psi_C \end{cases} : \ H \otimes H \otimes C \rightrightarrows H \otimes C \xrightarrow{I_H \otimes \varepsilon_C} H \longrightarrow 0$$

Recall from Corollary 8.7.7 that

$$\delta_H^{-1} \otimes I_C : \ H \otimes H \otimes C \longrightarrow H \otimes H \otimes C : \ h \otimes k \otimes c \mapsto \sum hS(k_{(1)}) \otimes k_{(2)} \otimes c$$

is an isomorphism. Therefore H is also the coequalizer of the maps

$$\begin{cases} \beta = (I_H \otimes \varepsilon_C \otimes I_C) \circ (\delta_H^{-1} \otimes I_C) \\ \alpha = (I_H \otimes \psi_C) \circ (\delta_H^{-1} \otimes I_C) \end{cases} : \ H \otimes H \otimes C \rightrightarrows H \otimes C \xrightarrow{I_H \otimes \varepsilon_C} H \longrightarrow 0$$

One easily verifies that

$$\begin{aligned} \alpha(h \otimes k \otimes c) &= \sum hS(k_{(1)}) \otimes (k_{(2)} \rightharpoonup c) \\ \beta(h \otimes k \otimes c) &= hS(k) \otimes c \end{aligned}$$

Now consider the map

$$\delta' : \ H \otimes H \otimes C \longrightarrow D \otimes H \otimes C : \ h \otimes k \otimes c \mapsto \sum (h \rightharpoonup c_{(1)}) \otimes k \otimes c_{(2)}$$

and observe that the diagram

$$\begin{array}{ccc}
H\otimes H\otimes C & \stackrel{\delta'}{\longrightarrow} & D\otimes H\otimes C \\
{\scriptstyle\beta}\downarrow\downarrow{\scriptstyle\alpha} & & {\scriptstyle\psi_D\otimes I_C}\downarrow\downarrow{\scriptstyle I_D\otimes\psi_C} \\
H\otimes C & \stackrel{\delta}{\longrightarrow} & D\otimes C \\
\downarrow{\scriptstyle I_H\otimes\varepsilon_C} & & \downarrow \\
H & & D\otimes_H C \\
\downarrow & & \downarrow \\
0 & & 0
\end{array}$$

commutes. Indeed,

$$\begin{aligned}
((\psi_D\circ I_C)\circ\delta)(h\otimes k\otimes c) &= \sum S(k)h\rightharpoonup c_{(1)}\otimes c_{(2)}\\
&= (\delta\circ\beta)(h\otimes k\otimes c)
\end{aligned}$$

and

$$\begin{aligned}
(\delta\circ\alpha)(h\otimes k\otimes c) &= \sum((hS(k_{(1)})k_{(2)})\rightharpoonup c_{(1)})\otimes(k_{(3)}\rightharpoonup c_{(2)})\\
&= \sum(h\rightharpoonup c_{(1)})\otimes(k\rightharpoonup c_{(2)})\\
&= ((I\otimes\psi_C)\circ\delta')(h\otimes k\otimes c)
\end{aligned}$$

Now δ and δ' are isomorphisms, and the two columns in the above diagram are exact. It therefore follows that δ descends to an isomorphism $H\cong D\otimes_H C$.
We still have to show that D is an H-Galois coobject. It is clear that D is faithfully flat as an R-module, since $D=C$ as an R-module, and it therefore suffices to show that

$$\delta_D:\ H\otimes D\longrightarrow D\otimes D:\ h\otimes d\mapsto\sum(S(h)\rightharpoonup d_{(2)})\otimes d_{(1)}$$

is an isomorphism.
We first show that δ is surjective. Take $d\otimes e\in D\otimes D$, and let $\delta_C^{-1}(e\otimes d)=\sum_i h_i\otimes c_i\in H\otimes C$. Then

$$\begin{aligned}
d\otimes e &= \sum_i c_{i_{(2)}}\otimes h_i\rightharpoonup c_{i_{(1)}}\\
&= \sum_i (S(h_{i_{(3)}})h_{i_{(2)}})\rightharpoonup c_{i_{(2)}}\otimes h_{i_{(1)}}\rightharpoonup c_{i_{(1)}}\\
&= \delta_D\Big(\sum_i h_{i_{(2)}}\otimes h_{i_{(1)}}\rightharpoonup c_i\Big)
\end{aligned}$$

Let us finally show that δ_D is injective. Suppose that

$$\delta_D(\sum_i \ell_i\otimes d_i)=\sum_i S(\ell_i)\rightharpoonup d_{i_{(2)}}\otimes d_{i_{(1)}}=0$$

Then

$$\begin{aligned}
0 &= \sum_i d_{i_{(1)}}\otimes S(\ell_i)\rightharpoonup d_{i_{(2)}}\\
&= \sum_i (\ell_{i_{(3)}}S(\ell_{i_{(2)}}))\rightharpoonup d_{i_{(1)}}\otimes S(\ell_{i_{(1)}})\rightharpoonup d_{i_{(2)}}\\
&= \delta_C\Big(\sum_i \ell_{i_{(2)}}\otimes S(\ell_{i_{(1)}})\rightharpoonup d_i\Big)
\end{aligned}$$

Now δ_C is injective, and therefore

$$\sum_i \ell_{i_{(2)}} \otimes S(\ell_{i_{(1)}}) \rightharpoonup d_i = 0$$

Applying $\tau \circ \Delta_H$ to the first factor, we obtain

$$\sum_i \ell_{i_{(3)}} \otimes \ell_{i_{(2)}} \otimes S(\ell_{i_{(1)}}) \rightharpoonup d_i = 0$$

Now let the second factor act on the third one. This yields that

$$\sum_i \ell_i \otimes d_i = 0$$

and it follows that δ_D is injective. □

It follows from Corollary 8.7.7, Proposition 10.1.1 and Proposition 10.1.2 that the set of isomorphism classes of H-Galois coobjects is a group under the operation induced by the tensor product over H. H is the unit element of this group. We call this group the *group of H-Galois coobjects*, and we denote it by $\mathrm{Gal}^{co}(R, H)$. We may also state this as follows: let $\underline{\underline{\mathrm{Gal}}}^{co}(R, H)$ be the category of H-Galois coobjects and H-module coalgebra isomorphisms. The tensor product over H is a product on this category, and $\mathrm{Gal}^{co}(R, H)$ is its Grothendieck group:

$$\mathrm{Gal}^{co}(R, H) = \mathrm{K}_0\underline{\underline{\mathrm{Gal}}}^{co}(R, H)$$

In the next Lemma, we compute the Whitehead group of $\underline{\underline{\mathrm{Gal}}}^{co}(R, H)$.

Lemma 10.1.3

$$\mathrm{K}_1\underline{\underline{\mathrm{Gal}}}^{co}(R, H) = G(H)$$

Proof From Proposition 10.1.2, it follows that H is a cofinal object of $\underline{\underline{\mathrm{Gal}}}^{co}(R, H)$. We therefore have that

$$\mathrm{K}_1\underline{\underline{\mathrm{Gal}}}^{co}(R, H) = \mathrm{Aut}_{R\text{-coalg},H\text{-mod}}(H)$$

Now $\mathrm{Aut}_{H\text{-mod}}(H) \cong \mathbb{G}_m(H)$, so we have to investigate when left multiplication by an invertible element g of H is comultiplicative. Let $f : H \to H$ be given by left multiplication by $g \in \mathbb{G}_m(H)$. Then

$$(f \otimes f)\Delta(h) = \sum gh_{(1)} \otimes gh_{(2)} \quad \text{and} \quad \Delta(f(h)) = \sum g_{(1)}h_{(1)} \otimes g_{(2)}h_{(2)}$$

for all $h \in H$. It follows immediately that f is comultiplicative if and only if g is grouplike, proving the Lemma. □

As usual, $\underline{\underline{\mathrm{Pic}}}(H)$ will be the category of invertible H-modules and H-linear isomorphisms, and $\mathrm{Pic}(H) = \mathrm{K}_0\underline{\underline{\mathrm{Pic}}}(H)$ is the Picard group of H. It is well-known that $\mathrm{K}_1\underline{\underline{\mathrm{Pic}}}(H) = \mathbb{G}_m(H)$.

From Proposition 10.1.2, it follows that an H-Galois coobject is invertible as an H-module. The functor F forgetting the coalgebra structure is therefore a functor

$$F:\ \underline{\underline{\mathrm{Gal}}}^{\mathrm{co}}(R,H) \longrightarrow \underline{\underline{\mathrm{Pic}}}(H)$$

it is clear that this functor is cofinal, and the exact sequence (C.8) takes the form

$$\mathrm{K}_1\underline{\underline{\mathrm{Gal}}}^{\mathrm{co}}(R,H) \longrightarrow \mathrm{K}_1\underline{\underline{\mathrm{Pic}}}(H) \xrightarrow{d} \mathrm{K}_0\underline{\Psi F} \xrightarrow{f} \mathrm{K}_0\underline{\underline{\mathrm{Gal}}}^{\mathrm{co}}(R,H) \longrightarrow \mathrm{K}_0\underline{\underline{\mathrm{Pic}}}(H) \qquad (10.1)$$

In the next Lemma, we will show that the middle term of this exact sequence consists of Harrison 2-cocycles.

Lemma 10.1.4 *With notations as above, we have*

$$\mathrm{K}_0\underline{\Psi F} \cong Z^2_{\mathrm{Harr}}(H,R,\mathbb{G}_m)$$

Proof Consider $x = [(C,\kappa)] \in \mathrm{K}_0\underline{\Psi F}$, with C an H-Galois coobject, and $\gamma:\ C \to H$ an H-linear isomorphism. We identify C and H using γ, that is, we put C equal to H as an H-module, but with a different comultiplication Δ_C. We can therefore write $x = [(C, I_H)]$, and we define

$$\alpha:\ \mathrm{K}_0\underline{\Psi F} \longrightarrow Z^2_{\mathrm{Harr}}(H,R,\mathbb{G}_m)$$

as follows:

$$\alpha(x) = \Delta_C(1)$$

Let us show that α is a well-defined map. We can view α as a map $\mathrm{K}_0\underline{\Psi F} \longrightarrow H\otimes H$. It is clear that α is multiplicative: if $x = [(C,I_H)]$ and $y = [(D,I_H)]$, then $xy = [(C\otimes_H D, I_H)]$, and

$$\alpha(xy) = \Delta_{C\otimes_H D}(1\otimes 1) = \Delta_C(1)\Delta_D(1) = \alpha(x)\alpha(y)$$

Since every element $x \in \mathrm{K}_1\underline{\Phi F}$ has an inverse, it now follows that $\alpha(x) \in \mathbb{G}_m(H\otimes H)$. Let us show that $u = \alpha(x)$ is a cocycle. Remark first that for all $c \in C$, we have that

$$\Delta_C(c) = \Delta_C(c.1) = (\sum c_{(1)}\otimes c_{(2)})\Delta_C(1) = \Delta_H(c)\Delta_C(1)$$

Write $u = \sum_i u_i \otimes v_i$. Then

$$\begin{aligned}
(\Delta_C \otimes I_C)(u) &= \sum_i \Delta_C(u_i)\otimes v_i \\
&= \sum_{i,j} u_{i_{(1)}}u_j \otimes u_{i_{(2)}}v_j \otimes v_i \\
&= \sum_{i,j} (u_{i_{(1)}} \otimes u_{i_{(2)}} \otimes 1)(u_j\otimes v_j\otimes 1) \\
(I_C\otimes \Delta_C)(u) &= \sum_i u_i \otimes \Delta_C(v_i) \\
&= \sum_{i,j} u_i \otimes v_{i_{(1)}}u_j \otimes v_{i_{(2)}}v_j \\
&= \sum_{i,j} (1\otimes v_{i_{(1)}} \otimes v_{i_{(2)}})(u_j\otimes v_j\otimes 1)
\end{aligned}$$

and it follows from the coassociativity of C that u is a cocycle.
We now define a map

$$\beta : Z^2_{\mathrm{Harr}}(H, R, \mathbb{G}_m) \longrightarrow \mathrm{K}_0\underline{\Psi F}$$

For $u \in Z^2_{\mathrm{Harr}}(H, R, \mathbb{G}_m)$, we take $C = H$ as an H-module, with comultiplication given by

$$\Delta_C(c) = u\Delta_H(h)$$

From the fact u is a cocycle, it follows that C is coassociative. Let us show that C has a counit. Since $u = \sum_i u_i \otimes v_i$ is a cocycle, we have that

$$\sum_{i,j} u_j u_{i_{(1)}} \otimes v_j u_{i_{(2)}} \otimes v_i = \sum_{i,j} u_i \otimes u_j v_{i_{(1)}} \otimes v_j v_{i_{(2)}}$$

Apply $I_H \otimes \varepsilon_H \otimes I_H$ to both sides to obtain

$$\sum_{i,j} u_j u_i \otimes \varepsilon(v_j) v_i = \sum_{i,j} u_i \otimes \varepsilon(u_j) v_j v_i$$

or

$$(1 \otimes \sum_j \varepsilon(v_j) u_j) u = (1 \otimes \sum_j \varepsilon(u_j) v_j) u$$

or, since u is invertible

$$\sum_j \varepsilon(v_j) u_j = \sum_j \varepsilon(u_j) v_j$$

Observe that $\sum_j \varepsilon(v_j) u_j$ is invertible in H, since $\varepsilon_H \otimes I_H$ is a multiplicative map. We now define

$$\varepsilon_C(c) = (\sum_j \varepsilon(v_j) u_j)^{-1} \varepsilon_H(c) = (\sum_j \varepsilon(u_j) v_j)^{-1} \varepsilon_H(c)$$

It is straightforward to show that ε_C is a counit, and it follows that C is a coalgebra. Left multiplication by elements of H makes C into an H-module coalgebra. Let us show that C is an H-Galois coobject. It suffices to show that

$$\delta_C : H \otimes C \longrightarrow C \otimes C : h \otimes c \mapsto \sum h u_i c_{(1)} \otimes v_i c_{(2)} = u \delta_H (h \otimes c)$$

is an isomorphism. This is obvious, since u is invertible. Now define β as follows:

$$\beta(u) = [(C, I_H)]$$

with C defined as above. It is straightforward to show that α and β are each others inverses, and this proves the Lemma. □

Lemma 10.1.5 *The diagram*

$$\begin{array}{ccc} \mathbb{G}_m(H) & \xrightarrow{\Delta_1} & Z^2_{\mathrm{Harr}}(H, R, \mathbb{G}_m) \\ \downarrow \cong & & \downarrow \beta \\ \mathrm{K}_1\underline{\underline{\mathrm{Pic}}}(H) & \xrightarrow{d} & \mathrm{K}_0\underline{\Psi F} \end{array}$$

is anticommutative.

Proof Take $g \in \mathbb{G}_m(H)$. The image of g in $\mathrm{K}_1\underline{\underline{\mathrm{Pic}}}(H)$ is represented by (H, g), and $d[(H,g)] = [(H,g)] = [(C, I_H)]$, where $C = H$ as an H-module, but with comultiplication given by

$$\Delta_C(h) = \sum g^{-1}(gh)_{(1)} \otimes g^{-1}(gh)_{(2)}$$

On the other hand, $\Delta_1(g) = (1 \otimes g)(\sum g_{(1)} \otimes g_{(2)})^{-1}(g \otimes 1)$, and therefore $\beta(\Delta_1(g)) = [(D, I_H)]$, where $D = H$ as an H-module, but with comultiplication given by

$$\Delta_D(h) = \sum (g \otimes g)\Delta_H(g)^{-1}\Delta_H(h)$$

and it follows that $\beta(\Delta_1(g)) = d[(H,g)]^{-1}$. □

Theorem 10.1.6 *Let H be a commutative faithfully flat Hopf algebra. Then we have exact sequences*

$$1 \longrightarrow G(H) \longrightarrow \mathbb{G}_m(H) \xrightarrow{\Delta_1} Z^2_{\mathrm{Harr}}(H, R, \mathbb{G}_m) \xrightarrow{\beta} \mathrm{Gal}^{\mathrm{co}}(R, H) \xrightarrow{\gamma} \mathrm{Pic}(H) \qquad (10.2)$$

and

$$1 \longrightarrow H^2_{\mathrm{Harr}}(H, R, \mathbb{G}_m) \xrightarrow{\beta} \mathrm{Gal}^{\mathrm{co}}(R, H) \xrightarrow{\gamma} H^1_{\mathrm{Harr}}(H, R, \mathrm{Pic}) \qquad (10.3)$$

Proof (10.2) follows immediately if we substitute the above results in the exact sequence (10.1). (10.3) follows if we can show that $\mathrm{Im}(\gamma) \subset H^1_{\mathrm{Harr}}(H, R, \mathrm{Pic})$, in other words, an H-Galois coobject C viewed as an invertible H-module is a Harrison cocycle.
Observe that, with notations as in Section 9.2,

$$\begin{aligned}\mathrm{Pic}(\varepsilon_0)(C) &= (H \otimes H) \otimes_{\varepsilon_0} C = H \otimes C \\ \mathrm{Pic}(\varepsilon_2)(C) &= (H \otimes H) \otimes_{\varepsilon_2} C = C \otimes H\end{aligned}$$

and therefore

$$\mathrm{Pic}(\varepsilon_0)(C) \otimes_{H \otimes H} \mathrm{Pic}(\varepsilon_2)(C) \cong C \otimes C$$

On the other hand

$$\mathrm{Pic}(\varepsilon_1)(C) = (H \otimes H) \otimes_{\varepsilon_1} C$$

is generated by monomials of the form

$$(h \otimes k) \otimes_{\varepsilon_1} c$$

subject to the relation

$$(h \otimes k) \otimes_{\varepsilon_1} l \cdot c = \sum (hl_{(1)} \otimes kl_{(2)}) \otimes_{\varepsilon_1} c$$

for all $h, k, l \in H$ and $c \in C$. Consider the map

$$\theta : (H \otimes H) \otimes_{\varepsilon_1} C \longrightarrow H \otimes C$$

given by

$$\theta((h \otimes k) \otimes_{\varepsilon_1} c) = \sum hS(k_{(1)}) \otimes k_{(2)}c$$

θ is well-defined, since

$$\begin{aligned}\theta(\sum(hl_{(1)} \otimes kl_{(2)}) \otimes_{\varepsilon_1} c) &= \sum hl_{(1)}S(l_{(2)})S(k_{(1)}) \otimes k_{(2)}l_{(3)}c \\ &= \sum hS(k_{(1)}) \otimes k_{(2)}l_{(3)}c \\ &= \theta((h \otimes k) \otimes_{\varepsilon_1} lc)\end{aligned}$$

θ is an isomorphism of R-modules. Its inverse is given by

$$\theta^{-1}(h \otimes c) = (h \otimes 1) \otimes_{\varepsilon_1} c$$

Indeed,

$$\begin{aligned}\theta^{-1}(\theta((h \otimes k) \otimes_{\varepsilon} c)) &= \sum(hS(k_{(1)}) \otimes 1) \otimes_{\varepsilon_1} k_{(2)}c \\ &= \sum(hS(k_{(1)})k_{(2)} \otimes k_{(3)}) \otimes_{\varepsilon_1} c \\ &= (h \otimes k) \otimes_{\varepsilon} c\end{aligned}$$

and

$$\theta(\theta^{-1}(h \otimes c)) = \theta((h \otimes 1) \otimes_{\varepsilon_1} c) = h \otimes c$$

It follows that

$$\delta_C \circ \theta : (H \otimes H) \otimes_{\varepsilon_1} C \longrightarrow C \otimes C$$

is an isomorphism of R-modules. $\delta_C \circ \theta$ can be described explicitly as follows:

$$\begin{aligned}(\delta_C \circ \theta)((h \otimes 1) \otimes_{\varepsilon_1} c) &= \delta_C(\sum hS(k_{(1)}) \otimes k_{(2)}c) \\ &= \sum hS(k_{(1)})k_{(2)}c_{(1)} \otimes k_{(3)}c_{(2)} \\ &= (h \otimes k)\Delta_C(c) \qquad (10.4)\end{aligned}$$

From (10.4), it follows easily that $\delta_C \circ \theta$ is $H \otimes H$-linear, and this finishes our proof. □

$\mathrm{Gal}^{\mathrm{co}}(R, H)$ is now described completely if we can add one more term to the long exact sequence (10.3). The obvious candidate for this next term is the third Harrison cohomology group $H^3_{\mathrm{Harr}}(H, R, \mathbb{G}_m)$. If H is finite, then this works: the exact sequence (10.3) (or at least a dual version of it), with the H^3-term added to it is then exact. This was shown independently by Early and Kreimer [81] and Yokogawa [204]. In the general case, we are only able to describe a subgroup of $\mathrm{Gal}^{\mathrm{co}}(R, H)$, that coincides with the full $\mathrm{Gal}^{\mathrm{co}}(R, H)$ if H is finite. This is what we will be doing in the sequel.

10.2 Galois coobjects with geometric normal basis

A Galois coobject C has normal basis if $C \cong H$ as an H-module, or, equivalently, if $[C] \in \mathrm{Ker}\,(\gamma)$ in Theorem 10.1.6. It follows from Theorem 10.1.6 that $\mathrm{Gal}^{\mathrm{co}}_{\mathrm{nb}}(R, H)$, the subgroup of $\mathrm{Gal}^{\mathrm{co}}(R, H)$ consisting of Galois objects with a normal basis, is isomorphic to the second Harrison cohomology group $H^2_{\mathrm{Harr}}(H, R, \mathbb{G}_m)$. This statement, which is much older then the exact sequence (10.3) is known as the *normal basis Theorem*, and goes back to several authors, cf. for example [16], [119] and [140]. We now introduce the following geometric version of Galois object with a normal basis.

Definition 10.2.1 *Let A be a commutative faithfully flat R-algebra, and H a commutative faithfully flat Hopf algebra. An invertible A-module I has a geometric normal basis if there exists a faithfully flat commutative R-algebra S such that $I \otimes S \cong A \otimes S$ as S-modules. An H-Galois coobject C has a geometric normal basis if it has a geometric normal basis as an invertible H-module.*

Obviously, the subsets of $\mathrm{Gal}^{\mathrm{co}}(R,H)$ and $\mathrm{Pic}(A)$ consisting of isomorphism classes of objects with geometric normal basis are subgroups. These subgroups will be denoted by $\mathrm{Gal}^{\mathrm{co}}_{\mathrm{gnb}}(R,H)$ and $\mathrm{Pic}_{\mathrm{gnb}}(R,H)$. We have the following inclusions:

$$\mathrm{Gal}^{\mathrm{co}}_{\mathrm{nb}}(R,H) \subset \mathrm{Gal}^{\mathrm{co}}_{\mathrm{gnb}}(R,H) \subset \mathrm{Gal}^{\mathrm{co}}(R,H)$$

$$\mathrm{Pic}_{\mathrm{gnb}}(R,A) \subset \mathrm{Pic}(A)$$

Lemma 10.2.2 *If A (resp. H) is faithfully projective as an R-module, then*

$$\mathrm{Gal}^{\mathrm{co}}_{\mathrm{gnb}}(R,H) = \mathrm{Gal}^{\mathrm{co}}(R,H)$$

and

$$\mathrm{Pic}_{\mathrm{gnb}}(R,A) = \mathrm{Pic}(A)$$

Proof Let I be an invertible H-module, and take $p \in \mathrm{Spec}(R)$. Then $H_p = H \otimes R_p$ is a finitely generated projective R_p-algebra and is therefore semilocal. Thus $I \otimes R_p$ is free of rank one as an H_p-module. A standard argument now shows that there is a Zariski covering $S = R_{f_1} \times \cdots \times R_{f_n}$ of R such that $I \otimes S$ is free of rank one as an $H \otimes S$-module. □

With notations as in Theorem 10.1.6, we have that

$$\mathrm{Gal}^{\mathrm{co}}_{\mathrm{gnb}}(R,H) = \gamma^{-1}(\mathrm{Pic}^{\mathrm{co}}_{\mathrm{gnb}}(R,H))$$

and

$$\mathrm{Im}\,(\beta) = \mathrm{Gal}^{\mathrm{co}}_{\mathrm{nb}}(R,H) \subset \mathrm{Gal}^{\mathrm{co}}_{\mathrm{gnb}}(R,H)$$

The exact sequence (10.3) therefore restricts to an exact sequence

$$1 \longrightarrow H^2_{\mathrm{Harr}}(H,R,\mathbb{G}_m) \xrightarrow{\beta} \mathrm{Gal}^{\mathrm{co}}_{\mathrm{gnb}}(R,H) \longrightarrow H^1_{\mathrm{Harr}}(H,R,\mathrm{Pic}_{\mathrm{gnb}}(R,\bullet))$$

Before extending this sequence, let us state the following technical Lemma.

Lemma 10.2.3 *Suppose that S is a faithfully flat R-algebra, and that C is an H-module coalgebra. If $S \otimes C$ is an $S \otimes H$-Galois coobject, then A is an H-Galois coobject.*

Proof C is a faithfully flat R-module, because $S\otimes C$ is a faithfully flat S-module, and S is a faithfully flat R-algebra. Furthermore

$$\delta_S:\ (S\otimes H)\otimes_S(S\otimes C)=S\otimes(H\otimes C)\longrightarrow(S\otimes C)\otimes_S(S\otimes C)=S\otimes(C\otimes C)$$

defined by

$$\delta_S(s\otimes(h\otimes c))=\sum s\otimes(h\rightharpoonup c_{(1)})\otimes c_{(2)}$$

is an isomorphism of S-modules. The fact that S is a faithfully flat R-algebra implies that

$$\delta:\ H\otimes C\longrightarrow C\otimes S$$

is an isomorphism of R-modules. □

Theorem 10.2.4 *Let H be a commutative faithfully flat Hopf algebra. Then we have the following exact sequence*

$$1\longrightarrow H^2_{\rm Harr}(H,R,\mathbb{G}_m)\xrightarrow{\beta}{\rm Gal}^{\rm co}_{\rm gnb}(R,H)\xrightarrow{\gamma}H^1_{\rm Harr}(H,R,{\rm Pic}_{\rm gnb}(R,\bullet))\xrightarrow{\delta}H^3_{\rm Harr}(H,R,\mathbb{G}_m) \tag{10.5}$$

Proof

Definition of the map δ.

Take a cocycle $C\in Z^1_{\rm Harr}(H,R,{\rm Pic}_{\rm gnb}(R,\bullet))$. We have an isomorphism

$$f:\ H^{\otimes 2}\otimes_{\varepsilon_1}C\longrightarrow C\otimes C$$

of $H^{\otimes 2}$-modules. Consider the following maps.

$$\hat{u}:\ C\longrightarrow C\otimes C$$

given by

$$\hat{u}(c)=f((1\otimes 1)\otimes_{\varepsilon_2}c)$$

and

$$\zeta_1,\ \zeta_2:\ H^{\otimes 2}\otimes C\longrightarrow C^{\otimes 3}$$

given by

$$\begin{aligned}\zeta_1(h\otimes k\otimes c)&=(h\otimes k\otimes 1)\rightharpoonup(((\hat{u}\otimes I_C)\circ\hat{u})(c))\\ \zeta_2(h\otimes k\otimes c)&=(h\otimes k\otimes 1)\rightharpoonup(((I_C\otimes\hat{u})\circ\hat{u})(c))\end{aligned}$$

It is clear that $\hat{u}$ makes C into a coassociative coalgebra if and only if $\zeta_1=\zeta_2$. Suppose for a moment that $C\cong H$ as an H-module. Then for all $h\in H$, we have

$$\begin{aligned}\hat{u}(h)&=f((1\otimes 1)\otimes_{\varepsilon_1}h)\\ &=f((\sum h_{(1)}\otimes h_{(2)})\otimes_{\varepsilon_1}1)\\ &=(\sum h_{(1)}\otimes h_{(2)})f((1\otimes 1)\otimes_{\varepsilon_1}1)\\ &=\hat{u}(1)\Delta_H(h)\end{aligned}$$

Now write

$$u=\hat{u}(1)=\sum u^1\otimes u^2=\sum U^1\otimes U^2$$

and let

$$f^{-1}(1\otimes 1)=v\otimes_{\varepsilon_2}1$$

with

$$v=\sum v^1\otimes v^2=\sum V^1\otimes V^2\in H\otimes H$$

Then $1\otimes 1=f(f^{-1}(1\otimes 1))=uv$, and $v=u^{-1}$. Observe next that the map

$$\alpha:\ H^{\otimes 3}\longrightarrow H^{\otimes 3}:\ h\otimes k\otimes l\mapsto \sum hl_{(1)}\otimes kl_{(2)}\otimes l_{(3)}$$

is bijective; the inverse of α is given by the formula

$$\alpha^{-1}(h\otimes k\otimes l)=\sum hS(l_{(2)})\otimes kS(l_{(1)})\otimes l_{(3)}$$

We now have that

$$\begin{aligned}
&\zeta_1(h\otimes k\otimes l)\\
&\quad=\sum(h\otimes k\otimes 1)(\hat{u}\otimes I_H)(h^1l_{(1)}\otimes u^2l_{(2)})\\
&\quad=\sum(h\otimes k\otimes 1)(U^1u^1_{(1)}l_{(1)}\otimes U^2u^1_{(2)}l_{(2)}\otimes u^2l_{(3)})\\
&\quad=\sum(U^1\otimes U^2\otimes 1)(u^1_{(1)}\otimes u^1_{(2)}\otimes u_{(2)})(hl_{(1)}\otimes kl_{(2)}\otimes l_{(3)})\\
&\quad=\varepsilon_3(u)\varepsilon_1(u)\alpha(h\otimes k\otimes l)
\end{aligned}$$

and, in a similar way,

$$\zeta_2(h\otimes k\otimes l)=\varepsilon_0(u)\varepsilon_2(u)\alpha(h\otimes k\otimes l)$$

Write $m(\varepsilon_i(u))$ for the map given by multiplication by $\varepsilon_i(u)$. Then

$$\begin{aligned}
\zeta_2&=m(\varepsilon_0(u))\circ m(\varepsilon_2(u))\circ\alpha\\
\zeta_1^{-1}&=\alpha^{-1}\circ m(\varepsilon_1(v))\circ m(\varepsilon_3(v))
\end{aligned}$$

and therefore

$$\zeta_2\circ\zeta_1^{-1}=m(\varepsilon_0(u))\circ m(\varepsilon_2(u))\circ m(\varepsilon_1(v))\circ m(\varepsilon_{(3)}(v))=m(\Delta_{(2)}(u))$$

is given by multiplication by the coboundary $\Delta_2(u)$.
We now return to the general case. Let S be a faithfully flat extension of R such that $S\otimes C\cong S\otimes H$ as $S\otimes H$-modules. The map

$$\zeta_1\otimes I_S:\ C^{\otimes 3}\otimes S\longrightarrow H^{\otimes 2}\otimes C\otimes S$$

is bijective (see above), and this implies that ζ_1 is also bijective (S is faithfully flat). Consider the map

$$\zeta_2\circ\zeta_1^{-1}:\ C^{\otimes 3}\xrightarrow{\zeta_1^{-1}}H^{\otimes 2}\otimes C\xrightarrow{\zeta_2}C^{\otimes 3}$$

Then the map $I_S\otimes(\zeta_2\circ\zeta_1^{-1})$ is given by multiplication by a coboundary in $B^3_{\rm Harr}(S\otimes H,S,\mathbb{G}_m)$, and is an isomorphism of $S\otimes H^{\otimes 3}$-modules. $\zeta_2\circ\zeta_1^{-1}$ is therefore an isomorphism of (rank one) $H^{\otimes 3}$-modules, and is given by multiplication by a unit $x\in\mathbb{G}_m(H^{\otimes 3})$. Moreover $1_S\otimes x\in B^3_{\rm Harr}(S\otimes H,S,\mathbb{G}_m)\subset Z^3_{\rm Harr}(S\otimes H,S,\mathbb{G}_m)$ is a cocycle, and thus x is a cocycle in $Z^3_{\rm Harr}(H,R,\mathbb{G}_m)$. We define $\delta([C])=[x]$. We leave it to the reader to show that δ is well-defined: if we repeat the above arguments with a different isomorphism $f' :\ H^{\otimes 2} \otimes_{\varepsilon_2} C \longrightarrow C^{\otimes 2}$ then we obtain a cocycle x' that is cohomologous to x.

Exactness at $H^1_{\rm Harr}(H, R, {\rm Pic}_{\rm gnb}(R, \bullet))$.

It is clear that $\delta \circ \gamma = 1$. If C is an H-Galois coobject, then we can choose the isomorphism

$$f :\ H^{\otimes 2} \otimes_{\varepsilon_2} C \longrightarrow C^{\otimes 2}$$

as follows:

$$f((h \otimes k) \otimes_{\varepsilon_2} c) = \sum h_{(1)}c_{(1)} \otimes k_{(2)}c_{(2)}$$

(see the end of the proof of Theorem 10.1.6). Now the map $\hat{u}$ defined above is nothing else then the comultiplication on Δ_C, and therefore $\zeta_1 = \zeta_2$ (C is coassociative), and $x = 1$.

Conversely, if $\delta([C]) = [x]$, with $x = \Delta_2(y^{-1})$ a coboundary, then we replace the isomorphism f by f' given by

$$f'((h \otimes k) \otimes_{\varepsilon_2} c) = yf((h \otimes k) \otimes_{\varepsilon_2} c)$$

Then it follows immediately that

$$\begin{aligned} \hat{u}'(c) &= y\hat{u}(c) \\ \zeta_1' &= m(\varepsilon_3(y)) \circ m(\varepsilon_1(y)) \circ \zeta_1 \\ \zeta_2' &= m(\varepsilon_0(y)) \circ m(\varepsilon_2(y)) \circ \zeta_1 \end{aligned}$$

and consequently

$$\zeta_2' \circ \zeta_1' = m(\Delta_2(y)) \circ \zeta_2 \circ \zeta_1 = 1$$

such that $\hat{u}'$ makes C into a coassociative coalgebra. Finally, observe that $S \otimes C$ is nothing else then $S \otimes H$ with comultiplication twisted by the Harrison cocycle $1_S \otimes \hat{u}'$. Therefore $S \otimes C$ is an $S \otimes H$-Galois coobject, and, by the previous proposition, C is an H-Galois coobject. □

We have already seen that an H-Galois coobject C is invertible as an H-module coalgebra, that is, there exists an H-module coalgebra D such that $C \otimes_H D \cong H$ as H-module coalgebras. For an H-module coalgebra with geometric normal basis, the converse also holds.

Corollary 10.2.5 *Let C and D be H-module coalgebras such that $C \otimes_H D \cong H$ as H-module coalgebras, and suppose that C (and therefore D) has a geometric normal basis. Then C and D are H-Galois coobjects.*

Proof From Proposition 10.2.3, it follows that we can assume that C and D have normal basis, that is, $C \cong D \cong H$ as H-modules. Write $\Delta_C(1) = u$ and $\Delta_D(1) = v$, and consider the canonical isomorphism $f :\ C \otimes_H D \to H :\ c \otimes d \mapsto cd$. The H-module coalgebra structure on $C \otimes_H D$ induces an H-module coalgebra structure on H. The new comultiplication is given by

$$\tilde{\Delta}(1) = uv$$

Let $\widetilde{H}$ be equal to H as an H-module with the new comultiplication $\tilde{\Delta}$. By assumption, $\widetilde{H}$ is isomorphic to H as an H-module coalgebra. From the exactness of the sequence (10.3), it follows that $uv \in B^2(H, R, \mathbb{G}_m)$ is a coboundary. It follows in particular that u and v are invertible. From the fact that C and D are coassociative, it follows that u and v are Harrison cocycles, and we already know that in this case C and D are H-Galois coobjects. □

We conclude this Section with the remark that the exact sequence (10.5) can be identified with the (first part of) the sequence (9.5) in the case where $S = R$ (see also Lemma 9.2.2), at least if we replace the functor $\underline{\underline{\mathrm{Pic}}}$ systematically by $\underline{\underline{\mathrm{Pic}}}_{\mathrm{gnb}}(R, \bullet)$. It suffices to find a map

$$\varphi:\ \mathrm{Gal}^{\mathrm{co}}_{\mathrm{gnb}}(R, H) \longrightarrow H^1_{\mathrm{Harr}}(H, R, \underline{\underline{\mathrm{Pic}}}(R, \bullet))$$

such that the diagram

$$\begin{array}{ccccc}
1 \longrightarrow & H^2_{\mathrm{Harr}}(H, R, \mathbb{G}_m) & \longrightarrow & \mathrm{Gal}^{\mathrm{co}}_{\mathrm{gnb}}(R, H) & \\
& \cong & & \downarrow \varphi & \\
1 \longrightarrow & H^2_{\mathrm{Harr}}(H, R, \mathbb{G}_m) & \longrightarrow & H^1_{\mathrm{Harr}}(H, R, \underline{\underline{\mathrm{Pic}}}_{\mathrm{gnb}}(R, \bullet)) & \\
\\
\longrightarrow & H^1_{\mathrm{Harr}}(H, R, \mathrm{Pic}_{\mathrm{gnb}}(R, \bullet)) & \longrightarrow & H^3_{\mathrm{Harr}}(H, R, \mathbb{G}_m) & \\
& \cong & & \cong & \\
\longrightarrow & H^1_{\mathrm{Harr}}(H, R, \mathrm{Pic}_{\mathrm{gnb}}(R, \bullet)) & \longrightarrow & H^3_{\mathrm{Harr}}(H, R, \mathbb{G}_m) & \longrightarrow \cdots
\end{array}$$

commutes. For $[C] \in \mathrm{Gal}^{\mathrm{co}}_{\mathrm{gnb}}(R, H)$, we define

$$\varphi([C]) = [(C, \delta_C \circ \theta)]$$

with δ_C and θ as in the proof of Theorem 10.1.6. Arguments similar to the ones in the proof of Theorem 10.2.4 show that φ is a well-defined homomorphism. From the Lemma of 5, we now deduce the following result.

Proposition 10.2.6 *Let H be a commutative faithfully flat Hopf algebra. Then*

$$\mathrm{Gal}^{\mathrm{co}}_{\mathrm{gnb}}(R, H) \cong H^1_{\mathrm{Harr}}(H, R, \underline{\underline{\mathrm{Pic}}}_{\mathrm{gnb}}(R, \bullet))$$

10.3 The group of Galois coobjects and Amitsur cohomology

Let H be a commutative, faithfully flat Hopf algebra, and let A and S be faithfully flat commutative R-algebras. We have a natural map

$$\mathrm{Gal}^{\mathrm{co}}(R, H) \longrightarrow \mathrm{Gal}^{\mathrm{co}}(S, S \otimes H)$$

which behaves functorially. We now define the following subgroups of $\mathrm{Gal}^{\mathrm{co}}(R,H)$:

$$\mathrm{Gal}^{\mathrm{co}}(S/R,H) = \mathrm{Ker}\left(\mathrm{Gal}^{\mathrm{co}}(R,H) \longrightarrow \mathrm{Gal}^{\mathrm{co}}(S,S\otimes H)\right)$$

$$\mathrm{Gal}^{\mathrm{s,co}}(R,H) = \bigcup_S \mathrm{Gal}^{\mathrm{co}}(S/R,H)$$

where the union is taken over all faithfully flat extensions S of R. $\mathrm{Gal}^{\mathrm{s,co}}(R,H)$ is called the *split part* of $\mathrm{Gal}^{\mathrm{co}}(R,H)$. Observe that

$$\mathrm{Gal}^{\mathrm{s,co}}(R,H) \subset \mathrm{Gal}^{\mathrm{co}}_{\mathrm{gnb}}(R,H)$$

and

$$\mathrm{Pic}_{\mathrm{gnb}}(R,A) = \bigcup_S \mathrm{Pic}(S\otimes A/A)$$

In this Section, we will describe $\mathrm{Gal}^{\mathrm{s,co}}(R,H)$ and $\mathrm{Pic}_{\mathrm{gnb}}(R,A)$ using Amitsur cohomology. To this end, we first need an H-module version of the Theorem of faithfully flat descent, cf. Theorems B.0.10 and B.0.13.

Proposition 10.3.1 *Suppose that, in the setting of Theorem B.0.10, M is an $A\otimes S$-module, and that $u:\ S\otimes M \to M\otimes S$ is an $A\otimes S^{\otimes 2}$-linear descent datum. Then the descended module N is an A-module, and is unique as an A-module up to A-module isomorphisms. If M has the structure of $H\otimes S$-module (co)algebra, and u is an $H\otimes S^{\otimes 2}$-module (co)algebra isomorphism, then N has a unique structure of H-module (co)algebra such that $\eta:\ N\otimes S\to M$ is an $H\otimes S$-module (co)algebra isomorphism.*

Proof We use the notations of Theorem B.0.10. Let $N=\{x\in M|x\otimes 1 = u(1\otimes x)\}$ and $\eta:\ N\otimes S\to M:\ n\otimes s\mapsto sn$ (cf. (B.4)). The only thing we have to show in order to prove the first part is that the A-action on M restricts to an A-action on N. This is easy: for every $x\in N$ and $a\in A$, we have that

$$u(1\otimes ax) = au(1\otimes x) = a(x\otimes 1) = ax\otimes 1$$

since u is A-linear. Hence $ax\in N$. The uniqueness is easy: it suffices to observe that the isomorphism ρ constructed in the proof of Theorem B.0.10 is A-linear.
If M is an H-module (co)algebra and u is an $H\otimes S^{\otimes 2}$-module (co)algebra isomorphism, then we know from Theorems B.0.13 and B.0.14 that N is a (co)algebra (and an H-module). Since the compatibility relation holds for M, it also holds for N and N is an H-module (co)algebra. □

As an application, we can now give the following cohomological interpretation of $\mathrm{Gal}^{\mathrm{co}}(S/R,H)$ and $\mathrm{Pic}(A\otimes S/S)$.

Theorem 10.3.2 *With notations as above, we have*

$$\mathrm{Pic}(A\otimes S/S) \cong H^1(S/R,\mathbb{G}_m(A\otimes \bullet)) \tag{10.6}$$

$$\mathrm{Gal}^{\mathrm{co}}(S/R,H) \cong H^1(S/R,G(H\otimes \bullet)) \tag{10.7}$$

Proof In the terminology of [114, Sec. II.8], the objects of $\mathrm{Pic}(A \otimes S/S)$ and $\mathrm{Gal}^{\mathrm{co}}(S/R, H)$ are the twisted forms of A and H, viewed respectively as an A-module and an H-module coalgebra, under the extension S. Applying the fact that the A-module automorphisms of A are the units of A, and the H-module coalgebra automorphisms of H are given by the grouplike elements $G(H)$ (cf. Lemma 10.1.3), we obtain the two isomorphisms above as pointed sets. A careful inspection of the connecting isomorphisms yields that they are abelian group isomorphisms. □

Lemma 10.3.3 *The presheaves $\mathbb{G}_m(A \otimes \bullet)$ and $G(H \otimes \bullet)$ are sheaves on R_{fl}.*

Proof Let $f : S \to S'$ be a covering in R_{fl}. Then $A \otimes S \to A \otimes S'$ is also a covering, and the sequence

$$1 \longrightarrow \mathbb{G}_m(A\otimes S) \longrightarrow \mathbb{G}_m(A\otimes S') \rightrightarrows \mathbb{G}_m\Big((A\otimes S')\otimes_{A\otimes S}(A\otimes S')\Big) = \mathbb{G}_m\Big(A\otimes(S'\otimes_S S')\Big)$$

is exact, because $\mathbb{G}_m$ is a sheaf. This proves the first statement. Now consider the commutative diagram

$$\begin{array}{ccccccc}
 & & 1 & & 1 & & 1 \\
 & & \downarrow & & \downarrow & & \downarrow \\
1 & \longrightarrow & G(H \otimes S) & \longrightarrow & G(H \otimes S') & \rightrightarrows & G\big(H \otimes (S' \otimes_S S')\big) \\
 & & \downarrow & & \downarrow & & \downarrow \\
1 & \longrightarrow & \mathbb{G}_m(H \otimes S) & \longrightarrow & \mathbb{G}_m(H \otimes S') & \rightrightarrows & \mathbb{G}_m\big(H \otimes (S' \otimes_S S')\big)
\end{array}$$

We already know that the bottomrow is exact. In order to show that the toprow is exact, we need to prove that $(I_H \otimes f)(x) \in G(H \otimes S')$ implies that $x \in G(H \otimes S)$. This can be seen as follows:

$$\begin{aligned}
&\big((I_H \otimes f) \otimes (I_H \otimes f)\big)(x \otimes x) \\
&= (I_H \otimes f)(x) \otimes (I_H \otimes f)(x) \\
&= \Delta_{H\otimes S'}\big((I_H \otimes f)(x)\big) \\
&= \big((I_H \otimes f) \otimes (I_H \otimes f)\big)\big(\Delta_{H\otimes S'}(x)\big)
\end{aligned}$$

and this implies that $\Delta_{H\otimes S}(x) = x \otimes x$, because

$$(I_H \otimes f) \otimes (I_H \otimes f) : (H \otimes S) \otimes_S (H \otimes S) \longrightarrow (H \otimes S') \otimes_{S'} (H \otimes S')$$

is faithfully flat. □

If we now take inductive limits over all faithfully flat R-algebras in 10.3.2, then we obtain the following result, using Proposition 5.6.9.

Corollary 10.3.4 *With notations as above, we have*

$$\begin{aligned}
\mathrm{Pic}_{\mathrm{gnb}}(R, A) &= H^1(R_{\mathrm{fl}}, \mathbb{G}_m(A \otimes \bullet)) \\
\mathrm{Gal}^{\mathrm{co,s}}(R, H) &= H^1(R_{\mathrm{fl}}, G(H \otimes \bullet))
\end{aligned}$$

10.4 The Picard group of a coalgebra

Let H be a cocommutative Hopf algebra. In view of the results of the three previous Sections, we could expect that the set of isomorphism classes of H-Galois objects forms a group under the operation induced by the cotensor product over H, and that this group can be described using Sweedler cohomology. This is what we will be discussing in the sequel. First, we need a dual version of the Picard group: for a cocommutative coalgebra C, we will introduce the Picard group $\mathrm{Pic}^{co}(C)$. Some technical difficulties arise here. For example, it turns out that the cotensor product is not always associative, and this is related to the fact that the tensor product over a commutative ring is not left exact. A possible escape is to restrict attention to the situation where the groundring R is a field, and this is what is usually done in the literature, see [183]. In this Section, we work over an arbitrary commutative ring. The theory will be saved by imposing the appropriate flatness conditions. First we state some sufficient conditions for the cotensor product to be associative.

Lemma 10.4.1 *Let C be a flat R-coalgebra, M a right C-comodule, N a (C,C)-bicomodule and P a left C-comodule. If M or P is flat as an R-module, then*

$$(M\square_C N)\square_C P \cong M\square_C(N\square_C P)$$

Proof Assume that P is flat; we leave the proof in the case where M is flat to the reader. By definition, the sequence

$$1\longrightarrow M\square_C N\longrightarrow M\otimes N\longrightarrow M\otimes C\otimes N$$

is exact, and therefore

$$1\longrightarrow (M\square_C N)\otimes P\longrightarrow M\otimes N\otimes P\longrightarrow M\otimes C\otimes N\otimes P$$

is exact, and it follows that

$$(M\square_C N)\otimes P\cong M\square_C(N\otimes P)$$

where the (left) C-coaction on $N\otimes P$ is induced by the C-coaction on N. Now look at the commutative diagram

$$\begin{array}{ccccccc}
 & & 1 & & 1 & & 1\\
 & & \downarrow & & \downarrow & & \downarrow\\
1 & \longrightarrow & (M\square_C N)\square_C P & \longrightarrow & M\square_C N\otimes P & \rightrightarrows & M\square_C N\otimes C\otimes P\\
 & & \downarrow & & \downarrow & & \downarrow\\
1 & \longrightarrow & (M\otimes N)\square_C P & \longrightarrow & M\otimes N\otimes P & \rightrightarrows & M\otimes N\otimes C\otimes P\\
 & & \downdownarrows & & \downdownarrows & & \downdownarrows\\
1 & \longrightarrow & (M\otimes C\otimes N)\square_C P & \longrightarrow & M\otimes C\otimes N\otimes P & \rightrightarrows & M\otimes C\otimes N\otimes C\otimes P
\end{array}$$

The three rows and the second and third colums are exact, because of the definition of the cotensor product and flatness of P and C. A diagram chasing shows that the first column is also exact, and this shows that $(M\square_C N)\square_C P\cong M\square_C(N\square_C P)$. □

Lemma 10.4.2 *Let C be an R-coalgebra, and P a left C-comodule. If C and P are flat as R-modules, the the functor $\bullet\square_C P:\ C\text{-comod}\to R\text{-mod}$ is left exact.*

Proof Let $N\to M$ be a C-colinear monomorphism of right C-comodules. Look at the commutative diagram

$$\begin{array}{ccccc}
 & & 0 & & 0\\
 & & \downarrow & & \downarrow\\
0 & \longrightarrow & N\square_C P & \longrightarrow & M\square_C P\\
 & & \downarrow & & \downarrow\\
0 & \longrightarrow & N\otimes P & \longrightarrow & M\otimes P\\
 & & \downarrow\downarrow & & \downarrow\downarrow\\
0 & \longrightarrow & N\otimes C\otimes P & \longrightarrow & M\otimes C\otimes P
\end{array}$$

The columns are exact, by the definition of the cotensor product. The second and the third row are exact, since C and P are flat as R-modules. It follows easily that the first row is also exact. Observe that we cannot conclude from this type of argument that $\bullet\square_C P$ is right exact. $\square$

It is possible to develop a Morita type theory for comodules over coalgebras. In the case where we work over a field, this was done in [107] and [177].

Theorem 10.4.3 *Let C and D be flat R-coalgebras, P an R-flat (C,D)-bicomodule and Q an R-flat (D,C)-bicomodule. Assume that*

$$h:\ P\square_D Q\longrightarrow C \text{ and } g:\ D\longrightarrow Q\square_C P$$

are respectively a C-bicolinear and D-bicolinear, and that the diagrams

$$\begin{array}{ccc}
P & \xrightarrow{\cong} & P\square_D D\\
\downarrow\cong & & \downarrow I_P\square g\\
C\square_C P & \xleftarrow{h\square I_P} & P\square_D Q\square_C P
\end{array}
\quad\text{and}\quad
\begin{array}{ccc}
Q & \xrightarrow{\cong} & D\square_D Q\\
\downarrow\cong & & \downarrow g\square I_Q\\
Q\square_C C & \xleftarrow{I_Q\square h} & Q\square_C P\square_D Q
\end{array}
\qquad(10.8)$$

commute. Then the functor

$$\bullet\square_D Q:\ \text{comod-}D\longrightarrow\text{comod-}C$$

is a right adjoint of

$$\bullet\square_C P:\ \text{comod-}C\longrightarrow\text{comod-}D$$

If h and g are bijective, then we have an equivalence of categories. Similar results apply to categories of left C-comodules.

Proof The proof is formally dual to the proof of Theorem 1.1.3. For $d\in D$, we will write

$$g(d)=\sum d^Q\otimes d^P\in Q\square_C P$$

From the commutativity of the diagrams (10.8), it follows that

$$\sum p_{(-1)}\otimes p_{(0)} = \sum h(p_{(0)}\otimes p^Q_{(1)})\otimes p^P_{(1)} \tag{10.9}$$
$$\sum q_{(0)}\otimes q_{(1)} = \sum q^Q_{(-1)}\otimes h(q^P_{(-1)}\otimes q_{(0)}) \tag{10.10}$$

for all $p\in P$ and $q\in Q$. Take $M\in$ comod-C and $N\in$ comod-D, and define two maps

$$\mathrm{Hom}^C(N\square_D Q, M)\underset{\Phi}{\overset{\Psi}{\rightleftarrows}}\mathrm{Hom}^D(N, M\square_C P)$$

as follows. For a C-colinear $v:\ N\square_D Q\to M$ and a D-colinear $u:\ N\to M\square_C P$, $\Psi(v)$ and $\Phi(u)$ are defined as the following compositions.

$$\Psi(v):\ N\xrightarrow{\cong}N\square_D D\xrightarrow{I_N\square g}N\square_D Q\square_C P\xrightarrow{v\square I_P}M\square_C P$$
$$\Phi(u):\ N\square_D Q\xrightarrow{u\square I_Q}M\square_C P\square_D Q\xrightarrow{I_M\square h}M\square_C C\xrightarrow{\cong}M$$

For $n\in N$, we write

$$u(n)=\sum n^M\otimes n^P\in M\square_D P$$

Then

$$\Psi(v)(n) = \sum v(n_{(0)}\otimes n^Q_{(1)})\otimes n^P_{(1)} \tag{10.11}$$
$$\Phi(u)(\sum_i n_i\otimes q_i) = \sum_i n_i^M\varepsilon_C(h(n_i^P\otimes q_i)) \tag{10.12}$$

for all $n\in N$ and $\sum_i n_i\otimes q_i\in N\square_D Q$. From the fact that u is C-colinear, we deduce that

$$\sum n^M\otimes(n^P)_{(0)}\otimes(n^P)_{(1)}=\sum n^M_{(0)}\otimes n^P_{(0)}\otimes n_{(1)} \tag{10.13}$$

for all $n\in N$. We are done if we can show that Ψ and Φ are each others inverses. First,

$$\begin{aligned}
\Psi(\Phi(u))(n) &= \sum \Phi(u)(n_{(0)}\otimes n^Q_{(1)})\otimes n^P_{(1)}\\
&= \sum n^M_{(0)}\varepsilon_C(h(n^P_{(0)}\otimes n^Q_{(1)}))\otimes n^P_{(1)}\\
\text{(by (10.13))}\quad &= \sum n^M\varepsilon_C(h((n^P)_{(0)}\otimes(n^P)^Q_{(1)}))\otimes(n^P)^P_{(1)}\\
\text{(by (10.9))}\quad &= \sum n^M\varepsilon_C((n^P)_{(-1)})\otimes(n^P)_{(0)}\\
&= \sum n^M\otimes n^P=u(n)
\end{aligned}$$

Now let $v:\ N\square_D Q\longrightarrow M$ C-colinear, and take $\sum_i n_i\otimes q_i\in N\square_D Q$. Then

$$\sum_i n_{i_{(0)}}\otimes n_{i_{(1)}}\otimes q_i=\sum_i n_i\otimes q_{i_{(-1)}}\otimes q_{i_{(0)}} \tag{10.14}$$

and

$$\begin{aligned}
\Phi(\Psi(v))(\sum_i n_i\otimes q_i) &= \sum_i v(n_{i_{(0)}}\otimes n^Q_{i_{(1)}})\varepsilon_C(h(n^P_{i_{(1)}}\otimes q_i))\\
\text{(by (10.14))}\quad &= \sum_i v(n_i\otimes q^Q_{i_{(-1)}})\varepsilon_C(h(q^P_{i_{(-1)}}\otimes q_{i_{(0)}}))\\
\text{(by (10.10))}\quad &= \sum_i v(n_i\otimes q_{i_{(0)}})\varepsilon_C(q_{i_{(1)}})\\
&= v(\sum_i n_i\otimes q_i)
\end{aligned}$$

This finishes the proof of the first part. Suppose now that the maps h and g are bijective. For any right D-comodule N, the map

$$\Psi(I_{N\Box_D Q}):\ N\longrightarrow N\Box_D Q\Box_C P$$

is given by

$$\begin{aligned}\Psi(I_{N\Box_D Q})(n) &= \sum n_{(0)}\otimes n^Q_{(1)}\otimes n^P_{(1)}\\ &= \sum n_{(0)}\otimes g(n_{(1)})\end{aligned}$$

This means that

$$\Psi(I_{N\Box_D Q}):\ N\xrightarrow{\rho_N} N\Box_D D\xrightarrow{I\Box g} N\Box_D Q\Box_C P$$

is a composition of isomorphisms, and it is therefore itself an isomorphism. In a similar way, we have for every left C-comodule M that the map

$$\Phi(I_{M\Box_C P}):\ M\Box_C P\Box_D Q\longrightarrow M$$

is given by

$$\Phi(I_{M\Box_C P})(\sum_i m_i\otimes p_i\otimes q_i)=\sum_i m_i\varepsilon_C(h(p_i\otimes q_i))$$

and

$$\Phi(I_{M\Box_C P}):\ M\Box_C P\Box_D Q\xrightarrow{I_M\otimes h} M\Box_C C\xrightarrow{I_M\otimes\varepsilon_C} M$$

is the composition of isomorphisms. We have shown that $\Psi(I_{N\Box_D Q})$ and $\Phi(I_{M\Box_C P})$ are isomorphism, and this is exactly what is needed to have an equivalence of categories. □

In the situation of Theorem 10.4.3, we call $(C, D, P, Q, f = h^{-1}, g)$ a *strict Morita-Takeuchi context*. If M is an invertible module over a commutative ring R, then we have a strict Morita context (R, R, M, M^*, f, g), and isomorphism classes of invertible R-modules correspond to isomorphism classes of Morita equivalences of R with itself. This observation leads to the following definition. Let C be a faithfully flat cocommutative coalgebra over a commutative ring R. From the fact that C is cocommutative, it follows that right C-comodules are also left C-comodules. The left C-coaction is given by $\rho^l(m)=\sum m_{(1)}\otimes m_{(0)}$. Thus a left or right C-comodule is automatically a C-bicomodule. By abuse of language, we call a strict Morita-Takeuchi context $P = (C, C, P, P', f, f')$, with P and Q left (or right) C-comodules, an *invertible C-comodule.* If R is a field, then the module P' can be described explicitly in terms of P. This description involves a "*co-hom*"-functor, but is much more complicated then the "*hom*"-functor used in the classical situations. For details, we refer the reader to [177].

Theorem 10.4.4 *Let C be a cocommutative faithfully flat R-coalgebra. The isomorphism classes of invertible C-comodules form an abelian group under the operation induced by the cotensor product over C. This group is called the Picard group of C, and we denote it by* $\mathrm{Pic}^{co}(C)$.

Proof Take two strict Morita-Takeuchi contexts $P = (C, C, P, P', f, f')$ and $Q = (C, C, Q, Q', g, g')$. Let us first show that $P\Box_C Q$ is flat as an R-module. Take a monomorphism $M \to N$ of R-modules. Then the map $M \otimes P \to N \otimes P$ is injective because P is flat. From Lemma 10.4.2, it follows that $(M \otimes P)\Box_C Q \to (N \otimes P)\Box_C Q$ is injective. From the associativity of the cotensor product (Lemma 10.4.1), it now follows that $M \otimes (P\Box_C Q) \to N \otimes (P\Box_C Q)$ is injective. Thus the tensor product over $P\Box_C Q$ is left exact, and this implies that $P\Box_C Q$ is flat. In a similar way, we find that $P'\Box_C Q'$ is flat, and it is straightforward to check that $P\Box_C Q = (C, C, P\Box_C Q, P'\Box_C Q', f\Box g, f'\Box g')$ is a strict Morita-Takeuchi context. From Lemma 10.4.1, it follows that the cotensor product of strict Morita-Takeuchi contexts is associative (all the underlying modules are flat). The neutral element is (C, C, C, C, I_C, I_C), and the inverse of $P = (C, C, P, P', f, f')$ is $P' = (C, C, P', P, f', f)$. Finally, the cotensor product of invertible C-comodules is commutative since C is cocommutative. □

Let S be a faithfully flat commutative R-algebra. We have a natural isomorphism of groups

$$\mathrm{Pic}^{\mathrm{co}}(R, C) \longrightarrow \mathrm{Pic}^{\mathrm{co}}(S, C \otimes S)$$

The kernel of this homomorphism will be denoted by

$$\mathrm{Pic}^{\mathrm{co}}(S/R, C)$$

and

$$\mathrm{Pic}^{\mathrm{co}}_{\mathrm{gnb}}(R, C) \subset \mathrm{Pic}^{\mathrm{co}}(R, C)$$

will be the union of the $\mathrm{Pic}^{\mathrm{co}}(S/R, C)$, where S runs over all faithfully flat R-algebras. We will now give a cohomological description of $\mathrm{Pic}^{\mathrm{co}}_{\mathrm{gnb}}(R, C)$. First, we need some technical preliminary results.

Proposition 10.4.5 (faithfully flat descent for comodules and comodule (co)algebras) *Let C be a faithfully flat cocommutative R-coalgebra, and suppose that, with notations and conventions as in Theorem B.0.10, M is a $C\otimes S$-comodule, and $u : S\otimes M \longrightarrow M\otimes S$ is a colinear descent datum. Then the descended R-module N is a C-comodule, and is unique up to C-comodule isomorphism.*
If C is a Hopf algebra, M has the structure of $C \otimes S$-comodule (co)algebra, and u is a $C \otimes S^{\otimes 2}$-comodule (co)algebra isomorphism, then N has a unique structure of C-comodule (co)algebra such that $\eta : N \otimes S \to M$ is a $C \otimes S$-comodule (co)algebra isomorphism.

Proof The proof is similar to the proof of Proposition 10.3.1. We only prove the first part, and leave the rest to the reader. As in Theorem B.0.10, let

$$N = \{x \in M | x \otimes 1 = u(1 \otimes x)\}$$

$$\eta : N \otimes S \to M : n \otimes s \mapsto sn$$

We will show that the $C \otimes S$-coaction on M restricts to a C-coaction on N, that is

$$\rho_M : M \longrightarrow M \otimes C \otimes S$$

restricts to

$$\rho_N:\ N \longrightarrow N\otimes C$$

The map

$$\begin{aligned} u\otimes(\tau\otimes I_S):\ &(S\otimes M)\otimes_{S^{\otimes 2}}(S\otimes(C\otimes S))\cong S\otimes(M\otimes_S(C\otimes S))\\ \longrightarrow\ &(M\otimes S)\otimes_{S^{\otimes 2}}((C\otimes S)\otimes S)\cong(M\otimes_S(C\otimes S))\otimes S \end{aligned}$$

is a descent datum for $M\otimes_S(C\otimes S)$, and the descended module is $N\otimes C$. Thus

$$\begin{aligned} N\otimes C=\{\sum_i x_i\otimes c_i\otimes s_i\in M\otimes C\otimes S\ \Big|\ &\sum_i u(1\otimes x_i)\otimes\tau_3(1\otimes c_i\otimes s_i)\\ &=\sum(x_i\otimes 1)\otimes(c_i\otimes s_i\otimes 1) \text{ in } (M\otimes S)\otimes_S(C\otimes S\otimes S)\} \end{aligned}$$

Now M is a $C\otimes S$-comodule via $\rho_M:\ M\longrightarrow M\otimes_S(C\otimes S)$. Therefore $S\otimes M$ is an $S\otimes C\otimes S$-comodule via

$$I_S\otimes\rho_M=\rho_{S\otimes M}:\ S\otimes M\longrightarrow(S\otimes M)\otimes_{S\otimes S}(S\otimes C\otimes S)$$

and $M\otimes S$ is a $C\otimes S\otimes S$-comodule via

$$\rho_M\otimes I_S=\rho_{M\otimes S}:\ M\otimes S\longrightarrow(M\otimes S)\otimes_{S\otimes S}(C\otimes S\otimes S)$$

The fact that u is colinear means that

$$(u\circ\tau_3)\rho_{S\otimes M}(s\otimes m)=\rho_{M\otimes S}(u(s\otimes m))$$

for all $s\in S$ and $m\in M$. Take $x\in N$, and write

$$\rho_M(x)=\sum_i x_i\otimes c_i\otimes s_i$$

Then

$$\begin{aligned} \sum_i u(1\otimes x_i)\otimes\tau_3(1\otimes c_i\otimes s_i) &= (u\otimes\tau_3)\big(\rho_{S\otimes M}(1\otimes x)\big)\\ &= \rho_{M\otimes S}(u(1\otimes x))\\ &= \rho_{M\otimes S}(x\otimes 1)\\ &= \sum_i(x_i\otimes 1)\otimes(c_i\otimes s_i\otimes 1) \end{aligned}$$

and it follows that $\rho_M(x)\in N\otimes C$. □

Lemma 10.4.6 *Let C be a cocommutative coalgebra. Then we have a ring isomorphism*

$$\alpha:\ \mathrm{End}^C(C),+,\circ\longrightarrow C^*,+,*$$

given by

$$\alpha(f)=\varepsilon\circ f \text{ and } \alpha^{-1}(c^*)=I_C*c^*$$

for all C-colinear $f:\ C\otimes C$ and $c^*\in C^*$. α restricts to a group isomorphism

$$\alpha':\ \mathrm{Aut}^C(C),\circ\longrightarrow\mathbb{G}_m(C^*),*$$

If $C=H$ is a Hopf algebra, then α restricts further to

$$\alpha'':\ \mathrm{Aut}^H_{R\text{-alg}}(H),\circ\longrightarrow\mathrm{Alg}(H,R)$$

Here $\mathrm{Aut}^C(C)$ stands for C-colinear automorphisms of C, $\mathrm{Aut}^H_{R\text{-alg}}(H)$ for H-colinear R-algebra automorphisms of H and $\mathrm{Alg}(H,R)$ for algebra maps from H to R.

Proof Take $c^* \in C^*$. We will first show that $I_C * c^*$ is C-colinear, that is, $(I_C * c^* \otimes I_C) \circ \Delta_C = \Delta_C \circ (I_C * c^*)$. For all $c \in C$, we have

$$\begin{aligned} \Delta_C\big((I_C * c^*)(c)\big) &= \Delta_C(\sum c_{(1)}\langle c^*, c_{(2)}\rangle) \\ &= \sum c_{(1)}\langle c^*, c_{(3)}\rangle \otimes c_{(2)} \\ &= \sum c_{(1)}\langle c^*, c_{(2)}\rangle \otimes c_{(3)} \\ &= (I_C * c^* \otimes I_C)(\sum c_{(1)} \otimes c_{(2)}) \\ &= (I_C * c^* \otimes I_C)(\Delta_C(c)) \end{aligned}$$

α and α^{-1} are each others inverses: for all $f \in \mathrm{End}^C(C)$, $c^* \in C^*$ and $c \in C$, we have

$$\begin{aligned} \alpha^{-1}(\alpha(f))(c) &= (I_C * (\varepsilon \circ f))(c) \\ &= \sum c_{(1)} \otimes \varepsilon(f(c_{(2)})) \\ &= (I_C * \varepsilon)(I_C \otimes f)(c) \\ &= (I_C * \varepsilon)\Delta_C(f(c)) \\ &= \sum f(c)_{(1)}\varepsilon(f(c)_{(2)}) = f(c) \end{aligned}$$

and

$$\begin{aligned} \langle \alpha(\alpha^{-1}(c^*)), c\rangle &= \varepsilon((I_C * c^*(c)) \\ &= \varepsilon(\sum c_{(1)}\langle c^*, c_{(2)}\rangle) = \langle c^*, c\rangle \end{aligned}$$

It is clear that α is R-linear. Let us next show that $\alpha(f \circ g) = \alpha(f) * \alpha(g)$, for all C-colinear $f, g : C \to C$.

$$\begin{aligned} (\varepsilon \circ f) * (\varepsilon \circ g) &= m \circ (\varepsilon \otimes \varepsilon) \circ (f \otimes g) \circ \Delta \\ &= m \circ (\varepsilon \otimes \varepsilon) \circ \tau(f \otimes I_C) \circ \tau \circ (g \otimes I_C) \circ \Delta \\ &= m \circ (\varepsilon \otimes \varepsilon) \circ \tau(f \otimes I_C) \circ \tau \circ \Delta \circ g \\ &= m \circ (\varepsilon \otimes \varepsilon) \circ \tau\Delta \circ f \circ g \\ &= (\varepsilon * \varepsilon) \circ f \circ g \\ &= \varepsilon \circ f \circ g \end{aligned}$$

It follows immediately that α' is a group isomorphism. Suppose finally that $C = H$ is a Hopf algebra. If f is an H-colinear algebra automorphism, then

$$\langle \alpha(f), hk\rangle = \varepsilon(f(hk)) = \varepsilon(f(h))\varepsilon(f(k)) = \langle \alpha(f), h\rangle\langle \alpha(f), k\rangle$$

and, conversely, if $h^* \in H^*$ is multiplicative, then

$$\begin{aligned} (I_H * h^*)(hk) &= \sum h_{(1)}k_{(1)}h^*(h_{(2)}k_{(2)}) \\ &= \sum h_{(1)}\langle h^*, h_{(2)}\rangle k_{(1)}\langle h^*, k_{(2)}\rangle \\ &= (I_H * h^*)(h)(I_H * h^*)(k) \end{aligned}$$

□

In the next Proposition, we will use $\mathbb{G}_m((C \otimes \bullet)^*)$ as a shorter notation for the sheaf sending S to $\mathbb{G}_m(\mathrm{Hom}_S(C \otimes S, S))$. Notice that there is a certain ambiguity in this notation.

Proposition 10.4.7 *Let C be a commutative faithfully flat R-coalgebra, and S a faithfully flat commutative R-algebra. Then*

$$\mathrm{Pic}^{\mathrm{co}}(S/R,C) \cong H^1(S/R, \mathbb{G}_m((C\otimes \bullet)^*))$$

$\mathbb{G}_m((C\otimes\bullet)^*)$ is a sheaf on R_{fl}, and, consequently,

$$\mathrm{Pic}^{\mathrm{co}}_{\mathrm{gnb}}(R,C) \cong H^1(R_{\mathrm{fl}}, \mathbb{G}_m((C\otimes \bullet)^*))$$

Proof The proof is similar to the proof of Proposition 10.3.2. Let us give some more detail this time. We will construct a map

$$\beta:\ \mathrm{Pic}^{\mathrm{co}}(S/R,C)\longrightarrow H^1(S/R, \mathbb{G}_m((C\otimes \bullet)^*))$$

Let $P=(C,C,P,P',f,f')$ be an invertible C-comodule representing an element of $\mathrm{Pic}^{\mathrm{co}}(S/R,C)$. Then we have a colinear isomorphism

$$\varphi:\ P\otimes S\longrightarrow C\otimes S$$

Define $f:\ S\otimes C\otimes S\longrightarrow C\otimes S\otimes S$ by commutativity of the following diagram of colinear isomorphisms:

$$\begin{array}{ccc} S\otimes P\otimes S & \stackrel{I_S\otimes\varphi}{\longrightarrow} & S\otimes C\otimes S\\ \Big\downarrow{\scriptstyle \tau\otimes I_S} & & \Big\downarrow{\scriptstyle f}\\ P\otimes S\otimes S & \stackrel{\varphi\otimes I_S}{\longrightarrow} & C\otimes S\otimes S \end{array} \tag{10.15}$$

From Lemma 10.4.6, it follows that

$$x=(\varepsilon_C\otimes I_S\otimes I_S)\circ f:\ S\otimes C\otimes S\longrightarrow S\otimes S$$

is convolution invertible. From the fact that f is a descent datum, it follows easily that $x\in \mathbb{G}_m((C\otimes S^{\otimes 2})^*)$ is a cocycle. We define $\beta([P])=[x]$.
Let us show that the map β is multiplicative, that is,

$$\beta([M\Box_C N])=\beta([M])\beta([N])$$

Observe first that the cotensor product behaves functorially, that is, if $f:\ M\to M'$ and $g:\ N\to N'$ are C-colinear maps, then $f\otimes g$ restricts to a map

$$f\Box g:\ M\Box_C N\longrightarrow M'\Box_C N'$$

Indeed, if $\sum_i m_i\otimes n_i\in M\Box_C N$, then

$$\begin{aligned}(\rho_{M'}\otimes I_{N'})\Big(\sum_i f(m_i)\otimes g(n_i)\Big) &= \sum_i (f\otimes I_C)(\rho_M(m_i))\otimes g(n_i)\\ &= \sum_i f(m_i)\otimes n_{i_{(1)}}\otimes g(n_{i_{(0)}})\\ &= \sum_i f(m_i)\otimes \tau(\rho_{N'}(g(n_i))\end{aligned}$$

Consider the special case $M = N = M' = N' = C$. Then we obtain a map

$$f\Box g:\ C\Box_C C \cong C \longrightarrow C \otimes_C C \cong C$$

given by

$$(f\Box_C g)(c) = \sum f(c_{(1)})\varepsilon(g(c_{(2)})) = \sum \varepsilon(f(c_{(1)}))g(c_{(2)})$$

The corresponding map $\alpha(f\Box g) \in C^*$ is

$$\alpha(f\Box g) = \varepsilon \circ (f\Box g) = (\varepsilon \circ f) * (\varepsilon \circ g)$$

Now let P and Q be invertible comodules split by S, and consider the data φ and f defined above, and let ψ and g be the corresponding maps for Q. The corresponding maps for $P\Box_C Q$ are now

$$\varphi\Box\psi:\ (P\Box_C Q)\otimes S \longrightarrow (C\Box_C C)\otimes S \cong C\otimes S$$

and

$$\begin{aligned} f\Box_{C\otimes S^{\otimes 2}} g\ :\ & (S\otimes C\otimes S)\Box_{S\otimes C\otimes S}(S\otimes C\otimes S) \cong S\otimes C\otimes S \\ & \longrightarrow (C\otimes S\otimes S)\Box_{C\otimes S\otimes S}(C\otimes S\otimes S) \cong C\otimes S\otimes S \end{aligned}$$

The corresponding cocycle is $\varepsilon \circ (f\Box_{C\otimes S^{\otimes 2}} g) = (\varepsilon\circ f) * (\varepsilon \circ g)$.
We will now show that β surjective. If $x \in Z^1(S/R, \mathbb{G}_m((C\otimes \bullet)^*))$ is a cocycle, then it follows from Lemma 10.4.6 that $\xi = I_{C\otimes S^{\otimes 2}} * x$ is a colinear isomorphism $S\otimes C\otimes S \longrightarrow C\otimes S\otimes S$, which is a descent datum for $C\otimes S$. From Proposition 10.4.5, it follows that $C\otimes S$ descends to a C-comodule P. Let y be the convolution inverse of x. Then $\eta = I_{C\otimes S^{\otimes 2}} * y$ is the composition inverse of ξ (see Lemma 10.4.6), and is also a descent datum for $C\otimes S$. We call the descended comodule P'. It is clear that P and P' are flat as R-modules. The map $\xi\Box\eta:\ S\otimes(C\Box_C C)\otimes S \to (C\Box_C C)\otimes S\otimes S$ is a descent datum, and the descended comodule is $P\Box_C P'$. Now the diagram

$$\begin{array}{ccc} S\otimes C\otimes S & \xrightarrow{I_S\otimes\Delta_C\otimes I_S} & S\otimes(C\Box_C C)\otimes S \\ \downarrow{\scriptstyle \tau\otimes I_S} & & \downarrow{\scriptstyle \xi\Box\eta} \\ C\otimes S\otimes S & \xrightarrow{\Delta_C\otimes I_S\otimes I_S} & (C\Box_C C)\otimes S\otimes S \end{array}$$

commutes. This follows from the fact that ξ and η are each others inverses. From Lemma B.0.11, it follows that $\Delta_C\otimes I_S$ descends to an isomorphism $f:\ C\to P\Box_C P'$. In a similar way, we obtain an isomorphism $f':\ C\to P'\Box_C P$. From the fact that

$$(C\otimes S, C\otimes S, C\otimes S, C\otimes S, \Delta_C\otimes I_S, \Delta_C\otimes I_S)$$

is a strict Morita-Takeuchi context (S is the groundring), it follows that $\mathbf{P} = (C, C, P, P', f, f')$ is a strict Morita-Takeuchi context. It is clear that $\beta([P]) = [x]$.
Let us next show that the presheaf $\mathbb{G}_m(\mathrm{Hom}_\bullet(C\otimes\bullet, \bullet))$ on R_{fl} is a sheaf. Let $S\to S'$ be a covering in R_{fl}. From faithfully flat descent of homomorphisms (cf. Corollary B.0.7), it follows that the sequence

$$1 \longrightarrow \mathrm{Hom}_S(C\otimes S, S) \longrightarrow \mathrm{Hom}_{S'}(C\otimes S', S') \rightrightarrows \mathrm{Hom}_{S'\otimes_S S'}(C\otimes S\otimes_S S', S'\otimes_S S')$$

is exact. Taking units, it follows that

$$1 \longrightarrow \mathbb{G}_m(\mathrm{Hom}_S(C \otimes S, S)) \longrightarrow \mathbb{G}_m(\mathrm{Hom}_{S'}(C \otimes S', S')) \rightrightarrows \mathbb{G}_m((\mathrm{Hom}_{S' \otimes_S S'}(C \otimes S \otimes_S S', S' \otimes_S S'))$$

is exact, and $\mathbb{G}_m(\mathrm{Hom}_\bullet(C \otimes \bullet, \bullet))$ is a sheaf.
Taking inductive limits over all faithfully flat R-algebras S, we obtain that

$$\mathrm{Pic}^{\mathrm{co}}_{\mathrm{gnb}}(R, C) \cong H^1(R_{\mathrm{fl}}, \mathbb{G}_m((C \otimes \bullet)^*))$$

□

Let us now show that the Picard group of coalgebras behaves functorially.

Lemma 10.4.8 $\mathrm{Pic}^{\mathrm{co}}(R, \bullet)$ *and* $\mathrm{Pic}^{\mathrm{co}}_{\mathrm{gnb}}(R, \bullet)$ *are contravariant functors from faithfully flat cocommutative* R*-coalgebras to abelian groups.*

Proof Suppose that $f: D \longrightarrow C$ is a morphism of coalgebras. A map

$$\mathrm{Pic}^{\mathrm{co}}(R, f): \mathrm{Pic}^{\mathrm{co}}(R, C) \longrightarrow \mathrm{Pic}^{\mathrm{co}}(R, D)$$

may be defined as follows: for an invertible C-comodule P, we define

$$\mathrm{Pic}^{\mathrm{co}}(R, f)([P]) = P \square_C D = [(D, D, P \square_C D, P' \square_C D, f \square I_D, f' \square I_D)]$$

where D is a C-comodule via f, that is, $\rho_D(d) = \sum d_{(1)} \otimes f(d_{(2)})$ for all $d \in D$. It is easy to show that $P \square_C D$ is an invertible D-comodule.
Suppose that P is split by a faithfully flat R-algebra S. Then we can easily show that $\mathrm{Pic}^{\mathrm{co}}(R, f)([P])$ is also split by S:

$$\begin{aligned} (P \square_C D) \otimes S &\cong (P \otimes S) \square_{C \otimes S} (D \otimes S) \\ &\cong (C \otimes S) \square_{C \otimes S} (D \otimes S) \cong D \otimes S \end{aligned}$$

as $D \otimes S$-comodules. Thus $\mathrm{Pic}^{\mathrm{co}}(R, f)$ restricts to a map

$$\mathrm{Pic}^{\mathrm{co}}_{\mathrm{gnb}}(R, f): \mathrm{Pic}^{\mathrm{co}}_{\mathrm{gnb}}(R, C) \longrightarrow \mathrm{Pic}^{\mathrm{co}}_{\mathrm{gnb}}(R, D)$$

It is straightforward to verify all the necessary functorial properties. □

It is well-known that, for an invertible R-module I, the natural map $R \to \mathrm{End}_R(I)$ is an isomorphism of R-algebras. In the next Proposition, we will prove a similar result for invertible C-comodules with a geometric normal basis.
For later use, we introduce the following notation, in the spirit of the convolution product. If A is an H-comodule algebra, $h^* \in H^*$, and $\alpha: A \to A$, then we define $h^* * \alpha: A \to A$ by the formula

$$(h^* * \alpha)(a) = \sum \langle h^*, a_{(1)} \rangle \alpha(a_{(0)})$$

for all $a \in A$.

Proposition 10.4.9 *Let C be a cocommutative faithfully flat R-coalgebra, and M an invertible C-comodule with geometric normal basis. Then the natural map*

$$\alpha : C^* \longrightarrow \mathrm{End}^C(M)$$

defined by

$$\alpha(c^*) = c^* * I_M$$

that is

$$\alpha(c^*)(m) = \sum \langle c^*, m_{(1)} \rangle m_{(0)}$$

is an isomorphism of R-algebras. α restricts to an isomorphism of groups $\mathbb{G}_m(C^*) \to \mathrm{Aut}^C(M)$.

Proof If $M \cong C$ as a C-comodule, then the result is true, see Proposition 10.4.6. Take a faithfully flat R-algebra S such that $M \otimes S \cong C \otimes S$ as $C \otimes S$-comodules. Then the result holds for $M \otimes S$, and the technique of the proof is to derive the result for M using descent theory.
It is easy to prove that α is a homomorphism of R-algebras, we leave verification to the reader. The most interesting part of the proof is to show that α is surjective. From Theorem 10.4.4 (see also Proposition 10.4.7), we know that there exists a C-comodule N such that $C \cong N \Box_C M$ as C-comodules. We write β for the connecting isomorphism.
Let $f : M \to M$ be C-colinear. Then $f \otimes I_S : M \otimes S \to M \otimes S$ is $C \otimes S$-colinear, and we know that there exists an S-linear map

$$g : C \otimes S \longrightarrow S$$

such that

$$f(m) \otimes s = (f \otimes I_S)(m \otimes s) = \sum m_{(0)} \otimes g(m_{(1)} \otimes 1)s \tag{10.16}$$

We will show that $g(c \otimes 1) \in R$, for all $c \in C$. From (10.16), it follows that

$$\begin{aligned} (f \otimes I_S \otimes I_S)(m \otimes 1 \otimes 1) &= \sum m_{(0)} \otimes g(m_{(1)} \otimes 1) \otimes 1 \\ &= \sum m_{(0)} \otimes 1 \otimes g(m_{(1)} \otimes 1) \end{aligned} \tag{10.17}$$

Now take $c \in C$, and let $\beta(c) = \sum n_i \otimes m_i \in N \Box_C M$. From (10.17), it follows that

$$\sum n_i \otimes m_{i_{(0)}} \otimes g(m_{i_{(1)}} \otimes 1) \otimes 1 = \sum n_i \otimes m_{i_{(0)}} \otimes 1 \otimes g(m_{i_{(1)}} \otimes 1)$$

and, using the fact that β is C-colinear,

$$\sum \beta(c_{(1)}) \otimes g(c_{(2)} \otimes 1) \otimes 1 = \sum \beta(c_{(1)}) \otimes 1 \otimes g(c_{(2)} \otimes 1)$$

Applying $\varepsilon_C \circ \beta^{-1}$ to the first factor of both sides, we obtain that

$$g(c \otimes 1) \otimes 1 = 1 \otimes g(c \otimes 1)$$

and from the faithfully flat descent Theorem B.0.6, it follows that $g(c \otimes 1) \in R$. We now have that

$$f(m) \otimes 1 = \sum g(m_{(1)} \otimes 1) m_{(0)} \otimes 1$$

and

$$f(m) = \sum g(m_{(1)} \otimes 1)m_{(0)}$$

and we have shown that $\alpha(g(\bullet \otimes 1)) = f$.
It is easy to show that α is injective: if $\alpha(c^*) = \alpha(d^*)$, then for all $m \in M$ and $s \in S$,

$$\sum \langle c^*, m_{(1)} \rangle m_{(0)} \otimes s = \sum \langle d^*, m_{(1)} \rangle m_{(0)} \otimes s$$

and it follows from Proposition 10.4.6 that $c^* \otimes I_S = d^* \otimes I_S$, and this implies that $c^* = d^*$. □

10.5 The group of Galois objects

From now on, H will be a faithfully flat, cocommutative Hopf algebra.

Proposition 10.5.1 *If A and B are H-Galois objects, then $A \square_H B$ is also an H-Galois object.*

Proof It follows from the proof of Theorem 10.4.4 that $A \square_H B$ is flat. We can also show directly that $A \square_H B$ is faithfully flat:

$$A \otimes (A \square_H B) \cong (A \otimes A) \square_H B \cong (A \otimes A) \square_H B \cong A \otimes (H \square_H B) \cong A \otimes B$$

is faithfully flat as an R-module. A is faithfully flat, so $A \square_H B$ is also faithfully flat. Observe that we used Lemma 10.4.1. From Theorem 8.1.6, we know that the maps

$$\gamma_A : A \otimes A \longrightarrow A \otimes H \quad : \quad a \otimes a' \mapsto \sum aa'_{(0)} \otimes a'_{(1)}$$
$$\gamma_B : B \otimes B \longrightarrow B \otimes H \quad : \quad b \otimes b' \mapsto \sum bb'_{(0)} \otimes b'_{(1)}$$

are isomorphisms. The map

$$\gamma_{A \square_C B} : (A \square_C B) \otimes (A \square_C B) \longrightarrow (A \square_C B) \otimes H$$

is the composition of the maps

$$\begin{aligned} (A \square_C B) \otimes (A \square_C B) &\cong (A \otimes A) \square_{C \otimes C} (B \otimes B) \\ &\xrightarrow{\gamma_A \square \gamma_B} (A \otimes H) \square_{H \otimes H} (A \otimes H) \\ &\cong (A \square_H H) \otimes H \end{aligned}$$

and is therefore an isomorphism. □

Theorem 10.5.2 *If A is an H-Galois object, then there exists an H-Galois object B such that $A \square_H B \cong H$ as H-comodule algebras. B is given by the following data: $B = A^{\mathrm{op}}$ as an R-algebra, and $\rho_B = (I_B \otimes S) \circ \rho_A$.*

Proof The proof is a formal dual of the proof of Theorem 10.1.2. Let $K = H^{\rm op}$ as an algebra, with H-coaction $\rho_K = (I \otimes S) \circ \Delta$:

$$\rho_K(h) = \sum h_{(1)} \otimes S(h_{(2)})$$

for all $h \in H$. Then the antipode $S: H \to K$ is an isomorphism of H-comodule algebras, and it suffices to show that

$$A\Box_H B \cong K$$

as H-comodule algebras. Since $A^{{\rm co}H} = R$, R is the equalizer of the maps

$$I_A \otimes \eta_H, \rho_A : R \longrightarrow A \rightrightarrows A \otimes H$$

Now

$$I_A \circ \gamma_H^{-1} : A \otimes H \otimes H \longrightarrow A \otimes H \otimes H : a \otimes h \otimes k \mapsto \sum a \otimes hS(k_{(1)}) \otimes k_{(2)}$$

is an isomorphism. Therefore H is also the equalizer of

$$\alpha = (I_A \circ \gamma_H^{-1}) \circ (\rho_A \otimes I_H),\ \beta = (I_A \circ \gamma_H^{-1}) \circ (I_A \otimes \eta_H) : A \otimes H \rightrightarrows A \otimes H \otimes H$$

An easy computation yields

$$\begin{aligned} \alpha(a \otimes h) &= \sum a_{(0)} \otimes a_{(1)}S(h_{(1)}) \otimes h_{(2)} \\ \beta(a \otimes h) &= \sum a \otimes S(h_{(1)}) \otimes h_{(2)} \end{aligned}$$

Consider the map

$$\gamma' : A \otimes H \otimes B \longrightarrow A \otimes H \otimes H : a \otimes h \otimes b \longrightarrow \sum ab_{(0)} \otimes h \otimes b_{(1)}$$

and observe that the diagram

$$\begin{array}{ccc} 0 & & 0 \\ \downarrow & & \downarrow \\ A\Box_H B & & H \\ \downarrow & & \downarrow \\ A \otimes B & \xrightarrow{\gamma} & A \otimes H \\ {\scriptstyle \rho_A \otimes I_B}\downdownarrows{\scriptstyle I_A \otimes \rho_B} & & {\scriptstyle \alpha}\downdownarrows{\scriptstyle \beta} \\ A \otimes H \otimes B & \xrightarrow{\gamma'} & A \otimes H \otimes H \end{array} \tag{10.18}$$

commutes. Indeed

$$\begin{aligned} (\beta \circ \gamma)(a \otimes b) &= \beta(\sum ab_{(0)} \otimes b_{(1)}) \\ &= \sum ab_{(0)} \otimes S(b_{(1)}) \otimes b_{(2)} \\ &= \gamma'(I_A \otimes \rho_B)(a \otimes b) \end{aligned}$$

and

$$\begin{aligned}(\alpha \circ \gamma)(a \otimes b) &= \alpha(\sum ab_{(0)} \otimes b_{(1)}) \\ &= \sum a_{(0)}b_{(0)} \otimes a_{(1)}b_{(1)}S(b_{(2)}) \otimes b_{(3)} \\ &= \sum a_{(0)}b_{(0)} \otimes a_{(1)} \otimes b_{(1)} \\ &= \gamma'(\sum a_{(0)} \otimes a_{(1)} \otimes b) \\ &= (\gamma' \circ (\rho_A \otimes I_B))(a \otimes b)\end{aligned}$$

for all $a \in A$ and $b \in B$.
δ and δ' are isomorphisms and the columns in (10.18) are exact. Thus $A\square_H B \cong H$, and we are done if we can show that B is an H-Galois object. It is clear that B is faithfully flat as an R-module, since $A \cong B$ as an R-module. It therefore suffices to show that

$$\gamma_B : B \otimes B \longrightarrow B \otimes H : b \otimes b' \mapsto \sum b'_{(0)}b \otimes S(b'_{(1)})$$

is bijective. Let us first show that γ_B is surjective. Take $b \otimes h \in B \otimes H$, and let

$$\gamma_A^{-1}(\sum b_{(0)} \otimes hb_{(1)}) = \sum_i a_i \otimes a'_i$$

then

$$\sum_i a_i a'_{i(0)} \otimes a'_{i(1)} = \sum b_{(0)} \otimes hb_{(1)}$$

Applying ρ_B to the first factor, we obtain

$$\sum a_{i(0)}a'_{i(0)} \otimes S(a_{i(1)}a'_{i(1)}) \otimes a'_{i(2)} = \sum b_{(0)} \otimes S(b_{(1)}) \otimes hb_{(2)}$$

Now switch and then multiply the two last factors. This gives us that

$$\sum a_{i(0)}a'_i \otimes S(a_{i(1)}) = b \otimes h$$

and

$$\gamma_B(\sum_i a'_i \otimes a_i) = \sum a_{i(0)}a'_i \otimes S(a_{i(1)}) = b \otimes h$$

Let us finally show that γ_B is injective. Suppose that

$$\gamma_B(\sum_i b_i \otimes b'_i) = \sum_i b'_{i(0)}b_i \otimes S(b'_{i(1)}) = 0$$

Applying ρ_A to the first factor, and switching and multiplying the two last factors yields

$$\begin{aligned}0 &= \sum_i b'_{i(0)}b_{i(0)} \otimes S(b'_{i(2)})b'_{i(1)}b_{i(1)} \\ &= \sum_i b'_i b_{i(0)} \otimes b_{i(1)} \\ &= \gamma_A(\sum b'_i \otimes b_i)\end{aligned}$$

From the fact that γ_A is injective, it follows that $\sum_i b'_i \otimes b_i = 0$ and $\sum_i b_i \otimes b'_i = 0$.
□

Theorem 10.5.3 *Let H be a cocommutative faithfully flat Hopf algebra. Then* $\mathrm{Gal}(R, H)$*, the set of isomorphism classes of H-Galois objects, forms a group under the operation induced by the cotensor product $\square_H$.*

Proof We know from Proposition 10.5.1 that the cotensor product of two H-Galois objects is again an H-Galois object. From Lemma 10.4.1, we know that the cotensor product of H-Galois objects is associative, since all the underlying comodules are flat. The neutral element is H viewed as an H-comodule algebra, and it follows from Theorem 10.5.2 that every H-Galois object has an inverse. □

Let A be an H-Galois object, and let B be as in Theorem 10.5.2. We have seen that the map γ restricts to a map $A\square_H B \to K$. We thus obtain an isomorphism of H-comodule algebras

$$f: \; A\square_H B \longrightarrow R\otimes H \subset A\otimes H: \; \sum_i a_i\otimes b_i \mapsto \sum_i a_i b_{i_{(0)}}\otimes S(b_{i_{(1)}})$$

Interchanging A and B, we obtain another isomorphism

$$f': \; B\square_H A \longrightarrow R\otimes H \subset B\otimes H: \; \sum_i b_i\otimes a_i \mapsto \sum_i b_i a_{i_{(0)}}\otimes a_{i_{(1)}}$$

Lemma 10.5.4 (H, H, A, B, f, f') *is a strict Morita-Takeuchi context.*

Proof We investigate the map

$$A\square_H B\square_H A \xrightarrow{f\square I_A} R\otimes H\square_H A \cong H\square_H A \xrightarrow{I_R\otimes\varepsilon_H\square I_A} R\otimes A = A$$

$$\begin{aligned}(I_R\otimes\varepsilon_H\square I_A)((f\square I_A)(\sum a_i\otimes b_i\otimes c_i)) &= (I_R\otimes\varepsilon_H\square I_A)(\sum a_i b_{i_{(0)}}\otimes S(b_{i_{(1)}})\otimes c_i)\\ &= \sum a_i b_{i_{(0)}}\otimes\varepsilon(b_{i_{(1)}})c_i\\ &= \sum_i a_i b_i c_i\end{aligned}$$

In a similar way, the composition

$$A\square_H B\square_H A \xrightarrow{I_A\square f'} A\square_H R\otimes H \cong A\square_H H \xrightarrow{I_R\otimes I_A\square\varepsilon_H} A\otimes R = A$$

also sends $\sum_i a_i\otimes b_i\otimes c_i$ to $\sum_i a_i b_i c_i$. We have shown that one of the two diagrams in the definition of a strict Morita-Takeuchi context is commutative. The proof of the commutativity of the other one is similar. Since the maps f and f' are isomorphisms, we have a strict Morita-Takeuchi context. □

Let $\underline{\underline{\mathrm{Gal}}}(R, H)$ and $\underline{\underline{\mathrm{Pic}}}^{\mathrm{co}}(R, H)$ be the categories of respectively H-Galois objects and invertible H-comodules with geometric normal basis. These categories are symmetric monoidal categories, the product is the cotensor product over H. H, considered respectively as an H-comodule algebra and an H-comodule, is a cofinal object of the two categories, and therefore the Grothendieck groups of these two categories are

$$K_0(\underline{\underline{\mathrm{Gal}}}(R,H)) = \mathrm{Gal}(R,H) \tag{10.19}$$
$$K_0(\underline{\underline{\mathrm{Pic}}}^{\mathrm{co}}(R,H)) = \mathrm{Pic}^{\mathrm{co}}(R,H) \tag{10.20}$$

and from Lemma 10.4.6, it follows that the Whitehead groups are

$$K_1(\underline{\underline{\mathrm{Gal}}}(R,H)) = \mathrm{Alg}(H,R) \tag{10.21}$$
$$K_1(\underline{\underline{\mathrm{Pic}}}^{\mathrm{co}}(R,H)) = \mathbb{G}_m(H^*) \tag{10.22}$$

We are again in the situation where we can apply [15, VIII.5], this time to the functor

$$F:\ \underline{\underline{\mathrm{Gal}}}(R,H) \longrightarrow \underline{\underline{\mathrm{Pic}}}^{\mathrm{co}}(R,H)$$

sending an H-Galois object H to the strict Morita-Takeuchi context (H,H,A,B,f,f') of Lemma 10.5.4 We therefore have an exact sequence

$$\mathrm{Alg}(H,R) \longrightarrow \mathbb{G}_m(H^*) \longrightarrow K_1\underline{\Phi F} \longrightarrow \underline{\underline{\mathrm{Gal}}}(R,H) \longrightarrow \underline{\underline{\mathrm{Pic}}}^{\mathrm{co}}(R,H) \tag{10.23}$$

We still have to compute $K_1\underline{\Phi F}$. Having Lemma 10.1.4 and the duality between Harrison and Sweedler cohomology, we expect that $K_1\underline{\Phi F}$ is the group of Sweedler 2-cocycles. This is exactly what we will prove in the next Lemma.

Lemma 10.5.5 *With notations as above, we have*

$$K_1\underline{\Phi F} \cong Z^2(H,R,\mathbb{G}_m)$$

Furthermore, the diagram

$$\begin{array}{ccc} \mathbb{G}_m(H^*) & \xrightarrow{\Delta_1} & Z^2(H,R,\mathbb{G}_m) \\ \downarrow\cong & & \downarrow\cong \\ K_1(\underline{\underline{\mathrm{Pic}}}^{\mathrm{co}}(R,H)) & \xrightarrow{d} & K_1\underline{\Phi F} \end{array} \tag{10.24}$$

commutes.

Proof Every $x \in K_1\underline{\Phi F}$ can be represented by a triple (A,γ,B), where A and B are H-Galois objects, and $\gamma:\ A \to B$ is an H-colinear isomorphism. We can take $A = H$, because H is a cofinal object. So take $x = [(H,\gamma,A)] \in K_1\underline{\Phi F}$. Identify A and H using γ. This comes down to saying that $A = H$ as an H-comodule, but with a different multiplication m_A. Then we can write $x = [(H,I_H,A)]$. Define

$$\alpha:\ K_1\underline{\Phi F} \longrightarrow Z^2(H,R,\mathbb{G}_m)$$

as follows:

$$\alpha(x) = \varepsilon \circ m_A:\ H \otimes H \xrightarrow{m_A} H \xrightarrow{\varepsilon} R$$

We can view α as a map $\alpha:\ K_1\underline{\Phi F}\longrightarrow(H\otimes H)^*$. Let us show that α is multiplicative. Take $x=[(H,I_H,A)]$ and $y=[(H,I_H,B)]$. Then $xy=[(H,\Delta_H,A\square_H B)]$ and we can describe $\alpha(xy):\ H\otimes H\longrightarrow R$ as follows.

$$\begin{aligned}\alpha(xy)(h\otimes k) &= \big(\varepsilon\circ m_{A\square_H B}\circ(\Delta_H\otimes\Delta_H)\big)(h\otimes k)\\ &= \sum\varepsilon_H(m_A(h_{(1)}\otimes k_{(1)}))\varepsilon(m_A(h_{(2)}\otimes k_{(2)}))\\ &= (\alpha(x)*\alpha(y))(h\otimes k)\end{aligned}$$

From the fact that every element of $K_1\underline{\Phi F}$ has an inverse, it follows that $\alpha(x)\in\mathbb{G}_m((H\otimes H)^*)$. Let us show that $f=\alpha(x)$ is a cocycle. We write $\cdot_A$ for the multiplication in A.
A is an H-comodule algebra, and thus we have, for all $a,b\in A=H$:

$$\rho(a\cdot_A b)=\sum(a\cdot_A b)_{(1)}\otimes(a\cdot_A b)_{(2)}=\sum a_{(1)}\cdot_A b_{(1)}\otimes a_{(2)}b_{(2)}$$

Applying $\varepsilon_H\otimes I_H$ to both sides, we obtain

$$a\cdot_A b=\sum\varepsilon_H(a_{(1)}\cdot_A b_{(1)})a_{(2)}b_{(2)}=f(a_{(1)}\otimes b_{(1)})a_{(2)}b_{(2)}$$

From the associativity of A, it follows that

$$(a\cdot_A b)\cdot_A c=\sum f(a_{(1)}\otimes b_{(1)})f(a_{(2)}b_{(2)}\otimes c_{(1)})a_{(3)}b_{(3)}c_{(2)}$$

equals

$$a\cdot_A(b\cdot_A c)=\sum f(b_{(1)}\otimes c_{(1)})f(a_{(1)}\otimes b_{(2)}c_{(2)})a_{(2)}b_{(3)}c_{(3)}$$

Applying ε to both sides, we find that

$$\sum f(a_{(1)}\otimes b_{(1)})f(a_{(2)}b_{(2)}\otimes c)=\sum f(b_{(1)}\otimes c_{(1)})f(a\otimes b_{(2)}c_{(2)})$$

which means that f is a cocycle.
We will now describe the inverse of the map α. If $f:\ H\otimes H\to R$ is a Sweedler 2-cocycle, then we define

$$\alpha^{-1}([f])=[(H,I_H,A)]$$

as follows: $A=H$ as an H-comodule, with multiplication

$$a\cdot_A b=\sum f(a_{(1)}\otimes b_{(1)})a_{(2)}b_{(2)}$$

The associativity of A follows easily from the fact that f is a cocycle. It is also easy to see that $A\cong H$ as an H-comodule algebra if f is a coboundary, and therefore α^{-1} is well-defined. It is straightforward to show that $u=f(1\otimes 1)^{-1}$ is a unit element in A (If we take the cocycle f normalized, then 1 is the unit element in A).
To show that A is an H-Galois object, it suffices to show that the map

$$\gamma_A:\ A\otimes A\longrightarrow A\otimes H:\ a\otimes b\mapsto\sum a\cdot_A b_{(1)}\otimes b_{(2)}=\sum f(a_{(1)}\otimes b_{(1)})a_{(2)}b_{(2)}\otimes b_{(3)}$$

is an isomorphism. A straightforward computation shows that the map

$$\gamma_A^{-1}:\ A\otimes H\longrightarrow A\otimes A:\ a\otimes h\mapsto\sum f^{-1}(a_{(1)}\otimes h_{(1)})a_{(2)}S(h_{(2)})\otimes h_{(3)}$$

is the inverse of γ_A. □

From (10.19-10.22) and Lemma 10.5.5, it follows that the exact sequence (10.23) may be written as

$$\mathrm{Alg}(H,R) \longrightarrow \mathbb{G}_m(H^*) \longrightarrow Z^2(H,R,\mathbb{G}_m) \longrightarrow \mathrm{Gal}(R,H) \longrightarrow \mathrm{Pic}^{\mathrm{co}}(R,H)$$

and we have an exact sequence

$$1 \longrightarrow H^2(H,R,\mathbb{G}_m) \longrightarrow \mathrm{Gal}(R,H) \longrightarrow \mathrm{Pic}^{\mathrm{co}}(R,H) \tag{10.25}$$

In view of the results of the previous Section, we expect that (10.25) can be extended to a long exact sequence. One expects that the third term in the sequence is the Sweedler cohomology group $H^1(H,R,\mathrm{Pic})$. If we look more carefully, then we see that there is a technical difficulty: the third term in the sequence (10.25) is the Picard group of H as a coalgebra, while the Sweedler cohomology group $H^1(H,R,\mathrm{Pic})$ is defined in terms of the "classical Picard group". An idea could be to consider Sweedler cohomology with values in $\mathrm{Pic}^{\mathrm{co}}(R,\bullet)$. But this makes not sense, since the functor $\mathrm{Pic}^{\mathrm{co}}(R,\bullet)$ is contravariant. To overcome this difficulty, we will introduce a new type of cohomology. It will turn out that this new cohomology is in some particular cases related to Sweedler cohomology. The definition is a formal dual version of the definition of Harrison cohomology.

Let H be a cocommutative Hopf algebra, C a cocommutative H-module coalgebra, and P a contravariant functor from cocommutative R-coalgebras to abelian groups. For $n \geq 0$, we consider the maps

$$\varepsilon_0, \varepsilon_1, \ldots, \varepsilon_n :\; H^{\otimes n+1} \otimes C \longrightarrow H^{\otimes n} \otimes C \tag{10.26}$$

defined by

$$\begin{aligned}
\varepsilon_0(h_1 \otimes \cdots \otimes h_{n+1} \otimes c) &= \varepsilon(h_1)h_2 \otimes \cdots \otimes h_{n+1} \otimes c \\
\varepsilon_i(h_1 \otimes \cdots \otimes h_{n+1} \otimes c) &= h_1 \otimes \cdots \otimes h_i h_{i+1} \otimes \cdots \otimes h_{n+1} \otimes c \\
\varepsilon_{n+1}(h_1 \otimes \cdots \otimes h_{n+1} \otimes c) &= h_1 \otimes h_2 \otimes \cdots \otimes (h_{n+1} \rightharpoonup c)
\end{aligned}$$

for all $h_1, h_2, \ldots h_{n+1} \in H$. Applying the contravariant functor P, we obtain maps

$$P(\varepsilon_i) :\; P(H^{\otimes n} \otimes C) \longrightarrow P(H^{\otimes n+1} \otimes C)$$

We define Δ_n as the alternating sum

$$\Delta_n = \sum_{i=0}^{n+1} (-1)^i P(\varepsilon_i)$$

It is now straightforward to show that we have a complex

$$\cdots \xrightarrow{\Delta_{n-1}} P(H^{\otimes n} \otimes C) \xrightarrow{\Delta_n} P(H^{\otimes n+1} \otimes C) \xrightarrow{\Delta_{n+1}} \cdots$$

The cohomology groups of this complex will be denoted by

$$H^n_{\mathrm{Sw}}(H,C,P) = \mathrm{Ker}\,(\Delta_n)/\mathrm{Im}\,(\Delta_{n-1}$$

In the following Lemma, we will give the relation with "classical" Sweedler cohomology.

Lemma 10.5.6 *Let H be a cocommutative Hopf algebra, and P a covariant functor from commutative R-algebras to abelian groups. Then*

$$H^n(H,R,P) \cong H^n_{\rm Sw}(H,R,P\circ(\bullet)^*)$$

Proof Compare the definitions of both cohomologies. □

Lemma 10.5.7 *Let S be a faithfully flat commutative R-algebra, and A an H-module coalgebra. If $S\otimes A$ is an $S\otimes H$-Galois object, then A is an H-Galois object*

Proof A is faithfully flat as an R-module, since $S\otimes A$ is a faithfully flat S-module, and S/R is faithfully flat.

$$\gamma_A\otimes I_S:\ S\otimes(A\otimes A)\cong(S\otimes A)\otimes_S(S\otimes A)\longrightarrow S\otimes(A\otimes H)\cong(S\otimes A)\otimes_S(S\otimes H)$$

is an isomorphism, and the fact that S/R is faithfully flat implies that γ_A is an isomorphism. It now suffices to apply Theorem 8.1.8. □

Definition 10.5.8 *An H-Galois object A has a geometric normal basis if there exists a faithfully flat commutative R-algebra S such that $A\otimes S$ is isomorphic to $H\otimes S$ as an $H\otimes S$-comodule. The subgroup of* ${\rm Gal}(R,H)$ *consisting of isomorphism classes of H-Galois objects with geometric normal basis is denoted by* ${\rm Gal}_{\rm gnb}(R,H)$.

We can now state our main Theorem.

Theorem 10.5.9 *Let H be a cocommutative faithfully flat Hopf algebra. The exact sequence (10.25) extends to a long exact sequence*

$$1\longrightarrow H^2(H,R,\mathbb{G}_m)\xrightarrow{\beta}{\rm Gal}_{\rm gnb}(R,H)\xrightarrow{\gamma}H^1_{\rm Sw}(H,R,{\rm Pic}^{\rm co}_{\rm gnb}(R,\bullet))\xrightarrow{\delta}H^3(H,R,\mathbb{G}_m) \tag{10.27}$$

Proof Let A be an H-Galois object with geometric normal basis. We have to show that A considered as an invertible H-comodule is a cocycle in $Z^1_{\rm Sw}(H,R,{\rm Pic}^{\rm co}_{\rm gnb}(R,\bullet))$. The maps $\varepsilon_1,\varepsilon_2,\varepsilon_3:\ H\otimes H\to H$ are given by the formulas

$$\varepsilon_0(h\otimes k)=\varepsilon(h)k,\ \ \varepsilon_1(h\otimes k)=hk,\ \ \varepsilon_2(h\otimes k)=\varepsilon(k)h$$

and

$$\begin{aligned}{\rm Pic}^{\rm co}_{\rm gnb}(R,\varepsilon_0)(A)&=A\square_{H,\varepsilon_0}(H\otimes H)\\ &=\{\textstyle\sum a_i\otimes(h_i\otimes k_i)\in A\otimes H\otimes K\ |\\ &\qquad\textstyle\sum a_i\otimes(h_i\otimes k_{i_{(1)}})\otimes k_{i_{(2)}}=\sum a_{i_{(0)}}\otimes(h_i\otimes k_i)\otimes a_{i_{(1)}}\}\\ &=H\otimes(H\square_H A)\cong A\otimes H\end{aligned}$$

In a similar way, we can show that

$${\rm Pic}^{\rm co}_{\rm gnb}(R,\varepsilon_2)(A)=H\otimes A$$

and

$$\begin{aligned}\mathrm{Pic}^{\mathrm{co}}_{\mathrm{gnb}}(R,\varepsilon_1)(A) &= A\square_{H,\varepsilon_1}(H\otimes H)\\ &= \{\sum a_i\otimes(h_i\otimes k_i)\in A\otimes H\otimes K \mid \\ &\qquad \sum a_i\otimes(h_{i_{(1)}}\otimes k_{i_{(1)}})\otimes h_{i_{(2)}}k_{i_{(2)}} = \sum a_{i_{(0)}}\otimes(h_i\otimes k_i)\otimes a_{i_{(1)}}\}\end{aligned}$$

Now $(A\otimes H)\square_{H\otimes H}(H\otimes A)\cong A\otimes A$, and it suffices to construct an $H\otimes H$-colinear isomorphism

$$A\otimes A\longrightarrow A\square_{H,\varepsilon_1}(H\otimes H)$$

Consider the map

$$\theta:\ A\otimes H\longrightarrow A\square_{H,\varepsilon_1}(H\otimes H)$$

defined by

$$\theta(a\otimes h)=\sum a_{(0)}\otimes(a_{(1)}S(h_{(1)})\otimes h_{(2)})$$

$\theta(a\otimes h)\in A\square_{H,\varepsilon_1}(H\otimes H)$ since

$$\sum a_{(0)}\otimes(a_{(1)}S(h_{(1)})\otimes h_{(2)})\otimes a_{(2)}S(h_{(3)})h_{(4)}=\sum a_{(0)}\otimes(a_{(1)}S(h_{(1)})\otimes h_{(2)})\otimes a_{(2)}$$

for all $a\in A$ and $h\in H$ (H is cocommutative!). θ is an isomorphism. Its inverse θ^{-1} is given by the formula

$$\theta^{-1}(\sum a_i\otimes(h_i\otimes k_i))=\sum a_i\otimes\varepsilon(h_i)k_i$$

it follows that

$$\varphi=\theta\circ\gamma_A:\ A\otimes A\longrightarrow A\square_{H,\varepsilon_1}(H\otimes H)$$

is an isomorphism of R-modules. It is straightforward to show that

$$\varphi(a\otimes b)=\sum a_{(0)}b_{(0)}\otimes a_{(1)}\otimes b_{(1)}$$

and it follows immediately that φ is $H\otimes H$-colinear.

Definition of the map δ.
Let $A\in Z^1_{\mathrm{Sw}}(H,R,\mathrm{Pic}^{\mathrm{co}}_{\mathrm{gnb}}(R,\bullet))$ be a cocycle. We then have an isomorphism

$$\phi:\ A\otimes A\longrightarrow A\square_{H,\varepsilon_1}(H\otimes H)$$

of $H\otimes H$-comodules. Consider the maps

$$\widehat{m}=(I_A\otimes\varepsilon_H\otimes\varepsilon_H)\circ\phi:\ A\otimes A\longrightarrow A$$

and

$$\zeta_1,\zeta_2:\ A^{\otimes 3}\longrightarrow A\otimes H^{\otimes 2}$$

given by

$$\begin{aligned}\zeta_1(a\otimes b\otimes c) &= \sum(\widehat{m}\circ(\widehat{m}\otimes I_A))(a\otimes b_{(0)}\otimes c_{(0)})\otimes b_{(1)}\otimes c_{(1)}\\ \zeta_2(a\otimes b\otimes c) &= \sum(\widehat{m}\circ(I_A\circ\widehat{m}))(a\otimes b_{(0)}\otimes c_{(0)})\otimes b_{(1)}\otimes c_{(1)}\end{aligned}$$

It is clear that $\widehat{m}$ makes A into an associative algebra if and only if $\zeta_1 = \zeta_2$. Suppose for a moment that $A \cong H$ as an H-comodule. Take $h, k \in H$, and write

$$\phi(h \otimes k) = \sum a_i \otimes (m_i \otimes n_i) \in A\Box_{H,\varepsilon_1}(H \otimes H)$$

Then

$$\sum a_{i_{(0)}} \otimes (m_i \otimes n_i) \otimes a_{i_{(1)}} = \sum a_i \otimes (m_{i_{(1)}} \otimes n_{i_{(1)}}) \otimes m_{i_{(2)}} n_{i_{(2)}} = \rho(f(h \otimes k))$$

and

$$\widehat{m}(h \otimes k) = \sum \varepsilon(m_i)\varepsilon(n_i)a_i$$

Now

$$\begin{aligned}
\sum \varepsilon(\widehat{m}(h_{(1)} \otimes k_{(1)}))h_{(2)} \otimes k_{(2)} &= \sum(\varepsilon^{\otimes 3} \otimes I_H)(f(h_{(1)} \otimes k_{(1)}) \otimes h_{(2)}k_{(2)}) \\
&= (\varepsilon^{\otimes 3} \otimes I_H)(\rho(\phi(h \otimes k))) \\
&= (\varepsilon^{\otimes 3} \otimes I_H)(\sum a_{i_{(0)}} \otimes (m_i \otimes n_i) \otimes a_{i_{(1)}}) \\
&= \sum \varepsilon(m_i)\varepsilon(n_i)a_i
\end{aligned}$$

and we have shown that

$$\widehat{m}(h \otimes k) = \sum f(h_{(1)} \otimes k_{(1)})h_{(2)}k_{(2)}$$

where $f = \varepsilon \circ \widehat{m} = \varepsilon^{\otimes 3} \circ \phi$.
We now claim that $f : H \otimes H \to R$ is convolution invertible. Define $g : H \otimes H \to R$ by

$$g(h \otimes k) = (\varepsilon \otimes \varepsilon)(\phi^{-1}(\sum h_{(1)} \otimes k_{(1)} \otimes h_{(2)}k_{(2)}))$$

then

$$\begin{aligned}
&(f * g)(h \otimes k) \\
&= \sum \varepsilon^{\otimes 5}\big(\phi(h_{(1)} \otimes k_{(1)}) \otimes \phi^{-1}(\sum h_{(2)}k_{(2)} \otimes h_{(3)} \otimes k_{(3)})\big) \\
&= \varepsilon^{\otimes 5}\big(\phi(\phi^{-1}(\sum h_{(1)}k_{(1)} \otimes h_{(2)} \otimes k_{(2)})_{(2)}) \otimes \phi^{-1}(\sum h_{(1)}k_{(1)} \otimes h_{(2)} \otimes k_{(2)})_{(1)}\big) \\
&\qquad (\phi^{-1} \text{ is } H \otimes H \text{ colinear}) \\
&= \varepsilon^{\otimes 5}\big(\phi(\phi^{-1}(\sum h_{(1)}k_{(1)} \otimes h_{(2)} \otimes k_{(2)})_{(1)}) \otimes \phi^{-1}(\sum h_{(1)}k_{(1)} \otimes h_{(2)} \otimes k_{(2)})_{(2)}\big) \\
&\qquad (H \text{ is cocommutative}) \\
&= \varepsilon^{\otimes 5}\big(((\phi \circ \phi^{-1})(\sum h_{(1)}k_{(1)} \otimes h_{(2)} \otimes k_{(2)}))_{(0)} \otimes ((\phi \circ \phi^{-1})(\sum h_{(1)}k_{(1)} \otimes h_{(2)} \otimes k_{(2)}))_{(1)}\big) \\
&\qquad (\phi \text{ is } H \otimes H \text{ colinear}) \\
&= \varepsilon^{\otimes 5}(\sum h_{(1)}k_{(1)} \otimes h_{(2)} \otimes k_{(2)} \otimes h_{(3)} \otimes k_{(3)}) \\
&= \varepsilon(h)\varepsilon(k)
\end{aligned}$$

Observe next that the map

$$\alpha : H^{\otimes 3} \longrightarrow H^{\otimes 3} : h \otimes k \otimes l \mapsto \sum hk_{(1)}l_{(1)} \otimes k_{(2)} \otimes l_{(2)}$$

is an isomorphism. The inverse α^{-1} is given by the formula

$$\alpha^{-1}(h \otimes k \otimes l) = \sum hS(k_{(1)}l_{(1)}) \otimes k_{(2)} \otimes l_{(2)}$$

We now compute easily that

$$\begin{aligned}\zeta_1(h\otimes k\otimes l) &= \sum f(h_{(1)}\otimes k_{(1)})f(h_{(2)}k_{(2)}\otimes l_{(1)})h_{(3)}k_{(3)}l_{(2)}\otimes k_{(4)}\otimes l_{(3)}\\ &= \alpha(f(h_{(1)}\otimes k_{(1)})f(h_{(2)}k_{(2)}\otimes l_{(1)})h_{(3)}\otimes k_{(3)}\otimes l_{(2)}\end{aligned}$$

hence

$$\zeta_1=\alpha\circ(\varepsilon_3(f)*\varepsilon_1(f)*I_{H^{\otimes 3}})$$

From Lemma 10.4.6, it follows that

$$\zeta_1^{-1}=(\varepsilon_3(g)*\varepsilon_1(g)*I_{H^{\otimes 3}})\circ\alpha^{-1}$$

In a similar way, we obtain

$$\zeta_2=\alpha\circ(\varepsilon_0(f)*\varepsilon_2(f)*I_{H^{\otimes 3}})$$

and, using Lemma 10.4.6, it follows that

$$\begin{aligned}\zeta_1^{-1}\circ\zeta_2 &= (\varepsilon_3(g)*\varepsilon_1(g)*I_{H^{\otimes 3}})\\ &\qquad \alpha^{-1}\circ\alpha\circ(\varepsilon_0(f)*\varepsilon_2(f)*I_{H^{\otimes 3}})\\ &= (\varepsilon_3(g)*\varepsilon_1(g)*I_{H^{\otimes 3}})\circ(\varepsilon_0(f)*\varepsilon_2(f)*I_{H^{\otimes 3}})\\ &= \Delta_2(f)*I_{H^{\otimes 3}} \qquad (10.28)\end{aligned}$$

Now we return to the general case. Let S be a faithfully flat extension of R such that $S\otimes A\cong S\otimes H$ as $H\otimes H$-comodules. We know from the above arguments that $\zeta_1\otimes I_S$ is an isomorphism of S-modules, and this implies that ζ_1 is an isomorphism of R-modules (S/R is faithfully flat). Now consider the map

$$\zeta_1^{-1}\circ\zeta_2:\ A^{\otimes 3}\xrightarrow{\zeta_2}A\otimes H^{\otimes 2}\xrightarrow{\zeta_1^{-1}}A^{\otimes 3}$$

$\zeta_1^{-1}\circ\zeta_2$ is an $H^{\otimes 3}$-colinear isomorphism of invertible $H^{\otimes 3}$-comodules with geometric normal basis. From Proposition 10.4.9, we know that there exists a convolution invertible $u\in(H\otimes H\otimes H)^*$ such that

$$\zeta_1^{-1}\circ\zeta_2=u*I_{H^{\otimes 3}}$$

From (10.28), it follows that u is a coboundary if $A\cong H$ as an H-comodule. Therefore $u\otimes I_S:\ H^{\otimes 3}\otimes S\to S$ is a coboundary in $B^3(S\otimes H,S,\mathbb{G}_m)\subset Z^3(S\otimes H,S,\mathbb{G}_m)$. From the fact that S is faithful as an R-algebra, it follows that u is a cocycle in $Z^3(H,R,\mathbb{G}_m)$, and we define $\delta([A])=[u]$. We leave it to the reader to show that δ is well-defined: if we start with a different isomorphism $\varphi':\ A^{\otimes 3}\to A\Box_{H,\varepsilon_1}H\otimes H$, then we obtain a cocycle u' that is cohomologous to u.

Exactness at $H^1_{\rm Sw}(H,R,{\rm Pic}^{\rm co}_{\rm gnb}))$
We first show that $\delta\circ\gamma=1$. If A is an H-Galois object, then we take the isomorphism

$$\varphi:\ \theta\circ\gamma_A:\ A\otimes A\longrightarrow A\Box_{H,\varepsilon_1}(H\otimes H):\ a\otimes b\mapsto\sum a_{(0)}b_{(0)}\otimes a_{(1)}\otimes b_{(1)}$$

from the first part of the proof. We see immediately that the map $\widehat{m}$ defined above is nothing else then the multiplication map on A, and therefore $\zeta_1=\zeta_2$, $\zeta_1^{-1}\circ\zeta_2=I_{A^{\otimes 3}}$ and $u=\varepsilon$.

Conversely, if $\delta([A]) = [\Delta_2(f)]$, with $f: H \otimes H \to R$ convolution invertible, then we replace the $H^{\otimes 2}$-colinear isomorphism

$$\varphi: A \otimes A \longrightarrow A \square_{H,\varepsilon_1} (H \otimes H)$$

by

$$\varphi': A \otimes A \longrightarrow A \square_{H,\varepsilon_1} (H \otimes H)$$

with

$$\varphi'(a \otimes b) = \sum g(a_{(1)} \otimes b_{(1)}) \varphi(a_{(0)} \otimes b_{(0)})$$

where g is the convolution inverse of f. We then compute easily that

$$\widehat{m}'(a \otimes b) = \sum g(a_{(1)} \otimes b_{(1)}) \widehat{m}(a_{(0)} \otimes b_{(0)})$$

and it follows that

$$\zeta_1'(a \otimes b \otimes c) = \sum g(a_{(1)} b_{(1)} \otimes c_{(1)}) g(a_{(2)} \otimes b_{(2)}) \zeta_1(a_{(0)} \otimes b_{(0)} \otimes c_{(0)}) \tag{10.29}$$

Using the notation introduced before Proposition 10.4.9, (10.29) can be rewritten as follows:

$$\zeta_1' = \zeta_1 \circ (\varepsilon_1(g) * \varepsilon_3(g) * I_{A^{\otimes 3}})$$

Using Proposition 10.4.9, we obtain

$$\zeta_1'^{-1} = (\varepsilon_1(f) * \varepsilon_3(f) * I_{A^{\otimes 3}}) \circ \zeta_1^{-1} \tag{10.30}$$

In a similar way,

$$\zeta_2' = \zeta_2 \circ (\varepsilon_0(g) * \varepsilon_2(g) * I_{A^{\otimes 3}}) \tag{10.31}$$

Now

$$\zeta_1^{-1} \circ \zeta_2 = \Delta_2(f) * I_{A^{\otimes 3}} \tag{10.32}$$

Combining (10.30-10.32) and using Proposition 10.4.9, we obtain

$$\begin{aligned} \zeta_1'^{-1} \circ \zeta_2' &= (\varepsilon_1(f) * \varepsilon_3(f) * I_{A^{\otimes 3}}) \circ (\Delta_2(f) * I_{A^{\otimes 3}}) \circ (\varepsilon_0(g) * \varepsilon_2(g) * I_{A^{\otimes 3}}) \\ &= \Delta_2(g) * \Delta_2(f) * I_{A^{\otimes 3}} = I_{A^{\otimes 3}} \end{aligned}$$

and it follows that $\widehat{m}'$ makes A into an associative algebra. Now $S \otimes A \cong S \otimes H$ as an $S \otimes H$-comodule, with multiplication twisted by the Sweedler cocycle $\varepsilon \circ \widehat{m}' \otimes I_S$. It follows that $S \otimes A$ is an $S \otimes H$-Galois object, and that A is an H-Galois object.
□

As one might expect, the exact sequence (10.27) can be extended to an exact sequence of infinite length. This sequence is similar the corresponding exact sequences for Amitsur cohomology and Harrison cohomology (see Theorems 6.1.3 and 9.2.1). We will give a brief sketch of the construction, and refer to [34, Sec. 5] for further details.

We need a formal dual of Harrison cohomology with values in the category $\underline{\underline{\mathrm{Pic}}}_{\mathrm{gnb}}(R, \bullet)$, that is, Sweedler cohomology with values in the category $\underline{\underline{\mathrm{DPic}}}^{\mathrm{co}}_{\mathrm{gnb}}(R, \bullet)$.

$\underline{\underline{\mathrm{DPic}}}^{\mathrm{co}}_{\mathrm{gnb}}(R, H)$ is the category of dual pairs of invertible H-comodules. We will see below that this category has the same Grothendieck and Whitehead group as

$\underline{\underline{\mathrm{Pic}}}^{\mathrm{co}}_{\mathrm{gnb}}(R,H)$. It also has the advantage that every object has a "natural" inverse in the category, and this is the reason why we work with the category of dual pairs of invertible comodules rather than the category of invertible comodules.

Objects: $\underline{M} = (M, M', \mu)$, where M and M' are invertible H-comodules with geometric normal basis, and $\mu: M'\square_H M \to H$ an H-colinear isomorphism.

Morphisms: $\underline{f} = (f, f'): \underline{M} \to \underline{N}$, with $f: M \to N$ and $f': M' \to N'$ H-colinear isomorphisms such that $\mu = \nu \circ (f\square f')$.

Product: The cotensor product induces a product on $\underline{\underline{\mathrm{Pic}}}_{\mathrm{gnb}}(R,H)$, making the category into a monoidal category:

$$\underline{M}\square_H\underline{N} = (M\square_H N, M'\square_H N', \mu\square\nu)$$

It is easy to see that

$$\begin{aligned} K_0\underline{\underline{\mathrm{DPic}}}^{\mathrm{co}}_{\mathrm{gnb}}(R,H) &\cong \mathrm{Pic}^{\mathrm{co}}_{\mathrm{gnb}}(R,H) \\ K_1\underline{\underline{\mathrm{DPic}}}^{\mathrm{co}}_{\mathrm{gnb}}(R,H) &\cong \mathbb{G}_m(H^*) \end{aligned} \tag{10.33}$$

The neutral element of $K_0\underline{\underline{\mathrm{DPic}}}^{\mathrm{co}}_{\mathrm{gnb}}(R,H)$ is represented by $\underline{H} = (H, H, I_H)$. The inverse $[\underline{M}]^{-1}$ of $[\underline{M}] \in K_0\underline{\underline{\mathrm{DPic}}}^{\mathrm{co}}_{\mathrm{gnb}}(R,H)$ is represented by

$$\underline{M}^{\tau} = (M', M, \mu\circ\tau)$$

where $\tau: M'\square_H M \to M\square_H M'$ is the switch map (well-defined because of the cocommutativity of H).

For a morphism $\underline{f}: \underline{M} \to \underline{N}$ in $\underline{\underline{\mathrm{DPic}}}^{\mathrm{co}}_{\mathrm{gnb}}(R,H)$, we define

$$\underline{f}^{\tau}: \underline{N}^{\tau} \to \underline{M}^{\tau}$$

by

$$\underline{f}^{\tau} = (f'^{-1}, f^{-1})$$

Now, we consider the maps

$$\varepsilon_0, \varepsilon_1, \ldots, \varepsilon_{n+1}: H^{\otimes n+1} \longrightarrow H^{\otimes n}$$

defined in (10.26) (with $C = R$). They induce $n+2$ functors

$$\epsilon_0, \epsilon_1, \ldots, \epsilon_{n+1}: \underline{\underline{\mathrm{DPic}}}^{\mathrm{co}}_{\mathrm{gnb}}(R, H^{\otimes n}) \longrightarrow \underline{\underline{\mathrm{DPic}}}^{\mathrm{co}}_{\mathrm{gnb}}(R, H^{\otimes n+1})$$

as follows:

$$\epsilon_i(M, M', \mu) = (H^{\otimes n+1}\square_{\varepsilon_i, H^{\otimes n}} M, H^{\otimes n+1}\square_{\varepsilon_i, H^{\otimes n}} M', I_{H^{\otimes n+1}}\square\mu)$$

Here the subscript ε_i means that we view $H^{\otimes n+1}$ as an $H^{\otimes n}$-comodule via ε_i. On the level of the morphisms, ϵ_i is defined by

$$\epsilon_i(f, f') = (I_{H^{\otimes n+1}}\square f, I_{H^{\otimes n+1}}\square f')$$

Now we define a functor,

$$\delta_n: \underline{\underline{\mathrm{DPic}}}^{\mathrm{co}}_{\mathrm{gnb}}(R, H^{\otimes n}) \longrightarrow \underline{\underline{\mathrm{DPic}}}^{\mathrm{co}}_{\mathrm{gnb}}(R, H^{\otimes n+1})$$

by

$$\delta_n(\underline{M}) = \epsilon_0(\underline{M})\Box_{H^{\otimes n+1}}\epsilon_1(\underline{M})^\tau\Box_{H^{\otimes n+1}}\cdots\Box_{H^{\otimes n+1}}\epsilon_{n+1}(\underline{M})^{(\tau)}$$
$$\delta_n(\underline{f}) = \epsilon_0(\underline{f})\Box\epsilon_1(\underline{f})^\tau\Box\cdots\Box\epsilon_{n+1}(\underline{f})^{(\tau)}$$

From the fact that the (modified) Sweedler complex is a complex, it follows that, for every $\underline{M} \in \underline{\underline{\mathrm{DPic}}}^{\mathrm{co}}_{\mathrm{gnb}}(R, H^{\otimes n})$, we have a natural map

$$\lambda_{\underline{M}}:\ (\delta_{n+1}\circ\delta_n)(\underline{M})\longrightarrow H^{\otimes n+2}$$

It is clear that δ_n is a cofinal, product preserving functor. Using (10.33), we find that the exact sequence (C.2) takes the form

$$\mathbb{G}_m(H^{*\otimes n})\xrightarrow{\Delta_n}\mathbb{G}_m(H^{*\otimes n+1})\xrightarrow{\kappa_n}K_0\Psi\underline{\delta_n}\xrightarrow{\theta_n}\mathrm{Pic}^{\mathrm{co}}_{\mathrm{gnb}}(R, H^{\otimes n})\xrightarrow{\Delta'_n}\mathrm{Pic}^{\mathrm{co}}_{\mathrm{gnb}}(R, H^{\otimes n+1}) \tag{10.34}$$

The map Δ_n comes from the Sweedler complex with values in $\mathbb{G}_m$. Δ'_n comes from the (modified) Sweedler complex with values in $\mathrm{Pic}^{\mathrm{co}}_{\mathrm{gnb}}(R,\bullet)$.
The elements of $K_0\Psi\underline{\delta_n}$ are represented by twotuples of the form $(\underline{M},\underline{\alpha})$, with $\underline{M} \in \underline{\underline{\mathrm{Pic}}}^{\mathrm{co}}_{\mathrm{gnb}}(R, H^{\otimes n+1})$ such that $\underline{\beta}\circ\delta_n(\underline{f}) = \underline{\alpha}$.
We define two more maps

$$D'_{n-1}:\ \mathrm{Pic}^{\mathrm{co}}_{\mathrm{gnb}}(R, H^{\otimes n-1}) \cong K_0\underline{\underline{\mathrm{DPic}}}^{\mathrm{co}}_{\mathrm{gnb}}(R, H^{\otimes n-1})\longrightarrow K_1\underline{\Phi\delta_n}$$

by

$$D'_{n-1}[\underline{M}] = [(\delta_{n-1}(\underline{M}), \lambda_{\underline{M}})] \tag{10.35}$$

and

$$D_n:\ K_1\underline{\Phi\delta_n}\to\mathbb{G}_m(H^{*\otimes n+2})$$

by

$$D_n[(\underline{M},\underline{\alpha})] = \varepsilon\circ\xi \tag{10.36}$$

where ξ is the first component of the composition

$$\underline{H}^{\otimes n+2}\xrightarrow{\lambda^{-1}_{\underline{M}}}\delta_{n+1}(\delta_n(\underline{M}))\xrightarrow{\delta_{n+1}(\alpha)}\underline{H}^{\otimes n+2}$$

One easily shows that $D_n\circ D'_{n-1} = 0$, and we define *modified Sweedler cohomology with values in the category of invertible comodules with geometric normal basis* as follows:

$$H^n_{\mathrm{Sw}}(H, R, \underline{\underline{\mathrm{DPic}}}^{\mathrm{co}}_{\mathrm{gnb}}(R,\bullet)) = \mathrm{Ker}\,(D_n)/\mathrm{Im}\,(D'_{n-1}) \tag{10.37}$$

Theorem 10.5.10 *With notations and assumptions as above, we have a long exact sequence of cohomology groups*

$$\begin{array}{llll}
0 & \longrightarrow\ H^1(H,R,\mathbb{G}_m) & \longrightarrow & H^0_{\mathrm{Sw}}(H,R,\underline{\underline{\mathrm{DPic}}}^{\mathrm{co}}_{\mathrm{gnb}}(R,\bullet))\\
 & \longrightarrow\ H^0_{\mathrm{Sw}}(H,R,\mathrm{Pic}^{\mathrm{co}}_{\mathrm{gnb}}(R,\bullet)) = 0 & \longrightarrow & H^2(H,R,\mathbb{G}_m)\\
 & \longrightarrow\ H^1_{\mathrm{Sw}}(H,R,\underline{\underline{\mathrm{DPic}}}^{\mathrm{co}}_{\mathrm{gnb}}(R,\bullet)) & \longrightarrow & H^1_{\mathrm{Sw}}(H,R,\mathrm{Pic}^{\mathrm{co}}_{\mathrm{gnb}}(R,\bullet))\\
 & \longrightarrow\ H^3(H,R,\mathbb{G}_m) & \longrightarrow & \cdots
\end{array} \tag{10.38}$$

Furthermore

$$H^0_{\mathrm{Sw}}(H, R, \underline{\underline{\mathrm{DPic}}}^{\mathrm{co}}_{\mathrm{gnb}}(R,\bullet)) \cong H^1(H, R, \mathbb{G}_m) \cong \mathrm{Alg}(H, R) \tag{10.39}$$

and

$$H^1_{\mathrm{Sw}}(H, R, \underline{\underline{\mathrm{DPic}}}^{\mathrm{co}}_{\mathrm{gnb}}(R,\bullet)) \cong \mathrm{Gal}(R, H) \tag{10.40}$$

Proof Similar to the proofs of Theorems 6.1.3 and 9.2.1. We omit the details. □

Proposition 10.5.11 *Let A and B be H-comodule algebras such that $A\Box_H B \cong H$ as H-comodule algebras. Suppose also that A (and B) have geometric normal basis. Then A and B are H-Galois objects.*

Proof From Proposition 10.5.7, it follows that we can assume that A and B have normal basis, that is, $A \cong B \cong H$ as H-comodules. Write m_A and m_B for the multiplication maps on A and B, and let $f = \varepsilon \circ m_A$, $g = \varepsilon \circ m_B$. Then for all $h, k \in H$

$$\Delta(m_A(h \otimes k)) = \sum m_A(h_1 \otimes k_1) \otimes h_2 k_2$$

(A is an H-comodule algebra). Applying $\varepsilon \otimes I_H$ to both sides, we obtain

$$m_A(h \otimes k) = \sum f(h_1 \otimes k_1) h_2 k_2$$

and we have

$$m_A = m_H \circ (f * I_H) \text{ and } m_B = m_H \circ (g * I_H)$$

Consider the canonical H-colinear isomorphism

$$\Delta : H \longrightarrow A\Box_H B : h \mapsto \Delta(h)$$

The multiplication on $A\Box_H B$ induces a new multiplication $\widetilde{m} = m_H \circ (f * g * I_H)$ on H. Let $\widetilde{H}$ be H as an H-comodule, and with $\widetilde{m}$ as multiplication. Then $\widetilde{H}$ and H are isomorphic as H-comodule algebras. From the exactness of the sequence (10.25), it follows that $f * g$ is a Sweedler coboundary, and is therefore invertible. From the fact that A and B are associative, it follows that f and g are Sweedler cocycles. Now A and B are isomorphic to H as H-comodules, with multiplication twisted by a Sweedler cocycle. We then know from the proof of Lemma 10.5.5 that A and B are H-Galois objects. □

10.6 The split part of the group of Galois objects

Let S be a faithfully flat commutative R-algebra. We have a natural map

$$\mathrm{Gal}_{\mathrm{gnb}}(R, H) \longrightarrow \mathrm{Gal}_{\mathrm{gnb}}(S, S \otimes H)$$

The kernel of this map is denoted by $\mathrm{Gal}(S/R, H)$, and we write

$$\mathrm{Gal}^{\mathrm{s}}(R, H) = \cup_S \mathrm{Gal}(S/R, H)$$

where the union runs over all faithfully flat commutative R-algebras S. We call $\mathrm{Gal}^{\mathrm{s}}(R, H)$ the split part of the group of Galois objects. We will now give cohomological interpretations of $\mathrm{Gal}(S/R, H)$ and $\mathrm{Gal}^{\mathrm{s}}(R, H)$.

Theorem 10.6.1 *Let H be a commutative faithfully flat R-algebra, and S a faithfully flat commutative R-algebra. Then*

$$\mathrm{Gal}(S/R,H) \cong H^1(S/R, \mathrm{Alg}(\bullet \otimes H, \bullet))$$

$\mathrm{Alg}(\bullet \otimes H, \bullet)$ is a sheaf on R_{fl}, and consequently

$$\mathrm{Gal}^s(R,H) \cong H^1(R_{\mathrm{fl}}, \mathrm{Alg}(\bullet \otimes H, \bullet))$$

Proof In the proof of Proposition 10.4.7, we associated a cocycle in $Z^1(S/R, \mathbb{G}_m(\mathrm{Hom}_\bullet(\bullet \otimes H, \bullet))$ to every invertible comodule split by S. If this comodule is an invertible H-comodule algebra, then the constructed cocycle is not only convolution invertible, but it is also an algebra map, by Lemma 10.4.6. To any H-Galois object split by S, we may therefore associate a cocycle in $Z^1(S/R, \mathrm{Alg}(\bullet \otimes H, \bullet))$, and we leave it to the reader to verify that it defines a well-defined isomorphism

$$\mathrm{Gal}(S/R,H) \longrightarrow H^1(S/R, \mathrm{Alg}(\bullet \otimes H, \bullet))$$

Let $S \to S'$ be a covering in R_{fl}. In the proof of Proposition 10.4.7, we have seen that the sequence

$$1 \longrightarrow \mathrm{Hom}_S(H \otimes S, S) \longrightarrow \mathrm{Hom}_{S'}(H \otimes S', S') \rightrightarrows \mathrm{Hom}_{S'\otimes_S S'}(H \otimes (S' \otimes_S S'), S' \otimes_S S')$$

is exact. It follows immediately that the sequence

$$1 \longrightarrow \mathrm{Alg}(H \otimes S, S) \longrightarrow \mathrm{Alg}(H \otimes S', S') \rightrightarrows \mathrm{Alg}(H \otimes (S' \otimes_S S'), S' \otimes_S S')$$

is also exact, and this means that $\mathrm{Alg}(\bullet \otimes H, \bullet)$ is a sheaf. The last statement now follows after we take inductive limits over all faithfully flat extensions of R. □

10.7 About the Picard invariant map

The map $\gamma : \mathrm{Gal}_{\mathrm{gnb}}(R,H) \longrightarrow H^1$ is often called the *Picard invariant map.* In the case where $H = RG$, G a finite abelian group, it can be shown easily that γ is surjective.

Proposition 10.7.1 *Let G be a finite abelian group. Then we have the following exact sequence*

$$1 \longrightarrow H^2(G, \mathbb{G}_m(R)) \xrightarrow{\beta} \mathrm{Gal}(R, RG) \xrightarrow{\gamma} H^1(G, \mathrm{Pic}(R)) \longrightarrow 1 \tag{10.41}$$

Proof Recall that $\mathrm{Gal}(R, RG)$ is the group of isomorphism classes of strongly G-graded rings with R as part of degree e. It suffices to show that γ is surjective. G is a finite abelian group and is the direct sum of cyclic subgroups

$$G = C_{n_1} \times C_{n_2} \times \cdots \times C_{n_r}$$

Let C_{n_i} be generated by u_i, and $f \in H^1(G, \mathrm{Pic}(R)) = \mathrm{Hom}(G, \mathrm{Pic}(R))$ is given by the images of $u_1, \ldots, u_k$. Let $f(u_i) = [I_i]$, where $[I_i]$ is an element of order n_i in the Picard group of R. Now for each i, consider the Rees ring

$$\check{R}[I_i] = \oplus_{k \in \mathbb{N}} I_i^{\otimes k}$$

and divide out the ideal generated by

$$x_1 \otimes x_2 \otimes \cdots \otimes x_{n_i} - \alpha_i(x_1 \otimes x_2 \otimes \cdots \otimes x_{n_i})$$

for $x_1, x_2, \ldots, x_{n_i} \in I_i$, and where $\alpha_i : I_i^{\otimes n_i} \to R$ is a fixed isomorphism. The quotient

$$S_i = R \oplus I_i \oplus I_i^{\otimes 2} \oplus \cdots \oplus I_i^{\otimes n_k - 1}$$

is strongly C_{n_i}-graded. The direct sum $\oplus_{i=1}^r S_i$ is G-strongly graded, and its image under γ is f. □

Even in the case where H is finite, commutative and cocommutative, no examples are known of a situation where the Picard invariant map γ is not surjective. L.N. Childs kindly informed the author that the Picard invariant map is surjective in the following particular situations:

1. $H = (RG)^*$, with the order of G invertible in R, and R containing an $\exp(G)$-th root of unity. This appeared first in [52]. In view of Theorem 7.1.10, this is a special case of Proposition 10.7.1.

2. $H = RG$, where G is a finite abelian group. This is Proposition 10.7.1 and appeared first in [58].

3. $H = RC_p^*$, with p a prime number, not a zero-divisor in R, and R containing a primitive p-th root of unity. For the proof, we refer to [54].

4. R is the ring of integers in an algebraic number field K, and H is such that $K \otimes H \cong (KC_p)^*$ (see [63]).

5. H is free of rank 2 (see [118]).

10.8 Pairings and noncommutative Galois objects

Let H be a commutative, cocommutative, faithfully flat Hopf algebra. In the previous Section, we have described the split part $\mathrm{Gal}^s(R, H)$ of $\mathrm{Gal}_{\mathrm{gnb}}(R, H)$. In this Section, we will study the quotient $\mathrm{Gal}_{\mathrm{gnb}}(R, H)/\mathrm{Gal}^s(R, H)$.

Proposition 10.8.1 *The H-Galois objects that can be split by a faithfully flat extension S of R are the commutative H-Galois objects.*

Proof If A is split by S, then $S \otimes A \cong S \otimes H$ as $S \otimes H$-comodule algebras. Since S and H are commutative, $S \otimes A \cong S \otimes H$ is commutative, and A is commutative. Conversely, if A is commutative, then the isomorphism

$$\gamma : A \otimes A \longrightarrow A \otimes H : a \otimes b \mapsto \sum ab_{(0)} \otimes b_{(1)}$$

is an isomorphism of $A \otimes H$-comodule algebras (here $A \otimes H$ is viewed as a Hopf algebra over A). Hence A is split by A itself. □

It follows immediately from Proposition 10.8.1 that the split Galois objects with normal basis are classified by symmetric Sweedler 2-cocycles. Let $H^2_{\text{symm}}(H, R, \mathbb{G}_m)$ be the subgroup of $H^2(H, R, \mathbb{G}_m)$ consisting of classes of symmetric cocycles.

Definition 10.8.2 *A map $g : H \otimes H \to R$ is called a pairing if the following conditions hold for all $h, k, l \in H$:*

$$\begin{aligned} g(1 \otimes h) &= g(h \otimes 1) = \varepsilon(h) & (10.42)\\ g(hk \otimes l) &= \sum g(h \otimes l_{(1)}) g(k \otimes l_{(2)}) & (10.43)\\ g(h \otimes kl) &= \sum g(h_{(1)} \otimes k) g(h_{(2)} \otimes l) & (10.44)\\ g(S(h) \otimes k) &= g(h \otimes S(k)) & (10.45) \end{aligned}$$

A pairing g is called skew if $g \circ \Delta = \varepsilon$, that is

$$\sum g(h_{(1)} \otimes h_{(2)}) = \varepsilon(h) \qquad (10.46)$$

for all $h \in H$.

Let $P(H, R)$ and $P_{\text{sk}}(H, R)$ be the sets of pairings and skew pairings from $H \otimes H$ to R. The convolution product makes $P(H, R)$ and $P_{\text{sk}}(H, R)$ into groups. We leave it to the reader to verify that the inverse of a pairing g is $g \circ (S \otimes I) = g \circ (I \otimes S)$.

Lemma 10.8.3 *If H is finite, then $P(H, R) \cong \text{Hopf}(H, H^*)$.*

Proof We have an isomorphism

$$F : \text{Hom}(H, H^*) \longrightarrow (H \otimes H)^*$$

given by

$$F(f)(h \otimes k) = \langle f(h), k \rangle$$

for all $h, k \in H$ and $f \in \text{Hom}(H, H^*)$. We will show that F restricts to an isomorphism

$$F : \text{Hopf}(H, H^*) \longrightarrow P(H, R)$$

Let $f : H \to H^*$ be a Hopf algebra map, and $g = F(f)$. Then f is multiplicative: $f(hk) = f(h) * f(k)$, for $h, k \in H$ and $f(1) = \varepsilon$. Therefore

$$\begin{aligned} g(hk \otimes l) &= \langle f(hk), l \rangle \\ &= \langle f(h) * f(k), l \rangle \\ &= \sum \langle f(h), l_{(1)} \rangle \langle f(k), l_{(2)} \rangle \\ &= \sum g(h \otimes l_{(1)}) g(k \otimes l_{(2)}) \end{aligned}$$

and

$$g(1 \otimes h) = \langle f(1), h \rangle = \varepsilon(h)$$

proving (10.43) and the first half of (10.42).

f is comultiplicative, and therefore

$$m^* \circ f = (f \otimes f) \circ \Delta \quad \text{and} \quad \eta^* \circ f = \varepsilon$$

The first identity gives

$$\begin{aligned} g(hk \otimes l) &= \langle f(h), kl \rangle \\ &= (m^*(f)(h))(k \otimes l) \\ &= (f \otimes f)(\Delta(h))(k \otimes l) \\ &= \sum (f(h_{(1)}) \otimes f(h_{(2)}))(k \otimes l) \\ &= \sum g(h_{(1)} \otimes k) g(h_{(2)} \otimes l) \end{aligned}$$

and this proves (10.44). The second identity gives us that

$$(\eta^* \circ f)(h) = f(h)(\eta(1)) = \varepsilon(h)$$

and

$$g(h \otimes 1) = \langle f(h), \eta(1) \rangle = \varepsilon(h)$$

proving the second half of (10.42).

f preserves the antipode: $f \circ S = S^* \circ f$. It follows that

$$g(S(h) \otimes k) = \langle f(S(h)), k \rangle = \langle S^*(f(h)), k \rangle = \langle f(h), S(k) \rangle = g(h \otimes S(k))$$

and this finishes the proof of the fact that $g \in P(H, R)$. A similar computation shows that, for $g \in P(H, R)$, $f = F^{-1}(g) \in \mathrm{Hopf}(H, H^*)$. □

Lemma 10.8.4 *If $H = RG$, with G an abelian group, then*

$$P(H, R) = \mathrm{Bil}(G, \mathbb{G}_m(R))$$

the group of all bilinear maps from $G \times G$ to $\mathbb{G}_m(R)$.

Proof Suppose that $f : G \times G \to \mathbb{G}_m(R)$ is bilinear. Define

$$\alpha(f) = g : RG \otimes RG \longrightarrow R$$

by

$$g(r\sigma \otimes s\tau) = rsf(\sigma, \tau)$$

A straightforward computation shows that $g \in P(RG, R)$ and that

$$\alpha : \mathrm{Bil}(G, \mathbb{G}_m(R)) \longrightarrow P(RG, R)$$

is an injective homomorphism. Consider $f \in P(RG, R)$. For all $\sigma, \tau \in G$, we have that

$$g(\sigma \otimes \tau) g(\sigma \otimes \tau^{-1}) = g(\sigma \otimes 1) = 1$$

so $g(\sigma\otimes\tau)\in \mathbb{G}_m(R)$. The map $f:\ G\times G\to \mathbb{G}_m(R)$ defined by $f(\sigma,\tau)=g(\sigma\otimes\tau)$ is bilinear, and $\alpha(f)=g$. Thus α is surjective. □

Consider the map

$$\beta:\ \mathbb{G}_m((H\otimes H)^*)\longrightarrow \mathbb{G}_m((H\otimes H)^*):\ f\mapsto f*(f^{-1}\circ\tau)$$

in other words

$$\beta(f)(h\otimes k)=\sum f(h_{(1)}\otimes k_{(1)})f^{-1}(k_{(2)}\otimes h_{(2)})$$

for all $f:\ H\otimes H\to R$ and $h,k\in H$.

Lemma 10.8.5 *If $f:\ H\otimes H\to R$ is a symmetric cocycle, then $\beta(f)=\varepsilon\otimes\varepsilon$. β induces a well-defined map, still denoted by β,*

$$\beta:\ H^2(H,R,\mathbb{G}_m)\longrightarrow P_{\rm sk}(H,R)$$

Proof The first statement is obvious. From the fact that every coboundary is symmetric, it follows that β is well-defined. We know from Lemma 9.1.2 that every class in $H^2(H,R)$ can be represented by a normalized cocycle, that is, a cocycle f satisfying the relation

$$f(1\otimes h)=f(h\otimes 1)=\varepsilon(h)$$

for all $h\in H$. Let $\beta([f])=g$. Then

$$g(1\otimes h)=g(h\otimes 1)=\varepsilon(h) \tag{10.47}$$

We claim that

$$g(hk\otimes l)=\sum g(h\otimes l_{(1)})g(h\otimes l_{(2)}) \tag{10.48}$$

for all $h,k,l\in H$. Indeed,

$$\begin{aligned}
&((f\otimes\varepsilon)*(g\circ(\mu\otimes I)))(h\otimes k\otimes l)\\
&= \sum f(h_{(1)}\otimes k_{(1)})f(h_{(2)}k_{(2)}\otimes l_{(1)})f^{-1}(l_{(2)}\otimes h_{(3)}k_{(3)})\\
&= \sum f(h_{(1)}\otimes k_{(1)}l_{(1)})f(k_{(2)}\otimes l_{(2)})f^{-1}(l_{(3)}\otimes h_{(2)}k_{(3)}) && (10.49)\\
&= \sum g(k_{(1)}\otimes l_{(1)})f(l_{(2)}\otimes k_{(2)})f(h_{(1)}\otimes k_{(3)}l_{(3)})f^{-1}(l_{(4)}\otimes h_{(2)}k_{(4)})\\
&= \sum g(k_{(1)}\otimes l_{(1)})f(h_{(1)}l_{(2)}\otimes k_{(2)})f(h_{(2)}\otimes l_{(3)})f^{-1}(l_{(4)}\otimes h_{(3)}k_{(3)}) && (10.50)\\
&= \sum g(k_{(1)}\otimes l_{(1)})f^{-1}(l_{(2)}\otimes h_{(1)})f(h_{(2)}\otimes l_{(3)})f(h_{(3)}\otimes k_{(2)}) && (10.51)\\
&= \sum g(k_{(1)}\otimes l_{(1)})g(h_{(1)}\otimes l_{(2)})f(h_{(2)}\otimes k_{(2)})
\end{aligned}$$

We used the cocycle relation three times: at step (10.49) (on the first two factors), at step (10.50) (on the second and the third factor) and at step (10.51) (on the second and the fourth factor). Taking the convolution product of both sides with $(f\otimes\varepsilon)^{-1}$, we obtain (10.48). A similar computation shows that

$$g(h\otimes kl)=\sum g(h_{(1)}\otimes k)g(h_{(2)}\otimes l) \tag{10.52}$$

Let us next show that

$$g(S(h) \otimes l) = g(h \otimes S(l)) \tag{10.53}$$

for all $h, l \in H$.

$$\begin{aligned}
&\sum f(h_{(1)} \otimes S(h_{(2)}))f(l_{(1)} \otimes S(l_{(2)})) \\
&= \sum f(h_{(1)} \otimes S(h_{(2)}))f(h_{(3)} \otimes S(l_{(1)}))f(h_{(4)}S(l_{(2)}) \otimes l_{(3)})f^{-1}(h_{(5)} \otimes 1) \quad \mathbf{(10.54)} \\
&= \sum f(h_{(1)}S(l_{(1)})l_{(2)} \otimes S(h_{(2)}))f(h_{(3)} \otimes S(l_{(3)}))f(h_{(4)}S(l_{(4)}) \otimes l_{(5)}) \\
&= \sum f(h_{(1)} \otimes S(l_{(1)}))f(h_{(2)}S(l_{(2)}) \otimes l_{(3)}S(h_{(3)}))f(l_{(4)} \otimes S(h_{(4)})) \quad (10.55) \\
&= \sum g(h_{(1)} \otimes S(l_{(1)}))f(S(l_{(2)}) \otimes h_{(2)})g^{-1}(S(h_{(3)}) \otimes l_{(3)})f(S(h_{(4)}) \otimes l_{(4)}) \\
&\qquad f(h_{(5)}S(l_{(5)}) \otimes l_{(6)}S(h_{(6)})) \\
&= \sum g(h_{(1)} \otimes S(l_{(1)}))g^{-1}(S(h_{(2)}) \otimes l_{(2)})f(S(l_{(3)}) \otimes h_{(3)}) \\
&\qquad f(h_{(4)}S(l_{(4)}) \otimes S(h_{(5)}))f(h_{(6)}S(l_{(5)})S(h_{(7)}) \otimes l_{(6)}) \quad (10.56) \\
&= \sum g(h_{(1)} \otimes S(l_{(1)}))g^{-1}(S(h_{(2)}) \otimes l_{(2)})f(S(l_{(3)}) \otimes h_{(3)}S(h_{(4)})) \\
&\qquad f(h_{(5)} \otimes S(h_{(6)}))f(S(l_{(4)}) \otimes l_{(5)}) \quad (10.57) \\
&= \sum g(h_{(1)} \otimes S(l_{(1)}))g^{-1}(S(h_{(2)}) \otimes l_{(2)})f(h_{(3)} \otimes S(h_{(4)}))f(l_{(3)} \otimes S(l_{(4)})) \\
&\qquad \qquad \mathbf{(10.58)}
\end{aligned}$$

Here we used the cocycle relation four times. Step (10.54): on the second factor; step (10.55): on the first and the third factor; step (10.56): on the fourth and the fifth factor; step (10.57): on the third and the fourth factor. At step (10.58), we used the fact that $f(S(h) \otimes h) = f(h \otimes S(h))$. This can be seen if we apply the cocycle relation with $h = l$ and $k = S(h)$.
(10.53) now follows if we take the convolution product of both sides with

$$\sum f^{-1}(h_{(1)} \otimes S(h_{(2)}))f^{-1}(l_{(1)} \otimes S(l_{(2)}))g(S(h_{(3)}) \otimes l_{(3)})$$

We have now shown that g is a pairing. It is clear that g is skew, since

$$\beta(f)(\Delta(h)) = \sum f(h_{(1)} \otimes h_{(2)})f^{-1}(h_{(3)} \otimes h_{(4)}) = \varepsilon(h)$$

and this finishes the proof of Lemma 10.8.5. □

It is clear that $\mathrm{Ker}\,(\beta) = H^2_{\mathrm{symm}}(H, R, \mathbb{G}_m)$, and we can summarize the above results as follows.

Theorem 10.8.6 *[33] If H is a faithfully flat, commutative and cocommutative Hopf algebra, then we have an exact sequence*

$$1 \longrightarrow H^2_{\mathrm{symm}}(H, R, \mathbb{G}_m) \longrightarrow H^2(H, R, \mathbb{G}_m) \xrightarrow{\beta} P_{\mathrm{sk}}(H, R) \tag{10.59}$$

Even in the case where H is a finite Hopf algebra, we do not know if the map β is surjective. No examples are known of situations where β fails to be surjective. If $H = RG$, with G a finite abelian group, then it is not difficult to show that β is surjective.

Theorem 10.8.7 *[65] Let G be a finite abelian group. Then we have a split exact sequence*

$$1 \longrightarrow H^2_{\text{symm}}(G, \mathbb{G}_m(R)) \longrightarrow H^2(G, \mathbb{G}_m(R)) \overset{\beta}{\longrightarrow} P_{\text{sk}}(RG, R) \longrightarrow 1$$

Proof Suppose that $g : G \times G \to \mathbb{G}_m(R)$ is a skew bilinear map. Since G is finite abelian, G is the direct sum of cyclic subgroups

$$G = C_{n_{(1)}} \times C_{n_{(2)}} \times \cdots \times C_{n_r}$$

Let C_{n_i} be generated by u_i, and define $f : G \times G \to \mathbb{G}_m(R)$ by

$$f(u_i^k, u_j^l) = \begin{cases} g(u_i^k, u_j^l) & \text{if } i > j; \\ 1 & \text{if } i \leq j. \end{cases}$$

We leave it to the reader to show that f is a cocycle and that $\beta(f) = g$. $\square$

Some more examples will be examined in the next Chapter. We will see each time that the map β is surjective and split.

Chapter 11

Some examples

11.1 Group algebras

Let $H = RG$, with G a finite abelian group. We can write

$$G = C_{n_1} \times C_{n_2} \times \cdots C_{n_r}$$

with C_{n_i} the cyclic group of order n_i. We will write σ_i as a generator for C_{n_i}.

Proposition 11.1.1 *We have an exact sequence*

$$1 \longrightarrow \prod_{i=1}^{r} \mathbb{G}_m(R)/\mathbb{G}_m(R)^{n_i} \longrightarrow \mathrm{Gal}^s(R, RG) \longrightarrow \mathrm{Hom}(G, \mathrm{Pic}(R)) \longrightarrow 1 \tag{11.1}$$

Proof For any faithfully flat R-algebra S, we have that

$$\mathrm{Alg}(SG, S) = \mathrm{Hom}(G(S), \mathbb{G}_m(S)) = \prod_{i=1}^{r} \mu_{n_i}(S)$$

where $G(\bullet)$ is the sheaf associated to the constant presheaf G on R_{fl}, in other words, $G(S)$ is the group of continuous functions from $\mathrm{Spec}(S)$ (with the Zariski topology) to G (with the discrete topology). We claim that the sequence

$$1 \longrightarrow \prod_{i=1}^{r} \mu_{n_i} \longrightarrow \prod_{i=1}^{r} \mathbb{G}_m \stackrel{\text{power}}{\longrightarrow} \prod_{i=1}^{r} \mathbb{G}_m \longrightarrow 1$$

is exact as a sequence of sheaves on R_{fl}. The map power is given by sending the i-th component to its n_i-th power. The only thing we have to show is that the map power is surjective as a map of sheaves: for any flat R-algebra S and $a \in \mathbb{G}_m(S)$, there exists a faithfully flat extension $S \to S'$ such that a has an n-th root in S': take

$$S' = S[x]/(x^n - a)$$

We know have a long exact sequence of cohomology groups

$$\begin{array}{ccccccc} 1 & \longrightarrow & \prod_{i=1}^r \mu_{n_i}(R) & \longrightarrow & \prod_{i=1}^r \mathbb{G}_m(R) & \xrightarrow{\text{power}} & \prod_{i=1}^r \mathbb{G}_m(R) \\ & \longrightarrow & H^1(R_{\text{fl}}, \prod_{i=1}^r \mu_{n_i}) & \cong & \text{Gal}^s(R, RG) & \longrightarrow & H^1(R_{\text{fl}}, \mathbb{G}_m)^r \\ & \cong & \text{Pic}(R)^r & \xrightarrow{\alpha_1} & H^1(R_{\text{fl}}, \mathbb{G}_m)^r & \cong & \text{Pic}(R)^r \end{array}$$

It is clear that $\text{Ker}\,(\alpha_1) = \text{Hom}(G, \text{Pic}(R))$, and the result follows. □

Remark 11.1.2 If $|G|$ is invertible in R, and, and R contains a primitive n_i-th root of unity for every i, then $RG \cong GR$, and the above results also hold for the group of G-Galois extensions. In particular, if $G = C_n$ is the cyclic group of order n, then the exact sequence (11.1) takes the form

$$1 \longrightarrow \mathbb{G}_m(R)/\mathbb{G}_m(R)^n \longrightarrow \text{Gal}(R, GR) \longrightarrow \text{Pic}_n(R) \longrightarrow 1$$

This sequence is called the Kummer exact sequence.

Corollary 11.1.3 *If G is a finite abelian group, then*

$$\text{Gal}(R, H)/\text{Gal}^s(R, H) \cong H^2(G, \mathbb{G}_m(R))/H^2_{\text{symm}}(G, \mathbb{G}_m(R))$$

Proof Compare the exact sequences (11.1) and (10.41), and take into account that $H^1(G, \text{Pic}(R)) = \text{Hom}(G, \text{Pic}(R))$. □

From Theorem 10.8.7, we know that it suffices to prove that $P_{\text{sk}}(RG, R) = 1$ in order to show that all Galois objects are commutative. This is the case if G is a cyclic group.

Proposition 11.1.4 *Let $H = RC_n$. Then we have an exact sequence*

$$1 \longrightarrow \mathbb{G}_m(R)/\mathbb{G}_m(R)^n \longrightarrow \text{Gal}(R, RC_n) \longrightarrow \text{Pic}_n(R) \longrightarrow 1 \qquad (11.2)$$

Proof The exact sequence (11.1) takes the form

$$1 \longrightarrow \mathbb{G}_m(R)/\mathbb{G}_m(R)^n \longrightarrow \text{Gal}^s(R, RC_n) \longrightarrow \text{Pic}_n(R) \longrightarrow 1$$

and it suffices to show that $\text{Gal}^s(R, RC_n) = \text{Gal}(R, RC_n)$, or, equivalently, that $P_{\text{sk}}(RC_n, R) = 1$.
Take $g \in P_{\text{sk}}(RC_n, R)$, and let σ be a generator for C_n. Then $g(\sigma, \sigma) = 1$, since g is skew, and therefore $g(\sigma^i, \sigma^j) = 1^{ij} = 1$ and g is trivial. □

Let us now study a slightly more complicated example.

Proposition 11.1.5 *Let $G = C_n \times C_n$, with n invertible in the connected ring R and suppose R contains a primitive n-th root of unity η. Then*

$$H^2(G, \mathbb{G}_m(R)) = \text{Gal}_n(R, RG) = \mathbb{G}_m(R)/(\mathbb{G}_m(R))^n \times \mathbb{G}_m(R)/(\mathbb{G}_m(R))^n \times C_n$$

Proof From Proposition 11.1.1, it follows that

$$H^2_{\text{symm}}(G, \mathbb{G}_m(R)) = \text{Gal}_n(R, RG) = \mathbb{G}_m(R)/(\mathbb{G}_m(R))^n \times \mathbb{G}_m(R)/(\mathbb{G}_m(R))^n$$

From Theorem 10.7.1, it now follows that it suffices to show that

$$P_{\text{sk}}(RG, R) = C_n$$

Write $G = \langle\sigma\rangle \times \langle\tau\rangle$, and define $\sigma^*, \tau^* \in G^*$ by

$$\langle\sigma^*, \sigma\rangle = \eta; \quad \langle\sigma^*, \tau\rangle = 1; \quad \langle\tau^*, \tau\rangle = \eta; \quad \langle\tau^*, \sigma\rangle = 1$$

Then $G^* = \langle\sigma^*\rangle \times \langle\tau^*\rangle$. Take $f : G \to G^*$, and suppose that

$$f(\sigma) = (\sigma^*)^i(\tau^*)^j; \quad f(\tau) = (\sigma^*)^k(\tau^*)^l$$

If f is skew, then we have for all integers α en β that

$$1 = \langle f(\sigma^\alpha\tau^\beta), \sigma^\alpha\tau^\beta\rangle = \eta^{\alpha(\alpha i+\beta j)+\beta(\alpha k+\beta l)}$$

and it follows that

$$n|\alpha\beta(j+k) + \alpha^2 i + \beta^2 j$$

If we take $\alpha = 1$ and $\beta = 0$, then it follows that $n|i$, so we may take $i = 0$. In a similar way, we may assume that $l = 0$. It now follows that $j + k$ is a multiple of n, and therefore the skew Hopf algebra maps from G to G^* are $f_0, f_1, \ldots, f_{n-1}$ defined by

$$f_j(\sigma) = (\tau^*)^j \text{ and } f_j(\tau) = (\sigma^*)^{n-j}$$

and this finishes our proof. □

Observe that the skew pairing d_j corresponding to the Hopf algebra map f_j described above satisfies the equation

$$d_j(u_\sigma \otimes u_\tau) = \langle f_j(\sigma), \tau\rangle = \eta^j$$

We can easily construct the noncommutative RG-Galois objects (that is, the G-strongly graded rings), corresponding to f_j: let A_j be given by the following data:

- $A_j = RG$ as a G-graded R-module;
- $(u_\sigma)^i = u_{\sigma^i}$; $(u_\tau)^i = u_{\tau^i}$; $u_\sigma u_\tau = u_{\sigma\tau}$; $u_\tau u_\sigma = \eta^j u_{\sigma\tau}$.

It is easy to see that

$$A_j \square_{RG} A_k \cong A_{j+k}$$

Remark 11.1.6 Let $G = C_q$. We have already remarked in Section 8.5 that a C_q-Galois extension in the "classical" sense is an RC_q^*-Galois object. If $q^{-1} \in R$, and if R contains a primitive q-th root of unity, then we can describe $\text{Gal}(R, RC_q^*)$ easily using Lemma 11.1.2. One can ask what happens if we drop one of the two conditions on R. The characteristic p case is discussed in Section 11.3.

Assume that p is invertible in R, but that R does not necessarily contain a primitive q-th root of unity. The procedure to compute $\mathrm{Gal}(R, RC_q^*)$ is then the following. Consider S_q, the q-th cyclotomic extension of R, and $\Gamma = \mathrm{Aut}_R(S_q)$. Lemma 11.1.2 can be applied to compute the cyclic Galois extensions of S_q, and one tries to recover the cyclic Galois extensions of R by descent (or by applying the Galois correspondence theorem). We refer to the literature for a more discussion, see e.g. [111] and [94].
A different, more conceptual, approach to this problem has been given by Janelidze in [109]. His result can be summarized as follows. Observe that $\mu_q(S_q)$, the group of q-th roots of unity in S_q, and $\mathbb{G}_m(S_q)$ can be viewed as Γ-sets. Then $\mathrm{Gal}(R, RC_q^*)$ is isomorphic to the subgroup of $\mathrm{Ext}_{\mathbb{Z}\Gamma}(\mu_q(S_q), \mathbb{G}_m(S_q))$ consisting of all extensions

$$1 \longrightarrow \mathbb{G}_m(S_q) \longrightarrow E \longrightarrow \mu_q(S_q) \longrightarrow 1$$

that have a section as a map of Γ. Apparently, the results in [111] can be derived from Janelidze's result.

11.2 Monogenic Larson orders

We now focus our attention to one-parameter Larson orders, as introduced at the end of Section 7.1. We recall the notations. Let R be a commutative ring, and suppose that p is not a zero-divisor in R. Write $q = p^d$, and consider ideals I and J of R such that $IJ^{p^{d-1}(p-1)} = pR$. Consider $K = \frac{1}{p} IJ^{p^{d-1}(p-1)-1} \subset R_p$. In Section 7.1 we have seen that

$$H_J = \check{R}[Ky]/\big(K^q((1+y)^q - 1)\big)$$

is a faithfully projective, commutative and cocommutative Hopf algebra. $u = 1+y$ is a grouplike element of H_J. The aim of this Section is to compute the group of Galois objects $\mathrm{Gal}(R, H_J)$. First observe that every H_J-Galois object is commutative: if A is an H_J-Galois object, then $R_p \otimes A$ is an R_pC_q-Galois object, and all R_pC_q-Galois objects are commutative. It will therefore be sufficient to compute

$$\mathrm{Gal}^s(R, H_J) = H^1(R_{\mathrm{fl}}, \mathrm{Alg}(\heartsuit \otimes H_J, \heartsuit))$$

First we will compute $\mathrm{Alg}(S \otimes H_J, S))$, for any flat R-algebra S. Observe that p is not a zero-divisor in S. Indeed, p is invertible in R_p and therefore in $S_p = R_p \otimes S$. Now $R \to R_p$ is monic, so $S \to S_p$ is monic (S is flat), and therefore p cannot be a zero-divisor in S_p.
It follows that $S \otimes H_J$ is also a one-parameter Larson order, but now over the groundring S, and starting from the S-ideal SJ:

$$S \otimes H_J = \check{S}[SKy]/\big(SK^q((1+y)^q - 1)\big)$$

Take $\rho \in \mathrm{Alg}(S \otimes H_J, S)$. Fix $m_i \in J$, $n_i \in K$ such that $\sum_i m_i n_i = 1$ (remember that $JK = R$). Then

$$\rho(u) = \rho(1+y) = 1 + \sum_i m_i \rho(n_i y) \in 1 + JS$$

since $\rho(n_i y) \in S$. Also $u^q = 1$, so

$$\rho(u) \in \mu_{q,J}(S) = \{u \in \mu_q(S) | u \equiv 1 \mod JS\}$$

Conversely, take $\eta \in \mu_{q,J}(S)$. The multiplicative map

$$\rho : S_p \otimes H_J \longrightarrow S_p$$

given by $\rho(u) = \eta$ restricts to

$$\rho : S \otimes H_J \longrightarrow S$$

Indeed, for all $k \in K$, $\rho(ky) = k\rho(u-1) = k(\eta - 1) \in KJS = S$.
For any ideal M of R, we introduce the notation

$$\mathbb{G}_{m,M}(S) = \{s \in \mathbb{G}_m(S) | s \equiv 1 \mod MS\}$$

Lemma 11.2.1 *If $s \in \mathbb{G}_{m,J}(S)$, then $s^q \in \mathbb{G}_{m,J^q}(S)$.*

Proof Write $s = 1 + t$, with $t \in JS$. Then

$$s^q - 1 = \sum_{i=1}^{q} \binom{q}{i} t^i$$

In the proof of Lemma 7.1.11, we have seen that there exists an ideal M of R such that

$$\binom{q}{i} R = MJ^{q-i}$$

hence

$$\binom{q}{i} t^i \in MJ^{q-i}J^i = MJ^q \subset J^q$$

and

$$s^q - 1 \in J^q$$

□

Lemma 11.2.2 *We have an exact sequence of sheaves on R_{fl}:*

$$1 \longrightarrow \mu_{q,J} \longrightarrow \mathbb{G}_{m,J} \xrightarrow{(\bullet)^q} \mathbb{G}_{m,J^q} \longrightarrow 1$$

Proof It is clear that $\mu_{q,J}(S) = \mathrm{Ker}\,(\mathbb{G}_{m,J}(S) \xrightarrow{(\bullet)^q} \mathbb{G}_{m,J^q}(S))$, so it suffices to show that $(\bullet)^q$ is surjective as a map of sheaves. Let S be a flat R-algebra, and take $s \in \mathbb{G}_{m,J^q}(S)$. Consider the Rees ring

$$\check{S}[KSt] = S \oplus KSt \oplus K^2St^2 \oplus \cdots$$

From Lemma 7.1.11 and the fact that $s - 1 \in J^q$, it follows that

$$K^q((1+t)^q - s) \subset \check{S}[KSt]$$

and

$$S' = \check{S}[KSt]/\Big(K^q((1+t)^q - s)\Big)$$

is a faithfully flat S-algebra. In S', we have that $(1+t)^q = s$, so $1+t$ is a q-th root of s. Also $t \in JS'$, and the result follows. □

We now have an exact sequence of cohomology groups

$$\begin{array}{ccccccc} 1 & \longrightarrow & \mu_{q,J}(R) & \longrightarrow & \mathbb{G}_{m,J}(R) & \xrightarrow{(\bullet)^q} & \mathbb{G}_{m,J^q}(R) \\ & \longrightarrow & H^1(R_{\mathrm{fl}}, \mu_{q,J}) = \mathrm{Gal}(R, H_J) & \longrightarrow & H^1(R_{\mathrm{fl}}, \mathbb{G}_{m,J}) & \longrightarrow & H^1(R_{\mathrm{fl}}, \mathbb{G}_{m,J^q}) \\ & \longrightarrow & \cdots & & & & \end{array} \tag{11.3}$$

Corollary 11.2.3 *Let* $\mathcal{K} = \mathrm{Ker}\Big(H^1(R_{\mathrm{fl}}, \mathbb{G}_{m,J}) \xrightarrow{(\bullet)^q} H^1(R_{\mathrm{fl}}, \mathbb{G}_{m,J^q})\Big)$. *We then have a short exact sequence*

$$1 \longrightarrow \mathbb{G}_{m,J^q}(R)/\mathbb{G}_{m,J}(R)^q \longrightarrow \mathrm{Gal}(R, H_J) \longrightarrow \mathcal{K} \longrightarrow 1 \tag{11.4}$$

Our aim is now to show that

$$\mathbb{G}_{m,J^q}(R)/\mathbb{G}_{m,J}(R)^q \cong H^2(H_J, R, \mathbb{G}_m)$$

To this end, we will construct a monomorphism

$$\mathcal{K} \longrightarrow H^1(H_J, R, \mathrm{Pic})$$

We will use that

$$H^1(H_J, R, \mathrm{Pic}) \cong H^1_{\mathrm{Sw}}(H_J, R, \mathrm{Pic}^{\mathrm{co}}) \cong H^1_{\mathrm{Sw}}(H_J, R, H^1(R_{\mathrm{fl}}, \mathrm{Aut}^{\heartsuit\otimes\diamondsuit}(\heartsuit \otimes \diamondsuit)))$$

We will always assume that $\heartsuit$ refers to Amitsur cohomology, and that $\diamondsuit$ refers to (modified) Sweedler cohomology.

Lemma 11.2.4 *We have a map*

$$\varphi : \mathbb{G}_{m,J}(R) \longrightarrow \mathrm{Aut}^{H_J}(H_J)$$

defined as follows. For any $s \in \mathbb{G}_{m,J}(R)$, $\varphi(s) = f$, *with*

$$f(u^j) = s^j u^j \tag{11.5}$$

for $j = 0, 1, 2, \ldots, q-1$.

Proof Obviously (11.5) defines a colinear automorphism

$$f : R_p C_q \longrightarrow R_p C_q$$

We have to show first that f restricts to a map

$$f : H_J \longrightarrow H_J$$

In view of Lemma 10.4.6, it suffices to show that $\varepsilon \circ f$ restricts to a map $H_J \to R$. This is ok if we can show that

$$(\varepsilon \circ f)(y^k) \in J^k$$

for $k = 1, 2, \ldots, q-1$. Now

$$\begin{aligned}
(\varepsilon \circ f)(y^k) &= (\varepsilon \circ f)((u-1)^k) \\
&= (\varepsilon \circ f)\Big(\sum_{i=0}^{k}(-1)^{k-i}\binom{k}{i}u^i\Big) \\
&= \varepsilon\Big(\sum_{i=0}^{k}(-1)^{k-i}\binom{k}{i}s^i u^i\Big) \\
&= \sum_{i=0}^{k}(-1)^{k-i}\binom{k}{i}s^i \\
&= (s-1)^k \in J^k
\end{aligned}$$

It is easy to show f is a bijection: the inverse of f is $\phi(s^{-1})$. □

Now let S be a commutative flat R-algebra. We can apply Lemma 11.2.4 to the S-Hopf algebra $S \otimes H_J$. This yields a map

$$\varphi(S):\ \mathbb{G}_{m,J}(S) \longrightarrow \mathrm{Aut}^{S\otimes H_J}(S \otimes H_J)$$

and

$$\varphi(S)(1_S \otimes u^j) = s^j \otimes u^j$$

Lemma 11.2.5 *We have a map between Amitsur complexes*

$$\begin{array}{ccccccc}
\cdots \xrightarrow{\Delta_{n-2}} & \mathbb{G}_{m,J}(S^{\otimes n}) & \xrightarrow{\Delta_{n-1}} & \mathbb{G}_{m,J}(S^{\otimes n+1}) & \xrightarrow{\Delta_n} \cdots \\
& \downarrow{\scriptstyle \varphi(S^{\otimes n})} & & \downarrow{\scriptstyle \varphi(S^{\otimes n+1})} & \\
\cdots \xrightarrow{\Delta_{n-2}} & \mathrm{Aut}^{S^{\otimes n}\otimes H_J}(S^{\otimes n} \otimes H_J) & \xrightarrow{\Delta_{n-1}} & \mathrm{Aut}^{S^{\otimes n+1}\otimes H_J}(S^{\otimes n+1} \otimes H_J) & \xrightarrow{\Delta_n} \cdots
\end{array}$$

and consequently $\varphi(S^{\otimes n+1})$ induces a monomorphism

$$\phi_n(S):\ H^n(S/R, \mathbb{G}_{m,J}) \longrightarrow H^n(S/R, \mathrm{Aut}^{\heartsuit\otimes H_J}(\heartsuit \otimes H_J))$$

Proof This follows immediately from the construction of the map $\varphi(S)$. It is clear that $s \in Z^n(S/R, \mathbb{G}_{m,J})$ is a coboundary if $\phi(S^{\otimes n+1})(s)$ is a coboundary in $Z^n(S/R, \mathrm{Aut}^{\heartsuit\otimes H_J})$, and this implies that $\phi_n(S)$ is a monomorphism. □

Lemma 11.2.6 *Suppose that S is a faithfully flat R-algebra, and let $[f] \in H^1(S/R, \mathrm{Aut}^{\heartsuit\otimes H_J}(\heartsuit \otimes H_J))$ be represented by*

$$f:\ S \otimes S \otimes H_J \longrightarrow S \otimes S \otimes H_J$$

Suppose that

$$f(1_S \otimes 1_S \otimes u^i) = f_i \otimes u^i \qquad (0 \le i \le q-1)$$

with $f_i \in \mathbb{G}_m(S\otimes S)$. *Then* $[f]$ *is a cocycle in* $Z^1_{\mathrm{Sw}}(H_J, R, H^1(S/R, \mathrm{Aut}^{\heartsuit\otimes\diamondsuit}(\heartsuit\otimes\diamondsuit)))$ *if and only if there exists an* $S\otimes H_J\otimes H_J$*-colinear automorphism*

$$g:\ S\otimes H_J\otimes H_J \longrightarrow S\otimes H_J\otimes H_J$$

such that

$$f_if_j = \begin{cases} (g_{ij}^{-1}\otimes g_{ij})f_{i+j} & \text{if } i+j<q \\ (g_{ij}^{-1}\otimes g_{ij})f_{i+j-q} & \text{if } i+j\geq q \end{cases} \tag{11.6}$$

where we wrote

$$g(1_S\otimes u^i\otimes u^j) = g_{ij}\otimes u^i\otimes u^j$$

for all $i,j\in\{0,1,\ldots,q-1\}$.

Proof Write

$$f(\diamondsuit) = H^1(S/R, \mathrm{Aut}^{\heartsuit\otimes\diamondsuit}(\heartsuit\otimes\diamondsuit))$$

Recall from Section 1 that we have three maps

$$\varepsilon_0,\ \varepsilon_1,\ \varepsilon_2:\ H_J^{\otimes 2}\longrightarrow H_J$$

given by the rules

$$\varepsilon_0(h_1\otimes h_2) = \varepsilon(h_1)h_2,\ \varepsilon_1(h_1\otimes h_2) = h_1h_2,\ \varepsilon_2(h_1\otimes h_2) = \varepsilon(h_2)h_1$$

For any H-comodule M, we have an isomorphism

$$\alpha:\ H\square_H M\longrightarrow M$$

given by

$$\alpha(\sum h_i\otimes m_i) = \sum \varepsilon(h_i)m_i \ \text{ and }\ \alpha^{-1}(m) = \sum m_{(1)}\otimes m_{(0)} \tag{11.7}$$

Let $H\square_{H,\varepsilon_i}(H\otimes H)$ be the cotensor product of H and $H\otimes H$, where $H\otimes H$ is considered as an H-comodule via ε_i, that is,

$$\rho(h\otimes k) = \sum h_1\otimes k_1\otimes \varepsilon_i(h_2\otimes k_2)$$

We can view the maps

$$F(\varepsilon_i)(f):\ S\otimes S\otimes (H\square_{H,\varepsilon_i}(H\otimes H))\longrightarrow S\otimes S\otimes (H\square_{H,\varepsilon_i}(H\otimes H))$$

as maps

$$F(\varepsilon_i)(f):\ S\otimes S\otimes H\otimes H\longrightarrow S\otimes S\otimes H\otimes H$$

For $i=0$, we obtain

$$\begin{aligned} & F(\varepsilon_0)(f)(1_S\otimes 1_S\otimes u^i\otimes u^j) \\ = \ & \big((I_{S^{\otimes 2}}\otimes\alpha)\circ(f\otimes I_{H^{\otimes 2}})\circ(I_{S^{\otimes 2}}\otimes\alpha^{-1})\big)(1_S\otimes 1_S\otimes u^i\otimes u^j) \\ = \ & (I_{S^{\otimes 2}}\otimes\alpha)\big((f\otimes I_{H^{\otimes 2}})(1_S\otimes 1_S\otimes u^j\otimes(u^i\otimes u^i))\big) \\ = \ & (I_{S^{\otimes 2}}\otimes\alpha)(f_j\otimes u^j\otimes(u^i\otimes u^i)) \\ = \ & f_j\otimes u^i\otimes u^j \end{aligned}$$

In a similar way, we obtain that

$$F(\varepsilon_2)(f)(1_S \otimes 1_S \otimes u^i \otimes u^j) = f_i \otimes u^i \otimes u^j$$

and it follows that

$$(F(\varepsilon_0) \circ F(\varepsilon_2))(f)(1_S \otimes 1_S \otimes u^i \otimes u^j) = f_i f_j \otimes u^i \otimes u^j \tag{11.8}$$

The isomorphism

$$\alpha : H\square_{H,\varepsilon_1} H \otimes H \longrightarrow H \otimes H$$

given by

$$\alpha(\sum h_i \otimes k_i \otimes l_i) = \sum \varepsilon(h_i) k_i \otimes l_i \quad \text{and} \quad \alpha^{-1}(k \otimes l) = \sum k_2 l_2 \otimes (k_1 \otimes l_1) \tag{11.9}$$

and therefore

$$\begin{aligned} & F(\varepsilon_1)(f)(1_S \otimes 1_S \otimes u^i \otimes u^j) \\ = & \Big((I_{S^{\otimes 2}} \otimes \alpha) \circ (f \otimes I_{H^{\otimes 2}}) \circ (I_{S^{\otimes 2}} \otimes \alpha^{-1})\Big)(1_S \otimes 1_S \otimes u^i \otimes u^j) \\ = & (I_{S^{\otimes 2}} \otimes \alpha)\Big((f \otimes I_{H^{\otimes 2}})(1_S \otimes 1_S \otimes u^{i+j} \otimes (u^i \otimes u^j)\Big) \\ = & \begin{cases} f_{i+j} \otimes u_i \otimes u_j & \text{if } i+j < q \\ f_{i+j-q} \otimes u_i \otimes u_j & \text{if } i+j \geq q \end{cases} \end{aligned} \tag{11.10}$$

Now $[f]$ is a Sweedler cocycle if $(F(\varepsilon_0) \circ F(\varepsilon_2))(f)$ is Amitsur cohomologous to $F(\varepsilon_1)(f)$. What is an Amitsur coboundary? Consider $g \in \mathrm{Aut}^{S\otimes H_J \otimes H_J}(S \otimes H_J \otimes H_J)$ and assume that

$$g(1_S \otimes u^i \otimes u^j) = g_{ij} \otimes u^i \otimes u^j \quad (0 \leq i, j < q)$$

Then we easily find that

$$\Delta_0(g) = (I_S \otimes g) \circ (g \otimes I_S)^{-1}$$

and

$$\Delta_0(g)(1_S \otimes 1_S \otimes u^i \otimes u^j = g_{ij}^{-1} \otimes g_{ij} \otimes u^i \otimes u^j \tag{11.11}$$

Now $[f]$ is a cocycle if and only if there exists a $g \in \mathrm{Aut}^{S\otimes H_J \otimes H_J}(S \otimes H_J \otimes H_J)$ such that

$$(F(\varepsilon_0) \circ F(\varepsilon_2))(f) = \Delta_0(g) \circ F(\varepsilon_1)(f) \tag{11.12}$$

If we evaluate both sides at $u^i \otimes u^j$, then we find (11.6), using (11.8,11.9) and (11.11). Conversely, if (11.6) holds, then (11.12) holds in the point $u^i \otimes u^j$, hence everywhere. □

Lemma 11.2.7 *Let $s \in \mathbb{G}_{m,J^q}(R)$. There exists a colinear isomorphism*

$$g : H_J \otimes H_J \longrightarrow H_J \otimes H_J$$

satisfying

$$g(u^i \otimes u^j) = \begin{cases} u^i \otimes u^j & \text{if } i+j < q; \\ s u^i \otimes u^j & \text{if } i+j \geq q; \end{cases}$$

Proof Let $s \in \mathbb{G}_{m,J^q}(R)$. Consider the commutative R-algebra

$$S' = \check{R}[Kt]/\big(K^q((1+t)^q - s)\big)$$

introduced in the proof of Lemma 11.2.2 (we replace S by R). As an R-module, $S' \cong H_J$ (map kt^i to ky^i, for all $k \in K$ and $i = 0, 1, \ldots, q-1$). The isomorphism induces an H_J-comodule structure on H_J, and, if we let $v = 1 + t$, then

$$\rho_{S'}(v^i) = v^i \otimes u^i$$

for $i = 0, 1, \ldots, q-1$. It is clear that $R_p \otimes S' = R_p[v]/(v^q - s)$ is an R_pC_q-comodule algebra, and therefore S' is an H_J-module algebra (the compatibility relation holds in $R_p \otimes S'$, and therefore in S'). The multiplication map on S' induces a map $m_{S'}:\ H_J \otimes H_J \to H_J$. Consider the map

$$\mu = \varepsilon \circ m_{S'}:\ H_J \otimes H_J \to R$$

Then

$$\mu(u^i \otimes u^j) = \begin{cases} \varepsilon(u^{i+j}) = 1 & \text{if } i+j < q \\ \varepsilon(su^{i+j-q}) = s & \text{if } i+j \geq q \end{cases}$$

In view of Lemma 10.4.6, the map

$$g = \mu * I_{H^{\otimes 2}}:\ H_J \otimes H_J \longrightarrow H_J \otimes H_J$$

satisfies all the requirements. □

Lemma 11.2.8 *Take* $[s] \in H^1(S/R, \mathbb{G}_{m,J})$. *If* $[s^q] = 1$ *in* $H^1(S/R, \mathbb{G}_{m,J^q})$, *then* $\phi_1(S)[s]$ *is a Sweedler cocycle in* $Z^1_{\mathrm{Sw}}(H, R, H^1(S/R, \mathrm{Aut}^{\heartsuit\otimes\diamondsuit}(\heartsuit \otimes \diamondsuit)))$.

Proof First suppose that $[s^q] = 1$ in $H^1(S/R, \mathbb{G}_{m,J^q})$. Then there exists $t \in \mathbb{G}_{m,J^q}(S)$ such that $s^q = t \otimes t^{-1}$. Let $f = \varphi(S \otimes S)$. With notations as above, we can write

$$f_i = s^i \quad (0 \leq i < q)$$

Consider

$$g:\ S \otimes H_J \otimes H_J \longrightarrow S \otimes H_J \otimes H_J$$

defined by

$$g(1_S \otimes u^i \otimes u^j) = \begin{cases} 1_S \otimes u^i \otimes u^j & \text{if } i+j < q \\ t \otimes u^i \otimes u^j & \text{if } i+j \geq q \end{cases}$$

It follows from Lemma 11.2.7 that g is well-defined. It follows that $\phi_1(S)$ satisfies (11.6): put $f = \phi_1(S)$. Then $f_i = s^i$ for $0 \leq i < q$ and, for $i + j < q$, we have

$$f_i f_j = s^{i+j} = f_{i+j} = (g_{ij}^{-1} \otimes g_{ij}) f_{i+j}$$

For $i + j \geq q$, we have

$$f_i f_j = s^{i+j} = s^{i+j-q} s^q = (t \otimes t^{-1}) f_{i+j-q} = (g_{ij}^{-1} \otimes g_{ij}) f_{i+j-q}$$

□

Corollary 11.2.9 *$\phi_1(S)$ restricts to a monomorphism*

$$\phi_1(S): \ \mathcal{K}_S = \mathrm{Ker}\left(H^1(S/R, \mathbb{G}_{m,J}) \to H^1(S/R, \mathbb{G}_{m,J^q})\right) \longrightarrow Z^1_{\mathrm{Sw}}(H, R, H^1(S/R, \mathrm{Aut}^{\heartsuit\otimes\diamondsuit}(\heartsuit \otimes \diamondsuit)))$$

Taking inductive limits over all faithfully flat extensions S of R, we find a monomorphism

$$\phi_1: \ \mathcal{K} \longrightarrow Z^1_{\mathrm{Sw}}(H_J, R, H^1(S/R, \mathrm{Aut}^{\heartsuit\otimes\diamondsuit}(\heartsuit \otimes \diamondsuit)))$$

Theorem 11.2.10 *Let R be a ring in which the prime number p is not a zero-divisor, and consider the Hopf algebra H_J. Then*

$$\mathbb{G}_{m,J^q}(R)/\mathbb{G}_{m,J}(R)^q \cong H^2(H_J, R, \mathbb{G}_m)$$

Proof We compare the sequences (11.4) and (10.27) and use Corollary 11.2.9

$$\begin{array}{ccccccccc}
1 & \to & \mathbb{G}_{m,J^q}(R)/\mathbb{G}_{m,J}(R)^q & \to & \mathrm{Gal}(R,H) & \to & \mathcal{K} & \to & 1 \\
 & & & & \cong & & \downarrow \phi_1 & & \\
1 & \to & H^2(H,R,\mathbb{G}_m) & \xrightarrow{\beta} & \mathrm{Gal}(R,H) & \xrightarrow{\gamma} & H^1(H,R,\mathrm{Pic}) & \xrightarrow{\delta} & H^3(H,R,\mathbb{G}_m)
\end{array}$$

A diagram chasing yields a monomorphism from $\mathbb{G}_{m,J^q}(R)/\mathbb{G}_{m,J}(R)^q$ to $H^2(H_J, R, \mathbb{G}_m)$ and from the fact that $\phi_1(S)$ is monic, it follows that this map is a monomorphism. □

Remark 11.2.11 In Lemma 7.1.13, we have mentioned that one can construct Hopf algebras as kernels of isogenies between n-dimensional polynomial formal groups. In [139], this approach is used to describe the group of Galois objects in the case where the groundring R is the valuation ring of a local field of characteristic 0. If R is local, then the Hopf algebra H_J can be obtained as the kernel of isogenies between one-dimensional formal groups (see [60]). It can be investigated that the description in [139] of the group of Galois objects is then equivalent to the one in Theorem 11.2.10 (see [60, Theorem 1.8]).
We also mention that Theorem 11.2.10 generalizes results in [105], [93] and [164]. In these three papers, the rank p case is studied, everytime with different techniques. In [188], Underwood gives a description of the group of Galois objects of an arbitrary Hopf algebra order in KC_{p^2} (in the case where the ground ring R is a principal ideal domain).

We now consider the case $d = 1$, that is, $q = p$ is a prime number. Suppose that I and J are ideals of R such that $IJ = (\eta - 1)R$. We know from Theorem 7.1.15 that H_J and H_I are each others dual. We will use the same notations as in Theorem 7.1.15:

$$H_J = \check{R}[Ky]/\left(K^p((1+y)^p - 1)\right) \ \text{and} \ H_I = \check{R}[Lt]/\left(L^p((1+t)^p - 1)\right)$$

$$K = \frac{1}{p} I^{p-1} J^{p-2} \ \text{and} \ L = \frac{1}{p} I^{p-2} J^{p-1}$$

$$u = 1 + y \text{ and } v = 1 + t$$

$$\langle u, v\rangle = 1 \text{ and } \langle y, t\rangle = \eta - 1$$

Our aim is now to compute $H^2(H_J \otimes H_I, R, \mathbb{G}_m)$. From Theorem 11.2.10 it follows that

$$\begin{aligned} H^2_{\text{symm}}(H_J \otimes H_I, R, \mathbb{G}_m) &\cong H^2(H_J, R, \mathbb{G}_m) \times H^2(H_I, R, \mathbb{G}_m) \\ &\cong \mathbb{G}_{m,J^p}(R)/\mathbb{G}_{m,J}(R)^p \times \mathbb{G}_{m,I^p}(R)/\mathbb{G}_{m,I}(R)^p \end{aligned}$$

We will prove that

$$\mathrm{P}_{\text{sk}}(H_J \otimes H_I, R) \cong C_p$$

We know that $R_p \otimes H_J \otimes H_I = R_p(C_p \times C_p)$, and also that

$$\mathrm{P}_{\text{sk}}(R_p(C_p \times C_p), R_p) \cong C_p$$

(see Proposition 11.1.5).

For $j = 0, 1, \ldots, p-1$, we consider the algebra A_j introduced at the end of Section 11.1. A_j can be described as follows:

$$A_j = \oplus_{m,n=0}^{p-1} R_p u^m v^n$$

with

$$vu = \eta^j uv \tag{11.13}$$

Now let

$$B_j = \oplus_{m,n=0}^{p-1} K^m x^m L^n t^{\subset} A_j$$

We claim that B_j is a subset of A_j. It suffices to show that $(lt)(ky) \in B_j$, for all $k \in K$ and $l \in L$. From (11.13), it follows that

$$1 + t + y + ty = \eta^j(1 + t + y + yt)$$

or

$$ty = \eta^j yt + (\eta^j - 1)(1 + y + y)$$

and

$$(lt)(ky) = \eta^j(ky)(lt) + (\eta^j - 1)kl(1 + y + y)$$

Now $(\eta - 1)KL = IJKL = R$, so $(\eta^j - 1)kl \in R$, and this implies that $(lt)(ky) \in B_j$. It is clear that B_j is an $H_J \otimes H_I$-Galois object with normal basis. Observe that we have a commutative diagram

$$\begin{array}{ccc} H^2(H_J \otimes H_I, R, \mathbb{G}_m) & \xrightarrow{\beta} & \mathrm{P}_{\text{sk}}(H_J \otimes H_I, R) \\ \downarrow \alpha & & \downarrow \gamma \\ H^2(R_p(C_p \times C_p), R_p, \mathbb{G}_m) & \xrightarrow{\beta} & \mathrm{P}_{\text{sk}}(R_p(C_p \times C_p), R_p) = C_p \end{array}$$

α and γ are induced by the base extension $R \to R_p$. It is clear that the map γ is injective: a skew pairing $H_J \otimes H_I \otimes H_J \otimes H_I \to R$ is always the restriction of a skew pairing $R_p(C_p \times C_p) \otimes R_p(C_p \times C_p) \to R_p$.

Consider one of the p skew pairings $d_j : R_p(C_p \times C_p) \otimes R_p(C_p \times C_p) \to R_p$ described at the end of Section 11. It can be verified directly that d_j restricts to a skew pairing

$$d'_j : H_J \otimes H_I \otimes H_J \otimes H_I \to R$$

and this shows that γ is also surjective. Another way to see this is the following: the algebra B_j is determined by a cocycle in $H^2(H_J \otimes H_I, R, \mathbb{G}_m)$, and the image of this cocycle under the map β is the skew pairing d'_j. This shows that the map β is surjective, and that

$$\mathrm{P_{sk}}(H_J \otimes H_I, R) \cong C_p$$

From the fact that

$$A_j \square_{R_p(C_p \times C_p)} A_k \cong A_{j+k}$$

it follows easily that the map β splits. We summarize our results in the next Theorem.

Theorem 11.2.12 *Let R be a connected ring of characteristic p in which p is not a zero-divisor, and containing a primitive p-th root of unity η. Let I and J be two ideals such that $IJ = (\eta - 1)R$. Then*

$$H^2(H_J \otimes H_I, R, \mathbb{G}_m) \cong \mathbb{G}_{m,J^p}(R)/\mathbb{G}_{m,J}(R)^p \times \mathbb{G}_{m,I^p}(R)/\mathbb{G}_{m,I}(R)^p \times C_p$$

Moreover the sequence (10.59) is split exact.

11.3 Examples in characteristic p

Throughout this Section, R will be a ring of characteristic p. Write $q = p^d$, and consider the Hopf algebra $H = R[x]/(x^q - x)$ with

$$\begin{aligned} \Delta(x) &= x \otimes 1 + 1 \otimes x \\ S(x) &= -x \\ \varepsilon(x) &= 0 \end{aligned}$$

Proposition 11.3.1 *Take $d = 1$. Then $H \cong RC_p^*$ is the dual of the group ring over the cyclic group of order p.*

Proof Let σ be a generator of the cyclic group C_p, and write $v = \sigma - 1 \in RC_p$. Then

$$0 = \sigma^p - 1 = (v+1)^p - 1 = v^p$$

and we can rewrite the group ring as follows:

$$RC_p = R[v]/(v^p)$$

with

$$\begin{aligned} \Delta(v) &= v \otimes v + v \otimes 1 + 1 \otimes v \\ \varepsilon(v) &= 0 \end{aligned}$$

Now define $f:\ RC_p \to R$ by

$$\langle f, v^i\rangle = \delta_{1i}$$

for $i = 0, 1, \ldots, p-1$. Then for all $n \geq 0$, we have that

$$\langle f, \sigma^n\rangle = \langle f, \sum_{i=0}^{n} \binom{n}{i} x^i\rangle = \binom{n}{1} = n$$

and

$$\langle f^p, \sigma^n\rangle = \langle f, \sigma^n\rangle^p = n^p = n$$

by Fermat's Theorem. We also have that

$$\langle \Delta(f), \sigma^n \otimes \sigma^m\rangle = \langle f, \sigma^{n+m}\rangle = n + m = \langle f \otimes \varepsilon + \varepsilon \otimes f, \sigma^n \otimes \sigma^m\rangle$$

and this implies that

$$\Delta f = f \otimes \varepsilon + \varepsilon \otimes f$$

and we have a Hopf algebra map $H \to RC_p^*$ sending x to f. It is clear that this map is injective, and we leave it to the reader to show that it is also surjective. □

We now return to the general case where $q = p^d$. Our aim is now to compute the group of Galois objects of H. We will first show that all H-Galois objects are commutative.

Proposition 11.3.2 *Let $H = R[x]/(x^q - x)$ be as above. Then $\mathrm{P_{sk}}(H, R) = 1$ and consequently*

$$H^2(H, R, \mathbb{G}_m) = H^2_{\mathrm{symm}}(H, R, \mathbb{G}_m)$$

Proof Let $g:\ H \otimes H \to R$ be a skew pairing. We will show that $g = \varepsilon \otimes \varepsilon$. From (10.42), it follows that

$$g(x^i \otimes 1) = g(1 \otimes x^i) = 0$$

for every $i > 0$. Observe next that, for any $r > 1$, we have

$$\Delta^{r-1}(x) = x \otimes 1 \otimes \cdots \otimes 1 + 1 \otimes x \otimes \cdots \otimes 1 + \cdots + 1 \otimes 1 \otimes \cdots \otimes x$$

and

$$\begin{aligned} g(x^r \otimes x) &= \sum g(x \otimes x_1) g(x \otimes x_2) \cdots g(x \otimes x_{r-1}) \\ &= (r-1) g(x \otimes 1)^{r-1} g(x \otimes x) = 0 \end{aligned}$$

This also implies that

$$g(x \otimes x) = g(x^q \otimes x) = 0$$

For a given integer r, consider the statement

$$g(x^i \otimes x^j) = 0 \qquad (11.14)$$

for any $i > 0$ and $0 \leq j \leq r$. We already know that (11.14) holds for $r = 0, 1$. Suppose that it holds for a given $r \geq 1$. We then have, for any $r > 0$:

$$\begin{aligned} g(x^i \otimes x^{r+1}) &= \sum g(x_1^i \otimes x^r) g(x_2^i \otimes x^r) \\ &= \sum_{k=0}^{i} \binom{i}{k} g(x^k \otimes x^r) g(x^{i-k} \otimes x) = 0 \end{aligned}$$

and this shows that (11.14) holds for any r. It follows that $g(x^i \otimes x^j) = 0$ for all $i > 0$ and $j \geq 0$. In a similar way, we can show that $g(x^i \otimes x^j) = 0$ for all $i \geq 0$ and $j > 0$. This finishes our proof. □

Now let S be a flat commutative R-algebra. It is clear that $\mathrm{char}(S) = p$. Let $\rho \in \mathrm{Alg}(H \otimes S, S)$. Then ρ is completely determined by $\rho(x) = \alpha$. Furthermore, given $\alpha \in S$, α is the image of x under some $\rho \in \mathrm{Alg}(H \otimes S, S)$ if and only if $\rho(x^q) = \alpha^q = \alpha = \rho(x)$. This shows that

$$\mathrm{Alg}(H \otimes S, S) = \{\alpha \in S | \alpha^q = \alpha\}$$

Proposition 11.3.3 *Let R be a ring of characteristic p, and let $H = R[x]/(x^q - x)$. We have a short exact sequence of sheaves on R_{fl}*

$$1 \longrightarrow \mathrm{Alg}(H \otimes \bullet, \bullet) \longrightarrow \mathbb{G}_a \xrightarrow{f} \mathbb{G}_a \longrightarrow 1 \tag{11.15}$$

where $f : \mathbb{G}_a \to \mathbb{G}_a$ is given by

$$f(S)(\alpha) = \alpha^q - \alpha$$

Proof It is clear from the above arguments that $\mathrm{Ker}\, f = \mathrm{Alg}(H \otimes \bullet, \bullet)$, so we only have to show that f is surjective as a map of sheaves. Take a flat R-algebra S and $\alpha \in S$. Write

$$S' = S[y]/(y^q - y - \alpha)$$

Then S' is a faithfully projective commutative S-algebra, and, in S', we have that $y^q - y = \alpha$. This means that the image of α in S' lies in $\mathrm{Im}\,(f(S'))$, and f is a surjective map of sheaves. □

Theorem 11.3.4 *Let R be a ring of characteristic p, and let $H = R[x]/(x^q - x)$. Then*

$$\mathrm{Gal}(R, H) = \mathrm{Gal}^s(R, H) = H^2(H, R, \mathbb{G}_m) = H^2_{\mathrm{symm}}(H, R, \mathbb{G}_m) = R/\mathcal{P}(R) \tag{11.16}$$

where $\mathcal{P}(R) = \{\alpha^q - \alpha | \alpha \in R\}$.

Proof From the exactness of (11.15), it follows that we have a long exact sequence

$$\begin{aligned} 1 &\longrightarrow \mathrm{Alg}(H, R) \longrightarrow R \longrightarrow R \\ &\longrightarrow H^1(R_{\mathrm{fl}}, \mathrm{Alg}(H \otimes \bullet, \bullet)) \cong \mathrm{Gal}^s(R, H) \longrightarrow H^2(H, R, \mathbb{G}_a) \longrightarrow \cdots \end{aligned} \tag{11.17}$$

From Proposition B.0.4, it follows that $H^n(H, R, \mathbb{G}_a) = 0$ for $n > 0$. Hence

$$\mathrm{Gal}^s(R, H) \cong R/\mathrm{Im}\, f(R) = R/\mathcal{P}(R)$$

The commutative H-Galois object that corresponds to $\alpha \in R$ is the *Artin-Schreier extension*

$$S = R[y]/(y^q - y - a)$$

S has normal basis, and therefore every commutative H-Galois object has normal basis. Thus

$$\mathrm{Gal}^s(R, H) \cong R/\mathcal{P}(R) \cong H^2_{\mathrm{symm}}(H, R, \mathbb{G}_m) \tag{11.18}$$

In Proposition 11.3.2, we have seen that

$$H^2(H, R, \mathbb{G}_m) = H^2_{\mathrm{symm}}(H, R, \mathbb{G}_m)$$

Finally, take an arbitrary H-Galois object A. We know that $A \otimes S$ has normal basis for some faithfully flat extension S of R. Applying Proposition 11.3.2 with S as groundring, we find that $A \otimes S$ is commutative. This implies that A is also commutative, and $[A] \in \mathrm{Gal}^s(R, H)$. We therefore have that

$$\mathrm{Gal}^s(R, H) = \mathrm{Gal}(R, H)$$

and this finishes our proof. □

Remark 11.3.5 If $q = p$ is prime, then $H = RC_p^*$, and the H-Galois objects are then nothing else then the cyclic Galois extensions of R. Let us describe the action of $C_p = \langle \sigma \rangle$ on $S = R[y]/(y^p - y - a)$. We use the notation of Proposition 11.3.1, that is, $H = R[x]/(x^p - x)$, and $\rho_S(y) = y \otimes 1 + 1 \otimes x$. Then

$$\begin{aligned} \sigma \rightharpoonup y &= \langle \sigma, 1 \rangle y + \langle \sigma, x \rangle \\ &= y + \langle v, x \rangle + \langle \varepsilon, x \rangle \\ &= y + 1 \end{aligned}$$

and

$$\begin{aligned} \sigma^i \rightharpoonup y &= y + i \\ \sigma^i \rightharpoonup y^j &= (y + i)^j \end{aligned} \tag{11.19}$$

Theorem 11.3.4 describes $\mathrm{Gal}(R, RC_p^*)$ in the case where R is a ring of characteristic p. This result can be generalized to cyclic groups of primary order.

Theorem 11.3.6 *Let R be a ring of characteristic p. Then*

$$\mathrm{Gal}(R, R_{p^d}) \cong \mathrm{W}_d(R)/\{\underline{x}^p \ominus \underline{x} | \underline{x} \in \mathrm{W}_d(R)\}$$

where $\mathrm{W}_d(R)$ is the ring of Witt-vectors of R, and $\ominus$ represents the subtraction in $\mathrm{W}_d(R)$.

Proof We refer to [94, Theorem VI.1.1]. □

Our next aim is to compute $H^2(RC_p \otimes RC_p^*, R, \mathbb{G}_m)$, the subgroup of $\mathrm{Gal}(RC_p \otimes RC_p^*)$ consisting of all Galois objects with normal basis. For the sake of simplicity, assume that R is connected; this is not really a restriction, since we will be able to apply our results to all the connected components of R. We will also assume that R contains no p-th roots of 1, except for 1 itself. This technical condition makes sure that RC_p^* has only one grouplike element. A typical example of such a ring R is a field of characteristic p. In this case, all Galois objects have normal basis, and our theory will give a description of the full Galois group.
Using Theorem 11.3.4 and Proposition 11.1.4, we find that

$$\begin{aligned} H^2_{\mathrm{symm}}(RC_p \otimes RC_p^*, R, \mathbb{G}_m) &= H^2_{\mathrm{symm}}(RC_p, R, \mathbb{G}_m) \times H^2_{\mathrm{symm}}(RC_p^*, R, \mathbb{G}_m) \\ &= \mathbb{G}_m(R)/\mathbb{G}_m(R)^p \times R/\mathcal{P}(R) \end{aligned}$$

Our first step is now to compute $\mathrm{P}_{\mathrm{sk}}(RC_p \otimes RC_p^*, R)$. Observe that

$$\begin{aligned} &\mathrm{P}(RC_p \otimes RC_p^*, R) \\ &\cong \mathrm{Hopf}(RC_p \otimes RC_p^*, RC_p^* \otimes RC_p) \\ &= \mathrm{Hopf}(RC_p, RC_p^*) \times \mathrm{Hopf}(RC_p, RC_p) \times \mathrm{Hopf}(RC_p^*, RC_p) \times \mathrm{Hopf}(RC_p^*, RC_p^*) \end{aligned}$$

Now RC_p^* has no nontrivial grouplike elements. This implies that $\mathrm{Hopf}(RC_p, RC_p^*)$ contains only one element: if $f : RC_p \to RC_p^*$ is a Hopf algebra map, then $f(\sigma) = \varepsilon$, and this implies that

$$f(h) = \varepsilon(h)\varepsilon$$

for all $h \in RC_p$. It also follows that

$$\mathrm{Hopf}(RC_p^*, RC_p) = \mathrm{Hopf}(RC_p, RC_p^*)^*$$

contains only one element, namely f^*, given

$$f^*(h^*) = \langle h^*, 1\rangle 1$$

for all $h^* \in H^*$. A Hopf algebra map $RC_p \to RC_p$ is completely determined by the image of σ, and this image should be grouplike, that is, a power of σ. Thus

$$\mathrm{Hopf}(RC_p, RC_p) = \{f_0, f_1, \ldots, f_{p-1}\}$$

where f_i is defined by

$$f_i(\sigma) = \sigma^i$$

A duality argument shows that

$$\mathrm{Hopf}(RC_p^*, RC_p^*) = \{f_0^*, f_1^*, \ldots, f_{p-1}^*\}$$

A Hopf algebra map $RC_p^* \to RC_p^*$ is completely determined by the image of x. Observe that

$$\langle f_i^*(x), \sigma^m\rangle = \langle x, \sigma^{mi}\rangle = im = i\langle x, m\rangle$$

and this proves that

$$f_i^*(x) = ix \tag{11.20}$$

Combining all the above arguments, we find that

$$\mathrm{Hopf}(RC_p \otimes RC_p^*, RC_p^* \otimes RC_p) = \{\tau \circ (f_i \otimes f_j^*) | i, j = 0, 1, \dots, p-1\}$$

where τ is the switch map. Let d_{ij} be the pairing in $\mathrm{P}(RC_p \otimes RC_p^*, R)$ corresponding to $\tau \circ (f_i \otimes f_j^*)$. We will now find out which of the d_{ij} are skew.
Write $\sigma^n x^m$ as a shorter notation for $\sigma^n \otimes x^m$. Then $\{\sigma^n x^m | n, m = 0, 1, \dots, p-1\}$ is basis of $RC_p \otimes RC_p^*$, and

$$\begin{aligned} \Delta(\sigma^n x^m) &= (\sigma^n \otimes \sigma^n)(x \otimes 1 + 1 \otimes x)^m \\ &= \sum_{i=0}^{m} \binom{m}{i} \sigma^n x^i \otimes \sigma^n x^{m-i} \end{aligned} \tag{11.21}$$

Recall that $v = \sigma - 1$ and $\langle x, v^j \rangle = \delta_{1j}$. It follows easily that

$$\begin{aligned} \langle x, \sigma^m \rangle &= \langle x, (v+1)^m \rangle \\ &= \langle x, \sum_{j=0}^{m} \binom{m}{j} v^j \rangle = m \end{aligned}$$

and

$$\langle x^k, \sigma^m \rangle = \langle x, \sigma^m \rangle^k = m^k \tag{11.22}$$

using the fact that σ is grouplike. Now

$$\begin{aligned} d_{kl}(\Delta(\sigma^n x^m)) &= \sum_{i=0}^{m} \binom{m}{i} \langle l^i x^i \otimes \sigma^{nk}, \sigma^n \otimes x^{m-i} \rangle \\ &= \sum_{i=0}^{m} \binom{m}{i} l^i n^i (nk)^{m-i} = n^m (k+l)^m \end{aligned} \tag{11.23}$$

Take $n = m = 1$ in (11.23). It follows that

$$d_{kl}(\Delta(\sigma x)) = k + l$$

Now $\varepsilon(\sigma x) = 0$, and it follows that d_{kl} is not skew if $k + l \neq p$. Assume that $l + k = p$. From (11.23), it follows that

$$d_{kl}(\Delta(\sigma^n x^m)) = n^m p^m = \delta_{m0} = \varepsilon(\sigma^n x^m)$$

and d_{kl} is skew. We have shown that

$$\mathrm{P}_{\mathrm{sk}}(RC_p \otimes RC_p^*, R) = \{d_{00}, d_{1,p-1}, d_{2,p-2}, \dots, d_{p-1,1}\}$$

We will now show that the map

$$\beta: \ H^2(RC_p \otimes RC_p^*, R, \mathbb{G}_m) \longrightarrow \mathrm{P}_{\mathrm{sk}}(RC_p \otimes RC_p^*, R)$$

is surjective and splits. For $k = 0, 1, \dots, p-1$, we consider the $RC_p \otimes RC_p^*$-comodule algebra A_k, defined as follows: as an $RC_p \otimes RC_p^*$-comodule, $A_k = RC_p \otimes RC_p^*$, with R-basis $\{\sigma^m x^n | m, n = 0, 1, 2, \dots, p-1\}$. The multiplication is given by the formula

$$x\sigma = \sigma(x + k) \tag{11.24}$$

or, more generally

$$x^n\sigma^m = \sigma^m(x+km)^n \tag{11.25}$$

It is clear that A_k is an associative algebra, and that it is an $RC_p \otimes RC_p^*$-Galois object with normal basis. Let $g_k \in Z^2(RC_p \otimes RC_p^*, R, \mathbb{G}_m)$ be the corresponding Sweedler cocycle. We claim that

$$\beta([g_k]) = d_{p-k,k} \tag{11.26}$$

To this end, it suffices to show that for all $a, b \in A_k = RC_p \otimes RC_p^*$:

$$a \cdot b = \sum d_{p-k,k}(a_{(1)} \otimes b_{(1)})b_{(0)})a_{(0)}) \tag{11.27}$$

where $\cdot$ refers to the multiplication in A_k. It is sufficient to check (11.27) for $a = x$ and $b = \sigma$. This can be done as follows:

$$\begin{aligned} & d_{p-k,k}(x \otimes \sigma)\sigma 1 + d_{p-k,k}(1 \otimes \sigma)\sigma x \\ = \; & \langle kx \otimes 1, \sigma \otimes \varepsilon\rangle\sigma + \langle \varepsilon \otimes 1, \sigma \otimes \varepsilon\rangle \sigma x \\ = \; & kx + \sigma x = x(\sigma + k) = \sigma x \end{aligned}$$

We now have that the map β is surjective. We will now show that β splits. We will show that $\{A_0, A_1, \ldots, A_{p-1}\}$ is a subgroup of $\mathrm{Gal}(R, RC_p \otimes RC_p^*)$. It suffices to show that

$$A_k \Box A_l \cong A_{kl} \tag{11.28}$$

where the cotensor product is taken over the Hopf algebra $RC_p \otimes RC_p^*$. We know that

$$A_k \Box A_l \cong RC_p \otimes RC_p^*$$

as $RC_p \otimes RC_p^*$-comodules. The connecting isomorphisms are $\varepsilon \otimes I$ and Δ. Let us compute the multiplication map on $RC_p \otimes RC_p^*$ induced by the multiplication on $A_k \Box A_l$. This is given by

$$\begin{aligned} x \cdot \sigma &= \Big((\varepsilon \otimes I) \circ m_{A_k \Box A_l} \circ \Delta\Big)(x \otimes \sigma) \\ &= (\varepsilon \otimes I)\Big(m_{A_k}(x \otimes \sigma) \otimes m_{A_l}(1 \otimes \sigma) + m_{A_k}(1 \otimes \sigma) \otimes m_{A_l}(x \otimes \sigma)\Big) \\ &= (\varepsilon \otimes I)\Big(\sigma(x+k) \otimes \sigma + \sigma \otimes \sigma(x+l)\Big) \\ &= k\sigma + \sigma(x+l) = \sigma(x+k+l) \end{aligned}$$

and this is exactly the multiplication rule on A_{k+l}. We summarize our results as follows.

Theorem 11.3.7 *Let R be a connected ring of characteristic p with no nontrivial p-th roots of 1. The sequence*

$$1 \longrightarrow H^2_{\mathrm{symm}}(RC_p \otimes RC_p^*, R, \mathbb{G}_m) \longrightarrow H^2(RC_p \otimes RC_p^*, R, \mathbb{G}_m) \xrightarrow{\beta} \mathrm{P}_{\mathrm{sk}}(RC_p \otimes RC_p^*, R) \longrightarrow 1$$

is split exact. Moreover

$$H^2(RC_p \otimes RC_p^*, R, \mathbb{G}_m) = R/\mathcal{P}(R) \times \mathbb{G}_m(R)/\mathbb{G}_m(R)^p \times C_p$$

Part III

The Brauer-Long group of a commutative ring

Chapter 12

H-Azumaya algebras

Throughout this Chapter, R will be a commutative ring. All the R-algebras that we consider have a unit element. H will be a faithfully projective, commutative and cocommutative Hopf algebra. Most of the theory can be generalized to the situation where H is an arbitrary Hopf algebra with bijective antipode, but then the technical details are much more complicated. A survey of this more recent theory is presented in Chapter 14.

12.1 Dimodules and dimodule algebras

Dimodules

An R-module M which is at once an H-module and an H-comodule is called an *H-dimodule* if for all $m \in M$ and $h \in H$, we have

$$\rho(h \rightharpoonup m) = \sum h \rightharpoonup m_{(0)} \otimes m_{(1)} \tag{12.1}$$

that is

$$\rho \circ \psi = (\psi \otimes I_H) \circ (I_H \otimes \rho)$$

The category of H-dimodules and H-dimodule homomorphisms (homomorphisms that are H-linear and H-colinear) will be denoted by R-dimod.

Proposition 12.1.1 *We have the following isomorphisms of categories*

$$H \otimes H^*\text{-comod} \cong H\text{-dimod} \cong H \otimes H^*\text{-mod}$$

Proof First define a functor $F:\ H \otimes H^*\text{-comod} \to H\text{-dimod}$ as follows: for an $H \otimes H^*$-comodule M, or, more precisely, an (H^*, H)-bicomodule, we define an H-dimodule structure on M as follows: if $\rho_{(H^*,H)}(m) = \sum m_{(-1)} \otimes m_{(0)} \otimes m_{(1)} \in H^* \otimes M \otimes H$, then we let

$$\rho_H(m) = \sum \langle m_{(-1)}, 1\rangle m_{(0)} \otimes m_{(1)} \quad \text{and} \quad h \rightharpoonup m = \sum \langle \varepsilon, m_{(1)}\rangle \langle m_{(-1)}, h\rangle m_{(0)}$$

It is straightforward to verify that these structures satisfy (12.1). From the fact that H is faithfully projective as an R-module, it follows that F is an isomorphism of categories. The second isomorphism may be handled in a similar way. □

In the sequel, we will sometimes view an H-dimodule as an (H^*, H)-bicomodule, and we will then use the notation

$$\rho_{(H^*,H)}(m) = \sum m_{(-1)} \otimes m_{(0)} \otimes m_{(1)} \in H^* \otimes M \otimes H$$

Proposition 12.1.2 *Suppose that M and N are H-dimodules. Then $M \otimes N$ and $\mathrm{Hom}_R(M, N)$ are again H-dimodules.*

Proof In Section 7.2, we have seen that $M \otimes N$ and $\mathrm{Hom}_R(M, N)$ are H-modules and H-comodules (cf. (7.31,7.32,7.38,7.39)). Let us show that these formulas satisfy (12.1). Take $h \in H$, $m \in M$, $n \in M$ and $f \in \mathrm{Hom}_R(M, N)$. Then

$$\begin{aligned} \rho(h\rightharpoonup(m \otimes n)) &= \rho(\sum h_{(1)}\rightharpoonup m \otimes h_{(2)}\rightharpoonup n) \\ &= \sum h_{(1)}\rightharpoonup m_{(0)} \otimes h_{(2)}\rightharpoonup n_{(0)} \otimes m_{(1)}n_{(1)} \\ &= h\rightharpoonup(m_{(0)} \otimes n_{(0)}) \otimes m_{(1)}n_{(1)} \end{aligned}$$

and this proves the assertion for the tensor product. Now observe that

$$\begin{aligned} \rho(h\rightharpoonup f)(m) &= \sum((h\rightharpoonup f)(m_{(0)}))_{(0)} \otimes ((h\rightharpoonup f)(m_{(0)}))_{(1)}S(m_{(1)}) \\ &= \sum(h_{(1)}\rightharpoonup f(S(h_{(2)})\rightharpoonup m_{(0)}))_{(0)} \otimes (h_{(1)}\rightharpoonup f(S(h_{(2)})\rightharpoonup m_{(0)}))_{(1)}S(m_{(1)}) \\ &= h_{(1)}\rightharpoonup f(S(h_{(2)})\rightharpoonup m_{(0)})_{(0)} \otimes f(S(h_{(2)})\rightharpoonup m_{(0)})_{(1)}S(m_{(1)}) \end{aligned}$$

and

$$\begin{aligned} &(\sum(h\rightharpoonup f_{(0)}) \otimes f_{(1)})(m) \\ &= \sum(h\rightharpoonup f_{(0)})(m) \otimes f_{(1)} \\ &= \sum h_{(1)}\rightharpoonup f_{(0)}(S(h_{(2)})\rightharpoonup m)) \otimes f_{(1)} \\ &= \sum h_{(1)}\rightharpoonup f((S(h_{(2)})\rightharpoonup m)_{(0)})_{(0)} \otimes f((S(h_{(2)})\rightharpoonup m)_{(0)})_{(1)}S(m_{(1)}) \\ &= h_{(1)}\rightharpoonup f(S(h_{(2)})\rightharpoonup m_{(0)})_{(0)} \otimes f(S(h_{(2)})\rightharpoonup m_{(0)})_{(1)}S(m_{(1)}) \end{aligned}$$

and this proves that $\mathrm{Hom}_R(M, N)$ is an H-dimodule. □

For an H-dimodule M, we will write

$$\begin{aligned} M^{H\mathrm{co}H} &= M^H \cap M^{\mathrm{co}H} \qquad (12.2) \\ &= \{m \in M | \rho(m) = m \otimes 1 \ \text{ and } \ h\rightharpoonup m = \varepsilon(h)m, \text{ for all } h \in H, m \in M\} \end{aligned}$$

Dimodule algebras

An R-algebra which is an H-dimodule, an H-module algebra and an H-comodule algebra is called an *H-dimodule algebra*.

Given two H-dimodule algebras A and B, we can form the smash product $A\#B$. Recall (see (7.45)) that the multiplication rule is given by

$$(a\#b)(c\#d) = \sum a(b_{(1)}\rightharpoonup c)\#b_{(0)}d \tag{12.3}$$

Proposition 12.1.3 *If A and B are dimodule algebras, then the tensor product $A\otimes B$ and the smash product $A\#B$ furnished with the dimodule structure on $A\otimes B$ are dimodule algebras. If M is an H-dimodule, then $\mathrm{End}_R(M)$ is an H-dimodule algebra.*

Proof In Section 7.2, we have seen that $A\otimes B$ is an H-module algebra and an H-comodule algebra. From Proposition 12.1.3, it follows that $A\otimes B$ is an H-dimodule algebra. A similar argument shows that $\mathrm{End}_R(M)$ is an H-dimodule algebra if M is an H-dimodule.
Let us show that $A\#B$ is an H-module algebra. For all $h\in H$, $a,c\in A$, $b,d\in B$, we have

$$\begin{aligned} h\rightharpoonup((a\#b)(c\#d)) &= \sum h\rightharpoonup(a(b_{(1)}\rightharpoonup c)\#b_{(0)}d) \\ &= \sum(h_{(1)}\rightharpoonup a)((h_{(2)}b_{(1)})\rightharpoonup c)\#(h_{(3)}\rightharpoonup b_{(0)})(h_{(4)}\rightharpoonup d) \\ &= \sum((h_{(1)}\rightharpoonup a)\#(h_{(3)}\rightharpoonup b))((h_{(2)}\rightharpoonup c)\#(h_{(4)}\rightharpoonup d)) \\ &= \sum(h_{(1)}\rightharpoonup(a\#b))(h_{(2)}\rightharpoonup(c\#d)) \end{aligned}$$

A similar argument shows that $A\#B$ is an H-comodule algebra, and it follows from Proposition 12.1.2 that $A\#B$ is an H-dimodule algebra. □

Suppose that M is a faithfully projective H-dimodule. For later use, we point out that the action and coaction of H on $\mathrm{End}_R(M)\cong M\otimes M^*$ may be described as follows (cf. (7.32, 7.40)). For $f=m\otimes m^*\in\mathrm{End}_R(M)$, $h\in H$, $p\in M$, we have

$$(h\rightharpoonup f)(p) = \sum\langle m^*, S(h_{(2)})\rightharpoonup p\rangle h_{(1)}\rightharpoonup m \tag{12.4}$$

$$\rho(f) = \sum f_{(0)}\otimes f_{(1)} = \sum m_{(0)}\otimes m^*_{(0)}\otimes m_{(1)}m^*_{(1)} \tag{12.5}$$

with (cf. 7.41)

$$\begin{aligned} \sum f_{(0)}(p)\otimes f_{(1)} &= \sum\langle m^*_{(0)}, p\rangle m_{(0)}\otimes m_{(1)}m^*_{(1)} \\ &= \sum\langle m^*, p_{(0)}\rangle m_{(0)}\otimes m_{(1)}S(p_{(1)}) \end{aligned} \tag{12.6}$$

The *H-opposite algebra* $\overline{A}$ of an H-dimodule algebra A is equal to A as an H-dimodule, but with multiplication structure given by

$$\overline{a}\cdot\overline{b} = \overline{\sum(a_{(1)}\rightharpoonup b)a_{(0)}} \tag{12.7}$$

We will now give some general properties of H-dimodule algebras, which are due to Long ([125]).

Proposition 12.1.4 *Let A, B and C be H-dimodule algebras. We have the following isomorphisms of H-dimodule algebras:*

$$(A\#B)\#C \cong A\#(B\#C) \tag{12.8}$$
$$A \cong \overline{\overline{A}} \tag{12.9}$$
$$\overline{B}\#\overline{A} \cong \overline{A\#B} \tag{12.10}$$

The connecting isomorphisms are given by

$$\begin{aligned} (a\#b)\#c &\mapsto a\#(b\#c) \\ a &\mapsto \sum a_{(1)}\rightharpoonup a_{(0)} \\ b\#a &\mapsto \sum b_{(1)}\rightharpoonup a\#b_{(0)} \end{aligned}$$

Proof (12.8) is straightforward. We leave it to the reader to show that the maps (12.9) and (12.10) are H-dimodule algebra homomorphisms and that their inverses are given respectively by

$$\begin{aligned} a &\mapsto \sum S(a_{(1)})\rightharpoonup a_{(0)} \\ a\#b &\mapsto \sum b_{(0)}\#S(b_{(1)})\rightharpoonup a \end{aligned}$$

□

Recall the following notation, for each $h \in H$.

$$\psi_h : M \to M$$

is defined by

$$\psi_h(m) = \psi(h \otimes m) = h\rightharpoonup m$$

We will now show that H acts and coacts trivially on ψ_h.

Lemma 12.1.5 *With notations as above, we have*

$$\rho(\psi_h) = \psi_h \otimes 1 \quad \text{and} \quad k\rightharpoonup\psi_h = \varepsilon(k)\psi_h$$

for all $h, k \in H$.

Proof From (7.39), it follows that, for all $m \in M$ and $h, k \in H$,

$$\begin{aligned} \rho(\psi_h)(m) &= \sum(\psi(h)(m_{(0)}))_{(0)} \otimes (\psi(h)(m_{(0)}))_{(1)}S(m_{(1)}) \\ &= \sum(h\rightharpoonup m_{(0)})_{(0)} \otimes (h\rightharpoonup m_{(0)})_{(1)}S(m_{(1)}) \\ &= \sum h\rightharpoonup m_{(0)} \otimes m_{(1)}S(m_{(2)}) \qquad \text{(by (12.1))} \\ &= (h\rightharpoonup m) \otimes 1 = \psi_h(m) \otimes 1 \end{aligned}$$

and

$$\begin{aligned} (k\rightharpoonup\psi_h)(m) &= \sum k_{(1)}\rightharpoonup\psi_h(S(k_{(2)})\rightharpoonup p) \\ \sum(k_{(1)}hS(k_{(2)}))\rightharpoonup m &= \varepsilon(k)\psi_h(m) \end{aligned}$$

□

Proposition 12.1.6 *(cf. [125, 3.7])*
Suppose that M is a faithfully projective H-dimodule, and that B is an H-dimodule algebra. Then the map

$$\Gamma : \ \mathrm{End}_R(M)\#B \longrightarrow \mathrm{End}_R(M)\otimes B$$

defined by

$$\Gamma(f\#b) = \sum (f\circ\psi_{b_{(1)}})\otimes b_{(0)}$$

is an H-dimodule algebra isomorphism.

Proof 1) Γ is an R-algebra homomorphism. For all $f,g\in \mathrm{End}_R(M)$ and $b,c\in B$, we have

$$\begin{aligned}\Gamma((f\#b)(g\#c)) &= \Gamma\Big(\sum f\circ\psi_{b_{(1)}}\circ g\circ\psi_{S(b_{(2)})}\#b_{(0)}c\Big)\\ &= \sum f\circ\psi_{b_{(1)}}\circ g\circ\psi_{S(b_{(2)})}\circ\psi_{b_{(3)}}\circ\psi_{c_{(1)}}\otimes b_{(0)}c_{(0)}\\ &= \Big(\sum f\circ\psi_{b_{(1)}}\otimes b_{(0)}\Big)\Big(\sum g\circ\psi_{c_{(1)}}\otimes c_{(0)}\Big)\\ &= \Gamma(f\#b)\Gamma(g\#c)\end{aligned}$$

2) Γ is H-linear. For all $h\in H$, $f\in \mathrm{End}_R(M)$ and $b\in B$, we have

$$\begin{aligned}\Gamma(h\rightharpoonup(f\#b)) &= \Gamma\Big(\sum \psi_{h_{(1)}}\circ f\circ\psi_{S(h_{(2)})}\#h_{(3)}\rightharpoonup b\Big)\\ &= \sum \psi_{h_{(1)}}\circ f\circ\psi_{S(h_{(2)})}\circ\psi_{b_{(1)}}\#h_{(3)}\rightharpoonup b_{(0)}\\ &= \sum \psi_{h_{(1)}}\circ f\circ\psi_{b_{(1)}}\circ\psi_{S(h_{(2)})}\#h_{(3)}\rightharpoonup b_{(0)}\\ &= h\rightharpoonup\Gamma(f\#b)\end{aligned}$$

3) Γ is H-colinear. For all $f\in \mathrm{End}_R(M)$ and $b\in B$, we have, using Lemma 12.1.5

$$\begin{aligned}\rho(\Gamma(f\#b)) &= \sum f_{(0)}\circ\psi_{b_{(1)}}\otimes b_{(0)}\otimes f_{(1)}b_{(1)}\\ &= (\Gamma\otimes I)(\rho(f\#b))\end{aligned}$$

4) Γ is bijective. It is straightforward to verify that the inverse of Γ is given by the formula

$$\Gamma^{-1}(f\otimes b) = \sum f\circ\psi_{S(b_{(1)})}\otimes b_{(0)}$$

□

Corollary 12.1.7 *If M and N are faithfully projective H-dimodules, then*

$$\mathrm{End}_R(M)\#\mathrm{End}_R(N)\cong \mathrm{End}_R(M)\otimes\mathrm{End}_R(N)$$

as H-dimodule algebras.

Proposition 12.1.8 *Suppose that M is a faithfully projective H-dimodule, and that B is an H-dimodule algebra. Then the map*

$$\Psi:\ B\#\mathrm{End}_R(M)\longrightarrow B\otimes \mathrm{End}_R(M)$$

defined by

$$\Psi(b\#(m\otimes m^*))=\sum(S(m_{(1)})\rightharpoonup b)\otimes(m_{(0)}\otimes m^*)$$

is an isomorphism of H-dimodule algebras. Here we identified $\mathrm{End}_R(M)$ *and* $M\otimes M^*$ *by their canonical isomorphism.*

Proof 1) Ψ is an R-algebra map. Recall that the product on $\mathrm{End}_R(M)=M\otimes M^*$ is given by

$$(m\otimes m^*)(n\otimes n^*)=\langle m^*,n\rangle m\otimes n^*$$

For all $b,c\in B,\ f=m\otimes m^*,\ g=n\otimes n^*\in \mathrm{End}_R(M)=M\otimes M^*$, we have

$$\begin{aligned}
&\Psi\Big((b\#(m\otimes m^*)(c\#(n\otimes n^*)\Big)\\
&= \Psi\Big(\sum b((m_{(1)}m^*_{(1)})\rightharpoonup c)\#\langle m^*_{(0)},n\rangle m_{(0)}\otimes n^*\Big)\\
&= \sum(S(m_{(1)})\rightharpoonup b)((\langle m^*_{(0)},n\rangle m^*_{(1)}\rightharpoonup c)\otimes(m_{(0)}\otimes n^*)\\
&= \sum(S(m_{(1)})\rightharpoonup b)((\langle m^*,n_{(0)}\rangle S(n_{(1)})\rightharpoonup c)\otimes(m_{(0)}\otimes n^*)\quad \text{(by (7.40)}\\
&= \Big(\sum(S(m_{(1)})\rightharpoonup b)\otimes(m_{(0)}\otimes m^*)\Big)\Big(\sum(S(n_{(1)})\rightharpoonup c)\otimes(n_{(0)}\otimes n^*)\Big)\\
&= \Psi(b\#(m\otimes m^*))\Psi(c\#(n\otimes n^*))
\end{aligned}$$

2) Ψ is H-colinear.

$$\begin{aligned}
&\Psi(\rho(b\#(m\otimes m^*)))=\Psi\Big(\sum(S(m_{(1)})\rightharpoonup b)\otimes(m_{(0)}\otimes m^*)\Big)\\
&= \sum(S(m_{(2)})\rightharpoonup b_{(0)})\otimes(m_{(0)}\otimes m^*_{(0)})\otimes b_{(1)}m_{(1)}m^*_{(1)}\\
&= (\Psi\otimes I)\Big(\sum(b_{(0)}\#m_{(0)}\otimes m^*_{(0)})\otimes b_{(1)}m_{(1)}m^*_{(1)}\Big)\\
&= (\Psi\otimes I)(\rho(b\#(m\otimes m^*)))
\end{aligned}$$

3) Ψ is H-linear.

$$\begin{aligned}
&\Psi(h\rightharpoonup(b\#(m\otimes m^*)))=\Psi\Big(\sum(h_1\rightharpoonup b)\#((h_2\rightharpoonup m)\otimes(m^*\circ\psi_{h_{(3)}}))\Big)\\
&= \sum(S(m_{(1)})h_{(1)}\rightharpoonup b)\#(h_2\rightharpoonup m_{(0)})\otimes(m^*\circ\psi_{h_{(3)}}))\\
&= h\rightharpoonup\Big(\sum S(m_{(1)})\rightharpoonup b\otimes(m_{(0)}\otimes m^*)\Big)\\
&= h\rightharpoonup\Psi(b\#(m\otimes m^*))
\end{aligned}$$

4) Ψ is bijective. The inverse of Ψ is given by the formula

$$\Psi^{-1}(b\otimes(m\otimes m^*))=\sum(m_{(1)}\rightharpoonup b)\#(m_{(0)}\otimes m^*)$$

□

Corollary 12.1.9 *If M is a faithfully projective H-dimodule and B is an H-dimodule algebra, then*

$$B\#\mathrm{End}_R(M) \cong \mathrm{End}_R(M)\#B$$

as H-dimodule algebras.

Proposition 12.1.10 *If M is a faithfully projective H-dimodule, then the map*

$$\sigma: \overline{\mathrm{End}_R(M)} \longrightarrow \mathrm{End}_R(M)^{\mathrm{op}}$$

defined by

$$\sigma(f)(p) = \sum (p_{(1)} \rightharpoonup f)(p_{(0)}) \tag{12.11}$$

for all $f: M \to M$ and $p \in M$ is an isomorphism of H-dimodule algebras.

Proof Using (12.4), we can rewrite (12.11) as follows.

$$\sigma(f)(p) = \sum \langle m^*, S(p_{(2)}) \rightharpoonup p_{(0)} \rangle (p_{(1)} \rightharpoonup m) \tag{12.12}$$

1) σ is an R-algebra homomorphism. Take $f = m \otimes m^*$, $g = n \otimes n^* \in \overline{\mathrm{End}_R(M)}$ and $p \in M$. We then obtain, using (12.4-12.6)

$$\begin{aligned} (f \cdot g)(p) &= \sum (f_{(1)} \rightharpoonup g)(f_{(0)}(p)) \\ &= \sum \big((m_{(1)} S(p_{(1)})) \rightharpoonup g\big)(\langle m^*, p_{(0)} \rangle m_{(0)}) \\ &= \langle m^*, p_{(0)} \rangle \langle n^*, (S(m_{(2)}) p_{(2)}) \rightharpoonup m_{(0)} \rangle (m_{(1)} S(p_{(1)})) \rightharpoonup n \end{aligned}$$

hence

$$\begin{aligned} \sigma(f \cdot g)(p) &= \sum p_{(1)} \rightharpoonup ((f \cdot g)(S(p_{(2)}) \rightharpoonup p_{(0)})) \\ &= \sum \langle m^*, S(p_{(2)}) \rightharpoonup p_{(0)} \rangle \langle n^*, (S(m_{(1)}) p_{(1)}) \rightharpoonup m_{(0)} \rangle (m_{(2)} \rightharpoonup n) \end{aligned}$$

From (12.12), we deduce

$$(\sigma(g) \circ \sigma(f))(p) = \sum \langle m^*, S(p_{(2)}) \rightharpoonup p_{(0)} \rangle \langle n^*, S(m_{(1)}) p_{(1)} \rightharpoonup m_{(0)} \rangle (m_{(2)} \rightharpoonup n)$$

and this shows that σ is indeed an R-algebra homomorphism.

2) σ is H-linear. For all $h \in H$, $f \in \mathrm{End}_R(M)$ and $p \in M$, we have

$$\begin{aligned} (h \rightharpoonup \sigma(f))(p) &= \sum h_{(1)} \rightharpoonup (\sigma(f)(S(h_{(2)}) \rightharpoonup p)) \\ &= \sum h_{(1)} \rightharpoonup ((p_{(1)} \rightharpoonup f)(S(h_{(2)}) \rightharpoonup p_{(0)})) \\ &= \sum (h p_{(1)} \rightharpoonup f)(p_{(0)}) \\ &= \sigma(h \rightharpoonup f)(p) \end{aligned}$$

3) σ is H-colinear. Take $f = m \otimes m^* \in \mathrm{End}_R(M)$ and $p \in M$. From (7.39), it follows that

$$\begin{aligned} \rho(\sigma(f))(p) &= \sum (\sigma(f)(p_{(0)}))_{(0)} \otimes (\sigma(f)(p_{(0)}))_{(1)} S(p_{(1)}) \\ &= \langle m^*, S(p_{(2)}) \rightharpoonup p_{(0)} \rangle (p_{(1)} \rightharpoonup m_{(0)}) \otimes m_{(1)} S(p_{(1)}) \end{aligned}$$

On the other hand, using (12.5), we have

$$\begin{aligned}((\sigma\otimes I)\circ\rho(f))(p) &= \sum\langle m^*_{(0)}, S(p_{(2)})\rightharpoonup p_{(0)}\rangle(p_{(1)}\rightharpoonup m_{(0)}\otimes m_{(1)}m^*_{(1)}\\ &= \langle m^*, S(p_{(2)})\rightharpoonup p_{(0)}\rangle(p_{(1)}\rightharpoonup m_{(0)})\otimes m_{(1)}S(p_{(1)})\end{aligned}$$

and this proves the colinearity.

4) σ is bijective. Its inverse σ^{-1} is given by the formula

$$\sigma^{-1}(f)(p)=\sum(S(p_{(1)})\rightharpoonup f)(p_{(0)})$$

□

Proposition 12.1.11 *If M is a faithfully projective H-dimodule, then*

$$\overline{\mathrm{End}_R(M)}\cong \mathrm{End}_R(M)^{\mathrm{op}}\cong \mathrm{End}_R(M^*)$$

as H-dimodule algebras.

Proof We only have to show that the isomorphism

$$\phi:\ \mathrm{End}_R(M)^{\mathrm{op}}\to \mathrm{End}_R(M^*):\ f\mapsto f^*$$

is H-linear and H-colinear.

1) ϕ is H-linear. Take $f=m\otimes m^*\in \mathrm{End}_R(M)$. Recall that the dual map f^* is given by the formula

$$f^*(p^*)(m)=\langle p^*,m\rangle m^*$$

for all $p^*\in M^*$. Also observe that it follows from (12.4) that the action of H on M^* is given by

$$\langle h\rightharpoonup p^*,p\rangle=\langle p^*,S(h)\rightharpoonup p\rangle$$

for all $h\in H, p^*\in M^*$ and $p\in M$. Now

$$\langle (h\rightharpoonup f)^*(p^*),p\rangle=\langle p^*,(h\rightharpoonup f)(p)\rangle=\sum\langle m^*,S(h_{(2)})\rightharpoonup p\rangle\langle p^*,h_{(1)}\rightharpoonup m\rangle$$

and

$$(h\rightharpoonup f^*)(p^*)=\sum h_{(1)}\rightharpoonup(f^*(S(h_{(2)})\rightharpoonup p^*))=\sum\langle S(h_{(2)})\rightharpoonup p^*,m\rangle(h_{(1)}\rightharpoonup m^*)$$

hence

$$\begin{aligned}\langle (h\rightharpoonup f^*)(p^*),p\rangle &= \sum\langle S(h_{(2)})\rightharpoonup p^*,m\rangle\langle h_{(1)}\rightharpoonup m^*,p\rangle\\ &= \langle p^*,h_{(2)}\rightharpoonup m\rangle\langle m^*,S(h_{(1)})\rightharpoonup p\rangle\end{aligned}$$

and this proves that ϕ is H-linear.

2) ϕ is H-colinear. Take $f=m\otimes m^*\in \mathrm{End}_R(M)$. Then $\rho(f)=\sum m_{(0)}\otimes m^*_{(0)}\otimes m_{(1)}m^*_{(1)}$, and for all $p^*\in M^*$, we have

$$(\phi\otimes I)(\rho(f))(p^*)=\sum\langle p^*,m_{(0)}\rangle m^*_{(0)}\otimes m_{(1)}m^*_{(1)}$$

Let us now compute $\rho(\phi(f)) = \rho(f^*)$. Using (7.39) and (12.6), we obtain

$$\begin{aligned}\rho(f^*)(p^*) &= \sum f^*_{(0)}(p^*)\otimes f^*_{(1)}\\ &= f^*(p^*_{(0)})_{(0)}\otimes f^*(p^*_{(0)})_{(1)}S(p^*_{(1)})\\ &= \langle p^*_{(0)}, m\rangle m^*_{(0)}\otimes m^*_{(1)}S(p^*_{(1)})\\ &= \langle p^*, m_{(0)}\rangle m^*_{(0)}\otimes m^*_{(1)}m_{(1)}\end{aligned}$$

and this shows that ϕ is H-linear. □

12.2 H-Azumaya algebras

Morita theory for H-dimodules

Let H be a faithfully projective, commutative and cocommutative Hopf algebra, and let A and B be H-dimodule algebras. An $(H, A\text{-}B)$-dimodule is an R-module M which is at once an H-dimodule and an A-B-bimodule such that

$$h \rightharpoonup (amb) = \sum (h_{(1)} \rightharpoonup a)(h_{(2)} \rightharpoonup m)(h_{(3)} \rightharpoonup b) \tag{12.13}$$

$$\rho(amb) = \sum a_{(0)}m_{(0)}b_{(0)} \otimes a_{(1)}m_{(1)}b_{(1)} \tag{12.14}$$

for all $a \in A$, $b \in B$ and $m \in M$. The category of $(H, A\text{-}B)$-bimodules and $(A\text{-}B)$-bilinear, H-linear, H-colinear maps will be denoted by $(H, A\text{-}B)$-dimod.
In a similar way, we define $(H, A\text{-}B)$-dimodules and $(H, A\text{-}B)$-comodules and the categories $(H, A\text{-}B)$-mod and $(H, A\text{-}B)$-comod.
A sextuple (A, B, P, Q, f, g) is called a *(strict) H-Morita context* if it is a (strict) Morita context and if A and B are H-dimodule algebras, P is an $(H, A\text{-}B)$-dimodule, Q is an $(H, B\text{-}A)$-dimodule and the maps f and g are H-linear and H-colinear.
We have the following (obvious) dimodule version of the Morita Theorems.

Theorem 12.2.1 *Let (A, B, P, Q, f, g) be a strict H-Morita context. Then the functors*

$$\begin{cases} P\otimes_B \bullet : & (H, B\text{-}R)\text{-dimod} \longrightarrow (H, A\text{-}R)\text{-dimod}\\ Q\otimes_A \bullet : & (H, A\text{-}R)\text{-dimod} \longrightarrow (H, B\text{-}R)\text{-dimod}\end{cases}$$

and

$$\begin{cases} \bullet\otimes_A P : & (H, R\text{-}A)\text{-dimod} \longrightarrow (H, R\text{-}B)\text{-dimod}\\ \bullet\otimes_B Q : & (H, R\text{-}B)\text{-dimod} \longrightarrow (H, R\text{-}A)\text{-dimod}\end{cases}$$

are two pairs of inverse equivalences.
The bimodule isomorphisms

$$P \cong \mathrm{Hom}_A(Q, A) \cong \mathrm{Hom}_B(Q, B)$$

$$Q \cong \mathrm{Hom}_B(P, B) \cong \mathrm{Hom}_A(P, A)$$

are isomorphisms in resp. the categories $(H, A\text{-}B)$-dimod *and* $(H, B\text{-}A)$-dimod, *and the isomorphisms*

$$A \cong \mathrm{End}_B(P) \cong \mathrm{End}_B(Q)^{\mathrm{op}}$$

$$B \cong \mathrm{End}_A(Q) \cong \mathrm{End}_A(P)^{\mathrm{op}}$$

are isomorphisms of H-dimodule algebras.

Proof It suffices to show that all the maps occurring in the proofs of Theorem 1.1.4 and Proposition 1.1.5 are H-linear and H-colinear. This is a straightforward verification, and we leave the details to the reader. □

Example 12.2.2 Let B be an H-dimodule algebra and P an $(H, R\text{-}B)$-bimodule that is a B-progenerator. Then the strict Morita context $(A = \mathrm{End}_B(P), B, P, Q = \mathrm{Hom}_B(Q, B), f, g)$ from Proposition 1.1.6 is a strict H-Morita context.

Proof We know from Proposition 12.1.3 that A is an H-dimodule algebra. Let us show that P is an $(H, A\text{-}B)$-bimodule. The A-action on P is given by the formula

$$a \cdot p = a(p)$$

for all B-linear $a: P \to P$. (12.13) is satisfied:

$$\sum (h_{(1)} \rightharpoonup a)(h_{(2)} \rightharpoonup p)(h_{(3)} \rightharpoonup b) = \sum h_{(1)} \rightharpoonup a\big(S(h_{(2)}) \rightharpoonup (h_{(3)} \rightharpoonup (pb))\big) = h \rightharpoonup (a \cdot pb)$$

and a duality argument shows that (12.14) also holds.
Q is an $(H, B\text{-}A)$-bimodule. The $(B\text{-}A)$-bimodule structure is given by the formula

$$(bqa)(p) = q(a(p))b$$

for all $a \in A$, $b \in B$, $q \in Q$, $p \in P$. To show that (12.13) holds, we proceed as follows:

$$\begin{aligned}
&\sum ((h_{(1)} \rightharpoonup b)(h_{(2)} \rightharpoonup q)(h_{(3)} \rightharpoonup a))(p) \\
&= \sum (h_{(2)} \rightharpoonup q)((h_{(3)} \rightharpoonup a)(p))(h_{(1)} \rightharpoonup b) \\
&= \sum \big(h_{(2)} \rightharpoonup q((S(h_{(2)})h_{(3)}) \rightharpoonup a(S(h_{(4)}) \rightharpoonup p))\big)(h_{(1)} \rightharpoonup b) \\
&= \sum h_{(2)} \rightharpoonup ((q \circ a)(S(h_{(3)}) \rightharpoonup p))(h_{(1)} \rightharpoonup b) \\
&= \sum h_{(1)} \rightharpoonup (q(a(S(h_{(2)}) \rightharpoonup p))b) \\
&= \sum h_{(1)} \rightharpoonup (bqa)(S(h_{(2)}) \rightharpoonup p) \\
&= (h \rightharpoonup bqa)(p)
\end{aligned}$$

Again, a duality argument shows that (12.14) is also satisfied. We leave it to the reader to show that f and g are H-linear and H-colinear. □

H-Azumaya algebras

Lemma 12.2.3 *Let A be an H-dimodule algebra, and consider the maps*

$$\begin{aligned}
F &: A\#\overline{A} \longrightarrow \mathrm{End}_R(A) \\
G &: \overline{A}\#A \longrightarrow \mathrm{End}_R(A)^{\mathrm{op}}
\end{aligned}$$

defined by

$$F(a\#\overline{b})(c) = \sum a(b_{(1)} \rightharpoonup c)b_{(0)} \tag{12.15}$$

$$G(\overline{a}\#b)(c) = \sum (c_{(1)} \rightharpoonup a)c_{(0)}b \tag{12.16}$$

F and G are H-dimodule algebra homomorphisms.

Before we prove the Lemma, we introduce the following notations:

$$\begin{aligned} A^{\rm e} &= A\otimes A^{\rm op}\\ A^{\#{\rm e}} &= A\#\overline{A}\\ {}^{\#{\rm e}}A &= \overline{A}\#A \end{aligned}$$

Proof Let us first show that F preserves the multiplication. The multiplication rule on $A\#\overline{A}$ is given by the formula

$$\begin{aligned} (a\#\overline{b})(c\#\overline{d}) &= \sum a(b_{(1)}\rightharpoonup c)\#\overline{b}_{(0)}\cdot\overline{d}\\ &= \sum a(b_{(2)}\rightharpoonup c)\#\overline{(b_{(1)}\rightharpoonup d)b_{(0)}} \end{aligned}$$

for all $a,b,c,d\in A$. Take $e\in A$. Then

$$\begin{aligned} F((a\#\overline{b})(c\#\overline{d}))(e) &= F(\sum a(b_{(2)}\rightharpoonup c)\#\overline{(b_{(1)}\rightharpoonup d)b_{(0)}})(e)\\ &= \sum a(b_{(3)}\rightharpoonup c)((d_{(1)}b_{(1)})\rightharpoonup e)(b_{(2)}\rightharpoonup d_{(0)})b_{(0)}\\ &= F(a\#\overline{b})(\sum c(d_{(1)}\rightharpoonup e)d_{(0)})\\ &= (F(a\#\overline{b})\circ F(c\#\overline{d}))(e) \end{aligned}$$

Let us next show that F is H-linear. For all $a,b,c\in A$ and $h\in H$, we have

$$\begin{aligned} F(h\rightharpoonup(a\#\overline{b}))(c) &= F\Big(\sum h_{(1)}\rightharpoonup a\#h_{(2)}\rightharpoonup b\Big)(c)\\ &= \sum(h_{(1)}\rightharpoonup a)(b_{(1)}\rightharpoonup c)(h_{(2)}\rightharpoonup b_{(0)}) \end{aligned}$$

and

$$\begin{aligned} (h\rightharpoonup F(a\#\overline{b}))(c) &= \sum h_{(1)}\rightharpoonup(F(a\#\overline{b})(S(h_{(2)})\rightharpoonup c))\\ &= \sum h_{(1)}\rightharpoonup a(b_{(1)}S(h_{(2)})\rightharpoonup c)b_{(0)}\\ &= \sum(h_{(1)}\rightharpoonup a)(b_{(1)}\rightharpoonup c)(h_{(2)}\rightharpoonup b_{(0)}) \end{aligned}$$

Similar computations show that G is an H-dimodule algebra homomorphism. □

The map F makes A into a left $A^{\#{\rm e}}$-module, and G makes A into a right ${}^{\#{\rm e}}A$-module. For a left $A^{\#{\rm e}}$-module M, we will write

$$M^A=\mathrm{End}_{A^{\#{\rm e}}}(A,M)=\{m\in M|(a\#\overline{1})\cdot m=(1\#\overline{a})\cdot m,\ \text{for all } a\in A\}\quad(12.17)$$

In a similar way, we denote for a right ${}^{\#{\rm e}}A$-module:

$${}^AM=\mathrm{End}_{{}^{\#{\rm e}}A}(A,M)=\{m\in M|m\cdot(\overline{a}\otimes 1)=m\cdot(\overline{1}\otimes a),\ \text{for all } a\in A\}\quad(12.18)$$

Observe that

$$\begin{aligned} {}^AA &= \{a\in A|\sum(b_{(1)}\rightharpoonup a)b_{(0)}=ba \text{ for all } b\in A\}\\ A^A &= \{a\in A|\sum(a_{(1)}\rightharpoonup b)a_{(0)}=ab \text{ for all } b\in A\} \end{aligned}$$

If $A^A=R$, then we call A *left H-central.* If ${}^AA=R$, then we call A *right H-central.*

Proposition 12.2.4 *Let A be an H-dimodule algebra. The following assertions are equivalent.*
1) A is faithfully projective as an R-module, and $F: A^{\#e} \to \mathrm{End}_R(A)$ is an isomorphism;
2) A is left H-central and a left $A^{\#e}$-progenerator;
3) the pair of adjoint functors

$$F_l: H\text{-dimod} \longrightarrow (H, A^{\#e}\text{-}R)\text{-dimod} \; : \; N \mapsto A \otimes N$$
$$G_l: (H, A^{\#e}\text{-}R)\text{-dimod} \longrightarrow H\text{-dimod} \; : \; M \mapsto M^A$$

are inverse equivalences. In this case, we call A a left H-Azumaya algebra .

Proof 1) $\Longleftrightarrow$ 2): similar to the proof of 2) $\Longleftrightarrow$ 3) in Proposition 2.2.9.
1) $\Longrightarrow$ 3) We have a strict H-Morita context $(A^{\#e} \cong \mathrm{End}_R(A), R, A, A^*, f, g)$, and 3) follows from the Morita Theorems 12.2.1, taken into account the fact that H-dimod $\cong (H, R\text{-}R)$-mod and

$$M^A = \mathrm{Hom}_{A^{\#e}}(A, M) = M \otimes_{A^{\#e}} A^*$$

(see 5) of Proposition 1.1.5).
3) $\Longrightarrow$ 1). From the fact that F_l and G_l are inverse equivalences, it follows that $R \cong G_l(F_l(R)) = A^A = \mathrm{End}_{A^{\#e}}(A)$, and we have a strict H-Morita context

$$(R = \mathrm{End}_{A^{\#e}}(A), A^{\#e}, A, \mathrm{Hom}_{A^{\#e}}(A, A^{\#e}), f, g)$$

From Proposition 1.1.5, it follows that A is faithfully projective as an R-module and that $A^{\#e} \cong \mathrm{End}_R(A)$. It can be verified directly that the connecting isomorphism is F. □

We have the following right handed version of Proposition 12.2.4

Proposition 12.2.5 *Let A be an H-dimodule algebra. The following assertions are equivalent.*
1) A is faithfully projective as an R-module, and $G: {}^{\#e}A \to \mathrm{End}_R(A)^{\mathrm{op}}$ is an isomorphism;
2) A is right H-central and a right ${}^{\#e}A$-progenerator;
3) The pair of adjoint functors

$$F_r: H\text{-dimod} \longrightarrow (H, R\text{-}{}^{\#e}A)\text{-dimod} \; : \; N \mapsto N \otimes A$$
$$G_r: (H, R\text{-}{}^{\#e}A)\text{-dimod} \longrightarrow H\text{-dimod} \; : \; M \mapsto {}^AM$$

are inverse equivalences.
In this case, we call A a right H-Azumaya algebra .

Proof Similar to the proof of Proposition 12.2.4 □

An H-dimodule algebra A that is at once a left and right H-Azumaya algebra is called an *H-Azumaya algebra.*

Proposition 12.2.6 *1) If M is a faithfully projective H-dimodule, then* $\mathrm{End}_R(M)$ *is an H-Azumaya algebra;*
2) if A and B are H-Azumaya algebra, then $A\#B$ is an H-Azumaya algebra;
3) if A is an H-Azumaya algebra, then $\overline{A}$ is an H-Azumaya algebra.

Proof 1) Using Corollary 12.1.7 and Proposition 12.1.10, we find that

$$\mathrm{End}_R(M)\#\overline{\mathrm{End}_R(M)} \cong \mathrm{End}_R(M)\otimes \mathrm{End}_R(M)^{\mathrm{op}}$$

as H-dimodule algebras. Now $\mathrm{End}_R(M)\otimes \mathrm{End}_R(M)^{\mathrm{op}} \cong \mathrm{End}_R(\mathrm{End}_R(M))$ as R-algebras. If R is a field, then it follows that $\mathrm{End}_R(M)\#\overline{\mathrm{End}_R(M)}$ is a simple R-algebra. Then the kernel of the map $F:\ \mathrm{End}_R(M)\#\overline{\mathrm{End}_R(M)} \to \mathrm{End}_R(\mathrm{End}_R(M))$ is necessarily zero. A count of dimensions tells us that F is bijective. Applying Nakayama's Lemma, we find that F is still bijective if we work over a local ring, and the general case follows from a local-global argument. Similar arguments show that G is bijective.
2) Applying Proposition 12.1.4 and Corollaries 12.1.7 and 12.1.9 and using the fact that A and B are H-Azumaya algebras, we obtain the following isomorphisms of H-Azumaya algebras:

$$\begin{aligned}(A\#B)\#\overline{(A\#B)} &\cong (A\#B)\#(\overline{B}\#\overline{A})\\ &\cong A\#(B\#\overline{B})\#\overline{A}\\ &\cong A\#\mathrm{End}_R(B)\#\overline{A}\\ &\cong A\#\overline{A}\#\mathrm{End}_R(B)\\ &\cong \mathrm{End}_R(A)\#\mathrm{End}_R(B)\\ &\cong \mathrm{End}_R(A\otimes B)\\ &\cong \mathrm{End}_R(A\#B)\end{aligned}$$

since $A\otimes B$ and $A\#B$ are equal as H-dimodules. We then proceed as in 1).
3) It follows immediately from the definition of left and right H-Azumaya algebras and Proposition 12.1.4 that $\overline{A}$ is right (left) H-Azumaya if and only if A is left (right) H-Azumaya. □

12.3 Separability conditions

In this Section, we will try to describe H-Azumaya algebras as a type of central separable algebras. An H-dimodule algebra A will be called *left H-separable* if A is projective as a left $A^{\#e}$-module. A will be called *right H-separable* if A is projective as a right ${}^{\#e}A$-module. An H-dimodule algebra that is at once left and right H-separable is called *H-separable*.

Lemma 12.3.1 *For an H-dimodule algebra A, the following assertions are equivalent.*
1) A is left H-separable;
2) the multiplication map $m_A:\ A^{\#e}\to A$ splits as a sequence of left $A^{\#e}$-modules;
3) there exists $e_l\in A^{\#e}$ such that $m_A(e_l)=1$ and $(a\#\overline{1})e=(1\#\overline{a})e$ for all $a\in A$.
e_l is necessarily an idempotent, and is called a left H-separability idempotent.

Proof Exercise. We also leave it to the reader to state the right handed version of the Lemma. □

Corollary 12.3.2 *Let A be a left (right) H-Azumaya algebra. Then A is left (right) H-central and left (right) H-separable.*

Proof This follows immediately from Propositions 12.2.4 and 12.2.5. □

Lemma 12.3.3 *Let A be a left H-separable H-dimodule algebra, with left H-separability idempotent e_l, and M a left $A^{\#e}$-module. Then $M^A = e_l \cdot M$.*
Similarly, if A is right H-separable, with right H-separability idempotent e_r, and M is a right ${}^{\#e}A$-module, then ${}^AM = m \cdot e_r$.

Proof We know that $M^A = \mathrm{Hom}_{A^{\#e}}(A, M)$. Let $m \in M^A$, and $f : A \to M$ the corresponding homomorphism. Then

$$m = f(1) = (f \circ m_A)(e) = (f \circ m_A)(e^2) = e \cdot (f \circ m_A)(e) = e \cdot m \in e \cdot M$$

Conversely, take $m \in M$. Then $(a\#\overline{1})e{\cdot}m = (1\#\overline{a})e{\cdot}m$, and consequently $em \in M^A$. The proof of the second assertion is similar. □

We now introduce a stronger version of H-separability. An H-dimodule algebra A is called *strongly H-separable* if the multiplication map $m_A : A^{\#e} \to A$ splits as a sequence in $(H, A^{\#e}\text{-}R)$-mod. In the next Lemma, we will see that this notion is equivalent to its right handed analog.

Lemma 12.3.4 *For an H-dimodule algebra A, the following assertions are equivalent.*
1) A is strongly H-separable;
2) there exists $e \in (A^{\#e})^{H\mathrm{co}H}$ such that $m_A(e) = 1$ and $(a\#\overline{1})e = (1\#\overline{a})e$ for all $a \in A$;
3) there exists $e \in (A^e)^{H\mathrm{co}H}$ such that $m_A(e) = 1$ and $(a \otimes 1)e = (1 \otimes a)e$ for all $a \in A$;
4) there exists $e \in ({}^{\#e}A)^{H\mathrm{co}H}$ such that $m_A(e) = 1$ and $e(\overline{a}\#1) = e(\overline{1}\#a)$ for all $a \in A$;
5) m_A splits as a sequence of $(H, A^e\text{-}R)$-dimodules;
6) m_A splits as a sequence of $(H, R\text{-}{}^{\#e}A)$-dimodules.

Proof Exercise. Observe that the separability idempotents in assertions 2), 3) and 4) are equal if we view them as elements of $A \otimes A$. □

It is clear that strong H-separability implies separability and H-separability. In fact, strong H-separability is equivalent to separability, left or right H-separability, with the additional condition that the separability idempotent has to be invariant under the H-action and the H-coaction. We will now see when separability, left,

right and strong H-separability are equivalent.
Recall that $x \in H$ is called a (left) *integral* if

$$hx = \varepsilon(h)x \tag{12.19}$$

for all $h \in H$. Larson and Sweedler's version of Maschke's Theorem [171, 5.18] states that a Hopf algebra over a field is semisimple if and only if there exists an integral $x \in H$ such that $\varepsilon(x) = 1$. In general, we call a Hopf algebra *semisimple-like* if there exists an integral $x \in H$ such that $\varepsilon(x) = 1$. H is called *cosemisimple-like* if there exists an integral $x^* \in H^*$ such that $x^*(1) = 1$. The proof of the following Proposition is based on the proof of [171, 5.18].

Proposition 12.3.5 *Let H be a semisimple-like Hopf algebra, and A an H-module algebra. Suppose that M and N are $(H, A\text{-}R)$-modules. If $f : M \to N$ is an H-linear A-linear epimorphism that has a right A-linear inverse g, then it has also an A-linear H-linear right inverse $\tilde{g}$.*

Proof Define $\tilde{g} : N \to M$ by the rule

$$\tilde{g}(n) = \sum x_{(1)} \rightharpoonup g(S(x_{(2)}) \rightharpoonup n)$$

for all $n \in N$. x is a left integral such that $\varepsilon(x) = 1$. Then

$$f(\tilde{g}(n)) = \sum x_{(1)} \rightharpoonup f(g(S(x_{(2)})) \rightharpoonup n) = \varepsilon(x)n = n$$

for all $n \in N$, and this shows that $\tilde{g}$ is a right inverse of f. We next show that $\tilde{g}$ is A-linear. For all $a \in A$ and $n \in N$, we have that

$$\begin{aligned}
\tilde{g}(an) &= \sum x_{(1)} \rightharpoonup g(S(x_{(2)}) \rightharpoonup (an)) \\
&= \sum x_{(1)} \rightharpoonup g((S(x_{(2)}) \rightharpoonup a)(S(x_{(3)}) \rightharpoonup n)) \\
&= \sum x_{(1)} \rightharpoonup ((S(x_{(2)}) \rightharpoonup a) g(S(x_{(3)}) \rightharpoonup n)) \\
&= \sum ((x_{(1)} S(x_{(3)})) \rightharpoonup a)(x_{(2)} \rightharpoonup g(S(x_{(3)}) \rightharpoonup n) \\
&= a\tilde{g}(n)
\end{aligned}$$

Finally, $\tilde{g}$ is H-linear. Observe first that, for all $h \in H$, we have

$$\sum h_{(1)}x \otimes h_{(2)} = \sum \varepsilon(h_{(1)})x \otimes h_{(2)} = x \otimes h$$

and

$$\sum h_{(1)}x_{(1)} \otimes h_{(2)}x_{(2)} \otimes h_{(3)} = \sum x_{(1)} \otimes x_{(2)} \otimes h$$

Now

$$\begin{aligned}
h \rightharpoonup \tilde{g}(n) &= \varepsilon(x)^{-1} \sum hx_{(1)} \rightharpoonup g(S(x_{(2)}) \rightharpoonup n) \\
&= \varepsilon(x)^{-1} \sum h_{(1)}x_{(1)} \rightharpoonup g((S(x_{(2)})\varepsilon(h_{(2)})) \rightharpoonup n) \\
&= \varepsilon(x)^{-1} \sum h_{(1)}x_{(1)} \rightharpoonup g((S(x_{(2)})S(h_{(2)})h_{(3)}) \rightharpoonup n) \\
&= \varepsilon(x)^{-1} \sum x_{(1)} \rightharpoonup g((S(x_{(2)})h) \rightharpoonup n) \\
&= \tilde{g}(h \rightharpoonup n)
\end{aligned}$$

□

We call H cosemisimple-like if H^* is semisimple-like. As a simple consequence, we have

Corollary 12.3.6 *If H is semisimple-like and cosemisimple-like, then every epimorphism in $(H, A^{\#e}\text{-}R)$-mod that splits in $A^{\#e}$-mod, splits also in $(H, A^{\#e}\text{-}R)$-mod. Similarly, every epimorphism in $(H, R\text{-}^{\#e}A)$-mod that splits in mod-$^{\#e}A$ splits in $(H, R\text{-}^{\#e}A)$-mod.*

Proof $H \otimes H^*$ is semisimple-like. Then apply the Proposition 12.3.5 with H replaced by $H \otimes H^*$ and A by $A^{\#e}$. □

Corollary 12.3.7 *If H is semisimple-like and cosemisimple-like, then the following assertions are equivalent for any H-dimodule algebra A.*
1) A is left H-separable;
2) A is right H-separable;
3) A is separable;
4) A is strongly H-separable.
In this case every H-Azumaya algebra A is H-central and strongly H-separable.

Proof It follows from Lemma 12.3.4 that 4) implies 1), 2) and 3). The converse implications follow directly from Corollary 12.3.6. □

An H-dimodule algebra A that is H-central and strongly H-separable is called a *strong H-Azumaya algebra.* The aim of the rest of this Section is to show that a strong H-Azumaya algebra is automatically H-Azumaya. First we need some preliminary results.

Lemma 12.3.8 *Let S be a commutative R-algebra. If A is a strongly H-separable H-dimodule algebra, then $S \otimes A$ is a strongly $S \otimes H$-separable $S \otimes H$-dimodule algebra.*

Proof If e is a separability idempotent for A that is invariant under the H-action and H-coaction, then $1_S \otimes e$ is a separability idempotent for $S \otimes H$, invariant under $S \otimes H$-action and coaction. □

Proposition 12.3.9 *Let A and B be H-dimodule algebras.*
1) If A is strongly H-separable, then so is $\overline{A}$;
2) if A is a strong H-Azumaya algebra, then so is $\overline{A}$;
3) if A and B are strongly H-separable, then so is $A\#B$;
4) if A and B are strong H-Azumaya algebras, then so is $A\#B$.

Proof 1) Suppose that $e = \sum x_i \otimes y_i \in (A \otimes A^{\rm op})^{H{\rm co}H}$ satisfies the third condition of Lemma 12.3.4. Take

$$f = \sum \overline{y}_{i_{(0)}} \otimes \overline{S(y_{i_{(1)}}) \rightharpoonup x_i} \in (\overline{A} \otimes \overline{A}^{\rm op})^{H{\rm co}H}$$

We will show that f meets the third condition of Lemma 12.3.4 (with A replaced by $\overline{A}$). First of all, we have that

$$\begin{aligned} m_{\overline{A}}(f) &= \sum \overline{y}_{i_{(0)}} \cdot \overline{S(y_{i_{(1)}})\rightharpoonup x_i} \\ &= \sum \overline{((y_{i_{(1)}}S(y_{i_{(2)}}))\rightharpoonup x_i)y_{i_{(0)}}} \\ &= \sum x_i y_i = 1 \end{aligned}$$

Using the fact that H acts and coacts trivially on e, we obtain that, for all $a \in A$

$$\begin{aligned} (\overline{a} \otimes 1)f &= \sum \overline{ay}_{i_{(0)}} \otimes \overline{S(y_{i_{(1)}})\rightharpoonup x_i} \\ &= \overline{(a_{(1)}\rightharpoonup y_{i_{(0)}})a_{(0)}} \otimes \overline{S(y_{i_{(1)}})\rightharpoonup x_i} \\ &= \overline{(a_{(1)}\rightharpoonup y_{i_{(0)}})a_{(0)}} \otimes \overline{(S(a_{(3)}y_{i_{(1)}})a_{(2)}\rightharpoonup x_i} \\ &= \overline{y_{i_{(0)}}a_{(0)}} \otimes \overline{S(a_{(1)}y_{i_{(1)}})\rightharpoonup x_i} \end{aligned}$$

and

$$\begin{aligned} (\overline{1} \otimes a)f &= \sum \overline{y}_{i_{(0)}} \otimes \overline{S(y_{i_{(1)}})\rightharpoonup x_i} \cdot \overline{a} \\ &= \sum \overline{y}_{i_{(0)}} \otimes \overline{(x_{i_{(1)}}\rightharpoonup a)(S(y_{i_{(1)}})\rightharpoonup x_{i_{(0)}})} \\ &= \sum \overline{y}_{i_{(0)}} \otimes \overline{\big((x_{i_{(1)}}y_{i_{(1)}}S(y_{i_{(2)}}))\rightharpoonup a\big)(S(y_{i_{(3)}})\rightharpoonup x_{i_{(0)}})} \\ &= \sum \overline{y}_{i_{(0)}} \otimes \overline{S(y_{i_{(1)}})\rightharpoonup ax_i} \end{aligned}$$

From the fact that

$$\sum ax_i \otimes y_i = \sum x_i \otimes y_i a$$

it follows that

$$\sum ax_i \otimes y_{i_{(0)}} \otimes y_{i_{(1)}} = \sum x_i \otimes y_{i_{(0)}}a_{(0)} \otimes y_{i_{(1)}}a_{(1)}$$

and

$$\sum y_{i_{(0)}} \otimes \big(S(y_{i_{(1)}})\rightharpoonup ax_i\big) = \sum y_{i_{(0)}}a_{(0)} \otimes S(y_{i_{(1)}}a_{(1)})\rightharpoonup x_i$$

and

$$(\overline{a} \otimes 1)f = (\overline{1} \otimes a)f$$

and this proves the first assertion.

2) Let $e = \sum x_i \otimes y_i$ be a separability idempotent for A, and suppose that e is invariant and coinvariant under the H-action and H-coaction. From 1), we know that

$$f = \sum \overline{\overline{y}}_{i_{(0)}} \# \overline{S(y_{i_{(1)}})\rightharpoonup x_i} \in \overline{\overline{A}\#\overline{A}}$$

is a (right) H-separability idempotent for $\overline{A}$. Using Lemma 12.3.3, we find that

$$\begin{aligned} {}^{\overline{A}}\overline{A} = \overline{A}f &= \{\sum \overline{a} \cdot \big(\sum \overline{\overline{y}}_{i_{(0)}} \# \overline{S(y_{i_{(1)}})\rightharpoonup x_i}\big) \mid a \in A\} \\ &= \{\sum \overline{a_{(1)}\rightharpoonup y_{i_{(0)}}} \cdot \overline{a}_{(0)} \cdot \overline{S(y_{i_{(1)}})\rightharpoonup x_i} \mid a \in A\} \\ &= \{\sum \overline{(a_{(1)}\rightharpoonup x_i)(y_{i_{(1)}}\rightharpoonup a_{(0)})(a_{(2)}\rightharpoonup y_{i_{(0)}})} \mid a \in A\} \\ &= \{\sum \overline{x_i(y_{i_{(1)}}\rightharpoonup a)y_{i_{(0)}}} \mid a \in A\} \\ &= \{\sum \overline{(x_i\#y_i) \cdot a} \mid a \in A\} \\ &= \overline{eA} = \overline{A^A} = R \end{aligned}$$

In a similar way, we can prove that

$$\overline{A}^{\overline{A}} = \overline{{}^{A}A} = R$$

and this proves the second assertion.

3) Suppose that $\sum x_i \otimes y_i \in A \otimes A^{\rm op}$ and $\sum u_j \otimes v_j \in B \otimes B^{\rm op}$ are separability idempotents for A and B, both invariant under H-action and H-coaction. We claim that

$$g = \sum (x_i \# u_{j_{(0)}}) \otimes (S(u_{j_{(1)}}) \rightharpoonup y_i \# v_j) \in (A\#B) \otimes (A\#B)^{\rm op} \tag{12.20}$$

is a separability idempotent for $A\#B$. It is clear that g is invariant under H-(co)action. Furthermore

$$\begin{aligned} m_{A\#B}(g) &= \sum x_i u_{j_{(1)}} (S(u_{j_{(2)}}) \rightharpoonup y_i) \# u_{j_{(0)}} v_j \\ &= \sum x_i y_i \# u_j v_j = 1\#1 \end{aligned}$$

Now take $a \in A$ and $b \in B$. Using the fact that $\sum x_i \otimes y_i$ is invariant under the H-action, we find that

$$\begin{aligned} \sum (h \rightharpoonup x_i) \otimes y_i &= \sum h_{(1)} \rightharpoonup x_i \otimes (S(h_{(2)}) h_{(3)}) \rightharpoonup y_i \\ &= \sum x_i \otimes S(h) y_i \end{aligned}$$

for all $h \in H$. We therefore have that

$$\begin{aligned} ((a\#b) \otimes (1\#1)) \cdot g &= \sum (a\#b) \cdot (x_i \# u_{j_{(0)}}) \otimes (S(u_{j_{(1)}}) \rightharpoonup y_i \# v_i) \\ &= \sum (a(b_{(1)} \rightharpoonup x_i) \# b_{(0)} u_{j_{(0)}}) \otimes (S(u_{j_{(1)}}) \rightharpoonup y_i \# v_i) \\ &= \sum (a x_i \# b_{(0)} u_{j_{(0)}}) \otimes (S(b_{(1)} u_{j_{(1)}}) \rightharpoonup y_i \# v_i) \end{aligned}$$

Now $\sum u_j \otimes v_j$ is invariant under the H-coaction. Hence

$$\begin{aligned} \sum u_j \otimes v_j \otimes 1 &= \sum u_{j_{(0)}} \otimes v_{j_{(0)}} \otimes u_{j_{(1)}} v_{j_{(1)}} \\ \sum u_{j_{(0)}} \otimes v_{j_{(0)}} \otimes v_{j_{(1)}} \otimes S(u_{j_{(1)}}) \otimes 1 &= \sum u_{j_{(0)}} \otimes v_{j_{(0)}} \otimes v_{j_{(1)}} \otimes S(u_{j_{(1)}}) \otimes u_{j_{(2)}} v_{j_{(2)}} \\ \sum u_{j_{(0)}} \otimes v_{j_{(0)}} \otimes v_{j_{(1)}} \otimes S(u_{j_{(1)}}) &= \sum u_j \otimes v_{j_{(0)}} \otimes v_{j_{(1)}} \otimes v_{j_{(2)}} \end{aligned}$$

and

$$\begin{aligned} ((1\#1) \otimes (a\#b)) \cdot g &= \sum (x_i \otimes u_{j_{(0)}}) \otimes (S(u_{j_{(1)}}) \rightharpoonup y_i \# v_j)(a\#b) \\ &= (x_i \otimes u_{j_{(0)}}) \otimes \big((S(u_{j_{(1)}}) \rightharpoonup y_i)(v_{j_{(1)}} \rightharpoonup a) \# (v_{j_{(0)}} b)\big) \\ &= (x_i \otimes u_j) \otimes (v_{j_{(2)}} \rightharpoonup y_i)(v_{j_{(1)}} \rightharpoonup a) \# (v_{j_{(0)}} b) \\ &= (x_i \otimes u_j) \otimes (v_{j_{(1)}} \rightharpoonup y_i a) \# (v_{j_{(0)}} b) \end{aligned}$$

From the fact that

$$\begin{aligned} \sum a x_i \otimes y_i &= \sum x_i \otimes y_i a \\ \sum b u_j \otimes v_j &= \sum u_j \otimes v_j b \end{aligned}$$

it follows that

$$\sum(ax_i\#bu_j)\otimes(y_i\otimes v_j)=\sum(x_i\#u_j)\otimes(y_ia\otimes v_jb)$$

Apply $(I\otimes S)\circ\rho_B$ to the second factor, and then let the third factor act on the fourth one. This yields

$$\begin{aligned}\sum(ax_i\#b_{(0)}u_{j(0)})\otimes(S(b_{(1)}u_{j(1)})\rightharpoonup y_i\#v_i) &= \sum(x_i\otimes u_{j(0)}\otimes((S(u_{j(1)})\rightharpoonup y_i)a)\#v_jb) \\ &= (x_i\otimes u_j)\otimes(v_{j(1)}\rightharpoonup y_ia)\#(v_{j(0)}b)\end{aligned}$$

and

$$((a\#b)\otimes(1\#1))\cdot g=((1\#1)\otimes(a\#b))\cdot g$$

proving the third assertion.
4) Suppose that A and B are left H-central. The left H-center $(A\#B)^{A\#B}$ of $A\#B$ is generated by elements of the form $g\cdot(a\#b)$, with $a\in A, b\in B$ and g as in 3), but viewed as an element of $A\#B\#\overline{A\#B}$. Using the fact that $\sum u_j\#v_j\in(B\#\overline{B})^{H\mathrm{co}H}$, we obtain that

$$\begin{aligned}g\cdot(a\#b) &= \sum\Big((x_i\#u_{j(0)})\#\overline{S(u_{j(1)})\rightharpoonup y_i)\#v_j}\Big)\cdot(a\#b) \\ &= \sum(x_i\#u_{j(0)})((y_{i(1)}v_{j(1)})\rightharpoonup a\#(y_{i(2)}v_{j(2)})\rightharpoonup b)(S(u_{j(1)})\rightharpoonup y_{i(0)}\#v_{j(0)}) \\ &= \sum\Big(x_i((u_{j(1)}y_{i(1)}v_{j(1)})\rightharpoonup a)((u_{j(2)}b_{(1)}S(u_{j(3)}))\rightharpoonup y_{i(0)})\Big)\#\Big(u_{j(0)}((y_{i(2)}v_{j(2)})\rightharpoonup b_{(0)})v_{j(0)}\Big) \\ &= \sum\Big(x_i(y_{i(1)}\rightharpoonup a)(b_{(1)}\rightharpoonup y_{i(0)})\Big)\#\Big(u_j((y_{i(2)}v_{j(2)})\rightharpoonup b_{(0)})v_{j(0)}\Big) \\ &= \sum\Big((x_i\#(b_{(1)}\rightharpoonup y_{i(0)}))\cdot a\Big)\#\Big((u_j\#v_j)\cdot(y_{i(1)}\rightharpoonup b_{(0)})\Big)\end{aligned}$$

Using again the fact that $\sum u_j\otimes v_j$ is invariant under the H-coaction, we obtain that

$$\sum(u_j\otimes v_j)\cdot(h\rightharpoonup b_{(0)})\otimes b_{(1)}=\sum\Big((u_j\otimes v_j)\cdot(h\rightharpoonup b)\Big)_{(0)}\otimes\sum\Big((u_j\otimes v_j)\cdot(h\rightharpoonup b)\Big)_{(1)}$$

Now $\sum(u_j\otimes v_j)\cdot(h\rightharpoonup b)\in B^B=R$, so

$$\sum(u_j\otimes v_j)\cdot(h\rightharpoonup b_{(0)})\otimes b_{(1)}=\sum(u_j\otimes v_j)\cdot(h\rightharpoonup b)\otimes 1$$

In a similar way, we find that

$$\begin{aligned}\sum(u_j\otimes v_j)\cdot(h\rightharpoonup b) &= h\rightharpoonup\sum(u_j\otimes v_j)\cdot b \\ &= \varepsilon(h)\sum(u_j\otimes v_j)\cdot b\end{aligned}$$

and we conclude that

$$g\cdot(a\#b)=\sum\Big((x_i\#y_i)\cdot a\Big)\#\Big((u_j\#v_j)\cdot b\Big)\in A^A\#B^B=R$$

and $A\#B$ is left H-central. In a similar way, we show that $A\#B$ is right H-central.
□

Lemma 12.3.10 *Let A be a strong H-Azumaya algebra. Then R embeds in A as a direct summand in H-dimod.*

Proof Let $e \in A^{\#e}$ be an H-separability idempotent that is invariant under H-action and H-coaction. Define $t: A \to R$ by $t(a) = e \cdot a \in A^A = R$. t is a left inverse of the inclusion $R \to A$, and t is H-linear and H-colinear since $e \in A^{H\mathrm{co}H}$. □

An H-dimodule algebra A is called *H-simple* if A has no proper two-sided ideals that are H-dimodules (we call such ideals H-ideals), or, equivalently, if A is a simple object in $(H, A^{\#e}\text{-}R)$-dimod or $(H, R\text{-}^{\#e}A)$-dimod.

Proposition 12.3.11 *A strong H-Azumaya algebra is H-simple if and only if the groundring R is a field.*

Proof First suppose that A is H-simple, and let I be a nonzero ideal of R. Then IA is an H-ideal of A, and this implies that $IA = A$. Let t be the trace map of Lemma 12.3.10. Then $I = t(IA) = t(A) = R$, and it follows that R has no nontrivial ideals. Consequently R is a field.
Now let A be a strong H-Azumaya algebra over a field k. Then A is a separable extension of k, and is semisimple artinian (see [114, Theorem III.3.1]). Let M be an H-ideal of A. Then there is a central idempotent $c \in A$ such that $M = cA = Ac$. We claim that $c \in A^{H\mathrm{co}H}$. Observe that $c \in A$ and that

$$cm = mc = m$$

for all $m \in M$. For all $h \in H$, we have that

$$\begin{aligned} h \rightharpoonup c &= (h \rightharpoonup c)c \\ &= \sum (h_{(1)} \rightharpoonup c)(\varepsilon(h_{(2)})c) \\ &= \sum h_{(1)} \rightharpoonup (c(S(h_{(2)}) \rightharpoonup c)) \\ &= \sum (h_{(1)}S(h_{(2)})) \rightharpoonup c = \varepsilon(h)c \end{aligned}$$

Write $c = c'$. Since $c \in M$, we have

$$\sum c' c_{(0)} \otimes c_{(1)} = \sum c_{(0)} \otimes c_{(1)}$$

and

$$\sum c'_{(0)} c_{(0)} \otimes c'_{(1)} c_{(1)} \otimes c_{(2)} = \sum c_{(0)} \otimes c_{(1)} \otimes c_{(2)}$$

Apply S to the third factor, and let the third factor act on the second one. We obtain that

$$\sum c'_{(0)} c \otimes c'_{(1)} = c \otimes 1$$

and

$$\rho(c) = c \otimes 1$$

Now for any $a \in A$, we have that

$$\sum a_{(0)}(a_{(1)} \rightharpoonup c) = ac = ca$$

and this implies that $c \in A^A = k$. Since c is an idempotent, $c = 1$, and $M = cA = A$. This proves that A is H-simple. □

Proposition 12.3.12 *Let A be a strong H-Azumaya algebra. Then for every maximal H-ideal M of A, $M \cap R = I$ is a maximal ideal of R, and $M = IA$.*

Proof A/M is a strong H/IH-Azumaya algebra (over the groundring R/I). A/M is H/IH-simple, hence R/I is a field, and I is a maximal ideal. Now A/IA is also a strong H/IH-Azumaya algebra. Thus A/IA is H/IH-simple, and this implies that IA is a maximal ideal of A. but $IA \subset M$, so $IA = M$. □

Proposition 12.3.13 *A strong H-Azumaya algebra is an H-Azumaya algebra.*

Proof From condition 2) of Proposition 12.2.4, it follows that we only need to show that A is a left $A^{\#e}$-generator and a right ${}^{\#e}A$-generator. Write

$$T = \{\sum f_i(x_i) | x_i \in A, f_i \in \mathrm{Hom}_{A^{\#e}}(A, A^{\#e})\}$$

Take $f \in \mathrm{Hom}_{A^{\#e}}(A, A^{\#e})$. Then $f(1) = m \in (A^{\#e})^A = eA^{\#e}$, where $e \in A^{\#e}$ is the separability idempotent of A (Lemma 12.3.3). For $x \in A$, we find that

$$f(x) = (x\#\overline{1})f(1) = (x\#\overline{1})m \in A^{\#e}eA^{\#e}$$

and $T \subset A^{\#e}eA^{\#e}$.
Taking $m = 1$ and $x = 1$, we obtain that $e \in T$. It is clear that T is a two-sided ideal of $A^{\#e}$, and therefore

$$T = A^{\#e}eA^{\#e}$$

Now assume that $T \neq A^{\#e}$. Then $T \subset M$, a maximal H-ideal of A. From Proposition 12.3.9, we know that $A^{\#e}$ is strongly H-Azumaya, and Proposition 12.3.12 tells us that

$$A^{\#e}eA^{\#e} = T \subset M = (M \cap R)A^{\#e}$$

Applying the multiplication map m_A to both sides, we obtain that $A \subset (M \cap R)A$ and

$$A = (M \cap R)A$$

Now apply the splitting map t from Lemma 12.3.10 to both sides. This gives

$$R = t(A) = (M \cap R)t(A) = M \cap R$$

and, by Proposition 12.3.12,

$$M = (M \cap R)A = RA = A$$

and M is not a proper ideal of A. This is a contradiction. □

Here is another characterization of strong H-Azumaya algebras.

Proposition 12.3.14 *Let A be an H-Azumaya algebra. A is a strong H-Azumaya algebra if and only if the inclusion map $R \to A$ splits as a map of H-dimodules.*

Proof In Lemma 12.3.10, we have already seen that $R \to A$ splits in H-dimod if A is a strong H-Azumaya algebra.
Conversely, let $t: A \to R$ be an H-linear H-colinear left inverse of the inclusion map. Since A is H-Azumaya, we know that

$$F: A^{\#e} \longrightarrow \mathrm{End}_R(A)$$

is an isomorphism. Take

$$e = \sum x_i \otimes \overline{y}_i = F^{-1}(t) \in A^{\#e}$$

This means that

$$t(a) = \sum x_i (y_{i_{(1)}} \rightharpoonup a) y_{i_{(0)}}$$

for all $a \in A$. From the fact that t and F are H-linear and H-colinear, it follows that $e \in (A^{\#e})^A$.
We claim that e is a left H-separability idempotent for A. To this end, we have to show that

$$(a\#\overline{1})e = (1\#\overline{a})e$$

for all $a \in A$, or

$$F((a\#\overline{1})e) = F(1\#\overline{a})F(e)$$

Now for all $b \in A$

$$\begin{aligned} F((a\#\overline{1})e)(b) &= F(\sum a x_i \# \overline{y}_i)(b) \\ &= \sum a x_i (y_{i_{(1)}} \rightharpoonup b) y_{i_{(0)}} \\ &= a t(b) \\ &= t(b) a \\ &= F(1\#\overline{a})(t(b)) \\ &= F(1\#\overline{a})(F(e)(a)) \end{aligned}$$

□

Theorem 12.3.15 *Let A be an H-dimodule algebra, and B and C strong H-Azumaya algebras. If $A\#B \cong C$ as H-dimodule algebras, then A is also a strong H-Azumaya algebra.*

Proof We have to show that A is strongly H-separable and H-central. From Proposition 12.3.14, we know that R embeds as an H-dimodule direct summand in B. Thus we have an H-linear H-colinear monomorphism $i: R \to B$ with a right inverse $p: B \to R$, also H-linear and H-colinear.
Using Propositions 12.1.4 and 12.1.6, we obtain that

$$\begin{aligned} (A\#B)^{\#e} &= (A\#B)\#\overline{(A\#B)} \\ &\cong A\#B\#\overline{B}\#\overline{A} \\ &\cong A\#\mathrm{End}_R(B)\#\overline{A} \\ &\cong A^{\#e} \otimes \mathrm{End}_R(B) \\ &\cong A^{\#e}\#B^{\#e} \end{aligned}$$

The following diagram is commutative, and all the maps are H-linear and H-colinear.

$$\begin{array}{ccccc} A^{\#e}\#B^{\#e} & \cong & (A\#B)^{\#e} & \xrightarrow{m_{A\#B}} & A\#B \\ \downarrow I_A\otimes m_B & & & & \downarrow I_A\otimes p \\ A^{\#e} & & \xrightarrow{m_A} & & A \end{array}$$

$m_{A\#B}$ splits in $(H, A^{\#e}\#B^{\#e}\text{-}R)$-mod, and therefore in $(H, A^{\#e}\text{-}R)$-mod. p splits in H-dimod, so $I_A \otimes p$ splits in $(H, A^{\#e}\text{-}R)$-mod. thus m_A splits in $(H, A^{\#e}\text{-}R)$-mod, and A is strongly H-separable.
To show that A is H-central, we proceed as follows: in the proof of part 4) of Proposition 12.3.9, we have seen that

$$(A\#B)^{A\#B} = A^A\#B^B$$

If A and $A\#B$ are left H-central, then $R = A^H\#R$, $A^A = R$, and A is left H-central. In a similar way, it follows that A is right H-central. □

12.4 Examples of H-Azumaya algebras

Azumaya algebras

Let A be an Azumaya algebra in the sense of Section 2.2. We give A the following (trivial) H-dimodule algebra structure.

$$\rho(a) = a\otimes 1 \tag{12.21}$$
$$h \rightharpoonup a = \varepsilon(h)a \tag{12.22}$$

for all $a \in A$ and $h \in H$. It is easy to check that $\overline{A} = A^{\mathrm{op}}$, $A^{\#e} \cong A^{e}$ and ${}^{\#e}A \cong A^{\mathrm{op}} \otimes A$. The map $F:\ A^{\#e}\longrightarrow \mathrm{End}_R(A)$ is the same as the map $F:\ A^{e}\longrightarrow \mathrm{End}_R(A)$ from Proposition 2.2.9, and is therefore an isomorphism. In a similar way, the map $G:\ {}^{\#e}A\longrightarrow \mathrm{End}_R(A)^{\mathrm{op}}$ is an isomorphism, and A is an H-Azumaya algebra.

H-module Azumaya algebras

Let A be an Azumaya algebra, and suppose that A has also the structure of H-module algebra. Let H coact trivially on A, as in (12.21). Then A is an H-dimodule algebra, and the same arguments as the ones above show that A is an H-Azumaya algebra. We call A an H-module Azumaya algebra.
The notion of H-module Azumaya algebra still makes sense in the case where the Hopf algebra H is cocommutative, but not necessarily commutative. We then define an H-module Azumaya algebra as an Azumaya algebra that is also an H-module algebra.
Suppose that H is cocommutative, but not necessarily commutative. We will now construct some examples of H-module Azumaya algebras. Let S be a (not necessarily commutative) H-Galois object. Using the fact that H is cocommutative, we

can easily show that S^{op} is also an H-Galois object. From Theorem 8.3.1, it follows that $S\#H^* \cong \mathrm{End}_R(S)$, the connecting isomorphism is given by

$$\phi(s\#h^*)(t) = \sum \langle h^*, t_{(1)}\rangle t_{(0)}s$$

Being isomorphic to an endomorphism ring, $S\#H^*$ is an Azumaya algebra, and an H-action on $S\#H^*$ is given by the formula

$$h\rightharpoonup(s\#h^*) = \sum \langle h^*_{(1)}, h\rangle s\#h^*_{(2)} \tag{12.23}$$

We claim that this action makes $S\#H^*$ into an H-module algebra. Recall first that the multiplication on $S\#H^*$ is given by the formula

$$(s\#h^*)(t\#k^*) = \sum st_{(0)}\#(t_{(1)}\rightharpoonup h^*) * k^* \tag{12.24}$$

We remark that the smash product that is used here is not the same smash product as the one that is used for H-dimodule algebras in the rest of this chapter. For all $h \in H$, $h^*, k^* \in H^*$ and $s, t \in S$, we have that

$$\begin{aligned} h\rightharpoonup\big((s\#h^*)(t\#k^*)\big) &= h\rightharpoonup(\textstyle\sum st_{(0)}\#(t_{(1)}\rightharpoonup h^*) * k^*)) \\ &= \textstyle\sum \langle h^*_{(1)} * k^*_{(1)}, h\rangle st_{(0)}\#(t_{(1)}\rightharpoonup h^*_{(2)}) * k^*_{(2)} \\ &= \textstyle\sum \langle h^*_{(1)}, h_{(1)}\rangle\langle k^*_{(1)}, h_{(2)}\rangle st_{(0)}\#(t_{(1)}\rightharpoonup h^*_{(2)}) * k^*_{(2)} \\ &= \big(\textstyle\sum \langle h^*_{(1)}, h_{(1)}\rangle s\#h^*_{(2)}\big)\big(\textstyle\sum \langle k^*_{(1)}, h_{(2)}\rangle t\#k^*_{(2)}\big) \\ &= \textstyle\sum (h_{(1)}\rightharpoonup(s\#h^*))(k_{(1)}\rightharpoonup(t\#k^*)) \end{aligned}$$

We have shown the following:

Proposition 12.4.1 *Let H be a commutative, cocommutative faithfully projective Hopf algebra. Then $S\#H^*$ with the H-action (12.23) is an H-module Azumaya algebra.*

H-comodule Azumaya algebras

An Azumaya algebra that is also an H-comodule algebra is called an H-comodule Azumaya algebra. If we give A the trivial action ((12.21)), then A becomes an H-dimodule algebra, and arguments similar to the ones given above show that A is an H-dimodule algebra. Observe also that H-comodule Azumaya algebras are nothing else then H^*-module Azumaya algebras, and that Proposition 12.4.1 allows us to construct H-comodule Azumaya algebras. The notion of H-comodule Azumaya algebra still makes sense in the case where H is not necessarily cocommutative.
An H-coaction on an algebra A is called *inner* if there exists $w \in \mathbb{G}_m(A \otimes H)$ such that

$$\rho(A) = w(a \otimes 1)w^{-1}$$

for all $a \in A$ (cf. [26], [22]).

Proposition 12.4.2 *Let H be a commutative Hopf algebra, and A an H-comodule Azumaya algebra. If $\mathrm{Pic}(H) = 1$, then H coacts innerly on A.*

Proof Let $\underline{H}$ be equal to H as an R-algebra (we forget the comultiplication on H). Then $\underline{H}$ is a commutative R-algebra, and $A \otimes \underline{H}$ is an Azumaya algebra over the groundring $\underline{H}$. The map

$$F: A \otimes \underline{H} \longrightarrow A \otimes \underline{H}: a \otimes h \mapsto \sum a_{(0)} \otimes a_{(1)} h$$

is an $\underline{H}$-algebra isomorphism of $A \otimes \underline{H}$ (its inverse is given by the formula $F^{-1}(a \otimes h) = \sum a_{(0)} \otimes S(a_{(1)})h$). By the Skolem-Noether Theorem, F is an inner automorphism, so we can find $w \in A \otimes H$ such that

$$F(a \otimes h) = w(a \otimes h)w^{-1}$$

Taking $h = 1$, we obtain that

$$\rho(a) = w(a \otimes h)w^{-1}$$

□

Similar statements hold for H-module algebras. An H-action on an algebra A is called *inner* if there exists a convolution invertible $u: H \to A$ such that

$$h \rightharpoonup a = \sum u(h_{(1)}) a v(h_{(2)})$$

where v is the convolution inverse of u. Inner actions on central simple algebras have been studied in [26] and [136]. For more results about inner actions on Azumaya algebras, we refer to [22], [24], [117] and [130]. If H is faithfully projective, then the H-action on an algebra A is inner if and only if the corresponding H^*-coaction is inner. As a direct consequence of Proposition 12.4.2, we obtain the following Corollary, which is a special case of a much more general result due to Masuoka [130].

Proposition 12.4.3 *Let H be a cocommutative faithfully projective Hopf algebra, and A an H-module Azumaya algebra. If* $\mathrm{Pic}(H^*) = 1$, *then H acts innerly on A.*

θ-Azumaya algebras

Fix a Hopf algebra map $\theta: H \to H^*$ (or, equivalently, a pairing $\theta: H \otimes H \to R$). Let M be an H-dimodule. The H-module structure on M defines an H^*-comodule structure on M, denoted by

$$\varphi: M \longrightarrow M \otimes H^*$$

We call M a θ-module if

$$\varphi = (I_M \otimes \theta) \circ \rho$$

or, equivalently, if

$$h \rightharpoonup m = \sum \langle \theta(m_{(1)}), h \rangle m_{(0)} \tag{12.25}$$

An H-dimodule algebra that is a θ-module is called a θ-algebra. A θ-Azumaya algebra is an H-Azumaya algebra that is a θ-algebra.
As a special case, consider the situation where $\theta: H \to H^*$ is given by the formula

$$\theta(h) = \varepsilon(h)\varepsilon$$

Then the H-action defined by (12.25) is trivial, and a θ-Azumaya algebra is nothing else then an H-comodule Azumaya algebra.

Example 12.4.4 Let R be a ring in which 2 is invertible, and take $H = RC_2$. Let $\theta: RC_2 \longrightarrow RC_2^* \cong RC_2^*$ be the canonical isomorphism. Let $A = RC_2$ as a graded C_2-algebra. The C_2-action on A induced by the grading is given by the formula

$$\sigma \rightharpoonup a = (-1)^{\deg(a)}a$$

for all homogeneous $a \in A$, and where we wrote $C_2 = \{1, \sigma\}$. A is a θ-algebra, and it can be seen easily that A is RC_2-Azumaya: it is easy to prove that $A^A = R$; furthermore, RC_2 is semisimple-like and cosemisimple-like, and A is a separable R-algebra. From Corollary 12.3.7 it follows that A is strongly RC_2-separable, and A is a (strong) RC_2-Azumaya algebra.
Observe that A is not a central R-algebra (A is commutative!). Thus we have our first example of an H-Azumaya algebra that is not an Azumaya algebra. This example already appeared in the work of Wall [200]. He called the non-central RC_2-Azumaya algebras "minus"-algebras, while the central ones were called "plus"-algebras.

Our next aim is to generalize Example 12.4.4 to an arbitrary Hopf algebra A. Let A be a not necessarily commutative H-Galois object with normal basis. From (10.25), we know that A is equal to H as an H-comodule, and that the multiplication on A is given by a Sweedler cocycle $f \in Z^2(H, R, \mathbb{G}_m)$, namely

$$h \cdot k = \sum f(h_{(1)} \otimes k_{(1)})h_{(2)}k_{(2)}$$

for all $h, k \in A = H$. The dot $\cdot$ represents the multiplication in A. Recall from Section 10.8 that we have a map $\beta: H^2(H, R, \mathbb{G}_m) \to P_{\mathrm{sk}}(H, R)$. We view $d = \beta(f)$ as a Hopf algebra map $d: H \to H^*$. From the definition of β, it may be deduced easily that the commutation rule in A is given by the formula

$$h \cdot k = \sum \langle d(h_{(2)}), k_{(2)} \rangle k_{(1)} \cdot h_{(1)} = \sum \langle h_{(2)}, d^*(k_{(2)}) \rangle k_{(1)} \cdot h_{(1)} \qquad (12.26)$$

Later on we will use the maps S_A, $S'_A: A \to A$ defined by

$$S_A(h) = \sum g(S(h_{(1)}) \otimes h_{(2)})S(h_{(3)}) \text{ and } S'_A(h) = \sum g(h_{(1)} \otimes S(h_{(2)}))S(h_{(3)})$$

for all $h \in A$. Here g is the convolution inverse of f. It may be verified immediately that

$$\sum S_A(h_{(1)}) \cdot h_{(2)} = \sum h_{(1)} \cdot S'_A(h_{(2)}) = \varepsilon(h) \qquad (12.27)$$

Now let $\theta: H \to H^*$ be a Hopf algebra map, and define an H-action on A by the formula

$$h \rightharpoonup k = \sum \langle \theta(k_{(2)}), h \rangle k_{(1)} \qquad (12.28)$$

for all $h, k \in H$. It is easy to show that A is a θ-algebra. We will investigate when A is an H-Azumaya algebra. To this end, we have to find out whether the maps

$$\begin{cases} F: & A\#\overline{A} \longrightarrow \mathrm{End}_R(A) \\ G: & \overline{A}\#A \longrightarrow \mathrm{End}_R(A)^{\mathrm{op}} \end{cases}$$

are isomorphisms. F and G are given by the formulas

$$\begin{aligned} F(a\#b)(k) &= \sum a \cdot (b_{(2)} \rightharpoonup k) \cdot b_{(1)} \\ &= \sum \langle \theta(k_{(2)}), b_{(2)} \rangle a \cdot k_{(1)} \cdot b_{(1)} \\ &= \sum \langle (\theta * d)(k_{(2)}), b_{(2)} \rangle a \cdot b_{(1)} \cdot k_{(1)} \qquad (12.29) \\ G(a\#b)(k) &= \sum (k_{(2)} \rightharpoonup a) \cdot k_{(1)} \cdot b \\ &= \sum \langle \theta(a_{(2)}), k_{(2)} \rangle a_{(1)} \cdot k_{(1)} \cdot b_{(1)} \\ &= \sum \langle (\theta * d)(a_{(2)}), k_{(2)} \rangle k_{(1)} \cdot a_{(1)} \cdot b_{(1)} \qquad (12.30) \end{aligned}$$

for all $a, b, k \in H$.

Theorem 12.4.5 *Take a Sweedler cocycle $f \in Z^2(H, R, \mathbb{G}_m)$ and a Hopf algebra map $\theta : H \to H^*$. Consider the Hopf algebra map $d = \beta(f) : H \to H^*$ and the θ-algebra A defined above. Then the following assertions are equivalent.*
1) $\theta * d$ is an isomorphism;
2) A is a left H-Azumaya algebra;
3) A is a right H-Azumaya algebra;
4) A is an H-Azumaya algebra.

Proof 1) $\Longrightarrow$ 2). We have to show that the map F is an isomorphism. It suffices to show that, for every prime ideal p of R, the localization

$$F_p : A_p\#\overline{A}_p \longrightarrow \mathrm{End}_{R_p}(A_p)$$

is an isomorphism. Thus we may restrict attention to the case where R is local. If we can show that F is surjective, then a rank argument shows that F is injective too.
Take $h \in H$ and $h^* \in H^*$ and let $\alpha \in \mathrm{End}_R(A)$ be defined by the formula $\alpha(k) = \langle h^*, k \rangle h$. We will be done if we can show that $\alpha \in \mathrm{Im}\,(F)$.
Recall from Section 8.6 that $\int_{H^*}^l$ is an invertible R-module. Since R is local, it is free of rank one, and we can find an integral x^* such that $\int_{H^*}^l = Rx^*$. Consider the isomorphism (cf. Lemma 8.6.2) $v : H \longrightarrow H^*$ given by

$$\langle v(h), k \rangle = \langle x^*, kS(h) \rangle$$

Recall also formula (8.29):

$$\sum \langle v(h), k_{(2)} \rangle k_{(1)} = \sum \langle v(h_{(1)}), k \rangle h_{(2)} \qquad (12.31)$$

for all $h, k \in H$. Now let $l = v^{-1}(h^*) \in H$. Observe that $\theta^* * d^* = (\theta * d)^*$ is bijective, and take

$$\Gamma = \sum h \cdot S_A(l_{(1)}) \cdot S_A((\theta^* * d^*)^{-1}(v(l_{(2)})_{(1)}))\#(\theta^* * d^*)^{-1}(v(l_{(2)})_{(2)})$$

Using (12.29), we obtain, for all $k \in A$,

$$\begin{aligned}
F(\Gamma)(k) &= \sum \langle(\theta * d)(k_{(2)}), (\theta^* * d^*)^{-1}(v(l_{(2)})_{(3)})\rangle \\
&\quad h \cdot S_A(l_{(1)}) \cdot S_A((\theta^* * d^*)^{-1}(v(l_{(2)})_{(1)})) \cdot (\theta^* * d^*)^{-1}(v(l_{(2)})_{(2)}) \cdot k_{(1)} \\
&= \sum \langle k_{(2)}, v(l_{(2)})\rangle h \cdot S_A(l_{(1)}) \cdot k_{(1)} \quad \text{(by (12.27))} \\
&= \sum \langle k_{(3)}, v(l)\rangle h \cdot S_A(k_{(2)}) \cdot k_{(1)} \quad \text{(by (12.31))} \\
&= \sum \langle h^*, k\rangle h = \alpha(k) \quad \text{(by (12.27)}
\end{aligned}$$

It follows that $\alpha \in \mathrm{Im}\,(F)$, and F is surjective.

1) $\Longrightarrow$ 3). This part is similar to the proof of 1) $\Longrightarrow$ 2). It suffices to show that G is surjective in the case where R is local. Let $\alpha \in \mathrm{End}_R(A)$, $h \in H$, $h^* \in H^*$ and $l = v^{-1}(h^*) \in H$ be as above. Take

$$\Delta = \sum (\theta * d)^{-1}(v(l_{(2)})_{(2)}) \# S'_A((\theta * d)^{-1}(v(l_{(2)})_{(1)})) \cdot S'_A(l_{(1)}) \cdot h$$

in $\overline{A}\#A$. For all $k \in A$, we obtain, using (12.30),

$$\begin{aligned}
G(\Delta)(k) &= \sum \langle(\theta * d)\big((\theta * d)^{-1}(v(l_{(2)})_{(3)})\big), k_{(2)}\rangle \\
&\quad k_{(1)} \cdot (\theta * d)^{-1}(v(l_{(2)})_{(2)}) \cdot S'_A((\theta * d)^{-1}(v(l_{(2)})_{(1)})) \\
&= \sum \langle k_{(2)}, v(l_{(2)})\rangle k_{(1)} \cdot S'_A(l_{(1)}) \cdot h \\
&= \sum \langle k, v(l_{(2)})\rangle l_{(3)} \cdot S'_A(l_{(1)}) \cdot h \\
&= \langle k, h^*\rangle h = \alpha(k)
\end{aligned}$$

and it follows that G is surjective.

2) $\Longrightarrow$ 1). We assume that the cocycle f is normalized, which is no restriction. As in the two previous parts of the proof, it suffices to show that $\theta * d$, or, equivalently, $\theta^* * d^*$ is surjective in the case where R is local. Suppose that F is surjective, and take an arbitrary $h^* \in H^*$. As above, we will write $h^* = v(l)$, with $l \in H$. Let $\alpha \in \mathrm{End}_R(A)$ be given by the formula $\alpha(k) = \langle h^*, k\rangle$, for all $k \in H$. We know that $\alpha \in \mathrm{Im}\,(F)$, and we write

$$\alpha(\sum_i a_i \# b_i) = \alpha$$

for some $a_i, b_i \in A$. We claim that the coaction of H on $h^* = v(l)$ is given by the formula

$$\rho(v(l)) = \sum v(l_{(1)}) \otimes S(l_{(2)}) \tag{12.32}$$

For all $k \in H$, it suffices to verify that

$$\sum \langle v(l_{(1)}), k\rangle S(l_{(2)}) = \sum \langle v(h), k_{(1)}\rangle S(k_{(2)})$$

(see (7.41)), and this follows immediately from (12.32). Now F is H-colinear, and therefore

$$\sum_i F(a_{i_{(1)}} \# b_{i_{(1)}}) \otimes a_{i_{(2)}} b_{i_{(2)}} = \sum (1 \otimes v(l_{(1)})) \otimes S(l_{(2)})$$

where we identified $\mathrm{End}_R(A)$ and $A \otimes A^*$. Applying the antipode to the second factor, we obtain that

$$\sum_i F(a_{i_{(1)}} \# b_{i_{(1)}}) \otimes S(a_{i_{(2)}} b_{i_{(2)}}) = \sum (1 \otimes v(l_{(1)})) \otimes l_{(2)}$$

For all $k \in H$, we obtain that

$$\sum_i F(a_{i_{(1)}}\#b_{i_{(1)}})(k) \otimes S(a_{i_{(2)}}b_{i_{(2)}}) = \sum \langle v(l_{(1)}), k\rangle 1 \otimes l_{(2)}$$

or, using (12.31) and (12.29)

$$\sum \langle (\theta * d)(k_{(2)}), b_{i_{(2)}}\rangle a_{i_{(1)}} \cdot b_{i_{(1)}} \cdot k_{(1)} \otimes S(a_{i_{(2)}}b_{i_{(3)}}) = \sum \langle v(l), k_{(1)}\rangle 1 \otimes k_{(2)} \quad (12.33)$$

Switching the two factors in (12.33), and applying the multiplication map in A, we obtain that

$$\sum \langle (\theta * d)(k_{(2)}), b_{i_{(3)}}\rangle (S(b_{i_{(2)}})S(a_{i_{(2)}})) \cdot a_{i_{(1)}} \cdot b_{i_{(1)}} \cdot k_{(1)} = \sum \langle v(l), k_{(1)}\rangle k_{(2)} \quad (12.34)$$

The left hand side of (12.34) amounts to

$$\begin{aligned}
&\sum \langle (\theta * d)(k_{(2)}), b_{i_{(4)}}\rangle f(S(b_{i_{(3)}}a_{i_{(4)}}) \otimes a_{i_{(3)}})S(b_{i_{(2)}})S(a_{i_{(2)}})a_{i_{(1)}} \cdot b_{i_{(1)}} \cdot k_{(1)} \\
&= \sum \langle (\theta * d)(k_{(2)}), b_{i_{(4)}}\rangle f(S(b_{i_{(3)}}a_{i_{(2)}}) \otimes a_{i_{(1)}})S(b_{i_{(2)}}) \cdot b_{i_{(1)}} \cdot k_{(1)} \\
&= \sum \langle (\theta * d)(k_{(2)}), b_{i_{(6)}}\rangle f(S(b_{i_{(5)}}a_{i_{(2)}}) \otimes a_{i_{(1)}})f(S(b_{i_{(4)}}) \otimes b_{i_{(3)}})S(b_{i_{(2)}})b_{i_{(1)}} \cdot k_{(1)} \\
&= \sum \langle (\theta * d)(k_{(2)}), b_{i_{(4)}}\rangle f(S(b_{i_{(3)}}a_{i_{(2)}}) \otimes a_{i_{(1)}})f(S(b_{i_{(2)}}) \otimes b_{i_{(1)}})k_{(1)}
\end{aligned}$$

Applying ε to both sides of (12.34), we now obtain that

$$\sum \langle (\theta * d)(k), b_{i_{(4)}}\rangle f(S(b_{i_{(3)}}a_{i_{(2)}}) \otimes a_{i_{(1)}})f(S(b_{i_{(2)}}) \otimes b_{i_{(1)}}) = \langle v(l), k\rangle$$

and

$$(\theta^* * d^*)\Big(\sum f(S(b_{i_{(3)}}a_{i_{(2)}}) \otimes a_{i_{(1)}})f(S(b_{i_{(2)}}) \otimes b_{i_{(1)}})b_{i_{(4)}}\Big) = v(l) = h^*$$

Thus $\theta^* * d^*$ is surjective.
The proof of 3) $\Longrightarrow$ 1) is similar to the proof of 2) $\Longrightarrow$ 1). 4) $\Longrightarrow$ 2) and 4) $\Longrightarrow$ 3) are trivial. Since 1) implies 2) and 3), we have that 1) $\Longrightarrow$ 4). □

Remark 12.4.6 Let $H = RG$ be a group ring, and write

$$\begin{aligned}
\phi(\sigma,\tau) &= \langle \theta(\tau), \sigma\rangle \\
\delta(\sigma,\tau) &= \langle d(\sigma), \tau\rangle = f(\sigma,\tau)f(\tau,\sigma)^{-1}
\end{aligned}$$

for all $\sigma, \tau \in G$. The Hopf algebra map $\theta * d$ is composition invertible if and only if the $|G| \times |G|$-matrix with entries $\phi(\sigma,\tau)\delta(\tau,\sigma)$ is invertible. At first glance, the conditions stated by Orzech (see [152, Prop. 2.8]) are more complicated. Orzech requires that the $|G| \times |G|$-matrices with entries

$$\phi(\sigma,\tau)f(\sigma^{-1},\tau)f(\sigma^{-1}\tau,\sigma) \quad \text{and} \quad \phi(\tau,\sigma^{-1})f(\sigma^{-1},\tau)f(\sigma^{-1}\tau,\sigma)$$

are invertible. Using the cocycle relations on f, we see that the regularity of both of these matrices are equivalent to the regularity of the matrix with entries $\phi(\sigma,\tau)g(\tau,\sigma)$. Indeed,

$$\begin{aligned}
\phi(\sigma,\tau)f(\sigma^{-1},\tau)f(\sigma^{-1}\tau,\sigma) &= \phi(\sigma,\tau)f(\sigma^{-1},\tau\sigma)f(\tau,\sigma) \\
&= \phi(\sigma,\tau)f(\sigma^{-1},\sigma\tau)f(\sigma,\tau)f(\sigma,\tau)^{-1}f(\tau,\sigma) \\
&= f(\sigma^{-1},\sigma)f(1,\tau)\phi(\sigma,\tau)\delta(\tau,\sigma)
\end{aligned}$$

and (replace σ^{-1} by σ in the second matrix)

$$\begin{aligned}
\phi(\tau,\sigma)f(\sigma,\tau)f(\sigma\tau,\sigma^{-1}) &= \phi(\tau,\sigma)f(\sigma,\tau)f(\tau,\sigma)^{-1}f(\tau,\sigma)f(\tau\sigma,\sigma^{-1}) \\
&= \phi(\tau,\sigma)g(\sigma,\tau)f(\tau,\sigma)f(\tau\sigma,\sigma^{-1})
\end{aligned}$$

Remark 12.4.7 Theorem 12.4.5 allows us to construct H-Azumaya algebras that are not separable as R-algebras and that are therefore non-strong H-Azumaya algebras. Let R be a commutative ring in which p is not a zero-divisor, containing a primitive root of unity η. Let I be an ideal of R such that $I^2 = (\eta - 1)R$, and consider the Tate-Oort order H_I. We have seen in Theorems 7.1.14 and 7.1.15 that there is a Hopf algebra isomorphism $\theta : H_I \to H_I^*$. Thus H_I is a θ-Azumaya algebra, and it is clear that H_I is not separable as an R-algebra, if we assume that p is not invertible in R.
We can easily construct a situation that meets all our requirements. For example, take $p = 2$, $R = \mathbb{Z}[\sqrt{2}]$ and $I = (\sqrt{2})$.

Remark 12.4.8 The data needed to determine the algebra A are the cocycle f and the Hopf algebra map θ. The corresponding data for the opposite algebra $A^{\rm op}$ and the H-opposite algebra $\overline{A}$ can be determined easily. The map θ stays the same in both cases. The cocycle describing the multiplication on $A^{\rm op}$ is $f \circ \tau$, and the corresponding skew pairing is d^*.
The multiplication on $\overline{A}$ is given by the formula

$$a \cdot b = \sum (a_{(2)} \rightharpoonup b) \cdot a_{(1)} = \sum \langle \theta(b_{(3)}), a_{(3)} \rangle f(b_{(2)} \otimes a_{(2)}) b_{(1)} a_{(1)}$$

The cocycle $\overline{f}$ describing the multiplication is therefore given by the formula

$$\overline{f}(a \otimes b) = \sum \langle \theta(b_{(2)}), a_{(2)} \rangle f(b_{(1)} \otimes a_{(1)})$$

and its convolution inverse $\overline{g}$ by

$$\overline{g}(a \otimes b) = \sum \langle \theta(S(b_{(2)})), a_{(2)} \rangle g(b_{(1)} \otimes a_{(1)})$$

for all $a, b \in A$. It follows that the corresponding skew pairing $\overline{d}$ is given by the formula

$$\begin{aligned} \langle \overline{d}(a), b \rangle &= \sum \langle \theta(b_{(1)}), a_{(1)} \rangle \langle (\theta \circ S)(a_{(2)}), b_{(2)} \rangle \langle d^*(a_{(3)}), b_{(3)} \rangle \\ &= \sum \langle \theta^*(a_{(1)}) * (\theta \circ S)(a_{(2)}) * d^*(a_{(3)}), b \rangle \end{aligned}$$

or

$$\overline{d} = \theta^* * (\theta \circ S) * d^* \tag{12.35}$$

Remark 12.4.9 Consider the case $H = RG$, with $G = C_n^k$ the direct sum of k copies of the cyclic group of order n. Assume that n is invertible in R, and that R contains a primitive n-th root of unity η. From Theorem 7.1.10, it follows that $(RG)^* \cong RG^*$. Let $\{\sigma_1, \ldots, \sigma_k\}$ be a free basis of G considered as a free C_n-module, and let $\{\sigma_1^*, \ldots, \sigma_k^*\}$ be the corresponding dual basis of G^*. We then have that

$$\langle \sigma_i^*, \sigma_j \rangle = \begin{cases} \eta & \text{if } i = j; \\ 1 & \text{if } i \neq j. \end{cases}$$

There is a one to one correspondence between Hopf algebra maps from RG to $(RG)^*$, and group homomorphisms (or C_n-module homomorphisms) from G to G^*. A C_n-module homomorphism from G to G^* is given by a $k \times k$-matrix with entries in $C_n \cong$

$\mathbb{Z}/n\mathbb{Z}$. The correspondence is the following: if $\theta : RG \to (RG)^*$ is a Hopf algebra map, and if the entries of the corresponding matrix $[\theta]$ are a_{ij} $(i,j = 1,\ldots,k)$, then

$$\theta(\sigma_i) = \prod_{j=1}^{k} (\sigma_j^*)^{a_{ij}}$$

It is easy to check that, for any two Hopf algebra maps $\theta, \psi : RG \to (RG)^*$, we have that

$$[\theta * \psi] = [\theta] + [\psi] \text{ and } [\theta^*] = [\theta]^t$$

Now we have an easy procedure to produce RG-Azumaya algebras: take a cocycle $f : G \times G \to \mathbb{G}_m(R)$, and take its corresponding pairing d, which is represented by a skew symmetric matrix $[d]$. Take an arbitrary invertible $k \times k$-matrix M, and let θ be represented by the matrix $[\theta] = M - [d]$. Then the θ-algebra A determined by f and θ is an H-Azumaya algebra. In the next two examples, respectively due to Orzech ([152]) and Tilborghs ([181]), we exploit this construction.

Example 12.4.10 (Orzech) Take $n = k = 2$, and write $\sigma_1 = \sigma$, $\sigma_2 = \tau$. Consider d and θ given by the matrices

$$[d] = \begin{pmatrix} 0 & 1 \\ 1 & 0 \end{pmatrix} \text{ and } [\theta] = \begin{pmatrix} 1 & 0 \\ 1 & 1 \end{pmatrix}$$

and let f be the inverse image of d under the map β as described in the proof of Theorem 10.8.7. It is clear that

$$[\theta * d] = [\theta] + [d] = \begin{pmatrix} 1 & 0 \\ 1 & 1 \end{pmatrix}$$

is invertible, and it follows that the corresponding θ-algebra A is an RC_2-Azumaya algebra. We give an explicit description of A. As an R-module, A is given by

$$A = R \oplus Ru_\sigma \oplus Ru_\tau \oplus Ru_{\sigma\tau}$$

The multiplication is given by the formulas

$$u_x^2 = 1 \;\; ; \;\; u_\sigma u_\tau = -u_\tau u_\sigma = u_{\sigma\tau}$$

for all $x \in G$. The RG-coaction (or better: the G-grading) is the natural one:

$$\deg(u_x) = x$$

and the RG-action (or better: the G-action) is induced by θ:

$$\sigma \rightharpoonup u_\sigma = \tau \rightharpoonup u_\sigma = -u_\sigma \;\; ; \;\; \sigma \rightharpoonup u_\tau = -(\tau \rightharpoonup u_\tau) = u_\tau$$

As an algebra, A is isomorphic to the matrix ring $M_2(R)$. This is clear, since A is in fact the quaternion algebra ${}^1R^1$. An explicit isomorphism is given by

$$\sigma \mapsto \begin{pmatrix} 1 & 0 \\ 0 & -1 \end{pmatrix} \text{ and } \tau \mapsto \begin{pmatrix} 0 & 1 \\ 1 & 0 \end{pmatrix}$$

Now we compute the matrix of the map $\overline{d}$ describing the RG-opposite algebra $\overline{A}$. Using (12.35) and the fact that $S = I_A$, we obtain

$$[\overline{d}] = [\theta^* * (\theta \circ S) * d^*] = [\theta^*] + [\theta] + [d^*] = \begin{pmatrix} 0 & 0 \\ 0 & 0 \end{pmatrix}$$

This means that $\overline{d}$ is the trivial map, and that $\overline{A}$ is commutative. We thus have an example of an H-Azumaya algebra that is an Azumaya algebra, while its H-opposite $\overline{A}$ is not Azumaya.

Example 12.4.11 (Tilborghs) Take $n = k = 3$, and let d and θ given by the matrices

$$[d] = \begin{pmatrix} 0 & 2 & 0 \\ 1 & 0 & 2 \\ 0 & 1 & 2 \end{pmatrix} \quad \text{and} \quad [\theta] = \begin{pmatrix} 0 & 0 & 1 \\ 1 & 0 & 0 \\ 0 & 1 & 0 \end{pmatrix}$$

As in the previous example, let f be the inverse image of d under the map β as described in the proof of Theorem 10.8.7. It can be checked easily that the determinant of the matrix

$$[\theta * d] = \begin{pmatrix} 0 & 2 & 1 \\ 2 & 0 & 2 \\ 0 & 2 & 2 \end{pmatrix}$$

is 2. Hence the corresponding algebra A is an RG-Azumaya algebra. We have seen above that the skew pairing corresponding to $A^{\rm op}$ is d^*. Now the matrix

$$[\theta * d^*] = \begin{pmatrix} 0 & 1 & 1 \\ 0 & 0 & 1 \\ 0 & 0 & 2 \end{pmatrix}$$

is not invertible, so $A^{\rm op}$ is not RG-Azumaya. We therefore have that opposite algebra of an H-Azumaya algebra is not always an H-Azumaya algebra.

H-Azumaya algebras that are endomorphism rings

H itself may be viewed as a faithfully projective H-dimodule, and obviously $\mathrm{End}_R(H)$ is an H-Azumaya algebra. We will now define an H-action and an H-coaction on H in such a way that the compatibility relation (12.1) is not satisfied. Thus H is not an H-dimodule under this action and coaction, but, as it happens, the endomorphism ring $\mathrm{End}_R(H)$ will turn out to be an H-dimodule algebra, and even an H-Azumaya algebra. In the case where H is a group ring, the construction presented below is due to Deegan [64] and Orzech [152]. The Hopf algebra case is due to the author [29].

Let $f:\ H \to H$ be a Hopf algebra automorphism, and let $g = I * (S \circ f)$, that is

$$g(h) = \sum h_{(1)} S(f(h_{(2)})) = \sum h_{(1)} f(S(h_{(2)}))$$

for all $h \in H$. Now let H_f be equal to H as an H-comodule, and consider the following H-action on H_f:

$$k \cdot h = g(k)h \tag{12.36}$$

for all $h, k \in H$. If f is different from the identity, then it can be seen easily that H_f is not an H-dimodule. As we will see, $A_f = \mathrm{End}_R(H_f)$ with the induced H-action and H-coaction is an H-Azumaya algebra.

Lemma 12.4.12 *With notations as above, A_f is an H-dimodule algebra.*

Proof We identify A_f and $H_f \otimes H_f^*$ in the canonical way. The action on H_f^* and A_f induced by multiplication in H will be denoted by $\rightharpoonup$. Thus we have that

$$\langle k \rightharpoonup h^*, h\rangle = \langle h^*, S(k)h\rangle$$

for all $h, k \in H$ and $h^* \in H^*$. The action induced by the new action (12.36) will be denoted by a dot $\cdot$. Thus

$$\langle k \cdot h^*, h\rangle = \langle h^*, S(k)\cdot h\rangle = \langle h^*, g(S(k))h\rangle = \langle g(k) \rightharpoonup h^*, h\rangle$$

and

$$k \cdot h^* = g(k) \rightharpoonup h^*$$

and

$$k \cdot (h \otimes h^*) = g(k) \rightharpoonup (h \otimes h^*)$$

for all $h, k \in H$ and $h^* \in H^*$. Using the fact that $\mathrm{End}_R(H)$ with the usual action and coaction is an H-dimodule, we obtain that

$$\begin{aligned}
\rho(k \cdot (h \otimes h^*)) &= \rho(g(k) \rightharpoonup (h \otimes h^*))\\
&= \sum (g(k) \rightharpoonup (h \otimes h^*)_{(0)}) \otimes (h \otimes h^*)_{(1)}\\
&= \sum (k \cdot (h \otimes h^*)_{(0)}) \otimes (h \otimes h^*)_{(1)}
\end{aligned}$$

It follows that A_f is an H-dimodule, and therefore an H-dimodule algebra, since we know already that A_f is an H-module algebra and an H-comodule algebra. □

Recall (7.33) that the H-action on $A_f = \mathrm{End}_R(H_f)$ may be written under the form

$$h \cdot a = \sum \psi_{h_{(1)}} \circ a \circ \psi_{S(h_{(2)})}$$

for $h \in H$ and $a \in A$. $\psi_h \in A_f$ is given by the formula

$$\psi_h(k) = h \cdot k = g(h)k$$

for all $k \in H$. With these notations, we claim that

$$\rho(\psi_h) = \sum \psi_{h_{(1)}} \otimes g(h_{(2)}) \tag{12.37}$$

We have to show that (7.40) is satisfied for all $k \in H$. The right hand side of (7.40) amounts to

$$\begin{aligned}
\sum \psi_h(k_{(1)})_{(1)} \otimes \psi_h(k_{(1)})_{(2)} S(k_{(2)}) &= \sum (g(h)k_{(1)})_{(1)} \otimes (g(h)k_{(1)})_{(2)} S(k_{(2)})\\
&= \sum g(h_{(1)})k_{(1)} \otimes g(h_{(2)})k_{(2)} S(k_{(3)})\\
&= \psi_{h_{(1)}}(k) \otimes g(h_{(2)})
\end{aligned}$$

which is the left hand side of (7.40).

Proposition 12.4.13 *With notations as above, A_f is an H-Azumaya algebra.*

Proof We have to show that the maps F and G defined in Lemma 12.2.3 are isomorphisms. It suffices to show that F and G are surjective, the injectivity then follows from the usual rank argument. Let us show that F is surjective. We know that A_f is an Azumaya algebra in the classical sense, and therefore the map F' : $A_f \otimes A_f^{\mathrm{op}} \to \mathrm{End}_R(A_f)$ given by

$$F'(a \otimes b)(c) = a \circ c \circ b$$

for all $a, b, c \in A_f$ is an isomorphism. Take $\alpha \in \mathrm{End}_R(A_f)$, and consider $\sum a_i \otimes b_i = F'^{-1}(\alpha) \in A_f \otimes A_f^{\mathrm{op}}$. We claim that

$$\Gamma = \sum a_i \circ \psi_{S(f^{-1}(b_{i_{(2)}}))} \# \psi_{f^{-1}(b_{i_{(1)}})} \circ b_{i_{(0)}}$$

is the inverse image of α under F in $A_f \# \overline{A}_f$. Using (12.37), we compute that

$$\begin{aligned}
\rho(\sum \psi_{f^{-1}(b_{i_{(1)}})} \circ b_{i_{(0)}}) &= \sum \psi_{f^{-1}(b_{i_{(2)}})} \circ b_{i_{(0)}} \otimes g(f^{-1}(b_{i_{(3)}}))b_{i_{(1)}} \\
&= \sum \psi_{f^{-1}(b_{i_{(2)}})} \circ b_{i_{(0)}}) \otimes f^{-1}(b_{i_{(3)}})S(f(f^{-1}(b_{i_{(4)}})))b_{i_{(1)}} \\
&= \sum \psi_{f^{-1}(b_{i_{(1)}})} \circ b_{i_{(0)}} \otimes f^{-1}(b_{i_{(2)}})
\end{aligned}$$

and using (12.15), we obtain that

$$\begin{aligned}
F(\Gamma)(c) &= \sum a_i \circ \psi_{S(f^{-1}(b_{i_{(3)}}))} \circ f^{-1}(b_{i_{(2)}} \cdot c) \circ \psi_{f^{-1}(b_{i_{(1)}})} \circ b_{i_{(0)}} \\
&= \sum a_i \circ \psi_{S(f^{-1}(b_{i_{(4)}}))} \circ \psi_{f^{-1}(b_{i_{(2)}})} \circ c \circ \psi_{S(f^{-1}(b_{i_{(3)}}))} \circ \psi_{f^{-1}(b_{i_{(1)}})} \circ b_{i_{(0)}} \\
&= \sum a_i \circ c \circ b_i = \alpha(c)
\end{aligned}$$

finishing the proof of the Proposition. □

Chapter 13

The Brauer-Long group of a commutative ring

13.1 The Brauer-Long group and its subgroups

Throughout this Chapter, R will be a commutative ring, and H a faithfully projective, commutative and cocommutative Hopf algebra. All R-algebras are assumed to have a unit element.
Two H-Azumaya algebras A and B are called H-equivalent if there exist faithfully projective H-dimodules M and N such that

$$A\#\mathrm{End}_R(M) \cong B\#\mathrm{End}_R(N)$$

as H-dimodule algebras. It is clear from the results in Section 12.2 that H-equivalence defines an equivalence relation on the set of H-isomorphism classes of H-dimodule algebras.

Proposition 13.1.1 *Two H-Azumaya algebras A and B are H-equivalent if and only if there exists a faithfully projective H-dimodule M such that*

$$A\#\overline{B} \cong \mathrm{End}_R(M)$$

as H-dimodule algebras.

Proof Suppose that A and B are H-equivalent. Then $A\#\overline{B}$ and $B\#\overline{B} \cong \mathrm{End}_R(B)$ are H-equivalent. Since $\mathrm{End}_R(B)$ is H-equivalent to R, there exist faithfully projective H-dimodules M and N such that

$$A\#\overline{B}\#\mathrm{End}_R(M) \cong \mathrm{End}_R(N)$$

From Proposition 12.1.8, we deduce that

$$A\#\overline{B}\#\mathrm{End}_R(M) \cong A\#\overline{B} \otimes \mathrm{End}_R(M) \cong \mathrm{End}_{A\#\overline{B}}((A\#\overline{B}) \otimes M)$$

as H-dimodule algebras. From Example 12.2.2, it follows that we have a strict H-Morita context

$$(\mathrm{End}_{A\#\overline{B}}((A\#\overline{B})\otimes M), A\#\overline{B}, (A\#\overline{B})\otimes M, Q, f, g)$$

and the categories $(H, A\#\overline{B}\#\mathrm{End}_R(M)\text{-}R)$-dimod and $(H, A\#\overline{B}\text{-}R)$-dimod are equivalent. Now the first of these two categories is isomorphic to $(H, \mathrm{End}_R(N)\text{-}R)$-dimod, which is equivalent to H-dimod. We find a category equivalence

$$T:\ (H, A\#\overline{B}\text{-}R)\text{-dimod}\longrightarrow H\text{-dimod}$$

$P = T(A\#\overline{B})$ is an R-progenerator, because $A\#\overline{B}$ is an $A\#\overline{B}$-progenerator. Finally

$$A\#\overline{B}\cong \mathrm{End}_{A\#\overline{B}}(A\#\overline{B})\cong \mathrm{End}_R(T(A)) = \mathrm{End}_R(P)$$

Conversely, if $A\#\overline{B}\cong \mathrm{End}_R(P)$, then

$$A\#\mathrm{End}_R(B^*)\cong A\#\mathrm{End}_R(B)^{\mathrm{op}}\cong A\#\overline{B}\#B\cong \mathrm{End}_R(P)\#B\cong B\#\mathrm{End}_R(P)$$

□

From Propositions 12.1.4, 12.2.4 and 12.2.5, we immediately obtain the following result.

Theorem 13.1.2 *H-equivalence of H-Azumaya algebras defines an equivalence relation on the set of H-dimodule algebra isomorphism classes of H-Azumaya algebras. The quotient is a group under the operation induced by the smash product. The inverse of a class $[A]$ is represented by $[\overline{A}]$. This group is called the Brauer-Long group of R and H, and is denoted* $\mathrm{BD}(R, H)$.

The split part of the Brauer-Long group

If $f:\ R\to S$ is a commutative R-algebra, the $H\otimes S$ is a Hopf algebra over the groundring S. If A is an H-Azumaya algebra, then $A\otimes S$ is an $H\otimes S$-Azumaya algebra, and we have a map

$$\mathrm{BD}(f, H):\ \mathrm{BD}(R, H)\longrightarrow \mathrm{BD}(S, H\otimes S):\ [A]\mapsto [A\otimes S]$$

We leave it to the reader to show that $\mathrm{BD}(\bullet, H)$ behaves functorially.

Proposition 13.1.3 *With notations as above,* $\mathrm{BD}(\bullet, H)$ *is a covariant functor from commutative R-algebras to abelian groups.*

$\mathrm{BD}(S/R, H)$ will denote the kernel of the map $\mathrm{BD}(R, H)\to \mathrm{BD}(S, H\otimes S)$. The *split part* is by definition the union of all the $\mathrm{BD}(S/R, H)$, where S runs through all faithfully flat R-algebras.

$$\mathrm{BD}(R, H) = \cup_S \mathrm{BD}^s(S/R, H)$$

Obviously the Brauer group $\mathrm{Br}(R)$ is a subgroup of the Brauer-Long group $\mathrm{BD}(R,H)$ (every Azumaya algebra is an H-Azumaya algebra, if we give it the trivial H-action and H-coaction. Every Azumaya algebra can be split in the Brauer group (and in the Brauer-Long group) by a faithfully flat extension of R. Thus

$$\mathrm{Br}(R) \subset \mathrm{BD}^s(R,H) \subset \mathrm{BD}(R,H)$$

Observe that these three groups coincide if we take $H = R$.

The strong part of the Brauer-Long group

From Proposition 12.3.9, it follows that

$$\mathrm{BDS}(R,H) = \{[A] \in \mathrm{BD}(R,H) | A \text{ is a strong } H\text{-Azumaya algebra}\}$$

is a subgroup of the Brauer-Long group. We call it the *strong part* of the Brauer-Long group. From Corollary 12.3.7, we obtain the following result.

Proposition 13.1.4 *If H is semisimple-like and cosemisimple-like, then*

$$\mathrm{BDS}(R,H) = \mathrm{BD}(R,H)$$

In particular, if $H = RG$, with $|G|$ invertible in R, then the strong part of the Brauer-Long group covers the full Brauer-Long group.

The Brauer group of θ-Azumaya algebras

Fix a Hopf algebra map $\theta : H \to H^*$, and recall that an H-dimodule M is called a θ-module if

$$h \rightharpoonup m = \sum \langle \theta(m_{(1)}), h \rangle m_{(0)} \tag{13.1}$$

for all $h \in H$ and $m \in M$. An H-dimodule algebra that is a θ-module is called a θ-algebra.

Proposition 13.1.5 *If M and N are θ-modules, then $M \otimes N$ and $\mathrm{Hom}_R(M,N)$ are also θ-modules. If A and B are θ-algebras, then $A\#B$, $A \otimes B$ and $\overline{A}$ are θ-algebras.*

Proof Take $m \in M$, $n \in N$ nd $h \in H$. Then

$$\begin{aligned} h \rightharpoonup (m \otimes n) &= \sum h_{(1)} \rightharpoonup m \otimes h_{(2)} \rightharpoonup n \\ &= \sum \langle \theta(m_{(1)}), h_{(1)} \rangle m_{(0)} \otimes \langle \theta(n_{(1)}), h_{(2)} \rangle n_{(0)} \\ &= \sum \langle \theta(m_{(1)}) * \theta(n_{(1)}), h \rangle m_{(0)} \otimes n_{(0)} \\ &= \sum \langle \theta(m_{(1)} n_{(1)}), h \rangle m_{(0)} \otimes n_{(0)} \end{aligned}$$

as needed. Next, take $f : M \to N$, and recall that

$$\sum f_{(0)}(m) \otimes f_{(1)} = \sum f(m_{(0)})_{(0)} \otimes f(m_{(0)})_{(1)} S(m_{(1)})$$

(see (7.40). Now

$$\begin{aligned}(h\rightharpoonup f)(m) &= \sum h_{(1)}\rightharpoonup f(S(h_{(2)})\rightharpoonup m)\\ &= \sum\langle\theta(m_{(1)}), S(h_{(2)})\rangle h_{(1)}\rightharpoonup f(m_{(0)})\\ &= \sum\langle S^*(\theta(m_{(1)})), h_{(2)}\rangle\langle f(m_{(0)})_{(1)}, h_{(1)}\rangle f(m_{(0)})_{(0)}\\ &= \sum\langle\theta(f(m_{(0)})_{(1)}S(m_{(1)})), h\rangle f(m_{(0)})_{(0)}\\ &= \sum\langle\theta, f_{(1)}\rangle f_{(0)}m\end{aligned}$$

as needed. The two other assertions are obvious. □

Corollary 13.1.6 *The subset*

$$\mathrm{B}_\theta(R,H) = \{[A]\in \mathrm{BD}(R,H)|A \text{ is a } \theta\text{-algebra}\}$$

is a subgroup of BD(R,H).

Special cases

1) Let $H = RC_2$, with $C_2 = \{0,1\}$ the cyclic group of order 2. Consider the bilinear map

$$\theta:\ C_2\times C_2 \longrightarrow R:\ (\sigma,\tau)\mapsto(-1)^{\sigma\tau}$$

If $2\neq 0$ in R, then θ represents the only non-trivial pairing $RC_2\to RC_2^*$. $\mathrm{B}_\theta(R) = \mathrm{BW}(R)$ is called the *Brauer-Wall group* of R. For R a field this invariant was introduced by Wall in [200]. For the case where R is a commutative ring, we refer to [169], [14] and [30].

2) Let $H = RG$, with G a finite abelian group. This case has been studied by Knus [112] in the case where R is a field, and by Childs, Garfinkel and Orzech in [57] and [53] in the the case where R is an arbitrary commutative ring.

3) The definition in the case where H is an arbitrary Hopf algebra was given first by Orzech in [154]. Consider now the special situation where $\theta:\ H\to H^*$ is trivial, that is,

$$\theta(h) = \varepsilon(h)\varepsilon$$

Then the H-action induced by the H-coaction is trivial, and a θ-Azumaya algebra is nothing else then an H-comodule algebra. The group $\mathrm{BC}(R,H) = \mathrm{B}_\theta(R,H)$ is called the Brauer group of H-comodule Azumaya algebras.

4) The Brauer group of H-module Azumaya algebras (sometimes called the *equivariant Brauer group*, cf. [87],[88]), can be introduced in a similar way: it is the subgroup of the Brauer-Long group consisting of equivalence classes of H-Azumaya algebras that may be represented by an H-module Azumaya algebra. Observe that

$$\mathrm{BM}(R,H)\cong \mathrm{BC}(R,H^*)$$

and

$$\mathrm{BM}(R,H)\cap \mathrm{BC}(R,H) = \mathrm{Br}(R)$$

We will now prove that the Brauer group itself may be viewed as a special case of the Brauer group of θ-Azumaya algebras.

Proposition 13.1.7 *Let $\theta : H \otimes H^* \to (H \otimes H^*)^* \cong H^* \otimes H$ be defined by the formula*

$$\theta(h \otimes h^*) = h^* \otimes \eta(\varepsilon(h))$$

Then $\mathrm{BD}(R, H) \cong \mathrm{B}_\theta(R, H \otimes H^*)$.

Proof Let θ-mod be the full subcategory of $H \otimes H^*$-dimod consisting of θ-modules. Forgetting the $H \otimes H^*$-action, we obtain a functor

$$\theta\text{-mod} \longrightarrow H \otimes H^*\text{-comod}$$

and it is clear that this functor is an equivalence of categories. We already know that the categories $H \otimes H^*$-comod and H-dimod are equivalent, so we obtain an equivalence between the categories θ-mod and H-dimod. We also obtain a correspondence between θ-algebras and H-dimodule algebras, and the result will follow if we can show the following.
Consider two θ-algebras A and B. Are the smash products $A\#B$ (with respect to the H-dimodule structure of A and B) and $A\#_\theta B$ (with respect to the $H \otimes H^*$-dimodule structure of A and B) equal?
We will consider an H-dimodule M as an (H^*, H)-bimodule, and we use the notations of Proposition 12.1.1.
In $A\#_\theta B$, we have

$$\begin{aligned}(a\#b)(c\#d) &= \sum a((b_{(1)} \otimes b_{(-1)}) \rightharpoonup c)\#b_{(0)}d \\ &= \sum a\theta(c_{(1)} \otimes c_{(-1)})(b_{(1)} \otimes b_{(-1)})c_{(0)}\#b_{(0)}d \\ &= \sum a\langle c_{(-1)}, b_{(1)}\rangle\langle b_{(-1)}, \eta(\varepsilon(c_{(1)}))\rangle c_{(0)}\#b_{(0)}d \\ &= \sum a\langle c_{(-1)}, b_{(1)}\rangle c_{(0)}\#b_{(0)}d \\ &= \sum a(b_{(1)} \rightharpoonup c)\#b_{(0)}d\end{aligned}$$

which is exactly the multiplication in $A\#B$. □

We have seen above that $\mathrm{BM}(R, H) \cong \mathrm{BC}(R, H^*)$. We will now show that $\mathrm{BD}(R, H)$ and $\mathrm{BD}(R, H^*)$ are anti-isomorphic. The following result is due to Tilborghs [181].

Theorem 13.1.8 (Tilborghs) *Let A be an H-Azumaya algebra. Then A^{op} is an H^*-Azumaya algebra. We have an anti-isomorphism*

$$\mathrm{BD}(R, H) \longrightarrow \mathrm{BD}(R, H^*) : \ [A] \mapsto [A^{\mathrm{op}}]$$

Proof As in the previous proof, we view an H-dimodule as an (H^*, H)-bicomodule. An H^*-dimodule may also be viewed as an (H^*, H)-bicomodule, so the categories H-dimod and H^*-dimod are equivalent. If A is an H-dimodule algebra, then it is also an H^*-dimodule algebra. Let A and B be H-dimodule algebras. Then the smash product $A\#B$ (with respect to the H-dimodule structure) has the following multiplication:

$$(a\#b)(c\#d) = \sum\langle c_{(-1)}, b_{(1)}\rangle ac_{(0)}\#b_{(0)}d$$

The multiplication rule on the smash product $A\#_*B$ (with respect to the H^*-dimodule structure) is the following:

$$(a\#_*b)(c\#_*d) = \sum\langle c_{(1)}, b_{(-1)}\rangle ac_{(0)}\#_*b_{(0)}d$$

Consider the map

$$f:\ (A\#B)^{\rm op}\longrightarrow B^{\rm op}\#_*A^{\rm op}:\ a\#b\mapsto b\#_*a$$

It is clear that f is an isomorphism of R-modules. Let us show that f is also multiplicative. In the sequel, $\cdot$ will denote the multiplication in $A^{\rm op}$ or $B^{\rm op}$.

$$\begin{aligned} f((a\#b)\cdot(c\#d)) &= f((c\#d)(a\#b)) \\ &= f(\sum\langle a_{(-1)}, d_{(1)}\rangle ca_{(0)}\#d_{(0)}b) \\ &= \sum\langle a_{(-1)}, d_{(1)}\rangle d_{(0)}b\#_*ca_{(0)} \\ &= \sum\langle d_{(1)}, a_{(-1)}\rangle b\cdot d_{(0)}\#_*a_{(0)}\cdot c \\ &= (b\#_*a)(d\#_*c) \\ &= f(a\#b)f(c\#d) \end{aligned}$$

In $\overline{A}$, the multiplication is given by the formula

$$m_{\overline{A}}(\overline{a}\otimes\overline{b}) = \sum\langle a_{(1)}, b_{(-1)}\rangle b_{(0)}a_{(0)}$$

In $\tilde{A}$, the H^*-opposite algebra of A, it is given by the formula

$$m_{\tilde{A}}(\tilde{a}\otimes\tilde{b}) = \sum\langle a_{(-1)}, b_{(1)}\rangle b_{(0)}a_{(0)}$$

It is clear that the identity is an algebra isomorphism $\overline{A}^{\rm op}\to\widetilde{A^{\rm op}}$. Indeed,

$$m_{\overline{A}^{\rm op}}(\overline{a}\otimes\overline{b}) = m_{\overline{A}}(\overline{b}\otimes\overline{a}) = \sum\langle b_{(1)}, a_{(-1)}\rangle a_{(0)}b_{(0)} = m_{\widetilde{A^{\rm op}}}(\tilde{a}\otimes\tilde{b})$$

Now consider the diagram

$$\begin{array}{ccc} (A\#\overline{A})^{\rm op} & \xrightarrow{F} & \mathrm{End}_R(A)^{\rm op} \\ \downarrow f & & \downarrow I \\ \widetilde{A^{\rm op}}\#A^{\rm op} & \xrightarrow{G^*} & \mathrm{End}_R(A)^{\rm op} \end{array}$$

where G^* is the map G from Lemma 12.2.3, but with H replaced by H^*, and A by $A^{\rm op}$. We claim that this diagram is commutative. Indeed, using our bicomodule notation, we find that, for all $a, b, c\in A$:

$$\begin{aligned} F(a\#\overline{b})(c) &= \sum a(b_{(1)}\rightharpoonup c)b_{(0)} \\ &= \sum\langle b_{(1)}, c_{(-1)}\rangle ac_{(0)}b_{(0)} \\ &= \sum\langle b_{(1)}, c_{(-1)}\rangle b_{(0)}\cdot c_{(0)}\cdot a \\ &= \sum(c_{(-1)}\rightharpoonup b)\cdot c_{(0)}\cdot a \\ &= G(\tilde{b}\#a)(c) \end{aligned}$$

We conclude that F is an isomorphism if and only if G_* is an isomorphism, and A is a left H-Azumaya algebra if and only if A^* is a right H^*-Azumaya algebra. In a similar way, we show that A is a right H-Azumaya if and only if A^* is a left H^*-Azumaya. This finishes the proof. □

13.2 The Brauer group of H-module Azumaya algebras

We have seen in Section 13.1 that the set of classes in $\mathrm{BD}(R,H)$ represented by H-module Azumaya algebras (that is, H-Azumaya algebras with trivial coaction) forms a subgroup $\mathrm{BM}(R,H)$ of $\mathrm{BD}(R,H)$. In the case where H is not necessarily commutative, we can still define $\mathrm{BM}(R,H)$. In this Section, we will see that $\mathrm{BM}(R,H)$ is closely related to the group of H-Galois objects $\mathrm{Gal}(R,H)$.

Theorem 13.2.1 (Beattie [17]) *Let H be a faithfully projective cocommutative Hopf algebra. We have a split exact sequence*

$$1\longrightarrow \mathrm{Br}(R)\longrightarrow \mathrm{BM}(R,H)\xrightarrow{\pi}\mathrm{Gal}(R,H)\longrightarrow 1$$

In the sequel, we will present a proof that follows closely the original proof in [16], [17]. In the special case where $H = RG$ is a group ring, the result has been proved first by Picco and Platzeck, cf. [157].
We will prove Theorem 13.2.1 in several steps. Let A be an H-module Azumaya algebra. From Proposition 12.2.4, we easily deduce that the pair of adjoint functors

$$F_l:\ H\text{-comod}\longrightarrow (H,A^e\text{-}R)\text{-comod}\ :\ N\mapsto A\otimes N$$
$$G_l:\ (H,A^e\text{-}R)\text{-comod}\longrightarrow H\text{-comod}\ :\ M\mapsto M^A$$

are inverse equivalences. Now $A\#H$ is an A^e-bimodule, the structure is given by the formula

$$(a\otimes b)\cdot(c\#h) = (a\#1)(c\#h)(b\#1)$$

Let H coact on $A\#H$ as follows

$$\rho(a\#h) = \sum a\#h_{(1)}\otimes h_{(2)}$$

It may be verified easily that $A\#H$ is an algebra in the category $(H,A^e\text{-}R)$-comod. Thus

$$G_l(A\#H) = (A\#H)^A = \{x\in A\#H \mid (a\#1)x = x(a\#1) \text{ for all } a\in A\}$$

is an algebra in H-comod, and this means that it is an H-comodule algebra. We will show that it is an H-Galois object. First of all, from the fact that A and H are faithfully flat R-modules, it follows that $A\#H\cong A\otimes(A\#H)^A$ is faithfully flat, and therefore $(A\#H)^A$ is faithfully flat.
Next, we have to show that the map

$$\gamma:\ (A\#H)^A\otimes(A\#H)^A\longrightarrow (A\#H)^A\otimes H\ :\ x\otimes y\mapsto \sum xy_{(0)}\otimes y_{(1)}$$

is an isomorphism. To this end, it suffices to show that $F_l(\gamma)$ is an isomorphism. Now

$$A\otimes(A\#H)^A\otimes(A\#H)^A\cong(A\#H)\otimes(A\#H)^A\cong(A\#H)\otimes_A(A\#H)$$

and

$$A \otimes (A\#H)^A \otimes H \cong (A\#H) \otimes H$$

so we may view $F_l(\gamma)$ as a map

$$F_l(\gamma): (A\#H) \otimes_A (A\#H) \longrightarrow (A\#H) \otimes H$$

Take $a, b \in A$ and $x = \sum a_i \otimes h_i,\ y = \sum b_j \# k_j \in (A\#H)^A$. Then

$$\begin{aligned} F_l(\gamma)(ax \otimes by) &= \sum axby_{(0)} \otimes y_{(1)} \\ &= \sum (aa_i\#h_i)(bb_j\#k_{i_{(1)}}) \otimes k_{i_{(2)}} \\ &= \sum \big(aa_i(h_{i_{(1)}} \rightharpoonup (bb_j))\#h_{i_{(2)}}k_{i_{(1)}}\big) \otimes k_{i_{(2)}} \end{aligned}$$

so

$$F_l(\gamma)((a\#h) \otimes (b\#k)) = \sum (a(h_{(1)} \rightharpoonup b)\#h_{(2)}k_{(1)}) \otimes k_{(2)}$$

for all $a, b \in A$ and $h, k \in H$. $F_l(\gamma)$ is bijective, and its inverse is given by the formula

$$F_l(\gamma)^{-1}((a\#h) \otimes k) = \sum (a\#hS(k_{(1)})) \otimes (1\#k_{(2)})$$

Indeed,

$$\begin{aligned} F_l(\gamma)^{-1}\big(F_l(\gamma)((a\#h) \otimes (b\#k))\big) &= F_l(\gamma)^{-1}\big(\sum (a(h_{(1)} \rightharpoonup b)\#h_{(2)}k_{(1)}) \otimes k_{(2)}\big) \\ &= \sum (a(h_{(1)} \rightharpoonup b)\#h_{(2)}k_{(1)}S(k_{(2)})) \otimes (1\#k_{(3)}) \\ &= (a\#h)(b\#1) \otimes (1\#k) \\ &= (a\#h) \otimes (b\#k) \end{aligned}$$

and

$$\begin{aligned} F_l(\gamma)\big(F_l(\gamma)^{-1}((a\#h) \otimes k)\big) &= F_l(\gamma)\big(\sum (a\#hS(k_{(1)})) \otimes (1\#k_{(2)})\big) \\ &= \sum \big(a((h_{(1)}S(k_{(2)})) \rightharpoonup 1)\#h_{(2)}S(k_{(1)})k_{(3)}\big) \otimes k_{(4)} \\ &= (a\#h) \otimes k \end{aligned}$$

and we have shown that $(A\#H)^A$ is an H-Galois object.

Lemma 13.2.2 *We have a homomorphism*

$$\pi: \mathrm{BM}(R, H) \longrightarrow \mathrm{Gal}(R, H): [A] \mapsto [(A\#H)^A]$$

Proof We will first show that

$$(A\#H)^A \square_H (B\#H)^B \cong ((A \otimes B)\square H)^{A \otimes B}$$

for any two H-Azumaya algebras A and B. Consider the map

$$\phi: ((A \otimes B)\square H)^{A \otimes B} \longrightarrow (A\#H)^A \square_H (B\#H)^B$$

defined by

$$\phi(\sum (a_i \otimes b_i)\#h_i) = \sum (a_i\#h_{i_{(1)}}) \otimes (b_i\#h_{i_{(2)}})$$

We leave it to the reader to show that ϕ is a well-defined H-comodule algebra homomorphism, and therefore an isomorphism, by Proposition 8.1.10.
Let us next show that π is well-defined. Let $A = \mathrm{End}_R(M)$, with M a faithfully projective H-module. Consider the map

$$\sigma :\ H \longrightarrow \mathrm{End}_R(M)\#H :\ h \mapsto \sum \psi_{S(h_{(1)})}\#h_{(2)}$$

(recall that $\psi_h(m) = \psi(h \otimes m) = h\rightharpoonup m$ for all $h \in H$ and $m \in M$). We see immediately that σ is multiplicative and H-colinear (the H-coaction on $\mathrm{End}_R(M)\#H$ is induced by the one on H). We also claim that $\mathrm{Im}\,(\sigma) \subset (A\#H)^A$. Indeed, for all $h \in H$ and $f \in A = \mathrm{End}_R(M)$, we have that

$$\begin{aligned}
\sigma(h)(f\#1) &= \sum(\psi_{S(h_{(1)})}\#h_{(2)})(f\#1)\\
&= \sum(\psi_{S(h_{(1)})} \circ (h_{(2)}\rightharpoonup f))\#h_{(3)}\\
&= \sum(\psi_{S(h_{(1)})} \circ \psi_{h_{(2)}} \circ f \circ \psi_{S(h_{(3)})})\#h_{(4)}\\
&= \sum f \circ \psi_{S(h_{(1)})}\#h_{(2)}\\
&= \sum(f\#1)(\sum \psi_{S(h_{(1)})}\#h_{(2)})\\
&= \sum(f\#1)\sigma(h)
\end{aligned}$$

We now have shown that $\sigma :\ H \longrightarrow (A\#H)^A$ is an H-comodule algebra homomorphism. Now H and $(A\#H)^A$ are H-Galois objects, and it follows from Proposition 8.1.10 that σ is an isomorphism. Thus $(A\#H)^A$ is trivial in $\mathrm{Gal}(R, H)$, and this proves that π is well-defined. □

Lemma 13.2.3

$$\mathrm{Ker}\,(\pi) = \mathrm{Br}(R)$$

Proof Suppose that A is an H-Azumaya algebra, with trivial H-action. Then $A\#H = A \otimes H$, and $(A\#H)^H = A^A \otimes H = R \otimes H = H$. This shows that $\mathrm{Br}(R) \subset \mathrm{Ker}\,(\pi)$. The inclusion map

$$i :\ \mathrm{Br}(R) \longrightarrow \mathrm{BM}(R)$$

has a left inverse j, given by forgetting the H-action on an H-Azumaya algebra. We will show that j restricts to an isomorphism

$$j :\ \mathrm{Ker}\,(\pi) \longrightarrow \mathrm{Br}(R)$$

To this end, it suffices to show that j is injective. So take $[A] \in \mathrm{Ker}\,(\pi) \cap \mathrm{Ker}\,(j)$. Then $[A] = 1$ in $\mathrm{Br}(R)$, and we have that $A = \mathrm{End}_R(M)$, with M a faithfully projective R-module.
From the fact that $[A] \in \mathrm{Ker}\,(\pi)$, we know that there exists an H-comodule algebra isomorphism $\sigma :\ H \longrightarrow (A\#H)^H$. $(I_A\#\varepsilon) \circ \sigma$ is an R-algebra homomorphism from H to $A = \mathrm{End}_R(M)$. Now we let H act on M as follows

$$h\rightharpoonup m = \psi_h(m) = ((I_A\#\varepsilon) \circ \sigma \circ S)(h)(m)$$

This action induces an H-action on $A = \mathrm{End}_R(M)$, and we claim that this action coincides with the original H-action on A.
Take $h \in H$, and write $\sigma(h) = \sum f_i \# h_i \in (A\#H)^A$. Then

$$\sum \sigma(h_{(1)}) \otimes h_{(2)} = \sum (f_i \# h_{i_{(1)}}) \otimes h_{i_{(2)}} \tag{13.2}$$

The fact that $\sigma(h) \in (A\#H)^A$ implies that

$$(f\#1)\sigma(h) = \sum (f \circ f_i)\#h_i = \sigma(h)(f\#1) = \sum f_i \circ (h_{i_{(1)}} \rightharpoonup f)\#h_{i_{(2)}}$$

Applying $I_A \# \varepsilon$ to both sides, we obtain that

$$\sum \varepsilon(h_i) f \circ f_i = \sum f_i \circ (h_{i_{(1)}} \rightharpoonup f_i) \tag{13.3}$$

Now

$$\begin{aligned} f \circ \big(((I_A\#\varepsilon)\circ\sigma)(h)\big) &= f \circ \big(\textstyle\sum \varepsilon(h_i) f_i\big) \\ \text{(by (13.3))} \quad &= \textstyle\sum f_i \circ (h_i \rightharpoonup f) \\ &= \textstyle\sum (I_A\#\varepsilon)(f_i \# h_{i_{(1)}} \circ (h_{i_{(2)}} \rightharpoonup f) \\ &= \textstyle\sum (I_A\#\varepsilon)(\sigma(h_{(1)})) \circ (h_{(2)} \rightharpoonup f) \end{aligned} \tag{13.4}$$

and we finally obtain that

$$\begin{aligned} \sum \psi_{h_{(1)}} \circ f \circ \psi_{S(h_{(2)})} &= \sum \psi_{h_{(1)}} \circ f \circ ((I_A\#\varepsilon)(\sigma(h_{(2)}))) \\ \text{(by (13.4))} \quad &= \sum \psi_{h_{(1)}} \circ ((I_A\#\varepsilon)(\sigma(h_{(2)}))) \circ (h_{(3)} \rightharpoonup f) \\ &= \sum \big(((I_A\#\varepsilon)\circ\sigma)(S(h_{(1)}))\big) \circ \big(((I_A\#\varepsilon)\circ\sigma)(h_{(2)})\big) \circ (h_{(3)} \rightharpoonup f) \\ &= \sum \big(((I_A\#\varepsilon)\circ\sigma)(S(h_{(1)})h_{(2)})\big) \circ (h_{(3)} \rightharpoonup f) \\ &\qquad ((I_A\#\varepsilon)\circ\sigma \text{ is an algebra homomorphism}) \\ &= \sum \big(((I_A\#\varepsilon)\circ\sigma)(1)\big) \circ (h \rightharpoonup f) \\ &= h \rightharpoonup f \end{aligned}$$

The original action on $A = \mathrm{End}_R(M)$ coincides with the action induced by the action on M, and $[A] = 1$ in $\mathrm{BM}(R, H)$. □

Our next aim is to show that π is surjective. Let S be an H-Galois object, and let $A = S\#H^*$, as in Section 12.4. Then A is an H-module algebra. The multiplication and H-action are given by the formulas

$$(s\#h^*)(t\#k^*) = \sum st_{(0)}\#(t_{(1)} \rightharpoonup h^*) * k^* \tag{13.5}$$

$$h \rightharpoonup (s\#h^*) = \sum \langle h^*_{(1)}, h\rangle s\#h^*_{(2)} \tag{13.6}$$

for all $s \in S$, $h \in H$ and $h^* \in H^*$. The proof of Theorem 13.2.1 will be finished after we have shown the following Lemma.

Lemma 13.2.4 *With notations as above, we have that*

$$\pi([A]) = [S]$$

Proof Observe that the multiplication in $(S\#H^*)\#H$ is given by the formula

$$\begin{aligned}((s\#h^*)\#h)((t\#k^*)\#k) &= \sum\langle k^*_{(1)}, h_{(1)}\rangle(s\#h^*)(t\#k^*_{(2)})\#h_{(2)}k \\ &= \sum\langle k^*_{(1)}, h_{(1)}\rangle(st_{(0)}\#(t_{(1)}\rightharpoonup h^*) * k^*_{(2)})\#h_{(2)}k\end{aligned}$$

Take $x = \sum(s_i\#h_i^*)\#h_i \in A\#H^*$. $x \in (A\#H^*)^A$ if and only if for all $t \in S$ and $k^* \in H^*$,

$$((t\#k^*)\#1)x = x((t\#k^*)\#1)$$

or, equivalently,

$$\sum(ts_{i_{(0)}}\#(s_{i_{(1)}}\rightharpoonup k^*) * h_i^*)\#h_i = \sum\langle k^*_{(1)}, h_{i_{(1)}}\rangle(s_it_{(0)}\#(t_{(1)}\rightharpoonup h_i^*) * k^*_{(2)})\#h_{i_{(2)}} \quad (13.7)$$

From (13.7), it follows that

$$\sum\langle k^*, s_{i_{(1)}}\rangle\langle h_i^*, 1\rangle\langle\varepsilon, h_i\rangle ts_{i_{(0)}} = \sum\langle k^*, h_i\rangle\langle h_i^*, t_{(1)}\rangle s_it_{i_{(0)}} \quad (13.8)$$

Now we define $\phi:\ (A\#H^*)^A \longrightarrow S$ by

$$\phi(\sum(s_i\#h_i^*)\#h_i) = \sum\langle h_i^*, 1\rangle\langle\varepsilon, h_i\rangle s_i$$

We claim that ϕ is multiplicative: take $x = \sum(s_i\#h_i^*)\#h_i$, $y = \sum(t_i\#k_i^*)\#k_i \in (A\#H^*)^A$. Applying (13.8) twice, we obtain that

$$\begin{aligned}\phi(xy) &= \sum\langle k_j^*, h_i\rangle\langle h_i^*, t_{j_{(1)}}\rangle\langle\varepsilon, k_j\rangle s_it_{j_{(0)}} \\ &= \sum\langle k_j^*, s_{i_{(1)}}\rangle\langle h_i^*, 1\rangle\langle\varepsilon, h_i\rangle\langle\varepsilon, k_j\rangle t_js_{i_{(0)}} \\ &= \sum\langle h_i^*, 1\rangle\langle\varepsilon, h_i\rangle\langle k_j^*, 1\rangle\langle\varepsilon, k_j\rangle s_it_j \\ &= \phi(x)\phi(y)\end{aligned}$$

To show that ϕ is H-colinear, it suffices to prove that ϕ is H^*-linear. Let $x \in (A\#H^*)^A$ be as above and take $k^* \in H^*$. Then

$$k^*\rightharpoonup x = \sum\langle k^*, h_{i_{(1)}}\rangle(s_i\#h_i^*)\#h_{i_{(2)}}$$

and, using (13.8),

$$\begin{aligned}k^*\rightharpoonup\phi(x) &= \sum\langle k^*, s_{i_{(1)}}\rangle\langle h_i^*, 1\rangle\langle\varepsilon, h_i\rangle s_{i_{(0)}} \\ &= \sum\langle k^*, h_i\rangle\langle h_i^*, 1\rangle s_i \\ &= \phi(k^*\rightharpoonup x)\end{aligned}$$

Being an H-comodule algebra homomorphism between H-Galois objects, ϕ is an isomorphism (cf. Proposition 8.1.10). This finishes the proof of the Lemma. □

If the action of H on A is inner, then we have the following description of $\pi([A])$, due to Beattie and Ulbrich [24].

Proposition 13.2.5 *Let H be an H-module Azumaya algebra, such that the action of H on A is induced by $u:\ H \to A$, with convolution inverse v. Then*

$$\pi([A]) = \{\sum v(h_{(1)})\#h_{(2)} | h \in H\} \subset A\#H \quad (13.9)$$

Proof Recall that $h \rightharpoonup a = \sum u(h_{(1)})av(h_{(2)})$, for all $a \in A$ and $h \in H$. Now

$$\begin{aligned}\left(\sum v(h_{(1)})\#h_{(2)}\right)(a\#1) &= \sum v(h_{(1)})(h_{(2)} \rightharpoonup a)\#h_{(3)} \\ &= \sum v(h_{(1)})u(h_{(2)})av(h_{(3)})\#h_{(4)} \\ &= \sum av(h_{(1)})\#h_{(2)} \\ &= (a\#1)\left(\sum v(h_{(1)})\#h_{(2)}\right)\end{aligned}$$

and it follows that $\sum v(h_{(1)})\#h_{(2)} \in \pi(A) = (A\#H)^A$. Conversely, suppose that $x = \sum_i a_i\#h_i \in \pi(A) = (A\#H)^A$. We will show that x may be written under the form $x = \sum v(h_{(1)})\#h_{(2)}$ for some $h \in H$. First, we claim that

$$\sum a_i u(h_i) \in Z(A) = R$$

Indeed, for all $b \in A$, we have

$$\begin{aligned}\sum_i ba_i\#h_i &= (b\#1)x = x(b\#1) = \sum_i a_i(h_{i_{(1)}} \rightharpoonup b)\#h_{i_{(2)}} \\ &= \sum_i a_i u(h_{i_{(1)}})bv(h_{i_{(2)}})\#h_{i_{(3)}}\end{aligned}$$

We now apply $m_A \circ (I \otimes u)$ to both sides, and we obtain that

$$\sum ba_i u(h_i) = \sum a_i u(h_i) b$$

Now $\rho(x) = \sum_i a_i\#h_{i_{(1)}} \otimes h_{i_{(2)}} \in \pi(A) \otimes H$, and

$$h = \sum_i a_i u(h_{i_{(1)}}) \otimes h_{i_{(2)}} \in R \otimes H = H$$

Now

$$\sum v(h_{(1)})u(h_{(2)}) = \sum a_i u(h_{i_{(1)}})v(h_{i_{(2)}})\#h_{i_{(3)}} = \sum_i a_i\#h_i = x$$

and this finishes our proof. □

Suppose now that H is commutative and cocommutative. A duality argument shows that we have a split exact sequence

$$1 \longrightarrow \mathrm{Br}(R) \longrightarrow \mathrm{BC}(R,H) \longrightarrow \mathrm{Gal}(R,H^*) \longrightarrow 1 \tag{13.10}$$

Recall from Section 13.1 that $\mathrm{BC}(R,H)$ is a special case of Orzech's subgroup $\mathrm{B}_\theta(R,H)$ consisting of equivalence classes of θ-Azumaya algebras. In [185], Ulbrich generalized the exact sequence (13.10) to $\mathrm{B}_\theta(R,H)$. He introduces a non-abelian group $\mathrm{D}(\theta,H)$ consisting of isomorphism classes of certain H^*-Galois biextensions of H, and shows that there is an exact sequence

$$1 \longrightarrow \mathrm{Br}(R) \longrightarrow \mathrm{B}_\theta(R,H) \longrightarrow \mathrm{D}(\theta,H) \tag{13.11}$$

Using Proposition 13.1.7, Ulbrich's construction can be applied to give a short exact sequence for the full Brauer-Long group. This exact sequence turns out to be consistent with the exact sequence that we will discuss in Section 13.9.

13.3 The Picard group of H-dimodules

In this Section, H is a faithfully projective, commutative and cocommutative Hopf algebra. The statements concerning the H-comodule structure only are also valid in the case where H is not necessarily commutative; in a similar way, the statements that only deal with the H-module structure are valid in the case where H is not necessarily cocommutative.
Let $\underline{\underline{\mathrm{PD}}}(R,H)$ be the category of invertible H-dimodules, that is the category of H-dimodules are projective of rank one as R-modules. The morphisms will be H-linear, H-colinear maps. In a similar way, we introduce the categories $\underline{\underline{\mathrm{PC}}}(R,H)$ of invertible H-comodules and H-colinear maps and $\underline{\underline{\mathrm{PM}}}(R,H)$ of invertible H-modules and H-linear maps. The tensor product over R is a product on these three categories. R is a cofinal object of the category, and it easily follows that the Whitehead groups of these three categories are all equal to $\mathbb{G}_m(R)$. For the Grothendieck groups, we introduce the following notation

$$\begin{aligned}
\mathrm{K}_0\underline{\underline{\mathrm{PD}}}(R,H) &= \mathrm{PD}(R,H)\\
\mathrm{K}_0\underline{\underline{\mathrm{PC}}}(R,H) &= \mathrm{PC}(R,H)\\
\mathrm{K}_0\underline{\underline{\mathrm{PM}}}(R,H) &= \mathrm{PM}(R,H)
\end{aligned}$$

They are respectively called the Picard groups of H-dimodules, H-comodules and H-modules. In the next Proposition, we will relate these three Picard groups to the classical Picard group $\mathrm{Pic}(R)$.

Proposition 13.3.1 *With notations as above, we have isomorphisms*

$$\mathrm{PC}(R,H) = \mathrm{Pic}(R)\times G(H) \tag{13.12}$$
$$\mathrm{PM}(R,H) = \mathrm{Pic}(R)\times G(H^*) \tag{13.13}$$
$$\mathrm{PD}(R,H) = \mathrm{Pic}(R)\times G(H)\times G(H^*) \tag{13.14}$$

Proof We define a map $F:\ \mathrm{Pic}(R)\times G(H)\to \mathrm{PC}(R,H)$ as follows: $F([I,g]) = [I(g)]$, where $I(g)$ is equal to I as an R-module, and with H-coaction given by $\rho(x) = x\otimes g$ for all $x\in I$. From the fact that g is grouplike, it follows easily that this defines an H-coaction on I. It is clear that F is an isomorphism of groups; the inverse of F can be constructed as follows. Take an invertible H-comodule I, and let $\underline{I}$ be equal to I as an R-module, but with trivial H-coaction. $\underline{I}$ is also an invertible H-comodule, and $J = \underline{I}^*\otimes I$ is isomorphic to R as an R-module (but not as an H-comodule). Identify J and R, and look at the image of 1 under the H-coaction on J:

$$\rho_J(1) = 1\otimes g$$

for some $g\in H$. We claim that $g\in G(H)$. Indeed,

$$1 = (1\otimes\varepsilon)\rho_J(1) = (1\otimes\varepsilon)(1\otimes g) = \varepsilon(g)$$

and

$$1\otimes\Delta(g) = (I\otimes\Delta)\rho_J(1) = (\rho_J\otimes I)\rho_J(1) = 1\otimes g\otimes g$$

Define $G: \mathrm{PC}(R,H) \to \mathrm{Pic}(R) \times G(H)$ by

$$G([I]) = ([\underline{I}], g)$$

It is straightforward to show that F and G are each others inverses, proving the first part of the Proposition.
The second and the third part follow immediately since $\mathrm{PM}(R,H) \cong \mathrm{PC}(R,H^*)$ and $\mathrm{PD}(R,H) \cong \mathrm{PC}(R, H \otimes H^*)$. For later use, we give an explicit description of the isomorphism

$$F: \mathrm{Pic}(R) \times G(H) \times G(H^*) \to \mathrm{PD}(R,H)$$

$F([I,g,g^*]) = [I(g,g^*)]$, where $I(g,g^*)$ is equal to I as an R-module, and with H-action and coaction given by the formulas

$$\rho(x) = x \otimes g \quad \text{and} \quad h \rightharpoonup x = \langle g^*, h\rangle x$$

for all $x \in I$ and $h \in H$. □

Remark 13.3.2 As we already observed at the beginning of the Section, (13.12) remains valid if H is not commutative; we can even drop the assumption that H is faithfully projective. (13.13) remains true if H is not cocommutative. If H is not faithfully projective, then the statement is replaced by

$$\mathrm{PM}(R,H) = \mathrm{Pic}(R) \times \mathrm{Alg}(H,R) \tag{13.15}$$

We now present a dimodule version of Proposition 1.3.1. We leave it to the reader to state the appropriate module and dimodule versions.

Proposition 13.3.3 *Let M and N be faithfully projective H-dimodules, and let*

$$\Phi: \mathrm{End}_R(M) \to \mathrm{End}_R(N)$$

be an isomorphism of H-dimodule algebras. Then there exists $I \in \underline{\mathrm{PD}}(R,H)$ and an H-linear H-colinear isomorphism $\varphi: M \otimes I \to N$ inducing Φ, that is

$$\Phi(\alpha) = \varphi \circ (\alpha \otimes I_I) \circ \varphi^{-1} \tag{13.16}$$

for all $\alpha \in \mathrm{End}_R(M)$. φ and I are unique up to isomorphism.

Proof The proof is an adaptation of the proof of Proposition 1.3.1. For completeness sake, we present an outline of the proof. Define an $\mathrm{End}_R(M)$-action on N as follows:

$$\alpha \cdot n = \Phi(\alpha)(n)$$

for all $n \in N$ and $\alpha \in \mathrm{End}_R(M)$. Now $(\mathrm{End}_R(M), R, M, M^*, f, g)$ is a strict H-Morita context (Example 12.2.2), and we have an isomorphism

$$N \cong M \otimes (M^* \otimes_{\mathrm{End}_R(M)} N)$$

of H-dimodules. A count of ranks shows that the rank of $I = M^* \otimes_{\mathrm{End}_R(M)} N$ is one, and $I \in \underline{\underline{PD}}(R, H)$. We have an $\mathrm{End}_R(M)$-linear isomorphism

$$\varphi : M \otimes I \to N$$

hence for all $n \in N$ and $\alpha \in \mathrm{End}_R(M)$

$$\varphi^{-1}(\alpha \cdot n) = (\alpha \otimes I_I)\varphi^{-1}(n)$$

and

$$\alpha \cdot n = \Phi(\alpha)(n) = (\varphi \circ (\alpha \otimes I_I) \circ \varphi^{-1})(n)$$

□

The map $\varphi : M \otimes I \to N$ is H-linear and H-colinear. Of course the map

$$\underline{\varphi} : M \otimes \underline{I} \to N$$

still induces Φ, but is no longer H-linear and H-colinear. In the sequel, we will often work with the map $\underline{\varphi}$. Let $([(\underline{I})], g, g^*)$ be the triple in $\mathrm{Pic}(R) \times G(H) \times G(H^*)$ corresponding to $[I]$ (see Proposition 13.3.1). We claim that the action and coaction of H on $\underline{\varphi}$ are given by the formulas

$$\rho(\underline{\varphi}) = \underline{\varphi} \otimes g \tag{13.17}$$

$$h \rightharpoonup \underline{\varphi} = \langle g^*, h\rangle \underline{\varphi} \tag{13.18}$$

for all $h \in H$. From the fact that φ is H-linear, it follows that, for all $h \in H, m \in M, x \in I$:

$$h \rightharpoonup \varphi(h \otimes x) = \sum \varphi(h_{(1)} \rightharpoonup m \otimes \langle g^*, h_{(2)}\rangle x)$$

and, using (7.32)

$$\begin{aligned} h \rightharpoonup \underline{\varphi}(h \otimes x) &= \sum h_{(1)} \rightharpoonup \underline{\varphi}(S(h_{(2)}) \rightharpoonup m \otimes x) \\ &= \sum h_{(1)} \rightharpoonup \underline{\varphi}(S(h_{(2)}) \rightharpoonup m \otimes \langle g^*, S(h_{(3)})\rangle \langle g^*, h_{(4)}\rangle x) \\ &= \langle g^*, h\rangle \underline{\varphi}(m \otimes x) \end{aligned}$$

and (13.18) follows. (13.17) can be obtained either directly using (7.40), or from (13.18) by a duality argument.

Now we consider two other faithfully projective H-dimodules P and Q and an H-linear H-colinear isomorphism

$$\Lambda : \mathrm{End}_R(P) \longrightarrow \mathrm{End}_R(Q)$$

induced by

$$\lambda : P \otimes J \longrightarrow Q$$

We identify J with the triple $(\underline{J}, l, l^*) \in \mathrm{Pic}(R) \times G(H) \times G(H^*)$. Using the isomorphism Γ defined in Proposition 12.1.6, we can define a map

$$\Theta : \mathrm{End}_R(M \otimes P) \longrightarrow \mathrm{End}_R(N \otimes Q)$$

by the commutativity of the diagram

$$\begin{array}{ccccc} \mathrm{End}_R(M)\#\mathrm{End}_R(P) & \xrightarrow{\Gamma} & \mathrm{End}_R(M)\otimes\mathrm{End}_R(P) & \cong & \mathrm{End}_R(M\otimes P) \\ \downarrow{\scriptstyle \Phi\otimes\Lambda} & & & & \downarrow{\scriptstyle \Theta} \\ \mathrm{End}_R(N)\#\mathrm{End}_R(Q) & \xrightarrow{\Gamma} & \mathrm{End}_R(N)\otimes\mathrm{End}_R(Q) & \cong & \mathrm{End}_R(N\otimes Q) \end{array}$$

Lemma 13.3.4 *With notations as above, Θ is induced by the map*

$$\theta:\ M\otimes P\otimes I\otimes J\longrightarrow N\otimes Q$$

given by

$$\theta=\varphi\otimes(\lambda\circ\psi_{g^*})=\langle g^*,l\rangle\varphi\otimes(\psi_{g^*}\circ\lambda)$$

with $\psi_{g^*}(p)=g^*\rightharpoonup p=\sum\langle g^*,p_{(1)}\rangle p_{(0)}$ for all $p\in P$.

Proof Take $\alpha\in\mathrm{End}_R(M)$ and $\beta\in\mathrm{End}_R(N)$. Recall from Proposition 12.1.6 that

$$\begin{aligned}\Gamma(\alpha\#\beta) &= \sum\alpha\circ\psi_{\beta_{(1)}}\otimes\beta_{(0)}\\ \Gamma^{-1}(\alpha\otimes\beta) &= \sum\alpha\circ\psi_{S(\beta_{(1)})}\#\beta_{(0)}\end{aligned}$$

Now

$$\begin{aligned}\Theta(\alpha\otimes\beta) &= \sum\Gamma\left(\Phi(\alpha\circ\psi_{S(\beta_{(1)})})\otimes\Lambda(\beta_{(0)})\right)\\ &= \sum\Phi(\alpha\circ\psi_{S(\beta_{(2)})})\circ\psi_{\beta_{(1)}}\otimes\Lambda(\beta_{(0)})\\ &= \sum\left(\underline{\varphi}\circ(\alpha\otimes I_{\underline{I}})\circ\psi_{S(\beta_{(2)})}\circ\underline{\varphi}^{-1}\circ\psi_{\beta_{(1)}}\right)\otimes\left(\underline{\lambda}\circ(\beta_{(0)}\otimes I_{\underline{J}})\circ\underline{\lambda}^{-1}\right)\\ &= \sum\left(\underline{\varphi}\circ(\alpha\otimes I_{\underline{I}})\circ(S(\beta_{(1)})\rightharpoonup\underline{\varphi}^{-1})\right)\otimes\left(\underline{\lambda}\circ(\beta_{(0)}\otimes I_{\underline{J}})\circ\underline{\lambda}^{-1}\right)\\ &= \sum\langle g^*,\beta_{(1)}\rangle\left(\underline{\varphi}\circ(\alpha\otimes I_{\underline{I}})\circ\underline{\varphi}^{-1}\right)\otimes\left(\underline{\lambda}\circ(\beta_{(0)}\otimes I_{\underline{J}})\circ\underline{\lambda}^{-1}\right)\end{aligned}$$

Now N is an H-dimodule, and therefore also an H^*-dimodule. The H^*-action on N induces an H^*-action on $\mathrm{End}_R(N)$. This H^*-coaction corresponds to the H-coaction on $\mathrm{End}_R(N)$. For all $h^*\in H^*$, we have that

$$h^*\rightharpoonup\beta=\sum\langle h^*,\beta_{(1)}\rangle\beta_{(0)}=\sum\psi_{h^*}\circ\beta\circ\psi_{S^*(h^*)}$$

If we take $h^*=g^*$, then we find that

$$\sum\langle g^*,\beta_{(1)}\rangle\beta_{(0)}=\sum\psi_{g^*}\circ\beta\circ(\psi_{g^*})^{-1}$$

and

$$\Theta(\alpha\otimes\beta)=\left(\underline{\varphi}\circ(\alpha\otimes I_{\underline{I}})\circ\underline{\varphi}^{-1}\right)\otimes\left((\underline{\lambda}\circ\psi_{g^*})\circ(\beta\otimes I_{\underline{J}})\circ(\psi_{g^*})^{-1}\circ\underline{\lambda}^{-1}\right)$$

and this proves that Θ is induced by θ. Observe finally that, for all $p\in P$.

$$\begin{aligned}(\psi_{g^*}\circ\underline{\lambda})(p) &= g^*\rightharpoonup\underline{\lambda}(p)\\ &= \sum\langle g^*,l\rangle\langle g^*,p_{(1)}\rangle\underline{\lambda}(p_{(0)})\\ &= \langle g^*,l\rangle(\underline{\lambda}\circ\psi_{g^*})(p)\end{aligned}$$

and this shows that

$$\psi_{g^*}\circ\underline{\lambda}=\langle g^*,l\rangle\underline{\lambda}\circ\psi_{g^*} \tag{13.19}$$

□

13.4 The cup product

Let H be a faithfully projective commutative cocommutative Hopf algebra, and consider a commutative H-Galois object U and a commutative H^*-Galois object T. We will view T as an H-Galois extension, that is, H acts on T. Now take the smash product $T\#U$. The multiplication rule is

$$\begin{aligned}(s\#u)(t\#v) &= \sum s(u_{(1)}\rightharpoonup t)\#u_{(0)}v \\ &= \sum \langle u_{(1)}, t^*_{(1)}\rangle st_{(0)}\#u_{(0)}v\end{aligned}$$

where we wrote $\rho(t) = \sum t_{(0)} \otimes t^*_{(1)} \in T \otimes H^*$.

Lemma 13.4.1 *With notations as above, $T\#U$ is an Azumaya algebra.*

Proof Take a faithfully flat extension S of R such that $S \otimes U \cong S \otimes H$ as an H^*-comodule algebra. From Theorem 8.3.1, it follows that

$$S \otimes (T\#U) \cong (S \otimes T)\#(S\#H) \cong \mathrm{End}_S(S \otimes T)$$

and, using Theorem 3.3.5, we obtain that $T\#U$ is an Azumaya algebra. □

Remark 13.4.2 It follows from the proof of Lemma 13.4.1 that $T\#U$ is also an Azumaya algebra in the situation where T is an H^*-Galois object, U is an H-Galois object and T or U is commutative. One of the two Galois objects (but noth both) is allowed to be noncommutative.

We now have a map

$$\psi:\ \mathrm{Gal}(S/R, H^*) \times \mathrm{Gal}(S/R, H) \longrightarrow \mathrm{Br}(S/R):\ ([T],[U]) \mapsto [T\#U]$$

We will now give a description of this map on the cocycle level. Let S be a faithfully flat R-algebra, and consider the map

$$\varphi:\ G(H \otimes S^{\otimes 2}) \times G(H^* \otimes S^{\otimes 2}) \longrightarrow \mathbb{G}_m(S^{\otimes 3})$$

given by

$$\varphi(g, g^*) = \langle g^* \otimes 1_S, 1_S \otimes g\rangle = \langle g^*_3, g_1\rangle \tag{13.20}$$

We recall that the indices here are defined as follows. If we have, for example $g = \sum g_i \otimes s_i \otimes s'_i \in H \otimes S^{\otimes 2}$, then

$$\begin{aligned}g_1 &= \sum g_i \otimes 1 \otimes s_i \otimes s'_i \\ g_2 &= \sum g_i \otimes s_i \otimes 1 \otimes s'_i \\ g_3 &= \sum g_i \otimes s_i \otimes s'_i \otimes 1\end{aligned}$$

in $H \otimes S^{\otimes 3}$. Also recall that, for $i \leq j$,

$$g_{ji} = g_{i,j+1}$$

Theorem 13.4.3 *φ induces a map*

$$\varphi:\ H^1(S/R, G(H\otimes\bullet))\times H^1(S/R, G(H^*\otimes\bullet))\longrightarrow H^2(S/R, \mathbb{G}_m)$$

such that the following diagram commutes

$$\begin{array}{ccccc} \mathrm{Gal}(S/R,H^*)\times \mathrm{Gal}(S/R,H) & \xrightarrow{\psi} & \mathrm{Br}(S/R) & \subset & \mathrm{Br}(R) \\ \cong \downarrow (\alpha,\alpha^*) & & & & \downarrow \subset \\ H^1(S/R,G(H\otimes\bullet))\times H^1(S/R,G(H^*\otimes\bullet)) & \xrightarrow{\varphi} & H^2(S/R,\mathbb{G}_m) & \hookrightarrow & H^2(R_{\mathrm{fl}},\mathbb{G}_m) \end{array} \tag{13.21}$$

Here the maps α and α^* are the ones defined in Theorem 10.6.1.

Proof Take Amitsur cocycles $g\in Z^1(S/R, G(H\otimes\bullet))$ and $g^*\in Z^1(S/R, G(H^*\otimes\bullet))$. Let us first show that $\varphi(g,g^*)$ is an Amitsur 2-cocycle.

$$\begin{aligned} \Delta_2(\varphi(g,g^*)) &= \Delta_2\langle g_3^*, g_1\rangle \\ &= \langle g_{31}^*, g_{11}\rangle\langle g_{32}^*, g_{12}\rangle^{-1}\langle g_{33}^*, g_{13}\rangle\langle g_{34}^*, g_{14}\rangle^{-1} \\ &= \langle g_{31}^* * g_{32}^*, g_{11}\rangle\langle g_{34}^*, g_{13}g_{14}^{-1}\rangle \\ &= \langle g_{14}^* * g_{24}^*, g_{11}\rangle\langle g_{34}^*, g_{21}g_{31}^{-1}\rangle \\ &= \langle (g_1^* * g_2^{*-1})_4, g_{11}\rangle\langle g_{34}^*, (g_2g_3^{-1})_1\rangle \\ &= \langle g_{34}^{*-1}, g_{11}\rangle\langle g_{34}^*, g_{11}\rangle = 1 \end{aligned}$$

Furthermore, if g or g^* is a coboundary, then $\varphi(g,g^*)$ is also a coboundary. Suppose for example that $g^* = k_1^* * (k_2^*)^{-1}\in B^1(S/R, G(H^*\otimes\bullet))$. Then

$$\begin{aligned} \Delta_1\langle k_2^*, g\rangle &= \langle k_{21}^*, g_1\rangle\langle k_{22}^*, g_2\rangle^{-1}\langle k_{23}^*, g_3\rangle \\ &= \langle k_{13}^*, g_1\rangle\langle k_{23}^*, g_2^{-1}g_3\rangle \\ &= \langle k_{13}^*, g_1\rangle\langle k_{13}^*, g_1^{-1}\rangle \\ &= \langle (k_1^* * (k_2^*)^{-1})_3, g_1\rangle \\ &= \langle g_3^*, g_1\rangle = \varphi(g,g^*) \end{aligned}$$

and this proves the first statement of the Theorem.
Before we are able to prove the commutativity of the diagram (13.21), we need some preparatory results. We remark first that the cocycles $\langle g_3^*, g_1\rangle$ and $\langle g_1^*, g_3\rangle^{-1}$ represent the same element in $H^2(S/R,\mathbb{G}_m)$. Indeed,

$$\begin{aligned} \Delta_1(\langle g^*, g\rangle) &= \langle g_1^*, g_1\rangle\langle g_2^*, g_2\rangle^{-1}\langle g_3^*, g_3\rangle \\ &= \langle g_1^*, g_1\rangle\langle g_1^* * g_3^*, g_1 * g_3\rangle^{-1}\langle g_3^*, g_3\rangle \\ &= \langle g_3^*, g_1\rangle\langle g_1^*, g_3\rangle \end{aligned} \tag{13.22}$$

From Theorem 8.3.1, we know that $H^*\#H$ is isomorphic to $\mathrm{End}_R(H^*)$. The connecting isomorphism j is given by the formula

$$j(h^*\#h)(k^*) = \sum\langle k_1^*, h\rangle k_2^* * h^*$$

for all $h^*, k^*\in H^*$ and $h\in H$.
Suppose now that $\Phi:\ H\to H$ and $\Psi:\ H^*\to H^*$ are respectively H-colinear

and H^*-colinear isomorphisms. From Lemma 10.4.6, we know that there exist $g^* \in G(H^*)$ and $g \in G(H)$ such that

$$\Phi = I_H * g^*, \quad \text{that is} \quad \Phi(h) = \sum \langle g^*, h_{(1)}\rangle h_{(2)}$$
$$\Psi = I_{H^*} * g, \quad \text{that is} \quad \Psi(h^*) = \sum \langle h^*_{(1)}, g\rangle h^*_{(2)}$$
$$\text{or} \quad \langle \Psi(h^*), h\rangle = \langle h^*, gh\rangle$$

for all $h \in H$ and $h^* \in H^*$. Now define $\Theta : \ \mathrm{End}_R(H^*) \to \mathrm{End}_R(H^*)$ by the commutativity of the following diagram

$$\begin{array}{ccc} H^*\#H & \xrightarrow{j} & \mathrm{End}_R(H^*) \\ \downarrow{\scriptstyle \Psi\otimes\Phi} & & \downarrow{\scriptstyle \Theta} \\ H^*\#H & \xrightarrow{j} & \mathrm{End}_R(H^*) \end{array}$$

We claim that θ is induced by the map $\theta : \ H^* \to H^*$ given by

$$\theta(k^*) = \sum \langle S^*(k_1^*), g\rangle g^* * k_2^*$$

for all $k^* \in H^*$. To this end, we have to show that, for all $\lambda \in \mathrm{End}_R(H^*)$

$$\Theta(\lambda) = \theta^{-1} \circ \lambda \circ \theta \tag{13.23}$$

and it suffices to show (13.23) for $\lambda = j(h^*\#h)$ with $h \in H$ and $h^* \in H^*$ arbitrary. Observe that the composition inverse θ^{-1} of θ is given by the formula

$$\theta^{-1}(k^*) = \sum \langle (g^*)^{-1} * k_1^*, g\rangle (g^*)^{-1} * k_2^*$$

Now for all $k^* \in H^*$, we have

$$\begin{aligned} (\theta^{-1} \circ \lambda \circ \theta)^{-1}(k^*) &= (\theta^{-1} \circ \lambda)\left(\sum \langle S^*(k_1^*), g\rangle g^* * k_2^*\right) \\ &= \theta^{-1}\left(\sum \langle S^*(k_1^*), g\rangle \langle g^* * k_2^*, h\rangle g^* * k_3^* * h^*\right) \\ &= \sum \langle S^*(k_1^*), g\rangle \langle g^* * k_2^*, h\rangle \langle (g^*)^{-1} * g^* * k_3^* * h_1^*, g\rangle (g^*)^{-1} * g^* * k_4^* * h_2^* \\ &= \sum \langle g^* * k_1^*, h\rangle \langle h_1^*, g\rangle k_2^* * h_2^* \end{aligned}$$

On the other hand

$$\begin{aligned} \Theta(\lambda)(k^*) &= (\Theta \circ j)(h^*\#h)(k^*) \\ &= j\left(\Psi(h^*) \otimes \Phi(h)\right)(k^*) \\ &= \sum \langle k_1^*, \Phi(h)\rangle k_2^* * \Psi(h^*) \\ &= \sum \langle g^*, h_1\rangle \langle k_1^*, h_2\rangle \langle h_1^*, g\rangle h_2^* * k_2^* \\ &= \sum \langle g^* * k_1^*, h\rangle \langle h_1^*, g\rangle k_2^* * h_2^* \end{aligned}$$

Let us now prove the commutativity of (13.21). We have an $H \otimes S$-colinear isomorphism $\sigma : \ U \otimes S \to H \otimes S$ and an $H^* \otimes S$-colinear isomorphism $\sigma^* : \ T \otimes S \to H^* \otimes S$. We define Φ and Ψ by the commutativity of the following diagrams (see also (10.15))

$$\begin{array}{ccc} S\otimes T\otimes S & \xrightarrow{I\otimes\sigma^*} & S\otimes H^*\otimes S \\ \downarrow{\scriptstyle \tau\otimes I} & & \downarrow{\scriptstyle \Phi} \\ T\otimes S\otimes S & \xrightarrow{\sigma^*\otimes I} & H^*\otimes S\otimes S \end{array} \qquad \begin{array}{ccc} S\otimes U\otimes S & \xrightarrow{I\otimes\sigma} & S\otimes H\otimes S \\ \downarrow{\scriptstyle \tau\otimes I} & & \downarrow{\scriptstyle \Psi} \\ U\otimes S\otimes S & \xrightarrow{\sigma\otimes I} & H\otimes S\otimes S \end{array}$$

Now we take the smash product of the two diagrams; we obtain the following diagram

$$\begin{array}{ccccc} S\otimes(T\#U)\otimes S & \longrightarrow & S\otimes(H^*\#H)\otimes S & \stackrel{\cong}{\longrightarrow} & \mathrm{End}_{S^{\otimes 2}}(S\otimes H^*\otimes S) \\ \downarrow{\scriptstyle \tau\otimes I} & & \downarrow{\scriptstyle \Phi\#\Psi} & & \downarrow{\scriptstyle \Theta} \\ (T\#Y)\otimes S\otimes S & \longrightarrow & (H^*\#H)\otimes S\otimes S & \stackrel{\cong}{\longrightarrow} & \mathrm{End}_{S^{\otimes 2}}(H^*\otimes S\otimes S) \end{array} \qquad (13.24)$$

Let $g^*\in G(H^*\otimes S^{\otimes 2})$ and $g\in G(H\otimes S^{\otimes 2})$ be the grouplike elements corresponding to Θ, as above (we work over the groundring $S^{\otimes 2}$). Then g^* and g are the Amitsur 1-cocycles corresponding respectively to U and T (see the proof of proposition 10.4.7 and Theorem 10.6.1). The map Θ is induced by $\theta:\ H^*\otimes S^{\otimes 2}\to H^*\otimes S^{\otimes 2}$ given by

$$\theta(k^*)=\sum\langle S^*(k^*_{(1)}),g\rangle g^* * k^*_{(2)}$$

for all $k^*\in H^*\otimes S^{\otimes 2}$. Now for all $l^*\in H^*\otimes S^{\otimes 3}$, we have

$$\begin{aligned}
&(\theta_2^{-1}\circ\theta_3\circ\theta_1)(l^*)\\
&= (\theta_2^{-1}\circ\theta_3)\left(\sum\langle S^*(k^*_{(1)}),g_1\rangle g_1^*k^*_{(2)}\right)\\
&= \theta_2^{-1}\left(\sum\langle S^*(k^*_{(1)}),g_1\rangle\langle S^*(g_1^*k^*_{(2)},g_3\rangle g_3^* * g_1^* * k^*_{(3)}\right)\\
&= \sum\langle S^*(k^*_{(1)}),g_1\rangle\langle S^*(g_1^*k^*_{(2)},g_3\rangle\langle (g_2^*)^{-1} * g_3^* * g_1^* * k^*_{(3)},g_2\rangle (g_2^*)^{-1} * g_3^* * g_1^* * k^*_{(4)}\\
&= \sum\langle k^*_{(1)},g_1^{-1}\rangle\langle (g_1^*)^{-1},g_3\rangle\langle k^*_{(2)},g_3^{-1}\rangle\langle k^*_{(3)},g_2\rangle k^*_{(4)}\\
&= \sum\langle g_1^*,g_3\rangle^{-1}\langle k^*_{(1)},g_1^{-1}g_3^{-1}g_2\rangle k^*_{(2)}\\
&= \langle g_1^*,g_3\rangle^{-1}k^*
\end{aligned}$$

We have shown that the image of $[T\#U]$ in $H^2(R_{\mathrm{fl}},\mathbb{G}_m)$ is represented by $\langle g_1^*,g_3\rangle^{-1}\in Z^2(S/R,\mathbb{G}_m)$. This cocycle is cohomologous to $\langle g_3^*,g_1\rangle=\varphi(g,g^*)$, and this proves the Theorem. □

Taking the inductive limit of (13.21) over all faithfully flat extensions S of R, we obtain the following commutative diagram

$$\begin{array}{ccc} \mathrm{Gal}^s(R,H^*)\times\mathrm{Gal}^s(R,H) & \stackrel{\psi}{\longrightarrow} & \mathrm{Br}(R) \\ \cong\downarrow{\scriptstyle (\alpha^*,\alpha)} & & \subset\downarrow{\scriptstyle \beta} \\ H^1(R_{\mathrm{fl}},G(H\otimes\bullet))\times H^1(R_{\mathrm{fl}},G(H^*\otimes\bullet)) & \stackrel{\varphi}{\longrightarrow} & H^2(R_{\mathrm{fl}},\mathbb{G}_m) \end{array} \qquad (13.25)$$

This diagram will play an important role in the next Section.

Remarks 13.4.4 1) The cup product map φ can be generalized as follows: let $r,s\geq 1$, and consider the map

$$\varphi:\ G(H\otimes S^{\otimes r+1})\times G(H^*\otimes S^{\otimes s+1})\longrightarrow\mathbb{G}_m(S^{\otimes r+s+1})$$

given by

$$\varphi(g,g^*)=\langle g^*\otimes 1_{S^{\otimes r}},1_{S^{\otimes s}}\otimes g\rangle$$

φ induces a map

$$\varphi:\ H^r(R_{\mathrm{fl}},G(H\otimes\bullet))\times H^s(R_{\mathrm{fl}},G(H^*\otimes\bullet))\to H^{r+s}(R_{\mathrm{fl}},\mathbb{G}_m)$$

2) Let $G = \langle\sigma\rangle$ be the cyclic group of order n, and suppose that T is a G-Galois extension of the commutative ring R. For $a \in \mathbb{G}_m(R)$, we can consider the crossed product $(T, G, f_a) = \bigoplus_{i=0}^{n-1} Tu_{\sigma^i}$ (see (6.18)). This crossed product may be viewed as a smash product: $U = R[x]/(x^n - a)$ is an RG-Galois object: if we put $\deg(x) = \sigma$, then U is a G-strongly graded ring. Furthermore the map

$$T\#U \longrightarrow (T, G, f_a) : \; t\#x \mapsto tu_\sigma$$

is an isomorphism.

3) Now we consider the following special case. Let $q = p^d$ be a primary number, K a field of characteristic different from p, and $H = KC_q$, with $C_q =< \sigma >$. Suppose that K contains a primitive q-th root of unity η. From Proposition 11.1.4, we know that

$$\mathrm{Gal}^s(R, H) \cong \mathrm{Gal}^s(R, H^*) \cong K^*/(K^*)^q$$

Take $a, b \in K^*$, and let U be the H-Galois object (that is, the strongly C_q-graded ring) corresponding to b, and T the H^*-Galois object (that is, the H-Galois extension) corresponding to a. We have

$$T = K[x]/(x^q - a) \quad \text{with} \quad \sigma \rightharpoonup x = \eta x$$
$$U = K[y]/(y^q - b) \quad \text{with} \quad \deg(y) = \sigma$$

Now write $x\#1 = t$, $1\#y = u$ in $T\#U$. Then

$$ut = (1\#y)(x\#1) = \eta x\#y = \eta tu$$

and we see that $T\#U$ is the cyclic algebra (a, b, η) (see for example [110, p. 56]). In the case where $q = 2$, $T\#U$ is a quaternion algebra.
Now we recall the definition of Milnor's K_2:

$$K_2^M(K) = K^* \otimes_{\mathbb{Z}} K^*/\langle a \otimes (1 - a) \mid a \neq 0, 1\rangle$$

The class represented by $a \otimes b$ in $K_2^M(K)$ is usually denoted by $\{a, b\}$. We have a map

$$R_{K,q} : \; K_2^M(K)/qK_2^M(K) \longrightarrow {}_q\mathrm{Br}(K)$$

given by

$$R_{K,q}(\{a, b\}) = (a, b, \eta)$$

This map is called the *norm resthomomorphism* (see [110, p. 77-78]). The celebrated Merkure'ev-Suslin Theorem [134] states that $R_{K,q}$ is an isomorphism. An outline of the proof may be found in [110]. Further applications of the Merkure'ev-Suslin Theorem were given by Merkure'ev in [132] and [133].
From the above arguments, it is clear that the cup product map ψ factors through the norm resthomomorphism $R_{K,q}$.

13.5 The split part of the Brauer-Long group

In the sequel, H will be a commutative, cocommutative, faithfully projective Hopf algebra, and S will be a commutative faithfully flat R-algebra. For the statements that involve the H-module structure, cocommutativity is not needed, and for statements that involve the H-comodule structure, commutativity is not needed.

Cohomology with values in the category of invertible H-dimodules

In Section 6.1 we introduced the cohomology groups $H^n(S/R, \underline{\underline{\mathrm{Pic}}})$ with values in the category $\underline{\underline{\mathrm{Pic}}}$. This construction mat be adapted easily to the situation where the category of invertible R-modules $\underline{\underline{\mathrm{Pic}}}$ is replaced by one of the following categories: $\underline{\underline{\mathrm{PM}}}(\bullet, H \otimes \bullet)$, $\underline{\underline{\mathrm{PC}}}(\bullet, H \otimes \bullet)$ or $\underline{\underline{\mathrm{PD}}}(\bullet, H \otimes \bullet)$. This leads to the definition of the cohomology groups

$$H^n(S/R, \underline{\underline{\mathrm{PM}}}(\bullet, H \otimes \bullet)),\ H^n(S/R, \underline{\underline{\mathrm{PC}}}(\bullet, H \otimes \bullet)) \text{ and } H^n(S/R, \underline{\underline{\mathrm{PD}}}(\bullet, H \otimes \bullet))$$

Arguments similar to the ones given in Section 6.1 show that we have long exact sequences

$$\begin{array}{lllll} 1 & \longrightarrow & H^1(S/R, \mathbb{G}_m) & \xrightarrow{\alpha_1} & H^0(S/R, \underline{\underline{\mathrm{PX}}}(\bullet, H \otimes \bullet)) & \xrightarrow{\beta_1} & H^0(S/R, \mathrm{PX}) \\ & \xrightarrow{\gamma_1} & H^2(S/R, \mathbb{G}_m) & \xrightarrow{\alpha_2} & H^1(S/R, \underline{\underline{\mathrm{PX}}}(\bullet, H \otimes \bullet))) & \xrightarrow{\beta_2} & H^1(S/R, \mathrm{PX}) \\ & \xrightarrow{\gamma_2} & \cdots & & & & \\ & \xrightarrow{\gamma_{n-1}} & H^n(S/R, \mathbb{G}_m) & \xrightarrow{\alpha_n} & H^{n-1}(S/R, \underline{\underline{\mathrm{PX}}}(\bullet, H \otimes \bullet))) & \xrightarrow{\beta_n} & H^{n-1}(S/R, \mathrm{PX}) \\ & \xrightarrow{\gamma_n} & \cdots & & & & \end{array} \tag{13.26}$$

where PX stands for PM, PC or PD.

Lemma 13.5.1 *For any $n \geq 0$, we have*

$$H^n(S/R, \underline{\underline{\mathrm{PM}}}(\bullet, H \otimes \bullet)) = H^n(S/R, G(H^* \otimes \bullet) \times H^n(S/R, \underline{\underline{\mathrm{Pic}}}) \tag{13.27}$$

$$H^n(S/R, \underline{\underline{\mathrm{PC}}}(\bullet, H \otimes \bullet)) = H^n(S/R, G(H \otimes \bullet) \times H^n(S/R, \underline{\underline{\mathrm{Pic}}}) \tag{13.28}$$

$$\begin{aligned} H^n(S/R, \underline{\underline{\mathrm{PD}}}(\bullet, H \otimes \bullet)) &= H^n(S/R, G(H \otimes \bullet) \times H^n(S/R, G(H^* \otimes \bullet) \\ &\quad \times H^n(S/R, \underline{\underline{\mathrm{Pic}}}) \end{aligned} \tag{13.29}$$

Proof We will prove the second statements, the proof of the other two statements is similar, or can be obtained by duality arguments. We will define an isomorphism

$$F:\ H^n(S/R, \underline{\underline{\mathrm{PC}}}(\bullet, H \otimes \bullet)) \longrightarrow H^n(S/R, G(H \otimes \bullet) \times H^n(S/R, \underline{\underline{\mathrm{Pic}}})$$

Take $(I, \alpha) \in Z^n(S/R, \underline{\underline{\mathrm{PC}}}(\bullet, H \otimes \bullet))$. Using Proposition 13.3.1, we can write $I = \underline{I}(g)$, with $\underline{I} = I$ as an $S^{\otimes n+1}$-module, but with trivial coaction. The coaction on $\underline{I}(g)$ is given by $\rho(x) = x \otimes g$, for all x. Here $g \in G(H \otimes S^{\otimes n+1})$ is grouplike. It is clear that $\underline{I} \in Z^n(S/R, \underline{\underline{\mathrm{Pic}}})$, and

$$\delta_{n-1}(I) = \delta_{n-1}(\underline{I}(g)) = \delta_{n-1}(\underline{I})(\Delta_{n-1} g)$$

Now $\delta_{n-1}(I) \cong S^{\otimes n+2}$ as an $H \otimes S^{\otimes n+2}$-comodule, so the coaction on $\delta_{n-1}(I)$ is the trivial one. This implies that $\Delta_{n-1} g = 1$, and g is a cocycle. We define

$$F[(I, \alpha)] = ([(\underline{I}, \alpha)], [g])$$

It is straightforward to show that F is well-defined. The inverse of F is given by the formula

$$F^{-1}\left([(\underline{I},\alpha)],[g]\right) = [(\underline{I}(g),\alpha)]$$

□

From Theorem 6.1.4, Proposition 13.3.1 and Lemma 13.5.1, it follows that

$$H^0(S/R, \underline{\underline{\mathrm{PX}}}(\bullet, H\otimes\bullet)) = \mathrm{PX}(R,H) \tag{13.30}$$

We also have the following property generalizing Theorem 6.2.1.

Theorem 13.5.2 *Let S be a faithfully flat commutative R-algebra. Let X stand for C or M (but not for D!). Then we have a monomorphism*

$$\mathrm{BX}(S/R,H) \hookrightarrow H^1(S/R, \underline{\underline{\mathrm{PX}}}(\bullet, H\otimes\bullet)) \tag{13.31}$$

which is an isomorphism if S is faithfully projective as an R-module

Proof The proof is an easy adaptation of the proof of Theorem 6.2.1. One has to rewrite this proof in the situation where one is dealing with the classical Brauer group $\mathrm{Br}(R)$ rather than $\mathrm{Br}'(R)$. The elementary algebras then become endomorphism rings. The cohomology groups with values in $\underline{\underline{\mathrm{Pic}}}$ or Pic have to be replaced systematically by cohomology groups with values in $\underline{\underline{\mathrm{PX}}}$ or PX. □

Taking inductive limits over all faithfully flat extensions S of R in Theorem 13.5.2, we obtain, using Lemma 13.5.1 and Theorem 6.1.4, an embedding

$$\beta:\ \mathrm{BC}^s(R,H) \hookrightarrow H^1(R_{\mathrm{fl}}, G(H\otimes\bullet)) \times H^2(R_{\mathrm{fl}}, \mathbb{G}_m)$$

We see that this result is consistent with Theorem 13.2.1, which implies that

$$\mathrm{BC}^s(R,H) \cong \mathrm{Gal}^s(R,H^*) \times \mathrm{Br}(R) \tag{13.32}$$

(see also Corollary 6.2.2 and Theorem 10.6.1). In a similar way, we have

$$\mathrm{BM}^s(R,H) \cong \mathrm{Gal}^s(R,H) \times \mathrm{Br}(R) \tag{13.33}$$

The relation between the maps defined in Theorem 13.2.1 and Theorem 13.5.2 is given in the next Proposition.

Proposition 13.5.3 *The following diagrams are commutative.*

$$\begin{array}{ccc}
\mathrm{BC}^s(R,H) & \xrightarrow{\pi\circ(\bullet)^{\mathrm{op}}} & \mathrm{Gal}^s(R,H^*) \\
\downarrow{\scriptstyle\beta} & & \downarrow{\scriptstyle\alpha} \\
H^1(R_{\mathrm{fl}}, G(H\otimes\bullet)) \times H^2(R_{\mathrm{fl}}, \mathbb{G}_m) & \xrightarrow{p_1} & H^1(R_{\mathrm{fl}}, G(H\otimes\bullet))
\end{array} \tag{13.34}$$

$$\begin{array}{ccc}
\mathrm{BM}^s(R,H) & \xrightarrow{\pi\circ(\bullet)^{\mathrm{op}}} & \mathrm{Gal}^s(R,H) \\
\downarrow{\scriptstyle\beta} & & \downarrow{\scriptstyle\alpha^*} \\
H^1(R_{\mathrm{fl}}, G(H^*\otimes\bullet)) \times H^2(R_{\mathrm{fl}}, \mathbb{G}_m) & \xrightarrow{p_1} & H^1(R_{\mathrm{fl}}, G(H^*\otimes\bullet))
\end{array} \tag{13.35}$$

α and α^* are the maps defined in theorem 10.6.1; p_1 is the projection onto the first component; π is the restriction of the map defined in Theorem 13.2.1.

Proof Recall that we have an isomorphism

$$\phi:\ H^*\#H\longrightarrow \mathrm{End}_R(H^*)$$

defined as follows:

$$\phi(h^*\#h)(k^*)=\sum\langle h,k^*_{(2)}\rangle h^**k^*_{(1)}$$

for all $h\in H$ and $h^*,k^*\in H^*$. We will define an H-coaction (or, equivalently, an H^*-action) on $\mathrm{End}_R(H^*)$ in such a way that ϕ is an H-comodule algebra isomorphism. Recall that the H-coaction on $H^*\#H$ is induced by the comultiplication on H. Now consider the following H^*-action on H^*:

$$h^*\rightharpoonup k^*=S(h^*)*k^*$$

The induced H^*-action on $\mathrm{End}_R(H^*)$ is given by

$$(h^*\rightharpoonup f)(k^*)=\sum S^*(h^*_{(1)})*f(h^*_{(2)}*k^*)$$

and this H^*-action induces an H-coaction. We claim that ϕ is H^*-linear (and therefore H-colinear). For all $h^*,k^*,l^*\in H^*$ and $h\in H$, we have

$$\begin{aligned}
&(l^*\rightharpoonup\phi(h^*\#h))(k^*)\\
&=\sum S^*(l^*_{(1)})*\phi(h^*\#h)(l^*_{(2)}*k^*)\\
&=\sum S^*(l^*_{(1)})\langle h,l^*_{(2)}*k^*_{(1)}\rangle h^**l^*_{(3)}*k^*_{(2)}\\
&=\sum\langle h_{(1)},l^*\rangle\langle h_{(2)},k^*_{(1)}\rangle h^**k^*_{(2)}\\
&=\phi(l^*\rightharpoonup(h^*\#h))
\end{aligned}$$

Now take an H^*-Galois object T, and suppose that T is split by a faithfully flat extension S of R. Let g be the Amitsur 1-cocycle corresponding to T. We are done if we can show that

$$(p_1\circ\beta)(T\#H^*)=S(g)$$

Consider (13.24) from the proof of Theorem 13.4.3, with $U=H$, the trivial H-Galois object. The map

$$\Theta:\ \mathrm{End}_{S^{\otimes 2}}(S\otimes H^*\otimes S)\longrightarrow\mathrm{End}_{S^{\otimes 2}}(H^*\otimes S\otimes S)$$

is induced by

$$\theta:\ H^*\otimes S^{\otimes 2}\longrightarrow H^*\otimes S^{\otimes 2}:\ k^*\mapsto\sum\langle S^*(k^*_{(1)}),g\rangle k^*_{(2)}$$

Observe that the Amitsur 1-cocycle corresponding to $U=H$ is the trivial ε. Our proof is finished if we can show that

$$\rho(\theta)=\theta\otimes S(g)$$

or, equivalently,

$$h^*\rightharpoonup\theta=\langle h^*,S(g)\rangle\theta$$

for all $h^* \in H^* \otimes S^{\otimes 2}$. Indeed, for all $h^*, k^* \in H^* \otimes S^{\otimes 2}$, we have

$$\begin{aligned}
(h^* \rightharpoonup \theta)(k^*) &= \sum h^*_{(1)} \rightharpoonup \theta(S^*(h^*_{(2)}) \rightharpoonup k^*) \\
&= \sum S^*(h^*_{(1)})\theta(h^*_{(2)} * k^*) \\
&= \sum S^*(h^*_{(1)})\langle S^*(h^*_{(2)} * k^*_{(1)}), g\rangle h^*_{(3)} * k^*_{(2)} \\
&= \sum \langle h^*, S(g)\rangle \langle S^*(k^*_{(1)}), g\rangle k^*_{(2)} \\
&= \sum \langle h^*, S(g)\rangle \theta(k^*)
\end{aligned}$$

□

Unfortunately, Theorem 13.5.2 does not apply to the split part of the Brauer-Long group $\mathrm{BD}^s(R,H)$, and this is what we are interested in. The reason is that the operation on $\mathrm{BD}^s(R,H)$ is the smash product, and not the tensor product. In order to obtain a description of $\mathrm{BD}^s(R,H)$, we now argue as follows. Consider the category $\mathcal{C}$ of H-Azumaya algebras that are split by a faithfully flat extension S of R, and observe that all the objects in the category are Azumaya algebras in the classical sense. As we know, two objects A and B in $\mathcal{C}$ are H-equivalent if and only if there exist faithfully projective H-dimodules M and N such that there is an H-linear H-colinear isomorphism

$$A\#\mathrm{End}_R(M) \cong B\#\mathrm{End}_R(N)$$

which is by Proposition 12.1.6 equivalent to

$$A \otimes \mathrm{End}_R(M) \cong B \otimes \mathrm{End}_R(N)$$

The set of equivalence classes is a group under the operation induced by the smash product. This group is the group that we are looking for, namely $\mathrm{BD}^s(R,H)$. Now this set is also a group under the operation induced by the tensor product. We denote this group by $\mathrm{BD}^{s\otimes}(R,H)$, and it is clear that the methods of Theorem 13.5.2 can be applied to describe this new group $\mathrm{BD}^{s\otimes}(R,H)$. We have that

$$\beta:\ \mathrm{BD}^{s\otimes}(S/R,H) \hookrightarrow H^1(S/R, \underline{\underline{\mathrm{PD}}}(\bullet, H\otimes\bullet))$$

In fact $\mathrm{BD}^{s\otimes}(R,H) \cong \mathrm{BD}^s(R, H\otimes H^*)$, and, as a set, $\mathrm{BD}^s(R,H)$ therefore equal to

$$\mathrm{Gal}^s(R,H^*) \times \mathrm{Gal}^s(R,H) \times \mathrm{Br}(R)$$

We have to investigate the multiplication rules. To this end, we have to go back to the construction of the map β, along the lines of the proof of Theorem 6.2.1. Let A be an H-Azumaya algebra split by S. We then have an H-linear H-colinear isomorphism

$$\rho:\ A\otimes S \longrightarrow \mathrm{End}_S(M)$$

for some faithfully projective $H\otimes S$-dimodule M. The map $\Phi:\ \mathrm{End}_{S\otimes S}(S\otimes M) \to \mathrm{End}_{S\otimes S}(M\otimes S)$ is defined by the commutativity of the diagram

$$\begin{array}{ccc}
S\otimes A\otimes S & \xrightarrow{\rho_2} & \mathrm{End}_{S\otimes S}(S\otimes M) \\
\big\downarrow{\scriptstyle \tau\otimes I_S} & & \big\downarrow{\scriptstyle \Phi} \\
A\otimes S\otimes S & \xrightarrow{\rho_1} & \mathrm{End}_{S\otimes S}(M\otimes S)
\end{array}$$

and is induced by some

$$\varphi : (S \otimes M) \otimes_{S\otimes S} I \longrightarrow M \otimes S$$

The image of $[A]$ in $H^1(S/R, \underline{\underline{\mathrm{PD}}}(\bullet, H \otimes \bullet))$ is represented by $(I, \varphi_2^{-1} \circ \varphi_3 \circ \varphi_1)$, corresponding to $((\underline{I}, \varphi_2^{-1} \circ \varphi_3 \circ \varphi_1), g, g^*)$ in

$$H^1(S/R, G(H \otimes \bullet) \times H^1(S/R, G(H^* \otimes \bullet) \times H^1(S/R, \underline{\underline{\mathrm{Pic}}})$$

(see (13.29)). Now let B be another H-Azumaya algebra split by S, and let $\sigma, P, \Lambda, \lambda, J$ be the data for B corresponding to $\rho, M, \Phi, \varphi, I$ for A. Look at the commutative diagram

$$\begin{array}{ccc}
\mathrm{End}_{S\otimes S}(S \otimes M)\#\mathrm{End}_{S\otimes S}(S \otimes P) & \xrightarrow{\Gamma} & \mathrm{End}_{S\otimes S}(S \otimes (M \otimes P)) \\
\downarrow{\scriptstyle \Phi\otimes\Lambda} & & \downarrow{\scriptstyle \Theta} \\
\mathrm{End}_{S\otimes S}(M \otimes S)\#\mathrm{End}_{S\otimes S}(P \otimes S) & \xrightarrow{\Gamma} & \mathrm{End}_{S\otimes S}((M \otimes P) \otimes S)
\end{array}$$

The map Γ is the one from Proposition 12.1.6. According to Lemma 13.3.4, Θ is induced by the map

$$\theta : (S \otimes (M \otimes P)) \otimes_{S\otimes S} I \otimes_{S\otimes S} J \longrightarrow (M \otimes P) \otimes S$$

with $\theta = \varphi \otimes (\lambda \circ \psi_{g^*}) = \langle g^*, l \rangle \varphi \otimes (\psi_{g^*} \circ \lambda)$. We have to compute $\theta_2^{-1} \circ \theta_3 \circ \theta_1$. Using (13.19), we obtain that

$$\begin{aligned}
&\left(\lambda \circ \psi_{g^*}\right)_2^{-1} \circ \left(\lambda \circ \psi_{g^*}\right)_3 \circ \left(\lambda \circ \psi_{g^*}\right)_1 \\
&= \langle g_2^*, l_2 \rangle \left(\psi_{g_2^*} \circ \lambda_2\right)^{-1} \circ \lambda_3 \circ \psi_{g_3^*} \circ \lambda_1 \circ \psi_{g_1^*} \\
&= \langle g_2^*, l_2 \rangle \lambda_2^{-1} \psi_{g_2^*}^{-1} \circ \lambda_3 \langle g_3^*, l_1 \rangle \lambda_1 \circ \psi_{g_3^* g_1^*} \\
&= \langle g_2^*, l_2 \rangle \langle (g_2^*)^{-1}, l_3 l_1 \rangle \langle g_3^*, l_1 \rangle \lambda_2^{-1} \circ \lambda_3 \circ \lambda_1 \circ \psi_{g_2^*}^{-1} \psi_{g_3^* g_1^*} \\
&= \langle g_2^*, l_2 l_3^{-1} l_1^{-1} \rangle \langle g_3^*, l_1 \rangle \lambda_2^{-1} \circ \lambda_3 \circ \lambda_1 \circ \psi_{g_2^{*-1} g_3^* g_1^*} \\
&= \langle g_3^*, l_1 \rangle \lambda_2^{-1} \circ \lambda_3 \circ \lambda_1
\end{aligned}$$

and this shows that the image of $[A\#B]$ is represented by

$$\left((\underline{I} \otimes_{S\otimes S} \underline{J}, \langle g_3^*, l_1 \rangle (\varphi_2 \otimes \lambda_2)^{-1} \circ (\varphi_3 \otimes \lambda_3) \circ (\varphi_1 \otimes \lambda_1), gl, g^* * l^*\right)$$

Observe that the image of the tensor product $[A \otimes B]$ is represented by

$$\left((\underline{I} \otimes_{S\otimes S} \underline{J}, (\varphi_2 \otimes \lambda_2)^{-1} \circ (\varphi_3 \otimes \lambda_3) \circ (\varphi_1 \otimes \lambda_1), gl, g^* * l^*\right)$$

The difference between the two multiplications is given by the factor $\langle g_3^*, l_1 \rangle$, representing the cup product of g^* and l (cf. (13.20))!). In fact we have shown the following result.

Theorem 13.5.4 *Let S be a faithfully flat commutative R-algebra. Then we have a monomorphism*

$$\mathrm{BD}(S/R, H) \hookrightarrow H^1(S/R, G(H \otimes \bullet) \times H^1(S/R, G(H^* \otimes \bullet) \times H^1(S/R, \underline{\underline{\mathrm{Pic}}})$$

where the multiplication on the right hand side is given by the formula

$$([g], [g^*], [x])([l], [l^*], [y]) = ([gl], [g^* * l^*], [x][y]\varphi([l], [g^*])) \tag{13.36}$$

This monomorphism is an isomorphism if S is faithfully projective as an R-module.

If we combine all our results, we obtain, using the fact that the cup product corresponds to the smash product on the algebra level (see Theorem 13.4.3), the following description of the split part of the Brauer-Long group.

Theorem 13.5.5

$$\mathrm{BD}^s(R,H) = \mathrm{Gal}^s(R,H^*) \times \mathrm{Gal}^s(R,H) \times \mathrm{Br}(R)$$

with multiplication rule

$$([S],[T],[A])([S'],[T'],[A']) = ([S\square_{H^*}S'],[T\square_H T'],[A\otimes A'\otimes (S'\#T)]) \qquad (13.37)$$

Theorem 13.5.5 can be applied to the subgroup $\mathrm{B}^s_\theta(R,H)$ of $\mathrm{BD}^s(R,H)$. Let $\theta : H \to H^*$ be a Hopf algebra map. θ induces maps

$$\begin{aligned}
\theta &: G(H)\longrightarrow G(H^*)\\
\theta &: H^1(R_{\mathrm{fl}},G(H\otimes \bullet))\longrightarrow H^1(R_{\mathrm{fl}},G(H^*\otimes \bullet))\\
\theta &: \mathrm{BC}(R,H)\longrightarrow \mathrm{BM}(R,H)
\end{aligned}$$

We therefore have a map between the split exact sequences

$$\begin{array}{ccccccccc}
1 & \longrightarrow & \mathrm{Br}(R) & \longrightarrow & \mathrm{BC}(R,H) & \longrightarrow & \mathrm{Gal}(R,H^*) & \longrightarrow & 1\\
 & & \downarrow\cong & & \downarrow\theta & & & & \\
1 & \longrightarrow & \mathrm{Br}(R) & \longrightarrow & \mathrm{BM}(R,H) & \longrightarrow & \mathrm{Gal}(R,H) & \longrightarrow & 1
\end{array}$$

and we obtain a map

$$\theta :\ \mathrm{Gal}(R,H^*)\longrightarrow \mathrm{Gal}(R,H)$$

Now take a θ-Azumaya algebra A, and view it as an H-Azumaya algebra. Let $\rho, M, \Phi, \varphi, I, g, g^*$ be as in the proof of Theorem 13.5.4. Then it is clear that M and I are θ-modules, and, as a consequence,

$$[g^*] = \theta([g])$$

On the algebra level, this can be restated as follows: the image of $[A]$ in $\mathrm{Gal}^s(R,H^*)\times \mathrm{Gal}^s(R,H)\times \mathrm{Br}(R)$ is of the form $([S],\theta([S]),[A])$. We therefore have the following result.

Corollary 13.5.6

$$\mathrm{B}^s_\theta(R,H) = \mathrm{Gal}^s(R,H^*) \times \mathrm{Br}(R)$$

with multiplication rule

$$([S],[A])([S'],[A']) = ([S\square_{H^*}S'],[A\otimes A'\otimes (S'\#\theta(S))]) \qquad (13.38)$$

13.6 A dimodule version of the Rosenberg-Zelinsky exact sequence

In the previous Section, we have given a cohomological description of the split part of the Brauer-Long group. Our next aim is to describe the cokernel of the inclusion $\mathrm{BD}^s(R,H) \to \mathrm{BD}(R,H)$. We will see in Section 13.7 that this cokernel embeds in a well-defined subgroup of $\mathrm{Aut}_{\mathrm{Hopf}}(H \otimes H^*)$. Our construction will be based on (a dimodule version of) the Skolem-Noether Theorem and its generalization, the Rosenberg-Zelinsky exact sequence. In the sequel, H will be a commutative, cocommutative, faithfully projective Hopf algebra, and A will be an H-Azumaya algebra. Let H-$\mathrm{Aut}(A)$ be the group of all H-linear, H-colinear R-automorphisms of A. We call $\alpha \in H$-$\mathrm{Aut}(A)$ *H-Inner* if there exists an invertible $u \in A$ such that

$$\alpha(x) = \sum (x_{(1)} \rightharpoonup u) x_{(0)} u^{-1} \tag{13.39}$$

for all $x \in A$. α is called *H-INNER* if both the action and coaction of H on u are trivial, or $u \in A^{H\mathrm{co}H}$. Recall that this means that $h \rightharpoonup u = \varepsilon(h)u$, for all $h \in H$, and $\rho(u) = u \otimes 1$. In this case, (13.39) takes the form

$$f(x) = \sum uxu^{-1} \tag{13.40}$$

The subgroup of H-$\mathrm{Aut}(A)$ consisting of H-inner, resp. H-INNER automorphisms will be denoted by H-$\mathrm{Inn}(A)$, resp. H-$\mathrm{INN}(A)$.

Theorem 13.6.1 (Dimodule version of the Rosenberg-Zelinsky exact sequence) *For any H-Azumaya algebra A, we have the following commutative diagram with exact rows.*

$$\begin{array}{ccccccc} 1 & \longrightarrow & H\text{-INN}(A) & \longrightarrow & H\text{-Aut}(A) & \xrightarrow{\Phi} & \mathrm{PD}(R,H) \\ & & \downarrow\subset & & \downarrow = & & \downarrow \\ 1 & \longrightarrow & H\text{-Inn}(A) & \longrightarrow & H\text{-Aut}(A) & \xrightarrow{\Psi} & \mathrm{Pic}(R) \end{array} \tag{13.41}$$

The image of $\alpha \in H$-$\mathrm{Aut}(A)$ is represented by

$$I_\alpha = \{a \in A \mid \sum (x_{(1)} \rightharpoonup a) x_{(0)} = \alpha(x)a, \text{ for all } x \in A\} \tag{13.42}$$

Proof The proof follows closely the proof of the classical version, see Section 2.3. It is based on the category equivalence between the categories H-dimod and $(H, A^{\#e}\text{-}R)$-dimod discussed in Proposition 12.2.4.
Observe first that A is an object of $(H, A^{\#e}\text{-}R)$-dimod, if we let $A^{\#e}$ act on A using the formula

$$(a\#\bar{b})c = \sum a(b_{(1)} \rightharpoonup c)b_{(0)}$$

For $\alpha, \beta \in H$-$\mathrm{Aut}(A)$, $A_{\alpha,\beta}$ will be the object of $(H, A^{\#e}\text{-}R)$-dimod that is equal to A as an H-dimodule and a right R-module, and with left $A^{\#e}$-action

$$(a\#\bar{b}) \cdot c = (\alpha(a)\#\overline{\beta(b)})c = \sum \alpha(a)(b_{(1)} \rightharpoonup c)\beta(b_{(0)})$$

It may be verified immediately that

$$\gamma : A_{\alpha,\beta} \longrightarrow A_{\gamma\circ\alpha,\gamma\circ\beta} \tag{13.43}$$
$$\alpha : A_{I,\alpha^{-1}} \longrightarrow A_{\alpha,I} \tag{13.44}$$

are isomorphisms in $(H, A^{\#e}\text{-}R)$-dimod. From these two formulas, it follows that

$$A_{\alpha,I} \otimes_A A_{\beta,I} \cong A_{\alpha\circ\beta,I} \tag{13.45}$$

in $(H, A^{\#e}\text{-}R)$-dimod. Indeed,

$$A_{\alpha,I} \otimes_A A_{\beta,I} \cong A_{\alpha\circ\beta,I} \cong A_{I,\alpha^{-1}} \otimes_A A_{\beta,I} \cong A_{\alpha\circ\beta,I} \cong A_{\beta,\alpha^{-1}} \cong A_{\alpha\circ\beta,I}$$

We now claim that α is H-inner if and only if $A_{\alpha,I}$ and A are isomorphic as $A^{\#e}$-modules, and that α is H-INNER if and only if $A_{\alpha,I}$ and A are isomorphic in $(H, A^{\#e}\text{-}R)$-dimod.
Suppose that $\alpha : A \to A_{\alpha,I}$ is a left $A^{\#e}$-linear homomorphism. Then

$$\nu(xy) = \alpha(x)\nu(y) \text{ and } \nu^{-1}(xy) = \alpha^{-1}(x)\nu^{-1}(y)$$

for all $x, y \in A$. With the help of these two equations, we can prove that $u = \nu(1)$ is invertible in A. Actually $u^{-1} = \alpha(\nu^{-1}(1))$, since

$$\alpha(\nu^{-1}(1))\nu(1) = \nu(\nu^{-1}(1)1) = 1$$

and

$$\nu(1)\alpha(\nu^{-1}(1)) = \alpha(\alpha^{-1}(\nu(1))\nu^{-1}(1)) = \alpha(\nu^{-1}(\nu(1)1)) = \alpha(1) = 1$$

Now for all $x \in A$, we have that

$$\begin{aligned} \nu(x) = \nu((1\#\overline{x})1) &= (1\#x)\cdot\nu(1) \\ &= (1\#x)u \\ &= \sum(x_{(1)}\rightharpoonup u)x_{(0)} \\ = \nu((x\#\overline{1})1) &= (x\#\overline{1})\cdot\nu(1) \\ &= (\alpha(x)\#\overline{1})u \\ &= \alpha(x)u \end{aligned}$$

and

$$\alpha(x) = \sum(x_{(1)}\rightharpoonup u)x_{(0)}u^{-1}$$

proving that α is H-inner. If ν is H-linear and H-colinear, then it is clear that $u = \nu(1) \in A^{H\mathrm{co}H}$, and α is H-INNER.
Conversely, suppose that α is H-inner, and define $\nu : A \to A_{\alpha,I}$ by

$$\nu(x) = \alpha(x)u$$

Then

$$\begin{aligned} \nu((a\#\overline{b})x) &= \nu(\sum a(b_{(1)}\rightharpoonup x)b_{(0)}) \\ &= \sum \alpha(a)(b_{(1)}\rightharpoonup\alpha(x))\alpha(b_{(0)})u \end{aligned}$$

and

$$\begin{aligned}(a\#\overline{b})\cdot\nu(x) &= (a\#\overline{b})\cdot(\alpha(x)u)\\ &= \sum\alpha(a)(b_{(1)}\rightharpoonup(\alpha(x)u))b_{(0)}\\ &= \sum\alpha(a)(b_{(2)}\rightharpoonup\alpha(x))(b_{(1)}\rightharpoonup u)b_{(0)}\\ &= \sum\alpha(a)(b_{(1)}\rightharpoonup\alpha(x))\alpha(b_{(0)})u\end{aligned}$$

and ν is an $A^{\#e}$-module isomorphism. If α is H-INNER, then ν is H-linear and H-colinear, and ν is an isomorphism in $(H, A^{\#e}\text{-}R)$-dimod.
Now take $\alpha\in H\text{-Aut}(A)$. From Proposition 12.2.4, it follows that

$$A_{\alpha,I}\cong F_l(G_l(A_{\alpha,I})) = A\otimes(A_{\alpha,I})^A \tag{13.46}$$

A count of ranks shows that $(A_{\alpha,I})^A$ is invertible as an R-module, so it is invertible as an H-dimodule. Also

$$\begin{aligned}(A_{\alpha,I})^A &= \{a\in A_{\alpha,I}|(x\#\overline{1})\cdot a = (1\#\overline{x})\cdot a,\ \text{for all } x\in A\}\\ &= \{a\in A_{\alpha,I}|\alpha(x)a = \sum(x_{(1)}\rightharpoonup a)x_{(0)},\ \text{for all } x\in A\} = I_\alpha\end{aligned}$$

It follows from (13.45) that the map

$$\Phi:\ H\text{-Aut}(A)\longrightarrow \mathrm{PD}(R,H):\ \alpha\mapsto[I_\alpha]$$

is a homomorphism. If $\alpha\in\mathrm{Ker}\,(\Phi)$, then $I_\alpha\cong A_{\alpha,I}$ as an H-dimodule, and $A_{\alpha,I}\cong A$ in $(H, A^{\#e}\text{-}R)$-dimod, and α is H-INNER. Conversely, if α is H-INNER, then $A_{\alpha,I}\cong A$ in $(H, A^{\#e}\text{-}R)$-dimod, and, by Proposition 12.2.4, $I_\alpha = G_l(A_{\alpha,I})\cong G_l(A)\cong R$ in H-dimod. Thus $\mathrm{Ker}\,\Phi = H\text{-INN}(A)$. The same argument proves that $\mathrm{Ker}\,\Psi = H\text{-Inn}(A)$. □

Corollary 13.6.2 *Suppose that $\alpha\in H\text{-Inn}(A)$. Then there exist $g\in G(H)$ and $g^*\in G(H^*)$ such that $\rho(u) = u\otimes g$ and $h\rightharpoonup u = \langle g^*, h\rangle u$, for all $h\in H$. Hence (13.39) can be rewritten as*

$$\alpha(x) = \sum\langle g^*, x_{(1)}\rangle u x_{(0)} u^{-1} \tag{13.47}$$

for all $x\in A$.

Proof From Theorem 13.6.1, we know that $I_\alpha = Ru$ for some invertible $u\in A$. I_α is an invertible H-dimodule that is isomorphic to R as an R-module. From the proof of Proposition 13.3.1, it follows that there exist $g\in G(H)$ and $g^*\in G(H^*)$ such that $\rho(x) = x\otimes g$ and $h\rightharpoonup x = \langle g^*, h\rangle x$, for all $h\in H$ and $x\in I_\alpha$. In particular, these statements hold for $x = u$, and this proves our corollary. □

13.7 A complex for the Brauer-Long group

Let H be a faithfully projective commutative and cocommutative Hopf algebra. We let $D = H\otimes H^*$, and $D^* = H^*\otimes H$. The switch map $t:\ D\otimes D^*$ is an isomorphism of Hopf algebras. For an H-Azumaya algebra A, we define a map

$$\rho = \rho_1\otimes\rho_2:\ D = H\otimes H^*\longrightarrow \mathrm{End}_R(A)$$

as follows:

$$\rho_1(h)(a) = h \rightharpoonup a \quad \text{and} \quad \rho_2(h^*)(a) = \sum \langle h^*, a_{(1)}\rangle a_{(0)}$$

for all $h \in H$, $h^* \in H^*$ and $a \in A$. ρ restricts to a map $\rho:\ G(D) \to H\text{-Aut}(A)$. Now define α_A to be the following composition:

$$\alpha_A:\ G(D) \xrightarrow{\rho} H\text{-Aut}(A) \xrightarrow{\Phi} \mathrm{PD}(R,H) \xrightarrow{p} G(D) \tag{13.48}$$

Here Φ is the map from the Rosenberg-Zelinsky exact sequence (13.41), and p is the projection of $\mathrm{PD}(R,H) = \mathrm{Pic}(R) \times G(D)$ onto $G(D)$ (see Section 13.3). We define $\beta_A:\ G(D) \to G(D)$ by

$$\beta_A(d) = d(\alpha_A(d))^{-1} \tag{13.49}$$

Consider $d = h \otimes h^* \in G(D) = G(H) \times G(H^*)$, and suppose that $\alpha_A(d) = k \otimes k^*$. Then $u \in A$ belongs to $I_{\rho(d)}$ if and only if

$$\sum (a_{(1)} \rightharpoonup u) a_{(0)} = \rho(h \otimes h^*)(a) u \tag{13.50}$$

or

$$\sum \langle k^*, a_{(1)}\rangle u a_{(0)} = \sum \langle h^*, a_{(1)}\rangle (h \rightharpoonup a_{(0)}) u \tag{13.51}$$

for all $a \in A$.

Lemma 13.7.1 *Let A and B be H-Azumaya algebras. Then $\beta_{A\#B} = \beta_A \circ \beta_B$.*

Proof It suffices to show that $\alpha_{A\#B} = \alpha_A \circ \beta_B$. Take $d = h \otimes h^* \in G(D)$, and let $\alpha_A(d) = k \otimes k^*$, $\beta_B(d) = hk^{-1} \otimes h^* * (k^*)^{-1} = p \otimes p^*$ and $(\alpha_A \circ \beta_B)(d) = g \otimes g^*$. Take $u \in I_{\rho(p\otimes p^*)}$ and $v \in I_{\rho(d)}$. For all $x \in H$, we have

$$\begin{aligned} x \rightharpoonup u &= \rho_1(x)(u) = \langle g^*, x\rangle u & \rho_A(u) &= u \otimes g \\ x \rightharpoonup v &= \rho_1(x)(v) = \langle k^*, x\rangle v & \rho_A(v) &= v \otimes k \end{aligned}$$

For all $a \in A$ and $b \in B$, we have

$$\begin{aligned} (\rho(h \otimes h^*)(a\#b))(u\#v) &= (\rho(h \otimes h^*)(a)\#\rho(h \otimes h^*)(b))(u\#v) \\ &= \sum \rho(h \otimes h^*)(a)(b_{(1)} \rightharpoonup u)\#\rho(h \otimes h^*)(b_{(0)})v \\ &= \sum \langle g^*, b_{(1)}\rangle \rho(p \otimes p^*)(\rho(k \otimes k^*)(a))u\#\rho(h \otimes h^*)(b_{(0)})v \\ &= \sum \langle g^*, a_{(1)}b_{(1)}\rangle u\rho(k \otimes k^*)(a_{(0)})\#\langle k^*, b_{(2)}\rangle v b_{(0)} \\ &= \sum \langle g^*, a_{(1)}b_{(1)}\rangle u(k \rightharpoonup a_{(0)})\#\langle k^*, a_{(2)}b_{(2)}\rangle v b_{(0)} \\ &= \sum (\langle g^*, a_{(1)}b_{(1)}\rangle u\#\langle k^*, a_{(2)}b_{(2)}\rangle v)(a_{(0)}\#b_{(0)}) \\ &= \sum ((a_{(1)}b_{(1)}) \rightharpoonup (u\#v))(a_{(0)}\#b_{(0)}) \end{aligned}$$

and it follows that $u\#v \in I_{\rho(d)}$, and $\alpha_{A\#B}(d) = gk \otimes g^* * k^* = (\alpha_A \circ \beta_B)(d)$, finishing the proof of the Lemma. □

Lemma 13.7.2 *Let P be a faithfully projective H-dimodule, and $A = \mathrm{End}_R(P)$. Then $\beta_A = I_{G(D)}$.*

Proof Recall (see (7.33)) that the action of H on $\mathrm{End}_R(P)$ can be written in the following way: for $h \in H$ and $f \in \mathrm{End}_R(P)$, $h \rightharpoonup f = \sum \psi_{h_{(1)}} \circ f \circ \psi_{S(h_{(2)})}$. For $h \in G(H)$, we have that $h \rightharpoonup f = \psi_h \circ f \circ \psi_{h^{-1}}$. From (7.40), it follows easily that $\rho(\psi_h) = \psi_h \otimes 1$. For all $k \in H$, we have that

$$k \rightharpoonup \psi_h = \sum \psi_{k_{(1)}} \circ \psi_h \circ \psi_{S(k_{(2)})} = \psi_{k_{(1)}hS(k_{(2)})} = \psi_h$$

It is now easy to check that $\psi_h \in I_{\rho(h\otimes\varepsilon)}$, and $\alpha(h \otimes \varepsilon) = 1 \otimes \varepsilon$. By a duality argument, it follows that $\alpha(1 \otimes h^*) = 1 \otimes \varepsilon$, and the Lemma follows. □

Lemma 13.7.3 *We have a well-defined map*

$$\beta : \ \mathrm{BD}(R, H) \longrightarrow \mathrm{Aut}(G(D)) : \ [A] \mapsto \beta_A$$

Proof Using the two previous Lemmas, we obtain that $\beta_A \circ \beta_{\overline{A}} = \beta_{\overline{A}} \circ \beta_A = I_{G(D)}$, and $\beta_A \in \mathrm{Aut}(G(D))$. It also follows that β is well-defined. □

Now we define a Hopf algebra map $\varphi : \ D \to D^*$ by

$$\varphi(h \otimes h^*) = h^* \otimes (\eta \circ \varepsilon)(h) \tag{13.52}$$

We identify H and H^{**} using the canonical isomorphism. The dual map φ^* of φ is then given by

$$\varphi^*(h \otimes h^*) = (\varepsilon^* \circ \eta^*)(h^*) \otimes h \tag{13.53}$$

The convolution $\varphi * \varphi^*$ is nothing else then the switch map t. The Hopf algebra maps φ, φ^* and t restrict to group homomorphisms $G(D) \to G(D)$. If $h \otimes h^*$ is grouplike, then $\varphi(h \otimes h^*) = h^* \otimes 1$ and $\varphi^*(h \otimes h^*) = \varepsilon \otimes h$.
Now we define $q : \ D \to R$ and $b : \ J \otimes J \to R$ by

$$q(h \otimes h^*) = \langle h^*, h\rangle \ \text{ and } \ b(h \otimes h^*, k \otimes k^*) = \langle h^*, k\rangle\langle k^*, h\rangle \tag{13.54}$$

Obviously $b(d \otimes e) = \langle d, t(e)\rangle$, and $q(d) = b(d \otimes d)$, for all $d, e \in D$. Now we introduce the following subgroups of $\mathrm{Aut}_{\mathsf{Hopf}}(D)$:

$$\mathrm{O}(R, H)_{\min} = \{f \in \mathrm{Aut}_{\mathsf{Hopf}}(D) | q = q \circ f\} \tag{13.55}$$

$$\begin{aligned} \mathrm{O}(R, H)_{\max} &= \{f \in \mathrm{Aut}_{\mathsf{Hopf}}(D) | b \circ (f \otimes f) = b\} \\ &= \{f \in \mathrm{Aut}_{\mathsf{Hopf}}(D) | f^* \circ t \circ f = t\} \end{aligned} \tag{13.56}$$

q and b restrict to maps

$$q : \ G(D) \to \mathbb{G}_m(R) \ \text{ and } \ b : \ G(D) \times G(D) \to \mathbb{G}_m(R)$$

b is a bilinear map and q is the associated quadratic form. We introduce the following "orthogonal" subgroups of $\mathrm{Aut}(G(D))$:

$$\mathrm{O}(G(D))_{\min} = \{f \in \mathrm{Aut}(G(D)) | q = q \circ f\} \tag{13.57}$$

$$\begin{aligned} \mathrm{O}(G(D))_{\max} &= \{f \in \mathrm{Aut}(G(D)) | b \circ (f \otimes f) = b\} \\ &= \{f \in \mathrm{Aut}(G(D)) | f^* \circ t \circ f = t\} \end{aligned} \tag{13.58}$$

Our terminology stems from the theory of quadratic forms. We now have the following result:

Proposition 13.7.4

$$\mathrm{Im}\,(\beta) \subset \mathrm{O}(G(D))_{\mathrm{min}} \subset \mathrm{O}(G(D))_{\mathrm{max}}$$

Proof Suppose that $f \in \mathrm{Im}\,(\beta)$. Then $f = \beta_A$ for some H-Azumaya algebra A, and $\alpha_A(d) = df(d)^{-1}$, for all $d \in D$. Take $d = h \otimes h^* \in D$, and suppose that $\alpha_A(d) = k \otimes k^*$. For $u \in I_{\rho(d)}$ and $x \in H$, we have that

$$\rho(u) = u \otimes k \ \text{ and } \ x \rightharpoonup u = \langle k^*, x\rangle u$$

Applying (13.51) with $a = u$, we obtain that

$$\langle k^*, k\rangle u^2 = \langle h^*, k\rangle\langle k^*, h\rangle u^2$$

If R is local, then the automorphism $\rho(d)$ is H-inner. Then we can take $u \in I_{\rho(d)}$ invertible in A, and it follows that

$$\langle k^*, k\rangle = \langle h^*, k\rangle\langle k^*, h\rangle \tag{13.59}$$

and a local-global argument shows that (13.59) also holds in the case where R is arbitrary. (13.59) is equivalent to the following assertions:

$$\begin{aligned}
&\langle k \otimes k^*, \varphi(k \otimes k^*)\rangle = \langle k \otimes k^*, t(h \otimes h^*)\rangle\\
&\langle \alpha_A(d), \varphi(\alpha_A(d))\rangle = \langle \alpha_A(d), t(d)\rangle\\
&\langle df(d)^{-1}, \varphi(df(d)^{-1})\rangle = \langle df(d)^{-1}, \varphi(d)\varphi^*(d)\rangle\\
&\langle df(d)^{-1}, \varphi(f(d)^{-1})\rangle = \langle df(d)^{-1}, \varphi^*(d)\rangle\\
&\langle d, \varphi(f(d))^{-1}\rangle\langle f(d)^{-1}, \varphi(f(d)^{-1})\rangle = \langle d, \varphi^*(d)\rangle\langle f(d)^{-1}, \varphi^*(d)\rangle\\
&\langle f(d)^{-1}, \varphi(f(d)^{-1})\rangle = \langle d, \varphi^*(d)\rangle\\
&\langle f(d), \varphi(f(d))\rangle = \langle d, \varphi(d)\rangle
\end{aligned}$$

and it follows that $f \in \mathrm{O}(G(D))_{\mathrm{min}}$.
Now take $f \in \mathrm{O}(G(D))_{\mathrm{min}}$. For all $d, e \in G(D)$, we have

$$\langle de, \varphi(de)\rangle = \langle f(de), \varphi(f(de))\rangle$$

or

$$\langle d, \varphi(de)\rangle\langle e, \varphi(de)\rangle = \langle f(d), \varphi(f(de))\rangle\langle f(e), \varphi(f(de))\rangle$$

or

$$\langle d, t(e)\rangle = \langle f(d), t(f(e))\rangle$$

and $f \in \mathrm{O}(G(D))_{\mathrm{max}}$. □

Proposition 13.7.5

$$\mathrm{BD}^s(R, H) \subset \mathrm{Ker}\,(\beta)$$

Proof Suppose that A is an H-Azumaya algebra such that $A \otimes S \cong \mathrm{End}_S(P)$ for some faithfully projective $S \otimes H$-dimodule P. By Lemma 13.7.2, $\beta_{S\otimes A} = I_{G(S\otimes D)}$. Now $\beta_A : G(D) \to G(D)$ is the restriction of the map $\beta_{S\otimes A}$, and is therefore trivial. □

We now have a complex

$$1 \longrightarrow \mathrm{BD}^s(R,H) \longrightarrow \mathrm{BD}(R,H) \xrightarrow{\beta} \mathrm{O}(G(D))_{\min}$$

In some cases, in particular when $H = RG$ is a group ring, with $|G|$ invertible in R and R having "enough" roots of unity, then this complex turns out to be an exact sequence. In general, however, the map $\beta_A : G(D) \to G(D)$ is too rough to describe $[A]$ as an element of $\mathrm{BD}(R,H)/\mathrm{BD}^s(R,H)$. In the sequel, we will extend the map β to a Hopf algebra automorphism $\beta_A : D \to D$. First we need some technical results.

13.8 The Hopf algebra $\mathcal{H} = \mathrm{Hom}_R(H,K)$

Let H be a Hopf algebra over the commutative ring R, and let S be a commutative R-algebra. Then $S \otimes H$ is a Hopf algebra over S. We will apply this in the following particular situation: let H be a Hopf algebra, and K a cocommutative coalgebra, and suppose that both H and K are faithfully projective as R-modules. Then K^* is a commutative R-algebra, and $K^* \otimes H$ is a Hopf algebra over K^*. We also have a natural isomorphism $\alpha : K^* \otimes H \longrightarrow \mathrm{Hom}_R(K,H)$ of R-modules, and α induces a Hopf algebra structure on $\mathcal{H} = \mathrm{Hom}_R(H,K)$. The structure maps are described in the next Proposition.

Proposition 13.8.1 *The Hopf algebra structure on $\mathcal{H} = \mathrm{Hom}_R(H,K)$ induced by the map α is the following*

$$\begin{aligned}
&m_{\mathcal{H}} : \mathcal{H} \otimes_{K^*} \mathcal{H} \longrightarrow \mathcal{H} :\ f \otimes g \mapsto f * g\\
&\Delta_{\mathcal{H}} : \mathcal{H} \longrightarrow \mathcal{H} \otimes_{K^*} \mathcal{H} \cong \mathrm{Hom}_R(K, H \otimes H) :\ f \mapsto \Delta_H \circ f\\
&\eta_{\mathcal{H}} : K^* \longrightarrow \mathcal{H} :\ k^* \mapsto \eta_H \circ k^*\\
&\varepsilon_{\mathcal{H}} : \mathcal{H} \longrightarrow K^* :\ f \mapsto \varepsilon_H \circ f\\
&S_{\mathcal{H}} : \mathcal{H} \longrightarrow \mathcal{H} :\ f \mapsto S_H \circ f
\end{aligned}$$

Proof We will first show that $m_{\mathcal{H}}(f \otimes g) = f * g$. Recall that $\alpha(k^* \otimes h)(k) = \langle k^*, k\rangle h$, and observe that it suffices to show the formula for $f, g \in \mathrm{Hom}_R(H,K)$ of the form $f = \alpha(k^* \otimes h)$ and $g = \alpha(k'^* \otimes h')$. For all $k \in K$, we have that

$$\begin{aligned}
m_{\mathcal{H}}(f \otimes g)(k) &= \alpha((k^* \otimes h)(k'^* \otimes h'))(k)\\
&= \alpha((k^* * k'^* \otimes hh'))(k)\\
&= \sum \langle k^*, k_{(1)}\rangle \langle k'^*, k_{(2)}\rangle hh'\\
&= (f * g)(k)
\end{aligned}$$

In a similar way, we find that

$$\Delta_{K^* \otimes H}(k^* \otimes h) = k^* \otimes \Delta_H(h)$$

and

$$\begin{aligned} \Delta_{\mathcal{H}}(f)(k) &= \alpha(k^* \otimes \Delta_H(h))(k) \\ &= \langle k^*, k\rangle \Delta_H(h) = \Delta_H(f(k)) \end{aligned}$$

and this proves the second statement. The other assertions may be proved in a similar way. □

Proposition 13.8.2 *With notations as above, we have that*

$$G(\mathcal{H}) = \{f \in \mathcal{H} | f \text{ is a coalgebra homomorphism}\}$$

Proof f is grouplike if and only if

$$\Delta_{\mathcal{H}}(f) = f \otimes_{K^*} f \quad (13.60)$$

$$\varepsilon_{\mathcal{H}}(f) = \varepsilon_K \quad (13.61)$$

and f is comultiplicative if and only if

$$\Delta_H \circ f = (f \otimes f) \circ \Delta_K \quad (13.62)$$

$$\varepsilon_H \circ f = \varepsilon_K \quad (13.63)$$

It is clear that (13.61) and (13.63) are equivalent. We claim that $f \otimes_{K^*} f$ viewed as an element of $\mathrm{Hom}_R(K, H \otimes H)$ is equal to $(f \otimes f) \circ \Delta_K$. Indeed, for all $f = \alpha(k^* \otimes h)$, we have that

$$(k^* \otimes h) \otimes_{K^*} (k^* \otimes h) = k^* * k^* \otimes h \otimes h$$

in $(K^* \otimes H) \otimes_{K^*} (K^* \otimes H) \cong K^* \otimes H \otimes H$, and

$$\begin{aligned} (f \otimes_{K^*} f)(k) &= \langle k^* * k^*, k\rangle h \otimes h \\ &= \sum \langle k^*, h_{(1)}\rangle \langle k^*, h_{(2)}\rangle h \otimes h \\ &= (f \otimes f)(\Delta_K(h)) \end{aligned}$$

□

Proposition 13.8.3 *With notations as above, we have that $\mathcal{H}^* \cong \mathrm{Hom}_R(K, H^*)$. The duality is given by the formula*

$$\langle \langle f^*, f \rangle l \rangle = \sum \langle f^*, l_{(1)}\rangle \langle f, l_{(2)}\rangle$$

Proof The upper index * is used for the dual of a K^*-module:

$$\mathrm{Hom}_{K^*}(M, K^*) = M^*$$

for any K^*-module M. For $m \in M$ and $m^* \in M^*$, we write

$$m^*(m) = \langle m^*, m \rangle$$

From the Hom-tensor relations, it follows immediately that

$$\begin{aligned}\mathcal{H}^* &\cong \mathrm{Hom}_{K^*}(K^* \otimes H, K^*) \cong \mathrm{Hom}_{K^*}(K^*, K^*) \otimes \mathrm{Hom}_R(H, R)\\ &\cong K^* \otimes H^* \cong \mathrm{Hom}_R(K, H^*)\end{aligned}$$

The duality between $K^* \otimes H^*$ and $K^* \otimes H$ is given by the formula

$$\langle k^* \otimes h^*, g^* \otimes h \rangle = \langle h^*, h \rangle k^* * g^*$$

Translating this into the duality between $\mathcal{H}^*$ and $\mathcal{H}$ using the map α, we obtain, for $f^* = \alpha(k^* \otimes h^*) \in \mathcal{H}^*$ and $f = \alpha(g^* \otimes h) \in \mathcal{H}$ that

$$\begin{aligned}\langle \langle f^*, f \rangle l \rangle &= \sum \langle h^*, h \rangle \langle k^*, l_{(1)} \rangle \langle k^*, l_{(2)} \rangle\\ &= \sum \langle f^*, l_{(1)} \rangle \langle f, l_{(2)} \rangle\end{aligned}$$

□

If M is an H-dimodule, then $\mathcal{M} = \mathrm{Hom}_R(K, M) \cong K^* \otimes M$ is an $\mathcal{H}$-dimodule. The structure maps $\psi_{\mathcal{M}} : \mathcal{H} \otimes_{K^*} \mathcal{M} \to \mathcal{M}$ and $\rho_{\mathcal{M}} : \mathcal{M} \to \mathcal{M} \otimes_{K^*} \mathcal{H} \cong \mathrm{Hom}_R(K, M \otimes H)$ are given by the formulas

$$\begin{aligned}\psi_{\mathcal{M}}(f \otimes \mu)(k) &= \sum f(k_{(1)}) \rightharpoonup \mu(k_{(2)})\\ \rho_{\mathcal{M}}(\mu) &= \rho_M \circ \mu\end{aligned}$$

for all $f \in \mathcal{H}$, $\mu \in \mathcal{M}$ and $k \in K$. The proof is similar to the proof of Proposition 13.8.1, so we will omit it here. If A is an H-dimodule algebra, then $\mathcal{A} = \mathrm{Hom}_R(K, A) \cong K^* \otimes A$ is an $\mathcal{H}$-dimodule algebra. The multiplication on $\mathcal{A}$ is the convolution.

Consider the canonical embedding

$$i : H \to K^* \otimes H \cong \mathcal{H}$$

For all $h \in H$, $i(h)(k) = \alpha(\varepsilon \otimes h)$, and this means that

$$i(h)(k) = \varepsilon(k)h$$

for all $h \in H$ and $k \in K$. Similarly, if A is an H-dimodule algebra, then $i : A \to \mathcal{A}$ is given by the formula

$$i(a)(k) = \varepsilon(k)a$$

for all $a \in A$ and $k \in K$.

13.9 A short exact sequence for the Brauer-Long group

We now return to the Brauer-Long group. As before, H is a faithfully projective commutative and cocommutative Hopf algebra, and $D = H \otimes H^*$. We will apply the above construction with the coalgebra K replaced by $D^{\otimes n}$. We will write

$$\mathcal{H}_n = \mathrm{Hom}_R(D^{\otimes n}, H) \cong D^{*\otimes n} \otimes H$$

In the sequel, we will use this construction in the cases $n = 1$ and $n = 2$. The index n will be omitted if $n = 1$. $\mathcal{H}_2$ will be used in the technical Lemmas 13.9.4 and 13.9.5, so, for simplicity, we recommend the reader to forget about the index n at a first reading of this Section. Observe that

$$\mathcal{H}_n^* \cong \mathrm{Hom}_R(D^{\otimes n}, H^*) \text{ and } \mathcal{D}_n = \mathcal{H}_n \otimes_{D^{*\otimes n}} \mathcal{H}_n^* \cong \mathrm{Hom}_R(D^{\otimes n}, D)$$

Now let A be an H-Azumaya algebra. Then

$$\mathcal{A}_n = \mathrm{Hom}_R(D^{\otimes n}, A) \cong D^{*\otimes n} \otimes A$$

is an $\mathcal{H}_n$-Azumaya $D^{*\otimes n}$-algebra. The construction preceding Lemma 13.7.1 gives us maps

$$\alpha_{\mathcal{A}_n}, \beta_{\mathcal{A}_n} : G(\mathcal{D}_n) \longrightarrow G(\mathcal{D}_n)$$

We will now give an explicit description of the maps $\alpha_{\mathcal{A}_n}$ and $\beta_{\mathcal{A}_n}$. We have

$$\alpha_{\mathcal{A}_n} : G(\mathcal{D}_n) \xrightarrow{R} \mathcal{H}_n\text{-}\mathrm{Aut}_{D^{*\otimes n}}(\mathcal{A}_n) \xrightarrow{\Phi} \mathrm{PD}(D^{*\otimes n}, \mathcal{H}_n) \longrightarrow G(\mathcal{D}_n)$$

where R is given by

$$R(\gamma)(f)(d) = \sum \rho(\gamma(d_{(1)}))(f(d_{(2)})) \tag{13.64}$$

for all $d \in D^{\otimes n}$, $\gamma \in G(\mathcal{D}_n)$ and $f \in \mathcal{A}_n$. $\beta_{\mathcal{A}_n}$ is given by the formula

$$\beta_{\mathcal{A}_n}(\gamma) = \gamma * (S \circ \alpha_{\mathcal{A}_n}(\gamma))$$

Recall from Section 13.6 that, for any $\mathcal{F} \in \mathcal{H}_n\text{-}\mathrm{Aut}_{D^{*\otimes n}}(\mathcal{A}_n)$, $\Phi(\mathcal{F})$ is represented by an invertible $\mathcal{H}_n$-dimodule $I_{\mathcal{F}}$, which is a submodule of $\mathcal{A}_n$. In the next Lemma, we will give an explicit description of $I_{R(\gamma)}$.

Lemma 13.9.1 *With notations as above, let $\gamma \in G(\mathcal{D}_n)$, $u \in \mathcal{A}_n$, and suppose that $\alpha_{\mathcal{A}_n}(\gamma) = \mu \otimes \mu^*$, with $\mu \in G(\mathcal{H}_n)$ and $\mu^* \in G(\mathcal{H}_n^*)$. Then $u \in I_{R(\gamma)}$ if and only if*

$$\sum \langle \mu^*(d_{(1)}), a_{(1)} \rangle u(d_{(2)}) a_{(0)} = \sum \rho(\gamma(d_{(1)}))(a) u(d_{(2)}) \tag{13.65}$$

for all $d \in D^{\otimes n}$ and $a \in A$.

Proof From (13.42), we know that $u \in I_{R(\gamma)}$ if and only if

$$\sum (f_{(1)} \rightharpoonup u) * f_{(0)} = R(\gamma)(f) * u$$

or

$$\sum \langle \mu^*, f_{(1)} \rangle * u * f_{(0)} = R(\gamma)(f) * u \tag{13.66}$$

for all $f \in \mathcal{A}_n$. Suppose now that $u \in I_{R(\gamma)}$, and apply (13.66) in the case $f = i(a)$, where $i : A \to \mathcal{A}$ is the natural embedding, and apply both sides of the equation to an arbitrary $d \in D^{\otimes n}$. The left hand side then amounts to

$$\sum \langle \langle \mu^*, i(a_{(1)}) \rangle, d_{(1)} \rangle u(d_{(2)}) i(a_{(0)})(d_{(3)}) = \langle \mu^*(d_{(1)}), a_{(1)} \rangle u(d_{(2)}) a_{(0)}$$

and the right hand side is

$$\begin{aligned} \sum R(\gamma)(i(a))(d_{(1)}) u(d_{(2)}) &= \sum \rho(\gamma(d_{(1)}))(i(a)(d_{(2)})) u(d_{(3)}) \\ &= \sum \rho(\gamma(d_{(1)}))(a) u(d_{(2)}) \end{aligned}$$

and (13.65) follows.
Conversely, suppose that (13.65) holds for all $d \in D^{\otimes n}$ and $a \in A$. Then (13.66) holds for all f of the form $f = i(a) = d^* * i(a)$, and, by $D^{*\otimes n}$-linearity, for all f of the form $f = d^* * i(a)$, and for arbitrary f. □

Our next Lemma shows that $\alpha_{\mathcal{A}_n}$ and $\beta_{\mathcal{A}_n}$ are known completely once we know $\alpha_{\mathcal{A}}(I)$ or $\beta_{\mathcal{A}}(I)$, where $I : D \to D$ is the identity map. Since I is comultiplicative, it is a grouplike element of the Hopf algebra $\mathcal{D}$.

Lemma 13.9.2 *With notations as above, we have, for all $\gamma \in G(\mathcal{D}_n)$, that*

$$\beta_{\mathcal{A}_n}(\gamma) = \beta_{\mathcal{A}}(I) \circ \gamma \quad \text{and} \quad \alpha_{\mathcal{A}_n}(\gamma) = \alpha_{\mathcal{A}}(I) \circ \gamma$$

Proof From (13.64), it follows that, for all $d \in D^{\otimes n}$, $\gamma \in G(\mathcal{D}_n)$ and $f \in \mathcal{A}$:

$$\begin{aligned} R(I)(f)(\gamma(d)) &= \sum \rho(\gamma(d_{(1)})) f(\gamma(d_{(2)})) \\ &= \sum \rho(\gamma(d)_{(1)}) f(\gamma(d)_{(2)}) \\ &= R(\gamma)(f \circ \gamma)(d) \end{aligned}$$

We used the fact that γ is grouplike and therefore comultiplicative. It follows that

$$R(I)(f) \circ \gamma = R(\gamma)(f \circ \gamma)$$

Take $u \in I_{R(I)}$ and suppose that $\alpha_{\mathcal{A}}(I) = \mu \otimes \mu^*$ Then for all $a \in A$ and $d \in D$, we have

$$\sum \langle \mu^*(d_{(1)}), a_{(1)} \rangle u(d_{(2)}) a_{(0)} = \sum \rho(d_{(1)})(a) u(d_{(2)})$$

Now replace d by $\gamma(d)$, with $d \in D^{\otimes n}$. Taking into account that γ is comultiplicative, we obtain that

$$\sum \langle (\mu^* \circ \gamma)(d_{(1)}), a_{(1)} \rangle (u \circ \gamma)(d_{(2)}) a_{(0)} = \sum \rho(\gamma(d_{(1)}))(a)(u \circ \gamma)(d_{(2)})$$

and from Lemma 13.9.1, it follows that $u \circ \gamma \in I_{R(\gamma)}$. Therefore

$$\alpha_{\mathcal{A}_n}(\gamma) = (\mu \circ \gamma) \otimes (\mu^* \circ \gamma)\alpha_{\mathcal{A}}(I) \circ \gamma$$

Finally

$$\beta_{\mathcal{A}_n}(\gamma) = \gamma * S_{\mathcal{D}_n}(\alpha_{\mathcal{A}_n}(\gamma)) = \gamma * S_D \circ \alpha_{\mathcal{A}}(I) \circ \gamma = (I * S_D \circ \alpha_{\mathcal{A}}(I)) \circ \gamma = \beta_{\mathcal{A}}(I) \circ \gamma$$

This finishes the proof of the Lemma. □

$\alpha_{\mathcal{A}}(I)$ and $\beta_{\mathcal{A}}(I)$ are the promised extensions of α_A and β_A to D. Let us first show that their restrictions to $G(D)$ are indeed α_A and β_A.

Lemma 13.9.3 *$\alpha_{\mathcal{A}}(I)_{|G(D)} = \alpha_A$ and $\beta_{\mathcal{A}}(I)_{|G(D)} = \beta_A$*

Proof Take $d \in G(D)$ and $u \in I_{R(I)}$. We will show that

$$I_{\rho(d)} = \{u(d)|u \in I_{R(I)} \subset \mathcal{A} = \mathrm{Hom}_R(D, A)\} \tag{13.67}$$

Take $u \in I_{R(I)}$, and let $\alpha_{\mathcal{A}}(I) = \mu \otimes \mu^*$. For all $a \in A$, we have, by Lemma 13.9.1,

$$\sum \langle \mu^*(d), a_{(1)} \rangle u(d) a_{(0)} = \rho(d)(a) u(d) \tag{13.68}$$

Furthermore

$$\rho_A(u(d)) = (\rho_{\mathcal{A}}(u))(d) = u(d) * \mu(d) \tag{13.69}$$

and, for all $h \in H$,

$$h \rightharpoonup u(d) = (i(h) \rightharpoonup u)(d) = \Big(\big\langle \mu^*, i(h) \big\rangle * u\Big)(d) = \langle \mu^*(d), h \rangle u(d) \tag{13.70}$$

It follows from (13.68) and (13.70) that $u(d) \in I_{\rho(d)}$, and we have that

$$\{u(d)|u \in I_{R(I)}\} \subset I_{\rho(d)} \tag{13.71}$$

If R is local, then D^* is semilocal and $\mathrm{Pic}(D^*) = 1$. Then $I_{R(I)}$ is free of rank one, and can be generated by an element $u \in \mathcal{A}$ that is invertible in $\mathcal{A}$. Let v be the inverse of u. Then $u * v = \varepsilon$, and $u(d)v(d) = \varepsilon(d) = 1$, since d is grouplike. But then $I_{\rho(d)}$ is generated by $u(d)$ as an R-module, and (13.71) is an equality. Thus (13.71) is an equality after we localize at an arbitrary $p \in \mathrm{Spec}(R)$, and this entails that (13.71) is an equality. Finally, the statement of the Lemma follows from (13.68) and (13.69). □

From now on, we will write β_A for the extended map $\beta_{\mathcal{A}}(I) : D \to D$. Now reconsider the maps φ, φ^* and $t : D \to D^*$ (see Lemmas 13.52 and 13.53). These maps extend to maps ϕ, ϕ^* and $T : \mathcal{D}_n \to \mathcal{D}_n$. The relations between φ, φ^*, t and ϕ, ϕ^*, T are given by the formulas

$$\phi(\mu) = \varphi \circ \mu \;\; ; \;\; \phi^*(\mu) = \varphi^* \circ \mu \;\; ; \;\; T(\mu) = t \circ \mu$$

for all $\mu \in \mathcal{D}_n$. We define $Q : \mathcal{D}_n \to D^*$ by the formula

$$Q(\mu) = \langle \phi(\mu), \mu \rangle$$

or

$$Q(\mu)(d) = \langle \varphi(\mu(d_{(1)})), \mu(d_{(2)}) \rangle$$

for all $\mu \in \mathcal{D}_n$ and $d \in D^{\otimes n}$.

If we apply Proposition 13.7.4 to the $\mathcal{H}_n$-Azumaya algebra $\mathcal{A}_n$, then it follows that

$$Q(\mu) = Q(\beta_{\mathcal{A}_n}(\mu)) \tag{13.72}$$

$$T = \beta^*_{\mathcal{A}_n} \circ T \circ \beta_{\mathcal{A}_n} \tag{13.73}$$

Lemma 13.9.4 *Let A be an H-Azumaya algebra. Then $\beta_A = \beta_{\mathcal{A}}(I) = f$ satisfies the identities*

$$q = q \circ f \text{ and } t = f^* \circ t \circ f$$

Proof From (13.72), it follows that $Q(I) = Q(f)$. Thus for all $d \in D$,

$$Q(I)(d) = \sum \langle d_{(1)}, \varphi(d_{(2)}) \rangle = q(d)$$

equals

$$Q(f)(d) = \sum \langle f(d_{(1)}), \varphi(f(d_{(2)})) \rangle = q(f(d))$$

and this shows that $q = q \circ f$.

In order to show that $t = f^* \circ t \circ f$, we will need the Hopf algebra $\mathcal{D}_n$ in the case $n = 2$. From (13.73) it follows that, for all $\mu, \nu \in G(\mathcal{D}_2)$,

$$\begin{aligned} \langle \beta_{\mathcal{A}_2}(\mu), (T \circ \beta_{\mathcal{A}_2})(\nu) \rangle &= \langle \mu, (\beta_{\mathcal{A}_2^*} \circ T \circ \beta_{\mathcal{A}_2})(\nu) \rangle \\ &= \langle \mu, T(\nu) \rangle = \langle \mu, t \circ \nu \rangle \end{aligned}$$

On the other hand

$$\langle \beta_{\mathcal{A}_2}(\mu), (T \circ \beta_{\mathcal{A}_2})(\nu) \rangle = \langle \mu, \beta_{\mathcal{A}_2^*}(T(\beta_{\mathcal{A}_2}(\nu))) \rangle$$

Now take $\mu = I \otimes \varepsilon$ and $\nu = \varepsilon \otimes I$. Then

$$\beta_{\mathcal{A}_2}(\mu) = \beta_{\mathcal{A}}(I) \circ (I \otimes \varepsilon) = \beta_{\mathcal{A}}(I) \otimes \varepsilon$$

and

$$\beta_{\mathcal{A}_2}(\nu) = \varepsilon \otimes \beta_{\mathcal{A}}(I)$$

We therefore have that

$$\langle I \otimes \varepsilon, t \circ (\varepsilon \otimes I) \rangle = \langle I \otimes \varepsilon, (f^* \circ t \circ f \circ I) \rangle$$

Apply both sides to $d \otimes e \in D \otimes D$. This gives

$$\langle j, t(k) \rangle = \langle j, (f^* \circ t \circ f)(k) \rangle$$

and it follows that $t = f^* \circ t \circ f$. □

$f = \beta_A$ is a grouplike element of $\mathcal{D}$, and it is therefore a coalgebra homomorphism (see Proposition 13.8.2). From the fact that $t = f^* \circ t \circ f$, it follows that f is an automorphism, its inverse is $t \circ f^* \circ t$. In the next two Lemmas, we will show that f also preserves the multiplication and the antipode. It will then follow that f is a Hopf algebra automorphism.

Lemma 13.9.5 *With notations as above, $f = \beta_A$ is an algebra automorphism of D.*

Proof Once more we need the Hopf algebra $\mathcal{D}_n$ in the case $n = 2$. Consider the grouplike elements $\varepsilon \otimes I$, m_D (the multiplication map) and $I \otimes \varepsilon$ of $\mathcal{D}_2$. From Lemma 13.9.2, it follows that

$$\begin{aligned}\beta_{A_2}(\varepsilon \otimes I) &= f \circ (\varepsilon \otimes I) = \varepsilon \otimes f\\ \beta_{A_2}(m_D) &= f \circ m_D\\ \beta_{A_2}(I \otimes \varepsilon) &= f \circ (I \otimes \varepsilon) = f \otimes \varepsilon\end{aligned}$$

Now $(\varepsilon \otimes I) * (I \otimes \varepsilon) = m_D$, so

$$\beta_{A_2}(\varepsilon \otimes I) * \beta_{A_2}(I \otimes \varepsilon) = \beta_{A_2}(m_D)$$

because β_{A_2} is a group automorphism. Applying both sides to $d \otimes e \in D \otimes D$, we obtain

$$\sum \varepsilon(d_{(1)}) f(e_{(1)}) f(d_{(2)}) \varepsilon(e_{(2)}) = f(de)$$

or

$$f(d)f(e) = f(de)$$

□

Lemma 13.9.6 *With notations as above, $f = \beta_A$ preserves the antipode.*

Proof We know that β_A preserves the restriction of the antipode $S_{\mathcal{D}}$ to $G(\mathcal{D})$. This means in fact that it preserves the inverse. Therefore

$$\beta_{\mathcal{A}}(S) = \beta_{\mathcal{A}}(S_{\mathcal{D}}(I)) = S_{\mathcal{D}}(\beta_{\mathcal{A}}(I)) = S \circ \beta_{\mathcal{A}}(I) = S \circ f$$

On the other hand

$$\beta_{\mathcal{A}}(S) = \beta_{\mathcal{A}}(I) \circ S = f \circ S$$

and this proves the Lemma. □

Corollary 13.9.7 *With notations as above, $f = \beta_A \in \mathrm{Aut}_{\mathrm{Hopf}}(D)$.*

Lemma 13.9.8 *If A and B are H-Azumaya algebras, then $\beta_{A\#B} = \beta_A \circ \beta_B$*

Proof Consider $\mathcal{A} = \mathrm{Hom}_R(H, A)$ and $\mathcal{B} = \mathrm{Hom}_R(H, B)$. Then

$$\mathcal{A}\#_{D^*}\mathcal{B} \cong D^* \otimes (A\#B) \cong \mathrm{Hom}_R(D, A\#B)$$

and

$$\beta_{A\#B} = \beta_{\mathcal{A}\#_{D^*}\mathcal{B}}(I) = \beta_{\mathcal{A}}(\beta_{\mathcal{B}}(I)) = \beta_{\mathcal{A}}(I) \circ \beta_{\mathcal{B}}(I) = \beta_A \circ \beta_B$$

On the way, we used Lemmas 13.7.1 and 13.9.2. □

Lemma 13.9.9 *If* $[A] \in \mathrm{BD}^s(R, H)$, *then* $\beta_A = I_D$.

Proof If $[A] \in \mathrm{BD}^s(R, H)$, then $\mathcal{A} \in \mathrm{BD}^s(D^*, \mathcal{H})$. By Proposition 13.7.5, $\beta_{\mathcal{A}}$ is the identity on $G(D)$. It follows that $\beta_A = \beta_{\mathcal{A}}(I_D) = I_D$. □

It follows from Lemma 13.9.9 that we have a well-defined map

$$\beta : \mathrm{BD}(R, H) \longrightarrow \mathrm{Aut}_{\mathrm{Hopf}}(D) : [A] \mapsto \beta_A$$

and that

$$\mathrm{BD}^s(R, H) \subset \mathrm{Ker}(\beta)$$

From Lemma 13.9.4, it follows that

$$\mathrm{Im}(\beta) \subset \mathrm{O}(R, H)_{\mathrm{min}} = \{f \in \mathrm{Aut}_{\mathrm{Hopf}}(D) | q = q \circ f\}$$

We therefore have a complex

$$1 \longrightarrow \mathrm{BD}^s(R, H) \longrightarrow \mathrm{BD}(R, H) \xrightarrow{\beta} \mathrm{O}(R, H)_{\mathrm{min}} \longrightarrow 1$$

and we hope that this complex is a short exact sequence!

Theorem 13.9.10 *With notations as above, we have that*

$$\mathrm{BD}^s(R, H) = \mathrm{Ker}(\beta)$$

Before we are able to prove Theorem 13.9.10, we need some Lemmas.

Lemma 13.9.11 *If* $[A] \in \mathrm{Ker}(\beta)$, *then* A *is* R*-central.*

Proof From the fact A is H-Azumaya, we know that

$$A^A = \{a \in A | xa = \sum (x_{(1)} \rightharpoonup a) x_{(0)} \text{ for all } x \in A\} = R$$

Now take $a \in Z(A)$. If we can show that the action of H on a is trivial, then we will have for all $x \in A$ that

$$xa = ax = \sum (x_{(1)} \rightharpoonup a) x_{(0)}$$

and this will imply that $a \in R$.
We will show that $\rho(d)(a) = \varepsilon(d)a$, for all $d \in D = H \otimes H^*$. Since $[A] \in \mathrm{Ker}(\beta)$, we have that

$$I_D = \beta_A = \beta_{\mathcal{A}}(I_D) = I_D * (\alpha_{\mathcal{A}}(I_D))^{-1} = I_D * \alpha_A^{-1}$$

(the inverses are the convolution inverses); it follows that

$$\alpha_A = \eta_D \circ \varepsilon_D = (\eta_H \circ \varepsilon_D) \otimes (\eta_{H^*} \circ \varepsilon_D)$$

From Lemma 13.9.1, it follows that

$$\sum\langle a_{(1)}, (\eta_{H^*} \circ \varepsilon_D)(d_{(1)})\rangle u(d_{(2)})a_{(0)} = \sum \rho(d_{(1)})(a)u(d_{(2)})$$

The left hand side is equal to $u(d)a$. Since $a \in Z(A)$, we therefore have

$$au(d) = u(d)a = \sum \rho(d_{(1)})(a)u(d_{(2)})$$

or

$$i(a) * u = R(I)(i(a)) * u$$

and it follows that

$$(i(a) - R(I)(i(a))) * I_{R(I)} = 0$$

Now $I_{R(I)}$ is an invertible, and therefore a faithfully flat D^*-module. This implies that

$$i(a) = R(I)(i(a))$$

We now have

$$\varepsilon(d)a = i(a)d = R(I)(i(a))(d) = \sum \rho(d_{(1)})(i(a)(d_{(2)})) = \rho(d)(a)$$

for all $d \in D$. This finishes the proof of the Lemma. □

Lemma 13.9.12 *If $[A] \in \mathrm{Ker}(\beta)$, and $u \in I_{R(I)}$, then*

$$\rho(e)u(d) = \varepsilon(e)u(d)$$

for all $d, e \in D$.

Proof First observe that the property holds if $d \in G(D)$. This follows from (13.67) and the fact that D acts trivially on $I_{\rho(d)}$. Thus we have that

$$\rho(e)(u(d)) = \varepsilon(e)u(d) \tag{13.74}$$

For all $e \in D$ and $d \in G(D)$. Of course (13.74) still holds if we apply a base extension.
It suffices to show that the Lemma holds in the case where R is local; the general case then follows from the usual local-global argument. If R is local, then D^* is semilocal, $\mathrm{Pic}(D^*) = 1$, and we may assume that $I_{R(I)}$ is free of rank one. Since $\alpha_A = \eta_D \circ \varepsilon_D$, we also have that $I_{R(I)}$ represents the trivial element of $\mathrm{PD}(D^*, \mathcal{D})$, and the action of D on A is inner; suppose that it is induced by the map $u: D \to A$, with convolution inverse v. Thus

$$\rho(d)(a) = \sum u(d_{(1)})av(d_{(2)})$$

for all $d \in D$ and $a \in A$. Now we apply a base extension; our new groundring will be $(D \otimes D)^*$. As before, we write $(D \otimes D)^* \otimes D \cong \mathrm{Hom}_R(D \otimes D, D) = \mathcal{D}_2$, $(D \otimes D)^* \otimes A \cong \mathrm{Hom}_R(D \otimes D, A) = \mathcal{A}_2$ etc. The action of $\mathcal{D}_2$ on $\mathcal{A}_2$ is now induced by the map

$$U = I_{(D\otimes D)^*} \otimes u: \ \mathcal{D}_2 \longrightarrow \mathcal{A}_2$$

Now take $\delta = \alpha(d^* \otimes d) \in \mathcal{D}_2$, with $d^* \in (D \otimes D)^*$ and $d \in D$. For all $e \in D^{\otimes 2}$, we now have

$$\begin{aligned} U(\delta)(e) &= \alpha(d^* \otimes u(d))(e) = \langle d^*, e\rangle u(d) \\ &= u(\langle d^*, e\rangle d) = u(\alpha(d^* \otimes d)(e)) \\ &= (u \circ \delta)(e) \end{aligned}$$

and we have shown that $U(\delta) = u \circ \delta$, for all $\delta \in \mathcal{D}_2$. Now take $\gamma, \delta \in G(\mathcal{D}_2)$, and apply (13.74). We obtain that

$$R(\gamma)(U(\delta)) = R(\gamma)(u \circ \delta) = U(\delta) = u \circ \delta$$

Now take $\gamma = I_D \otimes \varepsilon_D$ and $\delta = \varepsilon_D \otimes I_D$. γ and δ are clearly grouplike. For all $d, e \in D$, we therefore have that

$$R(\gamma)(u \circ \delta)(d \otimes e) = \sum \rho(\gamma(d_{(1)} \otimes e_{(1)})u(\delta(d_{(2)} \otimes e_{(2)})) = \rho(d)u(e)$$

equals

$$(u \circ \delta)(d \otimes e) = u(\varepsilon(d)e) = \varepsilon(d)u(e)$$

and this proves the Lemma. □

Corollary 13.9.13 *If* $[A] \in \mathrm{Ker}\,(\beta)$, *and* $u \in I_{R(I)}$, *then*

$$u(d)u(e) = u(e)u(d)$$

for all $d, e \in D$.

Proof Replace a by $u(k)$ in (13.65). Since $\mu^* = \varepsilon$, we have that

$$\sum \langle u(e)_{(1)}, \varepsilon(d_{(1)})\rangle u(d_{(2)})u(e)_{(0)} = \sum \rho(d_{(1)})(u(e))u(d_{(2)})$$

or

$$u(d)u(e) = u(e)u(d)$$

□

Recall that $\mathrm{BM}(R, H)$ is the subgroup of $\mathrm{BD}(R, H)$ consisting of classes of H-Azumaya algebras $[A]$ such that the coaction of H on A is trivial, that is, $\rho(a) = a \otimes 1$, for all $a \in A$. Now consider the restriction

$$\beta_{|\mathrm{BM}} : \mathrm{BM}(R, H) \longrightarrow \mathrm{Aut}_{\mathrm{Hopf}}(H \otimes H^*)$$

Lemma 13.9.14 $\mathrm{Ker}\,(\beta_{|\mathrm{BM}}) = \mathrm{BM}^s(R, H)$.

Proof Let $[A] \in \mathrm{Ker}\,(\beta_{|\mathrm{BM}})$. Then we know that there exists a faithfully flat R-algebra S such that $S \otimes H$ acts innerly on $S \otimes A$. If $[S \otimes A] \in \mathrm{BM}^s(S, S \otimes H)$, then

$[A] \in \mathrm{BM}^s(R,H)$. Therefore we may restrict attention to the situation where the action of H on A is inner. Now consider the map

$$\pi: \ \mathrm{BM}(R,H) \longrightarrow \mathrm{Gal}(R,H)$$

discussed in Section 13.2. In Proposition 13.2.5, we have seen that $\pi(A)$ may be described as follows: if the action of H on A is induced by the map $u: \ H \to A$ with convolution inverse v, then

$$\pi([A]) = \{\sum v(h_{(1)})\#h_{(2)} | h \in H\} \subset A\#H$$

We will show that $\pi([A])$ lies in the split part of the group of Galois objects. To this end, it suffices to show that $\pi([A])$ is commutative. Take $h,k \in H$. Using Lemma 13.9.12 and Corollary 13.9.13, we obtain that

$$\begin{aligned}\Big(\sum v(h_{(1)})\#h_{(2)}\Big)\Big(\sum v(k_{(1)})\#k_{(2)}\Big) &= \sum v(h_{(1)})(h_{(2)} \rightharpoonup v(k_{(1)})\#h_{(3)}k_{(2)} \\ &= \sum v(h_{(1)})v(k_{(1)})\#h_{(2)}k_{(2)} \\ &= \sum v(k_{(1)})v(h_{(1)})\#h_{(2)}k_{(2)} \\ &= \Big(\sum v(k_{(1)})\#k_{(2)}\Big)\Big(\sum v(h_{(1)})\#h_{(2)}\Big)\end{aligned}$$

and this proves the Lemma. □

Lemma 13.9.15 *If $[A] \in \mathrm{Ker}(\beta)$, then A is an Azumaya algebra.*

Proof It suffices to show that $A \otimes S$ is Azumaya, for some faithfully flat R-algebra S. Therefore, replacing R by S, we may assume that the action of H on A is inner, as in the previous proof. Let this action be induced by $u: \ H \to A$, with convolution inverse v. Now define $f: \ H \otimes H \to R$ by

$$f(h \otimes k) = \sum v(h_{(2)})v(k_{(2)})u(h_{(1)}k_{(1)})$$

From Corollary 13.9.13, it follows that f is symmetric. A straightforward computation shows that f is a Sweedler 2-cocycle. On A, we define a new multiplication in the following way:

$$m(a \otimes b) = \sum f(a_{(1)} \otimes b_{(1)})b_{(2)}a_{(2)}$$

We call A equipped with this new multiplication $A^{f\text{-op}}$. Now consider the map

$$j: \ A\#\overline{A} \longrightarrow A \otimes A^{f\text{-op}}: \ a\#b \mapsto \sum au(b_{(1)}) \otimes b_{(0)}$$

j is an R-algebra homomorphism. To see this, observe that

$$\begin{aligned} j((a\#b)(c\#d)) &= \sum j(a(b_{(1)} \rightharpoonup c)\#b_{(0)}d) \\ &= \sum j(au(b_{(1)})cv(b_{(2)})\#b_{(0)}d) \\ &= \sum au(b_{(1)})cv(b_{(2)})u(b_{(3)}d_{(1)}) \otimes b_{(0)}d_{(0)} \\ &= \sum au(b_{(1)})cu(d_{(1)})f(b_{(2)} \otimes d_{(2)}) \otimes b_{(0)}d_{(0)} \end{aligned}$$

and

$$\begin{aligned} j(a\#b)j(c\#d) &= \sum (au(b_{(1)}) \otimes b_{(0)})(cu(d_{(1)}) \otimes d_{(0)}) \\ &= \sum au(b_{(1)})cu(d_{(1)}) \otimes f(b_{(2)} \otimes d_{(2)})b_{(0)}d_{(0)} \end{aligned}$$

j is invertible, its inverse is given by the formula

$$j^{-1}(a \otimes b) = \sum av(b_{(1)})\#b_{(0)}$$

Now f is a symmetric cocycle, so there exists a faithfully flat R-algebra S such that $f \otimes 1_S$ is a coboundary. Replacing R by S, we obtain that $A^{f\text{-op}} \cong A^{\text{op}}$. But then we have that

$$A \otimes A^{\text{op}} \cong A \otimes A^{f\text{-op}} \cong A\#\overline{A} \cong \text{End}_R(A)$$

and this proves that A is an Azumaya algebra. □

We can now finish the proof of Theorem 13.9.10.

Proof of Theorem 13.9.10 Take $[A] \in \text{Ker}(\beta)$. By Lemma 13.9.15, $[A]$ is an Azumaya algebra. We have a D-action on A, so A is a D-module Azumaya algebra. From Lemma 13.9.14, it follows that $[A] \in \text{BM}^s(R, D)$. As sets, $\text{BM}^s(R, D)$ and $\text{BD}^s(R, H)$ are equal, the only difference is that the operation on $\text{BM}^s(R, D)$ is induced by the tensor product, while the operation on $\text{BD}^s(R, H)$ is induced by the smash product. We can conclude that $[A] \in \text{BD}^s(R, D)$. □

We summarize our results in the following Theorem.

Theorem 13.9.16 *([29]) Let H be a commutative, cocommutative faithfully projective Hopf algebra over a commutative ring R. Then we have a short exact sequence*

$$1 \longrightarrow \text{BD}^s(R, H) \longrightarrow \text{BD}(R, H) \xrightarrow{\beta} \text{O}(R, H)_{\text{min}} \tag{13.75}$$

13.10 Application to some particular cases

The big question that remains is the following: when is the map β surjective? This remains an open question, similar to the questions whether the Picard invariant map (see Section 10.7) and the map $\gamma:\ \text{Gal}(R, H) \to P_{\text{sk}}(H, R)$ (see Section 10.8) are surjective. There is an explicit relation between the maps β en γ, and this is what we will be discussing next.
To make our formalism more transparent, we first introduce the following notations. Let

$$\begin{aligned} f_{11}:\ H \longrightarrow H \qquad & f_{12}:\ H^* \longrightarrow H \\ f_{21}:\ H \longrightarrow H^* \qquad & f_{22}:\ H^* \longrightarrow H^* \end{aligned}$$

be Hopf algebra maps, and consider the Hopf algebra map $f:\ H \otimes H^* \to H \otimes H^*$ given by the formula

$$f(h \otimes h^*) = \sum f_{11}(h_{(1)})f_{12}(h^*_{(1)}) \otimes f_{21}(h_{(2)})f_{22}(h^*_{(2)})$$

We will represent f by the matrix

$$f = \begin{pmatrix} f_{11} & f_{12} \\ f_{21} & f_{22} \end{pmatrix}$$

Our notation is consistent with matrix multiplication:

$$\begin{pmatrix} f_{11} & f_{12} \\ f_{21} & f_{22} \end{pmatrix} \circ \begin{pmatrix} g_{11} & g_{12} \\ g_{21} & g_{22} \end{pmatrix} = \begin{pmatrix} f_{11} \circ g_{11} * f_{12} \circ g_{21} & f_{11} \circ g_{12} * f_{12} \circ g_{22} \\ f_{21} \circ g_{11} * f_{22} \circ g_{21} & f_{21} \circ g_{12} * f_{22} \circ g_{22} \end{pmatrix}$$

Moreover, every Hopf algebra map $f : H \otimes H^* \to H \otimes H^*$ has such a matrix representation: define f_{ij} by the formulas

$$\begin{aligned} f_{11}(h) &= ((I_H * \varepsilon_{H^*}) \circ f)(h \otimes \varepsilon) \\ f_{12}(h^*) &= ((I_H * \varepsilon_{H^*}) \circ f)(1 \otimes h^*) \\ f_{21}(h) &= ((\varepsilon_H * I_{H^*}) \circ f)(h \otimes \varepsilon) \\ f_{22}(h^*) &= ((\varepsilon_H * I_{H^*}) \circ f)(1 \otimes h^*) \end{aligned}$$

Then

$$\begin{aligned} \sum f_{11}(h_{(1)}) f_{12}(h^*_{(1)}) \otimes f_{21}(h_{(2)}) f_{22}(h^*_{(2)}) &= \\ \sum ((I_H * \varepsilon_{H^*}) \circ f)(h_{(1)} \otimes h^*_{(1)}) \otimes ((\varepsilon_H * I_{H^*}) \circ f)(h_{(2)} \otimes h^*_{(2)}) &= \\ \Big(((I_H * \varepsilon_{H^*}) \otimes (\varepsilon_H * I_{H^*})) \circ \Delta_{H \otimes H^*}\Big) f(h \otimes h^*) &= f(h \otimes h^*) \end{aligned}$$

The ontoness of β versus noncommutative Galois objects

Recall from Theorem 13.2.1 that we have a split exact sequence

$$1 \longrightarrow \mathrm{Br}(R) \longrightarrow \mathrm{BM}(R,H) \xrightarrow{\pi} \mathrm{Gal}(R,H) \longrightarrow 1$$

In Section 10.7, we discussed another exact sequence

$$1 \longrightarrow \mathrm{Gal}^s(R,H) \longrightarrow \mathrm{Gal}(R,H) \xrightarrow{\gamma} P_{\mathrm{sk}}(H,R)$$

Here $P_{\mathrm{sk}}(H,R) = \{g \in \mathrm{Hopf}(H,H^*) | \sum \langle h_{(1)}, g(h_{(2)}) \rangle = \varepsilon(h) \text{ for all } h \in H\}$. We now claim that we have the following diagram with exact rows:

$$\begin{array}{ccccccc} 1 & \longrightarrow & \mathrm{Gal}^s(R,H) & \longrightarrow & \mathrm{Gal}(R,H) & \xrightarrow{\gamma} & P_{\mathrm{sk}}(H,R) \\ & & \downarrow \subset & & \downarrow \subset & & \downarrow \cong \\ 1 & \longrightarrow & \mathrm{BM}^s(R,H) & \longrightarrow & \mathrm{BM}(R,H) & \longrightarrow & P_{\mathrm{sk}}(H,R) \\ & & \downarrow \subset & & \downarrow \subset & & \downarrow \subset \\ 1 & \longrightarrow & \mathrm{BD}^s(R,H) & \longrightarrow & \mathrm{BD}(R,H) & \longrightarrow & \mathrm{O}(R,H)_{\mathrm{min}} \end{array} \tag{13.76}$$

The embedding $\alpha : P_{\mathrm{sk}}(H,R) \to \mathrm{O}(R,H)_{\mathrm{min}}$ is defined as follows: for $g \in P_{\mathrm{sk}}(H,R)$, $\alpha(g) = f$ is defined by

$$f(h \otimes h^*) = \sum h_{(1)} \otimes (g(h_{(2)}) * h^*)$$

or, in matrix form,

$$f = \begin{pmatrix} I_H & \eta_H \circ \varepsilon_{H^*} \\ g & I_{H^*} \end{pmatrix} \tag{13.77}$$

The inverse of f is given by the formula

$$f^{-1}(h \otimes h^*) = \sum h_{(1)} \otimes (g(S(h_{(2)})) * h^*)$$

For all $h \otimes h^* \in D$, we have that

$$\begin{aligned} (q \circ f)(h \otimes h^*) &= q\Big(\sum h_{(1)} \otimes (g(h_{(2)}) * h^*)\Big) \\ &= \sum \langle h_{(1)}, g(h_{(3)}) \rangle \langle h_{(2)}, h^* \rangle \\ &= \langle h^*, h \rangle = q(h \otimes h^*) \end{aligned}$$

and it follows that $f \in \mathrm{O}(R,H)_{\min}$. We leave it to the reader to show that the above diagram is commutative.

Application to the group ring case

Let G be a finite abelian group, and suppose that $H = RG$. If $|G|$ is invertible in R and R contains an $\exp(G)$-th primitive root of unity, then $(RG)^* = RG^* \cong RG$ is also a group ring (see Theorem 7.1.10). In this case, it turns out that the map β is surjective.

Theorem 13.10.1 *([37, Theorem 3.4]) Suppose that G is a finite abelian group, and that R contains $|G|^{-1}$ and an $\exp(G)$-th primitive root of unity η. Then we have a short exact sequence*

$$1 \longrightarrow \mathrm{BD}^s(R,RG) \longrightarrow \mathrm{BD}(R,RG) \xrightarrow{\beta} \mathrm{O}(R,RG)_{\min} \longrightarrow 1 \tag{13.78}$$

Proof The only thing that we have to show is that β is surjective. First we reduce to the case where the base ring R is connected. Take $f \in \mathrm{O}(R,RG)_{\min}$. Then f is an automorphism of $R(G \times G^*)$. Write

$$G \times G^* = \{j_1, \ldots, j_n\}$$

and let $\{u_{j_1}, \ldots, u_{j_n}\}$ be a basis of $R(G \times G^*)$ as a free R-module. Now $f(u_{j_1})$ is a grouplike element of $R(G \times G^*)$, and this implies that we can find orthogonal idempotents $e_1, \ldots, e_n$ with sum 1 such that

$$f(u_{j_1}) = \sum_{i=1}^{n} e_i u_{j_i}$$

Take i such that $e = e_i \neq 0$. Then $f(eu_{j_1}) = eu_{j_i}$. Now

$$f(eu_{j_2}) = \sum_{k=1}^{n} f_k u_{j_k}$$

where now $f_1, \ldots, f_n$ are orthogonal idempotents with sum e. One of the e_k's is different from 0. Replacing e by this f_k, we obtain that $f(eu_{j_1}) = eu_{j_i}$ and $f(eu_{j_2}) = eu_{j_k}$. Continuing in this way, we obtain a nonzero idempotent e such that for all $j \in G \times G^*$, there exists a $k \in G \times G^*$ such that

$$f(eu_j) = eu_k \tag{13.79}$$

Applying the same argument to $R(1-e)$, and proceeding by induction, we obtain that we may write $R = \oplus_{i=1}^m Re_i$ such that (13.79) holds for $e = e_1, \ldots, e_m$. It now suffices to show that, for all $i = 1, \ldots, m$, there exists an $H \otimes Re_i$-Azumaya algebra A_i such that $\beta_{A_i} = f_{|Re_iG}$. Hence, we may assume that, for all $j \in G \times G^*$, there exists a $k \in G \times G^*$ such that $f(u_j) = u_k$. If we now write $f(j) = k$, then it turns out that we can consider f as an automorphism of $G \times G^*$.
By definition, we have that

$$f^* \circ t \circ f = t \tag{13.80}$$

and

$$\langle \varphi(j), j \rangle = \langle \varphi(f(j)), f(j) \rangle \tag{13.81}$$

for all $j \in G \otimes G^*$. Recall that, on $G \otimes G^*$, φ is defined as follows: $\varphi(g, g^*) = (g^*, 1)$. Now define $a:\ G^* \times G \to G \times G^*$ as follows:

$$a = (I * S \circ f) \circ t^{-1} = t^{-1} \circ (I * S \circ (f^*)^{-1}) \tag{13.82}$$

where S and $*$ are the restrictions of the antipode and the convolution. Then we also have that

$$a^* = (I * S \circ f^{-1}) \circ t^{-1} = t^{-1} \circ (I * S \circ f^*) \tag{13.83}$$

hence

$$f = I * S \circ a \circ t \qquad f^* = I * S \circ t \circ a^* \tag{13.84}$$

$$f^{-1} = I * S \circ a^* \circ t \qquad (f^*)^{-1} = I * S \circ t \circ a \tag{13.85}$$

(13.80) and (13.85) imply that

$$\begin{aligned} t^{-1} &= f^{-1} \circ t^{-1} \circ (f^*)^{-1} \\ &= (I * S \circ a^* \circ t) \circ t^{-1} \circ (I * S \circ t \circ a) \\ &= t^{-1} * S \circ a^* * S \circ a * a^* \circ t \circ a \end{aligned}$$

and

$$a * a^* * S \circ a^* \circ t \circ a = \varepsilon$$

Now we consider the map

$$d = a^* \circ \varphi \circ a * S \circ a :\ G^* \times G \to G \times G^*$$

Then $d * d^* = \varepsilon$; moreover, we have that $d \in P_{\mathrm{sk}}(R(G^* \times G), R)$. Indeed, we have

$$\begin{aligned} \varphi * S \circ f^* \circ \varphi \circ f &= \varphi * S \circ (I * S \circ t \circ a^*) \circ \varphi \circ (I * S \circ a \circ t) \\ &= t \circ a^* \circ \varphi * \varphi \circ a \circ t * S \circ t \circ a^* \circ \varphi \circ a \circ t \end{aligned}$$

and, using (13.81), we obtain

$$\begin{aligned} 1 &= \langle j, (\varphi * S \circ f^* \circ \varphi \circ f)(j)\rangle \\ &= \langle j, (t \circ a^* \circ \varphi)(j)\rangle \langle j, (\varphi \circ a \circ t * S \circ t \circ a^* \circ \varphi \circ a \circ t)(j)\rangle \\ &= \langle (\varphi^* \circ a)(t(j)), j\rangle \langle j, (\varphi \circ a * S \circ t \circ a^* \circ \varphi \circ a)(t(j))\rangle \\ &= \langle (\varphi^* \circ a * \varphi \circ a * t \circ S \circ a^* \circ \varphi \circ a)(t(j)), j\rangle \\ &= \langle t((S \circ d)(t(j))), j\rangle = \langle d(t(j)), t(j)\rangle^{-1} \end{aligned}$$

for all $j \in G \times G^*$. Let us next show that

$$\operatorname{Ker}(a) = \operatorname{Ker}(a^*) \tag{13.86}$$

To prove (13.86), it suffices to observe that the following statements are equivalent for all $k \in G \times G^*$.

$$\begin{array}{ll} a(k) = 1 & (t^{-1} \circ (I * S \circ (f^*)^{-1}))(k) = 1 \\ k = (f^*)^{-1}(k) & k = f^*(k) \\ (t^{-1} \circ (I * S \circ f^*))(k) = 1 & a^*(k) = 1 \end{array}$$

It now follows that the map d is trivial on $\operatorname{Ker}(a)$, and therefore d defines an element $d \in P_{\mathrm{sk}}(R((G^* \times G)/\operatorname{Ker}(a)), R)$. Applying Theorem 10.8.7, we obtain a cocycle

$$\gamma \in Z^2((G^* \times G)/\operatorname{Ker}(a), \mathbb{G}_m(R))$$

such that $d(j,k) = \gamma(j,k)\gamma(k,j)^{-1}$. We define A to be the following crossed product: $A = R((G^* \times G)/\operatorname{Ker}(a))$, with multiplication rule

$$u_j u_k = \gamma(j,k) u_{jk}$$

On A, we consider the following $G \times G^*$-grading:

$$\deg_{G \times G^*}(u_j) = a(j)$$

This grading is well-defined and makes A into an $R(G \times G^*)$-comodule algebra. The G^*-grading induces a G-action, making A into an RG-dimodule algebra. The G-action and G-grading are then the following:

$$\deg_G(u_j) = p_1(a(j)) \qquad \sigma \rightharpoonup u_j = \langle p_2(a(j)), \sigma\rangle u_j$$

where p_1 and p_2 are the natural projections of $G \times G^*$ on G and G^*.
Let us show that A is an RG-Azumaya algebra. A is separable as an R-algebra, because the order of the group $(G^* \times G)/\operatorname{Ker}(a)$ is invertible in R. In view of Corollary 12.3.7, it now suffices to show that A is RG-central. We will prove that $A^A = R$, and leave it to the reader to show that ${}^A A = R$. Take $c = \sum_j r_j u_j \in A^A$. Then for all $a \in A$, we have that

$$ac = \sum (a_{(1)} \rightharpoonup c) a_{(0)}$$

and for all $k \in (G^* \times G)/\mathrm{Ker}\,(a)$, we have that

$$\begin{aligned}
u_k(\sum_j r_j u_j) &= \sum_j r_j u_k u_j \\
&= \sum_j r_j(((p_1 \circ a)(k)) \rightharpoonup u_j)u_k \\
&= \sum_j r_j \langle (p_2 \circ a)(j), (p_1 \circ a)(k)\rangle u_j u_k \\
&= \sum_j r_j \langle a(j), (\varphi^* \circ a)(k)\rangle \langle j, d(k)\rangle u_k u_j \\
&= \sum_j r_j \langle j, (a^* \circ \varphi^* \circ a * d)(k)\rangle u_k u_j \\
&= \sum_j r_j \langle j, a(k)\rangle u_k u_j
\end{aligned}$$

In the last step, we used the fact that $a^* \circ \varphi^* \circ a * d = a$; this follows from

$$d * d^* = \varepsilon \text{ and } d^* = a^* \circ \varphi^* \circ a * S \circ a$$

Take homogeneous parts of both sides. We obtain, for all $k \in (G^* \times G)/\mathrm{Ker}\,(a)$ that

$$r_j u_k u_j = r_j \langle j, a(k)\rangle u_k u_j$$

or

$$r_j(1 - \langle j, a(k)\rangle) = 0$$

Now $\langle j, a(k)\rangle$ is a root of unity, and it follows from the proof of Theorem 7.1.10 that $1 - \langle j, a(k)\rangle$ is either 0, or invertible in R. Hence $r_j = 0$, or $\langle j, a(k)\rangle = 1$ for all $k \in (G^* \times G)/\mathrm{Ker}\,(a)$. So if $r_j \neq 0$, then $j \in \mathrm{Ker}\,(a^*) = \mathrm{Ker}\,(a)$, and $u_j = 1$, and $r_j u_j = r_j \in R$. It follows that $c \in R$.
Finally, let us show that $\beta_A = f$. This is equivalent to showing that, for all $j = (h, h^*) \in (G^* \times G)/\mathrm{Ker}\,(a)$:

$$(I * S \circ \alpha_A)(t(j)) = (I * a \circ t)(t(j))$$

or

$$\alpha_A(t(j)) = a(j)$$

We claim that

$$I_{\rho(h,h^*)} = Ru_j$$

Recall that $a \in I_{\rho(h,h^*)}$ if and only if

$$\sum (b_{(1)} \rightharpoonup a)b_{(0)} = \rho(h, h^*)(n)a \qquad (13.87)$$

for all $b \in A$. Take $k = (g^*, g) \in G^* \times G$. Then

$$\rho_A(u_k) = u_k \otimes (p_1 \circ a)(k) \text{ and } \sigma \rightharpoonup u_k = \langle (p_2 \circ a)(k), \sigma\rangle u_k$$

for all $\sigma \in G$. It suffices to show that (13.87) holds for $b = u_k$. We have

$$\begin{aligned}
\rho(h,h^*)(u_k)u_j &= \langle j, a(k)\rangle u_k u_j \\
&= \langle j, a(k)\rangle\langle k, d(j)\rangle u_j u_k \\
&= \langle k, (a^* * d)(j)\rangle u_j u_k \\
&= \langle k, (a^* \circ \varphi \circ a)(j)\rangle u_j u_k \\
&= \langle a(k), \varphi(a(j))\rangle u_j u_k \\
&= ((p_1 \circ a)(k) \rightharpoonup u_j) u_k \\
&= \sum (b_{(1)} \rightharpoonup u_j) b_{(0)}
\end{aligned}$$

Now $\deg_{G\times G^*}(u_j) = a(j)$, hence $\alpha_A(t(j)) = a$, and this finishes the proof of the Theorem. □

Proposition 13.10.2 *With notations as in Theorem 13.10.1, we have*

$$\begin{aligned}
\mathrm{Ker}\,(d) &= \mathrm{Ker}\,(I * S \circ \varphi \circ a) \oplus \mathrm{Ker}\,(a) \\
\mathrm{Ker}\,(d^*) &= \mathrm{Ker}\,(I * S \circ \varphi^* \circ a) \oplus \mathrm{Ker}\,(a)
\end{aligned}$$

Proof It is clear that $\mathrm{Ker}\,(I * S \circ \varphi \circ a) \cap \mathrm{Ker}\,(a) = \{e\}$. We also have that

$$\begin{aligned}
\mathrm{Ker}\,(a) &= \mathrm{Ker}\,(a^*) \subset \mathrm{Ker}\,(a^* \circ \varphi \circ a * S \circ a) = \mathrm{Ker}\,(d) \\
&= \mathrm{Ker}\,(I * S \circ \varphi \circ a) \subset \mathrm{Ker}\,(a^* \circ (I * S \circ \varphi \circ a)) = \mathrm{Ker}\,(d)
\end{aligned}$$

Take $j \in \mathrm{Ker}\,(d)$. Then $a^*((\varphi \circ a * S)(j)) = e$, hence $(\varphi \circ a * S)(j) \in \mathrm{Ker}\,(a^*) = \mathrm{Ker}\,(a)$, so $a((\varphi \circ a * S)(j)) = e$, and $(\varphi \circ a)((\varphi \circ a * S)(j)) = (\varphi \circ a * S)((\varphi \circ a)(j))e$. It follows that

$$(\varphi \circ a)(j) \in \mathrm{Ker}\,(\varphi \circ a * S) = \mathrm{Ker}\,(S \circ \varphi \circ a * I)$$

Finally

$$j = (\varphi \circ a)(j)(I * S \circ \varphi \circ a)(j) \in \mathrm{Ker}\,(I * S \circ \varphi \circ a) \oplus \mathrm{Ker}\,(a)$$

The second assertion can be proved in a similar way □

We will now give a description of the center of the H-Azumaya A constructed in the proof of Theorem 13.10.1. From Proposition 13.10.2, it follows that

$$Z(A) = R(\mathrm{Ker}\,(I * S \circ \varphi \circ a)) = R(\mathrm{Ker}\,(I * S \circ \varphi^* \circ a)) \tag{13.88}$$

Using the matrix notation

$$\alpha_A = \begin{pmatrix} a_{11} & a_{12} \\ a_{21} & a_{22} \end{pmatrix} \text{ and } \beta_A = \begin{pmatrix} b_{11} & b_{12} \\ b_{21} & b_{22} \end{pmatrix} \tag{13.89}$$

we easily compute that

$$I * S \circ \varphi \circ a = \begin{pmatrix} b_{22} & b_{21} \\ \eta_{H^*} \circ \varepsilon_H & I_H \end{pmatrix}$$

$$I * S \circ \varphi^* \circ a = \begin{pmatrix} I_{H^*} & \eta_H \circ \varepsilon_{H^*} \\ b_{12} & b_{11} \end{pmatrix}$$

The groups

$$\begin{aligned} K_1 &= \{(\sigma^*, 1) | b_{22}(\sigma^*) = \varepsilon_H\} \\ K_2 &= \{(\varepsilon, \sigma) | b_{11}(\sigma) = 1\} \end{aligned} \tag{13.90}$$

can therefore be used as bases for $Z(A)$.
If A and B are two H-Azumaya algebras such that $\beta_A = \beta_B$, then $Z(A) \cong Z(B)$. This can be seen as follows: take a faithfully flat extension S of R such that $A \otimes S$ and $B \otimes S$ represent the same element of $\text{BD}(S, S \otimes H)$. Then $Z(A \otimes S) \cong Z(B \otimes S)$ and a descent argument shows that $Z(A) \cong Z(B)$. We have therefore shown the following result.

Proposition 13.10.3 *Let R and G be as in Theorem 13.10.1. Let A be an H-Azumaya algebra, and use the matrix notation (13.89). The following are equivalent:*
1) A is R-central;
2) $b_{11} : G \longrightarrow G$ is an isomorphism;
3) $b_{22} : G^* \longrightarrow G^*$ is an isomorphism.

Now suppose that G is the direct sum of cyclic groups of prime order, and let L be a complement of K_2 in $\text{Ker}\,(I * S \circ \varphi^* \circ a)$. Then

$$A \cong RL \otimes RK_2 = RH \otimes Z(A)$$

Furthermore, RL is R-central, and separable, since the order of L is a unit in R. Thus the rank of RL is a square, and it follows that $Z(A)$ is of square rank if and only if A is of square rank.
Suppose that A and B are two H-Azumaya algebras such that $\beta_A = \beta_B$. Then the rank of A is a square if and only if the rank of B is a square. We therefore have the following result.

Proposition 13.10.4 *Let R and G be as in Theorem 13.10.1, and suppose that G is the direct sum of cyclic groups of prime order. For any H-Azumaya algebra A, A is of square rank if and only if $Z(A)$ is of square rank.*

Suppose now that the rank of the center of A is a big as possible. This happens if $b_{11} = \eta_H \circ \varepsilon_H$. We then also have that $b_{22} = \eta_{H^*} \circ \varepsilon_{H^*}$, since the groups K_1 and K_2 described above contain the same number of elements. The orthogonality condition then implies that

$$b_{12} = (b_{21}^*)^{-1}$$

and

$$\beta_A = \begin{pmatrix} \eta_H \circ \varepsilon_H & (b_{21}^*)^{-1} \\ b_{21} & \eta_{H^*} \circ \varepsilon_{H^*} \end{pmatrix}$$

and an easy computation shows that

$$a = \alpha_A \circ t = \begin{pmatrix} S \circ (b_{21}^*)^{-1} & I_H \\ I_{H^*} & S \circ b_{21} \end{pmatrix}$$

and

$$(\sigma^*, \sigma) \in \operatorname{Ker} a \iff \sigma^* = b_{21}^*(\sigma) = b_{21}(\sigma)$$

We can take $K_2 = \{u_{(\varepsilon,\sigma)} | \sigma \in G\}$ as a basis for $Z(A)$. The action and coaction of G on $Z(A)$ are the following:

$$\begin{aligned} \deg_G(u_{(\varepsilon,\sigma)}) &= \sigma \\ \tau \rightharpoonup u_{(\varepsilon,\sigma)} &= \langle S(b_{21}(\sigma)), \tau \rangle u_{(\varepsilon,\sigma)} \end{aligned}$$

and it follows that $Z(A)$ is an $S \circ b_{21}$-algebra. By Theorem 12.4.5, $Z(A)$ is an $S \circ b_{21}$-Azumaya algebra. $\beta_{Z(A)}$ will be computed in Theorem 13.10.12. We can also compute easily that

$$d = \begin{pmatrix} b_{12} * (S \circ b_{12}^*) & \eta_H \circ \varepsilon_{H^*} \\ \eta_{H^*} \circ \varepsilon_H & \eta_{H^*} \circ \varepsilon_{H^*} \end{pmatrix}$$

If $b_{12} = b_{12}^*$ is selfdual, then d is trivial, meaning that $A = Z(A)$ is commutative. In this case

$$\beta_A = \beta_{Z(A)} = \begin{pmatrix} \eta_H \circ \varepsilon_H & (b_{21}^*)^{-1} \\ b_{21} & \eta_{H^*} \circ \varepsilon_{H^*} \end{pmatrix}$$

and this is consistent with Theorem 13.10.12.

Deegan's subgroup of the Brauer-Long group

We have a natural embedding $i : \operatorname{Aut}_{\mathrm{Hopf}}(H) \to \mathrm{O}(R, H)_{\mathrm{min}}$. i is given by the formula

$$i(f) = f \otimes (f^{-1})^* = \begin{pmatrix} f & \eta_H \circ \varepsilon_{H^*} \\ \eta_{H^*} \circ \varepsilon_H & (f^*)^{-1} \end{pmatrix}$$

or

$$i(f)(h, h^*) = f(h) \otimes ((f^{-1})^*)(h^*)$$

Let us show that $i(f) \in \mathrm{O}(R, H)_{\mathrm{min}}$. For all $h \in H$ and $h^* \in H^*$, we have that

$$\begin{aligned} (q \circ i(f))(h \otimes h^*) &= \langle f(h), (f^{-1})^*(h^*) \rangle \\ &= \langle f^{-1}(f(h)), h^* \rangle \\ &= \langle h, h^* \rangle = q(h \otimes h^*) \end{aligned}$$

We will show that $i(\operatorname{Aut}_{\mathrm{Hopf}}(H)) \subset \operatorname{Im}(\beta)$. Furthermore, $\operatorname{Aut}_{\mathrm{Hopf}}(H)$ can be embedded as a subgroup in $\mathrm{BD}(R, H)$. This subgroup will be denoted by $\mathrm{BT}(R, H)$. In the case where H is a group ring, this subgroup of the Brauer-Long group was first described by Deegan in [64], and this is why we will call it *Deegan's subgroup of the Brauer-Long group* .

Let $\mathrm{BT}(R, H)$ be the subset of $\mathrm{BD}(R, H)$ consisting of classes of algebras represented by an R-algebra of the form $\operatorname{End}_R(M)$, where M is at once a faithfully projective H-module and H-comodule (but not an H-dimodule), and where the action and coaction of H on A are the once induced by the action and coaction of H on M.

Theorem 13.10.5 *$\mathrm{BT}(R,H)$ is a subgroup of $\mathrm{BD}(R,H)$. The map $\beta : \mathrm{BD}(R,H) \to \mathrm{O}(R,H)_{\min}$ restricts to an isomorphism*

$$\beta_t : \mathrm{BT}(R,H) \longrightarrow i(\mathrm{Aut}_{\mathrm{Hopf}}(H))$$

Consequently $i(\mathrm{Aut}_{\mathrm{Hopf}}(H)) \subset \mathrm{Im}\,(\beta)$.

Theorem 13.10.5 will follow from Lemmas 13.10.6, 13.10.7, 13.10.8 and 13.10.9.

Lemma 13.10.6 *If $[A] \in \mathrm{BT}(R,H)$, then $\beta_A = f \otimes (f^{-1})^*$ for some Hopf automorphism of H.*

Proof Recall the definition of $\alpha_{\mathcal{A}}$:

$$\alpha_{\mathcal{A}} : G(\mathcal{D}) \xrightarrow{R} \mathcal{H}\text{-}\mathrm{Aut}_{D^*}(\mathcal{A}) \xrightarrow{\Phi} \mathrm{PD}(D^*, \mathcal{H}) \longrightarrow G(\mathcal{D})$$

We have to find $\alpha_{\mathcal{A}}(I)$. Observe that

$$I_D = (I_H \otimes \varepsilon_{H^*}) * (\varepsilon_H \otimes I_{H^*})$$

We will first compute $\alpha_{\mathcal{A}}(I_H \otimes \varepsilon_{H^*})$. Define $u \in \mathcal{A} = \mathrm{Hom}_R(D, A)$ by

$$u(h \otimes h^*)(m) = \langle h^*, 1\rangle (h \rightharpoonup m)$$

for all $h \in H$, $h^* \in H^*$ and $m \in M$. We then have for all $\mu \in \mathcal{H} = \mathrm{Hom}_R(J, H)$ that

$$\mu \rightharpoonup u = \varepsilon_{\mathcal{H}^*}(\mu) * u \tag{13.91}$$

Indeed, for all $d = k \otimes k^* \in D$ and $m \in M$, we have that

$$\begin{aligned}
&(\mu \rightharpoonup u)(k \otimes k^*)(m)\\
= &\sum \big(\mu(k_{(1)} \otimes k^*_{(1)}) \rightharpoonup u(k_{(2)} \otimes k^*_{(2)})\big)(m)\\
= &\sum \big(\mu(k_{(1)} \otimes k^*_{(1)})_{(1)} \rightharpoonup u(k_{(2)} \otimes k^*_{(2)})\big)\big(S(\mu(k_{(1)} \otimes k^*_{(1)})_{(2)}) \rightharpoonup m\big)\\
= &\sum \langle k^*_{(2)}, 1\rangle \mu(k_{(1)} \otimes k^*_{(1)})_{(1)} \rightharpoonup \big(k_{(2)} \rightharpoonup (S(\mu(k_{(1)} \otimes k^*_{(1)})_{(2)} \rightharpoonup m)\big)\\
= &\sum \langle k^*_{(2)}, 1\rangle \big(\mu(k_{(1)} \otimes k^*_{(1)})_{(1)} k_{(2)} S(\mu(k_{(1)} \otimes k^*_{(1)})_{(2)})\big) \rightharpoonup m\\
= &\sum \varepsilon(\mu(k_{(1)} \otimes k^*_{(1)})) \langle k^*_{(2)}, 1\rangle (k_{(2)} \rightharpoonup m)\\
= &(\varepsilon_{\mathcal{H}^*}(\mu) * u)(m)
\end{aligned}$$

We now claim that

$$u \in I_{R(I_H \otimes \varepsilon_{H^*})} \tag{13.92}$$

By Lemma 13.9.1, it suffices to show that

$$u(k \otimes k^*)a = \sum \rho(k_{(1)} \otimes \varepsilon(k^*_{(1)}))(a) u(k_{(2)} \otimes k^*_{(2)})$$

or

$$u(k \otimes k^*)a = \sum (k_{(1)} \rightharpoonup a) u(k_{(2)} \otimes k^*)$$

for all $k \otimes k^* \in D$ and $a \in A$. The left hand side evaluated at m gives

$$\langle k^*, 1\rangle (k \rightharpoonup a(m))$$

while the right hand side is

$$\sum k_{(1)} \rightharpoonup a\big(S(k_{(2)}) \rightharpoonup (\varepsilon(k^*) k_{(3)} \rightharpoonup m)\big)$$

and it is clear that both sides are equal.
From (13.92) and (13.92), it follows that

$$\alpha_{\mathcal{A}}(I_H \otimes \varepsilon_{H^*}) = \lambda \otimes \varepsilon_{\mathcal{H}^*}$$

for some $\lambda \in G(\mathcal{H})$. From a duality argument, it follows that

$$\alpha_{\mathcal{A}}(\varepsilon_H \otimes I_{H^*}) = \varepsilon_{\mathcal{H}} \otimes \lambda'$$

for some $\lambda' \in G(\mathcal{H}^*)$. Therefore

$$\alpha_A = \alpha_{\mathcal{A}}(I) = \alpha_{\mathcal{A}}(I_H \otimes \varepsilon_{H^*}) \alpha_{\mathcal{A}}(\varepsilon_H \otimes I_{H^*}) = \lambda \otimes \lambda'$$

and $\beta_A = f \otimes f'$. From the fact that β_A is a Hopf algebra automorphism, it follows that f and f' are Hopf algebra automorphisms. $\beta_A \in \mathrm{O}(R, H)_{\min}$ implies that $f' = (f^{-1})^*$. □

Lemma 13.10.7 BT(R, H) *is a multiplicative subset of* BD(R, H).

Proof Take $A = \mathrm{End}_R(M) \cong M^* \otimes M$ and $B = \mathrm{End}_R(N) \cong N^* \otimes N$, representing elements of BT(R, H). We have to show that $A\#B \in \mathrm{BT}(R, H)$. In the sequel, we will identify

$$A \otimes B = \mathrm{End}_R(M) \otimes \mathrm{End}_R(N) = \mathrm{End}_R(M \otimes N)$$

As we have seen in Lemma 13.10.6, the actions of H on A and B are inner, and are induced by

$$u: H \longrightarrow A \;;\; u(h)(m) = h \rightharpoonup m$$
$$v: H \longrightarrow B \;;\; u(h)(n) = h \rightharpoonup n$$

for all $h \in H$, $m \in M$ and $n \in N$. Remark that u and v are multiplicative maps, and that H acts trivially on u and v:

$$h \rightharpoonup u(k) = \varepsilon(h) u(k) \text{ and } h \rightharpoonup v(k) = \varepsilon(h) v(k)$$

From the proof of Proposition 12.1.6, it follows that the map

$$\Gamma: A\#B \rightarrow A \otimes B: a\#b \mapsto \sum a u(b_{(1)}) \otimes b_{(0)}$$

is an H-module algebra isomorphism; Γ is not H-colinear, because M is not an H-dimodule. The inverse of Γ is given by the formula

$$\Gamma^{-1}(a\#b) = a u(S(b_{(1)})) \otimes b_{(0)}$$

We define $\text{End}_R(M \otimes N)_{(1)}$ to be $\text{End}_R(M \otimes N)$, with the H-dimodule algebra structure induced by Γ; then the H-action on $\text{End}_R(M \otimes N)_{(1)}$ is induced by the H-action on $M \otimes N$, but the same is not true for the H-coaction. The map

$$\Gamma: \ A\#B \to \text{End}_R(M \otimes N)_{(1)}$$

is then an H-dimodule algebra isomorphism.

In a similar way, it follows from Proposition 12.1.8 that we have an H-comodule algebra isomorphism

$$\Psi: \ A\#B \to A \otimes B: \ a\#b \mapsto \sum (S(n_{(1)}) \rightharpoonup a) \otimes n_{(0)} \otimes n^*$$

for all $a \in A$ and $b = n \otimes n^* \in B = N \otimes N^*$. We define $\text{End}_R(M \otimes N)_{(2)}$ to be $\text{End}_R(M \otimes N)$, with the H-dimodule algebra structure induced by Ψ; now the H-coaction on $\text{End}_R(M \otimes N)_{(2)}$ is induced by the H-coaction on $M \otimes N$, but the same is not true for the H-action. The map

$$\Psi: \ A\#B \to \text{End}_R(M \otimes N)_{(2)}$$

is an H-dimodule algebra isomorphism. Now consider the composition

$$\text{End}_R(M \otimes N)_{(1)} \xrightarrow{\Gamma^{-1}} A\#B \xrightarrow{\Psi} \text{End}_R(M \otimes N)_{(2)}$$

We already know that the coaction of H on $\text{End}_R(M \otimes N)_{(2)}$ is induced by the coaction on $M \otimes N$. If we can show also the action is induced by an action on $M \otimes N$, then we are done. Now the action of H on $\text{End}_R(M \otimes N)_{(1)}$ is induced by the action of H on $M \otimes N$; therefore it is induced by the map

$$\Lambda = (u \otimes v) \circ \Delta: \ H \longrightarrow \text{End}_R(M \otimes N)_{(1)}$$

The action of H on $\text{End}_R(M \otimes N)_{(1)}$ is therefore induced by $\Omega = \Psi \circ \Gamma^{-1} \circ \Lambda$; using the fact that H acts trivially on u, we obtain that

$$\Omega(h) = \sum u(h_{(1)} S(v(h_{(2)})_{(1)})) \otimes v(h_{(2)})_{(0)}$$

Define an action $\triangleright$ of H on $M \otimes N$ by

$$h \triangleright m = \Omega(h)(m \otimes n)$$

From the fact that u and v are multiplicative maps, it follows that Λ, and therefore Ω are multiplicative. But this means that the action $\triangleright$ makes $M \otimes N$ into an H-module. Moreover, this action induces the action of H on $\text{End}_R(M \otimes N)_{(2)}$, and this finishes the proof of the Lemma. □

From Lemma 13.10.7, it follows that β restricts to a homomorphism

$$\beta_t: \ \text{BT}(R, H) \longrightarrow i(\text{Aut}_{\text{Hopf}}(H))$$

Let us prove that β_t is surjective. Consider $f \in \text{Aut}_{\text{Hopf}}(H)$, and let H_f be equal to H as an H-comodule algebra, with H-action given by the formula

$$k \cdot h = g(k)h$$

where $g = I * (S \circ f)$. We have seen (Proposition 12.4.13) that $A_f = \text{End}_R(H_f)$ with the induced action and coaction is an H-Azumaya algebra, and it is clear that it represents an element of $\text{BT}(R, H)$. The surjectivity of β_t now follows from the next Lemma.

Lemma 13.10.8 *With notations as above, we have that*

$$\beta_{A_f} = f \otimes (f^{-1})^*$$

Proof We have already remarked that A_f represents an element of $\mathrm{BT}(R, H)$, and we can apply Lemma 13.10.6. Looking at the proof of Lemma 13.10.6, we conclude that it suffices to show that

$$\rho_{\mathcal{A}}(u) = u \otimes (g \otimes \eta_H \circ \varepsilon_{H^*})$$

where $u \in \mathcal{A}_f = \mathrm{Hom}_R(D, A_f)$ is given by the formula

$$u(k \otimes k^*)(h) = \langle k^*, 1\rangle (k \rightharpoonup h) = \langle k^*, 1\rangle g(k)h$$

Observe first that $\rho_{\mathcal{A}_f}(u) \in \mathcal{A}_f \otimes_{D^*} \mathcal{H}$ is equal to $\rho_{\mathcal{A}_f}(u) = \rho_{A_f} \circ u$. Take $d = k \otimes k^* \in D$ and $h \in H$. Then

$$\begin{aligned}
\big(\rho_{\mathcal{A}_f}(u)(k \otimes k^*)\big)(h) &= \big(\rho_{A_f}(u(k \otimes k^*))\big)(h) \\
&= \sum (u(d)(h_{(1)}))_{(0)} \otimes (u(d)(h_{(1)}))_{(1)} S(h_{(2)}) \\
&= \sum \langle k^*, 1\rangle g(k_{(1)}) h_{(1)} \otimes g(k_{(2)}) h_{(2)} S(h_{(3)}) \\
&= \sum \langle k^*, 1\rangle g(k_{(1)}) h \otimes g(k_{(2)})
\end{aligned}$$

On the other hand, we have

$$\begin{aligned}
& (u \otimes (g \otimes \eta_H \circ \varepsilon_{H^*}))(k \otimes k^*)(h) \\
= & \sum u(k_{(1)} \otimes k^*_{(1)}) \otimes g(k_{(2)}) \otimes \eta_H(\varepsilon_{H^*}(k^*_{(2)}))(h) \\
= & \sum g(k_{(1)}) \langle k^*, 1\rangle h \otimes g(k_{(2)})
\end{aligned}$$

□

The proof of Theorem 13.10.5 will be complete after the next Lemma.

Lemma 13.10.9 $\beta_t : \mathrm{BT}(R, H) \to i(\mathrm{Aut}_{\mathrm{Hopf}}(H))$ *is injective.*

Proof Take $[A = \mathrm{End}_R(M)] \in \mathrm{BT}(R, H)$, and suppose that $\beta_t([A]) = I_D$, or $\alpha_A = \varepsilon_D$. We will show that M is an H-dimodule, that is

$$\sum (h \rightharpoonup m)_{(0)} \otimes (h \rightharpoonup m)_{(1)} = \sum h \rightharpoonup m_{(0)} \otimes m_{(1)} \tag{13.93}$$

for all $h \in H$ and $m \in M$. From the previous Lemmas, we know that the action of H on A is induced by the map

$$H \longrightarrow A : \ h \mapsto \psi_h = (h \rightharpoonup \bullet)$$

Now $\rho_{\mathcal{A}}(u) = \rho_A \circ u = u \circ \varepsilon_D$, since $\alpha_A = \varepsilon_D$. Therefore

$$\rho_A \circ \psi_\bullet = u \otimes \varepsilon_H$$

or

$$\rho_A(\psi_h) = \sum \psi_{h_{(1)}} \otimes \varepsilon(h_{(2)}) = \psi_h \otimes 1$$

Recall that

$$\rho_A :\ A \to A \otimes H = \mathrm{End}_R(M) \otimes H \cong \mathrm{End}_R(M, M \otimes H)$$

so

$$\rho_A(\psi_h)(m) = (h \rightharpoonup m) \otimes 1 \tag{13.94}$$

On the other hand, for $a \in A$, $\rho_A(a)$ is defined by

$$\rho(a)(m) = \sum a(m_{(0)})_{(0)} \otimes a(m_{(0)})_{(1)} S(m_{(1)})$$

Taking $a = \psi_h$, we obtain

$$\rho_A(\psi_h)(m) = \sum (h \rightharpoonup m_{(0)})_{(0)} \otimes (h \rightharpoonup m_{(0)})_{(1)} S(m_{(1)}) \tag{13.95}$$

(13.93) now follows from (13.94) and (13.95). □

Let $\widetilde{H}_f$ be equal to H as an H-module, but with modified H-coaction

$$\rho(h) = \sum h_{(1)} \otimes g(h_{(2)})$$

It is straightforward to show that $\widetilde{A}_f = \mathrm{End}_R(\widetilde{H}_f)$ with the induced action and coaction is an H-dimodule algebra. Arguments similar to the ones given in the proof of Proposition 12.4.13 show that $\widetilde{A}_f$ is an H-Azumaya algebra, and it is clear that $\widetilde{A}_f$ represents an element of $\mathrm{BT}(R, H)$.

Proposition 13.10.10 *With notations as above, we have that* $\beta_{\widetilde{A}_f} = f \otimes (f^{-1})^*$. *Consequently* $[A_f] = [\widetilde{A}_f]$ *in* $\mathrm{BT}(R, H) \subset \mathrm{BD}(R, H)$.

Proof The proof is similar to the proof of Lemma 13.10.8. It suffices to show that

$$\rho_{\widetilde{A}_f}(u) = u \otimes (g \otimes \eta_H \circ \varepsilon_{H^*})$$

where now $u \in \widetilde{A}_f = \mathrm{Hom}_R(D, \widetilde{A}_f)$ is given by the formula

$$u(k \otimes k^*)(h) = \langle k^*, 1\rangle kh$$

for all $d = k \otimes k^* \in D$ and $h \in H$. The assertion follows from the following computation.

$$\begin{aligned}
\big(\rho_{\widetilde{A}_f}(u)(k \otimes k^*)\big)(h) &= \big(\rho_{\widetilde{A}_f}(u(k \otimes k^*))\big)(h) \\
&= \sum ((u(d)(h_{(1)}))_{(1)} \otimes g\big(((u(d)(h_{(1)}))_{(2)}\big) g(S(h_{(2)})) \\
&= \sum \langle k^*, 1\rangle k_{(1)} h_{(1)} \otimes g(k_{(2)} h_{(2)}) g(S(h_{(3)})) \\
&= \sum \langle k^*, 1\rangle k_{(1)} h_{(1)} \otimes g(k_{(2)}) \\
&= \sum \big(u(k_{(1)} \otimes k^*_{(1)}) \otimes g(k_{(2)}) \otimes \eta_H(\varepsilon_{H^*}(k^*_{(2)}))\big)(h) \\
&= \big(u \otimes (g \otimes \eta_H \circ \varepsilon_{H^*})\big)(h)
\end{aligned}$$

□

Orzech's subgroup of the Brauer-Long group

Theorem 13.10.11 *Let $\theta : H \to H^*$ be a Hopf algebra map. The exact sequence (13.75) can be restricted to an exact sequence*

$$1 \longrightarrow \mathrm{B}^s_\theta(R,H) \longrightarrow \mathrm{B}_\theta(R,H) \xrightarrow{\gamma} \mathrm{O}_\theta(R,H) \tag{13.96}$$

where

$$\mathrm{O}_\theta(R,H) = \{f \in \mathrm{Hopf}(H,H^*) | f * (S \circ f^* \circ \theta^* \circ f) \in \mathrm{P}_{\mathrm{sk}}(H^*,R)\}$$

with multiplication

$$f \cdot g = f * g * (f \circ (\theta * \theta^*) \circ g)$$

Proof (13.75) restricts to

$$1 \longrightarrow \mathrm{B}^s_\theta(R,H) \longrightarrow \mathrm{B}_\theta(R,H) \xrightarrow{\beta} \mathrm{O}(R,H)_{\mathrm{min}}$$

Let A be a θ-Azumaya algebra, and consider the map α_A. In matrix form, we can write

$$\alpha_A = \begin{pmatrix} a_{11} & a_{12} \\ a_{21} & a_{22} \end{pmatrix}$$

From the fact A is a θ-algebra, it follows that $\mathcal{A} = \mathrm{Hom}(D,A)$ is a Θ-algebra, with $\Theta : \mathcal{H} \to \mathcal{H}^*$ given by the formula

$$\Theta(\mu) = \theta \circ \mu$$

for all $\mu \in \mathcal{H} = \mathrm{Hom}(D,H)$. Now take $u \in I_{R(I)} \subset \mathcal{A} = \mathrm{Hom}_R(D,A)$. Then

$$\rho_{\mathcal{A}}(u) = (m_H \circ (a_{11} \otimes a_{12}))$$

and

$$\mu \rightharpoonup u = \big\langle m_{H^*} \circ (a_{21} \otimes a_{22}), \mu \big\rangle * u$$

for all $\mu \in \mathcal{H} = \mathrm{Hom}(D,H)$. Now $\mathcal{A}$ is a Θ-algebra, so we also have that

$$\begin{aligned} \mu \rightharpoonup u &= \big\langle \Theta(m_H \circ (a_{11} \otimes a_{12})), \mu \big\rangle * u \\ &= \big\langle \theta \circ m_H \circ (a_{11} \otimes a_{12}), \mu \big\rangle * u \end{aligned}$$

This holds for every $u \in I_{R(I)}$, and $I_{R(I)}$ is an invertible D^*-module. Hence it follows that

$$\theta \circ m_H \circ (a_{11} \otimes a_{12}) = m_{H^*} \circ (a_{21} \otimes a_{22})$$

Applying both sides to $h \otimes \varepsilon$ and $a \otimes h^*$, we find that

$$a_{21} = \theta \circ a_{11} \quad \text{and} \quad a_{22} = \theta \circ a_{12} \tag{13.97}$$

We will next show that a_{11} is known once we know a_{12}. The map

$$\vartheta : H \otimes H^* \to H \otimes H^* : h \otimes h^* \mapsto 1 \otimes \theta^*(h) * h^*$$

is comultiplicative, and it determines a grouplike element of $\mathcal{D} = \mathcal{H} \otimes_{D^*} \mathcal{H}^* = \mathrm{Hom}(D, D)$. For all $f \in \mathcal{A} = \mathrm{Hom}(D, A)$ and $h \otimes h^* \in D = H \otimes H^*$, we have, by (13.64),

$$\begin{aligned} R(I)(f)(h \otimes h^*) &= \sum \langle h^*_{(1)}, f(h_{(2)} \otimes h^*_{(2)})_{(1)}\rangle (h_{(1)} \rightharpoonup f(h_{(2)} \otimes h^*_{(2)})_{(0)}) \\ &= \sum \langle h^*_{(1)}, f(h_{(2)} \otimes h^*_{(2)})_{(1)}\rangle \langle h_{(1)}, \theta(f(h_{(2)} \otimes h^*_{(2)})_{(2)})\rangle f(h_{(2)} \otimes h^*_{(2)})_{(0)} \\ &= \sum \langle h^*_{(1)} * \theta^*(h_{(1)}), f(h_{(2)} \otimes h^*_{(2)})_{(1)}\rangle f(h_{(2)} \otimes h^*_{(2)})_{(0)} \\ &= R(\vartheta)(f)(h \otimes h^*) \end{aligned}$$

It follows that

$$R(I) = R(\vartheta) \text{ and } I_{R(I)} = I_{R(\vartheta)}$$

and, using Lemma 13.9.2,

$$\alpha_A = \alpha_A(I) = \alpha_A(\vartheta) = \alpha_A(I) \circ \vartheta = \alpha_A \circ \vartheta$$

Now

$$\begin{aligned} \alpha_A(h \otimes \varepsilon) &= \sum a_{11}(h_{(1)}) \otimes a_{21}(h_{(2)}) \\ &= \alpha_A(\vartheta(h \otimes \varepsilon)) \\ &= \alpha_A(1 \otimes \theta^*(h)) \\ &= \sum a_{12}(\theta^*(h_{(1)})) \otimes a_{22}(\theta^*(h_{(2)})) \end{aligned}$$

Applying ε_{H^*} to the second factor, we see that

$$a_{11} = a_{12} \circ \theta^* \tag{13.98}$$

From (13.97) and (13.98) it follows that α_A and β_A are completely determined by $a_{12} : H \to H^*$. In matrixform:

$$\alpha_A = \begin{pmatrix} a_{12} \circ \theta^* & a_{12} \\ \theta \circ a_{12} \circ \theta^* & \theta \circ a_{12} \end{pmatrix}$$

and

$$\beta_A = \begin{pmatrix} I_H & \eta_{H^*} \circ \varepsilon_H \\ \eta_H \circ \varepsilon_{H^*} & I_{H^*} \end{pmatrix} * \begin{pmatrix} S \circ a_{12} \circ \theta^* & S \circ a_{12} \\ S \circ \theta \circ a_{12} \circ \theta^* & S \circ \theta \circ a_{12} \end{pmatrix}$$

From Lemma 13.9.4, it follows that $q = q \circ \beta_A$. Recall that

$$q : H \otimes H^* \to R : h \otimes h^* \to \langle h^*, h\rangle$$

In particular, we have that

$$\begin{aligned} \langle h^*, 1\rangle &= q(1 \otimes h^*) \\ &= q(\beta_A(1 \otimes h^*)) \\ &= \sum \langle (I_{H^*} * S \circ \theta \circ a_{12})(h^*_{(1)}), (S \circ a_{12})(h^*_{(2)})\rangle \\ &= \sum \langle h^*_{(1)}, ((S \circ a_{12}) * (a^*_{12} \circ \theta^* \circ a_{12}))(h^*_{(2)})\rangle \end{aligned}$$

and it follows that $(S \circ a_{12}) * (a_{12}^* \circ \theta^* \circ a_{12})$ and $a_{12} * (S \circ a_{12}^* \circ \theta^* \circ a_{12})$ are skew. We define $\gamma([A]) = a_{12}$.
Finally, let B be another θ-Azumaya algebra, and suppose that $\gamma([B]) = b_{12}$. Then

$$\beta_B = \begin{pmatrix} I_H & \eta_{H^*} \circ \varepsilon_H \\ \eta_H \circ \varepsilon_{H^*} & I_{H^*} \end{pmatrix} * \begin{pmatrix} S \circ b_{12} \circ \theta^* & S \circ b_{12} \\ S \circ \theta \circ b_{12} \circ \theta^* & S \circ \theta \circ b_{12} \end{pmatrix}$$

We know that $\beta_{A\#B} = \beta_A \circ \beta_B$. A direct computation shows that the entry in the northeast corner of the matrix of $\beta_{A\#B}$ is $S \circ (a_{12} * b_{12} * (a_{12} \circ (\theta^* * \theta) \circ b_{12}))$ and it follows that

$$\gamma([A\#B]) = a_{12} * b_{12} * (a_{12} \circ (\theta^* * \theta) \circ b_{12})$$

□

Now consider the θ-Azumaya algebra A discussed in Theorem 12.4.5. Recall that $A = H$ as an H-comodule, with multiplication twisted by a Sweedler cocycle f. To f we may associate a skew pairing $d: H \otimes H^* \to R$. For all $h, k \in A$, we have that

$$h \cdot k = \sum \langle d(h_{(2)}), k_{(2)} \rangle k_{(1)} \cdot h_{(1)}$$

The action of H on A is given by the rule

$$h \rightharpoonup k = \sum \langle \theta(k_{(2)}), h \rangle k_{(1)}$$

and we have seen that A is H-Azumaya if and only if $\theta * d$ is an isomorphism. In the sequel, λ will be the composition inverse of $\theta * d$.

Theorem 13.10.12 *With notations as above, we have that*

$$\gamma([A]) = (\theta * d)^{-1} = \lambda$$

and, consequently

$$\beta_A = \begin{pmatrix} I_H & \eta_{H^*} \circ \varepsilon_H \\ \eta_H \circ \varepsilon_{H^*} & I_{H^*} \end{pmatrix} * \begin{pmatrix} S \circ \lambda \circ \theta^* & S \circ \lambda \\ S \circ \theta \circ \lambda \circ \theta^* & S \circ \theta \circ \lambda \end{pmatrix}$$

Proof Define $u \in \mathcal{A} = \mathrm{Hom}(D, A)$ and $\psi^* \in \mathcal{H}^* = \mathrm{Hom}(D, H^*)$ by

$$\begin{aligned} u(h \otimes h^*) &= \lambda(h^*)(\lambda \circ \theta^*)(h) \\ \psi^* &= \theta \circ u \end{aligned}$$

1) Viewed as a map $D \to A = H$, u is comultiplicative. Hence u is a grouplike element of $\mathcal{H}$, and $\rho_{\mathcal{A}}(u) = \Delta_{\mathcal{H}}(u) = u \otimes u$.
2) Take $\mu \in \mathcal{H}$. We claim that

$$\mu \rightharpoonup u = \langle \psi^*, \mu \rangle * u$$

For all $h \otimes h^* \in D$, we have that

$$\begin{aligned}(\mu\rightharpoonup u)(h \otimes h^*) &= \sum \mu(h_{(1)} \otimes h^*_{(1)})\rightharpoonup u(h_{(2)} \otimes h^*_{(2)}) \\ &= \sum \mu(h_{(1)} \otimes h^*_{(1)})\rightharpoonup(\lambda(h^*_{(2)})\lambda(\theta^*(h_{(2)}))) \\ &= \sum \langle \theta(\lambda(h^*_{(2)}))\theta(\lambda(\theta^*(h_{(2)}))), \mu(h_{(1)} \otimes h^*_{(1)})\rangle \lambda(h^*_{(3)})\lambda(\theta^*(h_{(3)})) \\ &= \sum \langle \psi^*(h_{(2)} \otimes h^*_{(2)}), \mu(h_{(1)} \otimes h^*_{(1)})\rangle u(h_{(3)} \otimes h^*_{(3)}) \\ &= (\langle \psi^*, \mu\rangle * u)(h \otimes h^*)\end{aligned}$$

3) We will now prove that $u \in I_{R(I)}$. In view of Lemma 13.9.1, we have to show that

$$\sum \langle h_{(1)}, \psi^*(k_{(1)} \otimes k^*_{(1)})\rangle u(k_{(2)} \otimes k^*_{(2)}) \cdot h_{(2)} = \sum \rho(k_{(1)} \otimes k^*_{(1)})(h) \cdot u(k_{(2)} \otimes k^*_{(2)}) \quad (13.99)$$

for all $h, k \in H$ and $k^* \in H^*$. The left hand side of (13.99) is

$$\begin{aligned}&\sum \langle \theta(\lambda(k^*_{(1)})), h_{(1)}\rangle \langle (\theta \circ \lambda \circ \theta^*)(k_{(1)}), h_{(2)}\rangle \lambda(k^*_{(2)})(\lambda \circ \theta^*)(k_{(2)})) \cdot h_{(3)} \\ = &\sum \langle \theta(\lambda(k^*_{(1)})), h_{(1)}\rangle \langle (\theta(\lambda(\theta^*(k_{(1)}))), h_{(2)}\rangle \langle d(\lambda(k^*_{(2)})), h_{(3)}\rangle \\ &\quad \langle (d \circ \theta \circ \lambda \circ \theta^*)(k_{(2)}), h_{(4)}\rangle h_{(5)} \cdot \lambda(k^*_{(3)})(\lambda \circ \theta^*)(k_{(3)}) \\ = &\sum \langle k^*_{(1)}, h_{(1)}\rangle \langle \theta(h_{(2)}), k_{(1)}\rangle h_{(3)} \cdot \lambda(k^*_{(2)})(\lambda \circ \theta^*)(k_{(2)})\end{aligned}$$

and this is exactly the right hand side of (13.99).
4) u is convolution invertible. Its convolution inverse is given by the formula

$$v(h \otimes h^*) = \lambda(S^*(h^*))(\lambda \circ \theta^* \circ S)(h)$$

and consequently $I_{R(I)} = D^*u$ is the free D^*-module generated by u. It now follows that

$$\begin{aligned}\alpha_A(h \otimes h^*) &= \alpha_A(I)(h \otimes h^*) \\ &= \sum \lambda(h^*_{(1)})(\lambda \circ \theta^*)(h_{(1)}) \otimes (\theta \circ \lambda)(h^*_{(2)})(\theta \circ \lambda \circ \theta^*)(h_{(2)})\end{aligned}$$

In matrixform, this may be rewritten as

$$\alpha_A = \begin{pmatrix} \lambda \circ \theta^* & \lambda \\ \theta \circ \lambda \circ \theta^* & \theta \circ \lambda \end{pmatrix}$$

The element in the northeast corner of the matrix of α_A is $\gamma([A]) = \lambda$. □

Remark 13.10.13 Consider the particular situation where d is trivial and $\theta : H \to H^*$ is selfdual. Then $\theta = \theta^*$, $\lambda = \theta^{-1}$ and

$$\beta_A = \begin{pmatrix} I_H * S \circ \theta^{-1} \circ \theta & S \circ \theta^{-1} \\ S \circ \theta \circ \theta^{-1} \circ \theta & I_{H^*} * (S \circ \theta \circ \theta^{-1}) \end{pmatrix} = \begin{pmatrix} \eta_H \circ \varepsilon_H & S \circ \theta^{-1} \\ S \circ \theta & \eta_{H^*} \circ \varepsilon_{H^*} \end{pmatrix}$$

and we can conclude the following: if $f : H \to H^*$ is a selfdual isomorphism, then the map

$$h \otimes h^* \mapsto f^{-1}(h^*) \otimes f(h)$$

lies in the image of β.

More subgroups of the Brauer-Long group

For two (or more) subgroups $\mathrm{BX}(R,H)$ and $\mathrm{BY}(R,H)$ of the Brauer-Long group, we will write $\mathrm{BXY}(R,H)$ for the subgroup generated by $\mathrm{BX}(R,H)$ and $\mathrm{BY}(R,H)$. For example, $\mathrm{BM^sC}(R,H)$ is the subgroup generated by $\mathrm{BM^s}(R,H)$ and $\mathrm{BC}(R,H)$. In this Section, we will characterize some of these subgroups. From Theorem 13.5.5, it follows immediately that

$$\mathrm{BD^s}(R,H) = \mathrm{BM^sC^s}(R,H) \tag{13.100}$$

Assume now that the map $\gamma : \mathrm{Gal}(R,H^*) \to \mathrm{P}_{sk}(H^*,R)$ discussed in Section 10.8 is surjective. This happens for example if $H = RG$, with the order of G invertible in R and R containing enough roots of unity. Other cases where γ is surjective have been discussed in Chapter 11. If γ is surjective, then the restriction of the map β to $\mathrm{BC}(R,H)$ fits into a short exact sequence

$$1 \longrightarrow \mathrm{BC^s}(R,H) \longrightarrow \mathrm{BC}(R,H) \xrightarrow{\beta} \left\{ \begin{pmatrix} I_H & g \\ \eta_{H^*}\circ\varepsilon_H & I_{H^*} \end{pmatrix} : g \in \mathrm{P}_{sk}(H^*,R) \right\}$$

As before, we will use the matrix notation (13.89). If $[A] \in \mathrm{BM^s}(R,H)$, then $\beta_A = I_{H\otimes H^*}$. If $[A] \in \mathrm{BC}(R,H)$, then

$$\beta_A = \begin{pmatrix} I_H & g \\ \eta_{H^*}\circ\varepsilon_H & I_{H^*} \end{pmatrix}$$

for some $g \in \mathrm{P}_{sk}(H^*,R)$. Consequently, if $[A] \in \mathrm{BM^sC}(R,H)$, then

$$b_{11} = I_H,\ b_{22} = I_{H^*},\ b_{21} = \eta_{H^*}\circ\varepsilon_H \tag{13.101}$$

Conversely, suppose that (13.101) holds. From the surjectivity of γ, it follows that there exists an H-comodule Azumaya algebra B such that

$$\beta_B = \begin{pmatrix} I_H & b_{12} \\ \eta_{H^*}\circ\varepsilon_H & I_{H^*} \end{pmatrix}$$

Then $\beta_{A\#\overline{B}} = \beta_A \circ \beta_B^{-1} = I_{H\otimes H^*}$, and $[A\#\overline{B}] \in \mathrm{Ker}\,(\beta) = \mathrm{BD^s}(R,H) = \mathrm{BM^sC^s}(R,H)$. It follows that $[A] \in \mathrm{BM^sC^sC}(R,H) = \mathrm{BM^sC}(R,H)$. We have shown the following.

Proposition 13.10.14 *Suppose that $\gamma : \mathrm{Gal}(R,H^*) \to \mathrm{P}_{sk}(H^*,R)$ is surjective. Then*

$$\mathrm{BM^sC}(R,H) = \{[A] \in \mathrm{BD}(R,H) \mid b_{11} = I_H,\ b_{22} = I_{H^*},\ b_{21} = \eta_{H^*}\circ\varepsilon_H\}$$

For $[A] \in \mathrm{BT}(R,H)$,

$$\beta_A = \begin{pmatrix} f & \eta_H\circ\varepsilon_{H^*} \\ \eta_{H^*}\circ\varepsilon_H & (f^*)^{-1} \end{pmatrix}$$

for some automorphism f of H (see Theorem 13.10.5). Consequently, if $[A] \in \text{BM}^s\text{CT}(R,H)$, then $b_{21} = \eta_{H^*} \circ \varepsilon_H$.
Conversely, assume that $b_{21} = \eta_{H^*} \circ \varepsilon_H$, that is,

$$\beta_A = \begin{pmatrix} b_{11} & b_{12} \\ \eta_{H^*} \circ \varepsilon_H & b_{22} \end{pmatrix}$$

Applying the orthogonality condition $\beta_A^* \circ t \circ \beta_A = t$ ($\beta_A \in \text{O}(R,H)_{\text{min}}$), we find easily that $b_{22^*} \circ b_{11} = I_H$, or $b_{22} = (b_{11}^*)^{-1}$. Using Theorem 13.10.5, we find a $[B] \in \text{BT}(R,H)$ such that

$$\beta_B = \begin{pmatrix} b_{11}^{-1} & \eta_H \circ \varepsilon_{H^*} \\ \eta_{H^*} \circ \varepsilon_H & b_{11}^* \end{pmatrix}$$

Now

$$\begin{aligned} \beta_{A\#B} = \beta_A \circ \beta_B &= \begin{pmatrix} b_{11} & b_{12} \\ \eta_{H^*} \circ \varepsilon_H & (b_{11}^*)^{-1} \end{pmatrix} \begin{pmatrix} b_{11}^{-1} & \eta_H \circ \varepsilon_{H^*} \\ \eta_{H^*} \circ \varepsilon_H & b_{11}^* \end{pmatrix} \\ &= \begin{pmatrix} I_H & b_{12} \circ b_{11}^* \\ \eta_{H^*} \circ \varepsilon_H & I_{H^*} \end{pmatrix} \end{aligned}$$

From the orthogonality condition, it follows that $b_{12} \circ b_{11}^* \in \text{P}_{\text{sk}}(H^*,R)$. It follows from Proposition 13.10.14 that $[A\#B] \in \text{BM}^s\text{C}(R,H)$, and consequently $[A] \in \text{BM}^s\text{CT}(R,H)$.

Proposition 13.10.15 *Suppose that* $\gamma :\ \text{Gal}(R,H^*) \to \text{P}_{sk}(H^*,R)$ *is surjective. Then*

$$\text{BM}^s\text{CT}(R,H) = \{[A] \in \text{BD}(R,H) \mid b_{21} = \eta_{H^*} \circ \varepsilon_H\}$$

Suppose now that $\text{Pic}(H^*) = 1$. This guarantees that the action of H on an H-module algebra A is inner (see Proposition 12.4.3). Take a central H-Azumaya algebra A. If we forget the H-coaction on A, then we obtain an H-module Azumaya algebra. Thus the action of H on A is inner, and it is induced by a map $u :\ H \to A$, with convolution inverse v. This means that

$$h \rightharpoonup a = \sum u(h_{(1)}) a v(h_{(2)})$$

for all $h \in H$ and $a \in A$. Now consider the map

$$\pi :\ \text{BM}(R,H) \longrightarrow \text{Gal}(R,H)$$

discussed in Section 13.2. In Proposition 13.2.5, we have seen that

$$\pi([A]) = \{\sum v(h_{(1)}) \# h_{(2)} | h \in H\} \subset A\#H$$

We also know that $\pi([A])$ is a Galois object with normal basis, since $\text{Pic}(H^*) = 1$. Thus $\pi([A])$ corresponds to a Sweedler cocycle (see Theorem 10.5.9). This Sweedler cocycle may be found as follows: for all $h,k\in H$, we have

$$\begin{aligned}
&\sum(v(h_{(1)})\#h_{(2)})(v(k_{(1)})\#k_{(2)})\\
=&\sum v(h_{(1)})(h_{(2)}\rightharpoonup v(k_{(1)}))\#h_{(3)}k_{(2)}\\
=&\sum v(h_{(1)})u(h_{(2)})v(k_{(1)})v(h_{(3)})\#h_{(4)}k_{(2)}\\
=&\sum v(k_{(1)})v(h_{(1)})u(h_{(2)}k_{(2)})\big(v(h_{(3)}k_{(3)})\#h_{(4)}k_{(4)}\big)
\end{aligned}$$

and it follows that the corresponding Sweedler cocycle is given by the formula

$$f(h\otimes k)=\sum v(k_{(1)})v(h_{(1)})u(h_{(2)}k_{(2)}) \tag{13.102}$$

f is called the *action cocycle* of A. It is clear that f does not depend on the choice of A in $[A]$.

Now suppose that $[A]\in \mathrm{BM^sCT}(R,H)$. If we forget the H-coaction, then we obtain an algebra representing a class in $\mathrm{BM^s}(R,H)$. The Galois object corresponding to $[A]$ is then commutative, and this means that the action cocycle of A is symmetric. Conversely, if A is central, and the action cocycle on A is symmetric, then A with trivial H-coaction represents an element of $\mathrm{BM^s}(R,H)$. Thus we have shown that $\mathrm{BM^sCT}(R,H)$ is the subgroup of $\mathrm{BD}(R,H)$ consisting of central H-Azumaya algebras with symmetric action cocycle.

Suppose now that the action cocycle of A is trivial. If B is A, but with trivial H-coaction, then $\pi([B])=1$, and $[B]\in \mathrm{Br}(R)$ (see Theorem 13.2.1). It follows that $[A]\in \mathrm{BCT}(R,H)$. Conversely, if $[A]\in\mathrm{BCT}(R,H)$, then A has trivial action cocycle. We summarize our results as follows.

Proposition 13.10.16 *Suppose that* $\gamma:\ \mathrm{BC}(R,H)\to \mathrm{Gal}(R,H^*)$ *is surjective, and that* $\mathrm{Pic}(H^*)=1$. *Then*

$$\begin{aligned}
\mathrm{BCT}(R,H)&=\{[A]\in\mathrm{BD}(R,H)\mid A\text{ has trivial action cocycle}\}\\
\mathrm{BM^sCT}(R,H)&=\{[A]\in\mathrm{BD}(R,H)\mid A\text{ has symmetric action cocycle}\}
\end{aligned}$$

Duality arguments imply the following result. We leave it to the reader to define the *coaction cocycle* of a central H-Azumaya algebra.

Proposition 13.10.17 *Suppose that the* $\gamma:\ \mathrm{BM}(R,H)\to\mathrm{Gal}(R,H)$ *is surjective. Then*

$$\begin{aligned}
\mathrm{BC^sM}(R,H)&=\{[A]\in\mathrm{BD}(R,H)\mid b_{11}=I_H,\ b_{22}=I_{H^*},\ b_{12}=\eta_H\circ\varepsilon_{H^*}\}\\
\mathrm{BC^sMT}(R,H)&=\{[A]\in\mathrm{BD}(R,H)\mid b_{12}=\eta_H\circ\varepsilon_{H^*}\}
\end{aligned}$$

If $\mathrm{Pic}(H)=1$, *then*

$$\begin{aligned}
\mathrm{BMT}(R,H)&=\{[A]\in\mathrm{BD}(R,H)\mid A\text{ has trivial coaction cocycle}\}\\
\mathrm{BC^sMT}(R,H)&=\{[A]\in\mathrm{BD}(R,H)\mid A\text{ has symmetric coaction cocycle}\}
\end{aligned}$$

It is clear that $\mathrm{BMT}(R,H)\cap\mathrm{BCT}(R,H)=\mathrm{Br}(R)\times\mathrm{BT}(R,H)$. From Propositions 13.10.16 and 13.10.17, we therefore find the following result. In the case where $H=RG$ is a group ring, it is originally due to Deegan ([64]).

Corollary 13.10.18 *Suppose that* $\mathrm{Pic}(H)=\mathrm{Pic}(H^*)=1$ *and that the maps* $\mathrm{BC}(R,H)\to\mathrm{Gal}(R,H^*)$ *and* $\mathrm{BM}(R,H)\to\mathrm{Gal}(R,H)$ *are surjective. Then* $\mathrm{Br}(R)\times\mathrm{BT}(R,H)$ *is the subgroup of* $\mathrm{BD}(R,H)$ *consisting of classes of central H-Azumaya algebras with trivial action and coaction cocycle.*

Proposition 13.10.19 *Suppose that the maps* $\mathrm{BC}(R,H)\to\mathrm{Gal}(R,H^*)$ *and* $\mathrm{BM}(R,H)\to\mathrm{Gal}(R,H)$ *are surjective. Then* $\mathrm{BMCT}(R,H)$ *contains all central H-Azumaya algebras.*

Proof Let A be a central H-Azumaya algebra, and let B be equal to A as an H-module algebra, but with trivial H-coaction. Then

$$A\#\overline{B}=A\otimes B^{\mathrm{op}}\cong\mathrm{End}_R(A)$$

as H-module algebras. As an element of $\mathrm{BM}(R,H)$, $A\#\overline{B}$ is trivial, so $A\#\overline{B}$ has trivial action cocycle. Thus $[A\#\overline{B}]\in\mathrm{BM^sCT}(R,H)$, and $A\in\mathrm{BMM^sCT}(R,H)=\mathrm{BMCT}(R,H)$. □

13.11 Computing $\mathrm{O}(R,H)_{\min}$

Cyclic group rings

Let R be a commutative ring. Assume that R contains a primitive n-th root of unity η, and that n is invertible in R. To simplify our formalism, we also assume that R is connected, although this is not really a restriction. Fixing the primitive root of unity η, we can identify RC_n and RC_n^*.

Proposition 13.11.1 *With assumptions as above,* $\mathrm{O}(R,RC_n)_{\min}$ *is the dihedral group on* $\mathbb{G}_m(\mathbb{Z}/n\mathbb{Z})$.

Proof We may restrict attention to the case where $n=p^e$ is a primary number. Indeed, if $n=n_1n_2$, with $(n_1,n_2)=1$, then $C_n=C_{n_1}\times C_{n_2}$, $\mathrm{Aut}(C_{n_1}\times C_{n_1}\times C_{n_2}\times C_{n_2})\cong\mathrm{Aut}(C_{n_1}\times C_{n_1})\times\mathrm{Aut}(C_{n_2}\times C_{n_2})$, and

$$\mathrm{O}(R,RC_n)_{\min}\cong\mathrm{O}(R,RC_{n_1})_{\min}\times\mathrm{O}(R,RC_{n_2})_{\min}$$

Write $C_{p^e}=\langle\sigma\rangle$ and $C_{p^e}^*=\langle\sigma^*\rangle$, with $\langle\sigma,\sigma^*\rangle=\eta$. We have isomorphisms

$$C_{p^e}\times C_{p^e}^*\cong(\mathbb{Z}/p^e\mathbb{Z})^2 \qquad (13.103)$$

$$\mathrm{End}_{\mathrm{Hopf}}(RC_{p^e}\otimes RC_{p^e}^*)\cong\mathrm{End}(C_{p^e}\times C_{p^e}^*)\cong\mathrm{M}_2(\mathbb{Z}/p^e\mathbb{Z}) \qquad (13.104)$$

Now suppose that

$$f=\begin{pmatrix}b_{11}&b_{12}\\b_{21}&b_{22}\end{pmatrix}\in\mathrm{O}(R,RC_{p^e})_{\min}$$

Using the identifications (13.103) and (13.104), we obtain, for all $x, y \in \mathbb{Z}/p^e\mathbb{Z}$, that

$$\begin{aligned} q(x,y) &= \langle \sigma^x, (\sigma^*)^y \rangle = \eta^{xy} \\ &= q(b_{11}x + b_{12}y, b_{21}x + b_{22}y) = \eta^{(b_{11}x+b_{12}y)(b_{21}x+b_{22}y)} \end{aligned}$$

or

$$(b_{11}x + b_{12}y)(b_{21}x + b_{22}y) \equiv xy \bmod p^e$$

or

$$b_{11}b_{21}x^2 + (b_{12}b_{21} + b_{22}b_{11} - 1)xy + b_{12}b_{22}y^2 \equiv 0 \bmod p^e$$

Let $v_{ij} = v_p(b_{ij})$ be the p-adic valuation of b_{ij}. Then

$$v_{11} + v_{21} \geq e \ ; \ v_{12} + v_{22} \geq e$$

and

$$\min\{v_{12} + v_{21}, v_{22} + v_{11}\} \leq v_p(b_{12}b_{21} + b_{22}b_{11}) = v(1) = 0$$

If $v_{22} + v_{11} = 0$, then $v_{22} = v_{11} = 0$, and this means that b_{11} and b_{22} are invertible in $\mathbb{Z}/p^e\mathbb{Z}$. Then $v_{21} \geq e$ and $v_{12} \geq e$, so $b_{21} = b_{12} = 0$. Thus

$$f = \begin{pmatrix} b_{11} & 0 \\ 0 & b_{22} \end{pmatrix}$$

and the orthogonality relations yields that $b_{22} = B_{11}^{-1}$. In this case $f = \beta_A$, for some $[A] \in \mathrm{BT}(R, RC_{p^e})$.
If $v_{12} + v_{21} = 0$, then $v_{12} = v_{21} = 0$, and this means that b_{12} and b_{21} are invertible in $\mathbb{Z}/p^e\mathbb{Z}$. Then $v_{11} \geq e$ and $v_{22} \geq e$, so $b_{11} = b_{22} = 0$. Using the orthogonality condition, we now find that

$$f = \begin{pmatrix} 0 & b_{12} \\ b_{12}^{-1} & 0 \end{pmatrix}$$

and $f = \beta_A$ for some θ-Azumaya algebra A. It now follows that

$$\mathrm{O}(R, RC_{p^e})_{\min} = \left\{ \begin{pmatrix} a & 0 \\ 0 & a^{-1} \end{pmatrix}, \begin{pmatrix} 0 & b \\ b^{-1} & 0 \end{pmatrix} \mid a, b \in \mathbb{G}_m(\mathbb{Z}/p^e\mathbb{Z}) \right\} \tag{13.105}$$

and this is isomorphic to the dihedral group D_{p^e} containing $2p^e$ elements. □

Corollary 13.11.2 *Suppose that R is a strictly Henselian ring in which the prime number p is invertible. For example, R is a separably closed field of characteristic different from p. Then*

$$\mathrm{BD}(R, RC_{p^e}) \cong \mathrm{D}_{p^e}$$

Remark 13.11.3 If p is not invertible in R (e.g. R is a field of characteristic p), or if R contains no primitive p^e-th root of unity, then the Hopf algebras RC_{p^e} and $(RC_{p^e})^*$ are no longer isomorphic. An adaptation of the above arguments yields that $\mathrm{O}(R, RC_{p^e}) = \mathrm{Aut}(C_{p^e}) = \mathbb{G}_m(\mathbb{Z}/p^e\mathbb{Z})$ in this case.

The group ring $H = R(C_p \times C_p)$

We assume again that the prime number p is invertible in R, and that R contains a primitive p-th root of unity η. Let $G = C_p \times C_p$. We take two generators σ and τ for G, and we write σ^* and τ^* for the corresponding two generators of G^*:

$$G = \langle\sigma\rangle \times \langle\tau\rangle \qquad G^* = \langle\sigma^*\rangle \times \langle\tau^*\rangle$$
$$\begin{aligned} \langle\sigma^*,\sigma\rangle &= \langle\tau^*,\tau\rangle = \eta \\ \langle\sigma^*,\tau\rangle &= \langle\sigma^*,\tau\rangle = 1 \end{aligned}$$

We can identify

$$\begin{aligned} G \times G^* &\cong (\mathbb{Z}/p\mathbb{Z})^4 \\ \mathrm{End}_{\mathrm{Hopf}}(RG) &\cong \mathrm{End}(G) \cong \mathrm{M}_4(\mathbb{Z}/p\mathbb{Z}) \end{aligned}$$

For $f \in \mathrm{M}_4(\mathbb{Z}/p\mathbb{Z})$, we write f_{ij} for the entry in the ij-position. We also write

$$f = \begin{pmatrix} b_{11} & b_{12} \\ b_{21} & b_{22} \end{pmatrix} = \begin{pmatrix} I - a_{11} & -a_{12} \\ -a_{21} & I - a_{22} \end{pmatrix}$$

where the b_{ij} and a_{ij} are 2×2-matrices. I represents the 2×2 identity matrix. We will also write

$$J = \begin{pmatrix} 0 & 1 \\ -1 & 0 \end{pmatrix}$$

We have seen in Section 11.1 that

$$\mathrm{Gal}(R, R(C_p\times C_p))/\mathrm{Gal}^s(R, R(C_p\times C_p)) \cong \mathrm{Gal}(R, R(C_p\times C_p)^*)/\mathrm{Gal}^s(R, R(C_p\times C_p)^*) \cong \mathbb{Z}/p\mathbb{Z}$$

If $[A] \in \mathrm{BM}(R,H)$ and $[B] \in \mathrm{BC}(R,H)$, then

$$\beta_A = \begin{pmatrix} I & 0 \\ aJ & I \end{pmatrix} \quad \text{and} \quad \beta_B = \begin{pmatrix} I & bJ \\ 0 & I \end{pmatrix}$$

for some $a, b \in \mathbb{Z}/p\mathbb{Z}$. Take

$$\mu = \begin{pmatrix} a & b \\ c & d \end{pmatrix} \in \mathrm{SL}_2(\mathbb{Z}/p\mathbb{Z})$$

and write

$$G_\mu = \begin{pmatrix} aI & bJ \\ -cJ & dI \end{pmatrix}$$

Using the fact that $J^2 = I$, we easily find that

$$G_\mu G_\nu = G_{\mu\nu}$$

for all $\mu, \nu \in \mathrm{SL}_2(\mathbb{Z}/p\mathbb{Z})$. Since $\mathrm{SL}_2(\mathbb{Z}/p\mathbb{Z})$ can be generated by elementary matrices, we find that the subgroup of $O(R, R(C_p \times C_p))$ generated by the images under β of $\mathrm{BM}(R, R(C_p \times C_p))$ and $\mathrm{BC}(R, R(C_p \times C_p))$ is

$$\{G_\mu | \mu \in \mathrm{SL}_2(\mathbb{Z}/p\mathbb{Z})\}$$

and

$$\mathrm{BMC}(R,H) = \{[A] \in \mathrm{BD}(R,H) | \beta_A = G_\mu \text{ for some } \mu \in \mathrm{SL}_2(\mathbb{Z}/p\mathbb{Z})\}$$

Theorem 13.11.4 (Beattie [20, Theorem 1.8])
With notations as above, $\mathrm{BMCT}(R, R(C_p \times C_p))$ *is the subgroup of* $\mathrm{BD}(R, H)$ *consisting of classes of algebras of square rank.*

Proof Suppose that A is of square rank, and that

$$\beta_A = f = \begin{pmatrix} b_{11} & b_{12} \\ b_{21} & b_{22} \end{pmatrix}$$

We have three possibilities.
1) $\mathrm{rk}(b_{11}) = 2$. Then A is central, by Proposition 13.10.3, and $[A] \in \mathrm{BMCT}(R, R(C_p \times C_p))$ by Proposition 13.10.19.
2) $\mathrm{rk}(b_{11}) = 1$. Then $Z(A)$ has rank p (see (13.90)). $Z(A)$ is not of square rank, so A is not of square rank, by Proposition 13.10.3. This is a contradiction.
3) $\mathrm{rk}(b_{11}) = 0$. Then $\mathrm{rk}(b_{22}) = 0$ (see (13.90)), and $b_{11} = b_{22} = 0$. From the orthogonality relations, it follows that

$$\beta_A = \begin{pmatrix} 0 & b_{12} \\ (b_{12}^t)^{-1} & 0 \end{pmatrix}$$

Take an H-Azumaya algebra B such that

$$\beta_B = \begin{pmatrix} 0 & J \\ J & 0 \end{pmatrix} = G_J$$

Then $B \in \mathrm{BMC}(R, H)$ and

$$\beta_A\beta_B = \beta_{A\#B} = \begin{pmatrix} b_{12}J & 0 \\ 0 & (b_{12}^t)^{-1}J \end{pmatrix}$$

Now $(J^{-1})^t = J$ and $(b_{12}^t)^{-1}J = ((Jb_{12})^t)^{-1} = ((b_{12}J)^t)^{-1}$, so $[A\#B] \in \mathrm{BD^sT}(R, H) = \mathrm{BM^sC^sT}(R, H)$. Consequently $[A] \in \mathrm{BMCM^sC^sT}(R, H) = \mathrm{BMCT}(R, H)$.
Conversely, $\mathrm{BMCT}(R, H)$ is generated by algebras of square rank, so it contains no algebras that have not a square rank. □

The image of $\mathrm{BMC}(R, RG)$ in $\mathrm{O}(R, RG)_{\min}$ is

$$\{G_\mu \mid \mu \in \mathrm{SL}_2(\mathbb{Z}/p\mathbb{Z})\}$$

For $\mu \in \mathrm{SL}_2(\mathbb{Z}/p\mathbb{Z})$, let

$$F_\mu = \begin{pmatrix} \mu & 0 \\ 0 & \mu^{-1} \end{pmatrix}$$

F_μ represents an element of $\mathrm{BT}(R, RG)$. In fact

$$\mathrm{BT}(R, RG) = \{I_x G_\mu \mid \mu \in \mathrm{SL}_2(\mathbb{Z}/p\mathbb{Z}), x \in \mathbb{G}_m(\mathbb{Z}/p\mathbb{Z})\}$$

where

$$I_x = \begin{pmatrix} xI & 0 \\ 0 & x^{-1}I \end{pmatrix}$$

It is also clear that the intersection of the images of BMC(R, RG) and BT(R, RG) in $O(R, RG)_{\min}$ is

$$\{I_x \mid x \in \mathbb{G}_m(\mathbb{Z}/p\mathbb{Z})\}$$

For all $\mu \in SL_2(\mathbb{Z}/p\mathbb{Z})$, we easily compute that

$$J\mu = \text{adj}(\mu)J = \mu^{-1}J$$

Using this identity, we easily verify that

$$F_\mu G_\nu = G_\nu F_\mu$$

For all $\mu, \nu \in SL_2(\mathbb{Z}/p\mathbb{Z})$. We therefore have shown:

Proposition 13.11.5 *The image of* BMCT(R, RG) *in* $O(R, RG)_{\min}$ *is isomorphic to the direct product of two copies of* $SL_2(\mathbb{Z}/p\mathbb{Z})$*:*

$$\beta(\text{BMCT}(R, R(C_p \times C_p))) = SL_2(\mathbb{Z}/p\mathbb{Z}) \times SL_2(\mathbb{Z}/p\mathbb{Z})$$

This group contains $p^2(p^2-1)^2$ *elements.*

If A is an RG-Azumaya algebra, then $A\#A$ has square rank. Therefore BMCT(R, RG) is normal of order 2 in the Brauer-Long group, and, applying β, we find that $SL_2(\mathbb{Z}/p\mathbb{Z}) \times SL_2(\mathbb{Z}/p\mathbb{Z})$ is normal of order 2 in $O(R, RG)_{\min}$. Consider

$$X = \begin{pmatrix} 0 & 0 & 1 & 0 \\ 0 & 1 & 0 & 0 \\ 1 & 0 & 0 & 0 \\ 0 & 0 & 0 & 1 \end{pmatrix}$$

It is straightforward to show that $X \in O(R, RG)_{\min}$. Let A be the inverse image of X under β constructed in the proof of Theorem 13.10.1.

$$\alpha_A = I - X = \begin{pmatrix} 1 & 0 & -1 & 0 \\ 0 & 0 & 0 & 0 \\ -1 & 0 & 1 & 0 \\ 0 & 0 & 0 & 0 \end{pmatrix}$$

has rank 1, so Ker (α_A) contains p^3 elements. A has rank p as an R-module. Clearly b_{11} has rank one, so $Z(A)$ is also of rank p and A is commutative. A can be described easily: $A = RC_p$ as an R-algebra, with the following action and grading:

$$\deg(u_{\sigma^i}) = (\sigma^i, 1) \ \text{ and } \ (\sigma^i, \tau^j) \rightharpoonup u_{\sigma^k} = \eta^{ik}$$

The rank of A is not a square, so X represents the quotient $O(R, RG)_{\min}/(SL_2(\mathbb{Z}/p\mathbb{Z})^2$. A straightforward computation shows that

$$XF_\mu = G_{\mu^{-1}}X$$

We summarize our results as follows:

Theorem 13.11.6 *Suppose that R is a connected commutative ring, containing p^{-1} and a primitive p-th root of unity η. Then*

$$\mathrm{O}(R, R(C_p \times C_p))_{\min} = \big(\mathrm{SL}_2(\mathbb{Z}/p\mathbb{Z}) \times \mathrm{SL}_2(\mathbb{Z}/p\mathbb{Z})\big) \times \mathbb{Z}/2\mathbb{Z}$$

contains $2p^2(p^2-1)^2$ elements. The multiplication rules are the following: the two copies of $\mathrm{SL}_2(\mathbb{Z}/p\mathbb{Z})$ commute, and, if X represents the nontrivial element in $\mathbb{Z}/2\mathbb{Z}$, then

$$XF_\mu = G_{\mu^{-1}}X$$

Remark 13.11.7 If R is a strictly Henselian ring containing p^e, then $\mathrm{BD}(R, R(C_p \times C_p)) = \mathrm{O}(R, R(C_p \times C_p))_{\min}$ is described completely by Theorem 13.11.6.

Tate-Oort algebras

Let R be a connected commutative ring in which the prime number p is not a zero-divisor. Assume that I and J are ideals of R such that

$$IJ^{p-1} = pR$$

As in Section 7.1, we write

$$K = \frac{1}{p} IJ^{p-2} \subset R_p$$

and we consider the Hopf algebra

$$H_J = \check{R}[Ky]/K^p((1+y)^p - 1)$$

Write $u = 1 + y$. Then we know that $G(H_J) = \{1, u, u^2, \dots, u^{p-1}\}$.
We will work with the Hopf algebra $H = H_J^{\otimes n}$. We will write

$$u_i = 1 \otimes \cdots \otimes u \otimes \cdots \otimes 1 \text{ and } y_i = 1 \otimes \cdots \otimes y \otimes \cdots \otimes 1$$

Lemma 13.11.8 *With assumptions as above, we have*

$$\mathrm{End}_{\mathrm{Hopf}}(H_J^{\otimes n}) \cong \mathrm{M}_n(\mathbb{Z}/p\mathbb{Z}) \text{ and } \mathrm{Aut}_{\mathrm{Hopf}}(H_J^{\otimes n}) \cong \mathrm{GL}_n(\mathbb{Z}/p\mathbb{Z})$$

Proof Let f be a Hopf algebra endomorphism of $H_J^{\otimes n}$. Then

$$f \otimes I_{R_p} : H_J^{\otimes n} \otimes R_p = R_p C_p^n \longrightarrow H_J^{\otimes n} \otimes R_p = R_p C_p^n$$

is a Hopf algebra map, induced by a group homomorphism of C_p^n. f is therefore represented by an $n \times n$-matrix with entries in $\mathbb{Z}/p\mathbb{Z}$. We are done if we can show that every Hopf algebra endomorphism f of $R_p C_p^n$ restricts to a Hopf algebra endomorphism of $H_J^{\otimes n}$.
Suppose that f is represented by the matrix

$$M = \begin{pmatrix} m_{11} & m_{12} & \cdots & m_{1n} \\ m_{21} & m_{22} & \cdots & m_{2n} \\ \vdots & \vdots & & \vdots \\ m_{n1} & m_{n2} & \cdots & m_{nn} \end{pmatrix}$$

This means that

$$f(u_i)=\prod_{j=1}^{m}u_j^{m_{ji}}$$

Now take $k\in K$. Then

$$\begin{aligned} f(ky_i) &= k(f(u_i)-1)=k(u_1^{m_{1i}}\cdots u_n^{m_{ni}}-1)\\ &= k\sum_{l=1}^{n}(u_l^{m_{li}}-1)\Big(\prod_{r=l+1}^{n}u_r^{m_{ri}}\Big)\end{aligned}$$

Now

$$k(u_l^{m_{li}}-1)=ky_l\sum_{s=0}^{m_{li}-1}u_l^s\in H_J^{\otimes n}$$

and it follows that $f(ky_i)\in H_J^{\otimes n}$, as needed. This shows that f restricts to an endomorphism of $H_J^{\otimes n}$.

If f is an automorphism, then its restriction is a monomorphism. Since we know that f^{-1} also restricts to a monomorphism of $H_J^{\otimes n}$, it follows that the restriction of f is composition invertible, and an automorphism. This proves the second assertion. □

Theorem 13.11.9 *Let R be a commutative ring in which p is not a zero-divisor. Assume that R contains a primitive p-th root of unity η, and that J is an ideal of R such that $J^2=(\eta-1)R$. Then*

$$\mathrm{O}(R,H_J^{\otimes n})_{\min}=\mathrm{O}(R,R_pC_p^n)_{\min}$$

Proof From Theorem 7.1.15, we know that H_J and H_J^* are isomorphic. An isomorphism $g:\ H_J\to H_J^*$ is given by the formula

$$\langle g(u),u\rangle=\eta \ \text{ or }\ \langle g(y),y\rangle=\eta-1$$

Identifying H_J and H_J^* through g, we see that

$$\mathrm{Aut}_{\mathrm{Hopf}}(H_J^{\otimes n}\otimes H_J^{\otimes n*})\cong\mathrm{Aut}_{\mathrm{Hopf}}(H_J^{\otimes 2n})\cong\mathrm{Aut}_{\mathrm{Hopf}}(H_J^{\otimes n}(R_pC_p^{2n}))\cong\mathrm{GL}_{2n}(\mathbb{Z}/p\mathbb{Z})$$

Suppose that an automorphism of $R_pC_p^{2n}$ lies in $\mathrm{O}(R,R_pC_p^n)_{\min}$. Then it is clear that its restriction in $\mathrm{Aut}_{\mathrm{Hopf}}(H_J^{\otimes n}\otimes H_J^{\otimes n*})$ is also orthogonal, hence it lies in $\mathrm{O}(R,H_J^{\otimes n})_{\min}$. This proves our result. □

Corollary 13.11.10 *Let R be as in Theorem 13.11.9. Then*

$$\begin{aligned}\mathrm{O}(R,H_J)_{\min} &\cong D_p\\ \mathrm{O}(R,H_J^{\otimes 2})_{\min} &\cong \big(\mathrm{SL}_2(\mathbb{Z}/p\mathbb{Z})\times\mathrm{SL}_2(\mathbb{Z}/p\mathbb{Z})\big)\times\mathbb{Z}/2\mathbb{Z}\end{aligned}$$

with the same multiplication rules as in Theorem 13.11.6. For $i=1,2$, the map $\beta:\ \mathrm{BD}(R,H_J^{\otimes i})\to\mathrm{O}(R,H_J^{\otimes i})_{\min}$ is surjective.

13.12 The multiplication rules

Up to the surjectivity of the map β, we know the Brauer-Long group as a set: it is the cartesian product of $\mathrm{BD}^s(R,H)$ and $\mathrm{O}(R,H)_{\mathrm{min}}$. To determine the structure of the Brauer-Long group completely, we have to investigate the commutation rules between the elements of $\mathrm{BD}^s(R,H)$ and $\mathrm{O}(R,H)_{\mathrm{min}}$. As far as we know, there is no general method to compute these commutation rules. We present some partial results in this Section. They will allow us to describe some of the subgroups of the Brauer-Long group. In some particular cases, for example if H is the group ring of a cyclic group, we can provide a description of the full Brauer-Long group.

Computing $\mathrm{BCT}(R,H)$

As a set

$$\mathrm{BCT}(R,H) \cong \mathrm{Aut}_{\mathrm{Hopf}}(H) \times \mathrm{Gal}(R,H^*) \times \mathrm{Br}(R)$$

$\mathrm{Br}(R)$, $\mathrm{Gal}(R,H^*)$ and $\mathrm{Aut}_{\mathrm{Hopf}}(H)$ embed as subgroups in $\mathrm{BCT}(R,H)$. We will consider H^*-Galois objects as H-Galois extensions, that is, H acts on $[S] \in \mathrm{Gal}(R,H^*)$. We recall from Section 13.2 that $\mathrm{Gal}(R,H^*)$ embeds in $\mathrm{BC}(R,H) \subset \mathrm{BCT}(R,H)$ as follows: the H-Azumaya algebra corresponding to an H-Galois extension S is $S\#H$, with trivial H-action, and H-coaction induced by the comultiplication on H:

$$\rho(s\#h) = \sum (s\#h_{(1)}) \otimes h_{(2)}$$

for all $s \in S$ and $h \in H$. Following (12.24), we find for the multiplication rule (H has to be replaced by H^*):

$$\begin{aligned}(s\#h)(u\#k) &= \sum su_{(0)}\#(u_{(1)}\rightharpoonup h)k \\ &= \sum \langle u_{(1)}, h_{(1)}\rangle su_{(0)}\#h_{(2)}k \\ &= \sum s(h_{(1)}\rightharpoonup t)\#h_{(2)}k\end{aligned}$$

for all $s, u \in S$ and $h, k \in H$.
Recall also from Theorem 13.10.5 that $\mathrm{Aut}_{\mathrm{Hopf}}(H) \cong \mathrm{BT}(R,H)$ embeds in $\mathrm{BCT}(R,H) \subset \mathrm{BD}(R,H)$ as follows: for $j \in \mathrm{Aut}_{\mathrm{Hopf}}(H)$, we take $A_j = \mathrm{End}_R(H_j)$, with $H_j = H$ as an H-comodule, with the modified action

$$k \cdot h = \sum k_{(1)} j(S(k_{(2)}))h$$

for all $h, k \in H$. A_j is then an H-Azumaya algebra, and it represents the class in $\mathrm{BD}(R,H)$ corresponding to j.
In the sequel, (j, S, A), with $j \in \mathrm{Aut}_{\mathrm{Hopf}}(H)$, S an H-Galois extension and A an Azumaya algebra, will be a shorter notation for the class in $\mathrm{BCT}(R,H)$ represented by $A_j\#(S\#H)\#A$.
We also recall the following notation: for $k \in H$, let

$$\psi_k : H_j \longrightarrow H_j : h \mapsto \psi_k(h) = k \cdot h$$

The action of H on A_j can be written as follows:

$$h \rightharpoonup f = \sum \psi_{h_{(1)}} \circ f \circ \psi_{S(h_{(2)})}$$

for all $f \in A_j$ and $h \in H$ (cf. (7.33)).
If T is an H-Galois object, and $j : H \to H$ is a Hopf algebra automorphism, then we define a new Galois object $T(j)$ as follows: $T(j) = T$ as an R-algebra, and the H-coaction of H on $T(j)$ is given by the formula

$$\rho_j(t) = \sum t_{(0)} \otimes j(t_{(1)})$$

for all $t \in T$. We leave it to the reader to show that T is an H-Galois object. If S is an H-Galois extension, then $S(j^*)$ is a new H-Galois extension, isomorphic to S as an R-algebra, and with H-action

$$h \cdot s = j(h) \rightharpoonup s$$

Theorem 13.12.1 *Let S be an H-Galois extension, and $j : H \to H$ a Hopf algebra automorphism. Then*

$$A_j\#(S\#H) \cong (S((j^*)^{-1})\#H)\#A_j \tag{13.106}$$

as H-dimodule algebras.

Proof We define

$$\Gamma :\ A_j\#(S\#H) \longrightarrow (S((j^*)^{-1})\#A_j = (S((j^*)^{-1})) \otimes A_j$$

as follows:

$$\Gamma(f \otimes (s\#h)) = \sum (s\#j(h_{(1)})) \otimes (f \circ \psi_{h_{(2)}})$$

for all $f \in A_j$, $s \in S$ and $h \in H$. We prove successively that Γ is multiplicative, H-colinear, H-linear and bijective.
1) Γ is multiplicative. For all $f, g \in A_j$, $s, u \in S$ and $h, k \in H$, we have that

$$\begin{aligned}
&\Gamma\Big((f \otimes (s\#h))(g \otimes (u\#k))\Big)\\
=\ &\Gamma\Big(\sum f \circ \psi_{h_{(1)}} \circ g \circ \psi_{S(h_{(1)})}\#(s(h_{(3)} \rightharpoonup u)\#h_{(4)}k)\Big)\\
=\ &\sum\Big(s(h_{(3)} \rightharpoonup u)\#j(h_{(4)}k_{(1)})\Big) \otimes \Big(f \circ \psi_{h_{(1)}} \circ g \circ \psi_{S(h_{(1)})} \circ \psi_{h_{(5)}} \circ \psi_{k_{(2)}}\Big)\\
=\ &\sum\Big(s(j^{-1}(j(h_{(2)})) \rightharpoonup u)\#j(h_{(3)})j(k_{(1)})\Big) \otimes \Big(f \circ \psi_{h_{(1)}} \circ g \circ \psi_{k_{(2)}}\Big)\\
=\ &\sum\Big((s\#j(h_{(2)})) \otimes (f \circ \psi_{h_{(1)}})\Big)\Big((u\#j(k_{(2)})) \otimes (g \circ \psi_{k_{(1)}})\Big)\\
=\ &\Gamma(f \otimes (s\#h))\Gamma(g \otimes (u\#k))
\end{aligned}$$

2) Γ is H-colinear. For all $f \in A_j$, $s \in S$ and $h \in H$, we have that

$$\begin{aligned}
\rho(\Gamma(f \otimes (s\#h))) &= \sum (s\#j(h_{(1)})) \otimes (f_{(1)} \circ \psi_{h_{(3)}}) \otimes j(h_{(2)})f_{(2)}h_{(4)}j(S(h_{(5)}))\\
&= \sum (s\#j(h_{(1)})) \otimes (f_{(1)} \circ \psi_{h_{(2)}}) \otimes f_{(1)}h_{(3)}\\
&= (\Gamma \otimes I_H)(\rho(f \otimes (s\#h)))
\end{aligned}$$

Here we used (12.36).
2) Γ is H-linear. For all $f \in A_j$, $s \in S$ and $h, k \in H$, we have that

$$\begin{aligned}\Gamma(k \rightharpoonup (f\#(s\#h)) &= \Gamma\big(\sum \psi_{k_{(1)}} \circ f \circ \psi_{S(k_{(2)})}\#(s\#h)\big)\\ &= \sum (s\#j(h_{(1)})) \otimes \psi_{k_{(1)}} \circ f \circ \psi_{S(k_{(2)})} \circ \psi_{h_{(2)}}\\ &= k \rightharpoonup \big(\sum (s\#j(h_{(1)})) \otimes f \circ \psi_{h_{(2)}}\big)\\ &= k \rightharpoonup \Gamma(f \oslash (s\#h))\end{aligned}$$

4) Γ is bijective. The inverse of Γ is given by

$$\Gamma^{-1}((s\#k) \otimes g) = \sum g \circ \psi_{S(k_{(1)})}\#(s\#j^{-1}(k_{(2)}))$$

□

Corollary 13.12.2 *Let H be a commutative, cocommutative, faithfully projective Hopf algebra over the commutative ring R. Then*

$$\mathrm{BCT}(R, H) \cong \mathrm{Aut}_{\mathrm{Hopf}}(H) \times \mathrm{Gal}(R, H^*) \times \mathrm{Br}(R)$$

with multiplication rules

$$(j, S, A)(j', S', A') = (jj', S(j'^*)S', AA') \tag{13.107}$$

We have implicitly introduced the following simplified notation: (j, S, A) stands for $(j, [S], [A]) \in \mathrm{Aut}_{\mathrm{Hopf}}(H) \times \mathrm{Gal}(R, H^*) \times \mathrm{Br}(R)$. In the right hand side of (13.107), jj' stands for $j \circ j'$, $S(j'^*)S'$ for $[S(j'^*)\square_{H^*}S'] = [S(j'^*)][S']$ and AA' for $[A \otimes A'] = [A][A']$. In the sequel, we will discuss formulas similar to (13.107), and we will use the same type of simplified notation.

Computing $\mathrm{BMT}(R, H)$

As a set

$$\mathrm{BMT}(R, H) = \mathrm{Aut}_{\mathrm{Hopf}}(H) \times \mathrm{Gal}(R, H) \times \mathrm{Br}(R)$$

Following Section 13.2, we have the following embedding of $\mathrm{Gal}(R, H)$ in $\mathrm{BM}(R, H) \subset \mathrm{BMT}(R, H)$: for an H-Galois object T, the corresponding H-module Azumaya algebra is $T\#H^*$, with trivial H-coaction, and H-action induced by the action on H^*:

$$h \cdot (t\#h^*) = \sum \langle h^*, h_{(1)}\rangle t\#h_{(2)}$$

and with multiplication rule

$$(t\#h^*)(v\#k^*) = \sum tv_{(0)}\#(v_{(1)} \rightharpoonup h^*) * k^*$$

for all $h \in H$, $h^*, k^* \in H^*$ and $t, v \in T$.
Now let (j, T, A) be a short notation for the class in $\mathrm{BMT}(R, H)$ represented by the H-Azumaya algebra $A_j\#(T\#H^*)\#A$.
Recall the definition of $\tilde{A}_j$ from Proposition 13.10.10.

Theorem 13.12.3 *Let T be an H-Galois object, and $j: H \to H$ a Hopf algebra automorphism. Then*

$$(T\#H^*)\#\widetilde{A}_j \cong \widetilde{A}_j\#(T(j^{-1})\#H^*) \tag{13.108}$$

as H-dimodule algebras. Consequently

$$[T\#H^*][A_j] = [A_j][T(j^{-1})\#H^*] \tag{13.109}$$

in BMT(R,H).

Proof T is an H-Galois object, and therefore an H^*-Galois extension. Applying Theorem 13.12.1, with H replaced by H^* and S by T^{-1}, we obtain that

$$A_{j^*}\#_*(T\#H^*) \cong (T(j^{-1})\#H^*)\#_*A_{j^*}$$

as H^*-dimodule algebras. We now apply Tilborghs' Theorem 13.1.8, and obtain that

$$(T^{-1}\#H^*)^{\rm op}\#A_{j^*}^{\rm op} \cong A_{j^*}^{\rm op}\#(T^{-1}(j^{-1})\#H^*)^{\rm op}$$

as H-dimodule algebras. Now $(T^{-1}\#H^*)^{\rm op} \cong T\#H^*$ and $(T^{-1}(j^{-1})\#H^*)^{\rm op} \cong T(j^{-1})\#H^*$. Furthermore, we claim that

$$A_{j^*}^{\rm op} \cong \widetilde{A}_j$$

The connecting isomorphism is given by the duality map

$$\widetilde{A}_j = {\rm End}_R(\widetilde{H}_j) \longrightarrow A_{j^*}^{\rm op} = {\rm End}_R(H_{j^*}^*)^{\rm op} : \ f \mapsto f^*$$

We leave it to the reader to verify that the H-action and H-coaction on $\widetilde{A}_j$ and $\widetilde{H}_j$ correspond to the H^*-action and H^*-coaction on $A_{j^*}^{\rm op}$ and $H_{j^*}^*$. This finishes the proof of (13.108). (13.109) follows from Proposition 13.10.10. □

Corollary 13.12.4 *Let H be a commutative, cocommutative, faithfully projective Hopf algebra over the commutative ring R. Then*

$$\text{BMT}(R,H) \cong \text{Aut}_{\text{Hopf}}(H) \times \text{Gal}(R,H) \times \text{Br}(R)$$

with multiplication rule

$$(j,T,A)(j',T',A') = (jj', T(j'^{-1})T', AA') \tag{13.110}$$

Computing $\text{BD}^s\text{T}(R,H)$

In Theorem 13.5.5, we have given the following description of the split part of the Brauer-Long group.

$$\text{BD}^s(R,H) = \text{Gal}^s(R,H^*) \times \text{Gal}^s(R,H) \times \text{Br}(R)$$

with multiplication rule

$$([S],[T],[A])([S'],[T'],[A']) = ([S\square_{H^*}S'],[T\square_H T'],[A\otimes A'\otimes |S'\#T|]) \qquad (13.111)$$

The H-module Azumaya algebra corresponding to an H-Galois object T is $T\#H^*$, and the H-comodule Azumaya algebra corresponding to an H-Galois extension S is $S\#H$. Here we are not completely accurate, since we have seen in Proposition 13.5.3 that the corresponding algebras are in fact the opposites of the ones we just mentioned. However, this does not affect the multiplication rule (13.111), since $S'\#T \cong (S')^{-1}\#T^{-1}$, $(T\#H^*)^{\rm op} \cong T^{-1}\#H^*$ and $(S\#H)^{\rm op} \sim S^{-1}\#H$. We use $|S'\#T|$ as a notation for $S'\#T$ equipped with trivial H-action and H-coaction.
If we combine (13.111) and Theorems 13.12.1 and 13.12.3, then we obtain the following description of $\mathrm{BD^sT}(R,H)$.

Theorem 13.12.5 *Let H be a commutative, cocommutative faithfully projective Hopf algebra over a commutative ring R. Then*

$$\mathrm{BD^sT}(R,H) \cong \mathrm{Aut_{Hopf}}(H)\times \mathrm{Gal^s}(R,H^*)\times \mathrm{Gal^s}(R,H)\times \mathrm{Br}(R)$$

with multiplication rules

$$(j,S,T,A)(j',S',T',A') = (jj',S(j'^*)S',T(j'^{-1})T',|S'\#T|AA') \qquad (13.112)$$

If $H = RC_{p^e}$, and if H is not isomorphic to its dual (that is, p is not invertible in R or R contains no primitive p-th root of unity), then $\mathrm{O}(R,H)_{\rm min} \cong \mathrm{Aut_{Hopf}}(H)$, and $\mathrm{BD^sT}(R,H) = \mathrm{BD}(R,H)$. In this situation, Theorem 13.12.5 describes the full Brauer-Long group. The same conclusion holds if H is a Tate-Oort algebra of rank p that is not isomorphic to its dual.

Computing $\mathrm{BD^s}\theta(R,H)$

Suppose that $\theta:\ H\to H^*$ is a self-adjoint Hopf algebra isomorphism. To θ, we may associate a symmetric pairing

$$\varphi:\ H\otimes H \longrightarrow R:\ h\otimes k\mapsto \varphi(h\otimes k) = \langle\theta(h),k\rangle$$

Observe that φ is a symmetric Sweedler 2-cocycle. H_θ will be equal to H as an H-comodule algebra, but with H-action

$$h\cdot k = \sum\langle\theta(k_{(1)}),h\rangle k_{(2)} = \sum\varphi(k_{(1)}\otimes h)k_{(2)} \qquad (13.113)$$

We have seen in Theorem 12.4.5 that H_θ is a θ-Azumaya algebra. The aim of this Section is to provide a full description of $\mathrm{BD^s}\theta(R,H)$, the subgroup of $\mathrm{BD}(R,H)$ generated by $\mathrm{BD^s}(R,H)$ and $[H_\theta]$.
We first introduce some notations. Let S be an H-Galois object, and $\alpha:\ H\to H^*$ a selfdual Hopf algebra map. Then

$$\rho_\alpha:\ H\otimes H\to R:\ h\otimes k\mapsto \langle\alpha(h),k\rangle = \rho_\alpha(h\otimes k)$$

is a Sweedler 2-cocycle. $S^\alpha = S$ as an H-comodule algebra, but with twisted multiplication

$$s \cdot t = \sum \rho_\alpha(s_{(1)} \otimes t_{(1)})s_{(0)}t_{(0)} \tag{13.114}$$

S^α is again an H-Galois object.
In the situation where S is commutative and α is an isomorphism, we let $S_\alpha = S$ as an H-comodule algebra, and with H-action

$$h \rightharpoonup s = \sum \rho_\alpha(h \otimes s_{(1)})s_{(0)}$$

S_α is an H-Azumaya algebra: after a faithfully flat base extension S_α becomes isomorphic to H_α, and we know from Theorem 12.4.5 that H_α is an H-Azumaya algebra. The multiplication formula for the H-opposite algebra $\overline{S}_\alpha$ is given by (13.114), and $\overline{S}_\alpha$ is isomorphic to S^α after we forget the H-action on $\overline{S}_\alpha$. We also have that

$$S^\alpha \cong S \square_H H^\alpha \quad \text{and} \quad (S^\theta)^{-1} = (S^{-1})^{\theta \circ S}$$

in $\mathrm{Gal}(R, H)$.
From now on, we fix the self-adjoint map $\theta : H \to H^*$. We have isomorphisms

$$\theta : \mathrm{Gal}(R, H) \to \mathrm{Gal}(R, H^*) \quad \text{and} \quad \theta : \mathrm{BC}(R, H) \to \mathrm{BM}(R, H)$$

defined as follows: for an H-Galois object S, $\theta(S) = S$ as an R-algebra, but with H-action given by a formula similar to (13.113). The other map can be defined in a similar way, and we obtain a commutative diagram of split exact sequences

$$\begin{array}{ccccccccc}
1 & \longrightarrow & \mathrm{Br}(R) & \longrightarrow & \mathrm{BM}(R,H) & \xrightarrow{\pi} & \mathrm{Gal}(R,H) & \longrightarrow & 1 \\
 & & \downarrow{=} & & \downarrow{\theta^{-1}} & & \downarrow{\theta} & & \\
1 & \longrightarrow & \mathrm{Br}(R) & \longrightarrow & \mathrm{BC}(R,H) & \xrightarrow{\pi} & \mathrm{Gal}(R,H^*) & \longrightarrow & 1
\end{array}$$

After we make the necessary identifications, we can describe $\mathrm{BD}^s(R, H)$ as the product of $\mathrm{Br}(R)$ and two copies of $\mathrm{Gal}(R, H)$:

$$\mathrm{BD}^s(R, H) = \mathrm{Gal}^s(R, H) \times \mathrm{Gal}^s(R, H) \times \mathrm{Br}(R)$$

with multiplication rule

$$([S], [T], [A])([S'], [T'], [A']) = ([S \square_H S'], [T \square_H T'], [A \otimes A' \otimes |S' \# T|]) \tag{13.115}$$

The smash product of two H-Galois objects is given by the multiplication rule

$$(s\#t)(u\#v) = \sum \varphi(t_{(1)} \otimes u_{(1)})st_{(0)} \otimes u_{(0)}v \tag{13.116}$$

For two commutative H-Galois objects S and T, we write $|S\#T|$, $(S\#T)_m$ and $(S\#T)_c$ for $S\#T$ equipped with respectively trivial action and coaction, trivial coaction and action induced by the action on T and trivial action and coaction induced by the coaction on T. The H-module Azumaya algebra and the H-comodule Azumaya algebra corresponding to an H-Galois object S are then precisely $(S\#H)_m$ and $(S\#H)_c$. Thus the isomorphism

$$\alpha : \mathrm{Gal}^s(R, H) \times \mathrm{Gal}^s(R, H) \times \mathrm{Br}(R) \longrightarrow \mathrm{BD}^s(R, H)$$

is given by

$$\alpha(S,T,A) = [(S\#H)_c][(T\#H)_m][A]$$

From (13.115), it follows in particular that

$$[(S\#H)_m][(T\#H)_c] = [(T\#H)_c][(S\#H)_m][|T\#S|] \tag{13.117}$$

Lemma 13.12.6 *Let S and T be commutative H-Galois objects. Then*

$$[|S\#S|][|T\#T|] = [|ST\#ST|] \tag{13.118}$$

in $\mathrm{Br}(R)$.

Proof We present two proofs. From (13.117), it follows that

$$\begin{aligned}
[(ST\#H)_m][(ST\#H)_c] &= [(ST\#H)_c][(ST\#H)_m][|ST\#ST|]\\
&= [(S\#H)_m][(T\#H)_m][(S\#H)_c][(T\#H)_c]\\
&= [(S\#H)_c][(T\#H)_c][(S\#H)_m][(T\#H)_m][|S\#S|][|T\#T|]
\end{aligned}$$

and the result follows. A cohomological argument is the following. Suppose that S and T are split by a faithfully flat extension U of R, and let $g,h \in Z^1(U/R, G(H\otimes \bullet))$ be the corresponding cocycle. The Amitsur 2-cocycles in $Z^2(U/R,\mathbb{G}_m)$ representing $|S\#S|$, $|T\#T|$ and $|ST\#ST|$ are respectively $\langle\theta(g_3),g_1\rangle$, $\langle\theta(h_3),h_1\rangle$ and $\langle\theta(g_3h_3),g_1h_1\rangle$ (notations as in Section 13.4). Now

$$\begin{aligned}
\langle\theta(g_3h_3),g_1h_1\rangle &= \langle\theta(g_3),g_1\rangle\langle\theta(h_3),h_1\rangle\langle\theta(g_3),h_1\rangle\langle\theta(h_3),g_1\rangle\\
&= \langle\theta(g_3),g_1\rangle\langle\theta(h_3),h_1\rangle\langle\theta(g_3),h_1\rangle\langle\theta(g_1),h_3\rangle
\end{aligned}$$

and $\langle\theta(g_3),h_1\rangle\langle\theta(g_1),h_3\rangle$ is a coboundary (see (13.22)). □

Now let S be a commutative H-Galois object, and consider the corresponding θ-Azumaya algebra S_θ. From Lemma 13.10.13, we know that

$$\beta_{H_\theta} = \begin{pmatrix} \eta_H\circ\varepsilon_H & S\circ\theta^{-1}\\ S^*\circ\theta & \eta_{H^*}\circ\varepsilon_{H^*}\end{pmatrix}$$

S_θ and H_θ are isomorphic after a faithfully flat base extension, so $\beta_{S_\theta} = \beta_{H_\theta}$ and

$$\beta_{S_\theta\#H_\theta} = \beta_{S_\theta}\circ\beta_{H_\theta} = \beta_{H_\theta}\circ\beta_{H_\theta} = I_{H\otimes H^*}$$

Consequently $[S_\theta\#H_\theta] \in \mathrm{BD}^s(R,H)$. We will compute the corresponding triple in $\mathrm{Gal}^s(R,H)\times\mathrm{Gal}^s(R,H)\times\mathrm{Br}(R)$. The underlying Azumaya algebra is $|S\#H|$ and is trivial in the Brauer group. The two other components are equal because $S_\theta\#H_\theta$ is a θ-algebra. After we forget the H-coaction on $S_\theta\#H_\theta$, we obtain an H-module Azumaya algebra, and it suffices to find its image under the map $\pi:\ \mathrm{BM}(R,H)\to\mathrm{Gal}(R,H)$.

Lemma 13.12.7 *With notations as above, we have that*

$$\pi([S_\theta\#H_\theta]) = [S^{\theta\circ S}]$$

Proof We define a map

$$\Gamma:\ S^{\theta\circ S}\longrightarrow((S_\theta\#H_\theta)\#H_\theta)^{S_\theta\#H_\theta}$$

by

$$\Gamma(s)=\sum(s_{(0)}\#S(s_{(1)}))\#s_{(2)}$$

step 1: Γ is well-defined.
For all $s,t\in S$ and $h\in H$, we have that

$$\begin{aligned}&\Big(\sum(s_{(0)}\#S(s_{(1)}))\#s_{(2)}\Big)((t\#h)\#1)\\ =&\ \sum\varphi(t_{(2)}h_{(1)}\otimes s_{(3)})\varphi(S(s_{(4)})\otimes t_{(1)})(s_{(0)}t_{(0)}\#S(s_{(1)})h_{(2)})\#s_{(2)}\\ =&\ \sum\varphi(h_{(1)}\otimes s_{(3)})(s_{(0)}t_{(0)}\#S(s_{(1)})h_{(2)})\#s_{(2)}\\ =&\ ((t\#h)\#1)\Big(\sum(s_{(0)}\#S(s_{(1)}))\#s_{(2)}\Big)\end{aligned}$$

step 2: Γ is multiplicative.
For all $s,t\in S$, we have that

$$\begin{aligned}\Gamma(s)\Gamma(t)&=\Big(\sum(s_{(0)}\#S(s_{(1)}))\#s_{(2)}\Big)\Big(\sum(t_{(0)}\#S(t_{(1)}))\#t_{(2)}\Big)\\ &=\varphi(s_{(3)}\otimes t_{(3)}S(t_{(4)}))\varphi(S(s_{(4)})\otimes t_{(5)})\sum(s_{(0)}t_{(0)}\#S(s_{(1)}t_{(1)}))\#s_{(2)}t_{(2)}\\ &=\sum\varphi(S(s_{(3)})\otimes t_{(3)})\sum(s_{(0)}t_{(0)}\#S(s_{(1)}t_{(1)}))\#s_{(2)}t_{(2)}\\ &=\Gamma(s\cdot t)\end{aligned}$$

step 3: Γ is H-colinear.
This is obvious, since the H-coaction on $((S_\theta\#H_\theta)\#H_\theta)^{S_\theta\#H_\theta}$ is induced by the comultiplication on the last factor.
step 4: Γ is an H-comodule algebra homomorphism between two H-Galois objects, and is an isomorphism, by Proposition 8.1.10. □

Corollary 13.12.8 *With notations as above, we have that*

$$\alpha(1,S^{\theta\circ S},S^{\theta\circ S})=[S_\theta\#H_\theta]$$

and

$$[S_\theta\#H_\theta]=[(S^{\theta\circ S}\#H)_c][(S^{\theta\circ S}\#H)_m]$$

in $\mathrm{BD}(R,H)$. *Consequently*

$$[H_\theta\#H_\theta]=[(H^{\theta\circ S}\#H)_c][(H^{\theta\circ S}\#H)_m]$$

in $\mathrm{BD}(R,H)$.

Corollary 13.12.9 *Let S and T be commutative H-Galois objects. With notations as above, we have that*

$$\begin{aligned}[\overline{H}_\theta\#S_\theta]&=[(S^{-1}\#H)_m][(S^{-1}\#H)_c]\\ &=[(S^{-1}\#H)_c][(S^{-1}\#H)_m][|S\#S|] \qquad (13.119)\\ [S_\theta\#T_\theta]&=\Big[\big((S^{\theta\circ S}\square_H T^{-1})\#H\big)_c\Big]\Big[\big((S^{\theta\circ S}\square_H T^{-1})\#H\big)_m\Big]\\ &\quad\Big[|T^{-1}\#(S^{\theta\circ S}\square_H T^{-1})|\Big] \qquad (13.120)\end{aligned}$$

in $\mathrm{BD}(R,H)$.

Proof Observing that $(S^\theta)^{\theta\circ S} = S^{\theta*\theta\circ S} = S$, and using (13.117), we obtain

$$\begin{aligned}[\overline{H}_\theta\#S_\theta] &= [\overline{S}_\theta\#H_\theta]^{-1}\\ &= [(S^\theta)_\theta\#H_\theta]^{-1}\\ &= \left([(S\#H)_c][(S\#H)_m]\right)^{-1}\\ &= [(S^{-1}\#H)_m][(S^{-1}\#H)_c]\\ &= [(S^{-1}\#H)_c][(S^{-1}\#H)_m][|S\#S|]\end{aligned}$$

proving (13.119). Furthermore

$$\begin{aligned}[S_\theta\#T_\theta] &= [S_\theta\#H_\theta][\overline{H}_\theta\#T_\theta]\\ &= [(S^{\theta\circ S}\#H)_c][(S^{\theta\circ S}\#H)_m][(T^{-1}\#H)_c][(T^{-1}\#H)_m][|T\#T|]\\ &= \left[\left((S^{\theta\circ S}\square_H T^{-1})\#H\right)_c\right]\left[\left((S^{\theta\circ S}\square_H T^{-1})\#H\right)_m\right][|T^{-1}\#S^{\theta\circ S}|][|T\#T|]\\ &= \left[\left((S^{\theta\circ S}\square_H T^{-1})\#H\right)_c\right]\left[\left((S^{\theta\circ S}\square_H T^{-1})\#H\right)_m\right]\left[|T^{-1}\#(S^{\theta\circ S}\square_H T^{-1})|\right]\end{aligned}$$

□

Remark 13.12.10 As an H-comodule algebra, $\overline{H}_\theta = H^\theta$, and the Azumaya algebra underlying $\overline{H}_\theta\#S_\theta$ is $|H^\theta\#S|$. It follows from Corollary 13.12.9 that

$$|H^\theta\#S| = |S\#S| \tag{13.121}$$

in $\mathrm{Br}(R)$. (13.121) can also be proved as follows:

$$[|H^\theta S^{-1}\#S|] = [|(S^{-1})^\theta\#S|] = |\overline{S}\#S| = 1$$

in $\mathrm{Br}(R)$, so

$$[|H^\theta\#S|][|S^{-1}\#S|] = 1$$

in $\mathrm{Br}(R)$, and our assertion follows.

Lemma 13.12.11 *Let S be a commutative H-Galois object. Then*

$$S_\theta\#(S\#H)_m \cong (S\#H)_m\#H_\theta$$

as H-dimodule algebras.

Proof We define

$$\Gamma:\ S_\theta\otimes(S\#H)_m = S_\theta\#(S\#H)_m \longrightarrow (S\#H)_m\#H_\theta$$

by

$$\Gamma(t\otimes(s\#h)) = \sum(s\#h)(t_{(0)}\#1)\#t_{(1)}$$

It is clear that Γ is H-linear and H-colinear. Let us prove that Γ is multiplicative. For all $s,t,u,v \in S$ and $h,k \in H$, we have that

$$\begin{aligned}
&\Gamma(t\otimes(s\#h))\Gamma(v\otimes(u\#k))\\
&= \Big(\sum(s\#h)(t_{(0)}\#1)\#t_{(1)}\Big)\Big(\sum(u\#k)(v_{(0)}\#1)\#v_{(1)}\Big)\\
&= \sum(s\#h)(t_{(0)}\#1)\big(t_{(2)}\rightharpoonup((u\#k)(v_{(0)}\#1))\big)\#t_{(1)}v_{(1)}\\
&= \sum(s\#h)(t_{(0)}\#1)\varphi(t_{(1)}\otimes k_{(1)})(u\otimes k_{(2)})(v_{(0)}\#1)\#t_{(2)}v_{(1)}\\
&= \sum\varphi(t_{(2)}\otimes k_{(1)})\varphi(h_{(1)}\otimes t_{(1)}u_{(1)})(st_{(0)}u_{(0)}\#h_{(2)}k_{(2)})(v_{(0)}\#1)\#t_{(3)}v_{(1)}\\
&= \sum\varphi(h_{(1)}k_{(1)}\otimes t_{(1)})\varphi(h_{(2)}\otimes u_{(1)})(st_{(0)}u_{(0)}\#h_{(3)}k_{(2)})(v_{(0)}\#1)\#t_{(2)}v_{(1)}\\
&= \sum\varphi(h_{(1)}\otimes u_{(1)})(su_{(0)}\#h_{(2)}k)(t_{(0)}v_{(0)}\#1)\#t_{(1)}v_{(1)}\\
&= \Gamma\Big(\sum tv\otimes\varphi(h_{(1)}\otimes u_{(1)})su_{(0)}\#h_{(2)}k\Big)\\
&= \Gamma((t\otimes(s\#h))(v\otimes(u\#k)))
\end{aligned}$$

In order to show that Γ is surjective, it suffices to show that Γ is surjective after we localize at an arbitrary prime ideal p of R. Thus it suffices to consider the case where R is local. But then S has normal basis, that is, $S \cong H$ as an H-comodule. Identifying S and H as an H-comodule, we see that

$$(s\#h)\#k = \Gamma\Big(\sum k_{(2)}\otimes(s\#h)(S(k_{(1)})\#1)\Big)$$

Now Γ is a surjective map between projective R-modules of equal rank, and Γ is necessarily injective. □

Corollary 13.12.12 *Let S be a commutative H-Galois object.*

$$[(S\#H)_m][H_\theta] = [H_\theta][(S^{-1}\#H)_c][|S\#S|]$$

in $\mathrm{BD}(R,H)$.

Proof

$$\begin{aligned}
[(S\#H)_m][H_\theta] &= [S_\theta][(S\#H)_m]\\
&= [H_\theta][\overline{H}_\theta\#S_\theta][(S\#H)_m]\\
&= [H_\theta][(S^{-1}\#H)_c][(S^{-1}\#H)_m][|S\#S|][(S\#H)_m]\\
&= [H_\theta][(S^{-1}\#H)_c][|S\#S|]
\end{aligned}$$

□

The preceding results allow to describe $\mathrm{BD}^s\theta(R,H)$ completely. As a set,

$$\mathrm{BD}^s\theta(R,H) = \mathbb{Z}/2\mathbb{Z}\times\mathrm{Gal}^s(R,H)\times\mathrm{Gal}^s(R,H)\times\mathrm{Br}(R)$$

The H-Azumaya algebra corresponding to $(-,1,1,1)$ is H_θ. The second factor $\mathrm{Gal}^s(R,H) = \theta^{-1}(\mathrm{Gal}^s(R,H^*))$ represents $\mathrm{BC}^s(R,H)$, and the third factor $\mathrm{Gal}^s(R,H)$ represents $\mathrm{BM}^s(R,H)$. The multiplication rules are now

$$(+,1,T,1)(+,S,1,1) = (+,S,T,|S\#T|) \tag{13.122}$$

(see (13.115)). From Corollary 13.12.8 , we know that

$$(-,1,1,1)(-,1,1,1) = (+, H^{\theta\circ S}, H^{\theta\circ S}, 1) \tag{13.123}$$

Using the fact that $[H^{\theta\circ S}] = [H^{\theta}]^{-1}$ in $\mathrm{Gal}(R,H)$, we obtain that

$$(-,1,1,1)^{-1} = (-, H^{\theta}, H^{\theta}, |H^{\theta}\#H^{\theta}|) \tag{13.124}$$

Corollary 13.12.12 can be restated as follows.

$$(+,1,T,1)(-,1,1,1) = (-, T^{-1}, 1, |T\#T|) \tag{13.125}$$

It follows that

$$(-,T^{-1},1,1) = (+,1,T,|T\#T^{-1}|)(-,1,1,1)$$

Multiply both sides to the left by $(-,1,1,1)^{-1}$ and to the right by $(-,1,1,1)$. Applying Lemma 13.12.10, this gives us that

$$\begin{aligned}
&(+,T^{-1},1,1)(-,1,1,1)\\
&= (-,H^{\theta},H^{\theta},|H^{\theta}\#H^{\theta}|)(+,1,T,|T\#T^{-1}|)(+,H^{\theta\circ S},H^{\theta\circ S},1)\\
&= (-,1,T,|H^{\theta}\#H^{\theta}||T\#T^{-1}||H^{\theta}\#H^{\theta\circ S}||T\#H^{\theta\circ S}|)\\
&= (-,1,T,|T\#H^{\theta\circ S}T^{-1}|) = (-,1,T,1)
\end{aligned}$$

Replacing T^{-1} by S, we obtain

$$(+,S,1,1)(-,1,1,1) = (-,1,S^{-1},1) \tag{13.126}$$

Theorem 13.12.13 *Let H be a commutative, cocommutative faithfully projective Hopf algebra, and let $\theta: H \to H^*$ be a self-adjoint Hopf algebra isomorphism. Then*

$$\mathrm{BD}^s\theta(R,H) = \mathbb{Z}/2\mathbb{Z} \times \mathrm{Gal}^s(R,H) \times \mathrm{Gal}^s(R,H) \times \mathrm{Br}(R)$$

with multiplication rules (13.122), (13.123), (13.125) and (13.126). These can be summarized as follows

$$(\pm,S,T,A)(+,S',T',A') = (\pm,SS',TT',|S'\#T|AA') \tag{13.127}$$

$$(+,S,T,A)(-,S',T',A') = (-,T^{-1}S',S^{-1}T',|S^{-1}\#S'||ST\#T|AA') \tag{13.128}$$

$$(-,S,T,A)(-,S',T',A') = (+,H^{\theta\circ S}T^{-1}S',H^{\theta\circ S}S^{-1}T',|T\#ST^2||S'S^{-1}\#S'|AA') \tag{13.129}$$

The cyclic group of order 2

We now consider the particular situation where $H = RC_2$, and we assume that $2^{-1} \in R$. Let $\theta: RC_2 \to RC_2^*$ be the canonical isomorphism. Then $H^{\theta} = C = R[x]/(x^2+1)$ (the "complex numbers" over R), and the group of Galois objects is 2-torsion. In this case, $\mathrm{BD}^s\theta(R,H) = \mathrm{BD}(R,H)$ is the full Brauer-Long group, and the multiplication rules (13.127), (13.128) and (13.129) simplify slightly. We obtain the following result.

Theorem 13.12.14 (DeMeyer-Ford [71]) *Let R be a commutative ring. Assume that 2 is invertible in R and that $C = R[x]/(x^2+1)$. Then*

$$\mathrm{BD}(R, RC_2) = \mathbb{Z}/2\mathbb{Z} \times \mathrm{Gal}(R, RC_2) \times \mathrm{Gal}(R, RC_2) \times \mathrm{Br}(R)$$

with multiplication rules

$$\begin{aligned}
(\pm, S, T, A)(+, S', T', A') &= (\pm, SS', TT', |S'\#T|AA') && (13.130)\\
(+, S, T, A)(-, S', T', A') &= (-, TS', ST', |S\#S'||ST\#T|AA') && (13.131)\\
(-, S, T, A)(-, S', T', A') &= (+, CTS', CST', |T\#S||S'S\#S'|AA') && (13.132)
\end{aligned}$$

Remarks 13.12.15 1) If R contains a squareroot of -1, then $C \cong RC_2$ is trivial in the Galois group of quadratic extensions, and the multiplication formulas simplify further.
2) If the two-torsion part of the Picard group of R is trivial, then every quadratic extension of R has normal basis, and $\mathrm{Gal}(R, RC_2) = \mathbb{G}_m(R)/\mathbb{G}_m(R)^2$. The description of the Brauer-Long group then becomes more explicit. Observe that, for $S = R[x]/(x^2-a)$, and $T = R[x]/(x^2-b)$, $|S\#T|$ is the quaternion algebra ${}^aR^b$.
3) Take $R = \mathbb{R}$. Then $\mathrm{Gal}(\mathbb{R}, \mathbb{R}C_2) = \{1, [\mathbb{C}]\} = \mathbb{Z}/2\mathbb{Z}$, $\mathrm{Br}(R) = \{1, [\mathbb{H}]\} = \mathbb{Z}/2\mathbb{Z}$, and $\mathrm{BD}(\mathbb{R}, \mathbb{R}C_2)$ consists of 16 elements. From Theorem 13.12.14, it follows that $\mathrm{BD}(\mathbb{R}, \mathbb{R}C_2) \cong D_8$, the dihedral group of order 16. This result is originally due to Long [124].
4) Theorem 13.12.14 is due to DeMeyer and Ford [71]. Let us make clear that the multiplication rules (13.130), (13.131) and (13.132) are equivalent to the ones that may be found in [71]. Put

$$[+, S, T, A] = (+, T, S, A) \quad\text{and}\quad [-, S, T, A] = (-, T, S, A|T\#T|)$$

From the multiplication rules (13.130), (13.131) and (13.132), it follows that

$$\begin{aligned}
[+, S, T, A][+, S', T', A'] &= [+, SS', TT', |S\#T'|AA']\\
[-, S, T, A][+, S', T', A'] &= [-, SS', TT', |ST'\#T'|AA']\\
[+, S, T, A][-, S', T', A'] &= [-, TS', ST', |T\#ST'|AA']\\
[-, S, T, A][-, S', T', A'] &= [+, CTS', CST', |T\#STT'|AA']
\end{aligned}$$

These are exactly the multiplication rules in [71, p.199] (take into account that $[|CS\#T'|] = [|ST'\#T'|]$ and $[|T\#CSTT'|] = [|T\#ST'|]$ in $\mathrm{Br}(R)$, by (13.121)).

As an application, we can compute the Brauer-Wall group of a commutative ring in which the number 2 is invertible: observe that $(+, S, T, A)$ may be represented by a θ-algebra if and only if $S = T$. We obtain the following description, originally due to Wall [200], in the case where R is a field, and to DeMeyer and Ford [71] in the general case.

Corollary 13.12.16 *Assume that $2^{-1} \in R$, and write $C = R[x]/(x^2+1)$. Then*

$$\mathrm{BW}(R) = \mathbb{Z}/2\mathbb{Z} \times \mathrm{Gal}(R, RC_2) \times \mathrm{Gal}(R, RC_2)$$

is the abelian group with multiplication rules

$$(\pm, S, A)(+, S', A') = (\pm, SS', |S\#S'|AA') \quad (13.133)$$
$$(-, S, A)(-, S', A') = (+, CSS', |SS'\#SS'|AA') \quad (13.134)$$

In particular, we obtain that $\mathrm{Br}(\mathbb{R}) = C_8$, the cyclic group of order 8. The multiplication rules (13.133) and (13.134) can be identified with the ones in [200] and [71] after we put $[+, S, A] = (+, S, A)$ and $[-, S, A] = (-, S, |S\#S|A)$.

Tate-Oort algebras of rank 2

Now let R be a commutative ring in which 2 is not a zero-divisor, and let I be an ideal of R such that $I^2 = 2R$. From Theorem 7.1.12, we know that the Hopf algebra

$$H_I = \check{R}\Big[\frac{Iy}{2}\Big]/\Big(\frac{y^2+2y}{2}\Big)$$

with

$$\Delta(y) = y \otimes y + y \otimes 1 + 1 \otimes y, \quad \varepsilon(y) = 0, \quad S(y) = y$$

is selfdual. The isomorphism $\theta : H_I \to H_I^*$ is given by the formula

$$\langle \theta(y), y \rangle = -2$$

Now

$$H_I^\theta = C = \check{R}\Big[\frac{It}{2}\Big]/\Big(\frac{t^2+2t+2}{2}\Big)$$

With this notation, we obtain the following result, as an immediate consequence of Theorem 13.12.13 .

Theorem 13.12.17 *With notations as above,*

$$\mathrm{BD}(R, H_I) = \mathbb{Z}/2\mathbb{Z} \times \mathrm{Gal}(R, H_I) \times \mathrm{Gal}(R, H_I) \times \mathrm{Br}(R)$$

with multiplication rules (13.130), (13.131) and (13.132). Furthermore

$$\mathrm{B}_\theta(R, H_I) = \mathbb{Z}/2\mathbb{Z} \times \mathrm{Gal}(R, H_I) \times \mathrm{Br}(R)$$

with multiplication rules (13.133) and (13.134).

Example 13.12.18 Let $R = \mathbb{Z}[\sqrt{2}]$, and $I = \sqrt{2}R$. Then

$$H_I = R\Big[\frac{y}{\sqrt{2}}\Big]/\Big[\frac{(y+1)^2-1}{2}\Big] = R[x]/(x^2+\sqrt{2}x)$$

with

$$\Delta(x) = \sqrt{2}x \otimes x + x \otimes 1 + 1 \otimes x; \quad \varepsilon(x) = 0; \quad S(x) = x$$

It can be shown that $\mathrm{Pic}(H_I) = 1$, and that every H_I-Galois object has normal basis. Consequently

$$\mathrm{Gal}(\mathbb{Z}[\sqrt{2}], H_I) = \mathbb{G}_{m,2}(R)/\mathbb{G}_{m,\sqrt{2}}(R)^2 = \{1, -1\}$$

The only nontrivial Galois object is

$$C = R\Big[\frac{t}{\sqrt{2}}\Big]/\Big[\frac{(t+1)^2+1}{2}\Big] = R[u]/(u^2+\sqrt{2}u+2) = \mathbb{Z}[\sqrt{2}, \frac{1+i}{\sqrt{2}}]$$

with $u = -(1+i)/\sqrt{2}$. The coaction is given by the formula

$$\rho(u) = \sqrt{2}u \otimes x + u \otimes 1 + 1 \otimes x$$

From (6.50), we know that $\mathrm{Br}(\mathbb{Z}[\sqrt{2}]) = \mathbb{Z}/2\mathbb{Z}$. We claim that the nontrivial element of $\mathrm{Br}(\mathbb{Z}[\sqrt{2}])$ is represented by $C\#C$. This can be seen as follows: 2 is invertible in $\mathbb{R}$, so $\mathbb{R} \otimes_R C = \mathbb{C}$, and $\mathbb{R} \otimes_R (C\#C) = \mathbb{C}\#\mathbb{C} = \mathbb{H}$ is nontrivial in $\mathrm{Br}(\mathbb{R})$. It follows also that $C\#C$ is the sub-$\mathbb{Z}[\sqrt{2}]$-algebra of $\mathbb{H}$ generated by $(1+i)/2$ and $(1+j)/2$. From Theorem 13.12.17, it now follows easily that

$$\mathrm{BD}(\mathbb{Z}[\sqrt{2}], H_I) = D_8 \quad \text{and} \quad \mathrm{B}_\theta(\mathbb{Z}[\sqrt{2}], H_I) = C_8$$

Remarks 13.12.19 1) $C\#C$ is the most elementary example of an Azumaya algebra that is not equivalent to a Galois crossed product in the classical sense. This follows easily from the fact that $\mathbb{Z}[\sqrt{2}]$ has no separable extensions, apart from the direct product of a finite number of copies of $\mathbb{Z}[\sqrt{2}]$, see [114, III.6.5]. However, it is clear that $C\#C$ is a crossed product, if one allows splittings by Galois objects over a Hopf algebra. In [55], it is investigated to what extent Azumaya algebras over quadratic number rings are equivalent to smash products of Galois objects over rank 2 Hopf algebras.
2) $\mathbb{Z}[\sqrt{2}]$ has no nontrivial separable extensions, so H_I is not separable as a $\mathbb{Z}[\sqrt{2}]$-algebra. In particular, $H_{I,\theta}$ is an H_I-Azumaya algebra that is not strong, because strong Azumaya algebras are separable (see Lemma 12.3.4). Moreover, any H_I-Azumaya algebra A that is equivalent to $H_{I,\theta}$ is not strong. Indeed, there exist faithfully projective H_I-dimodules P and Q such that

$$A \otimes \mathrm{End}_R(P) \cong A\#\mathrm{End}_R(P) \cong H_{I,\theta}\#\mathrm{End}_R(Q) \cong H_{I,\theta} \otimes \mathrm{End}_R(Q)$$

We used Proposition 12.1.8. If A is separable, then $A\otimes\mathrm{End}_R(P)$ and $H_{I,\theta}\otimes\mathrm{End}_R(Q)$ are separable. From Lemma 3.5.2, it the follows that $H_{I,\theta}$ is separable. This is a contradiction, so A is not separable, and therefore not strong.
We can conclude that $[H_{I,\theta}] \in \mathrm{BD}(R, H_I) \setminus \mathrm{BDS}(R, H)$, and we have an example of a situation where the strong part of the Brauer-Long group is not equal to the full Brauer-Long group.
3) Consider the subcategory $\mathcal{C}$ of the category H_I-dimod, consisting of θ-modules. This category is a symmetric monoidal category. The symmetry is the following

$$M \otimes N \longrightarrow N \otimes M : \; m \otimes n \mapsto \sum \langle \theta(n_{(1)}), m_{(1)} \rangle n_{(0)} \otimes m_{(0)}$$

In [156], the Brauer group of a symmetric monoidal category is introduced. In fact, two different versions $\mathcal{B}_1(\mathcal{C})$ and $\mathcal{B}_2(\mathcal{C})$ of the Brauer group are defined, and it is shown that we have a natural morphism

$$\xi:\ \mathcal{B}_2(\mathcal{C})\longrightarrow\mathcal{B}_1(\mathcal{C})$$

For $\mathcal{C}=R$-mod, we have that $\mathcal{B}_1(\mathcal{C})=\mathcal{B}_2(\mathcal{C})=\mathrm{Br}(R)$.
If we apply these definitions to our category $\mathcal{C}$ of θ-modules, then it turns out that

$$\mathcal{B}_1(\mathcal{C})=\mathrm{B}_\theta(\mathbb{Z}[\sqrt{2}],H_I)$$

and

$$\mathcal{B}_2(\mathcal{C})=\mathrm{B}_\theta(\mathbb{Z}[\sqrt{2}],H_I)\cap\mathrm{BDS}(\mathbb{Z}[\sqrt{2}],H_I)$$

Now $H_{I,\theta}$ represents an element of $\mathcal{B}_1(\mathcal{C})$, but not of $\mathcal{B}_2(\mathcal{C})$, so we have an example of the situation where the morphism ξ fails to be an isomorphism.

Computing $\mathrm{BD^sT^a}\theta(R,H)$

As before, $\theta:\ H\to H^*$ is a selfdual Hopf algebra isomorphism. Let $\mathrm{BD^sT^a}\theta(R,H)$ be the subgroup of $\mathrm{BD}(R,H)$ generated by $\mathrm{BD^s}(R,H)$, $[H_\theta]$ and

$$\{[A_j]\in\mathrm{Br}'(R,H)|\theta\circ j:\ H\to H^*\text{ is selfdual}\}$$

Otherwise stated, $\mathrm{BD^sT^a}\theta(R,H)$ is the the subgroup of $\mathrm{BD}(R,H)$ generated by $\mathrm{BD^s}(R,H)$ and

$$\{[H_\psi]|\psi:\ H\to H^*\text{ is a selfdual isomorphism}\}$$

Our aim is now to give a full description of $\mathrm{BD^sT^a}\theta(R,H)$. First, we will rewrite the relations (13.106) and (13.109) using the identification $\theta:\ H\to H^*$. Using this identification, we have that

$$(S\#H)_m=S\#H^*\ \text{ and }\ (S\#H)_c=\theta(S)\#H$$

(13.109) can be rewritten as

$$[(T\#H)_m][A_j]=[A_j][(T(j^{-1})\#H)_m] \tag{13.135}$$

Rewriting (13.106) is slightly more complicated. Let S be an H-Galois object, and suppose that $\theta\circ j$ is selfdual. Then we claim that

$$\theta(S(j))=\theta(S)(j^*) \tag{13.136}$$

Indeed, the H-action on $\theta(S)(j^*)$ is given by the formula

$$h\rightharpoonup s=j(h)\cdot s=\sum\langle\theta(s_{(1)}),j(h)\rangle s_{(0)}$$

and the H-action on $\theta(S(j))$ is given by

$$h\rightharpoonup s=\sum\langle\theta(j(s_{(1)})),h\rangle s_{(0)}$$

$\theta \circ j$ is selfdual, so $j^* \circ \theta = \theta \circ j$, and the two actions coincide.
From (13.106), we now find that

$$\begin{aligned} [(S\#H)_c][A_j] &= [\theta(S)\#H][A_j] \\ &= [A_j][\theta(S)(j^*)\#H] \\ &= [A_j][(S(j)\#H)_c] \end{aligned} \tag{13.137}$$

In order to describe $\mathrm{BD^sT^a}\theta(R,H)$, we have to investigate how $A_j = \mathrm{End}_R(H_j)$ multiplies with H_θ. This will result from the subsequent Lemmas.

Lemma 13.12.20 *Consider a Hopf algebra automorphism j of H, and a selfadjoint Hopf algebra isomorphism $\theta:\ H \to H^*$. Then*

$$A_j\#H_\theta \cong A_j \otimes H_{\theta\circ j^{-1}}$$

as H-dimodule algebras.

Proof Consider the map

$$\Omega:\ A_j\#H_\theta \longrightarrow A_j \otimes H_{\theta\circ j^{-1}}:\ f\#h \mapsto \sum f \circ \psi_{h_{(1)}} \otimes j(h_{(2)})$$

step 1: Ω is multiplicative.

$$\begin{aligned} &\Omega((f\#h)(g\#k)) \\ =\ &\Omega\Big(\sum f\circ\psi_{h_{(2)}}\circ g\circ\psi_{S(h_{(3)})}\#h_{(1)}k\Big) \\ =\ &\sum f\circ\psi_{h_{(3)}}\circ g\circ\psi_{S(h_{(4)})}\circ\psi_{h_{(1)}}\circ\psi_{k_{(1)}}\otimes j(h_{(2)}k_{(2)}) \\ =\ &\Big(\sum f\circ\psi_{h_{(1)}}\otimes j(h_{(2)})\Big)\Big(\sum g\circ\psi_{k_{(1)}}\otimes j(k_{(2)})\Big) \\ =\ &\Omega(f\#h)\Omega(g\#k) \end{aligned}$$

for all $f,g \in A_j$ and $h,k \in H$.
step 2: Ω is H-colinear. This follows trivially after we recall from Lemma 12.1.5 that

$$\rho(\psi_h) = \psi_h \otimes 1 \quad\text{and}\quad k\rightharpoonup\psi_h = \varepsilon(k)\psi_h$$

for all $h,k \in H$.
step 3: Ω is H-linear.
For all $f \in A_j$ and $h,k \in H$, we have that

$$\begin{aligned} &k\rightharpoonup\Omega(f\#h) \\ =\ &k\rightharpoonup\Big(\sum f\circ\psi_{h_{(1)}}\otimes j(h_{(2)})\Big) \\ =\ &\sum \psi_{k_{(1)}}\circ f\circ\psi_{h_{(1)}}\circ\psi_{S(k_{(2)})}\otimes\varphi(k_{(3)}\otimes j^{-1}(j(h_{(2)})))j(h_{(3)}) \\ =\ &\sum \Omega\Big(\psi_{k_{(1)}}\circ f\circ\psi_{S(k_{(2)})}\otimes\varphi(k_{(3)}\otimes h_{(1)})h_{(2)}\Big) \\ =\ &\Omega(k\rightharpoonup(f\#h)) \end{aligned}$$

step 4: Ω is bijective. Its inverse is given by the formula

$$\Omega^{-1}(f\otimes h) = \sum f\circ\psi_{S(h_{(1)})}\#j^{-1}(h_{(2)})$$

□

Lemma 13.12.21 *Consider a Hopf algebra automorphism j of H, and a selfadjoint Hopf algebra isomorphism $\theta : H \to H^*$. Then*

$$A_j \otimes H_{\theta\circ j^{-1}} \cong H_{\theta\circ j^{-1}}\#(A_j \otimes H_{\theta\circ j^{-1}})^{\mathrm{co}H}$$

as H-dimodule algebras.

Proof Consider the map

$$\Gamma : A_j \otimes H_{\theta\circ j^{-1}} \longrightarrow H_{\theta\circ j^{-1}}\#(A_j \otimes H_{\theta\circ j^{-1}})^{\mathrm{co}H} : f \otimes k \mapsto \sum f_{(1)}k\#(f_{(0)} \otimes S(f_{(2)}))$$

step 1: Γ is well-defined.
H is flat, so it suffices to show that

$$(I_H \otimes \rho)(\Gamma(f \otimes k)) = \Gamma(f \otimes k) \otimes 1$$

This is obvious.
step 2: Γ is H-linear.
For all $h, k \in H$ and $f \in A_j$, we have that

$$\begin{aligned}\Gamma(h\rightharpoonup(f \otimes k)) &= \Gamma\Big(\sum \psi_{h_{(1)}} \circ f \circ \psi_{S(h_{(2)})} \otimes \varphi(h_{(3)} \otimes j^{-1}(k_{(1)}))k_{(2)}\Big)\\ &= \sum \varphi(h_{(3)} \otimes j^{-1}(k_{(1)}))f_{(1)}k_{(2)}\#\Big(\psi_{h_{(1)}} \circ f_{(0)} \circ \psi_{S(h_{(2)})} \otimes S(f_{(2)})\Big)\end{aligned}$$

equals

$$\begin{aligned}h\rightharpoonup\Gamma(f \otimes k) = \sum \varphi(h_{(1)} \otimes j^{-1}(f_{(1)}k_{(1)}))f_{(2)}k_{(2)}\\ \#\Big(\psi_{h_{(2)}} \circ f_{(0)} \circ \psi_{S(h_{(3)})} \otimes \varphi(h_{(4)} \otimes j^{-1}(S(f_{(3)}))S(f_{(4)})\Big)\end{aligned}$$

step 3: It is obvious that Γ is H-colinear and multiplicative.
step 4: Γ is bijective. The inverse of Γ is given by the formula

$$\Gamma^{-1}(h\#\sum_i f_i \otimes k_i) = \sum f_i \otimes hk_i$$

Indeed,

$$\Gamma^{-1}(\Gamma(f \otimes k)) = \sum f_{(0)} \otimes f_{(1)}S(f_{(2)})k = f \otimes k$$

If $\sum_i f_i \otimes k_i$ is coinvariant under the H-coaction, then

$$\sum f_{i_{(0)}} \otimes k_{i_{(1)}} \otimes f_{i_{(1)}}k_{i_{(2)}} = \sum_i f_i \otimes k_i \otimes 1$$

If we apply $(I \otimes S) \circ \Delta$ to the second factor, and multiply the third and the fourth factor, then we obtain

$$\sum f_{i_{(0)}} \otimes k_i \otimes f_{i_{(1)}} = \sum f_i \otimes k_{i_{(1)}} \otimes S(k_{i_{(2)}}) \qquad (13.138)$$

Now

$$\begin{aligned}&\Gamma(\Gamma^{-1}(h\#\sum f_i \otimes k_i))\\ =\;& \Gamma(\sum f_i \otimes hk_i)\\ =\;& \sum f_{i_{(1)}}hk_i\#(f_{i_{(0)}} \otimes S(f_{i_{(2)}}))\\ =\;& \sum S(k_{i_{(2)}})hk_{i_{(1)}}\#(f_i \otimes k_{i_{(3)}})\\ =\;& h\#\sum f_i \otimes k_i\end{aligned}$$

□

Lemma 13.12.22 *$A = (A_j \otimes H_{\theta\circ j^{-1}})^{\mathrm{co}H}$ is a trivial Azumaya algebra.*

Proof We have an isomorphism of R-algebras

$$\Gamma : (A_j \otimes H_{\theta\circ j^{-1}})^{\mathrm{co}H} \longrightarrow A_j = \mathrm{End}_R(H_j) : \sum f_i \otimes h_i \mapsto \sum f_i \varepsilon(h_i)$$

The inverse of Γ is given by the formula

$$\Gamma^{-1}(f) = \sum f_{(0)} \otimes S(f_{(1)})$$

□

H acts on A, so A is an H-module Azumaya algebra. We will now compute $\pi([A])$.

Lemma 13.12.23 *The action of H on A is inner; it is induced by the map*

$$u : H \longrightarrow A : h \mapsto \sum \psi_{h_{(1)}} \circ \xi_{S(h_{(2)})} \otimes S(h_{(3)})$$

where the map $\xi_h : H \otimes H$ is given by

$$\xi_h(k) = \sum \varphi(h \otimes j^{-1}(k_{(1)}))k_{(2)}$$

Proof The convolution inverse v of u is given by the formula

$$v(h) = \sum \xi_{h_{(1)}} \circ \psi_{S(h_{(2)})} \otimes h_{(3)}$$

For all $h \in H$ and $\sum f_i \otimes k_i \in A$, we have

$$\begin{aligned} &\sum u(h_{(1)})(\sum f_i \otimes k_i)v(h_{(2)}) \\ =\ &\sum \psi_{h_{(1)}} \circ \xi_{S(h_{(2)})} \circ f_i \circ \xi_{h_{(3)}} \circ \psi_{S(h_{(4)})} \otimes S(h_{(5)})k_i h_{(6)} \\ =\ &\sum \psi_{h_{(1)}} \circ \xi_{S(h_{(2)})} \circ f_i \circ \xi_{h_{(3)}} \circ \psi_{S(h_{(4)})} \otimes k_i \end{aligned}$$

For all $l \in H$, we have that

$$\begin{aligned} &\sum \big(\xi_{S(h_{(1)})} \circ f_i \circ \xi_{h_{(2)}}\big)(l) \\ =\ &\sum \varphi(h_{(2)} \otimes j^{-1}(l_{(2)}))\varphi(S(h_{(1)}) \otimes j^{-1}(f_i(l_{(1)})_{(1)}))f_i(l_{(1)})_{(2)} \\ \text{(by (7.40))} \quad =\ &\sum \varphi\big(S(h_{(1)}) \otimes j^{-1}(S(l_{(2)})f_i(l_{(1)})_{(1)})\big)f_i(l_{(1)})_{(2)} \\ =\ &\sum \varphi(S(h_{(1)}) \otimes j^{-1}(f_{i_{(1)}}))f_{i_{(0)}}(l) \end{aligned}$$

and

$$\sum \xi_{S(h_{(1)})} \circ f_i \circ \xi_{h_{(2)}} = \sum \varphi(S(h_{(1)}) \otimes j^{-1}(f_{i_{(1)}}))f_{i_{(0)}}$$

Using (13.138), we now obtain that

$$\begin{aligned} &\sum u(h_{(1)})(\sum f_i \otimes k_i)v(h_{(2)}) \\ =\ &\sum \psi_{h_{(1)}} \circ \varphi\big(h_{(3)} \otimes j^{-1}(S(f_{i_{(1)}}))\big)f_{i_{(0)}} \circ \psi_{S(h_{(2)})} \otimes k_i \\ =\ &\sum \psi_{h_{(1)}} \circ f_{i_{(0)}} \circ \psi_{S(h_{(2)})} \otimes \varphi(h_{(3)} \otimes j^{-1}(k_{i_{(2)}}))k_{i_{(1)}} \\ =\ &h \rightharpoonup (\sum f_i \otimes k_i) \end{aligned}$$

□

Lemma 13.12.24 *With notations as above,*

$$\pi([A]) = [S^{\theta\circ j^{-1}}]$$

Proof According to Proposition 13.2.5, $\pi([A])$ is represented by

$$\{a_h = \sum v(h_{(1)})\#h_{(2)} | h \in H\} \subset A\#H$$

A straightforward computation shows that

$$\begin{aligned} a_h \cdot a_k &= \sum \varphi(h_{(1)} \otimes j^{-1}(k_{(1)}))a_{h_{(2)}k_{(2)}} \\ &= \sum \langle(\theta \circ j^{-1})(k_{(1)}), h_{(1)}\rangle a_{h_{(2)}k_{(2)}} \end{aligned}$$

and this is exactly the multiplication rule in $S^{\theta\circ j^{-1}}$. □

Corollary 13.12.25

$$[A_j\#H_\theta] = [H_{\theta\circ j^{-1}}][(H\#H^{\theta\circ j^{-1}})_m]$$

in BD(R, H).

Lemma 13.12.26 *With notations as above,*

$$\pi([A]) = [H^{\theta\circ j^{-1}}]$$

Proof According to Proposition 13.2.5, $\pi([A])$ is represented by

$$\{a_h = \sum v(h_{(1)})\#h_{(2)} | h \in H\} \subset A\#H$$

It is straightforward to verify that

$$\begin{aligned} a_h \cdot a_k &= \sum \varphi(h_{(1)} \otimes j^{-1}(k_{(1)}))a_{h_{(2)}}a_{k_{(2)}} \\ &= \sum \langle(\theta \circ j^{-1})(k_{(1)}), h_{(1)}\rangle a_{h_{(2)}}a_{k_{(2)}} \end{aligned}$$

and this shows that $\pi([A]) = [H^{\theta\circ j^{-1}}]$. □

We summarize our results as follows.

Theorem 13.12.27 *Let $\theta : H \to H^*$ be a selfdual Hopf algebra automorphism, and $j : H \to H$ a Hopf algebra automorphism. then*

$$[A_j][H_\theta] = [H_{\theta\circ j^{-1}}][(H^{\theta\circ j^{-1}}\#H)_m]$$

in BD(R, H).

We will now state a similar formula for $[H_\theta][A_j]$. We will apply a duality argument that is similar to the one used in the proof of Theorem 13.12.3. Recall Tilborghs' Theorem 13.1.8, stating that we have an anti-isomorphism

$$\mathrm{BD}(R, H^*) \longrightarrow \mathrm{BD}(R, H) : [A] \mapsto [A^{\mathrm{op}}]$$

We need the following Lemma

Lemma 13.12.28 *Let $\theta : H \to H^*$ be a selfdual Hopf algebra isomorphism. Then $\theta : H_\theta \to (H^*_{\theta^{-1}})^{\mathrm{op}}$ is an isomorphism of H-Azumaya algebras, and $\theta : H^\theta \to (H^*)^{\theta^{-1}}$ is an isomorphism of R-algebras.*

Proof H^* acts and coacts on $H^*_{\theta^{-1}}$ as follows

$$\begin{aligned} \rho(h^*) &= \sum h^*_{(1)} \otimes h^*_{(2)} \\ k^* \rightharpoonup h^* &= \sum \langle \theta^{-1}(h^*_{(2)}), k^* \rangle h^*_{(1)} \end{aligned}$$

This means that H acts and coacts on $(H^*_{\theta^{-1}})^{\mathrm{op}}$ as follows

$$\begin{aligned} h \rightharpoonup h^* &= \sum \langle h^*_{(2)}, h \rangle h^*_{(1)} \\ \rho(h^*) &= \sum h^*_{(1)} \otimes \theta^{-1}(h^*_{(2)}) \end{aligned}$$

θ is H-linear since

$$\begin{aligned} \theta(h \rightharpoonup k) &= \theta(\sum \langle \theta(k_{(2)}), h \rangle k_{(1)}) \\ &= \sum \langle \theta(k)_{(2)}, h \rangle \theta(k)_{(1)} \\ &= h \rightharpoonup \theta(k) \end{aligned}$$

and θ is H-colinear since

$$\begin{aligned} (\theta \otimes I_H)(\rho(h)) &= \sum \theta(h_{(1)}) \otimes h_{(2)} \\ &= \theta(h_{(1)}) \otimes \theta^{-1}(\theta(h_{(2)})) \\ &= \rho(\theta(h)) \end{aligned}$$

This proves the first assertion. $\theta : H^\theta \to (H^*)^{\theta^{-1}}$ is multiplicative since

$$\begin{aligned} \theta(h \cdot k) &= \theta\big(\sum \langle \theta(h_{(2)}), k_{(2)} \rangle h_{(1)} k_{(1)}\big) \\ &= \sum \langle \theta(h)_{(2)}, \theta^{-1}(\theta(k)_{(2)}) \rangle \theta(h)_{(1)} \theta(k)_{(1)} \\ &= \theta(h) \cdot \theta(k) \end{aligned}$$

□

Theorem 13.12.29 *Suppose that θ and $\theta \circ j : H \to H^*$ are selfdual Hopf algebra isomorphisms. Then*

$$[H_\theta][A_j] = [(H^{\theta \circ j \circ S} \# H)_c][H_{\theta \circ j}] \tag{13.139}$$

in $\mathrm{BD}(R, H)$.

Proof We apply Theorem 13.12.27, with H replaced by H^*, j by j^* and θ by θ^{-1}. It follows that

$$\begin{aligned}[A_{j^*}][H^*_{\theta^{-1}}] &= [H_{\theta^{-1}\circ(j^*)^{-1}}][((H^*)^{\theta^{-1}\circ(j^*)^{-1}}\#H^*)_m] \\ &= [H_{\theta^{-1}\circ(j^*)^{-1}}][(H^*)^{\theta^{-1}\circ(j^*)^{-1}}\#H]\end{aligned}$$

in $\mathrm{BD}(R,H^*)$. We will now apply Tilborghs' Theorem 13.1.8. But first note that, for any H-Galois object S,

$$(\theta(S)\#H)^{\mathrm{op}} \cong \theta(S^{-1})\#H = (S^{-1}\#H)_c$$

in $\mathrm{BD}(R,H)$. Using Proposition 13.10.10 and Lemma 13.12.28, we obtain that

$$\begin{aligned}[H_\theta][\tilde{A}_j] &= [H_\theta][A_j] \\ &= [(H^{j^*\circ\theta\circ S}\#H)_c][H_{j^*\circ\theta}] \\ &= [(H^{\theta\circ j\circ S}\#H)_c][H_{\theta\circ j}]\end{aligned}$$

□

Corollary 13.12.30 *Suppose that θ and $\theta\circ j:\ H\to H^*$ are selfdual Hopf algebra isomorphisms. Then*

$$[A_j][H_\theta] = [H_\theta][A_{j^{-1}}][(H^{\theta\circ j^{-1}\circ S}(j)H^{\theta\circ j^{-1}}\#H)_m] \tag{13.140}$$

in $\mathrm{BD}(R,H)$.

Proof Using (13.126) and Theorems 13.12.27 and 13.12.29, we obtain

$$\begin{aligned}[A_j][H_\theta] &= [H_{\theta\circ j^{-1}}][(H^{\theta\circ j^{-1}}\#H)_m] \\ &= [(H^{\theta\circ j^{-1}}\#H)_c][H_\theta][A_{j^{-1}}][(H^{\theta\circ j^{-1}}\#H)_m] \\ &= [H_\theta][(H^{\theta\circ j^{-1}\circ S}\#H)_m][A_{j^{-1}}][(H^{\theta\circ j^{-1}}\#H)_m] \\ &= [H_\theta][A_{j^{-1}}][(H^{\theta\circ j^{-1}\circ S}(j)H^{\theta\circ j^{-1}}\#H)_m]\end{aligned}$$

□

We now know the structure of $\mathrm{BD^sT^a}\theta(R,H)$ completely. We are able to write down the multiplication rules, as in Theorem 13.12.13. We will not do this, since the formulas that we obtain are too ugly and complicated. In the next Theorem, we will summarize our results by giving all the commutation rules. The reader can write down the multiplication rules if he wants to.

To simplify the formalism, we introduce the following notations. For Azumaya algebras A and B, we will write $A=[A]$ and $B=[B]$ for their classes in the Brauer-Long group $\mathrm{BD}(R,H)$. For $j\in \mathrm{Aut_{Hopf}}(H)$, we write $A_j=[A_j]$. Also write $X=[H_\theta]$ and $S_m=[(S\#H)_m]$ and $S_c=[(S\#H)_c]$ for any H-Galois object S.

Theorem 13.12.31 *Let H be a faithfully projective, commutative, cocommutative Hopf algebra, and suppose that $\theta : H \to H^*$ is a selfdual Hopf algebra isomorphism. Then*

$$\mathrm{BD^sT^a}\theta(R,H) = \mathbb{Z}/2\mathbb{Z} \times \mathrm{Aut_{Hopf}}(H) \times \mathrm{Gal^s}(R,H) \times \mathrm{Gal^s}(R,H) \times \mathrm{Br}(R)$$

with the following multiplication rules. Br(R) *lies in the center, and*

$$\begin{aligned} T_m S_c &= S_c T_m |T\#S| & (13.141)\\ S_m A_j &= A_j S(j^{-1})_m & (13.142)\\ S_c A_j &= A_j S(j)_c & (13.143)\\ X^2 &= (H^{\theta\circ S})_c (H^{\theta\circ S})_m & (13.144)\\ S_m X &= X(S^{-1})_c |S\#S| & (13.145)\\ S_c X &= X(S^{-1})_m & (13.146)\\ A_j X &= X A_{j^{-1}} (H^{\theta\circ j^{-1}\circ S}(j) H^{\theta\circ j^{-1}})_m & (13.147) \end{aligned}$$

Cyclic group rings

Let $q = p^d$ be a primary number, and consider the cyclic group ring $H = RC_q$. Assume that p is a unit in R, and that R contains a primitive q-th root of unity η. The Hopf algebra isomorphisms $H \to H^*$ correspond to the primitive q-th roots of unity, and they are all selfdual. It follows from Proposition 13.11.1 that $\mathrm{BD}(R,H) = \mathrm{BD^sT^a}\theta(R,H)$, and, consequently, the Brauer-Long group is described completely by Theorem 13.12.31.
Now assume that η has a squareroot in R. If p is odd, this is always the case, since $(q,2) = 1$. We will see that the multiplication rules then simplify.

Lemma 13.12.32 *If the primitive q-th root of unity η has a squareroot in R, then every bilinear cocycle $f : C_q \times C_q \to \mathbb{G}_m(R)$ is a coboundary.*

Proof Let σ be a generator for C_q, and write $f(\sigma,\sigma) = a$. Then a is a q-th root of 1, since $a^q = f(\sigma^q, \sigma) = 1$. From the fact that η has a squareroot, it follows that a also has a squareroot. Let b be a squareroot of a^{-1}, and define $g : C_q \to \mathbb{G}_m(R)$ by

$$g(\sigma^i) = b^{i^2}$$

Then

$$\begin{aligned} (\Delta_1 g)(\sigma^i, \sigma^j) &= g(\sigma^i) g(\sigma^j) g(\sigma^{i+j})^{-1}\\ &= b^{i^2+j^2-(i+j)^2} = b^{-2ij} = a^{ij}\\ &= f(\sigma,\sigma)^{ij} = f(\sigma^i, \sigma^j) \end{aligned}$$

and f is a coboundary. □

Corollary 13.12.33 *Suppose that p is invertible in R, and that the primitive q-th root of unity η has a squareroot in R. Let $\theta : H = RC_q \to RC_q^*$ be the corresponding selfdual Hopf algebra isomorphism. Then $[H^\theta] = 1$ in* Gal(R,H) *and* $[|S\#S|] = 1$ *in* Br(R) *for every H-Galois object S.*

Proof $[H^\theta]$ equals H as an H-comodule, with multiplication twisted by a bilinear cocycle. This cocycle is a coboundary, by Lemma 13.12.32, and the first assertion follows. It follows from Lemma 13.12.10 that

$$[|S\#S|] = [|S\#H^\theta|] = 1$$

in $\mathrm{Br}(R)$. □

We now easily deduce the following result from Theorem 13.12.31.

Theorem 13.12.34 (Beattie-Caenepeel [23]) *Suppose that p is invertible in R, and that the primitive q-th root of unity η has a squareroot in R. Then*

$$\mathrm{BD}(R, RC_q) = \mathbb{Z}/2\mathbb{Z} \times \mathrm{Aut}(C_q) \times \mathrm{Gal}(R, RC_q) \times \mathrm{Gal}(R, RC_q) \times \mathrm{Br}(R)$$

with multiplication rules

$$\begin{aligned} &(\pm, j, S, T, A)(+, j', S', T', A') \\ = \; &(\pm, jj', S(j')S', T((j')^{-1})T', AA'|S'\#T((j')^{-1})|) \end{aligned} \tag{13.148}$$

$$\begin{aligned} &(\pm, j, S, T, A)(-, j', S', T', A') \\ = \; &(\mp, j^{-1}j', T^{-1}(j')S', S^{-1}((j')^{-1})T', AA'|S^{-1}((j')^{-1})\#S'||S\#T|) \end{aligned} \tag{13.149}$$

Remarks 13.12.35 1) At first glance, (13.148) and (13.149) do not match completely with the multiplication rules in Theorem 3.8 of [23]. This is due to the fact that our $S(j)$ is not defined in the same way as the corresponding one in [23]: our $S(j)$ equals $S(j^{-1})$ in [23] (see [23, Theorem 2.7]).
2) Observe that $\mathrm{Aut}(C_q) = \mathbb{G}_m(\mathbb{Z}/q\mathbb{Z})$. If every RC_q-Galois object has normal basis (e.g. R is local), then

$$\mathrm{Gal}(R, RC_q) = \mathbb{G}_m(R)/\mathbb{G}_m(R)^q$$

Let S be a the RC_q-Galois object corresponding to $a \in \mathbb{G}_m(R)$, This means that

$$S = R[x]/(x^q - a), \quad \text{with } \rho(x) = x \otimes \sigma$$

with σ a fixed generator for C_q. Let $j : C_q \to C_q$ be the automorphism corresponding to $\alpha \in \mathbb{G}_m(\mathbb{Z}/q\mathbb{Z})$. this means that $j(\sigma) = \sigma^\alpha$. It is not difficult to show that the Galois object $S(j)$ then corresponds to $a^{\alpha^{-1}}$. This provides a more explicit description of the multiplication rules (13.148) and (13.149).

Corollary 13.12.36 *With assumptions as in Theorem 13.12.34, we have that*

$$\mathrm{B}_\theta(R, RC_q) = \mathbb{Z}/2\mathbb{Z} \times \mathrm{Gal}(R, RC_q) \times \mathrm{Br}(R)$$

with multiplication rules

$$(\pm, S, A)(+, S', A') = (\pm, SS', AA'|S'\#S|) \tag{13.150}$$

$$(\pm, S, A)(-, S', A') = (\mp, S^{-1}S', AA'|S'\#S^{-1}|) \tag{13.151}$$

Tate-Oort algebras of rank p

Let R be a connected commutative ring in which the prime number p is not a zero-divisor. Assume that R contains a primitive p-th root of unity η, and let J be an ideal of R such that $J^2 = (\eta - 1)$. Let K be the fractional ideal of R such that $JK = R$, and recall that $H_J = \check{R}[Ky]/(K^p((1+y)^p - 1)$, with $JK = R$ is a Hopf algebra. $u = 1 + y$ is a grouplike element of H_J, and we have a Hopf algebra isomorphism $\theta: H_J \to H_J^*$ such that

$$\langle \theta(y), y \rangle = \eta - 1$$

It follows from Corollary 13.11.10 that $\mathrm{BD^sT^a}\theta(R, H_J) = \mathrm{BD}(R, H_J)$, and $\mathrm{BD}(R, H_J)$ is described completely by Theorem 13.12.31. We will show that the multiplication rules simplify, as in the cyclic group ring case. First we need some Lemmas.

Lemma 13.12.37 *We have a one-to-one correspondence between the symmetric pairings $f: H_J \otimes H_J \to R$ in the sense of Section 10.8 and the p-th roots of unity $1, \eta, \eta^2, \ldots, \eta^{p-1}$.*

Proof Let f be a symmetric pairing, and let $a = f(u \otimes u)$. Then

$$a^p = f(u \otimes u)^p = f(u^p \otimes u) = f(1 \otimes u) = 1$$

so a is a p-th root of unity. Moreover, f is completely determined if we know $f(u \otimes u)$. To complete our proof, we have to show that there exists a symmetric pairing $f: H_J \otimes H_J \to R$ such that $f(u \otimes u) = a$, for every p-th root of unity a. If $a = 1$, then this is trivial: we take $f = \varepsilon \otimes \varepsilon$.

Let $a = \eta^i$ be a primitive root of unity. Remark that $(\eta^i - 1) = (\eta - 1)(1 + \eta + \eta^2 + \cdots + \eta^{i-1})$, and the principal ideals generated by $\eta - 1$ and $\eta^i - 1$ are equal, by Lemma 7.1.9. Thus $J^2 = (\eta^i - 1)$, and we can replace η by η^i in all our constructions. In particular, we obtain a Hopf algebra isomorphism $\theta_i: H_J \to H_J^*$ satisfying

$$\langle \theta_i(y), y \rangle = \eta^i - 1$$

Now define $f: H_J \otimes H_J \to R$ as follows:

$$f(h \otimes k) = \langle \theta_i(h), k \rangle$$

for all $h, k \in H_J$. f is a symmetric pairing, and

$$f(u \otimes u) = \langle \theta_i(y+1), y+1 \rangle = \eta^i - 1 + 1 = \eta^i = a$$

□

Lemma 13.12.38 *With notations as above, assume that p is odd. Then every symmetric pairing $f: H_J \otimes H_J \to R$ is a cocycle. Consequently $[H_J^\theta] = 1$ in* $\mathrm{Gal}(R, H_J)$, *and*

$$[|H_J^\theta \# S|] = [|S \# S|] = 1$$

in $\mathrm{Br}(R)$, *for every H_J-Galois object S.*

Proof Suppose that $f(u \otimes u) = a$, and let b be a squareroot of a^{-1} in R. b exists because p is odd. By Lemma 13.12.38, there exists a symmetric pairing $\varphi : H_J \otimes H_J \to R$ such that $\varphi(u \otimes u) = b$. Now consider $g = \varphi \circ \Delta : H_J \to R$. $g(u^i) = b^{i^2}$, and $\Delta_1 g(u^i \otimes u^j) = f(u^i \otimes u^j)$, as in the proof of Lemma 13.12.32. This shows that $\Delta_1 g = f$ after we make p invertible. Consequently $\Delta_1 g = f$ is a cocycle.
The rest of the proof is easy, see the proof of Corollary 13.12.33. □

As an easy consequence of Theorem 13.12.31, we now obtain the following result.

Theorem 13.12.39 *Let R be a connected commutative ring in which the prime number p is not a zero-divisor. Assume that R contains a primitive p-th root of unity η, and let J be an ideal of R such that $J^2 = (\eta - 1)$. Then*

$$\mathrm{BD}(R, H_J) = \mathbb{Z}/2\mathbb{Z} \times \mathrm{Aut}_{\mathrm{Hopf}}(H_J) \times \mathrm{Gal}(R, H_J) \times \mathrm{Gal}(R, H_J) \times \mathrm{Br}(R)$$

with multiplication rules (13.148) and (13.149). If $\theta : H_J \to H_J^*$ is one of the $p-1$ Hopf algebra isomorphisms between H_J and H_J^*, then

$$\mathrm{B}_\theta(R, H_J) = \mathbb{Z}/2\mathbb{Z} \times \mathrm{Gal}(R, H_J) \times \mathrm{Br}(R)$$

with multiplication rules (13.150) and (13.151).

Remarks 13.12.40 1) $\mathrm{Aut}_{\mathrm{Hopf}}(H_J) = \mathbb{G}_m(\mathbb{Z}/p\mathbb{Z})$. In fact, the $p-1$ Hopf algebra automorphisms $R_pC_p \to R_pC_p$ restrict to Hopf algebra automorphisms $H_J \to H_J$. This follows from the proof of Lemma 13.12.37.
If R is local, then $J = Rb$ is a principal ideal, and every H_b-Galois object has normal basis. Then

$$\mathrm{Gal}(R, H_J) = \mathbb{G}_{m,b^p}(R)/\mathbb{G}_{m,b}(R)^p$$

Let $a \in \mathbb{G}_{m,b^p}(R)$ correspond to the H_b-Galois object S, and let $j : H_J \to H_J$ correspond to $\alpha \in \mathbb{G}_m(\mathbb{Z}/p\mathbb{Z})$. Then $S(j)$ corresponds to $a^{\alpha^{-1}}$.
2) If $p = 2$, then we expect that every symmetric pairing is a coboundary, under the assumption that R contains a squareroot i of -1. It turns out that we need the additional assumption that $1 + i$ is invertible in R.
This can be seen as follows: the only possibly nontrivial pairing f is given by the formula $f(u \otimes u) = -1$. Now define $g : H_J \to H_J$ by $g(1) = 1$ and $g(y) = i-1$. Then $g(y) \in J$, since $1+i$ is a unit and $(i-1)(1+i) = -2 \in J^2$. Now $g(u) = g(y)+g(1) = i$, and $\Delta_1 g(u \otimes u) = -1 = f(u \otimes u)$. It follows that f is a coboundary.
We now have that $[H_J^\theta] = 1$ in $\mathrm{Gal}(R, H_J)$, and we can simplify the multiplication rules.

13.13 The Brauer-Long group of a scheme

One can define Brauer-Long group and its subgroups, as well of the group of Galois (co)objects of a scheme X and a quasi-coherent sheaf of Hopf algebras $\mathcal{H}$. We use the same methods as in Section 6.11, where we introduced the Brauer group of a

scheme X.
Many of the results of Parts II and III of this book can be generalized in a straightforward way. In this Section, we indicate briefly how this should be done.
Let X be a scheme, and $\mathcal{H}$ a quasi-coherent sheaf of $\mathcal{O}_X$-algebras. Suppose that we have some morphisms

$$\begin{aligned} \Delta_{\mathcal{H}} &: \mathcal{H} \longrightarrow \mathcal{H} \otimes_{\mathcal{O}_X} \mathcal{H} \\ \varepsilon_{\mathcal{H}} &: \mathcal{H} \longrightarrow \mathcal{O}_X \\ S_{\mathcal{H}} &: \mathcal{H} \longrightarrow \mathcal{H} \end{aligned}$$

such that the obvious coassociativity and counit properties hold, that is,

$$(\Delta_{\mathcal{H}} \otimes I_{\mathcal{H}}) \circ \Delta_{\mathcal{H}} = (I_{\mathcal{H}} \otimes \Delta_{\mathcal{H}}) \circ \Delta_{\mathcal{H}}$$

and

$$(I_{\mathcal{H}} \otimes \varepsilon_{\mathcal{H}}) \circ \Delta_{\mathcal{H}} = (\varepsilon_{\mathcal{H}} \otimes I_{\mathcal{H}}) \circ \Delta_{\mathcal{H}} = I_{\mathcal{H}}$$

This amounts to saying that $\Gamma(U, \mathcal{H})$ is a Hopf algebra over the groundring $\Gamma(U, \mathcal{O}_X)$, for every open affine subset U of X. Then we say that $\mathcal{H}$ together with the structure maps $\Delta_{\mathcal{H}}$, $\varepsilon_{\mathcal{H}}$ and $S_{\mathcal{H}}$ is an $\mathcal{O}_X$-Hopf algebra. $\mathcal{H}$ is called cocommutative if all the $\Gamma(U, \mathcal{H})$ are cocommutative.
Now assume that $\mathcal{H}$ is a commutative, cocommutative, locally free $\mathcal{O}_X$-Hopf algebra. A quasi-coherent sheaf $\mathcal{M}$ together with morphisms

$$\begin{aligned} \psi_{\mathcal{M}} &: \mathcal{H} \otimes_{\mathcal{O}_X} \mathcal{M} \longrightarrow \mathcal{M} \\ \rho_{\mathcal{M}} &: \mathcal{M} \longrightarrow \mathcal{M} \otimes_{\mathcal{O}_X} \mathcal{H} \end{aligned}$$

is called an $\mathcal{H}$-dimodule if $\Gamma(U, \mathcal{M})$ is a $\Gamma(U, \mathcal{H})$-dimodule for every open affine subset U of X. In the same style, we can define $\mathcal{H}$-modules, $\mathcal{H}$-comodules, $\mathcal{H}$-module algebras, $\mathcal{H}$-comodule algebras and $\mathcal{H}$-dimodule algebras. An $\mathcal{H}$-dimodule algebra $\mathcal{A}$ is called an $\mathcal{H}$-Azumaya algebra if $\Gamma(U, \mathcal{A})$ is a $\Gamma(U, \mathcal{H})$-Azumaya algebra over the groundring $\Gamma(U, \mathcal{O}_X)$ for every open affine subset U of X.
Given two $\mathcal{H}$-dimodule algebras $\mathcal{A}$ and $\mathcal{B}$, we can define the smash product $\mathcal{A}\#\mathcal{B}$. As a sheaf of $\mathcal{O}_X$-modules, $\mathcal{A}\#\mathcal{B} = \mathcal{A} \otimes_{\mathcal{O}_X} \mathcal{B}$. It suffices to define the multiplication map on open affine subsets U of X. This is done in the obvious way:

$$\Gamma(U, \mathcal{A}\#\mathcal{B}) = \Gamma(U, \mathcal{A})\#\Gamma(U, \mathcal{B})$$

Now let $\mathcal{M}$ be an $\mathcal{H}$-dimodule that is locally free of finite positive rank. It follows from (6.54), Proposition 12.2.6 and the definitions of a locally free sheaf of modules and of an $\mathcal{H}$-Azumaya algebra that $\mathcal{E}nd_{\mathcal{O}_X}(\mathcal{M})$ is an $\mathcal{H}$-Azumaya algebra. Two $\mathcal{H}$-Azumaya algebras $\mathcal{A}$ and $\mathcal{B}$ are called equivalent if

$$\mathcal{A}\#\mathcal{E}nd_{\mathcal{O}_X}(\mathcal{M}) \cong \mathcal{B}\#\mathcal{E}nd_{\mathcal{O}_X}(\mathcal{N})$$

for some locally free $\mathcal{H}$-dimodules $\mathcal{M}$ and $\mathcal{N}$ of finite positive rank. The equivalences classes of $\mathcal{H}$-Azumaya algebras form a group under the operation induced by the smash product. This group $\mathrm{BD}(X, \mathcal{H})$ is called the Brauer-Long group of X and $\mathcal{H}$. Using the same type of local-global arguments as above, one can generalize many of the results in this Chapter and the preceding one to our actual situation. We also mention that the same type of construction allows us to generalize the Galois theory developed in Part II of this book to the scheme theoretic situation. In particular, if $\mathcal{H}$ is cocommutative and if $\Gamma(U,\mathcal{H})$ is faithfully flat as a $\Gamma(U,\mathcal{O}_X)$-module, for every affine open subset U of X, then we can introduce the group of $\mathcal{H}$-Galois objects $\mathrm{Gal}(X,\mathcal{H})$. If $\mathcal{H}$ is locally free of finite positive rank, then we have a split exact sequence

$$1 \longrightarrow \mathrm{Br}(X) \longrightarrow \mathrm{BM}(X,\mathcal{H}) \longrightarrow \mathrm{Gal}(X,\mathcal{H}) \longrightarrow 1 \tag{13.152}$$

and this generalizes Theorem 13.2.1. We leave it to the reader to check all the details, along the lines of the proof of Theorem 13.2.1.

It is well-known that commutative Hopf algebras over a commutative ring correspond to affine group schemes (see e.g. [201, Sec. 1.4]). The reader might expect that the Brauer-Long group of a commutative ring R and a commutative and cocommutative Hopf algebra H (or: a commutative affine group scheme) can be generalized to the situation where R is replaced by a scheme and H by a commutative group scheme. Unfortunately, the author could not find such a generalization. We remark that the language of affine group schemes was used in [176], in order to discuss properties of H-Azumaya algebras and the Brauer-Long group.

Let us finally mention that some further generalizations are possible. One can replace the category of R-modules, or the category of quasi-coherent sheaves of $\mathcal{O}_X$-modules, by a monoidal symmetric closed category. This has been done by Villanueva and his collaborators. In [126], the theory of Galois objects over a faithfully projective Hopf algebra is generalized. The main result of [198] is the generalization of Beattie's exact sequence (13.152).

Chapter 14

The Brauer group of Yetter-Drinfel'd module algebras

In the previous Chapters, we have discussed the Brauer-Long group, and we have been able to compute this invariant in some explicit situations. However, the Brauer-Long group is defined only for a limited class of Hopf algebras, namely Hopf algebras that are commutative, cocommutative and faithfully projective.
The aim of this Chapter is to show how the definition can be extended to the situation where the only assumption on H is that its antipode S is bijective. We will present a survey of the results that have been obtained so far and we refer to [46], [47] and [194] for further details.

14.1 Yetter-Drinfel'd modules

Let k be a commutative ring. We will need the letter R for something else, and this is why we call our base ring k. Let H be a Hopf algebra with invertible antipode. A (left-right) *Yetter-Drinfel'd H-module M* is a k-module with an H-action and an H-coaction such that the following compatibility relation holds, for all $m \in M$ and $h \in H$.

$$\sum h_{(1)} \cdot m_{(0)} \otimes h_{(2)} \cdot m_{(1)} = \sum (h_{(2)} \cdot m)_{(0)} \otimes (h_{(2)} \cdot m)_{(1)} h_{(1)} \tag{14.1}$$

(14.1) is equivalent to

$$\rho_M(h \cdot m) = \sum h_{(2)} \cdot m_{(0)} \otimes h_{(3)} m_{(1)} S^{-1}(h_{(1)}) \tag{14.2}$$

Yetter-Drinfel'd modules were introduced in [203], and, implicitly, in [161]. Yetter-Drinfel'd modules are sometimes named *crossed modules*, and, occasionally, *Quantum Yang-Baxter modules*. If H is commutative and cocommutative, then (14.2) reduces to the equation

$$\rho_M(h \cdot m) = \sum h \cdot m_{(0)} \otimes m_{(1)} \tag{14.3}$$

and we see that Yetter-Drinfel'd modules generalize dimodules. Let us also remark that Yetter-Drinfel'd modules can also be viewed as special cases of Doi-Hopf modules (cf. [78] and [40]).
The tensor product of two Yetter-Drinfel'd modules M and N is again a Yetter-Drinfel'd module. The H-action and H-coaction on $M\otimes N$ are given by the formulas

$$h\cdot(m\otimes n) = \sum h_{(1)}\cdot m\otimes h_{(2)}\cdot n \tag{14.4}$$
$$\rho(m\otimes n) = \sum m_{(0)}\otimes n_{(0)}\otimes n_{(1)}m_{(1)} \tag{14.5}$$

The category of (left-right) Yetter-Drinfel'd H-modules and H-linear H-colinear maps is denoted by ${}_H\mathcal{YD}(H)^H$. The tensor product is coherent in the sense of MacLane [127], and ${}_H\mathcal{YD}(H)^H$ is a monoidal category. For $M,N\in {}_H\mathcal{YD}(H)^H$, we have an isomorphism

$$t_{M,N}:\ M\otimes N\longrightarrow N\otimes M:\ m\otimes n\mapsto \sum n_{(0)}\otimes n_{(1)}\cdot m \tag{14.6}$$

in ${}_H\mathcal{YD}(H)^H$. The following axioms are satisfied, for all $M,N,P\in {}_H\mathcal{YD}(H)^H$.

$$t_{M,N\otimes P} = (I_N\otimes t_{M,P})\circ(t_{M,N}\otimes I_P) \tag{14.7}$$
$$t_{M\otimes N,P} = (t_{M,P}\otimes I_N)\circ(I_M\circ t_{N,P}) \tag{14.8}$$

It follows that ${}_H\mathcal{YD}(H)^H$ is a *braided monoidal category* (see e.g. [137, p. 198]).
If H is faithfully projective as a k-module, then the category of Yetter-Drinfel'd modules is isomorphic to $D(H)$-mod, the category of modules over the *Drinfel'd double* $D(H)=H^{*\rm cop}\bowtie H$. As a k-coalgebra, $D(H)=H^{*\rm cop}\otimes H$. $D(H)$ is a Hopf algebra. To describe the multiplication, we recall that H acts on H^* from the left and from the right as follows.

$$h\rightharpoonup h^* = \sum\langle h^*_{(2)},h\rangle h^*_{(1)} \quad\text{and}\quad h^*\leftharpoonup h = \sum\langle h^*_{(1)},h\rangle h^*_{(2)} \tag{14.9}$$

In a similar way, H is a left and right H^*-module. Now let the multiplication on $D(H)$ be given by the formula

$$(h^*\bowtie h)(k^*\bowtie k) = \sum(h^* * (h_{(1)}\rightharpoonup k^*_{(3)}\leftharpoonup S^{-1}(h_{(1)}))\bowtie(S^{-1}(k^*_{(1)})\rightharpoonup h_{(3)}\leftharpoonup k^*_{(2)}) \tag{14.10}$$

for all $h,k\in H$ and $h^*,k^*\in H^*$. The identity element is $\varepsilon_H\bowtie 1_H$, and the antipode is given by the formula

$$S(h^*\bowtie h) = \sum(S^*(h^*_{(2)})\leftharpoonup h_{(1)})\bowtie(S(h_{(2)})\leftharpoonup S^*(h^*_{(1)})) \tag{14.11}$$

If H is commutative and cocommutative, then $D(H)=H\otimes H^*=D$ as introduced in Section 13.7.
Let M and N be faithfully Yetter-Drinfel'd H-modules. We can make $\mathrm{Hom}_k(M,N)$ into a Yetter-Drinfel'd H-module in two different ways. A first possibility is to take

$$(h\cdot f)(m) = \sum h_{(1)}\cdot f(S(h_{(2)})\cdot m) \tag{14.12}$$
$$\rho_1(f)(m) = \sum f(m_{(0)})_{(0)}\otimes S^{-1}(m_{(1)})f(m_{(0)})_{(1)} \tag{14.13}$$

for the structure maps. We will call this the *type 1* structure on $\mathrm{Hom}_k(M,N)$. The *type 2* structure is given by the formulas

$$(h\rightharpoonup f)(m) = \sum h_{(2)}\cdot f(S^{-1}(h_{(1)})\cdot m) \tag{14.14}$$
$$\rho_2(f)(m) = \sum f(m_{(0)})_{(0)}\otimes f(m_{(0)})_{(1)}S(m_{(1)}) \tag{14.15}$$

Yetter-Drinfel'd module algebras

A k-algebra that is a Yetter-Drinfel'd H-module, a left H-module algebra and a right H^{op}-comodule algebra is called a *Yetter-Drinfel'd module algebra*. If H is commutative and cocommutative, then a Yetter-Drinfel'd module algebra is nothing else then an H-dimodule algebra.
Given two Yetter-Drinfel'd module algebras A and B, we define the *braided product* $A\natural B$ as follows. $A\natural B = A \otimes B$ as a Yetter-Drinfel'd module, and the multiplication is given by the formula

$$(a\natural b)(c\natural d) = \sum ac_{(0)}\natural(c_{(1)} \cdot b)d \tag{14.16}$$

for all $a, c \in A$ and $b, d \in B$. The relation between the braided product and the smash product (see (12.3)) is the following. The switch map

$$\tau : \ (A\natural B)^{\mathrm{op}} \longrightarrow B^{\mathrm{op}}\#A^{\mathrm{op}} \tag{14.17}$$

is an isomorphism of k-algebras. At this point, we remark that the whole further theory can be developed using the smash product. There are some categorical arguments in favour of the braided product, and this is why we use the braided product from now on.
In a similar way, the braided H-opposite algebra $\tilde{A}$ is the Yetter-Drinfel'd module A with multiplication

$$a \circ b = \sum b_{(0)}(b_{(1)} \cdot a) \tag{14.18}$$

the relation with the H-opposite algebra $\overline{A}$ (see (12.7)) is that

$$\overline{A^{\mathrm{op}}} \cong \tilde{A}^{\mathrm{op}} \tag{14.19}$$

The following result is a generalization of Proposition 12.1.4.

Proposition 14.1.1 *Let A, B and C be Yetter-Drinfel'd module algebras. With notations as above, $\tilde{A}$ and $A\natural B$ are Yetter-Drinfel'd module algebras, and we have isomorphisms*

$$\begin{aligned}(A\natural B)\natural C \longrightarrow A\natural(B\natural C) &: (a\natural b)\natural c \mapsto a\natural(b\natural c)\\ \tilde{B}\natural\tilde{A} \longrightarrow \widetilde{A\natural B} &: b\natural a \mapsto \sum a_{(0)}\natural(a_{(1)} \cdot b)\end{aligned}$$

If $S^2 = I_H$, then we have an isomorphism

$$A \longrightarrow \tilde{\tilde{A}} : \ a \mapsto \sum a_{(1)} \cdot a_{(0)}$$

Proposition 12.1.11 generalizes to

Proposition 14.1.2 *Let M be a faithfully projective H-dimodule. Then* $\mathrm{End}_k(M)$ *(with structure maps (14.12) and (14.13)) and Then* $\mathrm{End}_k(M)^{\mathrm{op}}$ *(with structure maps (14.14) and (14.15)) are H-dimodule algebras. The map*

$$\Gamma : \ \overline{\mathrm{End}_k(M)} \longrightarrow \mathrm{End}_k(M)^{\mathrm{op}}$$

defined by

$$\Gamma(f)(m) = \sum f_{(0)}(f_{(1)} \cdot m)$$

for all $f \in \mathrm{End}_k(M)$ *and* $m \in M$*, is an isomorphism of Yetter-Drinfel'd module algebras. Furthermore*

$$\mathrm{End}_k(M)^{\mathrm{op}} \cong \mathrm{End}_k(M^*)$$

where the action and coaction on M^* *are given by (14.14) and (14.15).*

Proposition 14.1.3 *Let* M *and* N *be faithfully projective Yetter-Drinfel'd modules. The map*

$$\Phi : \mathrm{End}_k(M)\sharp\mathrm{End}_k(N) \longrightarrow \mathrm{End}_k(M \otimes N)$$

given by

$$\Phi(f\sharp g)(m \otimes n) = \sum f(m_{(0)}) \otimes (m_{(1)} \cdot g)(n)$$

for all $f \in \mathrm{End}_k(M)$*,* $g \in \mathrm{End}_k(N)$*,* $m \in M$ *and* $n \in N$*, is an isomorphism of Yetter-Drinfel'd module algebras. From (14.6), it follows that*

$$\mathrm{End}_k(M)\sharp\mathrm{End}_k(N) \cong \mathrm{End}_k(N)\sharp\mathrm{End}_k(M)$$

as Yetter-Drinfel'd module algebras.

We also have the following generalization of Corollary 12.1.9.

Proposition 14.1.4 *Let* M *be a faithfully projective Yetter-Drinfel'd module, and* B *a Yetter-Drinfel'd module algebra. Then*

$$B\sharp\mathrm{End}_k(M) \cong \mathrm{End}_k(M)\sharp B$$

as Yetter-Drinfel'd module algebras.

14.2 H-Azumaya algebras and the Brauer group

Let A and B be Yetter-Drinfel'd module H-algebras. The definition of $(H, A\text{-}B)$-dimodule (cf. Section 12.2) can easily be generalized, and we obtain the notion of Yetter-Drinfel'd $(H, A\text{-}B)$-module. $(H, A\text{-}B)\text{-}\mathcal{YD}$ will be the category of Yetter-Drinfel'd $(H, A\text{-}B)$-modules and left A-linear, right B-linear, H-linear, H-colinear maps.

Consider the maps

$$F : A\sharp\tilde{A} \longrightarrow \mathrm{End}_k(A) \qquad F(a\sharp b)(c) = \sum ac_{(0)}(c_{(1)} \cdot b)$$
$$G : \tilde{A}\sharp A \longrightarrow \mathrm{End}_k(A)^{\mathrm{op}} \qquad G(a\sharp b)(c) = \sum a_{(0)}(a_{(1)} \cdot c)b$$

It can be shown that F and G are homomorphisms of Yetter-Drinfel'd module algebras. F makes A into a left $A\sharp\tilde{A}$-module, and G makes A into a right $\tilde{A}\sharp A$-module. For a left $A\sharp\tilde{A}$-module M, we write

$$M^A = \mathrm{Hom}_{A\sharp\tilde{A}}(A, M)$$

For a right $\tilde{A}\sharp A$-module N, let

$$^{A}N = \mathrm{Hom}_{\tilde{A}\sharp A}(A, N)$$

We now have the following generalization of Proposition 12.2.4.

Proposition 14.2.1 *Let A be a Yetter-Drinfel'd H-module. The following are equivalent.*
1) A is faithfully projective as an k-module, and the map $F: A\sharp\tilde{A}\longrightarrow \mathrm{End}_k(A)$ is an isomorphism;
2) $A^A = k$, and A is a left $A\sharp\tilde{A}$-progenerator;
3) The pair of adjoint functors

$$\begin{aligned} F_l:\ {}_H\mathcal{YD}^H &\longrightarrow (H, A\sharp\tilde{A}\text{-}k)\text{-}\mathcal{YD} &:\ N \mapsto A\otimes N\\ G_l:\ (H, A\sharp\tilde{A}\text{-}k)\text{-}\mathcal{YD} &\longrightarrow {}_H\mathcal{YD}^H &:\ M \mapsto M^A \end{aligned}$$

are inverse equivalences. In this situation, A is called a left H-Azumaya algebra.

We have a similar generalization of Proposition 12.2.5, leading to the notion of right H-Azumaya algebra. If A is a left and right H-Azumaya algebra, then we say that A is an *H-Azumaya algebra.*
If M is a faithfully projective Yetter-Drinfel'd H-module, then $\mathrm{End}_k(M)$ is an H-Azumaya algebra. For two H-Azumaya algebras A and B, the braided product $A\sharp B$ and the braided opposite $\tilde{A}$ are H-Azumaya algebras. A and B are called H-equivalent if there exist faithfully projective Yetter-Drinfel'd H-modules M and N such that

$$A\sharp\mathrm{End}_k(M) \cong B\sharp\mathrm{End}_k(N)$$

or, equivalently, if

$$A\sharp\tilde{B} \cong \mathrm{End}_k(Q)$$

for some faithfully projective Yetter-Drinfel'd H-module Q.
The set of equivalence classes of H-Azumaya algebras forms a group under the operation induced by the braided product $\sharp$. The inverse of a class $[A]$ in this group is represented by the braided opposite $\tilde{A}$. We call this new Brauer group the *Brauer group of Yetter-Drinfel'd module algebras* , and we denote it by $\mathrm{BQ}(k, H)$.
If the antipode S is not involutory, then we do not know whether $A \cong \tilde{\tilde{A}}$ (cf. Proposition 14.1.1). However, we have that A and $\tilde{\tilde{A}}$ are equivalent H-Azumaya algebras, since we have isomorphisms

$$\begin{aligned} A\sharp\mathrm{End}_k(\tilde{A}) &\cong A\sharp\tilde{A}\sharp\tilde{\tilde{A}}\\ &\cong \mathrm{End}_k(A)\sharp\tilde{\tilde{A}}\\ &\cong \tilde{\tilde{A}}\sharp\mathrm{End}_k(A) \end{aligned}$$

Proposition 14.2.2 *Let H be a faithfully projective, commutative and cocommutative Hopf algebra. Then $\mathrm{BQ}(k, H) \cong \mathrm{BD}(k, H)^{\mathrm{op}}$. The connecting isomorphism is given by sending $[A]$ to $[A^{\mathrm{op}}]$.*

The proof of Proposition 14.2.2 follows from (14.17) and (14.19). If we would have developed our theory using the smash product instead of the braided product, then we would have obtained a group that is identical to the Brauer-Long group in the situation where H is commutative, cocommutative and faithfully projective.

14.3 The subgroups of $\mathrm{BQ}(k, H)$

As expected, the various subgroups of the Brauer-Long group that we discussed in the previous Chapter have their generalizations in our new theory. We discuss these generalizations briefly in this Section. Obviously, the Brauer group $\mathrm{Br}(k)$ is a subgroup of $\mathrm{BQ}(k, H)$. With the other subgroups, the situation is somewhat more complicated.

Quasitriangular Hopf algebras and $\mathrm{BM}(k, H, R)$

Our aim is to generalize the Brauer group of H-module Azumaya algebras. Consider an H-module M, and let H coact trivially on M. If H is cocommutative, then M is an H-dimodule (and a Yetter-Drinfel'd H-module). This is not the case if H is not cocommutative.
To be able to generalize the Brauer group of H-module algebras to the noncocommutative situation, we have to consider the so-called *quasitriangular Hopf algebras.* As we will see, these are generalizations of cocommutative Hopf algebras.

Definition 14.3.1 ([79],[80]) *A Hopf algebra H with bijective antipode is called almost cocommutative if there exists*

$$R = \sum R^1 \otimes R^2 = \sum r^1 \otimes r^2 \in \mathbb{G}_m(H \otimes H)$$

such that

$$\sum h_{(2)} \otimes h_{(1)} = R(\sum h_{(1)} \otimes h_{(2)})R^{-1} \tag{14.20}$$

for all $h \in H$. If, in addition,

$$\sum \Delta(R^1) \otimes R^2 = \sum R^1 \otimes r^1 \otimes R^2 r^2 \tag{14.21}$$

$$\sum R^1 \otimes \Delta(R^2) = \sum R^1 r^1 \otimes r^2 \otimes R^2 \tag{14.22}$$

then we say that (H, R) is a quasitriangular Hopf algebra. (H, R) is triangular if $R^{-1} = \sum R^2 \otimes R^1$.

We recall the following notations:

$$R_1 = \sum 1 \otimes R^1 \otimes R^2,\ R_2 = \sum R^1 \otimes 1 \otimes R^2,\ R_3 = \sum R^1 \otimes R^2 \otimes 1$$

If (H, R) is quasitriangular, then we have the following properties (see e.g. [80], [137]):

$$R_3R_2R_1 = R_1R_2R_3 \tag{14.23}$$
$$\sum \varepsilon(R^1)R^2 = \sum \varepsilon(R^2)R^1 = 1 \tag{14.24}$$
$$\sum S(R^1) \otimes R^2 = \sum R^1 \otimes S^{-1}(R^2) = R^{-1} \tag{14.25}$$
$$\sum S(R^1) \otimes S(R^2) = R \tag{14.26}$$

Equation (14.23) is known in the literature as the *quantum Yang-Baxter equation.* Now let M be a left H-module, and consider the following H-coaction on M.

$$\rho(m) = \sum R^2 \cdot m \otimes R^1 \tag{14.27}$$

An easy computation shows that M is now a Yetter-Drinfel'd H-module:

$$\begin{aligned} \rho(h \cdot m) &= \sum (R^2 h) \cdot m \otimes R^1 \\ &= \sum (R^2 h_{(3)}) \cdot m \otimes R^{(1)} h_{(2)} S^{-1}(h_{(1)}) \\ \text{(by (14.20))} \quad &= \sum h_{(2)} R^2) \otimes h_{(3)} R^1 S^{-1}(h_{(1)}) \end{aligned}$$

If A is a left H-module algebra, then the coaction (14.27) makes A into a Yetter-Drinfel'd module algebra. Now consider classes of H-Azumaya algebras that are represented by an algebra A satisfying (14.27). These classes form a group, and we will denote this group by $\mathrm{BM}(k, H, R)$. If H is cocommutative, then $(H, 1 \otimes 1)$ is quasitriangular, and $\mathrm{BM}(k, H, R) = \mathrm{BM}(k, H)$. Thus we have the generalization that we were looking for.

Coquasitriangular Hopf algebras and BC(k, H, R)

We now dualize the definition of $\mathrm{BM}(k, H, R)$.

Definition 14.3.2 ([129], [122], [167]) *Let H be a Hopf algebra with bijective antipode. H is called almost commutative if there exists a convolution invertible $\mathcal{R}: H \otimes H \to k$ such that*

$$\sum \mathcal{R}(h_{(1)} \otimes k_{(1)}) h_{(2)} k_{(2)} = \sum \mathcal{R}(h_{(2)} \otimes k_{(2)}) k_{(1)} h_{(1)} \tag{14.28}$$

for all $h, k \in H$. $(H, \mathcal{R})$ is called coquasitriangular if, in addition, $\mathcal{R}$ is a pairing in the sense of Definition 10.8.2. To this end, it suffices that

$$\mathcal{R}(h \otimes kl) = \sum \mathcal{R}(h_{(1)} \otimes k)\mathcal{R}(h_{(2)} \otimes l) \tag{14.29}$$
$$\mathcal{R}(hk \otimes l) = \sum \mathcal{R}(h \otimes l_{(1)})\mathcal{R}(k \otimes l_{(2)}) \tag{14.30}$$

Now let M be a right H-comodule, and define a left H-action as follows

$$h \cdot m = \sum \mathcal{R}(m_{(1)} \otimes h) m_{(0)} \tag{14.31}$$

M is a Yetter-Drinfel'd module, since

$$\begin{aligned}
&\sum h_{(2)} \cdot m_{(0)} \otimes h_{(3)} m_{(1)} S^{-1}(h_{(1)}) \\
&= \sum \mathcal{R}(h_{(2)} \otimes m_{(1)}) m_{(0)} \otimes h_{(3)} m_{(2)} S^{-1}(h_{(1)}) \\
\text{(by (14.28))} \quad &= \sum m_{(0)} \otimes \mathcal{R}(h_{(3)} \otimes m_{(2)}) m_{(1)} h_{(2)} S^{-1}(h_{(1)}) \\
&= \rho(\sum \mathcal{R}(m_{(1)} \otimes h) m_{(0)}) = \rho(h \cdot m)
\end{aligned}$$

If A is an $H^{\rm op}$-comodule algebra, then (14.31) makes A into a Yetter-Drinfel'd module algebra. $\mathrm{BC}(k, H, \mathcal{R})$ is the subgroup of $\mathrm{BQ}(k, H)$ consisting of classes of H-Azumaya algebras represented by an algebra A that satisfies (14.31).

Now suppose that $(H, \mathcal{R})$ is coquasitriangular and cocommutative. Then it follows immediately from (14.29) and the fact that $\mathcal{R}$ is convolution invertible, that H is also commutative.

If $\mathcal{R}$ is a pairing and H is commutative and cocommutative, then (14.28) is automatically satisfied and $(H, \mathcal{R})$ is coquasitriangular. Thus we have a one-to-one correspondence between coquasitriangular structures on a commutative and cocommutative Hopf algebra and pairings $H \otimes H \to R$. If H is faithfully projective as a k-module, then these correspond to Hopf algebra maps $H \to H^*$ (see Lemma 10.8.3). The map θ that corresponds to $\mathcal{R}$ is given by

$$\mathcal{R}(h \otimes k) = \langle \theta(h), k \rangle$$

for all $h, k \in H$. Yetter-Drinfel'd modules that satisfy (14.31) are then nothing else then θ-modules (see (13.1)). From Proposition 14.2.2, it follows immediately that

$$\mathrm{BC}(k, H, \mathcal{R}) = \mathrm{B}_\theta(k, H)^{\rm op} \tag{14.32}$$

In particular, we have that

$$\mathrm{BC}(k, H, \varepsilon \otimes \varepsilon) = \mathrm{BC}(k, H) \tag{14.33}$$

Now we return to the situation where H is not necessarily commutative or cocommutative. We assume that H is faithfully projective. Then (H, R) is quasitriangular if and only if (H^*, R) is coquasitriangular, and

$$\mathrm{BM}(k, H, R) \cong \mathrm{BC}(k, H^*, R) \tag{14.34}$$

It is known (see [80]) that the Drinfel'd double $D(H)$ is always quasitriangular. We have the following property, generalizing Proposition 13.1.7.

$$\mathrm{BQ}(k, H) \cong \mathrm{BM}(k, D(H), R) \cong \mathrm{BC}(k, D(H)^*, R) \tag{14.35}$$

The Picard group of a Hopf algebra

A Yetter-Drinfel'd H-module I is called invertible if there exists another Yetter-Drinfel'd H-module I' such that $I \otimes I' \cong k$ as Yetter-Drinfel'd modules. Isomorphism classes of invertible Yetter-Drinfel'd H-modules form the Picard group $\mathrm{PQ}(k, H)$. If (H, R) is quasitriangular, then $\mathrm{PM}(k, H, R)$ is introduced in the obvious way. In a similar way, we can define $\mathrm{PC}(k, H, \mathcal{R})$ for a coquasitriangular Hopf algebra $(H, \mathcal{R})$. It is easy to show that

$$\mathrm{PM}(k, H, R) \cong \mathrm{Pic}(k) \times \mathrm{Alg}(H, k) \tag{14.36}$$
$$\mathrm{PC}(k, H, \mathcal{R}) \cong \mathrm{Pic}(k) \times G(H) \tag{14.37}$$

Let

$$E(H) = \{g \bowtie \lambda \in G(H) \bowtie \mathrm{Alg}(H, k) | g(\lambda \rightharpoonup h) = (h \leftharpoonup \lambda)g, \text{ for all } h \in H\}$$

(see (14.9) for the definition of $\lambda \rightharpoonup h$ and $h \leftharpoonup \lambda$. $E(H)$ is an abelian group under the operation

$$(g \bowtie \lambda)(g' \bowtie \lambda') = g'g \bowtie \lambda * \lambda'$$

The unit is $1 \bowtie \varepsilon$, and we have

$$\mathrm{PQ}(k, H) = \mathrm{Pic}(k) \times E(H) \tag{14.38}$$

Remarks 14.3.3 1) $\mathrm{PC}(k, H, \mathcal{R})$ is commutative, and (14.37) implies that $G(H)$ is commutative if H admits a coquasitriangular structure. This follows also directly from (14.28). Similarly, if (H, R) is quasitriangular, then $\mathrm{Alg}(H, k)$ is commutative.
2) If H is faithfully projective, then $E(H) = G(D(H)^*)$.
3) Let H-Aut(A) be the group of Yetter-Drinfel'd module algebra isomorphisms of A. $f \in H$-Aut(A) is called H-INNER if $f(x) = axa^{-1}$ for some $a \in A^{H\mathrm{co}H}$, and H-INN(A) is the subgroup of H-Aut(A) consisting of H-INNER automorphisms. We have the following version of the Rosenberg-Zelinsky exact sequence.

$$1 \longrightarrow H\text{-INN}(A) \longrightarrow H\text{-Aut}(A) \xrightarrow{\Phi} PQ(k, H) \tag{14.39}$$

where

$$\Phi(f) = I_f = \{x \in A | \sum x_{(0)}(x_{(1)} \cdot a) = f(a)x \text{ for all } a \in A\} \tag{14.40}$$

The split part of BQ(k, H)

$\mathrm{BQ}(k, H)$ is functorial in k: if l is a commutative k-algebra, then we have a natural map $\mathrm{BQ}(k, H) \longrightarrow \mathrm{BQ}(l, l \otimes H)$. As usual, we write

$$\mathrm{BQ}(l/k, H) = \mathrm{Ker}\,(\mathrm{BQ}(k, H) \to \mathrm{BQ}(l, l \otimes H)$$

and

$$\mathrm{BQ}^s(k, H) = \bigcup \mathrm{BQ}(l/k, H)$$

where the union is taken over all faithfully flat commutative extensions l of k. Using the Rozenberg-Zelinsky exact sequence (14.39), we can show that there is a complex

$$\mathrm{BQ}^s(k, H) \longrightarrow \mathrm{BQ}(k, H) \xrightarrow{\beta} O(E(H)) \tag{14.41}$$

where

$$O(E(H)) = \{f \in \mathrm{Aut}(E(H)) | \omega \circ f = \omega\}$$

and $\omega(g \bowtie f) = \langle \lambda, g \rangle$. This generalizes the results of Section 13.7. A next step would be to extend β to a map $\beta : \mathrm{BQ}(k, H) \to D(H)$, as in Section 13.9. The author has no idea how this could be done. Another interesting problem that is still open is to provide a cohomological description of $\mathrm{BQ}^s(k, H)$.

Generalizing Deegan's subgroup of the Brauer-Long group

Let H be a faithfully projective Hopf algebra, and suppose that $\alpha : H \to H$ is a Hopf algebra isomorphism. Let $H_\alpha = H$ as an H-comodule, and let H coact on H_α as follows.

$$h \cdot x = \sum \alpha(h_{(2)}) x S^{-1}(h_{(1)})$$

for all $h \in H$ and $x \in H_\alpha$. H_α is not a Yetter-Drinfel'd module, but it can be shown that $A_\alpha = \mathrm{End}_k(H_\alpha)$ is a Yetter-Drinfel'd module algebra, and even an H-Azumaya algebra. The map

$$\pi : \mathrm{Aut}_{\mathrm{Hopf}}(H) \longrightarrow \mathrm{BQ}(k, H) : \alpha \mapsto [A_\alpha]$$

is an anti-homomorphism of groups. Surprisingly, it is not necessarily a monomorphism. In [194], it is shown that we have an exact sequence

$$1 \longrightarrow G(D(H)^*) \longrightarrow G(D(H)) \longrightarrow \mathrm{Aut}_{\mathrm{Hopf}}(H) \longrightarrow \mathrm{BQ}(k, H) \tag{14.42}$$

Remarks 14.3.4 1) As in the classical situation, we can define $\mathrm{BT}(k, H)$: consider H-Azumaya algebras of the form $\mathrm{End}_k(M)$, where M is a faithfully projective H-module and H-comodule, but not necessarily a Yetter-Drinfel'd H-module. $\mathrm{BT}(k, H)$ is the subset of $\mathrm{BQ}(k, H)$ consisting of classes represented by such algebras. We can show that $\mathrm{BT}(k, H)$ is a multiplicative subset containing the image of the map π, but we do not know whether $\mathrm{BT}(k, H)$ is a group.
2) Define $\Theta : \mathrm{Aut}_{\mathrm{Hopf}}(H) \longrightarrow \mathrm{Aut}(E(H))$ as follows: for a Hopf algebra automorphism $\alpha : H \to H$ and $g \bowtie \lambda \in E(H)$, let

$$\Theta(\alpha)(g \bowtie \lambda) = \alpha^{-1}(g) \bowtie \lambda \circ \alpha$$

Then Θ is well-defined, and the diagram

$$\begin{array}{ccc} \mathrm{Aut}_{\mathrm{Hopf}}(H) & \xrightarrow{\Theta} & \mathrm{Aut}(E(H)) \\ \downarrow{\scriptstyle \pi} & & \uparrow{\scriptstyle \beta} \\ \mathrm{BT}(k, H) & \hookrightarrow & \mathrm{BQ}(k, H) \end{array}$$

commutes.
3) Using the exact sequence (14.42), it is possible to construct examples of Hopf algebras H such that the Brauer group $\mathrm{BQ}(k, H)$ is not a torsion group. We refer to [194] for more details.

The strong part of $\mathrm{BQ}(k, H)$

A Yetter-Drinfel'd H-module algebra A is called *strongly H-separable* if it contains a separability idempotent on which H acts and coacts trivially. A is called a *strong H-Azumaya algebra* if A is strongly H-separable and $A^A = {}^{A} = k$. Strong H-Azumaya algebras are Azumaya, and the classes in $\mathrm{BQ}(k, H)$ represented by a strong H-Azumaya algebra form a subgroup $\mathrm{BQS}(k, H)$. If H is semisimplelike and cosemisimplelike, then

$$\mathrm{BQ}(k, H) = \mathrm{BQS}(k, H)$$

More details may be found in [47].

Explicit examples?

Take your favourite Hopf algebra H and ask yourself the following question: can I compute the Brauer group of Yetter-Drinfel'd module algebras. That this will be a complicated problem can be seen from the computations that we carried out in the two last sections of the previous chapter, where we computed the Brauer-Long group in the easiest possible situation. Even in the case where the groundring k is an algebraically closed field, almost nothing is known.
F. Van Oystaeyen and Y. Zhang told the author that they recently made some progress in computing the Brauer group of a group ring over a noncommutative abelian group. Their results will appear in [195] and [196].

Appendix A

Abelian categories and homological algebra

A.1 Abelian categories

Definition A.1.1 *An Abelian category is a category $\underline{C}$ satisfying the following axioms :*

1. *$\underline{C}$ has a zero object ;*
2. *$\underline{C}$ has finite products and direct sums ;*
3. *every morphism in $\underline{C}$ has a kernel and a cokernel ;*
4. *every monic is a kernel and every epic is a cokernel.*

For equivalent definitions and more details, we refer the reader to any standard work on category theory, for example Freyd [86], Faith [82] or Popescu [159]. The basic example of an abelian category is of course the category $\underline{\underline{Ab}}$ of Abelian groups. Grothendieck [95] introduced some additional conditions on Abelian categories, which have interesting properties :

Definition A.1.2 *An abelian category $\underline{C}$ satisfies condition Ab3 if it has arbitrary direct sums. It satisfies Ab3* if it has arbitrary direct products.*

Consider a category $\underline{C}$ which is *Ab*3 and *Ab*3*. Recall that, for $\{X_i | i \in I\}$ in $\underline{C}$, the direct sum $X = \coprod_{i \in I} X_i$ is the object in $\underline{C}$ which is universal with respect to maps $u_i | X_i \to X$, that is, if, for $C \in \underline{C}$, maps $f_i : X_i \to C$ are given, then there exists a map $f : X \to C$ such that $f \circ u_i = f_i$. Fix $j \in I$ and consider maps $f_i : X_i \to X_j$ defined as follows : $f_i = 0$ if $i \neq j$, and $f_j = I_{X_j}$. Then the universal property gives us a map $p_j : X \to X_j$ such that $p_j \circ u_j = I_{X_j}$ and $p_j \circ u_i = 0$ for $i \neq j$. p_j is called the canonical projection of X onto X_j. The product $X' = \prod_{j \in J} X_i$ is universal with respect to maps $q_i : X' \to X_i$, that is, for $C \in \underline{C}$ and morphisms $g_i : C \to X_i$ there exists a morphism $g : C \to X'$ with $g_i = q_i \circ g$. If we take $C = X = \coprod_{i \in I} X_i$, $g_i = p_i$, then it follows that we have a canonical map

$$t: X = \coprod_{i \in I} X_i \to X' = \prod_{i \in I} X_i$$

with $p_i = q_i \circ t$.

Definition A.1.3 *An Ab3-category satisfies condition Ab4 if for any set of monics $\{f_i: X_i \to Y_i | i \in I\}$ in $\underline{C}$, the direct sum morphism $\coprod_{i \in I} f_i: \coprod_{i \in I} X_i \to \coprod_{i \in I} Y_i$ is monic, or, equivalently, if the direct product of a set of short exact sequences in $\underline{C}$ is exact. Similarly, an Ab3*-category is called Ab4* if the direct product of a set of epics in $\underline{C}$ is epic, or, equivalently, if the direct product of a set of short exact sequences in $\underline{C}$ is exact.*

Using the remarks made above, we can give a sufficient condition for $\underline{C}$ to be *Ab4* :

Proposition A.1.4 *If $\underline{C}$ is an Ab3 and an Ab3* category, and if for any set $\{X_i | i \in I\}$ in $\underline{C}$, the canonical map*

$$t : X = \coprod_{i \in I} X_i \to X' = \prod_{i \in I} X_i$$

is monic, then $\underline{C}$ is Ab4.

Proof Let $f_i: X_i \to Y_i$ be monic, for $i \in I$. Then

$$\prod_{i \in I} f_i: \prod_{i \in I} X_i \to \prod_{i \in I} Y_i$$

is monic, since the product is always left exact. Now look at the diagram

$$\begin{array}{ccc} \coprod_{i \in I} X_i & \xrightarrow{\coprod f_i} & \coprod_{i \in I} Y_i \\ \downarrow t & & \downarrow t \\ \prod_{i \in I} X_i & \xrightarrow{\prod f_i} & \prod_{i \in I} Y_i \end{array}$$

to conclude that $\coprod_{i \in I} f_i$ is monic. □

Now let $\underline{C}$ be an *Ab3*-category. Using the fact that $\underline{C}$ has arbitrary direct sums and cokernels, we may show that $\underline{C}$ is category with arbitrary inductive limits, we refer to [159, Th. I.4.1]. Now consider a directed set of short exact sequences in $\underline{C}$:

$$0 \to X_i \to Y_i \to Z_i \to 0. \tag{A.1}$$

It is easy to show that the direct inductive limit is right exact, that is, the sequence

$$\varinjlim X_i \to \varinjlim Y_i \to \varinjlim Z_i \to 0$$

is exact.

Definition A.1.5 *An Ab3-category $\underline{C}$ is called Ab5 if the direct inductive limit is exact, that is, for every directed set of short exact sequences (A.1) in $\underline{C}$ we have a short exact sequence*

$$0 \to \varinjlim X_i \to \varinjlim Y_i \to \varinjlim Z_i \to 0$$

in $\underline{C}$.

An *Ab5*-category is automatically *Ab4* (cf. e.g. [159, II.8.9]). Recall that a set of objects $\{U_i : i \in I\}$ in $\underline{C}$ is called a set of generators if for any two distinct morphisms $f, g: X \to Y$ in $\underline{C}$, there exists $i_0 \in I$ and $h: U_{i_0} \to X$ such that $h \circ f \neq h \circ g$. If a set of generators consists of only one object, then we call this object a *generator* for the category $\underline{C}$. If $\underline{C}$ has arbitrary direct products, and a set of generators $\{U_i: i \in I\}$, then $\prod_{i \in I} U_i$ is a generator for $\underline{C}$. Indeed, let $h: U_{i_0} \to X$ be such that $h \circ f \neq h \circ g$, then the fact that q_{i_0} is epic implies that $(q_{i_0} \circ h) \circ f \neq (q_{i_0} \circ h) \circ g$.

Recall that an object X in a category $\underline{C}$ is called *injective* if for every monic $f: U \to V$ in $\underline{C}$, and for every $h: U \to X$, there exists $\overline{h}: V \to X$ such that $h = \overline{h} \circ f$. $\underline{C}$ has enough injective objects if for every U in $\underline{C}$, there exists a monic $U \to X$ in $\underline{C}$, with X injective. For example, for a commutative ring R, the category R-mod has enough injective objects, cf. [48, I.3.3].

Theorem A.1.6 *An Ab5-category with a generator has enough injective objects.*

Proof We refer to the literature, e.g. [86, 6.25] or [159, III.10.10]. □

Example We leave it to the reader to show that $\underline{\underline{Ab}}$ is an *Ab5* and *Ab4**-category. Clearly $\mathbb{Z}$ is a generator for $\underline{\underline{Ab}}$, hence $\underline{\underline{Ab}}$ has enough injective objects.

A.2 Derived functors

Let $\underline{C}$ be an abelian category having enough injective objects. For any A in $\underline{C}$, we can find a monic $A \to X_0$, with X_0 injective. We therefore obtain an exact sequence

$$0 \to A \to X^0 \to C^0 \to 0.$$

Now let $C^0 \to X^1$ be monic with X^1 injective. Then we may write down an exact sequence

$$0 \to A \to X^0 \to X^1 \to C^1 \to 0$$

Proceeding in this way, we obtain a long exact sequence

$$0 \to A \to X^0 \to X^1 \to X^2 \to \cdots$$

called an *injective resolution* of A. The object C^i in $\underline{C}$ is called the i-th *syzygy* of the resolution:

$$C^i = \mathrm{Ker}\,(X^i \to X^{i+1}) = \mathrm{Im}\,(X^{i-1} \to X^i)$$

for $i \geq 0$.
Let $\underline{C}$, $\underline{D}$ be abelian categories, and suppose that $\underline{C}$ has enough injectives. Consider a left exact functor $f: \underline{C} \to \underline{D}$. It is well-known that we have an essentially unique sequence of functors $R^i f: \underline{C} \to \underline{D}$, for $i \geq 0$, called the right derived functors of f. These functors satisfy the following properties :

1. $R^0 f = f$;
2. $R^i f(I) = 0$ if $i > 0$ and if I is an injective object of $\underline{C}$;
3. For any exact sequence
$$0 \to A \to B \to C \to 0$$
in $\underline{C}$, we have the following long exact sequence in $\underline{D}$:
$$0 \to f(A) \to f(B) \to f(C) \to R^1 f(A) \to R^1 f(B) \to R^1 f(C) \to R^2 f(A) \to \cdots$$

Moreover the association of this long exact sequence to the short exact sequence behaves functorially. The right derived functors $R^i f$ may be determined as follows : take an object $A \in \underline{C}$, and let
$$0 \to A \to X^0 \to X^1 \to X^2 \to \cdots$$
be an injective resolution of A in $\underline{C}$. Applying f to this resolution, we obtain the following complex in $\underline{D}$:
$$0 \to f(A) \to f(X^0) \to f(X^1) \to f(X^2) \to \cdots$$
$R^i f(A)$ may be defined as the i-th cohomology group of this complex. It is a straightforward exercise to show that $R^i f(A)$ is independent of the chosen resolution, and that it satisfies the three conditions given above. We may also show that $R^i f(A)$ is uniquely determined by these three conditions : let $R^i f(A)$, $S^i f(A)$ be two sequences of functors satisfying the three conditions. Take a monic $A \to B$ in $\underline{C}$, with B injective, and let C be the quotient B/A. Write the long exact sequence associated to the short exact sequence $0 \to A \to B \to C \to 0$. It then follows that $R^i f(A) \cong S^i f(A)$.
An object A of $\underline{C}$ is called *f-acyclic* if $R^i f(A) = 0$, for all $i > 0$.

Lemma A.2.1 *Let $f: \underline{C} \to \underline{D}$ be as above, and suppose that $\underline{I}$ is a class of objects in $\underline{C}$ such that*

1. *every object of $\underline{C}$ is a subobject of an object of $\underline{I}$;*
2. *if $A \oplus A' \in \underline{I}$, then $A \in \underline{I}$;*
3. *if $A, A' \in \underline{I}$, and $\alpha: A' \to A$ is a $\underline{C}$-morphism, then $A'' = \operatorname{Coker} \alpha \in \underline{I}$;*
4. *$f_{|\underline{I}}: \underline{I} \to \underline{D}$ is exact.*

Then $\underline{I}$ contains all injective objects of $\underline{C}$, and every element of $\underline{I}$ is f-acyclic. To compute the derived functors of f, we can therefore use resolutions by objects in $\underline{I}$.

Proof Let I be injective, and consider $A \in \underline{I}$ such that $0 \to I \to A$ is exact (use 1.). Since I is injective, it is a direct summand of A, hence $I \in \underline{I}$, by 2. Next, take $A \in \underline{I}$, and consider an injective resolution

$$0 \to A \to X^0 \to X^1 \to \cdots$$

and let $C^i = \mathrm{Im}\,(X^{i-1} \to X^i = \mathrm{Ker}\,(X^i \to X^{i+1})$ be the corresponding syzygies. Then we have short exact sequences

$$0 \longrightarrow A \longrightarrow X^0 \longrightarrow C^1 \longrightarrow 0$$

$$0 \longrightarrow C^i \longrightarrow X^i \longrightarrow C^{i+1} \longrightarrow 0$$

From 3., it follows that all the C^i are in $\underline{I}$, and, by 4.,

$$0 \longrightarrow f(C^i) \longrightarrow f(X^i) \longrightarrow f(C^{i+1}) \longrightarrow 0$$

is exact, so $R^i f(A) = 0$, for all $i > 0$. Finally, consider $A \in \underline{C}$, and let

$$0 \to A \to T^0 \to T^1 \to \cdots$$

be a resolution by objects in $\underline{I}$. Write

$$Z^i = \mathrm{Im}\,(T^{i-1} \to T^i = \mathrm{Ker}\,(T^i \to T^{i+1})$$

To the exact sequence

$$0 \longrightarrow Z^i \longrightarrow T^i \longrightarrow Z^{i+1} \longrightarrow 0$$

we may associate a long exact sequence in $\underline{C}$. Since $R^n f(T) = 0$ for $n > 0$, we have that

$$0 \longrightarrow f(Z^i) \longrightarrow f(T^i) \longrightarrow f(Z^{i+1}) \longrightarrow R^1 f(Z^i) \longrightarrow 0$$

and

$$R^f(Z^{i+1}) \cong R^{n+1} f(Z^i)$$

Therefore

$$\begin{aligned} R^n f(A) &= R^n f(Z^0) = R^{n-1} f(Z^1) = \cdots = R^1 f(Z^{n-1}) \\ &= \mathrm{Coker}\,(f(T^{n-1}) \to f(Z^n)) \\ &= \mathrm{Ker}\,(f(T^n) \to f(T^{n+1}))/\mathrm{Im}\,(f(T^{n-1}) \to f(T^n)) \end{aligned}$$

□

Theorem A.2.2 *Let $\underline{B}$, $\underline{C}$, $\underline{D}$ be abelian categories, and suppose that $\underline{B}$ and $\underline{C}$ have enough injectives. Suppose that $f: \underline{B} \to \underline{C}$ and $g: \underline{C} \to \underline{D}$ are left exact functors, such that f takes injective objects of $\underline{B}$ to g-acyclics. Consider $A \in \underline{B}$, and an injective resolution*

$$0 \to A \to X^0 \to X^1 \to X^2 \to \cdots \qquad \text{(A.2)}$$

with syzygies

$$C^q = \mathrm{Ker}\,(X^q \to X^{q+1}) = \mathrm{Im}\,(X^{q-1} \to X^q) \tag{A.3}$$

Then we have, for all $q \geq 0$, a long exact sequence

$$\begin{aligned} 0 \;&\to\; R^1g(f(C^q)) \to R^{q+1}(gf)(A) \to g(R^{q+1}f(A)) \\ &\to\; R^2g(f(C^q)) \to R^1(g)(f(C^{q+1})) \to R^1g(R^{q+1}f(A)) \\ &\to\; \cdots \\ &\to\; R^{p+1}g(f(C^q)) \to R^p(g)(f(C^{q+1})) \to R^pg(R^{q+1}f(A)) \to \cdots \end{aligned} \tag{A.4}$$

Combining the sequences $q = 0$ and $q = 1$, we obtain the following :

$$\begin{aligned} 0 \;&\to\; R^1g(f(A)) \to R^1(gf)(A) \to g(R^1f(A)) \\ &\to\; R^2g(f(A)) \to \mathrm{Ker}\,(R^2(gf)(A) \to g(R^2f(A))) \\ &\to\; (R^1g)(R^1f)(A) \to R^3g(f(A)) \end{aligned} \tag{A.5}$$

If g is an exact functor, then for all $A \in \underline{\mathcal{B}}$ and $q \geq 1$, we have that

$$R^q(gf)(A) \cong g(R^qf(A)) \tag{A.6}$$

If f is an exact functor, then for all $A \in \underline{\mathcal{B}}$ and $q \geq 1$, we have that

$$(R^qg)f(A) \cong R^q(g \circ f)(A) \tag{A.7}$$

Remark A.2.3 The proof that we present here is taken from [199]. The result is originally due to Grothendieck [95], and may be restated as follows : under the assumptions of theorem, we have a Leray spectral sequence

$$R^pgR^qf(A) \Rightarrow R^{p+q}(gf)(A)$$

For details concerning spectral sequences, the reader is referred to the literature, e.g. [102].

Proof Since f is left exact, we have that $f(C^q) = \mathrm{Ker}\,(f(X^q) \to f(X^{q+1}))$. Let $B^q = \mathrm{Im}\,(f(X^{q-1}) \to f(X^q))$; then we have an exact sequence

$$0 \to f(C^q) \to f(X^q) \to B^{q+1} \to 0 \tag{A.8}$$

in $\underline{\mathcal{C}}$. Applying g, we obtain a long exact sequence in $\underline{\mathcal{D}}$. Since $f(X^q)$ is acyclic, $R^n(f(X^q)) = 0$ for all $n > 0$, so we have :

$$0 \to g(f(C^q)) \to g(f(X^q)) \to g(B^{q+1}) \to R^1g(f(C^q)) \to 0 \tag{A.9}$$

and

$$R^pg(B^q) \cong R^{p+1}g(f(C^q)), \tag{A.10}$$

for $p \geq 1$. We also have the following exact sequence in $\underline{\mathcal{C}}$:

$$0 \to B^{q+1} \to f(C^{q+1}) \to R^{q+1}f(A) \to 0. \tag{A.11}$$

Applying g, we obtain the following exact sequence in $\underline{\mathcal{D}}$:

$$\begin{aligned} 0 &\to g(B^{q+1}) \to g(f(C^{q+1})) \to g(R^{q+1}f(A)) \\ &\to R^1g(B^{q+1})) \cong R^2g(f(C^q)) \to R^1g(f(C^{q+1})) \to \cdots \end{aligned}$$

We have a commutative diagram

$$\begin{array}{ccccccccc} & & gf(X^q) & & gf(X^q) & & & & \\ & & \downarrow & & \downarrow & & & & \\ 0 & \to & g(B^{q+1}) & \to & g(f(C^{q+1})) & \to & g(R^{q+1}f(A)) & \to & R^2g(f(C^q)) & \to & \cdots \\ & & \downarrow & & \downarrow & & & & \\ 0 & \to & R^1g(f(C^q)) & \to & W & \to & g(R^{q+1}f(A)) & \to & R^2g(f(C^q)) & \to & \cdots \\ & & \downarrow & & \downarrow & & & & \\ & & 0 & & 0 & & & & \end{array}$$

The first column is (A.9), and the second is just the definition of W as the cokernel of the composite $gf(X^q) \to g(B^{q+1}) \to g(f(C^q))$. Diagram chasing shows that the exactness of the first long sequence implies the exactness of the second one. Finally, we can identify W and $R^{q+1}(gf)(A)$; this may be seen as follows : $f(C^{q+1}) = \operatorname{Ker}(f(X^{q+1}) \to f(X^{q+2}))$, and, by the left exactness of g, $g(f(C^{q+1})) = \operatorname{Ker}(gf(X^{q+1}) \to gf(X^{q+2}))$, hence

$$\begin{aligned} &\operatorname{Coker}(gf(X^q) \to g(f(C^{q+1}))) \\ &= \operatorname{Ker}(gf(X^{q+1}) \to gf(X^{q+2}))/\operatorname{Im}(gf(X^q) \to gf(X^{q+1})) \\ &= R^{q+1}(gf)(A). \end{aligned}$$

Then the second row of the diagram above is the first long exact sequence of the theorem. We leave it to the reader to derive the second one.
If g is exact, then $R^ig = 0$ for $i \geq 1$, and it follows immediately that

$$R^{q+1}(gf)(A) \cong g(R^{q+1}f(A))$$

for $A \in \underline{\mathcal{B}}$ and $q \geq 0$.
If f is exact, then $R^if = 0$, for any $i \geq 1$. We then have

$$(R^1g)(f(C^q)) \cong R^{q+1}(gf)(A)$$

and

$$R^{m+1}g(f(C^R)) \cong R^m g(f(C^{r+1}))$$

for $q \geq 0$ and $m, r \geq 1$. Consequently

$$\begin{aligned} R^{q+1}(gf(A)) &\cong (R^1g)(f(C^q)) \\ &\cong (R^2g)(f(C^{q-1})) \\ &\cong \cdots \\ &\cong (R^{q+1}g)(f(C^0)) \\ &= (R^{q+1}g)(f(A)) \end{aligned}$$

□

Appendix B

Faithfully flat descent

Let R be a commutative ring and S a commutative R-algebra ; then many properties of modules M over R are inherited by the corresponding S-module $S \otimes_R M$, such as the properties of being finitely generated, free or projective. General descent theory is concerned with the inverse question: how to "descend" properties over S to properties over R ? In particular, when does an element, a module, an algebra over S "come from" a similar object over R ? Throughout, we will use the following conventions. The unadorned tensor product symbol stands for $\otimes_R$. As usual, we will write $S^{\otimes n}$ for the tensor product over R of n copies of S, and we will write $\otimes_n$ for the tensor product over $S^{\otimes n}$. For any $M, N \in R-$ mod, the switch map $M \otimes N \to N \otimes M$ will be denoted by t or τ. For $M \in R$-mod $(i = 1, \cdots, n)$, we define, for $j = 1, \cdots, n+1$:

$$\epsilon_j : M_1 \otimes \cdots \otimes M_n \to M_1 \otimes \cdots \otimes S \otimes \cdots M_n$$

by

$$\epsilon_j(m_1 \otimes \cdots \otimes m_n) = m_1 \otimes \cdots \otimes 1 \otimes \cdots \otimes m_n.$$

Take $M_1 = M_2 = ... = M_n = S$ and let $\partial_{n-1}: S^{\otimes n} \to S^{\otimes(n+1)}$ be the alternating sum

$$\epsilon_1 - \epsilon_2 + \cdots + (-1)^n \epsilon_{n+1},$$

then we obtain the *Amitsur complex* $C^+(S/R)$ as

$$0 \longrightarrow R \xrightarrow{\partial_0} S \xrightarrow{\partial_1} S^{\otimes 2} \xrightarrow{\partial_2} S^{\otimes 3} \xrightarrow{\partial_3} \cdots \tag{B.1}$$

A first key result is then the following :

Proposition B.0.4 *If S is a faithfully flat R-algebra, then, for any R-module N, the sequence $N \otimes C^+(S/R)$ is exact.*

Proof (cf. [114, Prop. II.2.1]) Since S is faithfully flat as an R-module, it suffices to show that the sequence $N \otimes C^+(S/R) \otimes S$ is exact. Let us show that this sequence is exact in $N \otimes S^{\otimes k} \otimes S = N \otimes S^{\otimes(k+1)}$, for every $k \geq 0$. Suppose

$$x = \sum_j n_j \otimes s_{1j} \otimes s_{2j} \otimes \cdots \otimes s_{k+1,j} \in \mathrm{Ker}\,(I \otimes \partial_{k-1} \otimes I).$$

Then

$$\sum_j n_j \otimes 1 \otimes s_{1j} \otimes s_{2j} \otimes \cdots \otimes s_{k+1,j} =$$

$$\begin{aligned} & \sum_j n_j \otimes s_{1j} \otimes 1 \otimes s_{2j} \otimes \cdots \otimes s_{k+1,j} \\ - & \sum_j n_j \otimes s_{1j} \otimes s_{2j} \otimes 1 \otimes \cdots \otimes s_{k+1,j} \\ + & \cdots \\ + & (-1)^{k+1} \sum_j n_j \otimes s_{1j} \otimes s_{2j} \otimes \cdots \otimes 1 \otimes s_{k+1,j}. \end{aligned}$$

Multiplying the first two factors, we obtain

$$x = (I \otimes \partial_{k-2} \otimes I)(\sum_j n_j s_{1j} \otimes s_{2j} \otimes \cdots \otimes s_{k+1,j}),$$

which proves the assertion. □

Proposition B.0.4 has the following interesting consequences :

Corollary B.0.5 *Let S be a faithfully flat R-algebra. Then, with notations as in I.3 :*

$$\mathrm{H}^0(S/R, \mathbb{G}_m) = \mathbb{G}_m(R) \text{ and } \mathrm{H}^0(S/R, \mathbb{G}_a) = \mathbb{G}_a(R) = R.$$

Consequently, $\mathbb{G}_m$ and $\mathbb{G}_a$ are sheaves on R_E.

Proof Take $N = R$ in B.0.4. The fact that $\mathbb{G}_m$ and $\mathbb{G}_a$ are sheaves then follows from I.3.5.3. □

From B.0.5 we immediately obtain the following statements.

Corollary B.0.6 (Faithfully Flat Descent of Elements) *Let S be a faithfully flat R-algebra, and N an R-module. Then $\sum_i n_i \otimes s_i \in N \otimes S$ can be written under the form $n \otimes 1$ for some $n \in N$ if and only if $\sum_i n_i \otimes 1 \otimes s_i = \sum_i n_i \otimes s_i \otimes 1$ in $N \otimes S \otimes S$.*

Corollary B.0.7 (Faithfully Flat Descent of Homomorphisms) *Let S be a faithfully flat R-algebra, and $N,\ N' \in R$-mod. Let $g\colon N \otimes S \to N' \otimes S$ be a homomorphism of S-modules. Then $g = f \otimes I$ for some $f\colon N \to N'$ if and only if $g_2 = g_3$ in $\mathrm{Hom}_{S\otimes S}(N \otimes S \otimes S, N' \otimes S \otimes S)$.*

Proof It suffices to show that, for all $n \in N$, $g(n \otimes 1) = n' \otimes 1$ for some $n' \in N'$. By B.0.6, this follows from

$$(\epsilon_2 \circ g)(n \otimes 1) = g_2(n \otimes 1 \otimes 1) = g_3(n \otimes 1 \otimes 1) = (\epsilon_3 \circ g)(n \otimes 1).$$

□

Corollary B.0.6 has the following interesting consequence.

Corollary B.0.8 *Let $R \to S$ be a commutative faithfully flat R-algebra, and consider $a_i,\ b_i \in S$. If $\sum_i a_i \otimes b_i \in R \otimes S$, then*

$$\sum_i a_i \otimes b_i = \sum_i 1 \otimes a_i b_i$$

in $S \otimes S$.

Proof $R \otimes S \to S \otimes S$ is a faithfully flat extension, and it follows from Corollary B.0.6 that $\sum_i a_i \otimes b_i \in R \otimes S$ if and only if

$$\sum (a_i \otimes b_i) \otimes (1 \otimes 1) = \sum (1 \otimes 1) \otimes (a_i \otimes b_i)$$

in $(S \otimes S) \otimes_{R \otimes S} (S \otimes S)$, or, equivalently

$$\sum a_i \otimes b_i \otimes 1 = \sum a_i \otimes 1 \otimes b_i$$

in $S \otimes S \otimes S$. We obtain our results after applying the multiplication map to the first two factors. □

Suppose that M is an S-module. When is $M = N \otimes S$ for some $N \in R$-mod? Consider an $S^{\otimes 2}$-homomorphism $u : M_1 \to M_2$. Let $\psi : M \otimes S \to M$ be given by $\psi(m \otimes s) = sm$, and define $\hat{u} : M \otimes M$ by $\hat{u} = \psi \circ u \circ \epsilon_1$. We call u a *descent datum* for M if $u_2 = u_3 \circ u_1$ and $\hat{u} = I_M$. We have the following characterization of descent data :

Proposition B.0.9 *An $S^{\otimes 2}$-homomorphism $u : M_1 \to M_2$ is a descent datum for M if and only if $u_2 = u_3 \circ u_1$ and u is an isomorphism.*

Proof (cf. [114, II.3.1]) For $m \in M$, we write

$$u(1 \otimes m) = \sum_i m_i \otimes s_i \quad \text{and} \quad u(1 \otimes m_i) = \sum_j m_{ij} \otimes s_{ij}$$

Then

$$\begin{aligned} (u_3 \circ u_1)(1 \otimes 1 \otimes m) &= \sum_{i,j} m_{ij} \otimes s_{ij} \otimes s_i \qquad \text{(B.2)} \\ &= u_2(1 \otimes 1 \otimes m) \\ &= \sum_i m_i \otimes 1 \otimes s_i \end{aligned}$$

With the same notations, we also have

$$\hat{u}(m) = \sum_i s_i m_i \quad \text{and} \quad u(1 \otimes \sum_i s_i m_i) = \sum_{i,j} m_{ij} \otimes s_{ij} s_i.$$

Now suppose that u is an isomorphism. Multiplying the second and third factor in (B.2), we obtain

$$\sum_{i,j} m_{ij} \otimes s_{ij} s_i = \sum_i m_i \otimes s_i$$

or

$$u(1 \otimes \sum_i s_i m_i) = u(1 \otimes m)$$

or

$$\hat{u}(m) = \sum_i s_i m_i = m.$$

Conversely, suppose $\hat{u} = I_M$. Then

$$\begin{aligned}(t \circ u \circ t \circ u)(1 \otimes m) &= (t \circ u)(\sum_i s_i \otimes m_i)\\ &= \sum_{ij} s_{ij} \otimes s_i m_{ij}\\ &= \sum_i 1 \otimes s_i m_i\\ &= 1 \otimes \hat{u}(m)\\ &= 1 \otimes m,\end{aligned}$$

where we used the identity that arises if we multiply the first and third factor in (B.2). In a similar way, we can prove that $u \circ t \circ u \circ t$ is the identity, hence u is invertible. □

Theorem B.0.10 (Faithfully Flat Descent for R-modules) *If $u: M_1 \to M_2$ is a descent datum for M, and if S is a faithfully flat R-algebra, then there exists an R-module N and an S-isomorphism $\eta: N \otimes S \to M$ such that the diagram below commutes :*

$$\begin{array}{ccc} N_{13} = S \otimes N \otimes S & \xrightarrow{I_S \otimes \eta} & M_1 = S \otimes M \\ \downarrow{\scriptstyle \tau \otimes I_S} & & \downarrow{\scriptstyle u} \\ N_{23} = N \otimes S \otimes S & \xrightarrow{\eta \otimes I_S} & M_2 = M \otimes S \end{array} \tag{B.3}$$

The pair (N, η) is unique up to isomorphism. A possible choice for N and η is the following:

$$N = \{x \in M | x \otimes 1 = u(1 \otimes x)\} \quad \text{and} \quad \eta(n \otimes s) = sn \tag{B.4}$$

Proof (cf. [114, II.3.2]) Let (N, η) be given by (B.4). We then have an exact sequence

$$0 \to N \to M \rightrightarrows M \otimes S$$

with $\epsilon_2, g \circ \epsilon_1: M \to M \otimes S$. Now consider the following diagrams of left S-modules (S always acts on the first factor) :

$$\begin{array}{ccccccc} 0 & \longrightarrow & S \otimes N & \longrightarrow & S \otimes M & \xrightarrow{I \otimes \epsilon_2} & S \otimes M \otimes S \\ & & & & \downarrow{\scriptstyle u} & & \downarrow{\scriptstyle u_3} \\ 0 & \longrightarrow & M & \longrightarrow & M \otimes S & \xrightarrow{\epsilon_3} & M \otimes S \otimes S \end{array}$$

and

$$\begin{array}{ccccccc} 0 & \longrightarrow & S \otimes N & \longrightarrow & S \otimes M & \xrightarrow{I \otimes u \circ \epsilon_1} & S \otimes M \otimes S \\ & & & & \downarrow{\scriptstyle u} & & \downarrow{\scriptstyle u_3} \\ 0 & \longrightarrow & M & \longrightarrow & M \otimes S & \xrightarrow{\epsilon_2} & M \otimes S \otimes S \end{array}$$

The first diagram is commutative, since $u_3 \circ (I \otimes \epsilon_2) = u_3 \circ \epsilon_3 = \epsilon_3 \circ u$. Also the second diagram is commutative: for all $s \in S$, $m \in M$, we have

$$\begin{aligned}(u_3 \circ (I \otimes u \circ \epsilon_1))(s \otimes m) &= u_3(s \otimes u(1 \otimes m)) \\ &= (u_3 \circ u_1)(s \otimes 1 \otimes m) \\ &= u_2(s \otimes 1 \otimes m) \\ &= (\epsilon_2 \circ u)(s \otimes m)\end{aligned}$$

Since u and u_3 are isomorphisms, u restricts to an isomorphism $\Phi: S \otimes N \to M$. For $n \in N$, we have that $u(1 \otimes n) = n \otimes 1$, and therefore

$$\Phi(1 \otimes n) = n = \eta(n \otimes 1)$$

and, for all $s \in S$, $n \in N$:

$$\Phi(s \otimes n) = \eta(n \otimes s)$$

and

$$\Phi = \eta \otimes t$$

so η is an isomorphism. Observe that, for all $n \in N$:

$$(u \circ \eta_1)(1 \otimes n \otimes 1) = u(1 \otimes n) = n \otimes 1 = (\eta_3 \circ \tau_3)(1 \otimes n \otimes 1)$$

hence the diagram (B.3) commutes. Let us finally show that (N, η) is unique : suppose that (K, κ) also meets the conditions of the theorem. Then we have an isomorphism $\rho = \kappa^{-1} \circ \eta: N \otimes S \to K \otimes S$, and a commutative diagram

$$\begin{array}{ccccc} S \otimes N \otimes S & \xrightarrow{\eta_1} & S \otimes M & \xleftarrow{\kappa_1} & S \otimes K \otimes S \\ \downarrow{\scriptstyle \tau_3} & & \downarrow{\scriptstyle u} & & \downarrow{\scriptstyle \tau_3} \\ N \otimes S \otimes S & \xrightarrow{\eta_3} & M \otimes S & \xleftarrow{\kappa_3} & K \otimes S \otimes S \end{array}$$

Hence $\rho_3 = \kappa_3^{-1} \circ \eta_3^{-1} = \tau_3 \otimes (\kappa_1^{-1} \circ \eta_1^{-1}) \otimes \tau_3 = \kappa_2^{-1} \circ \eta_2^{-1} = \rho_2$, so, according to (B.0.7), $\rho = \alpha \otimes I$ for some isomorphism $\alpha: N \to K$. Finally $\kappa \circ (\alpha \otimes I) = \eta$, so (K, κ) and (N, η) are isomorphic. □

Remark B.0.11 Let $u : M_1 \to M_2$ and $v : M_1' \to M_2'$ be descent data for the S-modules M and M', and write N and N' for the descended modules. Let $f : M \to M'$ be an S-linear map such that the diagram

$$\begin{array}{ccc} S \otimes M & \xrightarrow{I_S \otimes f} & S \otimes M' \\ \downarrow{\scriptstyle u} & & \downarrow{\scriptstyle v} \\ M \otimes S & \xrightarrow{f \otimes I_S} & M' \otimes S \end{array} \tag{B.5}$$

commutes. Then the map f restricts to a map $N \to N'$.
Indeed, if $x \in N$, then $x \otimes 1 = u(1 \otimes x)$, according to (B.4). Then

$$\begin{aligned} f(x) \otimes 1 &= ((f \otimes I_S) \circ u)(1 \otimes x) \\ &= (v \circ (I \otimes f))(1 \otimes x) = v(1 \otimes f(x)) \end{aligned}$$

and it follows that $f(x) \in N'$.

Theorem B.0.10 and Lemma B.0.11 can be restated in a categorical way. Let $\mathcal{D}(S)$ be the category of descent data. The objects are couples (M,u), where M is an S-module, and $u: M_1 \to M_2$ is a descent datum. A morphism between two descent data (M,u) and (M',v) is a map $f: M \to M'$ such that (B.5) commutes. We have functors

$$\begin{aligned} F: R\text{-mod} \longrightarrow \mathcal{D}(S) &: N \mapsto (N \otimes S, \tau \otimes I_S) \\ G: \mathcal{D}(S) \longrightarrow R\text{-mod} &: M \mapsto \{x \in M | x \otimes 1 = u(1 \otimes x)\} \end{aligned}$$

Proposition B.0.12 *G is a right adjoint of F. F and G are inverse equivalences if and only if S is faithfully flat as an R-module.*

Proof Take $N \in R$-mod, and $(M,u) \in \mathcal{D}(S)$. Define

$$\alpha: \operatorname{Hom}_R(N, G(M,u)) \longrightarrow \operatorname{Hom}_{\mathcal{D}(S)}((N \otimes S, \tau_3), (M,u))$$

by

$$\alpha(f)(n \otimes s) = sf(n)$$

and

$$\beta: \operatorname{Hom}_{\mathcal{D}(S)}((N \otimes S, \tau_3), (M,u)) \longrightarrow \operatorname{Hom}_R(N, G(M,u))$$

by

$$\beta(g)(n) = g(n \otimes 1)$$

for all $f \in \operatorname{Hom}_R(N, G(M,u))$, $g \in \operatorname{Hom}_{\mathcal{D}(S)}((N \otimes S, \tau_3), (M,u))$, $s \in S$ and $n \in N$. It is easy to check that $\beta(g)$ is well-defined. We also have to prove that f is a homomorphism of descent data. Indeed,

$$\begin{aligned} (u \circ (I_S \circ \alpha(f)))(s \otimes n \otimes t) &= u(s \otimes tg(n)) \\ &= (s \otimes t)u(1 \otimes g(n)) \\ &= (s \otimes t)(g(n) \otimes 1) \\ &= ((\alpha(f) \otimes I_S) \circ \tau_3)(s \otimes n \otimes t) \end{aligned}$$

α and β are each others inverses, since

$$(\beta \circ \alpha)(f)(n) = \alpha(f)(n \otimes 1) = f(n)$$

and

$$(\alpha \circ \beta)(g)(n \otimes s) = s(\beta(g)(n)) = sg(n \otimes 1) = g(n \otimes s)$$

If S is faithfully flat as an R-module, then it follows essentially from Theorem B.0.10 that F and G are inverse equivalences. The converse is obvious. □

Theorem B.0.13 (Faithfully Flat Descent for R-algebras) *If in the setting of B.0.10, M is an S-algebra and u is an $S^{\otimes 2}$-algebra isomorphism, then N has a unique structure of R-algebra such that $\eta: N \otimes S \to M$ is an isomorphism of S-algebras. If M is associative, commutative, unitary, then so is N.*

Proof Recall from (B.4) that $N = \{x \in M | x \otimes 1 = u(1 \otimes x)\}$. Take $x, y \in N \subset M$. Then $xy \in N$, since

$$xy \otimes 1 = (x \otimes 1)(y \otimes 1) = u(1 \otimes x)u(1 \otimes y) = u(1 \otimes xy).$$

Clearly if M is commutative and/or associative, then so is N. Suppose that M has a unit element 1_M. Then $1_M \otimes 1 = u(1 \otimes 1_M)$, since u has to preserve unit elements. Therefore $1_M \in N$. □

Theorem B.0.14 (Faithfully Flat Descent for R-coalgebras) *If in the setting of B.0.10, M is an S-coalgebra and u is an $S^{\otimes 2}$-coalgebra isomorphism, then N has a unique structure of R-coalgebra such that $\eta: N \otimes S \to M$ is an isomorphism of S-coalgebras. If M is coassociative or cocommutative, then so is N. If M has a counit, then N also has a counit.*

Proof First observe that

$$N \otimes N \otimes S \cong (N \otimes S) \otimes_S (N \otimes S) \cong M \otimes_S M$$

and therefore the map

$$u \otimes u: \; S \otimes M \otimes_S M \cong (S \otimes M) \otimes_S (S \otimes M) \longrightarrow M \otimes_S M \otimes M \cong (M \otimes S) \otimes_S (M \otimes S)$$

is a descent datum for $N \otimes N$. It follows that

$$N \otimes N = \{\sum_i x_i \otimes y_i \in M \otimes_S M$$
$$\mid \sum_i u(1 \otimes x_i) \otimes u(1 \otimes y_i) = \sum_i u(x_i \otimes 1) \otimes u(y_i \otimes 1) \in (M \otimes S) \otimes_S (M \otimes S)\}$$

Now take $x \in N \subset M$. M is an S-coalgebra, and $S \otimes M$ is an $S^{\otimes 2}$-coalgebra. Now

$$\begin{aligned}
&\sum_i u(x \otimes x_1) \otimes u(x \otimes x_2) \\
&= (u \otimes u)\big(\Delta_{S \otimes M}(1 \otimes x)\big) \\
&= \Delta_{M \otimes S}(u(1 \otimes x)) \\
&\qquad (u \text{ is comultiplicative}) \\
&= \Delta_{M \otimes S}(x \otimes 1) \\
&= \sum (x_1 \otimes 1) \otimes (x_2 \otimes 1)
\end{aligned}$$

and it follows that $\Delta(x) \in N \otimes N$, and N is a coalgebra. It is clear that N is coassociative and/or cocommutative if M is coassociative and/or cocommutative. Suppose finally that M has a counit ε_M. We will show that $\varepsilon_M(x) \in R$ for all $x \in N$. The counit maps for $S \otimes M$ and $M \otimes S$ are respectively $I_S \otimes \varepsilon_M$ and $\varepsilon_M \otimes I_S$. Using the fact that u is comultiplicative, we obtain that

$$\begin{aligned}
\varepsilon_M(x) \otimes 1 &= (\varepsilon_M \otimes I_S)(x \otimes 1) \\
&= (\varepsilon_M \otimes I_S)(u(1 \otimes x)) \\
&= (I_S \otimes \varepsilon_M)(1 \otimes x) \\
&= 1 \otimes \varepsilon_M(x)
\end{aligned}$$

and if follows from Corollary B.0.6 that $\varepsilon_M(x) \in R$. □

Appendix C

Elementary algebraic K-theory

In this appendix, we will recall the definition of the Grothendieck and the Whitehead group of a category with product $\mathcal{C}$, and state some elementary properties. We will give only very brief sketches of the proofs, for full detail, we refer to the literature, e.g. [14], [15], [173]. We will look more closely at the special situation where every element in the Grothendieck group is represented by an object of the category, this is the situation that we will encounter several times in this monograph.
Let $\mathcal{C}$ be a category, and assume that the isomorphism classes of $\mathcal{C}$ form a set. A *product* on $\mathcal{C}$ is a functor

$$\perp: \mathcal{C}\times\mathcal{C}\longrightarrow\mathcal{C}$$

which is "coherently associative and commutative" in the sense of [127]. This means that there are natural isomorphisms

$$\perp\circ(I_{\mathcal{C}}\times\perp)\cong\perp\circ(\perp\times I_{\mathcal{C}}):\ \mathcal{C}\times\mathcal{C}\times\mathcal{C}\longrightarrow\mathcal{C}$$

and

$$\perp\circ\tau\cong\perp:\ \mathcal{C}\times\mathcal{C}\longrightarrow\mathcal{C}$$

where τ is the switch functor.
The *Grothendieck group* $\mathrm{K}_0\mathcal{C}$ is the the abelian group generated by the isomorphism classes $[C]$ of objects of $\mathcal{C}$ modulo the relations

$$[C_1\perp C_2]=[C_1][C_2]$$

One can show easily that every $x\in\mathrm{K}_0\mathcal{C}$ can be written under the form $x=[C_1][C_2]^{-1}$ with $C_1,C_2\in\mathcal{C}$. Moreover, $[C_1]=[C_2]$ in $\mathrm{K}_0\mathcal{C}$ if and only if there exists a $C\in\mathcal{C}$ such that $C_1\perp C\cong C_2\perp C$.
We now introduce a new category with product $\Omega\mathcal{C}$, as follows:
<u>objects</u>: (C,α), with $C\in\mathcal{C}$ and α a $\mathcal{C}$-isomorphism of C.
<u>morphisms</u>: A morphism $f:\ (C_1,\alpha_1)\to(C_2,\alpha_2)$ consists of a morphism $f:\ C_1\to C_2$ such that $f\circ\alpha_1=\alpha_2\circ f$.
<u>product</u>: $(C_1,\alpha_1)\perp(C_2,\alpha_2)=(C_1\perp C_2,\alpha_1\perp\alpha_2)$.
$\Omega\mathcal{C}$ is called the *loop category* of $\mathcal{C}$. The *Whitehead group* $\mathrm{K}_1\mathcal{C}$ is $\mathrm{K}_0\Omega\mathcal{C}$ modulo the subgroup generated by elements of the form

$$[(C,\alpha\circ\beta)]-[(C,\alpha)]-[(C,\beta)]$$

The Grothendieck and Whitehead groups may also be defined by a universal property (see e.g. [15, Ch. VII]). They behave functorially in the following sense: if $F : \mathcal{C} \to \mathcal{D}$ is a product preserving functor (this means that for all objects C_1 and C_2 in $\mathcal{C}$, we have a natural isomorphism $F(C_1 \perp C_2) \cong F(C_1) \perp F(C_2)$), then we have maps

$$\mathrm{K}_0F : \mathrm{K}_0\mathcal{C} \to \mathrm{K}_0\mathcal{D} \quad \text{and} \quad \mathrm{K}_1F : \mathrm{K}_1\mathcal{C} \to \mathrm{K}_1\mathcal{D}$$

These maps are defined in the obvious way.
The functor $F : \mathcal{C} \to \mathcal{D}$ is called *cofinal* if, for every D in $\mathcal{D}$, there exist $\widetilde{D} \in \mathcal{D}$ and $C \in \mathcal{C}$ such that $D \perp \widetilde{D} \cong F(C)$. Let $F : \mathcal{C} \to \mathcal{D}$ be such a functor. We define a new category with product $\underline{\Phi F}$ as follows.
objects: triples (C_1, α, C_2), with $C_1, C_2 \in \mathcal{C}$ and $\alpha : F(C_1) \to F(C_2)$ an isomorphism in $\mathcal{D}$.
morphisms: $(f_1, f_2) : (C_1, \alpha, C_2) \to (\tilde{C}_1, \tilde{\alpha}, \tilde{C}_2)$ consists of two isomorphisms $f_1 : C_1 \to \tilde{C}_1$ and $f_2 : C_2 \to \tilde{C}_2$ such that

$$F(f_2) \circ \alpha = \tilde{\alpha} \circ F(f_1)$$

product: $(C_1, \alpha, C_2) \perp (\tilde{C}_1, \tilde{\alpha}, \tilde{C}_2) = (C_1 \perp \tilde{C}_1, \alpha \perp \tilde{\alpha}, C_2 \perp \tilde{C}_2)$.
$\mathrm{K}_1\underline{\Phi F}$ will be the abelian group $\mathrm{K}_0\underline{\Phi F}$ modulo the subgroup generated by elements of the form

$$[(C_1, \beta \circ \alpha, C_3)][(C_1, \alpha, C_2)]^{-1}[(C_2, \beta, C_3)]^{-1} \tag{C.1}$$

With these notations, we have following result.

Theorem C.0.15 *Let $F : \mathcal{C} \to \mathcal{D}$ be a cofinal product preserving functor between two categories with a product. Then we have an exact sequence*

$$\mathrm{K}_1\mathcal{C} \xrightarrow{\mathrm{K}_1F} \mathrm{K}_1\mathcal{D} \xrightarrow{d} \mathrm{K}_1\underline{\Phi F} \xrightarrow{f} \mathrm{K}_0\mathcal{C} \xrightarrow{\mathrm{K}_0F} \mathrm{K}_0\mathcal{D} \tag{C.2}$$

Proof The proof consists of a long but straightforward computation. For full detail, we refer to [15, Ch. VII] or to [173]. We will restrict ourselves here to giving the definitions of the maps d and f.
Consider $[(D, \alpha)] \in \mathrm{K}_1\mathcal{D}$. Then there exist $\widetilde{D} \in \mathcal{D}$ and $C \in \mathcal{C}$ such that $D \perp \widetilde{D} \cong F(C)$. We can write

$$[(D, \alpha)] = [(D \perp \widetilde{D}, \alpha \perp I_{\widetilde{D}})] = [(F(C), \beta)]$$

in $\mathrm{K}_1\mathcal{D}$. We define $d[(D, \alpha)] = [(C, \beta, C)]$.
For $[(C_1, \alpha, C_2)] \in \mathrm{K}_1\underline{\Phi F}$, we define $f[(C_1, \alpha, C_2)] = [C_2][C_1]^{-1}$. □

Let $\mathcal{C}$ be category with product. A full subcategory with product $\mathcal{C}_0$ is called a *cofinal subcategory* if the inclusion functor $i : \mathcal{C}_0 \to \mathcal{C}$ is cofinal. Let $\mathcal{C}_0$ be a full cofinal subcategory of $\mathcal{C}$. It is not difficult to show that every $x \in \mathrm{K}_0\mathcal{C}$ can be written under the form

$$x = [C][B]^{-1}$$

with $C \in \mathcal{C}$ and $B \in \mathcal{C}_0$. For two objects $C_1, C_2 \in \mathcal{C}$, we also have that $[C_1] = [C_2]$ in $\mathrm{K}_0\mathcal{C}$ if and only if

$$C_1 \perp B \cong C_2 \perp B \tag{C.3}$$

for some $B \in \mathcal{C}_0$.
Similar, but more complicated, conditions hold for the Whitehead group. One can show that

$$[(C_1, \alpha_1)] = [C_2, \alpha_2)] \text{ in } \mathrm{K}_1\mathcal{C}$$

if and only if

$$(C_1, \alpha_1) \perp (B_1, \beta_1) \perp (B_2, \beta_2 \circ \tilde{\beta}_2) \perp (B_3, \beta_3) \perp (B_3, \tilde{\beta}_3) \tag{C.4}$$

$$\cong (C_2, \alpha_2) \perp (B_1, \beta_1) \perp (B_2, \beta_2) \perp (B_2, \tilde{\beta}_2) \perp (B_3, \beta_3 \circ \tilde{\beta}_3)$$

for some $(B_1, \beta_1), (B_2, \beta_2), (B_2, \tilde{\beta}_2), (B_3, \beta_3), (B_3, \tilde{\beta}_3) \in \Omega\mathcal{C}_0$.

Proposition C.0.16 *Let $\mathcal{C}_0$ be a full cofinal subcategory with product of the category with product $\mathcal{C}$. Then*

$$\begin{aligned} &\mathrm{K}_1 i: \ \mathrm{K}_1\mathcal{C}_0 \xrightarrow{\cong} \mathrm{K}_1\mathcal{C} \\ &\mathrm{K}_0 i: \ \mathrm{K}_0\mathcal{C}_0 \hookrightarrow \mathrm{K}_0\mathcal{C} \\ &\mathrm{K}_1(\underline{\Phi i}) = 1 \end{aligned} \tag{C.5}$$

Proof It follows immediately from (C.3) and (C.4) that $\mathrm{K}_0 i$ and $\mathrm{K}_1 i$ are injective. In the proof of Theorem C.0.15, we have seen that every $[(C, \alpha)] \in \mathrm{K}_1\mathcal{C}$ can be written under the form

$$[(C, \alpha)] = [(B, \beta)]$$

with $B \in \mathcal{C}_0$. This implies that $\mathrm{K}_1 i$ is surjective. The last statement follows from the Lemma of 5. □

We will now examine the special situation where every element in $\mathrm{K}_0\mathcal{C}$ is represented by an object of the category $\mathcal{C}$. Let the unit element 1 of $\mathrm{K}_0\mathcal{C}$ be represented $E \in \mathcal{C}$. For every $C \in \mathcal{C}$, $[C]$ is represented by an object of $\mathcal{C}$ that we will denote by C'. It follows easily that the set of isomorphism classes of objects in $\mathcal{C}$ is a group under the operation induced by the product $\perp$, and that the Grothendieck group is exactly this group. Conversely, if the isomorphism classes of objects in $\mathcal{C}$ form a group, then every element in the Grothendieck group is represented by an object of the category. We also have an easy description of the Whitehead group. The full subcategory $\{E\}$ of $\mathcal{C}$ is cofinal, and therefore

$$\mathrm{K}_1\mathcal{C} \cong \mathrm{K}_1\{E\}$$

Let α, β be $\mathcal{C}$-automorphisms of E. Then $f \perp g$ is an automorphism of $E \perp E \cong E$, and

$$\alpha \perp \beta = (I_E \circ \alpha) \perp (\beta \circ I_E) = (I_E \perp \beta) \circ (\alpha \perp I_E) = \beta \circ \alpha$$

In a similar way, we can show that

$$\alpha \perp \beta = \alpha \circ \beta$$

It follows that $\mathrm{Aut}_{\mathcal{C}}(E)$ is an abelian group and that

$$\mathrm{K}_1\mathcal{C} \cong \mathrm{Aut}_{\mathcal{C}}(E) \tag{C.6}$$

A typical example is the situation where $\mathcal{C} = \underline{\underline{\mathrm{Pic}}}(R)$, the category of invertible modules over a commutative ring R, and $\perp = \otimes_R$. In this case

$$\mathrm{K}_0\underline{\underline{\mathrm{Pic}}}(R) = \mathrm{Pic}(R)$$

the Picard group of R, and

$$\mathrm{K}_1\underline{\underline{\mathrm{Pic}}}(R) = \mathrm{Aut}_R(R) = \mathbb{G}_m(R)$$

Consider two categories with product $\mathcal{C}$ and $\mathcal{D}$, and suppose that every element in $\mathrm{K}_0\mathcal{C}$ and $\mathrm{K}_0\mathcal{D}$ is represented by an object in the category. Let $F: \mathcal{C} \to \mathcal{D}$ be a cofinal product preserving functor. We will give an easy description of $\mathrm{K}_1\underline{\Phi F}$. From (C.1), it follows that, for every $C \in \mathcal{C}$, we have

$$[(C, I_{F(C)}, C)]^2 = [(C, I_{F(C)}, C)]$$

and

$$[(C, I_{F(C)}, C)] = 1$$

in $\mathrm{K}_1\underline{\Phi F}$. We also have that

$$[(C_1, \alpha, C_2)][(C_2, \alpha^{-1}, C_1)] = [(C_1, I_{F(C_1)}, C_1)] = 1$$

in $\mathrm{K}_1\underline{\Phi F}$. This implies that every element in $\mathrm{K}_1\underline{\Phi F}$ is represented by an object (C_1, α, C_2) of $\underline{\Phi F}$. Let C_2' represent the inverse of C_2 in $\mathrm{K}_0\mathcal{C}$. Then

$$\begin{aligned}
[(C_1, \alpha, C_2)] &= [(C_1, \alpha, C_2)][(C_2', I_{F(C_2')}, C_2')] \\
&= [(C_1 \perp C_2', \alpha \perp I_{F(C_2')}, C_2 \perp C_2')] \\
&= [(C_1 \perp C_2', \alpha \perp I_{F(C_2')}, E)]
\end{aligned}$$

and we have shown that every element of $\mathrm{K}_1\underline{\Phi F}$ can be represented by a triple of the form (C, β, E). We can redefine $\mathrm{K}_1\underline{\Phi F}$ as follows: let $\underline{\Psi F}$ be the category given by the following data:
objects: (C, α), with $C \in \mathcal{C}$ and $\alpha: F(C) \to F(E)$ an isomorphism in $\mathcal{D}$.
morphisms: isomorphisms $f: C \to \tilde{C}$ in $\mathcal{C}$ such that

$$\tilde{\alpha} = F(f) \circ \alpha$$

product: $(C, \alpha) \perp (\tilde{C}, \tilde{\alpha}) = (C \perp \tilde{C}, \alpha \perp \tilde{\alpha})$.
It is now straightforward to see that

$$\mathrm{K}_1\underline{\Phi F} \cong \mathrm{K}_0\underline{\Psi F} \tag{C.7}$$

The exact sequence (C.2) now takes the form

$$\mathrm{Aut}_{\mathcal{C}}(E) \longrightarrow \mathrm{Aut}_{\mathcal{D}}(F(E)) \xrightarrow{d} \mathrm{K}_1\underline{\Phi F} \cong \mathrm{K}_0\underline{\Psi F} \xrightarrow{f} \mathrm{K}_0\mathcal{C} \xrightarrow{\mathrm{K}_0 F} \mathrm{K}_0\mathcal{D} \tag{C.8}$$

The maps d and f are given by the following formulas. For a $\mathcal{D}$-automorphism $\alpha: F(E) \to F(E)$, we have

$$d(\alpha) = [(E, \alpha)] \tag{C.9}$$

For $(C,\alpha)] \in K_1\underline{\Phi F} \cong K_0\underline{\Psi F}$, we have

$$f([C,\alpha]) = [C] \tag{C.10}$$

We use the exact sequence (C.8) at several places in this monograph. It is not difficult to show directly that this sequence is exact.
We have mentioned above the typical example $\mathcal{C} = \underline{\underline{\mathrm{Pic}}}(R)$. In this case, there is an additional property that turns out to be useful sometimes: the functor

$$(\bullet)^* : \underline{\underline{\mathrm{Pic}}}(R) \longrightarrow \underline{\underline{\mathrm{Pic}}}(R)$$

satisfies the property

$$[I^*] = [I]^{-1}$$

in $K_0\underline{\underline{\mathrm{Pic}}}(R)$. In other words, every object $I \in \underline{\underline{\mathrm{Pic}}}(R)$ has a "natural" inverse I^*. Of course we do not have this nice property for every category $\mathcal{C}$ satisfying the property that every element of its Grothendieck group is represented by an object of the category.
Given such a category $\mathcal{C}$, we construct a new category $D\mathcal{C}$, such that $D\mathcal{C}$ has the same Grothendieck and Whitehead groups as $\mathcal{C}$, and every element in $D\mathcal{C}$ has a "natural" inverse. $D\mathcal{C}$ is defined as follows:
objects: triples $\underline{C} = (C, C', \gamma)$, with $\gamma : C \perp C' \to E$ a $\mathcal{C}$-isomorphism.
morphisms: $\underline{f} = (f, f') : \underline{C} \to \underline{D}$ consists of two isomorphisms $f : C \to D$ and $f' : C' \to D'$ in $\mathcal{C}$ such that $\delta \circ (f \perp f') = \gamma$.
product: $(C, C', \gamma) \perp (D, D', \delta) = (C \perp D, C' \perp D', \gamma \perp \gamma')$.
It is clear that

$$K_0 D\mathcal{C} \cong K_0\mathcal{C} \quad \text{and} \quad K_1 D\mathcal{C} \cong K_1\mathcal{C}$$

The functor

$$D\mathcal{C} \longrightarrow D\mathcal{C} : (C, C', \gamma) \mapsto (C', C, \gamma \circ \tau)$$

sends an element of $K_0 D\mathcal{C}$ to its inverse. So $D\mathcal{C}$ has all the required properties.

Bibliography

[1] E. Abe, Hopf Algebras, Cambridge University Press, Cambridge, 1977.

[2] P. N. Áhn, L. Márki, Morita equivalence for rings without identity, *Tsukuba J. Math.* **11** (1987), 1-16.

[3] J.N. Alonso Alvarez, J.M. Fernández Vilaboa, E. Villanueva Novoa, Quantum Yang-Baxter H-Azumaya monoids in closed categories, *Appl. Categorical Structures*, to appear.

[4] S. Amitsur, Simple algebras and cohomology groups of arbitrary fields, *Trans. Amer. Math. Soc.* **90** (1959), 73-112.

[5] S. Amitsur, On central division algebras, *Israel J. Math.* **12** (1972), 408-420.

[6] E. Artin, C.J. Nesbitt, R.M. Thrall, Rings with minimum condition, University of Michigan Press, Ann Arbor, 1944.

[7] M. Artin, Grothendieck Topologies, Harvard University Lecture Notes, Cambridge, Massachussets, 1962.

[8] M. Artin, On the joins of Hensel rings, *Adv. in Math.* **7** (1971), 282-296.

[9] M. Artin, A. Grothendieck, J.-L. Verdier, Théorie des topos et cohomologie étale des schémas, *Lecture Notes* in Math. **269**, **270**, **305**, Springer Verlag, Berlin, 1972-1973.

[10] B. Auslander, The Brauer group of a ringed space, *J. Algebra* **4** (1966), 220-273.

[11] M. Auslander, A. Brumer, Brauer groups of discrete valuation rings, *Indag. Math.* **71** (1968), 286-296.

[12] M. Auslander, O. Goldman, The Brauer group of a commutative ring, *Trans. Amer. Math. Soc.* **97** (1960), 367-409.

[13] G. Azumaya, On maximally central algebras, *Nagoya Math. J.* **2** (1951), 119-150.

[14] H. Bass, Lectures on topics in algebraic K-theory, Tata Institute of Fundamental research, Bombay, 1967.

[15] H. Bass, Algebraic K-theory, Benjamin, New York, 1968.

[16] M. Beattie, Brauer groups of H-module and H-dimodule algebras, thesis, Queens University, Kingston, Ontario, 1976.

[17] M. Beattie, A direct sum decomposition for the Brauer group of H-module algebras, *J. Algebra* **43** (1976), 686-693.

[18] M. Beattie, The Brauer group of central separable G-Azumaya algebras, *J. Algebra* **54** (1978), 516-525.

[19] M. Beattie, Automorphisms of G-Azumaya algebras, *Canad. J. Math.* **37** (1985), 1047-1058.

[20] M. Beattie, Computing the Brauer group of graded Azumaya algebras from its subgroups, *J.Algebra* **101** (1986), 339-349.

[21] M. Beattie, The subgroup structure of the Brauer group of RG-dimodule algebras, in "Ring Theory Antwerp 1985", F. Van Oystaeyen (Ed.), *Lecture Notes in Math.* **1197**, Springer Verlag, Berlin, 1987.

[22] M. Beattie, Inner Gradings and Galois extensions with normal basis, *Proc. Amer. Math. Soc.* **107** (1989), 881-886.

[23] M. Beattie, S. Caenepeel, The Brauer-Long group of $\mathbb{Z}/p^t\mathbb{Z}$-dimodule algebras, *J. Pure and Appl. Algebra* **61** (1989), 219-236.

[24] M. Beattie, K. H. Ulbrich, A Skolem-Noether Theorem for Hopf algebra actions, *Comm. Algebra* **18** (1990), 3713-3724.

[25] G. Bergman, Everybody knows what a Hopf algebra is, *Contemporary Math.* **43** (1985), 25-48.

[26] R.J. Blattner, M. Cohen, S. Montgomery, Crossed Products and Inner Actions of Hopf Algebras, *Trans. Amer. Math. Soc.* **298** (1986), 671-711.

[27] N. Bourbaki, Commutative Algebra, Herman, Paris, 1965.

[28] S. Caenepeel, Computing the Brauer-Long group of a Hopf algebra I : the Cohomological theory, *Israel J. Math.* **72** (1990), 38-83.

[29] S. Caenepeel, Computing the Brauer-Long group of a Hopf algebra II : the Skolem-Noether theory, *J. Pure and Appl. Algebra* **84** (1993), 107-144.

[30] S. Caenepeel, A cohomological interpretation of the Brauer-Wall group, in "Proceedings of the Second Belgian-Spanish week on Algebra and Geometry", E. Villanueva (Ed.), *Alxebra Santiago de Compostela* **54** (1990), 31-46.

[31] S. Caenepeel, A note on Inner Actions of Hopf Algebras, *Proc. Amer. Math. Soc.* **113** (1991), 31-39.

[32] S. Caenepeel, A note on noncommutative Galois objects over a Hopf algebra, *Bull. Soc. Math. Belg.-Tijdschr. Belg. Wisk. Gen.* **43A** (1991), 15-31.

[33] S. Caenepeel, Harrison cohomology and the group of Galois coobjects, in "Proceedings of the seventh Contact Franco-Belge en Algèbre", J. Alev (Ed.), Collection SMF, SMF, Paris, 1996.

[34] S. Caenepeel, A variation of Sweedler's complex and the group of Galois objects of an infinite Hopf algebra, *Comm. Algebra* **24** (1996), 2991-3015.

[35] S. Caenepeel, Kummer theory for monogenic Larson orders, in "Algebraic and categorical methods in ring theory", S. Caenepeel and A. Verschoren (Eds.), Marcel Dekker, New York, to appear.

[36] S. Caenepeel, The Brauer-Long group revisited, preprint.

[37] S. Caenepeel, M. Beattie, A cohomological approach to the Brauer-Long group and the groups of Galois extensions and strongly graded rings, *Trans. Amer. Math. Soc.* **324** (1991), 747-775.

[38] S. Caenepeel, G. Elencwajg, M. Ojanguren, Etale algebras, in preparation.

[39] S. Caenepeel, F. Grandjean, A note on Taylor's Brauer group, *Pacific J. Math.*, to appear.

[40] S. Caenepeel, G. Militaru, S. Zhu, Crossed modules and Doi-Hopf modules, *Israel J. Math.*, to appear.

[41] S. Caenepeel, G. Militaru, S. Zhu, Doi-Hopf modules, Yetter-Drinfel'd modules and Frobenius type properties, *Trans. Amer. Math. Soc.*, to appear.

[42] S. Caenepeel, G. Militaru, S. Zhu, A Maschke-type Theorem for Doi-Hopf modules and applications, *J. Algebra*, **187** (1997), 388-412.

[43] S. Caenepeel, Ş. Raianu, Induction functors for the Doi-Koppinen unified Hopf modules, in "Abelian groups and Modules", A. Facchini and C. Menini (Eds.), Kluwer Academic Publishers, Dordrecht, 1995, 73-94.

[44] S. Caenepeel, Ş. Raianu, F. Van Oystaeyen, Induction and Coinduction for Hopf Algebras. Applications, *J. Algebra* **165** (1994), 204-222.

[45] S. Caenepeel, F. Van Oystaeyen, Brauer groups and the cohomology of graded rings, *Monographs and Textbooks in Pure and Appl. Math.* **121**, Marcel Dekker, New York, 1988.

[46] S. Caenepeel, F. Van Oystaeyen, Y. Zhang, Quantum Yang-Baxter module algebras, *K-theory* **8** (1994), 231-255.

[47] S. Caenepeel, F. Van Oystaeyen, Y. Zhang, The Brauer group of Yetter-Drinfel'd module algebras, *Trans. Amer. Math. Soc.*, to appear.

[48] H. Cartan, S. Eilenberg, Homological Algebra, Princeton University Press, Princeton, 1956.

[49] S. Chase, D. Harrison, A. Rosenberg, Galois theory and cohomology of commutative rings, *Mem. Amer. Math. Soc.* **52** (1965), 1-19.

[50] S. Chase, A. Rosenberg, Amitsur cohomology and the Brauer group, *Mem. Amer. Math. Soc.* **52** (1965), 20-45.

[51] S. Chase, M.E. Sweedler, Hopf algebras and Galois theory, *Lect. Notes in Math.* **97**, Springer Verlag, Berlin, 1969.

[52] L.N. Childs, Abelian Galois extensions of rings containing roots of unity, *Illinois J. Math.* **15** (1971), 273-280.

[53] L.N. Childs, The Brauer group of graded Azumaya algebras II : graded Galois extensions, *Trans. Amer. Math. Soc.* **204** (1975), 137-160.

[54] L.N. Childs, The group of unramified Kummer extensions of prime degree, *Proc. London Math. Soc.* **35** (1977), 407-422.

[55] L.N. Childs, Representing classes in the Brauer group of quadratic number rings as smash products, *Pacific J. Math.* **129** (1987).

[56] L.N. Childs, Galois extensions over local number rings, in "Rings, Extensions and Cohomology", A. Magid (Ed.), Marcel Dekker, New York, 1994, p. 41-65.

[57] L.N. Childs, G. Garfinkel, M. Orzech, The Brauer group of graded Azumaya algebras, *Trans. Amer. Math. Soc.* **175** (1973), 299-326.

[58] L.N. Childs, A. Magid, The Picard invariant map of a principal homogeneous space, *J. Pure Appl. Algebra* **4** (1974), 273-286.

[59] L.N. Childs, D. Moss, Hopf algebras and local Galois module theory, in "Advances in Hopf Algebras", J. Bergen and S. Montgomery (Eds.), Marcel Dekker, New York 1995, p. 41-65.

[60] L.N. Childs, C. Greither, D. Moss, J. Sauenberg, K. Zimmerman, Hopf algebras, formal groups and Raynaud orders, Monograph, to appear.

[61] M. Cohen, Smash Products, inner action and quotient rings, *Pacific J. Math.* **125** (1986), 45-66.

[62] P. M. Cohn, Algebra 1, John Wiley, New York, 1982.

[63] M. Cox-Paul, The image of the Picard invariant map for Hopf Galois extensions, thesis, State University of New York, Albany, New York, 1994.

[64] A. P. Deegan, A subgroup of the generalized Brauer group of G-Azumaya algebras, *J. London Math. Soc.* **2** (1981), 223-240.

[65] F. DeMeyer, Galois theory in separable algebras over commutative rings, *Ill. J. Math.* **10** (1966), 287-295.

[66] F. DeMeyer, The Brauer group of some separably closed rings, *Osaka J. Math.* **3** (1966), 201-204.

[67] F. DeMeyer, The Brauer group of polynomial rings, *Pac. J. Math.* **59** (1975), 391-398.

[68] F. DeMeyer, The Brauer group of a ring modulo an ideal, *Rocky Mountain J. Math.* **6** (1976), 191-198.

[69] F. DeMeyer, The Brauer group of affine curves, in "Brauer groups, Evanston, 1975", *Lecture Notes in Math.* **549**, D. Zelinsky (Ed.), Springer Verlag, Berlin, 1976.

[70] F. DeMeyer, T. Ford, On the Brauer group of surfaces, *J. Algebra* **86** (1984), 259-271.

[71] F. DeMeyer, T. Ford, Computing the Brauer-Long group of $\mathbb{Z}/2$-dimodule algebras, *J. Pure Appl. Algebra* **54** (1988), 197-208.

[72] F. DeMeyer, T. Ford, Nontrivial, locally trivial Azumaya algebras, *Contemp. Math.* **124** (1992), 39-50.

[73] F. DeMeyer, T. Ford, On the Brauer group of toric varieties, *Trans. Amer. Math. Soc.* **335** (1993), 559-577.

[74] F. DeMeyer, T. Ford, R. Miranda, The cohomological Brauer group of a toric variety, *J. Algebraic Geometry* **2** (1993), 137-154.

[75] F. DeMeyer, E. Ingraham, Separable algebras over commutative rings, *Lecture Notes in Math.* **181**, Springer Verlag, Berlin, 1971.

[76] F. DeMeyer, M.A. Knus, The Brauer group of a real curve, *Proc. Amer. Math. Soc.* **57** (1976), 227-232.

[77] F. DeMeyer, K. Reignier, Etale cohomology of toric varieties defined by infinite fans, *J. Pure Appl. Algebra* **107** (1996), 207-217.

[78] Y. Doi, Unifying Hopf modules, *J. Algebra* **153** (1992), 373-385.

[79] V.G. Drinfel'd, Quantum groups, *Proc. Int. Cong. Math. Berkeley* **1** (1986), 789-820.

[80] V.G. Drinfel'd, On almost cocommutative Hopf algebras, *Leningrad Math. J.* **1** (1990), 321-342.

[81] T. E. Early, H.F. Kreimer, Galois algebras and Harrison cohomology, *J. Algebra* **58** (1979), 136-147.

[82] C. Faith, Algebra: Rings, modules and categories I, Springer Verlag, Berlin, 1973.

[83] T. Ford, Every finite abelian group is the Brauer group of a ring, *Proc. Amer. Math. Soc.* **82** (1981), 315-321.

[84] T. Ford, On the Brauer group of a Laurent polynomial ring, *J. Pure Appl. Algebra* **51** (1988), 111-117.

[85] T. Ford, On the Brauer group and the cup product map, in "Perspectives in Ring Theory", L. Le Bruyn, F. Van Oystaeyen (Eds.), Kluwer Academic Publishers, Dordrecht, 1988.

[86] P. Freyd, Abelian Categories, Harper Row, New York, 1964.

[87] A. Fröhlich, C. T. C. Wall, Equivariant Brauer groups in algebraic number theory, *Bull. Soc. Math. France* **25** (1971), 91-96.

[88] A. Fröhlich, C. T. C. Wall, Generalizations of the Brauer group I, preprint.

[89] O. Gabber, Some Theorems on Azumaya algebras, in "Groupe de Brauer", M. Kervaire and M. Ojanguren (Eds.), *Lecture Notes in Math.* **844**, Springer Verlag, Berlin, 1981.

[90] J. Gamst, K. Hoechstman, Quaternions généralisés, *C.R. Acad. Sci. Paris* **269** (1969), 560-562.

[91] J. Giraud, Cohomologie non abélienne, Springer-Verlag, Berlin, 1971.

[92] F. Grandjean, Les Algèbres d'Azumaya sans unité, thesis, Catholic University of Louvain, Louvain-la-Neuve, 1997.

[93] C. Greither, Extensions of finite group schemes, and Hopf Galois theory over a discrete valuation ring, *Math. Z.* **210** (1992), 37-67.

[94] C. Greither, Cyclic Galois extensions of commutative rings, *Lecture Notes in Math.* **1534**, Springer Verlag, Berlin, 1992.

[95] A. Grothendieck, Sur quelques points d'algèbre homologique, *Tohôku Math. J.* **9** (1957), 119-221.

[96] A. Grothendieck, Technique de Descente I, *Sém. Bourbaki*, exp. **190** (1959-1960).

[97] A. Grothendieck, Le groupe de Brauer I, II, III in "Dix Exposés sur la cohomologie des schémas", North Holland, Amsterdam, 1968.

[98] D. Harrison, Abelian extensions of arbitrary fields, *Trans. Amer. Math. Soc.* **106** (1963), 230-235.

[99] D. Harrison, Abelian extensions of commutative rings, *Mem. Amer. Math. Soc.* **52** (1965), 66-79.

[100] R. Hartshorne, Algebraic Geometry, *Graduate texts in Math.* **52**, Springer Verlag, Berlin, 1977.

[101] I. N. Herstein, Noncommutative rings, *Carus Math. Monographs* **15**, Math. Ass. of America, Chicago, 1968.

[102] P. Hilton, U. Stammbach, A course in homological algebra, *Grad. Texts in Math.* **4**, Springer Verlag, Berlin, 1971.

[103] R. Hoobler, When is Br(X) = Br′(X)?, in "Brauer groups in ring theory and algebraic geometry", F. Van Oystaeyen and A. Verschoren (Eds.), *Lecture Notes in Math.* **917**, Springer Verlag, Berlin, 1982, 231-245.

[104] R. Hoobler, Functors of graded rings, in "Methods in ring theory", F. Van Oystaeyen (Ed.), Reidel, Dordrecht, 1984, 161-170.

[105] S. Hurley, Galois objects with normal bases for free Hopf algebras of prime degree, *J. Algebra* **109** (1987), 292-318.

[106] E. Ingraham, On the existence and conjugacy of inertial subalgebras, *J. Algebra* **31** (1974), 547-556.

[107] B. I-peng Lin, Morita's theorem for coalgebras, *Comm. Algebra* **1** (1974), 311-344.

[108] N. Jacobson, Finite dimensional division algebras over fields, Springer Verlag, Berlin, 1996.

[109] G. Janelidze, On abelian extensions of commutative rings (Russian), *Bull. Acad. Sci. Georgia* **108** (1982), 477-480.

[110] I. Kersten, Brauergruppen von Körpern, *Aspekte der Mathematik* **D6**, Vieweg, Braunschweig/Wiesbaden 1990.

[111] I. Kersten, J. Mičhalicek, Kummer theory without roots of unity, *J. Pure Appl. algebra* **50** (1988), 21-72.

[112] M.A. Knus, Algebras graded by a group, in "Category theory, homology theory and their applications II", P. Hilton (Ed.), *Lecture Notes in Math.* **92**, Springer Verlag, Berlin 1969, 117-133.

[113] M.A. Knus, A Teichmüller cocycle for finite extensions, Preprint.

[114] M.A. Knus, M. Ojanguren, Théorie de la descente et algèbres d'Azumaya, *Lecture Notes in Math.* **389**, Springer Verlag, Berlin, 1974.

[115] M.A. Knus, M. Ojanguren, A Mayer-Vietoris sequence for the Brauer group, *J. Pure and Appl. Algebra* **5** (1974), 345-360.

[116] M.A. Knus, M. Ojanguren, Cohomologie étale et groupe de Brauer, in "Groupe de Brauer", M. Kervaire and M. Ojanguren (Eds.), *Lecture Notes in Math.* **844**, Springer Verlag, Berlin, 1981.

[117] M. Koppinen, A Skolem-Noether theorem for Hopf algebra measurings, *Arch. Math. (Basel)* **57** (1991), 34-40.

[118] H. F. Kreimer, Quadratic Hopf algebras and Galois extensions, *Contemporary Math.* **13** (1981), 353-361

[119] H. F. Kreimer, P. M. Cook II, Galois theories and normal bases, *J. Algebra* **43** (1976), 115-121.

[120] E.E. Kummer, Über die Ergänzungssätze zu den allgemeinen Reciprocitätsgesetzen, *J. Reine Angew. Math.* **44** (1852), 93-146.

[121] R. G. Larson, Hopf algebras defined by group valuations, *J. Algebra* **38** (1976), 414-452.

[122] R. G. Larson, J. Towber, Two dual classes of bialgebras related to the concepts of "quantum group" and "quantum Lie algebra", *Comm. Algebra* **19** (1991), 3295-3345.

[123] Hongnian Li, Brauer groups over affine normal surfaces, *Proc. of Symposia in Pure Math.* **58.2** (1995), 309-317.

[124] F. Long, A generalization of the Brauer group of graded algebras, *Proc. London Math. Soc.* **29** (1974), 237-256.

[125] F. Long, The Brauer group of dimodule algebras, *J. Algebra* **30** (1974), 559-601.

[126] M.P. Lopez Lopez, Objetos de Galois sobre un algebra Hopf finita, *Alxebra Santiago de Compostela* **25** (1980).

[127] S. MacLane, Categories for the working mathematician, Springer Verlag, Berlin, 1971.

[128] A. Magid, Brauer groups of linear algebraic groups with characters, *Proc. Amer. Math. Soc.* **71** (1978), 164-168.

[129] S. Majid, Doubles of quasitriangular Hopf algebras, *Comm. Algebra* **19** (1991), 3061-3073.

[130] A. Masuoka, Coalgebra actions on Azumaya algebras, *Tsukuba J. of Math.* **14** (1990), 107-112.

[131] H. Matsumura, Commutative Algebra, Second edition, Benjamin/Cummings, Reading, Mass., 1980.

[132] A.S. Merkure'ev, Brauer groups of a fields, *Comm. Algebra* **11** (1983), 2611-2624.

[133] A.S. Merkure'ev, On the structure of the Brauer group of a field, *Math. Izvestiya* **27** (1986), 141-157.

[134] A.S. Merkure'ev, A.A. Suslin, K-cohomology of Severi-Brauer varieties and the norm rest homomorphism, *Math. Izvestiya* **21** (1983), 308-340.

[135] J.S. Milne, Etale cohomology, Princeton University Press, Princeton, 1980.

[136] S. Montgomery, Inner Actions of Hopf Algebras, in "Ring Theory 1989", L. Rowen (Ed.), The Weizmann Science Press of Israel, Jerusalem, 1989, p. 141-149.

[137] S. Montgomery, Hopf algebras and their actions on rings, American Mathematical Society, Providence, 1993.

[138] K. Morita, Duality for modules and its application to the theory of rings with minimum condition, *Science Reports of the Jokyo Kyoiku Daigaku Sect. A* **150** (1958), 84-142.

[139] D. Moss, Kummer theory of formal groups, *Amer. J. Math.* **118** (1996), 301-318.

[140] A. Nakajima, On generalized Harrison cohomology and Galois object, *Math. J. Okayama Univ.* **17** (1975), 135-148.

[141] A. Nakajima, Some results on H-Azumaya algebras, *Math. J. Okayama Univ.* **19** (1977), 101-110.

[142] A. Nakajima, Free algebras and Galois objects of rank 2, *Math. J. Okayama Univ.* **23** (1981), 181-187.

[143] D. Northcott, A note on polynomial rings, *J. London Math. Soc.* **33** (1958), 36-39.

[144] C. Năstăsescu, Ş. Raianu, F. Van Oystaeyen, Modules graded by G-sets, *Math. Z.* **203** (1990), 605-627.

[145] C. Năstăsescu, M. Van den Bergh, F. Van Oystaeyen, Separable functors applied to graded rings, *J. Algebra* **123** (1989), 397-413.

[146] C. Năstăsescu, F. Van Oystaeyen, Graded ring theory, Library of Math. 28, North Holland, Amsterdam, 1982.

[147] C. Năstăsescu, F. Van Oystaeyen, L. Shaoxue, Graded modules over G-sets II, *Math. Z.* **207** (1991), 341-358.

[148] H. Opolka, F. Van Oystaeyen, Brauer groups and projective representations of finite groups, Springer Verlag, Berlin, to appear.

[149] J. Osterburg, D. Quinn, A Skolem-Noether theorem for group graded rings, *J. Algebra* **113** (1988), 483-490.

[150] J. Osterburg, D. Quinn, Addendum to "A Skolem-Noether theorem for group graded rings", *J. Algebra* **120** (1989), 414-415.

[151] M. Orzech, A cohomological description of abelian Galois extensions, *Trans. Amer. Math. Soc.* **137** (1969), 481-499.

[152] M. Orzech, On the Brauer group of algebras having a grading and an action, *Can. J. Math.* **28** (1976), 533-552.

[153] M. Orzech, Correction to "On the Brauer group of algebras having a grading and an action", *Can. J. Math.* **32** (1980), 1523-1524.

[154] M. Orzech, Brauer groups of graded algebras, *Lecture Notes in Math.* **549**, Springer Verlag, Berlin 1976, 134-147.

[155] M. Orzech, C. Small, The Brauer group of a commutative ring, Marcel Dekker, Inc., New York 1975.

[156] B. Pareigis, The Brauer group of a symmetric monoidal category, in "Brauer groups, Evanston 1975", D. Zelinsky (Ed.), *Lecture Notes in Math.* **549**, Springer Verlag, Berlin, 1976.

[157] D.J. Picco, M.I. Platzeck, Graded algebras and Galois extensions, *Bol. Un. Mat. Argentina* **25** (1971), 401-415.

[158] R. S. Pierce, Associative Algebras, Springer Verlag, Berlin, 1982.

[159] N. Popescu, Abelian categories with applications to rings and modules, Academic Press, London and New York, 1973.

[160] I. Raeburn, J.L. Taylor, The Bigger Brauer group and étale cohomology, *Pacific J. Math.* **119** (1985), 445-463.

[161] D.E. Radford, The structure of Hopf algebras with a projection, *J. Algebra* **92** (1985), 322-347.

[162] M. Raynaud, Anneaux locaux Henséliens, *Lecture Notes in Math.* **169**, Springer Verlag, Berlin, 1971.

[163] M. Raynaud, Schémas en groupe de type $(p, p, \cdots, p)$, *Bull. Soc. Math. France* **102** (1974), 241-280.

[164] L. Roberts, The flat cohomology of group schemes of rank p, *Amer. J. Math.* **95** (1973), 688-702.

[165] A. Rosenberg, D. Zelinsky, Automorphisms of separable algebras, *Pac. J. Math.* **11** (1961), 1107-117.

[166] D. J. Saltman, The Brauer group is torsion, *Proc. Amer. Math. Soc.* **81** (1981), 385-387.

[167] P. Schauenburg, On coquasitriangular Hopf algebras and the quantum Yang-Baxter equation, *Algebra Berichte* **67** (1992), Verlag Uni-Druck/Verlag Reinhard Fischer, Mnchen, 1992.

[168] J.-P. Serre, Local Fields, Springer Verlag, Berlin, 1979.

[169] C. Small, The Brauer-Wall group of a commutative ring, *Trans. Amer. Math. Soc.* **156** (1971), 455-491.

[170] R.G. Swan, On seminormality, *J. Algebra* **67** (1980), 210-229.

[171] M. E. Sweedler, Hopf algebras, Benjamin, New York, 1969.

[172] M. E. Sweedler, Cohomology of algebras over Hopf algebras, *Trans. Amer. Math. Soc.* **133** (1968), 205-239.

[173] J. Sylvester, Introduction to algebraic K-theory, Chapman and Hall, London and New York, 1981.

[174] J. T. Tate, F. Oort, Group Schemes of prime order, *Ann. Scient. Ec. Norm. Sup. (4ième série)* **3** (1970), 1-21.

[175] J.L. Taylor, A bigger Brauer group, *Pacific J. Math.* **103** (1982), 163-203.

[176] M. Takeuchi, The #-product of group sheaf extensions applied to Long's theory of dimodule algebras, *Algebra Berichte* **34** (1977), Verlag Uni-Druck/Verlag Reinhard Fischer, München, 1977.

[177] M. Takeuchi, Morita theorems for categories of comodules, *J. Fac. Sc. Univ. Tokyo, Sec. IA* **24** (1977), 629-644.

[178] M. Takeuchi, On Villamayor and Zelinsky's long exact sequence, *Mem. Amer. Math. Soc.* **249** (1981), 1-178.

[179] J.P. Tignol, On the corestriction of central simple algebras, *Math. Z.* **194** (1987), 267-274.

[180] F. Tilborghs, The Brauer group of R-algebras which have compatible G-action and $\mathbb{Z} \times G$-grading, *Comm. Algebra* **18** (1990), 3351-3379.

[181] F. Tilborghs, An anti-homomorphism for the Brauer-Long group, *Math. J. Okayama Univ.* **32** (1990), 43-52.

[182] F. Tilborghs, F. Van Oystaeyen, Brauer-Wall algebras graded by $\mathbb{Z}/2 \times \mathbb{Z}$, *Comm. Algebra* **16** (1988), 1457-1478.

[183] B. Torrecillas, Y.H. Zhang, The Picard group of coalgebras, *Comm. Algebra* **24** (1996), 2235-2247.

[184] B. Torrecillas, F. Van Oystaeyen, Y.H. Zhang, The Brauer group of a cocommutative coalgebra, *J. Algebra*, to appear.

[185] K. H. Ulbrich, An exact sequence for the Brauer group of dimodule Azumaya algebras, *Math. J. Okayama Univ.* **35** (1993), 63-88.

[186] R. Underwood, R-Hopf algebra orders in KC^{p^2}, *J. Algebra*, **169** (1994), 418-440.

[187] R. Underwood, The valuative condition and R-Hopf algebras in KC_{p^3}, *Amer. J. Math.* **118** (1996), 701-743.

[188] R. Underwood, The group of Galois extensions over orders in KC_{p^2}, *Trans. Amer. Math. Soc.*, to appear.

[189] E.M. Vitale, The Brauer and Brauer-Taylor groups of a symmetric monoidal category, *Cahiers de Topologie et Géometrie Différentielle Catégoriques* **37** (1996), 91-122.

[190] F. Van Oystaeyen, On Brauer groups of arithmetically graded rings, *Comm. algebra* **9** (1981), 1873-1892.

[191] F. Van Oystaeyen, A. Verschoren, Relative invariants of rings I, *Monographs and Textbooks in Pure and Appl. Math.* **79**, Marcel Dekker, New York, 1983.

[192] F. Van Oystaeyen, A. Verschoren, Relative invariants of rings Part II, *Monographs and Textbooks in Pure and Appl. Math.* **86**, Marcel Dekker, New York, 1984.

[193] F. Van Oystaeyen, Y. Zhang, The Brauer group of a braided monoidal category, *J. Algebra*, to appear.

[194] F. Van Oystaeyen, Y. Zhang, Embedding the Brauer group into the Hopf automorphism group, *Bull. Canad. Math. Soc.*, to appear.

[195] F. Van Oystaeyen, Y. Zhang, Brauer Groups of Actions, in "Algebraic and geometric methods in ring theory", S. Caenepeel and A. Verschoren (Eds.), Marcel Dekker, New York, to appear.

[196] F. Van Oystaeyen, Y. Zhang, The Brauer Group of a group, in preparation.

[197] A. Verschoren, Brauer groups and class groups for Krull domains: a K-theoretic approach, *J. Pure Appl. Algebra* **33** (1984), 219-224.

[198] J.M. Fernandez Vilaboa, Grupos de Brauer y de Galois de un Algebra de Hopf en una categoría cerrada, *Alxebra Santiago de Compostela* **42** (1985).

[199] O.E. Villamayor, D. Zelinsky, Brauer groups and Amitsur cohomology for general commutative ring extensions, *J. Pure Appl. Algebra* **10** (1977), 19-55.

[200] C.T.C. Wall, Graded Brauer groups, *J. Reine Angew. Math.* **213** (1964), 187-199.

[201] W.C. Waterhouse, Introduction to affine group schemes, *Graduate texts in Math.* **66**, Springer Verlag, Berlin, 1979.

[202] E. Witt, Schiefkörper über diskret bewerteten Körper, *J. Reine Angew. Math.* **176** (1937), 153-156.

[203] D.N. Yetter, Quantum groups and representations of monoidal categories, *Math. Proc. Cambridge Phil. Soc.* **108** (1990), 261-290.

[204] K. Yokogawa, The cohomological aspects of Hopf Galois extensions over a commutative ring, *Osaka J. Math.* **18** (1981), 75-93.

[205] K. Yokogawa, Hopf-Galois extensions and smash products, *J. Algebra* **107** (1987), 138-152.

[206] S. Yuan, Reflexive modules and algebra class groups over noetherian integrally closed domains, *J. Algebra* **32** (1974), 405-417.

[207] O. Zariski, P. Samuel, Commutative Algebra Vol. 1, *Grad. Texts in Math.* **28**, Springer Verlag, Berlin, 1958.

[208] O. Zariski, P. Samuel, Commutative Algebra Vol. 2, *Grad. Texts in Math.* **29**, Springer Verlag, Berlin, 1958.

[209] D. Zelinsky, Long exact sequences and the Brauer group, in "Brauer groups, Evanston 1975", D. Zelinsky (Ed.), *Lecture Notes in Math.* **549**, Springer Verlag, Berlin, 1976.

Index